Frank Rösch (Ed.)

Nuclear- and Radiochemistry

Also of Interest

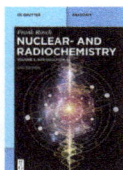

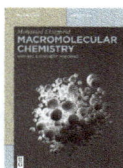

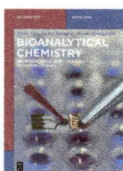

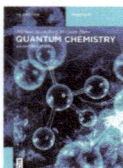

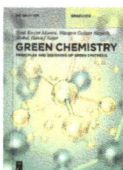

Nuclear- and Radiochemistry

Volume 2: Modern Applications

Edited by
Frank Rösch

2nd, extended edition

DE GRUYTER

Editor
Prof. Dr. Frank Rösch
Institute of Nuclear Chemistry
Johannes Gutenberg University Mainz
Fritz-Strassmann-Weg 2
55128 Mainz, Germany
E-mail: frank.roesch@uni-mainz.de

ISBN 978-3-11-074267-1
e-ISBN 978-3-11-074270-1
e-ISBN (EPUB) 978-3-11-074279-4

Library of Congress Control Number: 2022938896

Bibliographic information published by the Deutsche Nationalbibliothek
The Deutsche Nationalbibliothek lists this publication in the Deutsche Nationalbibliografie;
detailed bibliographic data are available on the Internet at http://dnb.dnb.de.

© 2022 Walter de Gruyter GmbH, Berlin/Boston
Cover image: ismagilov/iStock/Getty Images Plus
Typesetting: Integra Software Services Pvt. Ltd.
Printing and binding: CPI books GmbH, Leck

www.degruyter.com

Preface

Very shortly after the discovery of "radioactivity", and mainly because of the impact the newly discovered radioelements radium and polonium had on natural sciences at the beginning of the 20th century, Marie Curie, Pierre Curie and Ernest Rutherford individually published the first books about radioactivity. In the preface to a book summarizing his lectures (Curie P. *Œuvres de Pierre Curie. Publiés par les soins de la Société française de physique.* Paris: Gauthier–Villars; 1908, published posthumously), Pierre Curie noted

> [...] that radioactivity, although mainly connected to physics and chemistry, is not alien to other areas of science and is gaining there increasing importance. The phenomena of radioactivity are so manifold, its occurrence as miscellaneous and as prevalent in the universe, that it must be considered in all branches of natural science, and in particular in physiology, therapy, meteorology and geology.

More than a century later, this prediction and vision definitely has become true! Radiation emitted from naturally occurring or manmade unstable radionuclides, routinely produced at large scale, has become a vital and in many cases indispensable tool in research and development, in technology, in industry and in medicine.

This textbook is entitled NUCLEAR- AND RADIOCHEMISTRY, VOLUME 2: MODERN APPLICATIONS and illustrates the most important directions; structured into several groups. The first group involves three chapters on material and geological **analytics** (neutron activation analyses, nuclear mass spectroscopy, and nuclear dating). The second group covers four chapters on **chemistry** (radioelement speciation, radiochemical separations, radioactive elements). The third group is on **nuclear energy** and nuclear energy plants. Group four in three chapters covers applications of radioactivity in **life sciences** (tritium and carbon-14 labeling for *in vitro* studies, molecular diagnostic imaging *in vivo*, and endo-radiation therapy).

The application of radioactivity for various purposes is paralleled by the use of an impressive number of different radionuclides with – in many cases – extraordinarily high radioactivity, either addressing their production, separation, analytics or applications. Consequently, there is an urgent demand for adequate radiation measurement and for understanding biological effects of radiation, ultimately expressed in terms of radiation doses delivered to personnel and patients. Thus, the book starts with one further group of two individual chapters on **radiation measurement** and **radiation dosimetry**.

All the implementations of radioactivity today represent the essential impact radioactivity has on so many important fields. Nuclear power, radioisotope production, diagnostic and therapeutic use of radiopharmaceuticals have in many cases become everyday routine and are indispensable. Radioanalytical technologies occupy a unique position at the final frontier of low-sensitivity elemental, material, geological, and

https://doi.org/10.1515/9783110742701-202

astrophysical investigations. One may argue that there is no longer a field of "radio-chemistry" or "nuclear chemistry" but radio- and nuclear "chemistries".

Consequently, an impressive and steadily increasing number of chemists, physicists, physicians, technologists, radiation safety personnel, and many other professionals are handling a variety of radionuclides for very different purposes. This is the reason I believe that the present teaching book may serve as a source for many people – and for students interested in the fascinating topic of "modern applications" of radioactivity!

This VOLUME 2: MODERN APPLICATIONS completes the teaching book on NUCLEAR- AND RADIOCHEMISTRY, with VOLUME 1: INTRODUCTION published by De Gruyter in 2014. Also in 13 chapters, Volume 1 discussed the atom's structure, binding energies and shell structures of nucleons, concept and mathematics of unstable nuclei, primary transformations (α, β, spontaneous fission), secondary transformations and post-effects, and, finally, nuclear reactions as related to radionuclide production. The reader of the present volume is referred to Volume 1 for understanding the phenomena of radioactivity, in particular the origin of radioactive emissions, which in turn constitute the basic tool for MODERN APPLICATIONS of radioactivity ...

Modern methods and applications of radio- and nuclear chemistry are extremely versatile and thus each field deserves special attention. For me it is therefore obvious that (in contrast to earlier textbooks on radio- and nuclear chemistry) the complex matter cannot be covered adequately by a single author. Instead, it is the philosophy of this textbook that the 13 chapters are written by experts in each particular field. Fortunately, not all, but the vast majority of these authors have one thing in common: they are (for chapters 3, 7, 9, and 11) or were (chapters 6, 10, and 11) staff members or PhD students or are (for chapters 4 and 5) long-time collaborators of my institute, the Institute of Nuclear Chemistry of the Johannes Gutenberg University, Mainz, Germany. I am extremely grateful to all authors for their contributions and patience.

Vichuquen, November 2015 | Mainz, July 2016
Frank Rösch

Preface for the 2nd edition

Nuclear and Radiochemistry is, compared to the many other chemistries, a relatively young branch of this fascinating science. When historically referred to Marie Curiés identification of radium and the wording radioactivity as introduced by her, nuclear and radiochemistry is about 120 years old (or young, if you wish). Within this period, significant knowledge has been gained and since the first edition of this teaching book, no fundamental changes occurred in respect to the basic content of unstable nuclei and their transformations.

In contrast, the many applications of radioactivity and unstable nuclei are continuously increasing – this is permanent "work in progress". Accordingly, this 2nd edition represents an extended version of many chapters. Several corrections have been made as well. I am grateful to all the students and colleagues who added valuable comments.

Vichuquen, December 2021 | Berlin, March 2022
Frank Rösch

https://doi.org/10.1515/9783110742701-203

Contents

List of contributing auhtors

Cathy S. Cutler
Brookhaven National Laboratory
Upton, New York
USA

Christoph E. Düllmann
Institute of Nuclear Chemistry
University of Mainz
Mainz, Germany

Klaus Eberhardt
Institute of Nuclear Chemistry
Johannes Gutenberg University
Mainz, Germany

Norman M. Edelstein
Lawrence Berkeley National Laboratory,
Berkeley, USA

Alberto del Guerra
Department of Physics "E. Fermi"
University of Pisa
Pisa, Italy

Heinz Gäggeler
University of Bern
Bern, Switzerland
and
Paul Scherrer Institut
Villigen, Switzerland

Steffen Happel
Tristem International
3 Rue des Champs Géons ZAC de L'Éperon,
35170 Bruz
France

Gert Langrock
Areva GmbH
Erlangen, Germany

Thomas Moenius
Novartis AG
Basel, Switzerland

Lester R. Morss
Argonne National Laboratory
Argonne, USA

Daniele Panetta
Institute of Clinical Physiology
National Research Council
Pisa, Italy

Frank Roesch
Institute of Nuclear Chemistry
University Mainz
Mainz, Germany

Tobias L. Ross
Hannover Medical School
Hannover, Germany

Werner Rühm
Institute of Radiation Protection
Helmholtz Zentrum München
München, Germany

Michael Steppert
Institute for Radioecology and Radiation
Protection
Leibniz University Hannover
Hannover, Germany

Sönke Szidat
University of Bern
Bern, Switzerland

Norbert Trautmann
Institute of Nuclear Chemistry, University
Mainz, Mainz, Germany

Clemens Walther
Institute for Radioecology and Radiation
Protection
Leibniz University Hannover
Hannover, Germany

https://doi.org/10.1515/9783110742701-205

Klaus Wendt
Institute of Physics
Johannes Gutenberg University
Mainz, Germany

Peter Zeh
Areva GmbH
Erlangen, Germany

Alberto Del Guerra and Daniele Panetta

1 Radiation measurement

Aim: The interactions of radiation with matter are the basic processes underlying
the working principle of every radiation detection apparatus. This chapter focuses
on the methods and techniques for the detection of ionizing radiation. First, the different types of ionizing radiations (relevant in the context of radiochemistry) will be
introduced. In order to understand how a certain radiation can be qualitatively
and/or quantitatively assessed, it is important to figure out how it interacts with the
matter it traverses. In most cases, the result of such interaction is measured as an
electric signal by means of dedicated electronics, and then displayed on some suitable screen or printer. Such "active" detectors allow a real-time measurement.
Other detectors do not provide a direct reading, as they have to be exposed to the
radiation field for a given period of time: these detectors are called "passive". Another way to distinguish among the various types of detectors is based on the physical and chemical properties of the substance that constitutes the sensitive volume
of the detector itself. They can be gaseous, liquid or solid. The choice of one or another type of detector is mainly based on its final purpose: some are suitable as portable survey meters, others are better suited for personal monitoring, or for probing
the radiochemical purity of a radiopharmaceutical or detecting it for the purpose of
molecular imaging.

1.1 Physical basis of radiation detection

1.1.1 Types of radiation

The term *radiation* is applicable to every physical entity that is able to transport energy through space and time, away from the place where it originated (commonly
referred to as the *source* of the radiation). In the present context, radiation is emitted in the course of transformations of unstable nuclei and interacts with matter.
This interaction is at the basis of radiation measurement. Several types of radiation
can be distinguished.[1] In the field of radiochemistry and nuclear chemistry we will
only deal with electromagnetic radiation and charged particles that can ionize the
matter they interact with. For this reason, these types of radiation are called *ionizing
radiation*. Table 1.1 classifies the relevant types.

---

[1] This depends on the nature of the energy that is being transported. For instance, sound transports mechanical energy through longitudinal waves in an elastic medium: in this case, we refer to
acoustic waves or acoustic radiation.

https://doi.org/10.1515/9783110742701-001

Tab. 1.1: General classification scheme for radiation based on its ability to ionize matter.

Ionizing		Nonionizing
directly ionizing (electrically charged)	indirectly ionizing (electrically neutral)	
α, β⁺/β⁻ and atomic electrons, p, δ and fissions fragments	X, γ, hard UV, and n	electromagnetic waves and particles with <10 eV energy (e.g., infrared radiation, visible light, soft UV)

1.1.2 Definitions

In this context "ionizing radiation" refers to radiation emitted from the primary (such as beta electrons, α-particles, neutrons and fission fragments), secondary or post-effect (such as internal conversion electrons or AUGER and COSTER–KRONIG electrons, γ and X-ray photons and bremsstrahlung) transformations of an unstable nuclide. The *source* might have various dimensions, ranging from point-like or monoatomic layers to volumes of milliliters, liters or (in industrial processes) even cubic meters, to solids of various chemical and physical state and size, and to gases. The radiation emitted from a source distributes in all directions (4π geometry). Thus it creates a *radiation field*. A radiation *detector* represents a well-defined state of condensed matter, which translates the effect of radiation interaction in that particular matter into a quantitative and qualitative *signal*.

The energy transported by ionizing radiation is either represented electromagnetically or associated to the kinetic energy of fast nuclear or elementary particles after they have originated from a source. All particles with an electric charge can directly ionize atoms when they interact with matter through multiple collisions driven by coulomb forces: they are called *directly ionizing radiation*, cf. Vol. I, Chapter 12. Examples of directly ionizing radiation are α-particles, protons, light or heavy ions, muons and other elementary particles with sufficient kinetic energy. Neutral particles and electromagnetic waves (e.g., photons) do not have electric charge; they can ionize atoms through few, sparse interactions. These latter interactions give rise to *secondary charged particles* that, in turn, can continue to ionize the surrounding atoms. Secondary electrons with sufficiently high recoil kinetic energy to travel into the matter producing further ionization on atoms that are far away from the site of interaction with primary radiation are called *delta rays*. Thus, neutral particles can ionize the matter indirectly: for this reason this is called *indirectly ionizing radiation*. Commons examples of indirectly ionizing radiation are neutrons, X-rays and γ-photons.[2]

---

2 A photon is a quantum of electromagnetic energy, i.e., the elementary unit of the energy transported by an electromagnetic wave (in analogy, the electron charge is a quantum of electric

1.1.3 Parameters describing a radiation field

In order to allow a quantitative description of radiation fields, some basic definitions could be useful in the following sections. One could think of a radiation field
as a portion of three-dimensional space in which several particles move along individual trajectories. At a given point P in the field there is a *particle fluence* (or *flux*),
Φ, defined as the number N of particles passing in a time interval Δt through a
sphere of cross-sectional area s centered in P, when s becomes infinitely small.
More formally:

$$\Phi = \frac{\mathrm{d}N}{\mathrm{d}s}\ [\mathrm{m}^{-2}].\tag{1.1}$$

Because Φ has the dimension of the inverse of an area, it is measured in m^{-2}. Of
course, the number N is a stochastic quantity; thus, it is intended in the above
equation that the expectation value of N must be considered for the computation of
Φ in a sufficiently long time interval.

The time derivative of eq. (1.1) gives another quantity expressing the number of
particles passing through P in the unit time. It is called *fluence rate* (or *flux density*)
and is denoted by ϕ:

$$\phi = \frac{\mathrm{d}}{\mathrm{d}t}\left(\frac{\mathrm{d}N}{\mathrm{d}s}\right)\ [\mathrm{m}^{-2}\mathrm{s}^{-1}].\tag{1.2}$$

The unit of measure of the particle fluence rate is $\mathrm{m}^{-2}\,\mathrm{s}^{-1}$. The fluence and fluence
rate, as described above, only provide information about the *number* of particles
crossing a certain portion of space. Sometimes it is useful to have information not
only about the number of particles, but also about the total *energy* transported by
them. Let us denote by R the total kinetic energy of the N particles through P:

$$R = \sum_i E_i n_i\ \ i = 1, \ldots, M,\tag{1.3}$$

where n_i is the number of particles having energy E_i, M is the total number of energies
available in the field, and $\Sigma_i\, n_i = N$ is the total number of particles. The distribution of
particles n_i in each energy level in the field is called the *energy spectrum* of the radiation. For simplicity, eq. (1.3) supposes that only a finite number of energies are

---

charge). Because of the wave-particle duality of quantum mechanics, the photon can be treated as
a particle with its own energy $\varepsilon = h\upsilon$ in the context of radiation detection. Among the types of electromagnetic radiation, not only X-rays and γ-photons can ionize atoms, but also ultraviolet (UV)
light with sufficient energy. In fact, the electromagnetic spectrum can be conventionally partitioned
into ionizing and nonionizing at the energy of first ionization of the hydrogen atom, i.e., 13.6 eV
(approximately correspondingly to an electromagnetic radiation with a wavelength $\lambda = 90$ nm, i.e.,
below the "vacuum" UV light band).

present among all the N particles: in this case there is a *discrete* energy spectrum. Examples of radiation fields with a discrete spectrum are the photon field originating from a γ-emitting radionuclide (e.g., ^{99m}Tc, widely used in nuclear medicine, with main gamma emission of 140.51 keV) or the particle field originating from an α-emitter (such as ^{210}Po, with $E_\alpha = 5.3$ MeV, often used in polonium-beryllium alloys to create neutron sources). In the most general case, the energy of the particles in a field can have an arbitrary value in a continuous interval between 0 and E_{max}. In this case, the summation in eq. (1.3) is replaced by an integral, and the radiation field is referred to as a field with a *continuous* energy spectrum. Examples of radiation fields with continuous spectra are the β-electron field emitted by, e.g., ^{14}C, the positron field emitted by, e.g., ^{11}C (widely used for radiotracer labeling in Position Emission Tomography) or the photon field originating from bremsstrahlung.

In analogy to the definitions of particle fluence and fluence rate in eqs. (1.1) and (1.2), we define the *energy fluence* ψ and ene*rgy fluence rate* ψ, measured in the SI in J m^{-2} and J m^{-2} s^{-1}, respectively, as:

$$\Psi = \frac{dR}{ds}\left[\text{J m}^{-2}\right], \tag{1.4}$$

$$\psi = \frac{d}{dt}\left(\frac{dR}{ds}\right)\left[\text{J m}^{-2}\text{s}^{-1}\right]. \tag{1.5}$$

1.2 Interactions of charged particles with matter

When a radiation field or an individual particle hits a *target* material (e.g., inside a radiation detector), it can change some measurable physical parameter of the target by transferring part (or all) of the transported energy to its atoms. Such modifications are exploited within all radiation detectors to build a measureable signal (most often, an electric signal). The interaction mechanisms by which the radiation energy is deposited in matter are diverse,[3] depending on (a) the type of radiation (i.e., charged or neutral, heavy or light particles), (b) the energy spectrum of the radiation field, and (c) the chemical and physical properties of the traversed matter.

1.2.1 Stopping power

Charged particles, such as protons, heavy ions or fast electrons and positrons interact with electrons and nuclei of the atoms in the target via electric (COULOMB) forces.

---

3 Cf. Chapter 13 of Vol. I.

The energy loss (dE) by the incident particles per unit length (dx) in the target material is described quantitatively by means of the stopping power, S:

$$S = -\frac{dE}{dx} \, [J/m] \tag{1.6}$$

The SI unit for stopping power is J/m; nevertheless, in the practice of nuclear measurements, a more common unit of measure is keV/µm.[4] Stopping power consists of two components: one relative to the energy lost by COULOMB *collisions*, denoted by S_c, and another due to *radiative* losses (i.e., *bremsstrahlung*), denoted by S_r. Thus,

$$S = S_c + S_r. \tag{1.7}$$

Stopping power is a function of several variables, and its expression is slightly different for heavier particles such as protons, α-particles and larger ions, compared to lighter particles such as electrons. For heavier particles, the *mass* stopping power (i.e., the ratio of S to the material density ρ) has the following dependence on the relevant variables:

$$\frac{S}{\rho} \approx \begin{cases} \dfrac{Z_p^2 Z_a}{\beta^2} & \text{for} \quad \beta < 0.96 \\ Z_p^2 Z_a ln(\beta^2) & \text{for} \quad 0.96 < \beta < 1, \end{cases} \tag{1.8}$$

where Z_p is the atomic number of the primary particle, Z_a is the atomic number of the target material and $\beta = v/c$ is the particle velocity. The radiative component is negligible at low kinetic energy for heavy charged particles. In the field of radiochemistry, the heavy particles of interest have low kinetic energy compared to their mass, hence most of the energy losses are due to interactions with atomic electrons which cause the slowing down of the particle along trajectories that are roughly rectilinear and aligned to the original direction[5] (see Fig. 1.1).

In contrast, for light charged particles, the radiative component of the stopping power becomes important, especially for target materials with high atomic number. A *critical* energy, E_c, is defined as the energy for which the collision and radiative stopping power are equal, i.e., $S_c = S_r$. For energies $E < E_c$, the energy lost by collision is dominant. A good approximation for the critical energy, e.g., electrons is given by eq. (1.9):

$$E_c \approx \frac{800}{Z_a} \, [MeV]. \tag{1.9}$$

---

4 1 J/m = 6.242 · 10⁹ keV/µm.
5 Collisions with nuclei, such as those observed in the pioneering experiments of Rutherford, are less frequent and are responsible for particle scattering at large angles.

HEAVY CHARGED PARTICLES

LIGHT CHARGED PARTICLES

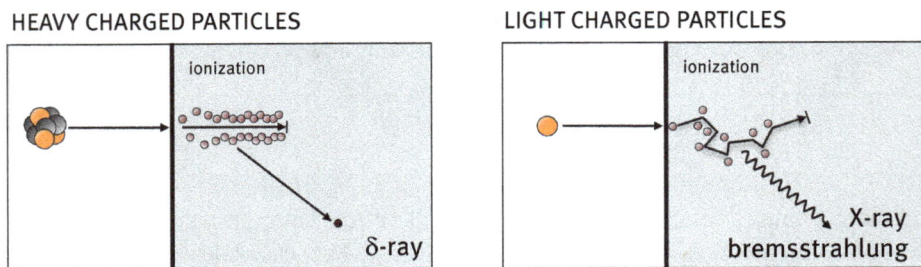

Fig. 1.1: Mechanism of energy transfer from charged ionizing particles to matter. Heavy charged particles such as α-particles slow down following a roughly rectilinear path until they reach the thermalization energy. Some knock-on collisions with atomic electrons can give rise to an energetic δ-ray (i.e., a fast electron e⁻). For light charged particles such as β-particles, the path in the target material is more irregular. The production of X-rays from "braking" radiation (bremsstrahlung) is more likely to occur in this case than for heavy particles.

The collision and radiative mass stopping power for fast electrons and positrons can be approximated as shown in eq. (1.10), where E_{kin} is the kinetic energy of the particle, and mc^2 is its rest mass:

$$\frac{S_c}{\rho} \approx \frac{Z_a}{\beta^2} \qquad \text{mass collision stopping power for light charged particles}$$

$$\frac{S_r}{\rho} \approx Z_a^2 \beta^2 \qquad \text{mass radiative stopping power for light charged particles}$$
$$\text{(approximation for } E_{\text{kin}} \gg mc^2\text{).} \tag{1.10}$$

After having lost most of its kinetic energy traveling through a medium, a charged particle could reach thermal equilibrium with surrounding atoms. The mean value of the path traveled by each particle of a charged particle beam in a target material is defined as the *range* of the particle in that medium and, as for the stopping power, is a function of the energy and charge of the particle and of the atomic number and density of the target material. The range of particles in different media are often described by means of empirical formulae (range-energy relations) derived by fitting experimental data from physical measurements. For instance, the range of α-particles in air can be calculated from the approximated formula:

$$R_\alpha \, [\text{cm}] = (0.005 \cdot E_{\text{kin}} + 0.285) E_{\text{kin}}^{3/2} \quad 4 \, \text{MeV} < E_{\text{kin}} < 15 \, \text{MeV}$$

and a good approximation for the maximum range of β-particles is

$$R_\beta \, [\text{mg/cm}^2] = 412 \cdot E_{\text{kin}}^{1.265 - 0.0954 \ln\left(E_{\text{kin}}\right)} \quad 0.01 \, \text{MeV} < E_{\text{kin}} < 3 \, \text{MeV}.$$

This latter formula, due to Katz and Penfold (1952), is independent of the target material density, as the range is expressed in terms of areal density (mg/cm²).

1.2.2 Interactions of X-rays and γ-photons with matter

Radiation without electric charge such as photons and neutrons can still ionize matter, although only indirectly. When a photon strikes an atom of the target (i.e., detector) material, it can interact with the atomic electrons (or, in certain circumstances, with the nucleus) with various mechanisms depending primarily on the photon energy $\varepsilon = h\upsilon$ and the Z_a of the material. In the photon energy range relevant for radiochemistry (i.e., from few keV to tens of MeV), mainly five types of photon-matter interactions can occur (see Tab. 1.2).

Tab. 1.2: Classes of interactions between photons and matter.

RAYLEIGH scattering
COMPTON scattering
Photoelectric absorption
Production of electron-positron pairs
Photonuclear interactions

RAYLEIGH scattering, often referred to as *coherent* scattering, is a process in which an incident low energy photon interacts with the target atom as a whole. This interaction causes a temporary excitation of the atomic electrons that are subsequently de-excited through the emission of another (scattered) photon, with the same wavelength as the incident one (cf. Fig. 1.2). Because the scattered photon has the same wavelength, and hence the same energy as the incident one, no net transfer of energy to matter arises from this type of interaction. Nevertheless, the scattered photon is emitted in a narrow cone with respect to the original direction, so that it is removed from the primary beam. Photons can have sufficient energy to ionize the target atom when colliding with its electrons; when interacting with internal electrons of binding energy E_{bind} of medium-to-high Z atoms, the photon is completely absorbed by the atom while the electron is ejected according to eq. (1.11):

$$E_{eject} = h\upsilon - E_{bind}, \tag{1.11}$$

where E_{eject} is the resulting kinetic energy of the ejected electron (referred to as a photoelectron) and E_{bind} is the binding energy of the electron orbital before the collision (cf. Fig. 1.3). This process is called *photoelectric* effect (or *photoelectric absorption*). It is evident from eq. (1.11) that for this process to occur it must be $h\upsilon > E_{bind}$.

If the binding energy of the electron involved in the collision is negligible with respect to the energy of the incident photon, there is a low probability that photoelectric absorption occurs and the interaction can be modeled as a relativistic collision between a photon and an unbound, stationary electron. This usually happens for interactions with electrons of external orbitals, and the process is called *incoherent*

primary
photon
atom
excited
atom
scattered
photon

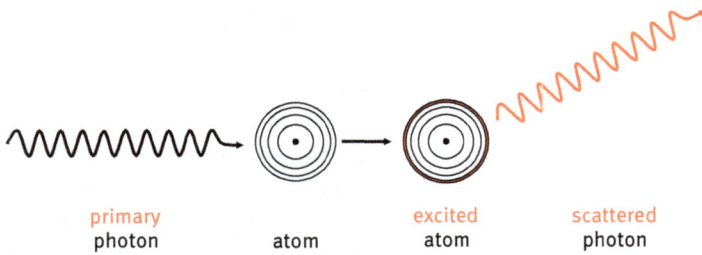

Fig. 1.2: Interactions of photons with matter: Rayleigh scattering.

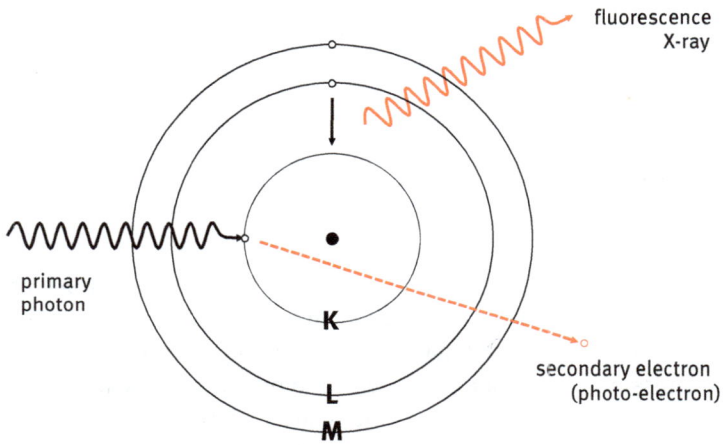

fluorescence
X-ray

primary
photon

K
L
M

secondary electron
(photo-electron)

Fig. 1.3: Interactions of photons with matter: photoelectric absorption.

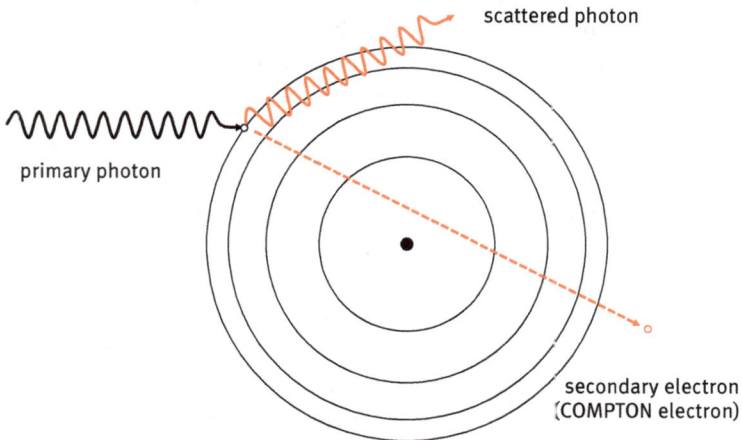

scattered photon

primary photon

secondary electron
(COMPTON electron)

Fig. 1.4: Interactions of photons with matter: Compton scattering.

(or Compton) *scattering* (cf. Fig. 1.4). The term "scattering" suggests that the incident photon is not absorbed, but instead it is scattered in a different direction. It is also called "incoherent" because each scatter event involves a single electron instead of the atom as a whole, and hence the photons coming from each event have no coherence (scattered photons have different wavelengths, and those emitted with the same wavelength have different phases). By applying conservation laws for the energy and relativistic momentum in the collision, it is possible to derive the energy $E^{scatter} = hv'$ of scattered photon as a function of the initial energy hv of the photon and the scattering angle θ.

$$hv' = \frac{hv}{1 + \frac{hv}{m_e c^2}(1 - \cos\theta)}. \tag{1.12}$$

It is also possible for an incident photon to interact with the electric field of the atomic nucleus, leading to the production of an *electron-positron pair* (cf. Fig. 1.5). This process is in agreement with the Einstein equation for mass-energy equivalence, i.e., $E = mc^2$. The production of the e^+e^- pair is only possible if the initial energy hv of the photon is greater than the sum of the rest energies of the two particles (*threshold energy* for e^+e^- pair production):

$$hv > m_{e^+}c^2 + m_{e^-}c^2 = 2\, m_e c^2. \tag{1.13}$$

For energies lower than this threshold, there is not enough (mass-)energy in the system to "materialize" the two particles; thus, pair production is not possible for $hv < 1022$ keV.[6] All the excess energy above the threshold is shared between the electron and the positron being created. In addition, the nucleus also has to share a very small fraction of recoil energy, in such a way that the fundamental conservation laws for energy and momentum are met.

At higher energies (above ca. 10 MeV), the photon can directly interact with the nucleus itself by nuclear reaction. Depending on the incident photon energy and on the target nucleus, several types of photonuclear interactions[7] are possible, such as elastic (γ, γ) or inelastic (γ, γ) scattering, photoproton (γ, p) or photoneutron (γ, n) emissions, or photofission (γ, f) (see Fig. 1.6).

---

6 To be more precise, the conservation of momentum imposes that the threshold energy for pair production must be slightly greater than 1022 keV.

7 These types of interactions are especially important at energies generally higher than those obtainable through radioactive transformation of unstable nuclides and most of them are observed in high-energy physics experiments, or in cosmic ray interaction with the atmosphere. Nevertheless, in the medical environment one must also account for (γ, n) reactions with megavoltage X-ray beams of radiation therapy linear accelerators (LINAC). Such reactions are responsible for the production of short-lived radioisotope, and especially of fast neutrons that require special attention as for the shielding design of the treatment room.

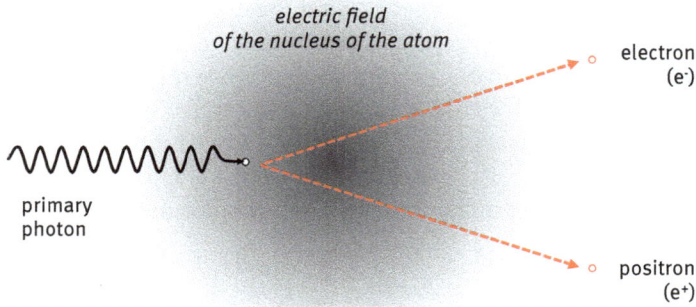

Fig. 1.5: Interactions of photons with matter: pair production.

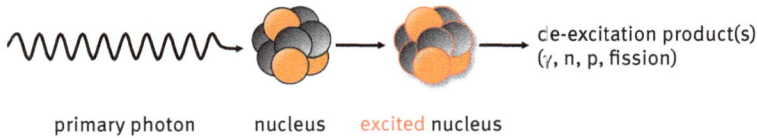

Fig. 1.6: Interactions of photons with matter: photonuclear reaction.

From a macroscopic point of view, it is often necessary to quantify the attenuation of a photon beam through a known thickness of material. The atomic or nuclear[8] cross section, σ, represents the probability for a photon to undergo that interaction on a single atom (or nucleus). The overall probability of interactions of a photon with a single atom or its nucleus is the sum of the cross sections for all the possible interactions:

$$\sigma_{\text{tot}} = \sum_i \sigma_i \, [\text{m}^2].\tag{1.14}$$

The cross section is expressed in m^2, although a common unit is the barn[9] ($1\,\text{b} = 10^{-28}\,\text{m}^2$). By multiplying the above total cross section with the number of atoms in the unit volume, one gets the *linear attenuation coefficient* μ (N_A is AVOGADRO's number, A is the mass number, and ρ the density of the chemical species).

---

8 See Chapter 13 in Vol. II for "Cross section" in nuclear reactions.
9 American physicists working on the atomic bomb during the Second World War used to say the uranium nucleus was "as big as a barn". For this reason, the term "barn" was adopted to denote an area of $10^{-24}\,\text{cm}^2$, which is approximately equal to the cross-sectional size of a uranium nucleus.

$$\mu = \frac{N_A \rho}{A} \sigma_{tot} \left[m^{-1} \right].$$ (1.15)

The linear attenuation coefficient has the dimension of the inverse of a length, so it is expressed in m^{-1} in the SI. Just like atomic and nuclear cross sections, the parameter μ depends on the incident photon energy and on Z_a and ρ of the target material.

Let us now suppose that we have a monoenergetic photon field with fluence Φ_0 impinging onto a homogeneous slab of material with linear attenuation coefficient μ and thickness x; it can be shown that the fluence Φ of all those photons that did not undergo any interaction when traversing the slab is related to the initial fluence Φ_0 by the *exponential attenuation law for photons*, cf. eq. (1.16):

$$\Phi = \Phi_0 e^{-\mu x}.$$ (1.16)

1.2.3 Interactions of neutrons with matter

Neutrons do not have electric charge but nevertheless they are able to ionize atoms in an indirect way. Instead of interacting with atomic electrons, neutrons are more likely to interact with atomic nuclei via elastic or inelastic nuclear *scattering*[10] as depicted in Figs. 1.7 and 1.8. Again, the atomic cross section of each type of interaction strongly depends on the energy of the incident neutron, as well as on the target material. For neutrons of kinetic energy in the range of keV or higher (referred to as *fast* neutrons[11]), the most important interaction is scattering; in case of *elastic* scattering (n,n), indirect ionization arises from the kinetic energy transferred from the incident neutron to the recoil nucleus. If the collision is inelastic (n,n'), the neutron enters the target nucleus and is then emitted at a certain angle with respect to the original direction; the target nucleus is now in an excited state, and the subsequent de-excitation is accompanied by the emission of a *prompt* γ-photon. Elastic collision is the main mechanism of energy loss for fast neutrons. By applying laws of conservation for energy and momentum to the reaction, it is easy to show that the highest fraction of neutron energy will be lost through elastic collision with particles of roughly the same mass (i.e., hydrogen nuclei). In general, the average energy loss for elastic collision is given by

$$\Delta E = \frac{2 E_{kin} A}{(A+1)^2},$$ (1.17)

---

10 The target nucleus may even **absorb** the neutron, leading to a nuclear reaction that can give rise to various kinds of secondary radiations (cf. Fig. 1.9).
11 Compared to thermal neutrons of 0.025 eV or cold neutrons of 10^{-3} eV or ultracold neutrons of $3 \cdot 10^{-7}$ eV. Their velocities are 2187 m/s, 437 m/s, and 7.58 m/s, respectively.

where A is the mass number of the target atom and E_{kin} is the initial kinetic energy of the neutron. It follows from eq. (1.17) that the most effective materials for slowing down neutrons (or to *moderate* them) are those with high hydrogen content.

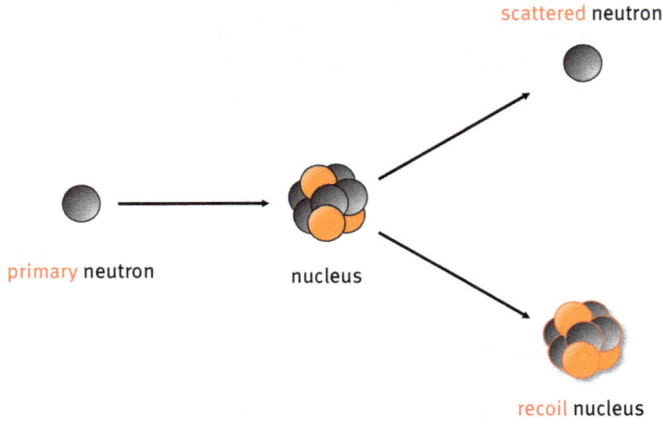

Fig. 1.7: Interaction of neutrons with matter: elastic scattering.

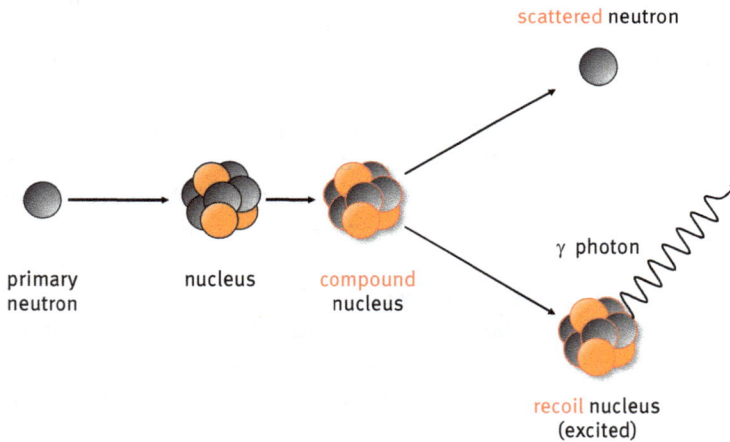

Fig. 1.8: Interaction of neutrons with matter: inelastic scattering.

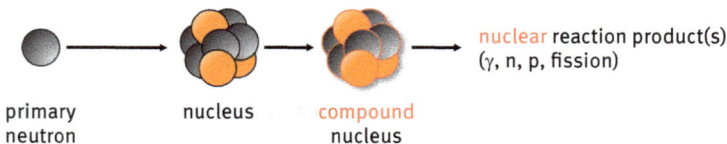

Fig. 1.9: Interaction of neutrons with matter: absorption.

For kinetic energies close to the thermal energy at room temperature (i.e., approximately $k_B T \approx 0.025$ eV for $T = 300$ K, where k_B is the BOLTZMANN constant), the absorption cross section can dominate the scattering cross section in some materials. The absorption of such thermal neutrons can lead to nuclear reactions such as the ejection of heavy charged particles from the target nuclei by spallation or fission, or transmutation of the target into a radioactive nuclide. The absorption cross section of *thermal* neutrons is particularly high for Boron-10. Nuclear reactions like $^{10}B(n,\alpha)^7Li$ or $^3He(n,p)^3H$ are often exploited in neutron detectors. The capture cross section is even higher for Cadmium-113 in the $^{113}Cd(n,\gamma)^{114}Cd$ reaction. The detection of secondary radiation (i.e., α, β, γ) emitted following neutron absorptions is one of the methods for thermal neutron detection, cf. Fig. 1.10.

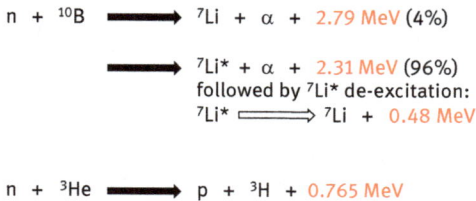

$$n \; + \; ^{10}B \; \longrightarrow \; ^7Li \; + \; \alpha \; + \; 2.79 \text{ MeV } (4\%)$$

$$\longrightarrow \; ^7Li^* \; + \; \alpha \; + \; 2.31 \text{ MeV } (96\%)$$
followed by $^7Li^*$ de-excitation:
$$^7Li^* \; \Longrightarrow \; ^7Li \; + \; 0.48 \text{ MeV}$$

$$n \; + \; ^3He \; \longrightarrow \; p \; + \; ^3H \; + \; 0.765 \text{ MeV}$$

Fig. 1.10: Examples of pathways of neutron capture reactions useful for the detection of thermal neutrons. The star (*) denotes an excited nuclear status.

Fast neutrons are registered by a two-step process: first, the fast neutrons are moderated with light materials (e.g., polyethylene), then the resulting slow neutrons are detected by means of absorption reactions as explained above (see Fig. 1.11). From a macroscopic point of view, the overall change of the neutron flux when it traverses a slab of material with thickness x can be described with a decreasing exponential law, pretty similar to that used for photon attenuation (*exponential attenuation law for neutrons*):

$$\Phi = \Phi_0 e^{-N\sigma_{tot} x}, \tag{1.18}$$

where $N = \rho(N_A/M)$ is the atomic density in the material occupying the active volume of the detector and σ_{tot} is the total atomic cross section, ρ is its mass density (kg/m^3), M is its molar mass (kg/mol), and N_A is AVOGADRO's number.

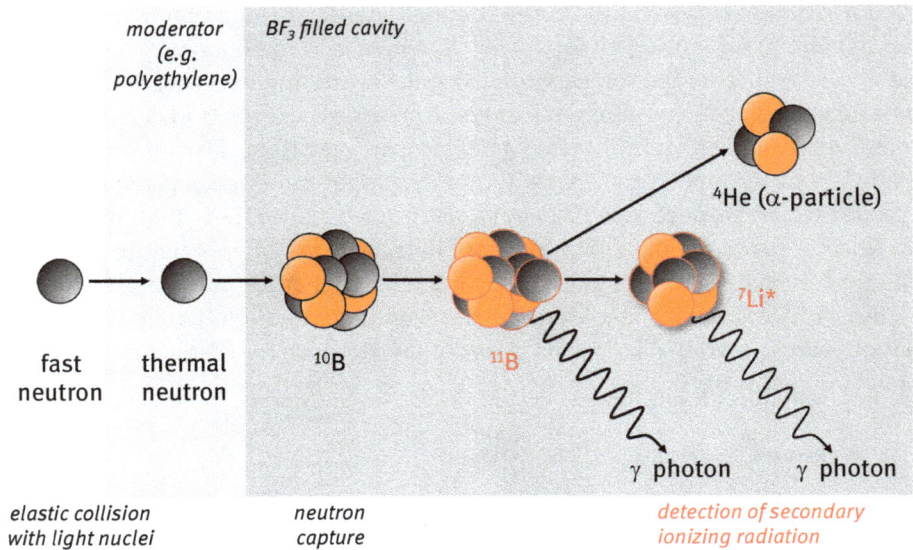

moderator (e.g. polyethylene)   *BF₃ filled cavity*

^{4}He (α-particle)

^{7}Li*

fast neutron   thermal neutron ^{10}B ^{11}B

γ photon γ photon

elastic collision with light nuclei   *neutron capture*   *detection of secondary ionizing radiation*

Fig. 1.11: Principles of detecting initial neutron radiation via secondary α-radiation as induced through neutron capture nuclear reactions. This figure refers to the particular case a of gas-filled BF₃ detector, exploiting the boron-neutron capture reaction.

1.3 Radiation detectors

1.3.1 Generalities about radiation detectors

The interaction mechanisms described in the previous section are exploited in radiation detectors to obtain a measurable signal that can be related to some physical parameters of the radiation field under study, mainly the kind of radiation, its energy and intensity. Many different radiation detection systems are in use. In fact, some detectors can be very efficient in revealing a given type of radiation and in a given energy range, but at the same time they can be completely insensitive to others. For instance, thin scintillator layers coupled to silicon photodiodes are widely used to detect low to medium energy X-rays used in diagnostic imaging, but they are almost completely insensitive to high energy γ-rays emitted from radionuclides such as ^{137}Cs or ^{60}Co. Yet, some kind of detectors may have excellent performance in measuring a given physical parameter of the radiation field under study, but they are unable to measure other parameters of the same type of radiation. As an example, a semiconductor detector (such as hyperpure germanium detectors – HPGe) is the best choice for measuring energy spectra of γ-radiation, but one must use ion chambers to make an absolute measurement of the energy transferred to the matter from that radiation. In other words, there are no detectors that could register all the physical parameters of a radiation field at the same time.

This section aims at giving an introduction to the various systems for detecting ionizing radiations, so as to figure out which system is better suited for a specific application. Important parameters that are common to all radiation detection systems are listed in Tab. 1.3.

Tab. 1.3: Parameters most relevant for the choice of radiation detectors.

Detection efficiency	Fraction of detected particles over all particles passing through the detector, depending on the type of radiation and on its energy.
Detector dead time	Fraction of time in which the detector is insensitive to radiation, due to physical processes following the detection of an ionizing particle (relevant for detectors operating in pulse mode, or counting mode).
Energy resolution capability	The ability of a radiation detector to discriminate the energy of the incoming radiation, especially important in spectrometers.
Energy dependence	This parameter takes into account the change of detector efficiency for different energy of incoming radiation. A flat energy response is desired in radiation dosimeters.

1.3.2 Detection efficiency: Intrinsic vs. geometric vs. absolute

Ideally, a "perfect" detector embedded in a radiation field should be able to gather all the particles passing through its sensitive volume.[12] This is not possible for all detectors, especially in case of sparsely interacting radiation such as photons and neutrons. The *intrinsic* efficiency of a detector, ε_{int}, is the ratio of the number of particles detected to the total number of particles passing through the detector volume:

$$\varepsilon_{int} = \frac{\text{number of particles detected}}{\text{number of particles passing through the detector}}. \tag{1.19}$$

In most cases, it is desirable to use a detector with the highest possible intrinsic efficiency. However, there are also cases in which the intrinsic efficiency should be low, for instance when the process of radiation measurement must give a negligible

---

12 By "sensitive volume", or "active volume", we refer to that specific portion of the detector where the energy of the incident radiation field is converted to some different form of energy, which is then used to produce a measurable signal. For instance, many types of portable radiation detectors come with handles and LCD screens, but those parts are not in their sensitive volume (which may be a gas filled cavity or a piece of crystal, as we will see later). The term "active area" is sometimes used to describe the cross-sectional area of the active volume, which is typically determined by the size and shape of the so-called "window" of the detector.

perturbation to the radiation field being measured. The intrinsic efficiency of a detector is an important parameter that must be accurately determined prior to the real measurements. This is commonly done by means of calibration sources, i.e., radioactive sources with suitable geometry and type of emission (depending upon the type of detector to calibrate) whose activity is known with good accuracy. Using such sources, the "ideal" response of the detector can be calculated theoretically and then compared to the experimental one to estimate the intrinsic efficiency of the detector.

Suppose that only a small fraction of the particles being emitted by a source are directed to the active area of the detector: in such a situation, it is clear that even a detector with an intrinsic efficiency of 100% would not be able to record all the particles emitted by the source. To take into account such an effect, the *geometric* efficiency of a detector, ε_{geom}, represents the ratio of the number of particles passing through the detector to the total number of particles emitted by a source:

$$\varepsilon_{geom} = \frac{\text{number of particles passing through the detector}}{\text{number of particles emitted by the source}}. \tag{1.20}$$

For a given experimental setup, the geometric efficiency can be derived by analytical calculations. As an example, let us suppose that a given detector is seen from an isotropically emitting point-source of radiation with a solid angle Ω; in this case, we have $\varepsilon_{geom} = \Omega/(4\pi)$ (remember that an isotropic source has an emission solid angle of 4π steradians). By multiplying the geometric by the intrinsic efficiency, we obtain the *absolute* efficiency of a radiation detector, eq. (1.21):

$$\varepsilon_{abs} = \varepsilon_{int} \cdot \varepsilon_{geom}. \tag{1.21}$$

1.3.3 Dead time

Most of the considerations given so far regarding radiation measurements seem to be strictly connected to the task of "counting" the particles of a radiation field, i.e., measuring the passage of each particle as an individual event. All radiation detectors designed to accomplish this task are referred to as *pulse mode* detectors. There are also cases (especially when the rate of interaction of the radiation within the detector is so high) that it is no longer important to consider the single collision of each particle with atoms in the active volume as an individual event. In these cases, the so-called *current mode* detectors work in such a way that the output is an integrated signal built upon the integrated energy released by the field to the detector.

In detectors operating in pulse mode, after each collision (or "hit") recorded there is always a certain amount of time in which the detector does not respond to any other event. Such a time interval is defined as the *dead time* of the detector. The dead time is somehow related to the "duration" of the signal generated by the detector as a consequence of the detected event. This duration is a complex function of both the detector and the particle to be detected. Whenever rather high activities, i.e., high particle fluences need to be measured, the time of nonresponsiveness of the detector can constitute a high fraction of the total measurement time. As a consequence, the detector reports just a fraction of the real activity. A typical example occurs when using GEIGER–MÜLLER counters (described in Section 1.4.1) near radioactive sources of high activity. The dead time of a detector thus must be determined in specially designed calibration experiments.

1.3.4 Energy resolution

Some detectors are designed to measure the energy spectrum of the radiation field. This allows analyzing the radionuclide composition of a radioactive material. In this context, a radiation detector is referred to as a spectrometer. In most cases, the α-radiation or X-ray and γ-radiation emitted by radionuclides in terms of energy and abundance provides sufficient information to determine each species present in a radioactive source. Beside the efficiency, already described above, the most important parameter to evaluate the performance of a spectrometer is the energy resolution. It is defined as the minimum energy difference that two signals[13] must have in order to be distinguished by the spectrometer. The energy resolution (R_{en}) is usually defined as the percentage ratio of the measured line width (expressed as the full width at half of its maximum, or FWHM) to the absolute energy of the line (E_0):

$$R_{en} = \frac{\text{FWHM}}{E_0} \cdot 100\%. \tag{1.22}$$

1.3.5 Energy dependence

In almost all radiation detectors, there is a complex relation between the energy carried by the radiation, the energy released within the detector volume and the detector reading (i.e., the output value of a detector measurement). There are situations in which it would be very important to have a good direct proportionality (i.e., linearity) of the detector reading to the energy released within the detector volume. This is especially important in the field of radiation dosimetry, for instance, for radiation protection purposes. In this context, radiation detectors are called *dosimeters*. In all radiation dosimeters the actual reading depends on the energy of the radiation under study. As explained later in Section 1.7 about individual monitoring, there are several "dosimetric quantities" that can be measured

---

13 The word for a signal obtained from a spectrometer is typically "peak" or "line".

by mean of dosimeters, depending on whether we need, e.g., to estimate the risk of an individual exposed to a radiation field, or just measure the absolute amount of radiant energy deposited per unit mass.[14] A common problem in dosimetry is related to the dependency of detector calibration on the energy of the incident radiation. Supposing a monoenergetic radiation field of a given energy E is measured, let us denote by $G(E)$ and $M(E)$ the true value of the dosimetric quantity (i.e., as measured by a perfect, ideal dosimeter) and the actual dosimeter reading, respectively. The *energy response* of a detector (R_G) here is the ratio between the two above values, as a function of energy E:

$$R_G = \frac{M(E)}{G(E)}. \tag{1.23}$$

Appropriate dosimeters offer a "flat" energy response, i.e., the weakest possible dependence on the energy. Detectors used in radiation dosimetry are often calibrated at a specific energy (for photons this is usually the energy of the gamma emission of ^{137}Cs at 662 keV).

1.4 Active systems for radiation detection

Radiation detection systems use the ability of ionizing particles to interact with matter in order to produce a measurable signal. The following two sections provide an overview of the main classes of radiation detectors, each of which differs from the others for the particular choice of the "active" material (i.e., the material where the conversion of the radiation energy into some other form of directly measurable signal takes place) and for other design considerations (e.g., geometry, mode of operation, entrance window, and so on) that can make them suitable for a specific application, radiation type and intensity. In short, there is no single radiation detector in the world that can be said to be "the best" for everything.

This section focuses on detectors that are able to generate an output in realtime (active detectors). The next section will give an overview of the most common types of detector that need a special processing after their exposure to radiation to produce a readable output (passive detectors).

---

14 The main dosimetric quantity, called *absorbed dose*, D, is defined by the International Commission on Radiological Protection (ICRP) as the expectation value of the energy imparted by the radiation into a target volume, divided by its mass. The special SI unit of the absorbed dose is the Gray (Gy), corresponding to 1 J/kg. Dosimetry is discussed in detail in Chapter 2.

1.4.1 Gas detectors

In most cases, radiation detection is based on the measurement of the electric charge or current induced by collective ionization of the atoms in the active volume of the detector, by both the primary particles of the incident radiation field and by its secondary radiation. When the detecting material is gas, each ionization event produces an electron-ion pair. Positive and negative charges can be then separated by an external electric field and collected on the electrodes, thus producing an electric current. This is the working principle of a gas detector, which could be represented schematically as a simple electrical circuit with a resistor and a capacitor (hence called RC circuit) with an amperometer in series (cf. Fig. 1.12). The "sensitive" element of this detector is the gas volume that is placed between the capacitor electrodes. Whenever the gas atoms are ionized by the incident radiation, the electrons and the ions are accelerated by the electric field towards the electrode of opposing sign. An electric current is "injected" into the circuit that can be measured by the amperometer. Depending on their operational design and experimental configuration, gas detectors allow absolute measurements of the energy transferred by the radiation to the gas atoms in the sensitive volume. W_e denotes the average energy required to produce an electron-ion pair in the detector gas. If each electrode collects a total electric charge Q during a measurement, then the total energy transferred by the radiation E_{tr} is given by

$$E_{tr} = nW_e = Q\frac{W_e}{e}, \qquad (1.24)$$

where $n = Q/e$ is the number of electron-ion pairs produced by ionization, and $e = 1.6 \cdot 10^{-19}$ C is the absolute value of the electron charge. The value of W_e can be measured in each type of gas: it is slightly greater than the average ionization potential for the atoms of the gas itself. As an example, in air at standard temperature and pressure, W_e is 34 eV per ion pair.

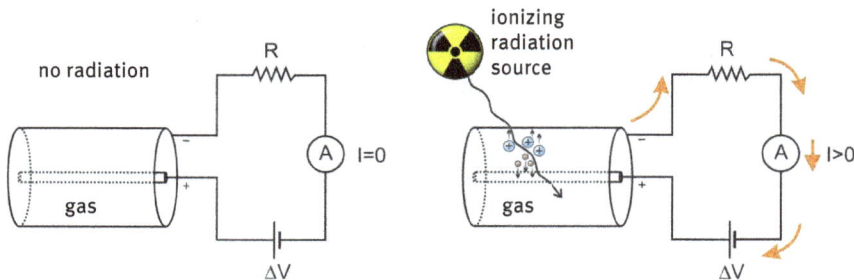

Fig. 1.12: Principle of a gas detector. Left: In absence of radiation, no current flows in the circuit. Right: An ionizing particle enters the sensitive volume of the detector and several electron-ion pairs are produced (depending on the energy lost by the particle inside the detector). Electric charges are then collected on the electrodes and a current starts flowing in the circuit.

The electric field between the electrodes must be strong enough to avoid the recombination of electrons and ions in their path along the electric lines of force. In practice, a small recombination fraction is always present; nevertheless, when the potential difference between the two electrodes reaches a sufficiently high value (which depends mainly on the detector geometry and on the type and pressure of the gas), the recombination fraction becomes negligible with respect to the total amount of ionization charge: the detector is said to be in saturation regime. The saturation regime is the working regime of the *ionization chamber*.

By increasing the voltage between the two electrodes, and hence the strength of the electric field, the electrons produced by ionization are likely to acquire sufficient kinetic energy in their path to produce further ionization inside the gas. This is the mechanism of charge multiplication that can be exploited to amplify the electrical signal at the electrodes (also called *Townsend avalanche*). The range of electrode bias voltage in which the charge multiplication is linear is called "proportional region"; this mode of operation is used in the *proportional* counters. Conversely to *ionization* chambers, which are usually operated in *current* mode, proportional counters are operated in *pulse* mode (as the name "counter" suggests). The linearity of the charge multiplication process allows discriminating the energy of the incident particle. Moreover, particles with different range inside the gas lead to different shapes of the electric pulse at the electrode, thus allowing for pulse-shape discrimination among different types of radiations.

When the bias voltage is increased further, the linearity of the charge multiplication is first reduced and then completely lost, reaching a regime in which a single ionization event in the gas produces a continuous discharge (the GEIGER *discharge*). Gas detectors operating in the Geiger region are referred to as GEIGER–MÜLLER counters (or GM counters).[15] The main advantage of detectors operating in the GEIGER region is the very high amount of electric charge collected at the electrode, leading to less demanding circuitry for signal amplification. On the other hand, after the GEIGER discharge is initiated the GM counter can detect no other particle until the discharge is terminated. Typical dead times for these detectors are in the order of 100 μs that reduce the counting efficiency at high count rates, thus suggesting their applicability to radiation fields with low fluence rates.

---

15 The GEIGER discharge is a complex phenomenon, which involves also the generation of UV photons from the gas molecules within the avalanche; these photons could be reabsorbed by other molecules, thus creating other free electrons and, in turn, other discharges. The above chain reaction must be avoided for a GM counter to work properly; for this purpose, the gas inside the detector is mixed with a component allowing for the quenching of the discharge.

1.4.2 Scintillation detectors

Some materials have the peculiarity to emit visible light, or to scintillate,[16] when struck by ionizing radiation. This phenomenon is widely used to detect radiation in an indirect way. Detecting radiation with scintillators is basically a two-step process. First, several visible light photons are produced by the interaction of the radiation field with the scintillating material, in a quantity that is proportional to the energy deposited by the primary radiation. In the second step, the visible light is converted into an electric signal by a dedicated instrument, such as a photomultiplier tube (PMT) or a solid-state photodetector, optically coupled to the scintillator. Figure 1.13 shows a typical scintillator detector that can be commonly found in a radiochemistry department.

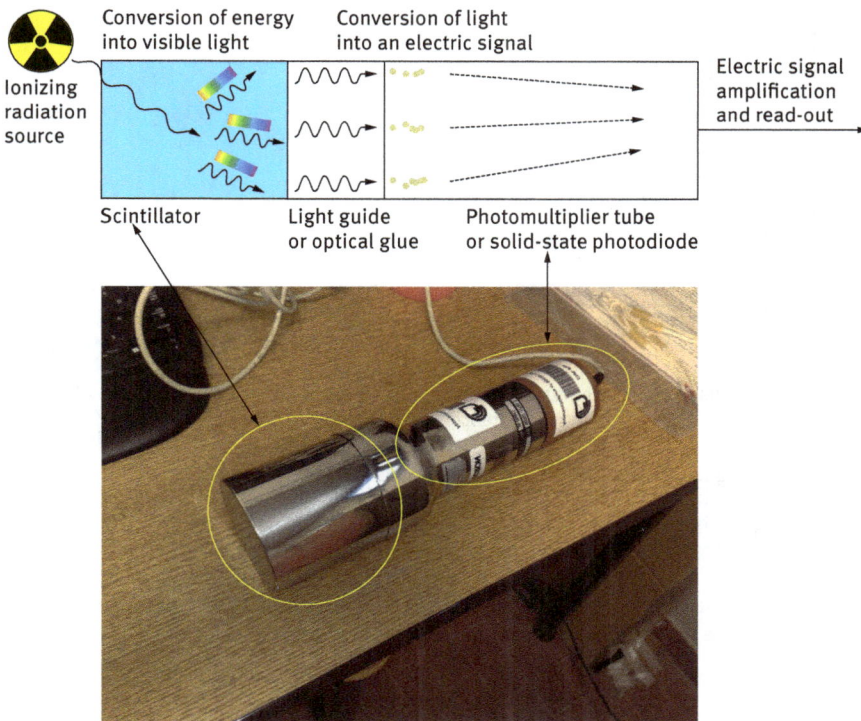

Fig. 1.13: General configuration and working principle of a scintillator detector. The photograph shows a widely used 3″ × 3″ NaI:Tl detector with a PMT (Ortec, Oak Ridge TN, USA). Both the crystal and the light guide are enclosed inside a light-tight envelope. In some case, a solid-state photodiode is used instead of the PMT to convert the visible light into an electrical signal.

---

16 Scintillation by definition is the emission of a flash of light after the passage of an ionizing particle.

Scintillator materials can be organic or inorganic. In organic materials, the scintillation process begins with the release of enough energy from the primary radiation to promote (at least) one valence electron to a higher energy level in the π-molecular orbitals that characterize these compounds. A typical energy level structure of an organic scintillator is shown in Fig. 1.14. A key property of the π-molecular orbitals is the presence of a fine structure, related to different vibrational states of each energy level. The de-excitation of the electron to the ground state energy level usually involves an excited vibrational level. Thus, the energy of the visible photon associated with this de-excitation is lower than the excitation energy. As a result, the material is transparent to the visible photons emitted in the scintillation process. Organic scintillators can be crystalline solids, or they can be dissolved in an organic solvent that, in turn, can be either solid (plastics) or liquid.

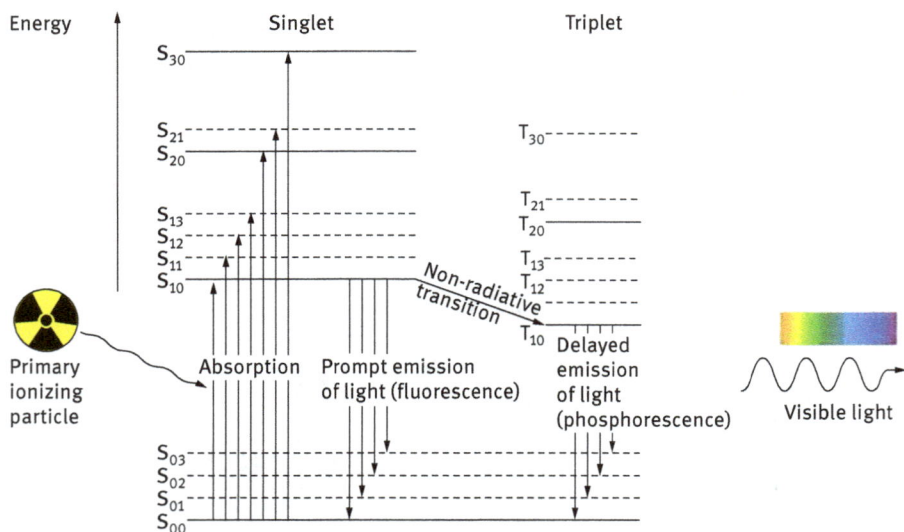

Fig. 1.14: Typical scheme of the energy level structure of an organic scintillator. Each energy level has a fine structure due to the different vibrational levels (drawn as dashed lines).

Inorganic scintillators are crystalline solids in which a small amount of a suitable impurity (or activator) can be added to obtain better light yield and time response. The scintillation mechanism can be understood by looking at the energy band configuration of a crystalline solid with activator impurity (Fig. 1.15). The excitation of several electrons from the valence band to the conduction band by the incident radiation leads to the creation of an equal number of electron-hole pairs. Before recombination, an electron-hole pair can migrate to an activator site where the hole can ionize the activator atom. After that, the ground state of the activator is left empty. At the same time, a free electron in the conduction band can fall nonradiatively into an excited state of the same activator atom in such a way that it can de-excite to the ground level with the emission of a photon whose frequency is typically within the visible spectrum. Because both the ground and excited state of the activator are within the forbidden band of the host crystal, the energy difference is lower than the band gap and hence the crystal cannot reabsorb the emitted photon.

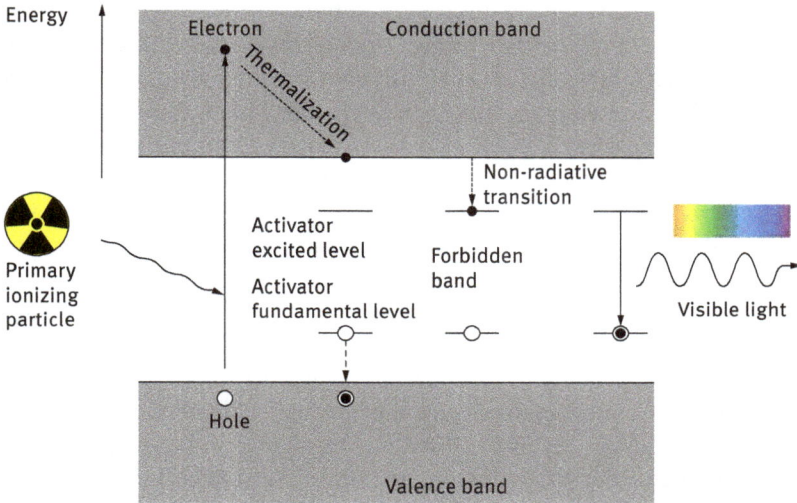

Fig. 1.15: Typical scheme of the energy band structure of an inorganic scintillator with activator impurities. The activator levels lie within the forbidden band.

Important parameters of choice for scintillating materials are: (1) the time shape of the light pulse produced in a scintillation event; (2) the number of visible photons produced per unit energy deposited by the primary radiation; and (3) the stopping power (for charged particles) or attenuation coefficient (for photons) of the crystal. Also the spectral characteristics of the visible light produced and the refractive index of the material can be important, as they should match those of the PMT or the photodetector to avoid an undesirable loss of counting efficiency. Ideally, the scintillation light would be produced by an instantaneous "burst" of visible photons, immediately after the primary particle has interacted with an atom inside the detector.

A drawback of organic scintillators is the relatively low atomic number, and hence a low linear attenuation coefficient that hampers the detection efficiency of medium to high-energy γ-photons. Conversely, inorganic crystals such as thallium-activated sodium iodide (NaI:Tl), sodium-activated or thallium-activated cesium iodide (CsI:Na, CsI:Tl) or pure bismuth germanate (BGO) contain atoms with higher atomic number which lead to higher detection efficiency. In general, inorganic crystals have slower response, in the order of tens to hundreds of nanoseconds for the fluorescence component. Moreover, they can be hygroscopic (NaI is a common example of a highly hygroscopic crystal), with practical problems in terms of manufacturing and shaping.

1.4.3 Semiconductor detectors

Some materials have intermediate characteristics in term of electrical conductivity, so that they can be considered neither conductors nor insulators: these materials are called semiconductors. Like other crystalline solids, the energetic configuration of the electrons in semiconductors is structured in "bands" instead of discrete levels, pretty similar to that shown in Fig. 1.15 for scintillating inorganic crystals. The width of the forbidden energy band in such materials (e.g., silicon or germanium are good examples of semiconductors) is such that, even at room temperature, a small number of electrons are in the conduction band and thus they are free to move when exposed to an electrical field. As a consequence, an equal number of holes, acting as positive charge carriers with lower mobility than electrons, remain in the valence band. These holes can also move in the semiconductor (even though they move less easily than free electrons of the conduction band) as other electrons moving across the lattice can recombine with them. A nice advantageous property of these materials is the possibility of "modulating" the overall number of available positive and negative charge carriers by adding impurities (called dopants in this context) in the bulk crystal. Dopants leading to an excess of electrons give rise to an n-type (or, equivalently, n-doped) semiconductor, whereas a dopant leading to an excess of holes will produce a p-type semiconductor. The huge developments seen in the electronic components industry in the last decades led to a corresponding increase in the refinement of manufacturing processes for semiconductors (of which solid-state transistors are made). For the purpose of radiation detection, the most important characteristics of semiconductors is that when placing n-type and p-type crystals in contact an electric field is formed at the p-n junction that causes all the available charge carriers to drift away from the junction (see Fig. 1.16). The so-formed depletion region (i.e., the region in which no free charge carriers are present at the steady state) can be used as a solid-state variant of an ionization chamber. In fact, the atoms in the depletion region play a role similar to that of the gas atoms in an ionization chamber (electrically neutral when no radiation is coming through it). When an ionizing particle with sufficient energy strikes an atom in the depletion region, an electron-hole pair is then created (in gases, we would have an electron-ion pair); the electric charge created is then collected on electrodes placed at both sides of the p-n junction. Because the silicon p–n junction is at the base of the solid-state diode, widely used in electronics, detectors operating with the principle described above are called silicon diode detectors. Unlike diodes commonly used in solid-state circuitry, diode radiation detectors are operated in reverse bias (i.e., the n-type side is placed at a positive electric potential with respect to the

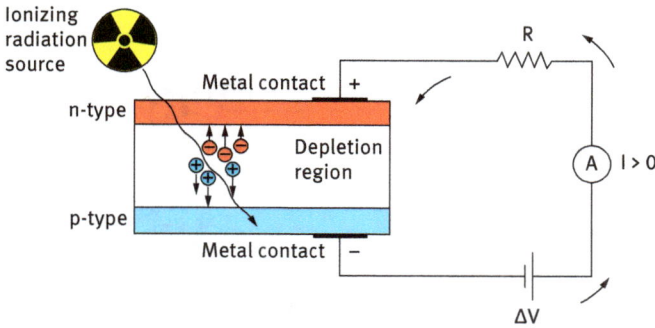

Fig. 1.16: Working principle of a semiconductor detector. See text for details.

p-type side). Reverse biasing increases the depth of the depletion region and re-duces the probability of electron-hole recombination.

There are many advantages of using silicon diodes instead of gas detectors. Among them, the higher density of solids can improve the detection efficiency by several orders of magnitude as compared to gases. Moreover, the energy required to form an electron-hole pair in common semiconductors is substantially lower than that required to create an electron-ion pair in gases (roughly 3.6 eV in Si and 2.9 eV in Ge). As a consequence, the number of charge carriers produced for a given amount of energy transferred by the radiation to the detector can be very large in semiconductors, thus leading to high performances in terms of energy resolution.[17]

Besides silicon diodes, another widely used semiconductor detector is the high-purity germanium (or HPGe) detector. In HPGe, larger volumes of depleted semiconductors are achieved by (1) manufacturing germanium crystals with very low concentrations of impurity, and (2) by placing the crystal at liquid nitrogen temperature (77 K) in order to reduce the number of free electrons in the conduction band, too high in number at room temperature. The electrodes are then created by adding very thin layers of n-type and p-type Ge at opposing edges of the pure crystal. The large depleted volume created in this way, added to the higher atomic number and density of germanium, is such that the detection efficiency for high-energy γ-photons is higher than that of silicon diodes (but lower than that of high Z scintillators). The most important characteristic of the HPGe detector is the out-standing energy resolution, so that they are used primarily in γ-spectroscopy. On the other hand, they have the drawback that they cannot be operated at room temperature, so a cryostat with a reservoir of liquid nitrogen in a Dewar must be constantly refilled in normal operative conditions (see Fig. 1.17).

---

17 $R_{en} \propto n_c^{-1/2}$, where R_{en} is the energy resolution, cf. eq. (1.22), and n_c is the number of charge carriers.

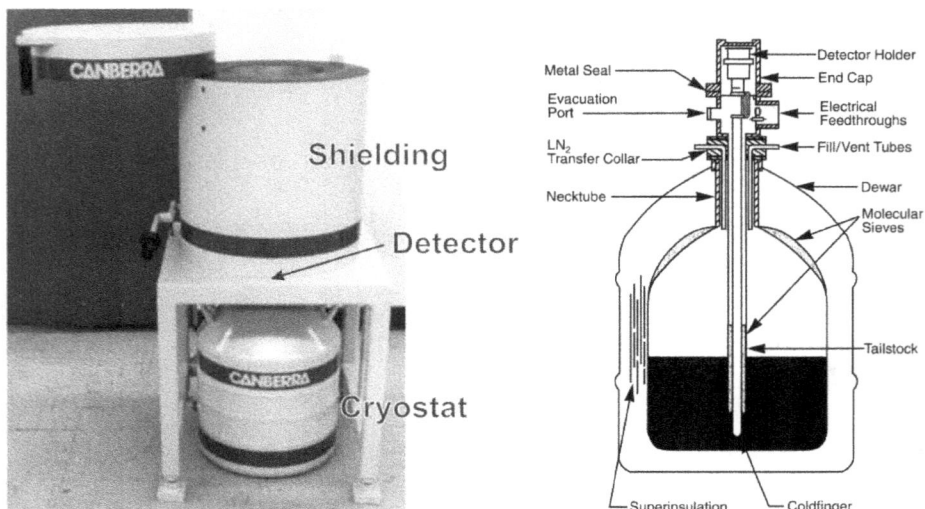

Fig. 1.17: A typical configuration of an HPGe detector with liquid nitrogen cooling. (Canberra Industries Meriden CT, USA; adapted from www.canberra.com – last access on June 4, 2012).

1.5 Passive systems for radiation detection

1.5.1 Thermoluminescent (TL) detectors

Unlike scintillator detectors, emitting most of the visible light within a short time after the interaction with the ionizing particle, there are also materials that can "store" the information about the absorbed energy without the emission of any measurable signal until they are properly stimulated. This allows including small pellets of such materials in miniaturized badges, which will integrate the absorbed energy over relatively long periods of time (usually a few months) before reading them with dedicated instrumentation. Inorganic crystals such as lithium fluoride (LiF) or calcium fluoride (CaF) can be doped with a suitable impurity (mostly manganese or magnesium) in order to create a spurious energy level within the forbidden gap. These levels act as "traps" for electron or holes, depending on whether they are close to the conduction band or to the valence band, respectively. Similarly to scintillators, the energy transferred by ionizing radiation can excite several electrons to the conduction band, thus creating several electron-hole pairs. Both electrons and holes can migrate to an impurity site in the crystal, where they can fall in trap levels by nonradiative processes, cf. Fig. 1.18. The trap level must be deep enough that an electron (or a hole) inside the trap has a low probability of escaping at room temperature. At the end of the measurement time, the crystal can be placed in a special reader where its temperature is increased in a controlled "sequence"; in this way, each trapped electron can be re-excited to the conduction band by the increased thermal energy. When the electron encounters a hole trap, it recombines with the hole by emitting a visible photon. The visible light emitted during the heating process (referred to as thermoluminescence) can be collected using a photomultiplier tube, which converts the light in an electric signal that can be digitized, displayed, recorded and quantified.

A relevant property of thermoluminescent detectors is the linearity of response with respect to the absorbed energy integrated over the measurement time. For this reason, and for the possibility of manufacturing them as small chips, they are widely used as dosimeters for individual monitoring.

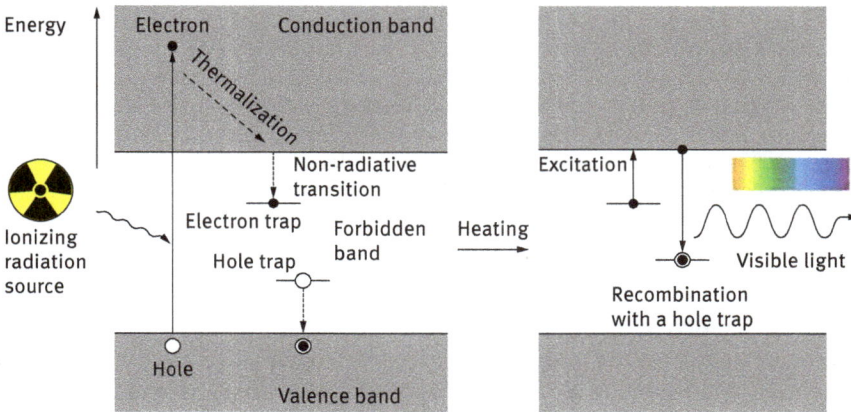

Fig. 1.18: Working principle of a thermoluminescent detector. See text for details.

1.5.2 Nuclear track detectors (NDT)

The high ionization density produced by heavy charged particles, such as α-particles or fission fragments, can leave microscopic damages on some dielectric materials along their path. These dielectric damages are associated with the rupture of chemical bonds, which can occur in dielectric solids with crystalline or polymeric structure. By means of a strong acid or base attack, referred to as an etching[18] process, the latent and hardly visible track left by the nuclear particle in the solid (of few nanometers in diameter)

---

18 The etching time can vary from few minutes to more than one hour, depending on the material used. There is not yet a comprehensive and fully satisfactory theory of the physical mechanisms that lead to the etchability of a latent track. The preferential etchability of the material along the latent track is thought to be due to local defects caused by thermal spikes generated along the track by electron collision cascades, or directly by atomic displacements due to locally intense electric fields arising from the ionization burst near the particle path (the so-called "ion explosion spike").

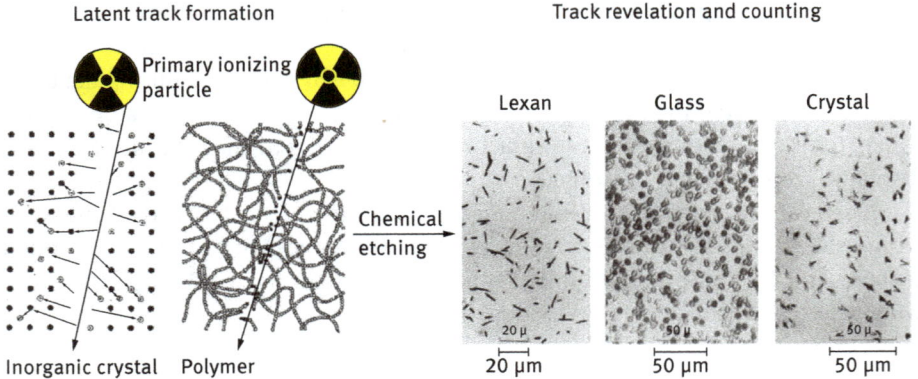

Fig. 1.19: Radiation detection by solid state nuclear track detectors (SSNTD). Left: the mechanisms of latent track formation are depicted for both inorganic crystals and polymers. Right: three photographs obtained with an optical microscope show different track patterns and shapes from fission fragments of ^{252}Cf. This type of detector is also used for indirect neutron detection through (n,p), (n,α), or (n,f) reactions. (Adapted from Fleisher et al. Nuclear Tracks in Solids. University of California Press, Berkley, USA, 1975).

can be significantly enlarged. Subsequently, the track becomes detectable by optical microscopes or, in some cases, the naked eye. This process is illustrated in Fig. 1.19.

Regardless of the microscopic model of damage formation, any suitable solid material for nuclear track recording must be (1) homogeneous, (2) dielectric, and (3) with a low atomic mobility. Common materials used as solid-state nuclear track detectors (SSNTD) are inorganic crystals (such as quartz or mica) glasses, or polycarbonate plastics (such as Lexan™ or Mylar™). The allyl diglicol polycarbonate, or CR-39 (already used to manufacture eyeglass lenses) has gained success as an SSNTD in the last decade for its good detection performance. The mass stopping power S/ρ of the SSNTD must be greater than a threshold (specific of each material) in order that the nuclear particle can be detected. For instance, CR-39 is very sensitive of the detection of α-particles since it has a threshold of $0.05\,\text{MeV}\,\text{mg}^{-1}\,\text{cm}^{-2}$ which is about two orders of magnitude lower than that of Lexan. Moreover, the etching process will be effective only if the entrance angle of the particle with respect to the detector's surface is greater than an angle θ_c (referred to as the "critical angle"). Both the detection threshold and the critical angle depend on the material type.

After the etching process, several techniques can be used to count, analyze and record nuclear tracks on SSNTDs. Optical microscopes are widely used for this purpose in all cases in which the etched track diameter is few micrometers or bigger. Modern microscopes with digital imaging sensors allow a direct conversion and storage on PC of the observed image, and the subsequent analysis with dedicated software. Alternative (and more complex) track revelation methods, more suitable for track diameters of tens to hundreds of nanometers, are scanning electron microscopy (SEM), transmission electron microscopy (TEM) or small angle X-ray scattering (SAXS).

1.6 Applications of radiation detectors

1.6.1 Absolute determination of activity

Perhaps the most important application of radiation detectors in radio- and nuclear chemistry is the accurate and reproducible measurement of the amount of radioactive substance contained in a sample. This task can be accomplished by absolute methods, but in most cases it is done by means of relative methods in which the detector output given by the sample is compared to the detector output of a known (calibrated) activity. In general, some a priori information is required such as the type of radionuclide that is being measured, along with its transformation scheme.

Activity measurements are performed very often with detectors operating in counting mode. Because the activity is defined as the number of nuclear transformations per unit time, and because each single transformation can lead to the emission of one or more ionizing particles of identical or different type (e.g., α and/or β and/or γ, etc.), one could relate the detection rate of such particles to the activity in the sample. The detecting rate or count rate[19] is expressed in counts per second (cps). It is very important to know with the highest accuracy the absolute efficiency of the detector, as well as the ratio of a specific emission X of the particle that is actually detected by the instrument chosen. If C is the detection rate (or count rate), we will have $C = \acute{A}\,\varepsilon_{abs} \cdot \varepsilon_{emis}$ where ε_{emis} is the relative abundance (or yield) of the emission under consideration; thus, it is straightforward to derive the activity $\acute{A}$ via eq. (1.25):

$$\acute{A} = \frac{C}{\varepsilon_{abs} \cdot \varepsilon_{emis}} . \tag{1.25}$$

The above equation allows the absolute measurement of the absolute activity in a sample. As already stated, in most cases the actual detection rate or count rate C is just related to the count rate C_{cal} of a source with a known activity $\acute{A}_{cal}$. It is very important that the radiation emitted by the calibration source has a similar energy spectrum to the measured sample, and also a comparable activity (not always possible in practice), to avoid inaccuracies due to possible nonlinearity of the detector. In this case, one can write

---

19 Cf. Chapter 4 in Vol. I for the relationship between relative activities to absolute activities, Fig. 5.11 and eq. (5.11).

$$\acute{A} = C \frac{\acute{A}_{cal}}{C_{cal}}. \tag{1.26}$$

For example, determination of absolute activities of, e.g., a source of ^{131}I would proceed as follows:

1. Identify the most relevant γ-energies of ^{131}I and their abundances (data are listed in Tab. 1.4).[20]
2. Determine the detector efficiency at some distances between source and detector for the relevant energies. For example, a given Ge detector, a calibration provided an energy dependence of ε_{abs} as shown in Fig. 1.20.
3. Determine the count rate C of the source under consideration, being careful to reproduce the same source-detector geometry used for the calibration of the detector.
4. Use eq. (1.25) to calculate the absolute activity.

Let us suppose that, using a Ge detector, we have measured a count rate $C = 1250$ cps for the most abundant gamma emission at $E_\gamma = 364$ keV. Based on the detector efficiency reported in Tab. 1.4, by using eq. (1.25) we obtain the following activity for the ^{131}I source:

$$\acute{A} = \frac{1250\,\text{cps}}{0.002\,\frac{\text{cps}}{\text{photon/s}} \cdot 0.82\,\frac{\text{photon/s}}{\text{Bq}}} \approx 762\,\text{kBq} \approx 20.6\,\mu\,\text{Ci}. \tag{1.27}$$

Tab. 1.4: Main photon emissions of I-131 and their relative abundances ε_{emis}, with examples of detection efficiency of a Ge detector at a fixed source-to-detector distance.

Line	E_γ (keV)	Relative abundance ε_{emis}	Detector efficiency* ε_{abs}
1	284	0.054	0.004
2	364	0.82	0.002
3	637	0.068	0.001

*for a given source-detector distance.

---

[20] Cf. also Chapter 11 on "Secondary and post-processes of transformations" in Vol. I and Figs. 11.11 and 11.12 showing electromagnetic multipole transitions for ^{131}I and a resulting γ-spectrum, respectively.

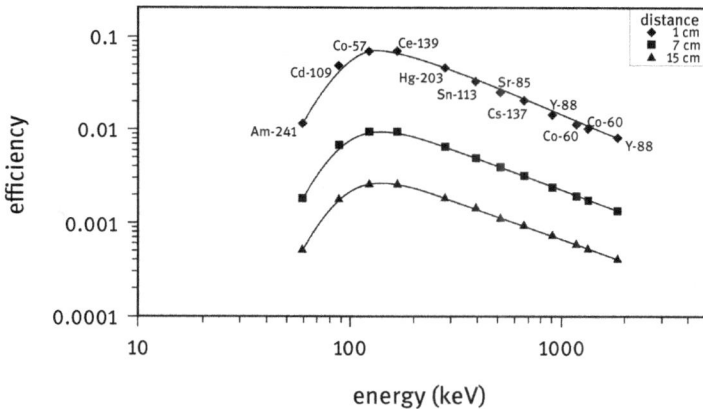

Fig. 1.20: Example of efficiency calibration curve of a p-type Ge detector of 20% relative efficiency for three source-to-detector distances.

1.6.2 Calibration sources

A detector with strong energy dependence must be recalibrated each time the energy range of the radiation field differs from the calibration energy. Even with a good flatness of the energy response function, some radiation detectors may have a response that changes over time (these weak fluctuations in the detector response are called "drifts"). A common cause of recalibration is, e.g., the dependence of the detector response on the room conditions (e.g., temperature and pressure). For applications requiring a good accuracy and reproducibility of the measurement output, such as the measurement of radionuclide activity for medical applications, it is important to establish an appropriate program of quality controls (QC) of radiation detectors. An important step of these QC is the employment of radioactive sources whose activity is certified by the manufacturer/vendor in such a way that its value can be referred directly or indirectly to a physical measurement performed on a "standard" source, only available in national metrology institutes (e.g., NIST in the US, NPL in the UK, or ENEA-INMRI in Italy). By definition, a primary radioactivity standard is a radioactive source whose activity cannot be calibrated as they represent the starting point for the subsequent chain of calibrations. A calibration source must be provided with a certificate reporting the chain of measurement by which the source activity has been compared with the activity of a primary standard: when this information is available, and the uncertainty associated with each measurement is known, the source is said to be "traceable" to the primary radioactivity standard.

Calibration sources are sold with diverse form factors and total activities for several radionuclides. Depending on their use, they can be rod-shaped, disc-shaped, point-like, but they typically come in a sealed form in such a way that they cannot release radioactive material in the environment, thus preserving both operator safety (avoiding contaminations) and the validity of their calibration. For example, Fig. 1.21 shows a sealed, vial-shaped [133]Ba gamma-emitting source used for calibration of 4π well counter detectors, along with its certificate of calibration; Fig. 1.22 shows a point [22]Na source embedded in a small acrylic cube used for calibration of Positron Emission Tomography (PET) scanners. Besides the absolute activity, important information on the calibration certificate includes the date of calibration and its accuracy. Calibration date must be considered at each usage because the radioactive decay at the date of each measurement must be taken into account. Calibration accuracy, on the other hand, is important as it influences the overall uncertainty associated with each measurement, which can be calculated by using error propagation theory. Most vendors provide calibrated activity with accuracy ranging from ± 2% to ± 5%.

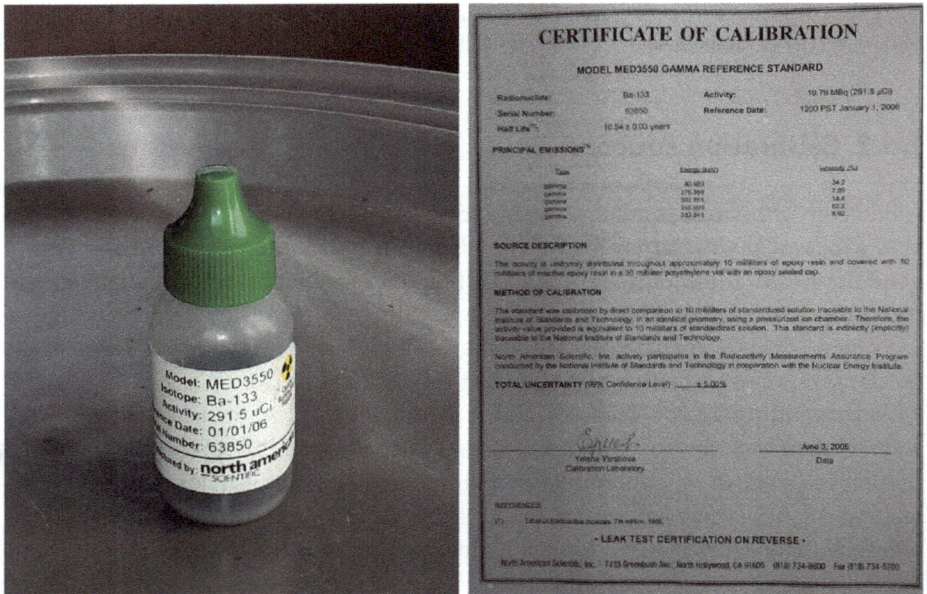

Fig. 1.21: Example of a sealed Ba-133 calibration source for well-type detectors.

1.6.3 Description of common radiation detection systems for radioactivity measurement

Scintillator detectors like NaI are widely used for radioactivity measurement due to their good intrinsic efficiency, but in most cases they are manufactured in cylindrical shape with 3″ or 5″ diameters, without cavities, so that their acceptance solid angle (and hence the geometric efficiency) could be rather low. Both the

Fig. 1.22: Example of a sealed point Na-22 source for calibration of PET scanners.

sample and the active volume of the detector must be placed on a shielded enclosure of high thickness in order to reduce undesired counts from the background activity in the room. This is also true for other detector configurations, as shown in Fig. 1.17 for the case of HPGe. By using gas-based proportional counters instead of scintillators, it is possible to realize detection systems with high acceptance angles, close to 4π steradians. In all cases, the sensitive volume of the detector is separated from the outside world by a "window" of various thicknesses and materials depending on the specific detector.

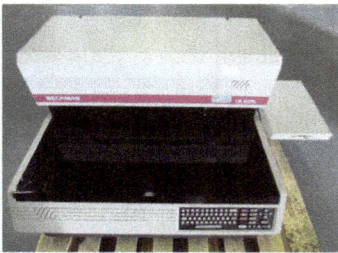

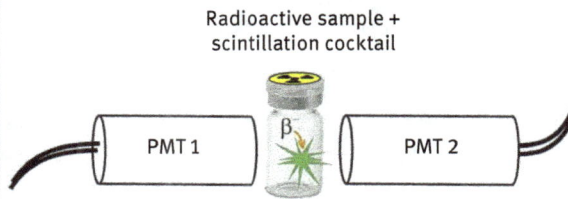

Fig. 1.23: A commercial liquid scintillation beta counter (Beckman Coulter, Brea CA, USA) (left) widely used in radioimmunoassay (RIA) laboratories. The working principle is shown at right: the scintillation cocktail is directly diluted into the radioactive sample and two photomultiplier tubes in coincidence read the visible light coming from the vial. (Photograph at left downloaded from http://www.labwrench.com/1equipment.view/equipmentNo/5915/Beckman-Coulter/LS-6500/, last access on June 10, 2012).

It is important to understand that the entrance window could affect the energy response of the detector itself, and this is especially true for alpha and beta counting, or for the detection of very low-energy X-rays. The measurement of the activity of commonly used β^--emitters such as ^{14}C, ^3H, or ^{32}P cannot be carried out efficiently by placing the sample in a test tube and then counting the sample with an external detector. As mentioned above, these particles are weakly penetrating; thus, very few of them would be able to travel beyond the test tube walls and the detector window. This problem is addressed by incorporating the sample under study in a solution containing a liquid scintillator. If both the solution and the vial are transparent, the scintillation light produced within the solution can be detected by means of photomultiplier tubes. In order to increase the sensitivity and to reduce noise, two opposing PMTs operating in coincidence are used (cf. Fig. 1.23). This means that a light signal in one of the two PMTs is rejected if the other PMT does not record another pulse at the same time. In this way, "spurious" signals due to background noise can be avoided. If the two PMTs are close enough, this configuration leads to an almost 4π geometry. There are basically three main ingredients in the solution used for liquid scintillation beta counting (very often referred to as the "scintillation cocktail"): an organic scintillator solute, a solvent, and a wavelength shifter. This last component absorbs the light emitted by the scintillator, and then re-emits it at a longer wavelength. This ensures that the scintillation light can match the spectral response of the PMTs that, in most cases, have higher efficiencies at longer wavelength.

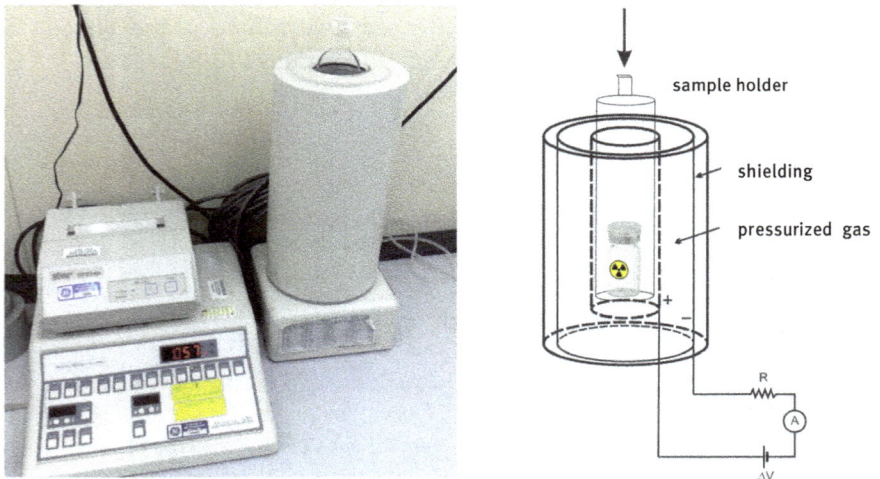

Fig. 1.24: A well-type ionization chamber used for dose calibration in a nuclear medicine department (Biodex Medical Systems, Shirley NY, USA). The ion chamber has an almost 4π geometry which maximizes the geometric efficiency. The user can select the type of radionuclide, which allows the correct conversion from exposure (C/kg) to activity (Bq), see eq. (1.28).

The 4π configuration is frequently used in the radiochemistry environment for gas-based activity meters operating in current mode (ion chambers) dose calibrators, cf. Fig. 1.24. The advantage of the current mode of operation is especially important at high levels of activity, where pulse-mode detectors would suffer decreased efficiency and loss of linearity due to dead times. Ion chamber dose calibrators are particularly well suited for the measurement of activity of γ-emitters. For radioactive emission, the relation between the exposure X (i.e., the electric charge produces in the gas by the indirect ionization from photons, per unit mass) measured at a distance r from the source and the activity A is given by the exposure rate constant, T, defined by:

$$\frac{dX}{dt} = \frac{\Gamma A}{r^2} \left[C\,kg^{-1}s^{-1} \right]. \tag{1.28}$$

Exposure X and exposure rate dX/dt are quantities that can be directly measured with an ionization chamber; thus, the knowledge of T for the radionuclides under study allows the absolute measurement of the sample activity.

1.6.4 Radiation spectrometry

Different to the case in which the radionuclide under study is known and just its activity must be determined, there are situations in which the type of radionuclide (or the types of radionuclides in a mixture) is unknown. Let us first consider a radioactive sample of β- and/or γ-emitters; it generates a mixed $\beta + \gamma$ field where the various gamma components have a characteristic discrete energy spectrum. The energy spectrum allows the identification of each radionuclide indirectly by the analysis of its lines in the gamma ray spectrum. It simultaneously provides the absolute activity of each radionuclide and thus finally analyzes the composition of the mixture.

In practice, the measured spectrum is somewhat different from a series of sharp lines; as explained in Section 1.3.4, each detector has a finite energy resolution, so the measured spectrum is actually made up of bell-shaped "peaks" of fine width. Moreover, a photon can interact within the detector without transferring all of its energy (for instance, when it undergoes COMPTON scattering), cf. Fig. 1.25. In this case, the photon energy "seen" by the detector is just the fraction that was transferred to the scattered electron. Because of the kinetic of COMPTON scattering, this fraction varies continuously from zero up to a maximum given by

$$E_{\text{edge}} = h\nu \left(1 - \frac{1}{1 + 2h\nu/(m_e c^2)} \right)$$
$$= h\nu - E_c, \tag{1.29}$$

where

$$E_c = \frac{h\nu}{1 + 2h\nu/(m_e c^2)}.$$ (1.30)

Figure 1.25 shows a comparison between a real gamma spectrum (for the very simple case of monoenergetic radiation) and the corresponding measured spectrum.

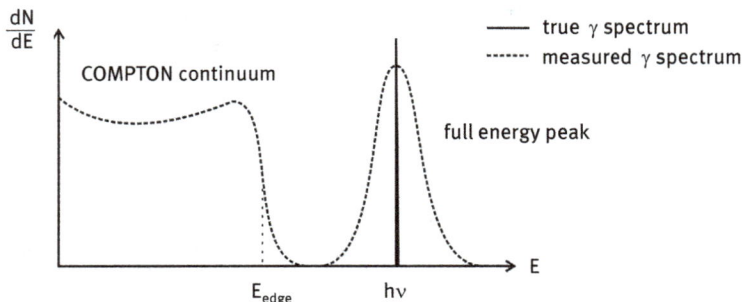

Fig. 1.25: Comparison between a true gamma spectrum of a hypothetical γ-emitter (solid vertical line) and the corresponding measured spectrum (dashed line). Note that this graph is still an exemplification, as several types of noise are present in the measured spectrum in the real case.

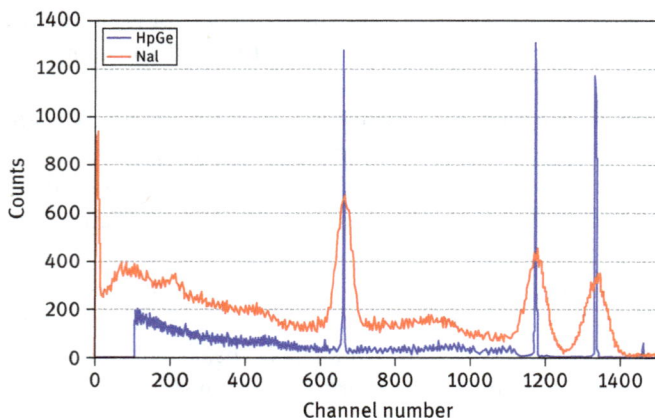

Fig. 1.26: Comparison of gamma-ray spectra from a mixed source of ^{137}Cs (E_γ = 662 keV) and ^{60}Co ($E_{\gamma,1}$ = 1173 keV; $E_{\gamma,2}$ = 1332 keV), as measured by HPGe (blue line) and NaI (purple line). The superior energy resolution of HPGe is evident. The peak at very low channel numbers is due to electronic noise and must be neglected. Note that the channel numbers have been calibrated in order to give a reading in keV; in this case, 1 channel = 1 keV. (From http://www.personal.psu.edu/elb5061/hpgedetector.html – last access on June 4, 2012).

Detector materials with high density and high atomic number are preferred in order to keep the relative height of the COMPTON continuum low with respect to the full-energy peak. Probably the most widely used radiation detector for gamma-ray spectroscopy is the NaI:Tl scintillator, coupled to a PMT and multichannel analyzer for pulse-height discrimination. The relatively high effective atomic number[21] ($Z_{eff} = 50.2$) and density ($\rho = 3.67$ g cm^{-3}), along with its high light yield makes it a good material for gamma-ray spectrometry in terms of both detection efficiency and intrinsic energy resolution ($R_{en} = 5\text{--}10\%$ at the energy of ^{137}Cs, i.e., $E_\gamma = 662$ keV). Higher energy resolutions can be achieved with semiconductor detectors, such as reverse biased Si diodes or HPGe. For example, Fig. 1.26 shows γ-spectra of a mixed source of ^{137}Cs and ^{60}Co, where three γ-lines of 662 keV, 1173 keV and 1332 keV are visible. The FHWM of those lines are about 2 keV for a state-of-the-art Ge-detector (<0.3% at 662 keV of the ^{137}Cs emission), but it may be up to 80 keV (>10% at 662 keV) for a NaI-detector.

For gamma-ray spectrometry, HPGe is the best choice in terms of energy resolution. Higher energy resolutions mean higher capability of revealing the fine structure of spectral lines of a radiation field, and hence a higher capability of radionuclide identification. Nevertheless, the relative height of the full-energy peak with respect to the COMPTON edge is lower in germanium than in NaI because of its lower atomic number ($Z = 32$ for Ge). While HPGe is appropriate for spectrometry with medium- to high-energy gamma rays (i.e., hundreds of keV or higher), Si detectors appear to be more suitable for low energy γ-photons or X-rays. The better performance of silicon for such low energy radiation is due to its low atomic number ($Z = 14$), allowing soft photons (usually fluorescence X-rays) to enter the sensitive volume of the detector and then releasing their energy. Silicon detectors can be also used for spectrometry of α-particles; a detector configuration referred to as surface barrier detectors (SBD), having a very thin dead layer allows the entrance of the (very weakly penetrating) alpha particles into the sensitive volume.

1.6.5 Radiochromatography

An important application of radiation detectors is to assess the radiochemical purity (RCP) of a radiolabeled compound or a radiopharmaceutical. The European Pharmacopoeia defines the RCP as the percentage ratio of the radionuclide activity of the radiopharmaceutical in its stated chemical form to the total activity of the radiopharmaceutical preparation. Radiochromatographic techniques are used to separate the

---

21 For a compound material (e.g., water) or a mixture of materials (e.g., organic tissues), the effective atomic number, Z_{eff}, can be defined as the mass-weighted average of the atomic numbers of all the elements present in the material. Several alternative and more or less complicated definitions have been proposed for Z_{eff}, to better describe the interaction of radiation with those composite materials.

different chemical substances containing the radionuclide, such as gas chromatography (GC), high performance liquid chromatography (HPLC) and thin layer chromatography (TLC).[22] What is important to discuss here is that in all the above-mentioned techniques, the measurement of the radionuclide activity at various points in a GC or HPLC or other column (called radio-GC and radio-HPLC, etc.) or plate (radio-TLC) leads to the development of a "chromatogram". The peaks in the chromatogram are then analyzed by means of radiation detectors and associated to the different chemical compounds in the sample.

In radio-GC, the gas effluent flowing in the column is detected at some point by a radiation detector. The choice of the detector is mainly based on the type of radionuclide present in the effluent. For β-emitters such as ^{14}C and ^{3}H, it is common to use windowless gas flow proportional counters, where the radioactive gas is mixed with the counter gas. This configuration gives a high sensitivity, because the radioactive source can be counted internally, thus avoiding the attenuation due to walls and windows. For γ-emitters such as ^{131}I (main γ-emission at 364 keV) or ^{59}Fe (main γ-emissions at 1099 keV and 1292 keV), NaI:Tl scintillators are preferred due to their higher efficiency for photons, as shown in Fig. 1.27 for the case of radio-HPLC.

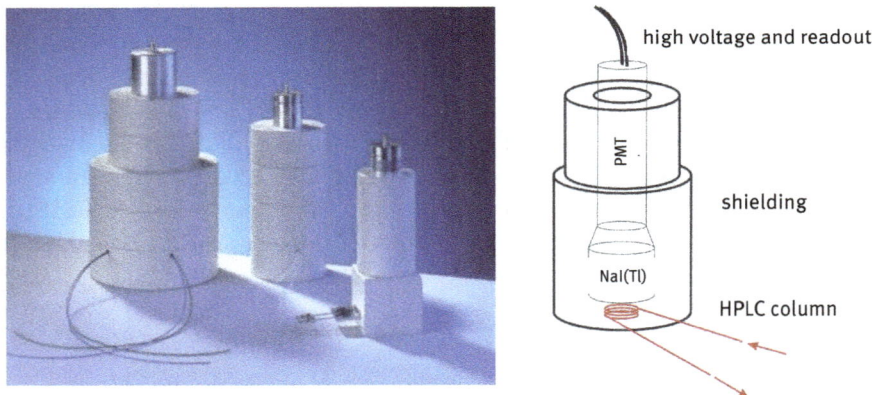

Fig. 1.27: Example of gamma radiation detector for radio-HPLC (raytest Isotopenmessgeräte GmbH, Straubenhardt, Germany). A thick lead shield protects the scintillator detector against undesired events from the background activity. The HPLC column passes through two small holes in the shield walls at the base of the apparatus. A small loop in the column provides a local increase of the measured activity (and hence an increased sensitivity) nearly at contact with the detector window, leading to an almost 2π geometry. (Photograph from www.raytest.de, last access on June 9, 2012).

---

22 The full description of the chromatographic techniques is beyond the goal of this chapter. For aspects of the radiochemical separation of chemically different radioactive species or the separation of one radioactive species from nonradioactive bulk materials cf. Chapter 7 on "Radiochemical separations".

Either photomultipliers tubes or solid-state photodiodes can be used to detect the scintillation light. An alternative way to measure the β-radiation in liquid systems is based on the detection of the CERENKOV light. CERENKOV photons are emitted in a transparent medium when an energetic particle travels inside it at a speed that is greater than the speed of light in that medium (i.e., at a speed greater than $v = c/n$, where n is the refractive index of the medium). The emitted radiation has a continuous spectrum, positioned between the visible light and the UV range. In this case, the effluent in the column must be mixed with additives that increase the wavelength of the CERENKOV light (referred to as wavelength shifters, often used also in liquid scintillator cocktails) in order to match the spectral response of most photomultiplier tubes.

A radio-TLC starts from a TLC plate with a one- or two-dimensional distribution of radioactive species. The radiochromatogram is obtained from a spatially resolved measurement of the radioactivity distribution by scanning the TLC plate. Collimated scintillator detectors are used, cf. Fig. 1.28. A motorized linear translator (for 1D-TLC) or x-y translator (for 2D-TLC) it capable of a complete and automated scan of the TLC plate. This latter solution is particularly well suited for radio-TLC with high-energy γ-emitters. Alternatively, imaging techniques such as autoradiography with photostimulable phosphor plates appear to be a practical concept, because of the improved spatial resolution. Some autoradiography techniques have the disadvantage of long exposure times required to reach a sufficient sensitivity. Phosphor imagers of various compositions are commercially available, with different sensitivity depending on the type of radionuclide in use.

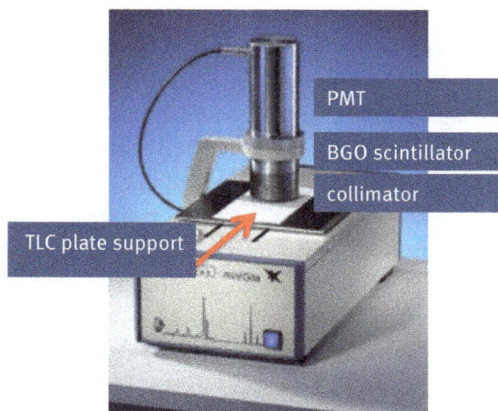

Fig. 1.28: A radio-TLC system for high energy emitting radionuclides (raytest Isotopenmessgeräte GmbH, Straubenhardt, Germany). Once the TLC plate is properly mounted on its support, a linear translator starts the scanning sequence of the chromatogram. (Modified from www.raytest.de, last access on June 9, 2012).

1.7 Environmental and individual monitoring

Whenever ionizing radiation is used, radiation detectors should monitor individuals and workplaces, thus ensuring the safety of people working with radioactive sources. Detectors used for this purpose must be calibrated in terms of relevant quantities for radiation protection. There are basically two pathways by which ionizing radiation can lead to a potential harm for workers: external or internal exposure. When working with sealed radioactive sources or X-ray generators contamination is excluded and workers could be struck by radiation exclusively from outside the body. In this case, the challenge lies in the quantification of those parameters of the radiation field that can be directly related to an increased risk for the "reference man"[23] to incur radiation-induced diseases in the short term and/or long term. The main operational quantities for radiation protection are defined by the International Committee for Radiological Units (ICRU) as the personal dose equivalent (H_p) and *ambient dose equivalent* (H^*), both measured in Sievert (Sv). Without entering in the detailed description of such quantities, it is enough to say that they are directly linked with the average energy imparted by the radiation at the point of observation, in a tissue-equivalent phantom.

Several types of active detectors can be used for this task. Ion chambers operated in current mode are often preferred for environmental dosimetry, because of the possibility of performing absolute measurement. Nevertheless, other detectors operating in pulse mode such as proportional, GM and scintillator counters can be employed after a suitable calibration, cf. Fig. 1.29. These latter detectors have a direct reading in counts-per-seconds (cps), but they can be calibrated in terms of dose equivalent rate (Sv/h) with some a priori information about the radiation field (e.g., the type of radionuclide). For the purpose of personal monitoring, passive detector systems such as TLD pellets (see Fig. 1.30) have the advantage that they can be easily placed inside wearable dosimeter badges. Although other passive detectors are used as personal dosimeters (such as radiographic films), TLD also have the advantage that they can be reused after the reading process. Moreover, the good equivalence of the LiF crystal to soft tissue ensures good dosimetric performances in terms of energy response.

The assessment of contamination risks in the presence of unsealed radioactive sources is in general more delicate and less straightforward. Radionuclides dispersed in liquids or dusts can adhere more or less strongly to nearby surfaces and objects; by touching them, workers could become contaminated or they could propagate the contamination to other surfaces. If the radionuclide is in a gaseous state,

---

23 "Reference man" was first defined in the Report n. 23 (1975) of the International Commission of Radiation Protection (ICRP), with the purpose of standardizing the main anatomical features and chemical composition of organs of adult males and females, thus facilitating the definition and calibration of dosimetric quantities related to individuals; cf. also the following chapter on "Radiation dosimetry".

GM counter

real-time monitor

Fig. 1.29: An environmental radiation monitor based on GM counters (Comecer S.p.A., Castel Bolognese, Italy). The LCD display shows information on ambient dose measured by several detectors in the laboratory.

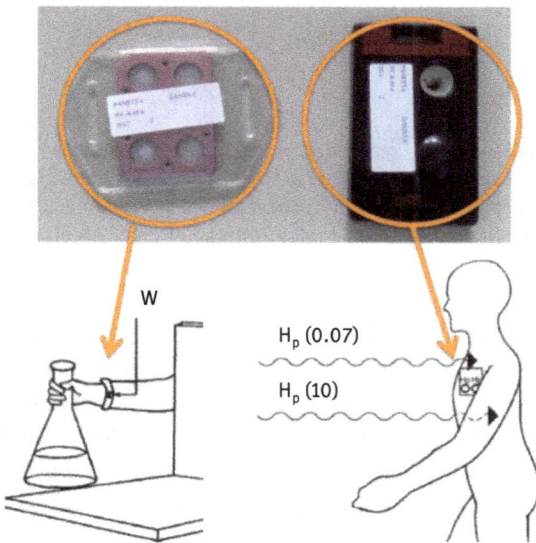

W

$H_p (0.07)$

$H_p (10)$

Fig. 1.30: TLD-based dosimeter badges for wrist (left) and body (right) for individual monitoring of β/γ radiation. (Adapted from "Practical Radiation Technical Manual", IAEA, Vienna, 2004).

individuals can breathe it in, thus leading to an internal exposure. The same type of contamination can occur if liquid or gaseous radioactive samples come in contact with foods or beverages, which can be ingested by individuals (or by animals, thus entering in the food chain).

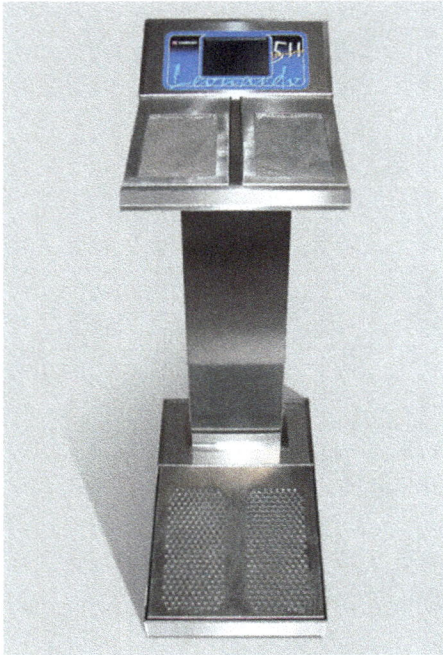

Fig. 1.31: A typical hand-and-feet contamination meter (Comecer S.p.A., Castel Bolognese, Italy). Four plastic scintillators (two for the hands, two for the feet) are read by PMTs and the readout is shown in real-time on the LCD display, after calibration in Bq/m^2 for several types of radionuclides (user-selectable). (From http://www.centroditaratura.it/servizi-taratura/strumenti/mani-piedi-vesti/ – last access on June 9, 2012).

The risk assessment in the presence of unsealed sources is based on the measurement of activity concentrations in all the surfaces, objects, foods, beverages and environments (or, in a single term, in all the "matrices") potentially contaminated. A smear-test (or wipe test) is often performed on potentially contaminated surfaces in order to measure the fraction of radioactive substance that is susceptible to be transferred to other matrices by contact. Well-counters based on scintillators in low background shielding are very often used to measure the amount of radioactive material removed by the smears. This test is generally performed along with a direct measurement of the contaminated surface, by mean of GM counters or portable scintillator detectors, calibrated in term of Bq/m^2 for several different radionuclides (in this context, these detectors are called "contamination meters"). The combined measurement by contamination meters and smear tests allows the assessment of both the total and the removable contamination on a surface. Contamination meters can be assembled in platforms with multiple probes for a fast contamination check on hands, feet and clothing (see Fig. 1.31).

For the assessment of airborne contamination, filtered air pumps can be used to estimate the probability of internal intake of radionuclides by respiration. At the end of the observation period, the filters are removed from the pumps and then the residual activity concentration is measured similarly to smears used for superficial measurements. In this case, the absolute activity of the air filter (expressed in Bq) is used to estimate the average activity concentration in air (in Bq/m^3) during the observation period. The information from both the superficial and airborne contamination can then be processed by means of biological models for the radionuclide biodistribution inside the body of the reference man, thus obtaining estimates for H_p and H^*.

1.8 Outlook

Detecting radiation is something that we learnt by mimicking Mother Nature. Human senses already have a great capability of detecting some kinds of radiation, such as visible light. Let us think of our eyes: no digital camera to date has the ability to adapt its response to such a wide range of light intensity (approximately 10 decades), from almost total darkness to bright sunlight. Nonetheless, the very reason why it took so much time for mankind to understand the fundamental laws governing the Universe at the atomic and subatomic level is that our senses cannot feel (not even barely) all the types of radiation that we defined as "ionizing" at the beginning of this chapter. The amazingly fast development of nuclear physics in the first half of the twentieth century, triggered by the discovery of X-rays by RÖNTGEN[24] and subsequently by BEC-QUEREL and Pierre and Marie CURIE in 1896, has led to a fast acceleration in the development and improvement of the related technology. This progress made it possible to conduct more and more complex nuclear physics experiments to confirm (or, in most cases, to retract) theories and hypothesis. All of these experiments where somehow related to the detection of ionizing radiation. As an example, RUTHERFORD's experiment on α-particle scattering on thin gold foils in 1911 necessitated a ZnS scintillator, which emitted tiny, localized flashes of visible light when struck by scattered α-particles.

Radiation detectors play a fundamental role in the medical field, being the key component of most medical scanners such as gamma cameras, Computed Tomography (CT), and Positron Emission Tomography (PET) systems. In CT, for instance, the transition from early pressurized Xenon gas detectors to modern solid-state detectors (such as Gadox scintillators) for the detection of X-rays allowed to reduce the patient dose by more than 20% (on a per-scan basis) due to the higher detection efficiency of solid detectors.

The largest and most complex detector ever built to date is the ATLAS detector (www.atlas.ch) at the Large Hadron Collider (LCH) at CERN, Geneva, Switzerland (along with the detectors of other LCH projects like CMS, ALICE, TOTEM, LHCf, and

---

24 By means of a scintillator screen of $BaPt(CN)_4$ – even though RÖNTGEN did not know about the scintillation mechanism in 1895.

LHCb), which is 45 meters long, more than 25 meters high and weighs about 7000 tons. This cylindrical detector, whose axis is a segment of the LHC accelerator (more precisely, its center is the place where opposing beams of protons strike each other, or "collide"), is made up of an inner part of silicon detectors (more than 80 million channels) and Xenon gas detectors (350,000 channels) able to track the path of the particles produced in the proton-proton collisions (tracker), immersed in a magnetic field of 2 Tesla. Moving outwards, a mid-part of the system comprises liquid Argon detectors and plastic scintillators alternated to metal absorbers to measure the energy of those particles (calorimeters); the outer part consists of gas-mixture detectors immersed in a strong magnetic field acting as a spectrometer for charged particles called muons (muon spectrometer). In the spring of 2012, this detector (and almost in parallel also the CMS detector) was able to acquire enough data to confirm for the first time the existence of the so-called "God particle", i.e., the HIGGS boson, proposed in the 1964 by three theoretical physicists (Peter HIGGS and François ENGLERT, and the late Robert BRUT).[25] The discovery of the Higgs boson has proved the validity of the "Standard Model" at least for collision energies up to 8 TeV ($= 8 \cdot 10^{12}$ eV). More experiments are planned at LHC starting from late 2015 with collision energies up to 14 TeV, searching for other answers that might increase our knowledge on the Universe and on its origin, such as the existence of extra dimensions of space, the unification of fundamental forces, and evidence for dark matter candidates.

For future experiments it is mandatory that new and more sophisticated detectors will be built, once more making ionizing detectors the fundamental bricks in the progress of particle physics.

References

Katz L, Penfold AS. Range-energy relations for electrons and the determination of beta-ray endpoint energies by absorption. Rev Mod Phys 1952, 24, 28–44.

Suggested reading

Attix HF. Introduction to radiological physics and radiation dosimetry. John Wiley and Sons, 2004.
Fleisher RL, et al. Nuclear tracks in solids. Principles and applications. University of California Press, 1974.
Knoll GF. Radiation detection and measurement, 4th ed. John Wiley and Sons, 2010.
Leo WR. Techniques for nuclear and particle physics experiments, 2nd ed. Springer-Verlag, 1994.
Stabin MG. Radiation protection and dosimetry. An introduction to health physics. Springer, 2008.

25 For their prediction, HIGGS and ENGLERT received the Nobel Prize in Physics in 2013.

Werner Rühm

2 Radiation dosimetry

Aim: The various interactions of radiation with matter, for example, emitted in the course of transformation of unstable nuclides, constitute the fundamentals of radiation measurement. In this context, any ionizing effects are governed by the physico-chemical parameters of the corresponding matter. Additionally, interaction of ionizing radiation with living organisms constitutes a separate field of science. Its aim is to understand the consequences of exposure to ionizing radiation for cellular and subcellular components, in particular the DNA. In this context, radiation dose is a measure of exposure to ionizing radiation.

This chapter discusses the most important dose concepts that are currently in use in radiation research and radiation protection and provides the scientific background behind these concepts. It shows how effective dose can be calculated after external exposure to cosmic radiation and finally describes the procedure to calculate effective dose after internal exposure to, as an example, incorporated radioactive cesium.

2.1 Introduction

The concept of "dose" was developed long before Wilhelm Conrad RÖNTGEN (1845–1923) discovered X-rays in 1895. In fact, it was the great physician and philosopher PARACELSUS (1493–1541) who coined the phrase *"dosis sola facit venenum"* – the dose makes the poison. In the context of his research, PARACELSUS meant the dose to be the amount of exposure to a certain substance, for example, a chemical or medicine. This idea is in essence still to date a major concept in modern toxicology. The term "poison" points towards any negative effect of this substance on the human body.

While negative effects can often be easily identified for chemicals, it was – at least at the beginning of radiation research – far from being clear what the negative consequences of a certain exposure to ionizing radiation were. It is true that quite soon after RÖNTGEN's pioneering first experiments with X-rays,[1] physicists and physicians who worked to investigate the properties of the newly discovered radiation suffered from certain health effects such as, for example, skin burns at their exposed hands. It was observed that these effects occurred quite soon after high exposures, but did not occur after only occasional low exposure to the radiation. Moreover, the severity of the observed effects appeared to be proportional to the amount of exposure. Radiation protection measures focused therefore on shielding of the most exposed parts of the

---

1 (First) Nobel Prize in Physics in 1901 ". . . in recognition of the extraordinary services he has rendered by the discovery of the remarkable rays subsequently named after him".

https://doi.org/10.1515/9783110742701-002

Tab. 2.1: Termini used in radiation dosimetry.

Parameter	Meaning	Units	
		SI	Conventional
Absorbed dose	Energy E deposited per mass m	J/kg	Gy
Equivalent dose	Absorbed dose multiplied by a radiation weighting factor defined by the International Commission on Radiological Protection (ICRP), to account for the biological effectiveness of a certain type of ionizing radiation	J/kg	Sv
Effective dose	Sum over equivalent doses for certain organs of the human body multiplied by tissue weighting factors as defined by ICRP to account for the different radiosensitivity of these organs	J/kg	Sv
Organ dose	Equivalent dose for a certain organ	J/kg	Sv
LD_{50} dose	Dose that will be lethal for 50% of an exposed group of individuals	J/kg	Gy
Relative biological effectiveness (RBE)	Factor that takes into account that different radiation types may show different biological effects		
Radiation weighting factor	Factor suggested by ICRP to take into account that different radiation types may show different biological effects		
Ambient dose equivalent	Dose equivalent at a depth of 10 mm within the ICRU (International Commission on Radiation Units and Measurements) sphere at a radius opposing the direction of an aligned field	J/kg	Sv
Directional dose equivalent	Dose equivalent at a certain depth in the ICRU sphere with respect to a specified direction	J/kg	Sv
Personal dose equivalent	Dose equivalent at a certain depth in the human body at the position where an individual dosimeter is worn	J/kg	Sv
Fluence-to-dose conversion coefficient	Coefficient that converts the fluence of a certain particle into dose	$Sv\,cm^2$	
Ionization	Process of removing an electron from the atomic shell		
Single-strand break (SSB)	One break in a DNA strand		

Tab. 2.1 (continued)

Parameter	Meaning	Units	
		SI	**Conventional**
Double-strand break (DSB)	Break in both DNA strands, with the breaks being not separated		
Linear Energy Transfer (LET)	Energy lost by charged particles in a medium per track length	keV/µm	
Specific effective energy (SEE)	Equivalent dose in target organ T per radioactive transformation in source organ S	Sv per transformation	
Biological half-life	Time it takes for the human body to excrete half of an acutely incorporated activity	s	

human body. Low doses of ionizing radiation seemed to be harmless, and many low-dose applications of radiation were readily adopted such as radioactive tooth pastes that promised brilliantly white teeth. The tragic events that happened after the two nuclear explosions over Hiroshima and Nagasaki in 1945 changed this view. A few years after the bombings, first reports were published that leukemia mortality increased among those who survived the bombings. Cataracts were also found to be increased several years after exposure, a finding which was confirmed by studies on physicists who worked at cyclotrons and who also showed an increase in cataracts. Today, one knows that – beside the so-called *tissue reactions* (or "deterministic effects" as they were called earlier), which are characterized by the fact that a major fraction of cells of a certain tissue must be affected by the radiation, and that a dose threshold exists below which such reactions do not occur – so-called *stochastic effects* such as leukemia or cancer also occur at lower doses. These health effects are characterized by the fact that (a) there might be no radiation dose threshold and (b) radiation dose does not govern the severity of the effect, but the *probability* that the effect occurs. The most relevant termini used in radiation dosimetry, which will be discussed in detail throughout this chapter, are listed in Tab. 2.1.

2.2 Radiation dose concepts

2.2.1 Ionizing radiation

When radiation hits an atom, various processes occur that govern the interaction of the incoming radiation with matter (see Chapter 1). For biological tissue, the process of ionization is the most important because it affects the electron shell of the atoms that comprise these molecules and, as a result, may disturb the chemical binding within the molecules of the tissue. Basically, an ionization occurs if the energy of the impinging radiation is high enough to remove an electron from the atomic shell, and a positively charged atom, i.e., an ion, remains. Due to the properties of the elements, at least 4–5 eV are required to ionize an atom (this energy is needed to remove the outermost electron from the atomic shell of alkaline elements; for other elements or for electrons in lower shells a higher energy is required). Radiation below that energy range such as radiowaves, microwaves, etc., is not dealt with here, in contrast for example to X-rays and gamma radiation which show energies well above the 4–5 eV mentioned above.[2]

Principally, ionizing radiation that hits the DNA in the nucleus of a cell may remove an electron from the DNA, and as a result, a covalent bond in the DNA molecule may break. This process is called *direct action of ionizing radiation*. Alternatively, the radiation may ionize a water molecule in the cell which then will form an OH radical that in turn may remove an electron from the DNA. This process is called *indirect action of ionizing radiation*. A break in the sugar-phosphate backbone of the DNA leads to a single-strand break (SSB). If two SSSBs are close together, a DSB may occur which – if not repaired – may lead to cell death or – if not repaired correctly – may lead to a chromosome aberration. Cell experiments indicate that exposure to an absorbed dose of 4 Gy of gamma radiation leads to the induction of about 200 DSBs in a cell. If one keeps in mind that exposure to natural and artificial sources of ionizing radiation results for example in a mean annual effective dose of about 4 mSv to the German population (see Tabs. 2.7 and 2.8), that the human body typically consists of about 10^{14} cells, and that a year has about $3 \cdot 10^7$ s, then one can easily estimate that very roughly about $10^5 - 10^6$ DSBs per second occur in the human body, due to these sources of ionizing radiation. Obviously, there must be efficient repair mechanisms that allow the cell to repair the damage and the human organism to live for 70–80 years or more. While the cell can repair an SSB for example by using the undamaged strand as a basis, it is more challenging for the cell to repair a DSB, and specific mechanisms such as nonhomologous end-joining and homologous recombination (where a sister chromatid or a homologous part of the DNA are used as a template) are required to repair the damage.

2.2.2 Protection quantities

2.2.2.1 Absorbed radiation dose

In radiation research, the term "radiation dose" (or "dose" as it will be called from now on) was initially defined on pure physical grounds. The simple idea behind

---

2 In fact it needs more energy to form a double-strand break (DSB) in the DNA (10–15 eV) or to ionize a molecule in air (about 34 eV).

this was that the more energy that was absorbed by an atom or deposited for example in a cell or in tissue by the radiation, the more effects on atomic, cellular or tissue levels are expected.[3] This led to the concept of *"absorbed dose"* D which is defined as the amount of energy dE deposited per mass dm, at any point in the material of interest (eq. (2.1)):

$$D = \frac{\mathrm{d}E}{\mathrm{d}m}.$$ (2.1)

Accordingly, D_T is the *mean absorbed dose* in a certain organ or tissue. The SI unit of absorbed dose is J/kg. It is instructive to realize that an energy of the order of one or several J per kg is quite usual in everyday life.[4]

On the other hand, it is well known (e.g., from studies on the atomic bomb survivors in Hiroshima and Nagasaki) that an absorbed dose to the whole body of 3–4 J/kg of ionizing radiation will cause about half of the affected individuals to die from radiation-induced tissue reactions. This is the so-called LD_{50} dose ("LD_{50}" stands for "lethal dose, 50%").[5] This pronounced biological effectiveness of ionizing radiation justifies the introduction of the special name Gray (Gy) for the unit of absorbed dose for applications in this field.[6]

----

3 For the interaction of the ionizing radiation with atoms this would be equivalent to the concept of radiation measurement, cf. Chapter 1: The higher the radioactivity of a source, the higher the readout of the radiation detector.
4 For example, consider walking down the step of a staircase with height h. Then the gravitational energy released can be calculated as $m \cdot g \cdot h$, where m is your body mass, and g is the acceleration due to gravity which is 9.81 m/s^2. Let your body mass be 70 kg and the height of the step 10 cm, then the gravitational energy released when you walk down one step is 70 kg $\cdot$ 0.1 m $\cdot$ 9.81 m/s^2 = 68.7 kg $\cdot$ m/s^2 = 68.7 J. Thus, the gravitational energy corresponds to almost 1 J per kg body mass. Or assume that you drink a cup of tea in the morning. If the tap water you used to prepare the tea had a temperature of say 20 °C, and if the mass of water m you need to fill your cup is 100 g (corresponding to 100 ml), then the amount of energy ΔQ required to heat the water can be calculated as $\Delta Q = c \cdot m \cdot \Delta T$, where c is the specific heat of water which is 4.2 J K^{-1} g^{-1}. If, for drinking, the tea has a temperature of 60 °C, you end up with a net energy of at least ΔQ = 4.2 J K^{-1} g^{-1} $\cdot$ 100 g $\cdot$ (60 – 20) K = 16.8 kJ you had to spend to heat your tea (boiling at 100 °C would require about twice as much). You can estimate the temperature increase you may feel after you drank the tea by assuming that an energy ΔQ = 4.2 J K^{-1} g^{-1} $\cdot$ 100 g $\cdot$ (60 – 37) K = 9.7 kJ is released to and distributed homogenously over your body. Again assuming a body mass of about 70 kg you finally end up with a temperature increase of ΔT = 9.7 kJ/(4.2 J K^{-1} g^{-1} $\cdot$ 70 kg) = 0.03 K. It is quite obvious that if you drink the tea carefully this will not cause any harm to your body, although you had incorporated about 9700 J / 70 kg, i.e., more than 100 J per kg of body mass.
5 A quantity which is also used in toxicology.
6 The International Commission on Radiological Protection (ICRP) and the International Committee on Radiation Units and Measurements (ICRU) decided to introduce "Gray" as the special name of the unit of absorbed dose (Gy), in appreciation of the major contributions of Louis Harold GRAY (1905–1965), a British radiation physicist, to radiation research.

2.2.2.2 Equivalent dose

The biological effectiveness of ionizing radiation suggests that the energy transferred to tissue after exposure to ionizing radiation – although a useful measure for the presence of radiation and its harmful effects to the human body – is not the only parameter needed to describe and understand the effects of ionizing radiation on biological systems. In fact, it is the local distribution of the energy absorbed which also plays an important role. In contrast to the above example of drinking tea where the heat was assumed to be distributed homogeneously over the human body, the energy released by ionizing radiation is absorbed very locally within the body, for example through induction of double-strand breaks in the DNA of the cell nucleus. This feature is unique to ionizing radiation, and it explains another observation that cannot be understood just based on the concept of absorbed energy: The biological effect of different types of ionizing radiation may be different even if the energy released by these radiation types per mass of biological material, i.e., the absorbed dose, is the same.

Assume an experiment where a certain number of cells is irradiated with a specified absorbed dose of gamma radiation and where then the irradiated cells are grown on a surface in a medium afterwards. After about two weeks, the number of colonies formed is counted and one finds that say 30% of the initially irradiated cells were able to form a colony. Obviously, the remaining 70% had lost their reproducibility or died. If one repeats this experiment with cells of the same type but irradiated at different absorbed doses, one will be able to establish a so-called survival curve which quantifies cell survival (given as a fraction relative to unexposed cells) as a function of absorbed dose (given in Gy) from gamma radiation. One can of course perform a similar experiment with a different type of ionizing radiation, for example with ions such as protons or alpha particles, and establish a second survival curve as a function of absorbed dose from these ions. Comparison of the two curves reveals major differences: It appears that many more cells can survive after irradiation with gamma radiation than after exposure to ions, although the absorbed dose was the same (Fig. 2.1).

The reason for this obvious difference is the way in which gamma radiation and charged particles release their energy in matter: While the mean loss of energy of ions along their path through matter, described by the so-called *linear energy transfer* (LET), is quite high (e.g., about 27 keV/μm for protons with an energy of 1 MeV in water), it is much lower for gamma radiation (e.g., about 0.2 keV/μm for photons from ^{60}Co with a mean energy of 1.25 MeV in water). As a result, ionization events that occur after irradiation within a cell are quite close together for charged particles and quite far apart for gamma radiation, on the scale of a cell nucleus. For illustration, Fig. 2.2 sketches two cells that are irradiated by these two types of radiation. The left cell gets four ionizations due to low-LET gamma radiation along its path through the cell, where energy is absorbed, while the right cell gets many more ionization events due to high-LET ionizing charged particles. It is intuitively clear even without addressing details of cellular damage and mechanisms that allow cells to

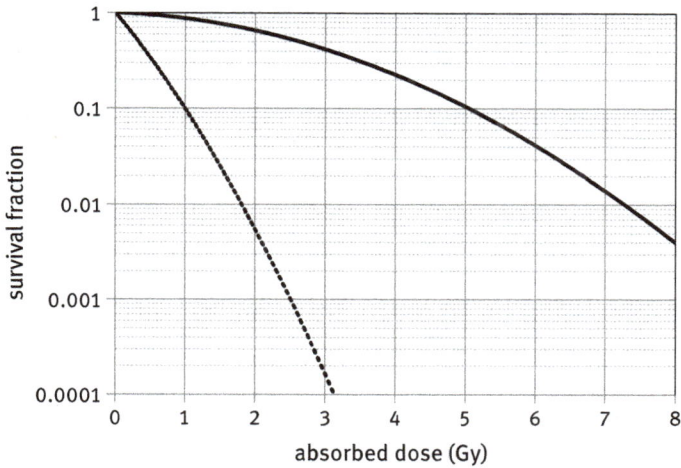

Fig. 2.1: Sketch of typical cell survival curves for high-LET ions such as protons or alpha particles (dashed line) and low-LET radiation such as gamma rays (solid line).

repair this damage, that repair will be more effective if the ionization events are far apart rather than if they are close together. This is the reason why the survival curve for gamma radiation shows a so-called shoulder (which describes the curvature of the survival curve after exposure to gamma radiation; solid line in Fig. 2.1) at low absorbed dose (which means that the repair mechanisms of the cells are quite effective at low doses), while the survival curve for ions does not show this shoulder (because even at these low doses repair is difficult since the ionization sites are too close together).

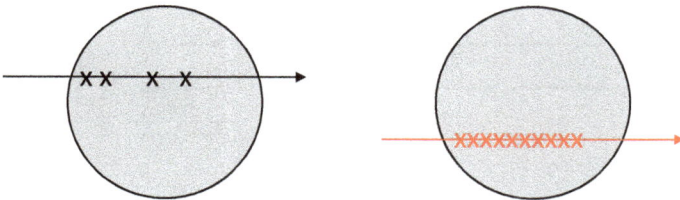

Fig. 2.2: Sketch of a cell irradiated with low-LET gamma radiation (left) or high-LET ions (right); crosses represent ionizations along the track through the cell.

In the example given in Fig. 2.1, a survival fraction of 0.1 is reached at an absorbed dose of about 1 Gy for ions, while it is reached at about 5 Gy for γ-radiation. Thus, for ions at an absorbed dose of 1 Gy, the so-called *relative biological effectiveness* (RBE), which is defined as the ratio of absorbed doses of the reference radiation (e.g., gamma radiation from, e.g., ^{60}Co or X-rays) and the investigated radiation that both produce the same biological effect is 5 in the given example.

In order to take the different biological effectiveness of different types of radiation into account – which has its origin in the different local distribution of ionizations – the concept of absorbed dose was extended, and the *equivalent dose* H_T was introduced. This quantity can be calculated by multiplying the mean absorbed dose D_T in tissue with the radiation weighting factor w_R (eq. (2.2)):

$$H_T = w_R D_T \tag{2.2}$$

In the example shown in Fig. 2.1, the RBE of ions with an absorbed dose of 1 Gy is 5.0 (corresponding to a survival fraction of 0.1). Note, however, that the RBE value depends on the absorbed dose considered: Because the shoulder of the survival curve at low absorbed doses is much more pronounced for γ-rays than for ions, the RBE is only 4 for ions with an absorbed dose of 1.8 Gy in the given example (corresponding to a survival fraction of 0.01), cf. Fig. 2.1.

For the purpose of radiation protection, International Commission on Radiological Protection (ICRP) specified numerical values of the radiation weighting factor w_R for various radiation types incident on the human body. The choice of these values is based on scientific literature available on the RBE of a certain radiation type including *in vitro*, animal, and human data, and on expert judgment. Per definition, w_R is set at 1 for low-LET ionizing radiation such as γ-radiation, X-rays, or electrons, while it is set for example at 20 for α-particles. For these particles, w_R is assumed to be independent of the energy of the radiation considered to keep the system of radiation protection as simple as possible (Tab. 2.2). Only for neutrons is a continuous curve as a function of neutron energy recommended for w_R, which varies between 2.5 and 20, cf. Fig. 2.3.

Tab. 2.2: Radiation weighting factors recommended by ICRP (2007); for external exposure, these values relate to the radiation incident on the body, while for internal exposure, they relate to radiation emitted from the incorporated radioactive sources.

Radiation type	W_R
Photons, electrons, muons	1
Protons, charged pions	2
Alpha particles, fission fragments, heavy ions	20
Neutrons	Continuous function, cf. Fig. 2.3

Because w_R has no dimension, the SI unit of equivalent dose is the same as that of absorbed dose, i.e., J/kg. However, to distinguish both the unit for absorbed dose and that for equivalent dose,[7] the unit of equivalent dose was given the special name "Sievert" (Sv).

---

7 And to appreciate the great achievements of the Swedish physicist Rolf SIEVERT (1896–1966) in radiation research.

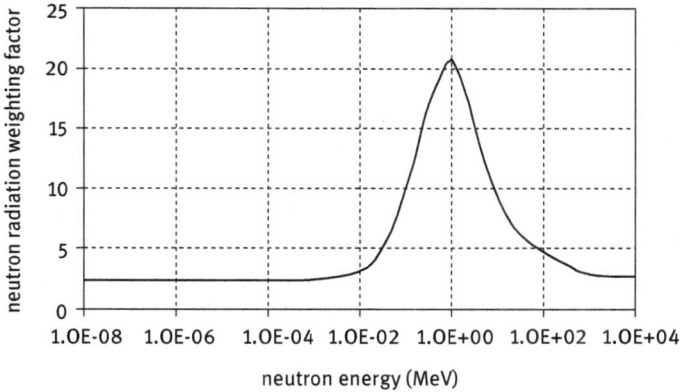

Fig. 2.3: Radiation weighting factor for neutrons as recommended by ICRP (2007).

2.2.2.3 Effective dose

Besides absorbed dose and radiation type, there are a number of other parameters that influence the effect of ionizing radiation on cellular and tissue levels. These include, for example, fractionation (i.e., the dose is delivered in fractions with dose-free time intervals in between), dose rate, oxygen concentration in the medium surrounding a cell during irradiation, and others. Among these, the cell type is of particular relevance, and it turned out that certain tissues are more sensitive to ionizing radiation than others. These include for example cells that proliferate quicker than other cells, due to a shorter cell cycle.

On the level of the human organism, the results obtained from the atomic bomb survivors indeed suggest that some organs are more sensitive to ionizing radiation than others. In fact, soon after the nuclear bombings of Hiroshima and Nagasaki on August 6 and August 9, 1945, efforts were made to initiate a large epidemiological study on this cohort. In 1950, a cohort of about 120,000 survivors was established, and since then, the cohort is followed up and the incidence and mortality of various diseases among this cohort has been quantified as part of the so-called "Life Span Study" (LSS). Dose estimates are available for each individual survivor taking into account individual location and shielding at the time of bombing. In order to account for the mixed gamma and neutron radiation field the survivors were exposed to, doses are given in terms of the weighted absorbed dose which is calculated as $D_\gamma + 10 \cdot D_n$, where D_γ is the gamma radiation absorbed dose and D_n is the neutron absorbed dose (a mean quality factor of 10 was chosen to account for the high RBE of neutrons – see also Section 2.2.2.2).

These data yielded a so-called *dose-effect curve* with the health data on the y-axis and the dose data on the x-axis. Figure 2.4 gives, as an example, the dose-effect curve that was obtained when the number O of all-cancer deaths observed in the LSS cohort of the atomic bomb survivors in the follow-up period 1950–2003 was compared with the number E expected without any radiation exposure from the bombing, normalized to this expected number (this corresponds to the so-called "Excess Relative Risk" ERR $= (O - E)/E$ and then plotted against the weighted absorbed dose to the colon. The slope of the curve, i.e., the ERR per dose, is called "risk coefficient" and is a measure of how a specific dose of ionizing radiation acts on the human body, in terms of all-cancer mortality. Among 86,611 survivors and in the follow-up period 1950–2003, 5235 deaths from solid tumors were observed, and 527 of those are believed to be radiation-induced (Ozasa et al. 2012). Due to the relatively large number of cases, this cohort allows to deduce even organ-specific risk estimates (see Fig. 2.5).

Figure 2.5 indicates that among the atomic bomb survivors, most of the investigated organs show a risk coefficient significantly greater than zero, and certain organs such as the lung or the female breast exhibit significantly higher risk coefficients than other organs. Based on these and other findings, ICRP defined so-called *tissue weighting factors* w_T for radiation-induced stochastic effects and introduced the concept of effective dose.

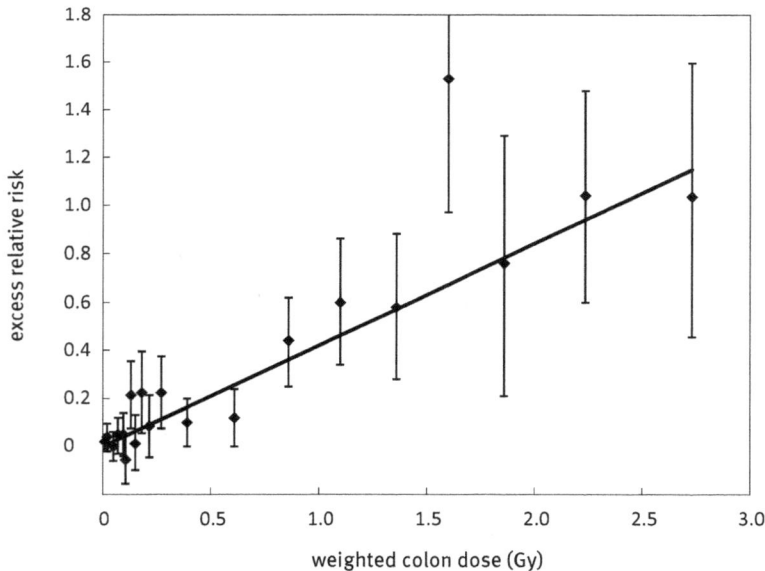

Fig. 2.4: Excess relative risk (ERR) for all solid cancers from LSS cohort of atomic bomb survivors and a linear fit through the data (figure modified from Ozasa et al. (2012)). Weighted colon dose means here that the neutron absorbed dose was multiplied by a factor of 10, to take into account the high relative biological effectiveness of neutrons, and then added to the absorbed dose from gamma radiation. Error bars represent 95% confidence intervals.

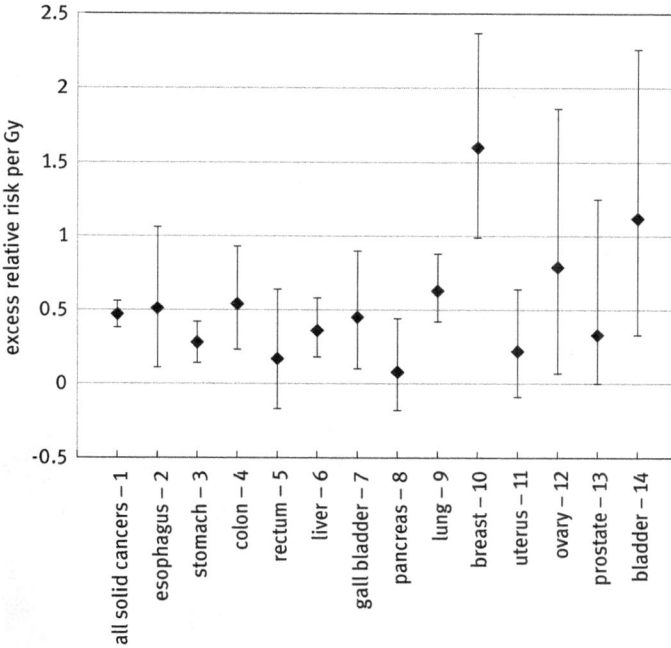

Fig. 2.5: Risk coefficients for different organs of atomic bomb survivors (Ozasa et al. 2012): Error bars represent 95% confidence intervals.

The quantity *effective dose* is defined according to eq. (2.3):

$$E = \sum_T w_T H_T = \sum_T w_T \sum_R w_R D_{R,T}, \tag{2.3}$$

where w_R is the radiation weighting factor for radiation type R, w_T is the weighting factor for tissue T, and $D_{R,T}$ is the mean absorbed dose from radiation of type R in tissue T. Numerical values of the tissue weighting factors as recommended by ICRP are given in Tab. 2.3.

Tab. 2.3: ICRP tissue weighting factors (ICRP 2007); remainder includes 14 additional organs/tissues.

Organ/tissue	W_T	$\sum W_T$
Red bone marrow, colon, lung, stomach, breast, remainder	0.12	0.72
Gonads	0.08	0.08
Bladder, liver, esophagus, thyroid	0.04	0.16
Bone surface, skin, brain, salivary gland	0.01	0.04
Total		**1.0**

In order to calculate $D_{R,T}$, the absorbed dose after external or internal exposure due to radiation type R in tissue T, realistic models of the human body are needed to take self-absorption of the radiation by the body and production of secondary particles[8] within the body into account. For this purpose, ICRP has recommended the use of two voxel phantoms as a reference for male and female individuals. These voxel phantoms are based on CT (computed tomography) scans of real persons and were then adapted to an ICRP reference male and female (including corresponding organ sizes and masses). Figure 2.6 shows the recommended ICRP voxel phantoms, which were developed at the Helmholtz Zentrum München, Germany.

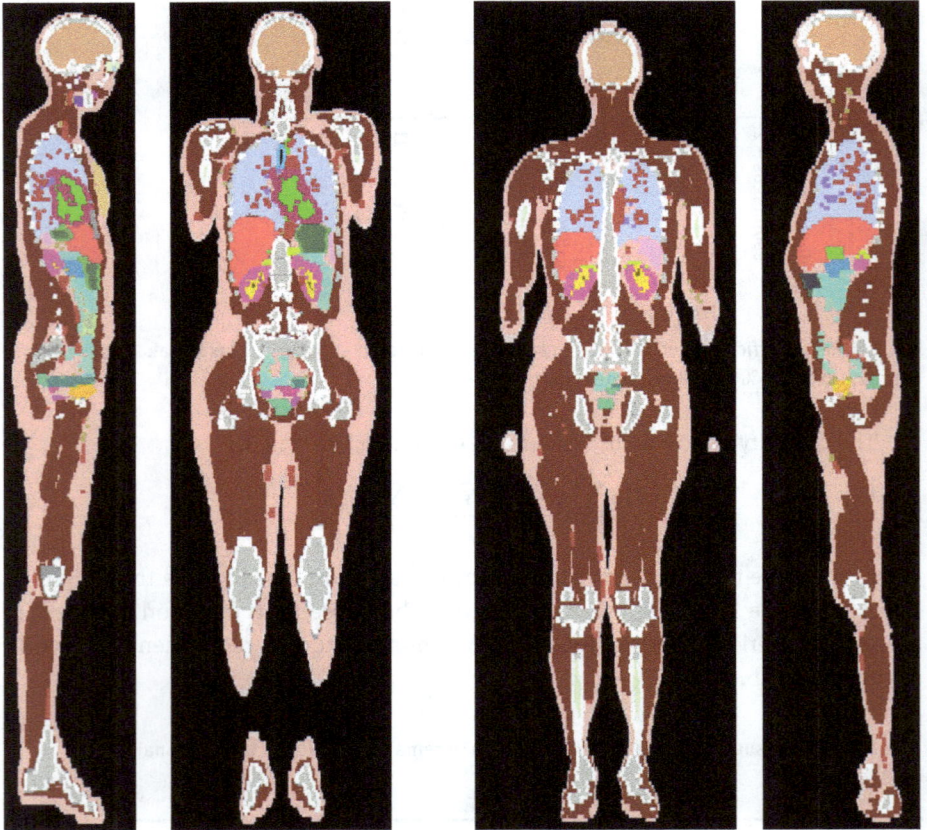

Fig. 2.6: Coronal and sagittal images (slices) of the ICRP reference voxel phantoms of the human body; left: female phantom; right: male phantom; both with additional details on various internal organs (ICRP 2009; pictures with permission from ICRP).

---

8 Secondary particles are for example photoelectrons or Compton electrons that are produced when photons enter the body and interact with tissue atoms, cf. Chapter 1.

Typically, the voxel sizes are $5 \times 1.875 \times 1.875 \, \text{mm}^3$ for the female phantom and $8 \times 2.08 \times 2.08 \, \text{mm}^3$ for the male phantom. Different tissue types (i.e., organs) are described with their specific elemental composition.

Once the $D_{R,T}$ values have been calculated for a certain exposure situation by use of these ICRP male and female voxel phantoms, for all organs and radiation types involved, then multiplication with the w_R (Tab. 2.2) and w_T values (Tab. 2.3) allows, together with eq. (2.3), calculation of the corresponding effective dose. The overall concept to calculate effective doses is summarized in Fig. 2.7.

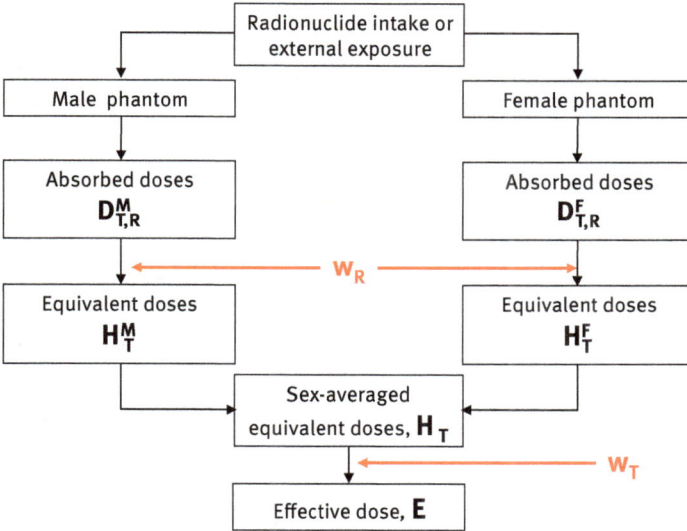

Fig. 2.7: ICRP concept to calculate effective doses.

Note that "effective dose" is an artificial quantity. It includes values based on literature reviews and expert judgment (w_R and w_T) and is meant to be valid for a reference person only (i.e., it is independent of sex, individual body size, and body mass and age). As such, effective dose cannot be measured. Although the value of effective dose does consider the individual exposure situation, it does not include the individual properties of the exposed person. Therefore, it should not be used to estimate individual risks after exposure to ionizing radiation. It is, however, useful to quantify radiation protection measures (i.e., whether effective dose decreased once an improved shielding was installed at a workplace) or to compare different radiation exposure scenarios (e.g., internal versus external exposures; see examples given below). Effective dose is also used for dose recording and to set limits and reference levels for workers and the public exposed to ionizing radiation.

2.2.3 Operational quantities

The *protection* quantities, equivalent and effective dose, are related to the human body, and basic dose limits in radiation protection are given in terms of these quantities. They, however, cannot be measured and hence for measurements in cases of external radiation fields the following "*operational* quantities" have been defined by ICRP and ICRU:

(a) ambient dose equivalent $H^*(10)$
(b) directional dose equivalent $H'(d, \Omega)$
(c) personal dose equivalent $H_p(d)$.

These quantities have been defined for area and individual monitoring and for controlling effective dose, skin equivalent dose, and equivalent dose to the eye lens (see Tab. 2.4). The measured values of the operational quantities provide – in most cases – conservative estimates of the protection quantities.

Tab. 2.4: Operational quantities for external exposure.

Monitoring of	Area	Individual
Effective dose	Ambient dose equivalent, $H^*(10)$	Personal dose equivalent, $H_p(10)$
Equivalent dose to skin and extremities (hands, feet)	Directional dose equivalent, $H'(0.07, \Omega)$	Personal dose equivalent, $H_p(0.07)$
Equivalent dose to the eye lens	Directional dose equivalent, $H^*(3, \Omega)$	Personal dose equivalent, $H_p(3)$

For whichever operational quantity is to be used, point of origin is again the absorbed dose D, i.e., the energy deposited per mass of tissue or material; eq. (2.1). As was noted earlier, on the biological level, different types of radiation (e.g., α-, β-, γ-radiation, neutrons, and protons) produce different radiation effects. The LET, which is the energy lost by charged particles in a medium per track length, is a useful quantity because it allows definition of a quality factor Q, which can be used to describe the biological effectiveness of ionizing radiation relative to photons (cf. Fig. 2.8) and which has to be applied to the absorbed dose at the point of interest.

Multiplication of the absorbed dose D and the quality factor Q defines the dose equivalent H; eq. (2.4) with L being the LET:

$$H = \int_L Q(L) \frac{dD}{dL} dL. \tag{2.4}$$

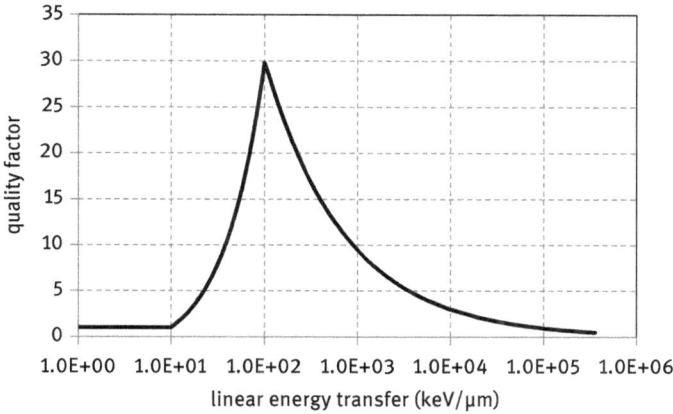

Fig. 2.8: Quality factor Q(LET) as a function of linear energy transfer (LET) in water, as defined by ICRP and ICRU (see, e.g., ICRP (1991)). Water instead of tissue is used for simplification. Similar to the radiation weighting factor w_R, the function (LET) was obtained based on biological data. For example, protons at an energy of 1 MeV show an Q(LET) of about 27 keV/μm, and accordingly, their quality factor is about 6.5. In contrast, 1 MeV electrons show an LET of about 0.19 keV/μm, and their quality factor is 1.

2.2.3.1 Ambient dose equivalent, $H^*(10)$

For area monitoring, ICRP/ICRU use a phantom that is much simpler than the phantoms used by ICRP to model the human body (Fig. 2.6), i.e., the so-called *ICRU sphere*. The ICRU sphere has a diameter of 30 cm and is made of tissue-equivalent material with a density of 1 g/cm³ (elemental composition: 76.2% oxygen, 11.1% carbon, 10.1% hydrogen, and 2.6% nitrogen). The quantity $H^*(10)$ is defined by the dose equivalent at a depth of 10 mm within the sphere at a radius opposing the direction of an aligned field. This definition implies that the real radiation field at the point of interest is hypothetically aligned and expanded over the area of the sphere, i.e., that the energy distribution and fluence are the same as that at the point of interest in the real field (Fig. 2.9). This definition considers that dosimeters for area monitoring should have an isotropic response.[9]

2.2.3.2 Directional dose equivalent, $H'(0.07, \Omega)$

To specify the dose equivalent in the ICRU sphere with respect to a specified direction Ω, the directional dose equivalent $H'(d, \Omega)$ is to be used at a certain depth, d,

---

9 "Isotropic" meaning that their response does not depend on the direction of the impinging radiation.

on the radius that points towards the direction Ω. For a unidirectional field, one can specify the direction Ω by the angle α between this direction and the direction opposing the incident field (Fig. 2.9). For the specific case of $\alpha = 0°$, $H'(d, 0°)$ is equal to $H^*(d)$. For low-penetrating radiation, a depth of 0.07 mm is recommended, cf. Fig. 2.9.

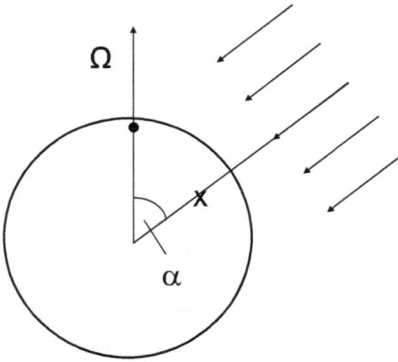

Fig. 2.9: ICRU sphere (diameter: 30 cm) with point of interest × at a depth of 10 mm on a radius opposing the incident aligned and expanded field (definition of $H'(10)$), and with point of interest • at a depth of 0.07 mm at an angle α between a specified direction Ω and the radius opposing the incident aligned and expanded field (definition of $H'(0.07, \alpha)$) (schematic view).

2.2.3.3 Personal dose equivalent, $H_p(d)$

For radiation protection purposes, a measure for the exposure of an individual is required, instead of the dose at a certain point in a radiation field discussed above. This is the personal dose equivalent $H_p(d)$ defined at a certain depth d in the human body at the position where an individual dosimeter is worn. Because radiation scattered within the body may contribute to the individual dose, $H_p(d)$ may be different for different individuals (characterized by different body mass and geometry), and it may also vary depending on where the dosimeter is worn on the body. For an assessment of effective dose, dosimeters are usually worn in front of the breast. For calibration of individual dosimeters, instead of the human body, a *simplified tissue-substitute phantom*, the ICRU slab with dimensions 30 cm × 30 cm × 15 cm is used, simulating the backscattering of the human body. The dosimeters are positioned in front of the slab and calibrated in terms of $H_p(10)$ for estimation of effective dose, and $H_p(0.07)$ for estimation of skin equivalent dose or the dose to the extremities.

It is noted that operational quantities are defined for external exposures only. In case of internal exposures by incorporated radionuclides, estimates of the protection quantities are based on the determination of incorporated activities and application of proper biokinetic models (see Section 2.3.2).

2.3 Examples

While the previous sections described the basic concepts currently in use in dosimetry and radiation protection, the following sections are meant to demonstrate how these concepts are being used in certain exposure scenarios. The section deals with calculation of effective dose after external exposure to secondary cosmic radiation during air travel – a situation which is characterized by a very complex radiation field – while another addresses calculation of effective dose after incorporation of radioactive cesium – an example which highlights the specific problems that arise in internal exposure situations.

2.3.1 Effective dose: External exposure from cosmic radiation at flight altitudes

Primary cosmic radiation consists of a galactic and a solar component (mainly high-energetic protons and alpha particles). When it hits the atmosphere of the Earth (mainly composed of nitrogen and oxygen nuclei), various nuclear reactions occur and secondary particles are produced including, among others, protons, neutrons, electrons and positrons, photons, and positive and negative muons. These particles, which are often called "secondary cosmic radiation", form a cascade that has its highest fluence rate (i.e., the number of particles per cm^2 and second) at an altitude of about 15–20 km, and that leads to an appreciable exposure of pilots,

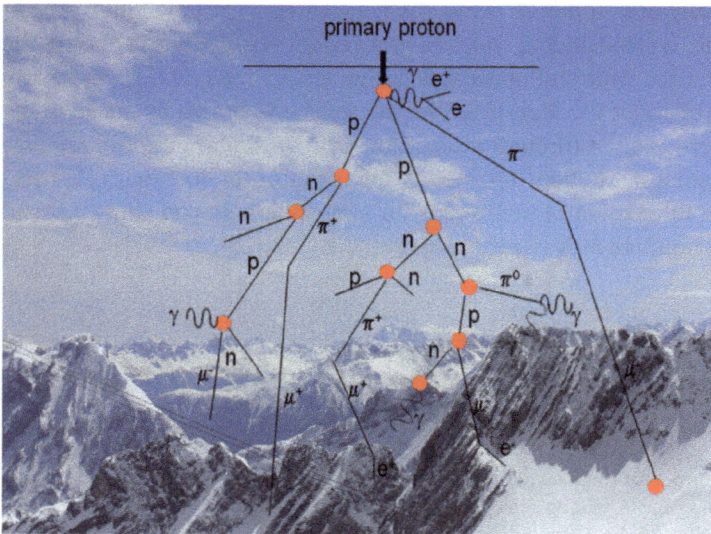

Fig. 2.10: Sketch of the hadron, electron, photon, and muon showers created by the interaction of primary cosmic radiation with nuclei in the atmosphere (photo: B. Rühm).

cabin crew, and air passengers to ionizing radiation at flight altitudes. Figure 2.10 schematically shows the proton-induced cascade of secondary cosmic radiation in the atmosphere.

Calculation of the associated effective doses is a complicated task which requires

(a) simulation of the fluence rates as a function of particle energy at every point in the atmosphere for all dose-relevant particles;

(b) calculation of fluence-to-effective-dose conversion coefficients for all particles and energies involved, including all organs of the human body relevant for the calculation of effective dose and use of proper radiation and tissue weighting factors as provided by ICRP (see Section 2.2.2); and

(c) combining these fluence rates and fluence-to-effective-dose conversion coefficients.

These topics are worked out in more detail below.

2.3.1.1 Secondary neutrons of cosmic radiation in the atmosphere

The complex field of *secondary cosmic radiation particles* produced in the atmosphere can be simulated by means of Monte Carlo methods.[10] Figure 2.11 shows the calculated spectral fluence rate distribution, $d\Phi/dE$, for secondary neutrons close to the geomagnetic poles, the situation at minimum solar activity[11] that occurred around 2009, and a height above sea level of 10.4 km. The so-called *lethargy representation* is used, where $d\Phi/d\ln(E)$ is plotted on the linear y-axis (note that $d\Phi/d\ln(E)$ corresponds mathematically to $E \cdot d\Phi/dE$) versus the neutron energy plotted on the logarithmic x-axis. In this representation, the area at a certain energy interval below the curve is a measure for the neutron fluence rate in that energy interval. The maximum at a neutron energy of about 1 MeV results from neutrons evaporated by nuclei in the atmosphere that were excited through interaction with secondary cosmic ray particles, while the maximum at about 100 MeV is the result of the fact that cross sections describing the interaction of neutrons with atmospheric nuclei show a broad minimum at several hundred MeV neutron energy.

---

10 For example, as an input for these simulations, primary cosmic radiation proton and alpha particle spectra were taken from NASA, and the GEANT4 toolkit developed at CERN was used to calculate the secondary radiation field (including neutrons, protons, electrons, positrons, muons (positive and negative), photons, pions) at various locations in the atmosphere.

11 Note that a minimum in solar activity corresponds to a maximum in fluence rate of cosmic particles in the atmosphere of the Earth, and vice versa.

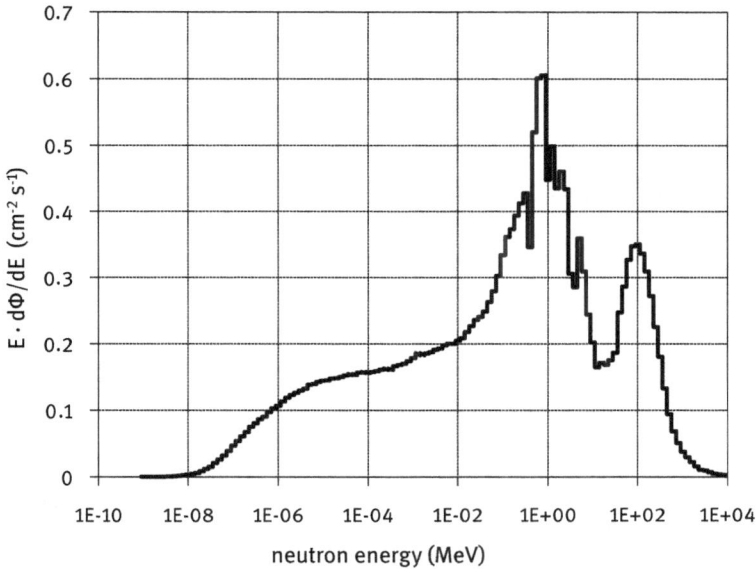

Fig. 2.11: Fluence rate distribution of secondary neutrons at an altitude of 10.4 km close to the geomagnetic poles and for minimum solar activity as calculated with GEANT4 (courtesy: C. Pioch, HMGU).

2.3.1.2 ICRP fluence-to-effective-dose conversion coefficients for neutrons

In order to calculate the effective dose contribution after external exposure to neutrons with energies shown in Fig. 2.11, the absorbed dose D deposited in the organs of the human body per incident neutron per cm^2 is required. This must be known for all relevant neutron energies – in the present case from meV up to some GeV, and can be calculated by means of Monte Carlo (MC) codes that allow simulation of particle transport within the human body. Obviously, a realistic model of the human body (cf. Fig. 2.6) is needed to take self-absorption of the radiation by the body and production of secondary particles in the body into account. By means of MC codes, an irradiation of these phantoms with neutrons can be simulated in the range between 1 meV and 10 GeV, and the absorbed energy per fluence of incident neutrons in various organs within the phantoms can be scored as a function of neutron energy. The resulting values can then be multiplied with the energy-dependent ICRP radiation weighting factor w_R for neutrons (Tab. 2.2), and the resulting organ equivalent doses multiplied with the ICRP tissue weighting factors w_T (Tab. 2.3) and summed up, see eq. (2.3). Based on such calculations, which also include an average over female and male values, cf. see Fig. 2.7, ICRP recommends values for the neutron effective dose per fluence (in units of pSv cm^2). Figure 2.12 shows the effective dose per neutron fluence recommended by ICRP for an isotropic exposure situation with neutrons.

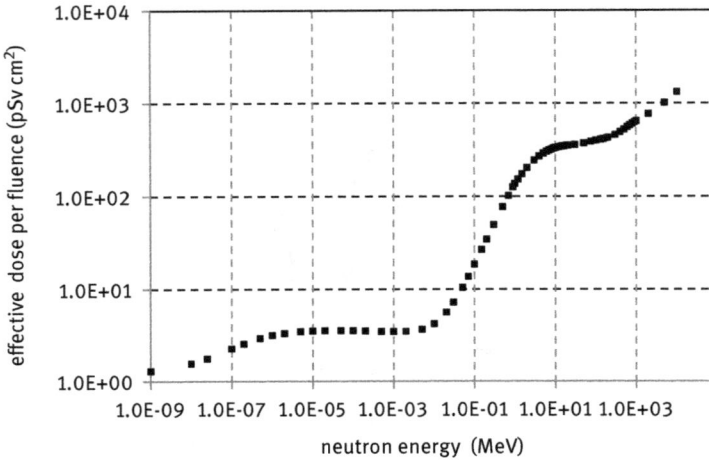

Fig. 2.12: Fluence-to-effective-dose conversion coefficients for neutrons for isotropic irradiation of the human body (ICRP 2010). Isotropic irradiation with a neutron fluence of 1/cm² at 1 meV results in an effective dose of about 1 pSv, while at 1 GeV, it results in an effective dose of about 640 pSv.

2.3.1.3 Effective dose from secondary cosmic radiation

If one is interested in the effective dose per unit of time from neutrons at an altitude of 10.4 km close to solar minimum activity and close to the geomagnetic poles, one has to fold the spectral fluence rate distribution shown in Fig. 2.11 with the ICRP fluence-to-effective-dose conversion coefficients for neutrons shown in Fig. 2.12. This results in an effective dose rate of 2.47 µSv/h.

Because the field of secondary cosmic radiation does not only consist of neutrons, the procedure discussed must also be applied to the other particles (protons, photons, electrons and positrons, positive and negative muons, charged pions) involved. Thus, the fluence distributions in energy of these particles must also be calculated at this altitude, and folded with the corresponding fluence-to-dose conversion coefficients recommended by ICRP for these particles (Tab. 2.5). Taken together, the total effective dose rate from secondary cosmic radiation at an altitude of 10.4 km close to solar minimum and close to the geomagnetic poles is about 4.3 µSv/h. Because there are some differences for the various particles when effective dose (a protection quantity) and ambient dose equivalent ($H^*(10)$) (an operational quantity) values are compared, the corresponding total ambient dose equivalent rate of 5.46 µSv/h is slightly larger than the total effective does rate. Thus it provides a reasonable and conservative upper estimate of effective dose rate.

Tab. 2.5: Effective dose per hour at 10.4 km altitude for various particles of secondary cosmic radiation, close to the geomagnetic poles, and for minimum solar activity; for comparison, the corresponding ambient dose equivalent per hour is also given (values are based on ICRP weighting factors (ICRP 2007)).

Particle	n	p	γ	μ^+, μ^-	e^+, e^-	π^+, π^-	Total
dE/dt (µSv/h)	2.47	0.82	0.46	0.16	0.36	0.003	4.27
$dH^*(10)/dt$ (µSv/h)	3.67	0.65	0.22	0.15	0.77	0.004	5.46

2.3.1.4 Determination of effective dose of air crew and flight passengers

In a similar way, use of the EPCARD code (European Program Package for Calculation of Aviation Route Doses) allows calculation of effective doses of pilots and cabin crew from cosmic radiation, for any flight route and time.[12] For example, for a flight from Munich, Germany, to New York, USA, on August 1, 2012, at a flight altitude of 10,400 m, the total route effective dose would be of 55 µSv (assuming a total flight time of 9 h including a duration of ascent and descent of 0.5 h each).

2.3.2 Effective dose: Internal exposure from incorporated radionuclides

Radioactivity is a natural phenomenon, cf. Chapter 3 of Vol. I. Due to their long physical half-lives, for example, primordial radionuclides such as ^{238}U ($t_{1/2} = 4.5 \cdot 10^9$ a), ^{235}U ($t_{1/2} = 0.7 \cdot 10^9$ a), or ^{40}K ($t_{1/2} = 1.3 \cdot 10^9$ a) have been present in the lithosphere since the earth was formed, while cosmogenic radionuclides such as ^{14}C ($t_{1/2} = 5730$ a) have a shorter half-life but are continuously being created by the interaction of cosmic radiation with the atmosphere. Additionally, a number of manmade (anthropogenic) radionuclides can also be found in the environment such as ^{137}Cs ($t_{1/2} = 30.17$ a) which is produced, for example, as a fission product of ^{235}U in nuclear reactors. Whatever origin these radionuclides may have, if they are present in the environment, they may – depending on their chemical properties – reach the food chain and, as a result, can be incorporated daily.

It is important to note that effective dose after radionuclide incorporation cannot be measured but must be calculated. The contribution of an incorporated radionuclide to effective dose is difficult to quantify, and quite a number of factors need

---

12 The code was developed at the Helmholtz Zentrum München, Germany, and is used by many airlines in Germany and other countries for dose monitoring. A simplified version of the code is available at www.helmholtz-muenchen.de/epcard.

to be considered that crucially govern the energy absorbed by the human body and the resulting effective dose once the incorporated radionuclide transforms. Some of these factors are the type of incorporation, the type of radioactive transformation and the biokinetics of these radionuclides and their progeny in the human body.

2.3.2.1 Type of incorporation

A radionuclide present in gaseous form will be inhaled, and the lung is the first and most important organ to be exposed. A typical example is ^{222}Rn ($t\frac{1}{2} = 3.8$ d), an α-emitting radioactive isotope of the noble gas radon which is part of the "natural" ^{238}U transformation chain.[13] If ^{222}Rn transforms in the lung, a considerable fraction of the progenies formed (e.g., ^{218}Po, ^{214}Bi) remain in the lung and their subsequent transformation adds to an exposure of the lung with α-particles. Depending on the solubility that is determined by the chemical form of the inhaled radionuclide, the radionuclide generated may remain in the lung for a long time (if the solubility is low), or it is transferred from the lung tissue to the blood plasma rather quickly (if the solubility is high).[14]

If ingested, the gastrointestinal tract is the first organ to be considered before the radionuclide is further distributed within the body. Here, uptake of the radionuclide from the gastrointestinal tract to the blood plasma is described by an f_A parameter, with a value of 1 indicating complete absorption. Ingestion is the most important pathway of incorporation for primordial ^{40}K, which is – together with stable potassium[15] – present in plants through root-uptake from soil. Milk that contains about 1.5 g potassium per liter (which is a typical value) includes 47 Bq of ^{40}K per liter. Finally, radionuclides may also be incorporated through the skin, in particular, in the presence of a wound.

2.3.2.2 Type of radioactive transformation

Primary and secondary transformation of a radionuclide[16] is mainly accompanied by the emission of α, β, and/or γ, X-ray, and electron radiation. Depending on the energy

---

13 Cf. Chapter 3 of Vol. I.
14 Because ^{238}U is naturally present in the lithosphere as a trace element, ^{222}Rn is being produced continuously in soil. As a noble gas it enters a house quite easily, for example through the cellar or through exhalation of building materials. As a result, mean ^{222}Rn indoor concentration for example in German dwellings is about 50 Bq per cubic meter of air. In regions where the natural concentration of ^{238}U in soil is very high, however, indoor radon concentrations may reach 10 000 Bq/m^3 or more.
15 The isotopic abundance of ^{40}K in potassium is 0.0117%.
16 Cf. Vol. I, Chapters 5–12.

involved, the range of the emitted particles in tissue may be considerably different. For example, α-particles emitted after the transformation of ^{222}Rn show energies of several MeV, and accordingly, their range in tissue is of the order of several μm. Thus, inhaled ^{222}Rn results in a very local exposure of lung tissue cells with α-particles that are highly biologically effective (see Section 2.2 and Tab. 2.2). In contrast, the energy of photons emitted after the transformation of ^{40}K is 1.461 MeV. Because their biological effectiveness is lower than that of α-particles (Tab. 2.2), and because their range in human tissue is so long that a major fraction can even leave the human body, the transformation of one ^{40}K nucleus in the human body will result in a smaller contribution to effective dose than transformation of one ^{222}Rn nucleus.

2.3.2.3 Biokinetics

Distribution of an incorporated radionuclide within the human body depends on its chemical characteristics.[17] Moreover, effective dose after incorporation of a radionuclide depends on whether the radionuclide in question is excreted by the human body quickly or retained for a long time. In case of ^{137}Cs, the physical half-life of 30.17 years is not very critical for dose assessment because the so-called *biological half-life* (which is the time needed for the body to excrete half of an acutely incorporated activity) is only about 100 days. The situation is different for ^{90}Sr which has a similar physical half-life as ^{137}Cs but which is built into the bone matrix and stays there for a very long time.

2.3.2.4 Calculation of effective dose per activity intake

If incorporation of an activity of 1 Bq of a certain radionuclide results – at time t after incorporation – in an activity $A_S(t)$ in a certain source organ S, the transformation of this radionuclide in that organ will result in emission of radiation type R and exposure of a neighboring target organ T (Fig. 2.13). The equivalent dose $H(T \leftarrow S, t)$ is the energy absorbed in organ T multiplied by the corresponding radiation weighting factor w_R and divided by the target organ mass, eq. (2.5). It depends on $A(t)$, type and energy of the radiation emitted, distance between S and T, organ sizes, and tissue composition involved:

---

17 While some radionuclides become concentrated in certain parts of the human body – for example ^{131}I ($t_{1/2}$ = 8.02 d) in its chemical form of iodide which is known to accumulate in the thyroid, or ^{89}Sr ($t_{1/2}$ = 50.57 d) in its chemical form of the Sr^{2+} cation which is incorporated in the bone matrix – others such as ^{137}Cs and ^{40}K (again as cations) are distributed quite homogenously over the whole body.

$$H(T \leftarrow S, R, t) = \text{SEE}(T \leftarrow S, R) \cdot A_s(t). \tag{2.5}$$

$H(T \leftarrow S, R, t)$ is the equivalent dose in target organ T due to particles of radiation type R emitted by source organ S. The *specific effective energy* SSE corresponds to the equivalent dose in target organ T per radioactive transformation in source organ S. $A_S(t)$ is the activity in source organ S at a certain time t.

Incorporation of 1 Bq of the radionuclide of interest may also result in the contamination of additional organs. Moreover, the radionuclide may emit several types of radiation. Thus, to get the total equivalent dose $H(T)$ for target organ T, one has to sum over all types of radiation involved and over all contaminated source organs S, and integrate over t, eq. (2.6). Equations (2.5) and (2.6) imply that, in fact, two models of the human body must be used for dose calculation, one to calculate $\text{SEE}(T \leftarrow S, R)$ and the other to calculate $A_S(t)$:

$$H(T) = \int \sum_S \sum_R H(T \leftarrow S, R, t) dt. \tag{2.6}$$

The first model includes anthropomorphic phantoms such as those shown in Fig. 2.6 that allow – in combination with particle transport calculations – simulation of (a) the field of particles generated in the human body[18] of a radionuclide in the various source organs S, and (b) the absorption of these particles in target organ T or their escape. This model produces the $\text{SEE}(T \leftarrow S, R)$ values for various source organ and target organ combinations. ICRP suggests the use of the voxel phantoms shown in Fig. 2.6 for this purpose. Once a certain source organ S in those phantoms is virtually contaminated by a radionuclide, use of particle transport codes such as MCNPX, GEANT4, or PHITS allows calculation of the energy absorbed in any other target organ T of the phantom. The resulting SEE values are tabulated for various combinations of source and target organs. Table 2.6 shows as an example SEE values as suggested by ICRP, calculated for the adult male phantom for ^{137}Cs. For example, one transformation of ^{137}Cs in the thyroid would result in an equivalent dose of $6.15 \cdot 10^{-16}$ Sv in the brain, while it would result in $2.14 \cdot 10^{-12}$ Sv in the thyroid itself (in this case the source and target organ is identical).

The second model is a compartment model of the human body that allows modeling of the activity $A_S(t)$ in the various source organs as a function of time. In this model, the human body is assumed to consist of a number of compartments, each representing a certain type of organ or tissue. Transport of a radionuclide from one organ to another is modeled through transfer parameters that describe the translocation of the radionuclide between the various compartments as a function of time. Figure 2.14 shows a model that describes the biokinetics of radiocesium (^{137}Cs or ^{134}Cs) in the human body as proposed by ICRP. This case is quite simple,

---

18 In the context of dose, the most relevant radiative emissions are α-particles, β⁻ and β⁺ particles, IC and Auger electrons, and electromagnetic radiation (X-rays and photons).

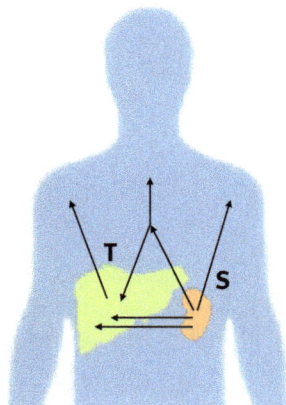

Fig. 2.13: Contaminated source organ *S*; when the radionuclides in the organ decay, they emit particles (e.g., photons, electrons, and alpha particles) which in turn may release their energy in target organ *T* where they contribute to absorbed dose.

Tab. 2.6: SEE values for an adult (i.e., human hermaphrodite with male dimensions as used by ICRP) and ^{137}Cs, given in Sv per transformation, for various source (columns) and target (rows) organs (courtesy: D. Nosske, BfS, Germany).

	Brain	Breasts	Kidneys	Liver	Lung	Muscle	Thyroid
Brain	3.88E − 14	2.53E − 17	3.45E − 18	1.09E − 17	5.81E − 17	1.09E − 16	6.15E − 16
Breasts	2.53E − 17	1.30E − 13	1.24E − 16	3.25E − 16	9.73E − 16	2.07E − 16	1.66E − 16
Kidneys	3.45E − 18	1.24E − 16	1.50E − 13	1.13E − 15	2.97E − 16	4.18E − 16	3.46E − 17
Liver	1.09E − 17	3.25E − 16	1.13E − 15	3.01E − 14	7.66E − 16	3.18E − 16	5.58E − 17
Lung	5.81E − 17	9.73E − 16	2.97E − 16	7.67E − 16	4.44E − 14	3.83E − 16	3.70E − 16
Muscle	1.09E − 16	2.07E − 16	4.18E − 16	3.18E − 16	3.83E − 16	1.89E − 15	4.84E − 16
Ovaries	7.54E − 19	2.83E − 17	3.21E − 16	1.83E − 16	4.37E − 17	5.80E − 16	5.30E − 18
Pancreas	9.00E − 18	2.99E − 16	1.96E − 15	1.39E − 15	6.70E − 16	5.04E − 16	5.28E − 17
Red Marrow	3.89E − 16	2.78E − 16	7.41E − 16	3.86E − 16	4.90E − 16	4.03E − 16	3.42E − 16
Skin	2.16E − 16	3.74E − 16	2.07E − 16	1.89E − 16	1.96E − 16	2.84E − 16	2.18E − 16
Spleen	1.16E − 17	2.30E − 16	2.64E − 15	3.12E − 16	6.45E − 16	4.29E − 16	5.01E − 17
Testes	1.64E − 19	0.00E + 00	3.47E − 17	2.04E − 17	6.72E − 18	4.41E − 16	1.13E − 18
Thymus	6.21E − 17	1.19E − 15	1.07E − 16	2.64E − 16	1.13E − 15	4.48E − 16	6.48E − 16
Thyroid	6.15E − 16	1.66E − 16	3.46E − 17	5.58E − 17	3.70E − 16	4.84E − 16	2.14E − 12
Uterus	6.99E − 19	3.20E − 17	2.99E − 16	1.57E − 16	3.49E − 17	5.76E − 16	4.99E − 18

and the differential equations that can be deduced from the model can be solved easily. After oral intake, a small fraction of ^{137}Cs is removed from the gastrointestinal tract by fecal excretion, while the major fraction is transferred to the blood ("transfer compartment") from where it is distributed throughout the body. Experimental data on human urinary excretion suggest that 10% of ^{137}Cs is transferred to the kidneys from where it is excreted through urine quite quickly, while the remaining 90% of ^{137}Cs is transferred to body tissue (muscle and organ tissues) where it remains much longer before urinary excretion takes place. Based on available data on human excretion of ^{137}Cs, ICRP estimated the fast component to be characterized

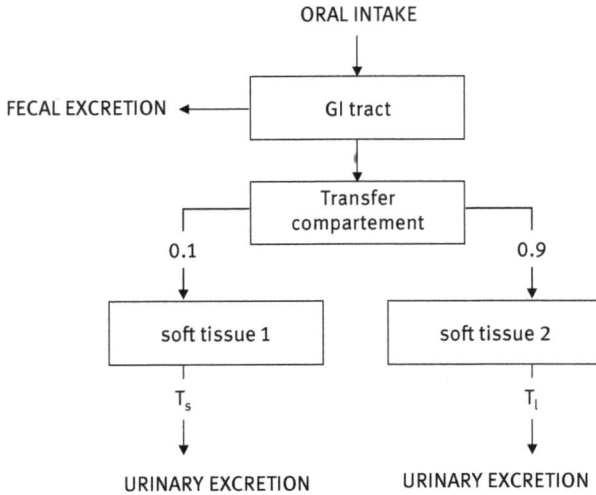

Fig. 2.14: Sketch of ICRP biokinetic model of radiocesium.

by a short half-life t_s of 2 days, while the slow component to be characterized by a öpmger half-life t_l of 110 days.

If the biokinetic model for a certain radionuclide is more complicated than that for ^{137}Cs and includes more compartments, then it becomes almost impossible to provide accurate numbers for all the transfer parameters that are needed to describe the biokinetics in the human body. For this reason, although as much knowledge as possible on human physiological parameters is used, many of the parameters have to be estimated based on animal experiments and expert judgment. It is for this reason that biokinetic modeling involves appreciable uncertainties, and the results should be interpreted with care (NCRP 2009).

Figure 2.15 demonstrates the predictions of the recommended biokinetic ICRP model for ^{137}Cs both in terms of daily urinary excretion and whole-body retention. The left panel of Fig. 2.15 suggests that after an acute intake of 1 Bq of ^{137}Cs, about 15 mBq are excreted the first day. A maximum occurs after the second day (23 mBq), followed by a continuous decrease thereafter which includes a fast ($t < 10$ d) and a slow ($t > 10$ d) component. Accordingly, after one day, almost 99% of the incorporated ^{137}Cs is still retained in the human body (right panel of Fig. 2.15). This fraction decreases continuously thereafter. Note that about 100 days after intake, about 50% of the incorporated ^{137}Cs is still in the human body. The model thus predicts a biological half-life for ^{137}Cs of about 100 days.

Simulations using the anthropomorphic phantom and radiation transport (i.e., calculation of SEE($T \leftarrow S, R$) values) and those using biokinetic models (i.e., calculation of $A_S(t)$) finally allow calculation of equivalent doses to each organ of the human body, by means of eq. (2.5). Once these equivalent doses are multiplied by

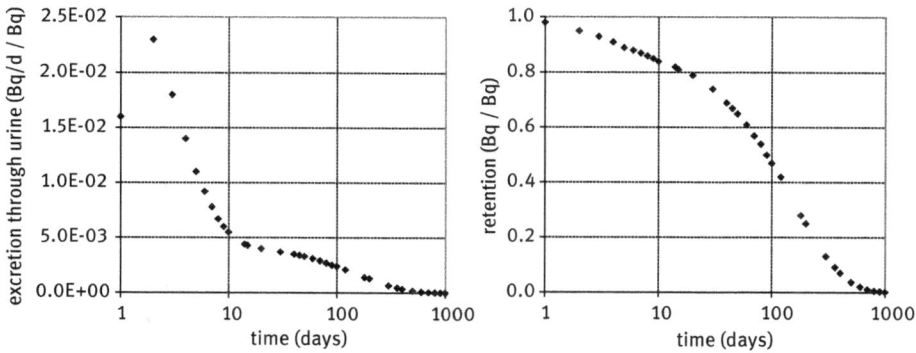

Fig. 2.15: Results of the ICRP biokinetic model for ^{137}Cs, after acute intake of 1 Bq of ^{137}Cs: Left panel – daily urinary excretion; right panel – whole-body retention (BMU 2007).

the proper tissue weighting factors (Tab. 2.3), one can finally calculate the effective dose that results from incorporation of 1 Bq of the investigated radionuclide. Based on these calculations, as far as ^{137}Cs is concerned, a single acute intake of 1 Bq would result in an overall effective dose of $1.3 \cdot 10^{-8}$ Sv.[19]

2.3.3 Determination of effective dose after intake of ^{137}Cs from forest products such as mushrooms

The previous section introduced how effective dose can be calculated after intake of 1 Bq of a certain radionuclide. However, the question of how to quantify the activity that might have been incorporated has not yet been addressed. Figure 2.15 suggests at least two options available – excretion analysis (left panel in the figure) and whole-body counting (right panel in the figure).

2.3.3.1 Excretion analysis

The activity of the radionuclide of interest could be determined in 24-h urine samples. If, for example 10 days after a single ingestion in the 24-h urine sample of an individual, a ^{137}Cs activity of A Bq was measured, then from Fig. 2.15 (left panel), it follows that a ^{137}Cs activity of $A/0.005$ Bq $= 200 \cdot A$ Bq was incorporated.

---

19 Such values are tabulated and can be found for all relevant radionuclides, for different intake scenarios (acute, chronic), contamination pathways (inhalation, ingestion, wound uptake), chemical speciation (soluble, nonsoluble compounds), and aerosol sizes (relevant for the inhalation pathway) (ICRP 1994).

2.3.3.2 Whole-body counting

In a similar way, if the whole-body activity of an individual was measured 50 days after single ingestion by means of a whole-body counter to be B Bq, then it follows from Fig. 2.15 (right panel) that initially a ^{137}Cs activity of $B/0.65$ Bq $= 1.5 \cdot B$ Bq was incorporated.

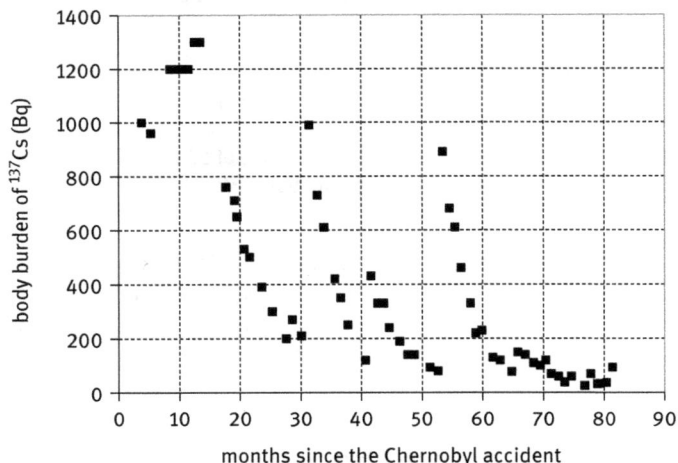

Fig. 2.16: Monthly ^{137}Cs whole-body activity of an individual living in the South of Germany after the Chernobyl accident; peaks 31, 41, and 53 months after the accident are due to ingestion of forest products such as mushrooms (figure modified from Rühm et al. (1999)).

This procedure was followed after the Chernobyl accident when large areas of Western Europe were contaminated by ^{137}Cs. Figure 2.16 shows the result of a series of monthly measurements performed with a whole-body counter on one individual.[20]

The time series depicted in Fig. 2.16 indicates a maximum in ^{137}Cs body burden of about 1300 Bq about one year after the accident followed by a continuous decrease thereafter. Obviously, 30 months after the accident a new intake of about 800 Bq occurred, 10 months later another one of about 300 Bq, and 12 months still another one of about 800 Bq. These intakes occurred in autumn and were due to the ingestion of forest products such as mushrooms that were contaminated with ^{137}Cs. If these incorporations were due to single intakes, then one can estimate that the first and third peak resulted in an additional effective dose of

---

20 In this case the whole-body counter was located at the Federal Office of Radiation Protection, Neuherberg, Germany. It consisted of 4 NaI scintillation counters that were placed above and below the individual who lay on a stretcher.

Tab. 2.7: Natural sources of ionizing radiation that contribute to global mean public exposure. Numbers are given as annual effective dose in mSv (UNSCEAR 2010).

Natural sources	Effective dose (mSv per year)	
	Average	Range
Total	2.4	1.0–13
Terrestrial radiation	0.48	0.3–1.0[a]
Cosmic radiation (sea level)	0.39	0.3–1.0[b]
Inhalation of radon	1.26	0.2–10.0[a]
Ingestion of radionuclides	0.29	0.2–1.0[c]

[a]Depending on radionuclides present in soil and building materials.
[b]Range from sea-level to high ground-level altitudes.
[c]Depending on radionuclides in food and drinking water.

about $1.3 \cdot 10^{-8}$ Sv/Bq $\cdot$ 800 Bq $\approx$ 10 µSv, while the second peak resulted in an additional effective dose of about $1.3 \cdot 10^{-8}$ Sv/Bq $\cdot$ 300 Bq $\approx$ 4 µSv.[21]

2.4 Effective doses from natural and artificial sources

Due to the presence of primordial, cosmogenic, and anthropogenic radionuclides in the environment, we all are exposed to ionizing radiation. There are at least four major natural sources of ionizing radiation which result in a total annual effective dose of the general population of about 2 mSv. Table 2.7 summarizes the contributions of these sources of ionizing radiation to annual effective dose as estimated by the United Nations Scientific Committee on the Effects of Atomic Radiation (UNSCEAR). Note that the numbers given "should be seen only as reference values and not as specific to any particular place" (UNSCEAR 2010).

Additionally, there are artificial sources of ionizing radiation that might also contribute to annual effective dose of the general population. The most important contribution comes from diagnostic medical procedures, which was estimated to be about 1.8 mSv per year, in Germany. Interestingly, doses from these procedures to the population are continuously increasing in many industrialized countries because the number of CT scans is continuously increasing. Note that a typical CT

---

21 For comparison, typically our body contains about 60 Bq of ^{40}K per kg body mass (for a body mass of 70 kg our body thus contains about 4000 Bq ^{40}K). This corresponds – using the tabulated ICRP dose coefficient for ^{40}K of 3 µSv/a per Bq/kg body mass for adults – to a yearly effective dose of 3 µSv $\cdot$ kg/(a $\cdot$ Bq) $\cdot$ 60 Bq/kg = 180 µSv per year, from ^{40}K.

scan of the chest results in an effective dose of the order of 10 mSv or more. Compared to that, in most cases, all other artificial sources of ionizing radiation such as exposures from the atmospheric nuclear tests in the 1950s and early 1960s, fallout from Chernobyl, emissions from running nuclear installations, and air travel, are negligible. Table 2.8 summarizes the contributions of artificial sources of ionizing radiation to annual effective dose[22] for the German population. Note that the numbers given represent mean values averaged over the whole population. Thus, individual effective doses for example of patients who require many X-ray images or CT scans might be much higher than the average number given in the table. The numbers given in Tab. 2.8 do not include doses from radiotherapy. Here, the local dose to a tumor is very high and of the order of 60 Gy or more. Because healthy tissue is also exposed, albeit to a much lesser extent, a patient may receive – depending on the treatment modality – a considerable effective dose from the radiation. For obvious reasons, however, radiation protection issues do not play a major role and tumor control is most important. Note also that for certain professions, individual annual effective doses might also be high: for example, the mean annual effective dose of pilots and cabin crew monitored in Germany due to their exposure with cosmic radiation was of the order of 2.4 mSv, in 2009.

Tab. 2.8: Artificial sources of ionizing radiation that contribute to mean public exposure. Numbers are given for Germany 2010, and as annual effective dose in mSv (BMU 2012).

Artificial sources	Mean annual effective dose
Total	1.8 mSv
Medical diagnostics	1.8 mSv
Applications in research, technology, household	<0.01 mSv
Nuclear installations	<0.01 mSv
Global weapons fallout	<0.01 mSv
Chernobyl fallout	<0.012 mSv

2.5 Outlook

Radiation research focuses on the interaction of ionizing radiation with matter and – in particular – with biological tissues. Understanding the effects that occur on a molecular, cellular, and organism level is a prerequisite for protecting humans against ionizing radiation without unduly preventing them from its beneficial use.

---

22 As estimated by the German Ministry for Environment, Nature Conservation and Reactor Safety (BMU).

- Is an exposure to 2–4 mSv per year dangerous (Tabs. 2.7 and 2.8)?
- What is the risk of 20 mSv per year which is the dose limit proposed by ICRP for radiation workers?

A reliable answer to these questions requires epidemiological studies on human cohorts that were exposed to such dose rates, in particular with respect to leukemia and cancer incidence and mortality. In this respect, the *LSS* on the atomic bomb survivors of Hiroshima and Nagasaki provides the most important results to date because of the long follow-up (since 1950), the size of the cohort (initially about 120,000 individuals), and its age and sex distribution (which is typical for a population including males and females, and children, adolescences, and adults). Even if data on all tumor types are combined, however, a statistically significant radiation-induced excess of solid cancer cases can only be found above a colon equivalent dose of about 100 mSv (Fig. 2.4). Accordingly, if one is interested in the radiation risk below such a dose one has to extrapolate, although it is far from being clear how this should be done.

For example, for radiation protection purposes, ICRP suggests use of a linear extrapolation to lower doses and assumes that no threshold exists in this dose region (linear-no-threshold (LNT) model). While this model appears justified for a robust and simple concept that needs to be used in daily radiation protection work, recent research in radiation biology points at a number of processes that are nonlinear at low doses. For example, irradiation of single cells with, e.g., alpha particles showed that neighboring cells that were not hit at all by any alpha particle (i.e., were not exposed to any radiation dose) went to apoptosis, i.e., initiated their own cell death. This so-called *bystander effect* may result – at least on a cellular level – in a modified dose effect at low doses. Genomic instability is another such example because it was demonstrated that irradiated cells survived, but several cell cycles (generations) later, progeny cells that were not at all exposed became unstable and died. In contrast, the so-called *adaptive response* of cells suggests that low doses of ionizing radiation might stimulate cellular repair mechanisms, and accordingly, cells exposed to low doses may be more resistant to a higher dose later than cells that were not exposed to the low dose before. In the extreme case, exposure to low doses of ionizing radiation may even be beneficial because it may stimulate the immune system of an organism, a speculation, which is called *hormesis*.

Whether these effects – that have mostly been shown to exist on a cellular level – are of any relevance to the development of cancer among humans is currently unknown, and research programs have been initiated worldwide to tackle this problem. Other open issues include the observation that there appears to be a statistically significant radiation-induced excess of noncancer diseases such as diseases of the cardio-vascular system among the atomic bomb survivors and some other epidemiological cohorts, which poses the question of whether, and if yes, how to include this in the current radiation protection system. Finally, it is well known that exposure of cell cultures to a certain dose at a high dose rate reduces cell survival compared to exposure to the same dose at low dose rate. It is for this reason that risk coefficients deduced from the Japanese atomic bomb survivors (who were exposed within a short period of time – high dose rate) are reduced by a certain factor before they are applied for radiation protection purposes (where exposure to a low dose rate is typical). In contrast, recent epidemiological studies on cancer incidence and mortality of nuclear workers hint at a similar risk to that deduced from the atomic bomb survivors – an observation that questions application of the above-mentioned factor.

All these effects and observations demonstrate that much more research is needed to bridge the gap between radiobiological studies performed on molecular and cellular levels and epidemiological studies performed on the human organism.

References

Federal Ministry for Environment, Nature Conservation and Reactor Safety (BMU). Umweltradioaktivität und Strahlenbelastung im Jahr 2010. Unterrichtung durch die Bundesregierung (in German), 2012.

Federal Ministry for Environment, Nature Conservation and Reactor Safety (BMU). Richtlinie für die physikalische Strahlenschutzkontrolle zur Ermittlung der Körperdosis, Teil 2: Ermittlung der Körperdosis bei innerer Strahlenexposition (Inkorporationsüberwachung) (§§ 40, 41 und 42 StrlSchV) (in German), 2007.

International Commission on Radiological Protection (ICRP). Conversion Coefficients for Radiological protection quantities for external radiation exposures. ICRP Publication 116, Ann ICRP 2010, 40(2–5), pp. 1–257.

International Commission on Radiological Protection (ICRP). Adult reference computational phantoms. ICRP Publication 110, Ann ICRP 2009 39(2), pp. 1–164.

International Commission on Radiological Protection (ICRP). Recommendations of the International Commission on Radiological Protection. ICRP Publication 103, Ann ICRP 2007, 37 (2–4), pp. 1–332.

International Commission on Radiological Protection. Human respiratory tract model for radiological protection. ICRP Publication 66, Ann ICRP 1994 24(1–3), 1–482.

International Commission on Radiological Protection (ICRP). Recommendations of the International Commission on Radiological Protection. ICRP Publication 60, Ann ICRP 1991, 21, 1–201.

National Council on Radiation Protection and Measurements (NCRP). Uncertainties in internal radiation dose assessment, Report No. 164, 2009.

Ozasa K, Shimizu Y, Suyama A, Kasagi F, Soda M, Grant EJ, Sakata R, Sugiyama H, Kodama K. Studies of the mortality of atomic bomb survivors, Report 14, 1950–2003: An Overview of cancer and noncancer diseases. Radiat Res 2012, 177, 229–243, and Radiat Res 2013, 179, e0040–0041.

Rühm W, König K, Bayer A. Long-term follow-up of the ^{137}Cs body burden of a reference group after the Chernobyl accident – A means for the determination of biological half-lives of individuals. Health Phys 1999, 77, 373–382.

United Nations Scientific Committee on the Effects of Atomic Radiation (UNSCEAR). Sources and effects of ionizing radiation. Report to the General Assembly with scientific annexes. United Nations, New York, 2010.

Klaus Eberhardt

3 Elemental analysis by neutron activation: NAA

Aim: This chapter introduces the principle and various applications of Neutron Activation Analysis (NAA) as a valuable tool for elemental analysis for a variety of different sample matrices. Its simple idea is to (i) convert a stable atom of a given chemical element to be analyzed into a corresponding radioactive isotope (i.e., "activate" it) in order to (ii) quantify its absolute radioactivity which is equivalent to the number of stable atoms of the chemical element of interest. Because the projectiles used to induce the corresponding nuclear reaction are neutrons, the method is referred to as "Neutron Activation Analysis".

The neutrons themselves are of mostly thermal, but sometimes also higher kinetic energy. The neutron absorption reaction is of the type $^AB(n,\gamma)^{A+1}C$, whereby the neutron-rich nucleus undergoes further radioactive transformation. The "analytics" is to measure typically the characteristic electromagnetic radiation as emitted by the nuclear reaction products. This characteristic γ-radiation is like a fingerprint for a given nucleus and thus allows the determination of the concentration of the corresponding element in the sample.

Although NAA is a mature analytical method, it offers excellent multi-elemental capabilities and remains an important tool in trace elemental analysis due to its high sensitivity that allows the determination of constituents present in very small concentrations. NAA is often used as a reference method for the certification of materials. In many cases no chemical pre-treatment of the sample is required, and NAA thus represents a "noninvasive" technology.

3.1 History

3.1.1 The neutron

In 1932 James CHADWICK directed α-particles as emitted from a polonium source into a beryllium foil. The α-particles interacted with ^9Be nuclei forming a ^{12}C nucleus and a highly energetic particle carrying no electric charge. The particles released were detected through their interaction with paraffin resulting in the formation of protons with a maximum kinetic energy of 5.7 MeV that could easily be observed with an ionization chamber. It was obvious that only a particle having almost the same mass as a hydrogen atom could produce proton radiation of such energy. The

https://doi.org/10.1515/9783110742701-003

experimental results showed that a collision of α-particles with beryllium nuclei releases massive neutral particles, which CHADWICK named neutrons.[1]

Figure 3.1 schematically shows the ^{9}Be(α,n)^{12}C reaction as used in CHADWICK's experiments as a source for neutrons with an average kinetic energy of 5.5 MeV. Since neutrons have no electrical charge they have the ability to penetrate deep layers of materials and easily interact (or "activate") with the positively charged nuclei of the target atoms, inducing so-called neutron capture reactions.

The activation could be carried out with small neutron sources consisting, e.g., of beryllium powder mixed with ^{226}Ra as the α-particle source (see Fig. 3.1). Here, about $2 \cdot 10^7$ neutrons per second are emitted per gram radium used.

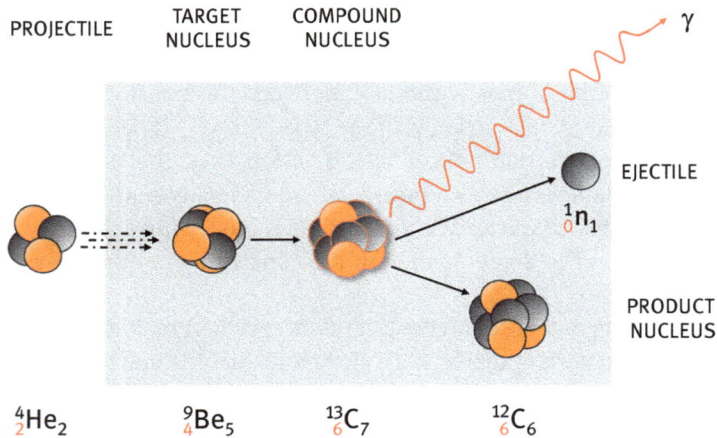

Fig. 3.1: Formation of neutrons via the reaction ^{9}Be(α,n)^{12}C induced by α-particles emitted from, e.g., ^{226}Ra. The neutron is ejected with an average kinetic energy of 5.5 MeV.

3.1.2 The neutron capture nuclear reaction

Neutron capture reactions turned out to be an efficient method for the production of radioisotopes, such as ^{128}I ($t^{1/2}$ = 25 min) via the ^{127}I(n,γ) reaction. Already in 1934 FERMI and coworkers described the "production" of artificial radioactivity by means of neutron bombardment of stable nuclei. The FERMI group showed that the activation yields could be increased substantially with neutrons slowed down (moderated), e.g., in a hydrogen containing material.[2]

---

1 The Nobel Prize in Physics 1935 awarded to James CHADWICK *"for the discovery of the neutron"*.
2 For protons as a moderator material approximately 20 collisions are required to slow down neutrons from several MeV to thermal energies. For this, shielding of the source with a 20-cm thick paraffin layer is adequate. Moderation is accomplished within a time window of a millisecond.

In their early experiments with rare earth elements George HEVESY and Hilde LEVI observed that some of the chemical elements – in particular dysprosium, europium, and holmium – became highly radioactive when exposed to neutrons. HEVESY recognized that this method in turn could be used for the qualitative and quantitative detection of the rare earth elements, and they were the first to report on the new method of "Activation Analysis", i.e., "Neutron Activation Analysis" (NAA), and its application for the analysis of dysprosium in various rare earth element salts (Hevesy and Levi 1935). Table 3.1 comprises some of the results they obtained on the intensity of the induced radioactivity of rare earth elements after neutron bombardment.[3]

Tab. 3.1: Intensity of induced radioactivity of rare earth elements after neutron bombardment, as obtained by Hevesy and Levi (1935). The nuclear reactions on, e.g., the stable isotopes of europium, dysprosium, and holmium are $^{151}Eu(n,\gamma)^{152m2}Eu$, $^{164}Dy(n,\gamma)^{165}Dy$, and $^{165}Ho(n,\gamma)^{166}Ho$, respectively. The activity values are given in arbitrary units, normalized to the radiation intensity resulting from the activation of dysprosium.

Element (analyte)	Product isotope		Induced radioactivity (arbitrary units)
	Half-life	Isotope	
Y	70 h	^{90}Y	0.5
La	1.9 d	^{140}La	–
Ce	–	–	–
Pr	19 h	^{142}Pr	4.5
Nd	1 h	^{149}Nd	0.04
Sa	40 min	^{155}Sm	0.6
Eu	9.2 h	$^{152m2}Eu$	80
Gd	8 h	^{159}Gd	very low
Tb	3.9 h	–	–
Dy	2.5 h	^{165}Dy	100
Ho	35 h	^{166}Ho	20
Er	12 h	^{169}Er	0.35
Yb	3.5 h	^{177}Yb	0.25
Lu	6 d/4 h	$^{177g}Lu/^{176m}Lu$	1.4/1

After these pioneering experiments the new method of activation analysis was not much used. This was mainly due to the lack of stronger neutron sources.[4] The

3 At that time, element discrimination was mainly based on differences in neutron capture cross sections and half-life of the corresponding isotopes rather than the energy of the emitted radiation.
4 Neutron sources are available through, e.g., ^{226}Ra-derived α-particle induced nuclear reactions on, e.g., Be, from spontaneous fission nuclides such as ^{252}Cf (cf. Chapter 10 in Vol. I) or from nuclear reactors (cf. Chapter 13 in Vol. I). All differ in terms of the maximum kinetic energy of the

developing reactor technology in the 1950s and later the availability of nuclear reactors specially dedicated to research and education strongly promoted the application of NAA and a rapid growth occurred.[5] In addition, new types of radiation detectors were developed and the proportional counters used so far were replaced by energy-resolving γ-ray spectrometers like scintillation counters and later by high-resolution semiconductor detectors coupled to multichannel pulse-high analyzers and convenient PC-based software packages for data acquisition and storage, cf. Chapter 1.

3.2 Principle of neutron activation analysis

3.2.1 Formation of nuclear reaction products

The basic analytical concept of NAA is to induce a nuclear reaction by means of neutrons as projectiles on a stable nuclide of a chemical element existing in unknown (and typically very low concentration) within the sample to be studied. The neutron capture reaction typically is of type $^A B_N(n,\gamma)^{A+1}C_{N+1}$, i.e., an isotope of the same chemical element is formed, cf. Chapter 13 of Vol. I. If this – now neutron-rich – isotope (A_{N+1}) is unstable, it stabilizes by means of primary β^--transformation, cf. Fig. 13.25 in Vol. I and Fig. 3.2 below. In both cases, a secondary transformation to a product nucleus often results in the emission of γ-rays characteristic for the particular isotope, which can be used for qualitative and quantitative determination of the elements of interest (often referred to as "analyte"). Besides single neutron capture double neutron capture is also possible as well as, e.g., (n,p) and (n,α) reactions (see Section 3.2.2 for a more detailed description of neutron capture reactions).

The basic principle of NAA is schematically shown in Fig. 3.3, exemplified for Fe as the analyte to be determined. The initial neutron absorption process creates a ^{59}Fe compound nucleus in an energetically excited state that transforms into its ground state by the emission of one or more γ-rays. These are called "prompt" because they appear almost immediately.[6] The compound nucleus is neutron rich and thus will undergo β-transformation followed by the emission of one or more γ-rays from the excited daughter nucleus (here ^{59}Co). The latter process is referred to as "delayed"

---

neutrons generated, the either discrete or continuous energy distribution, and the neutron flux per cm^2 and second.

5 A research reactor with a thermal power in the order of 10–20 MW can provide thermal neutron fluxes which are about a factor of 10^7 higher compared to Ra/Be sources, resulting in much stronger activation, cf. Chapter 11.

6 Normally the half-lives of excited nuclear states are around 10^{-12} s, though there are quite a few exceptions of longer-lived isomeric states with measurable half-lives, cf. Chapter 11 in Vol. I.

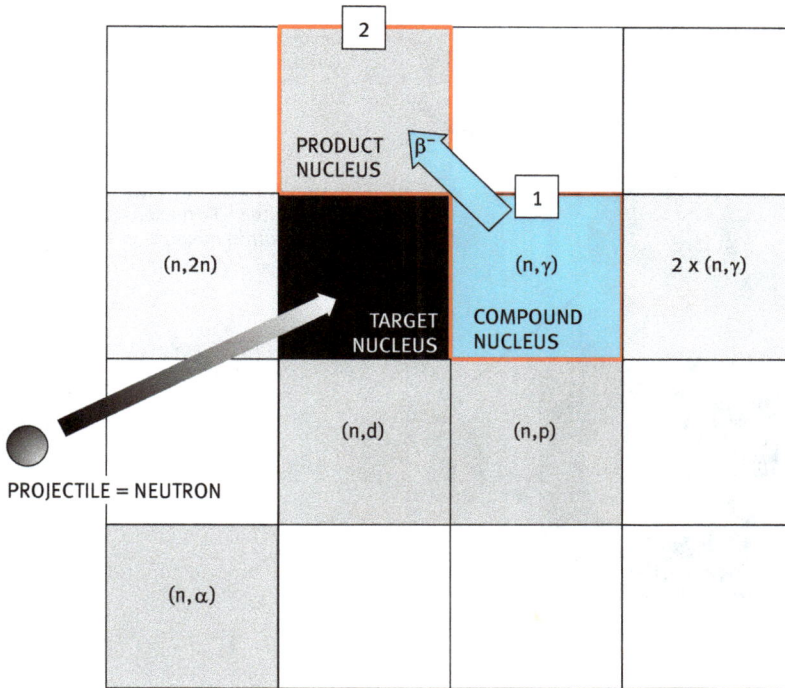

Fig. 3.2: Single neutron absorption reaction leading to a neutron-rich compound nucleus (1) that undergoes radioactive transformation into the product nucleus (2).

since the delay time corresponds to the β-transformation half-life. Both the prompt and delayed γ-rays possess unique energies that are characteristic for the particular excited nucleus and thus allow the identification and quantitative determination of the nuclides of the elements of interest. Due to the correlation between the radioactivity and the number of the target nuclei, the concentration of the target element can be derived.[7]

3.2.2 Neutron absorption cross sections

The neutron absorption cross section strongly depends on neutron energy. Neutrons that are slowed down to energies that are comparable with thermal energies at ambient temperature are called *thermal neutrons* and possess a kinetic energy of about 0.025 eV that corresponds to the most probable velocity in a MAXWELLian

---

7 Cf. Fig. 5.6 in Chapter 5.

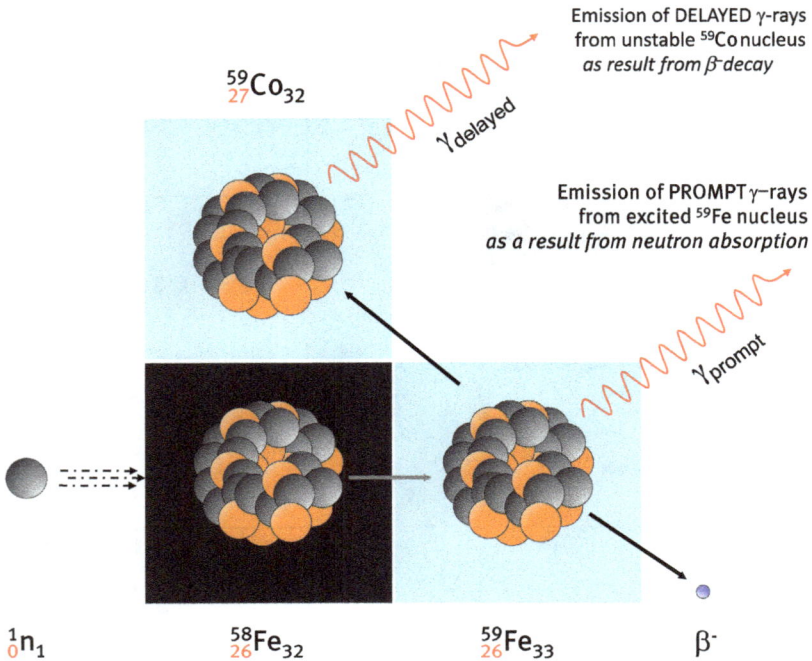

Fig. 3.3: The basic principle of NAA exemplified for the quantification of stable iron (^{58}Fe) through its neutron-capture nuclear reaction product ^{59}Fe (via ^{58}Fe(n,γ)^{59}Fe) and ^{59}Co (via ^{59}Fe → β$^-$ → ^{59}Co).

distribution at 20 °C. Neutrons in the energy range of 1 eV to 100 keV are called *epithermal* neutrons.[8,9]

Neutron absorption cross sections exhibit strong fluctuations in the energy range of 1 eV up to a few hundred eV. This allows selective activation of certain nuclides. Thus, a special form of NAA called *epithermal* NAA can be performed by covering the sample with, e.g., a Cd shield with a thickness of about 1 mm that completely blocks thermal neutrons due to absorption. Figure 3.4 shows the absorption cross section for ^{107}Ag as a function of neutron energy. For energies below about 1 eV the cross section σ is proportional to $(1/\sqrt{E_n})$, where E_n is the neutron energy. For higher energies, large fluctuations with energy are observed, called "resonances". These resonances are due to the nuclear shell structure and lead to a drastic increase of σ whenever the neutron energy corresponds to a certain neutron energy state in the target nucleus.

---

8 Cf. Figure 13.27 in Vol. I.

9 The ratio between the flux of thermal and fast neutrons depends on the kind of neutron source and the surrounding shielding material.

All neutrons with energies above 0.1 MeV (fast neutrons) contribute very little to (n,γ) reactions. They rather can induce nuclear reactions where one or more nuclear particles are ejected, e.g., in (n,p), (n,d), (n,α), (n,n′), and even (n,2n) reactions (see Fig. 3.2), that all have a certain energy threshold.[10]

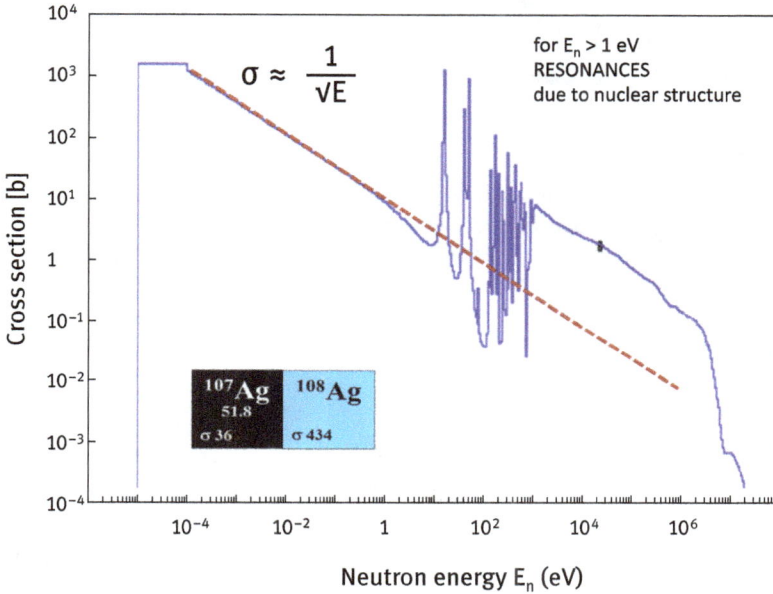

Fig. 3.4: Neutron absorption cross section for ^{107}Ag as a function of neutron energy (taken from IAEA 2014).

NAA with fast neutrons is a very selective technique; however, it can lead to particular interferences in trace element analysis. One example is the determination of traces of Na in an Al matrix. Here, ^{24}Na is basically produced via the reaction ^{23}Na (n,γ)^{24}Na with thermal neutrons. Simultaneously, it is produced via the (n,α) reaction on ^{27}Al (the bulk material) and hence hinders the determination of the initial sodium concentration. Interferences of the same kind can occur with Mg via the reaction ^{27}Al(n,p)^{27}Mg (cf. Fig. 3.5). All reactions occur simultaneously during the irradiation of an Al sample in a neutron field as delivered from a nuclear reactor and thus might hinder the elemental determination of Na or Mg, respectively.

---

10 It should be noted that charged particle emission is not restricted to reactions with fast neutrons. It can also occur in slow neutron capture with light Z-nuclei. One prominent example is the reaction ^{10}B(n,α)^{7}Li with a cross section as high as 3837 b for thermal neutrons. This reaction is often applied in neutron detectors.

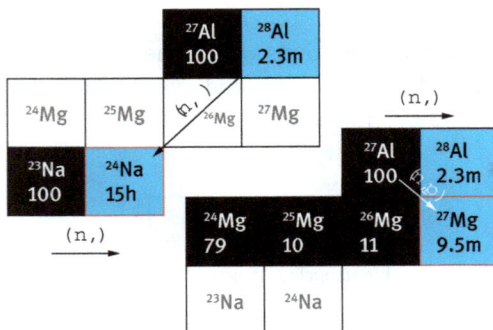

Fig. 3.5: Formation of ^{24}Na through the (n,γ) reaction on ^{23}Na (stable) and via the (n,α) reaction on ^{27}Al (stable), respectively. ^{27}Mg is produced through the (n,p) reaction on ^{27}Al. All reactions occur simultaneously during the irradiation of an Al sample in a neutron field as delivered from a nuclear reactor.

3.2.3 The different forms of NAA

In NAA, usually γ-rays emitted from excited nuclei are used for determination of the activation products (see also Fig. 3.3 and corresponding explanation). However, a special form of activation analysis used to determine the content of fissile material in a sample makes use of so-called *delayed neutrons* as emitted from highly excited fission product nuclei.

If the prompt γ-rays are detected, the technique is referred to as "Prompt Gamma Neutron Activation Analysis" (PGNAA).[11] If the delayed γ-rays are utilized, the acronym DGNAA is used to denote "Delayed Gamma Neutron Activation Analysis". Since DGNAA is much more widely used, the acronym NAA usually refers to DGNAA.

PGNAA is generally performed by using a neutron beam extracted through a reactor beam port. The sample is placed at the exit aperture of the port with the counting equipment adjacent to it, perpendicular to the neutron beam axis. PGNAA is performed in real time, that means that exposing the sample to a neutron flux and data acquisition and analysis are performed in coincidence. Since the neutron flux in a reactor beam port is normally orders of magnitude lower compared to in-core neutron flux densities, PGNAA is limited to nuclides with large neutron absorption cross sections or to nuclides that transform too rapidly and thus are not accessible off-line by DGNAA.

---

11 Reactor-based PGNAA was first applied in 1966 at the Institut Laue-Langevin in Grenoble (France).

In contrast, DGNAA is flexible in time: the sample is placed in an in-core irradiation position while the irradiation time can be adjusted over a wide range if (a) one desires to identify elements at very low concentrations, (b) the nuclides show only low absorption cross sections, or (c) the resulting radioactive isotopes have long half-lives. Subsequent to irradiation, the sample is removed from the irradiation position and, if necessary, allowed to transform until safe to handle or to improve sensitivity for a long-lived activation product that suffers from interference from a short-lived nuclide by waiting for the latter to transform. The International Atomic Energy Agency (IAEA) provides nuclear data databases as well as data evaluation software for NAA, including PGNAA (IAEA 2011).

A special form of NAA is the so-called Delayed Neutron Activation Analysis (DNAA). DNAA is applied for the fast determination of fissile nuclides, such as ^{233}U, ^{235}U, and ^{239}Pu, in samples of various chemical composition. DNAA uses the counting of β-delayed neutrons[12] emitted from neutron-rich fission products as obtained by irradiation of fissile material like ^{235}U or ^{239}Pu. Since neutrons are detected, this method is practically background-free and there are no interferences from γ-emitting isotopes. From the nuclides produced in nuclear fission, about 110 are known precursors of β-delayed neutron emission with half-lives ranging from milliseconds to minutes (Tomlinson 1973). Delayed neutrons emitted by the long-lived fission fragment isotopes ^{87}Br, ^{88}Br, and ^{137}I are generally measured. Furthermore, this technique can also be applied for the determination of thorium in samples containing uranium. ^{232}Th fissions only with neutrons of an energy of 1 MeV and above and, therefore, a cadmium cover for the absorption of thermal neutrons is used for eliminating delayed neutrons from thermal neutron induced uranium fission since in natural samples uranium is a common contaminant. Typically, the samples are irradiated for 2 min and after a delay time of 15–20 s, they are transferred to a detector arrangement with ^{3}He proportional tubes in a circular arrangement and counted for 1 min.[13] The delayed neutrons are emitted with energies in the order of MeV, are thermalized by paraffin and polyethylene, and finally are detected with an efficiency of about 30%. Figure 3.6 shows a schematic view of a detector setup as used for the detection of delayed neutrons in DNAA at the TRIGA-type research reactor of the University of Mainz, Germany (Eberhardt and Trautmann 2007).

In general, NAA (here DGNAA) can be divided into two categories, depending on whether sample analysis is performed with or without any chemical separation procedure:

---

12 cf. Chapter 10 in Vol. I.

13 For such a short irradiation cycle typically a fast pneumatic transfer system (also called "rabbit system") is used for sample transfer to and from the irradiation position. Such systems are installed in many research reactors and allow sample transfer times of the order of a few seconds.

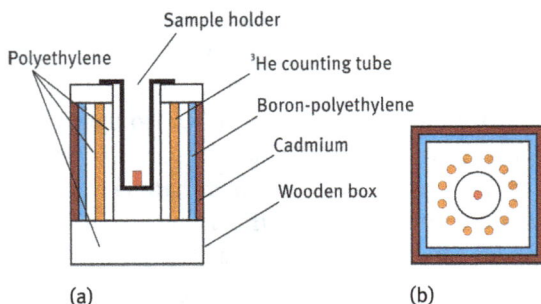

Fig. 3.6: Vertical (a) and horizontal (b) cross section view of the neutron detector arrangement used for DNAA.

Instrumental Neutron Activation Analysis (INAA): Sample irradiation time, transformation times, and counting times are optimized such that no chemical treatment of the sample is necessary.

Radiochemical Neutron Activation Analysis (RNAA): Chemical separation is involved in sample preparation for enrichment of the analyte prior to irradiation to gain sensitivity or to overcome matrix effects that interfere with the detection.

DNAA is applied both with and without sample pre-treatment; see Eberhardt and Trautmann (2007) and references therein.

3.2.4 High resolution γ-ray spectroscopy

Since nuclear excitation energies are specific for a certain nuclide, the electromagnetic radiation emitted from an excited nucleus is characteristic for this particular isotope. Hence, this unique radiation pattern serves like a "fingerprint" for determination of the isotopic composition of the irradiated sample and thus for elemental analysis using γ-ray spectroscopy, cf. Chapter 1. The equipment needed for high-resolution γ-ray counting consist of a semiconductor detector, which is either a high-purity germanium-crystal (HPGe) or a lithium-drifted germanium-crystal (GeLi), with a high-voltage supply (HV) and a preamplifier. This system is connected to an amplifier, a multichannel pulse-height analyzer (MCA), and a computer-based data acquisition system to link the hardware of the detector system to the data storage memory of the computer. The software will record the spectral parameters in combination with important physical data like, e.g., starting time, sample cooling time, energy calibration data, and user information on irradiation parameters as well as a sample description. For NAA, a detector with high resolution (about 1.8 keV at 1332 keV), high efficiency (>35%), and a high peak-to-Compton ratio (>60:1) is desirable. The detector should be shielded from

background radiation by 10 cm of lead lined with copper as absorber for secondary lead X-rays.

Figure 3.7 shows as an example the γ-ray spectrum of a copper sample irradi-
ated for 6 h at a thermal neutron flux of $7 \cdot 10^{11} \, cm^{-2} \, s^{-1}$. Concentrations of silver,
antimony, and selenium in this sample are in the order of 3–11 ppm. Table 3.2 comprises the relevant activation products, their half-life, and the corresponding element concentration.

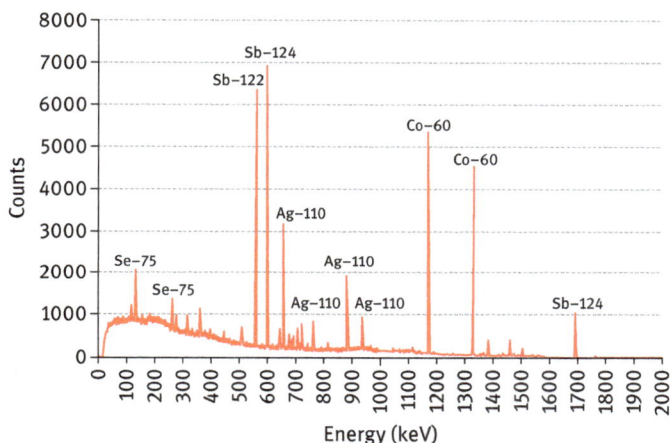

Fig. 3.7: Spectrum of a copper sample irradiated for 6 h with a thermal neutron flux
of $7 \cdot 10^{11} \, cm^{-2} \, s^{-1}$.

The copper matrix was dissolved in 6 M nitric acid. An aliquot of 2 ml containing
about 200 mg of copper was then irradiated in a standard irradiation position of the
100 kW TRIGA-type research reactor at the University of Mainz, Germany. The sample
was allowed to transform for 15 days to get rid of a number of short-lived nuclides
that interfere with the measurement of the long-lived activation products. The sample
was then measured for 24 h with a high-purity germanium detector. Finally, neutron
capture products such as ^{60}Co ($t\frac{1}{2} = 5.5$ a), ^{75}Se ($t\frac{1}{2} = 120$ d), ^{110}Ag ($t\frac{1}{2} = 250$ d), ^{122}Sb
($t\frac{1}{2} = 2.7$ d), and ^{124}Sb ($t\frac{1}{2} = 60$ d) were identified. Their activity is used to determine
the concentration of the four elements present in the Cu sample.

3.2.5 Activation equation

In neutron activation, radioactive nuclides are first produced and subsequently
transformed according to their particular transformation constants. The nuclear reaction rate R denotes the number of radioactive nuclei formed per unit time in a
reaction:

Tab. 3.2: Long-lived activation products, corresponding half-lives, and resulting element concentrations (in ppm) as deduced from a γ-spectroscopic analysis of an irradiated Cu sample. See text for further details.

Reaction	Cross section (b)	Half-life (d)	Concentration (ppm)
^{59}Co(n,γ)^{60}Co	37	1924	11
^{74}Se(n,γ)^{75}Se	50	120	3
^{109}Ag(n,γ)^{110}Ag	4	250	11
^{122}Sb(n,γ)^{122}Sb	6	2.7	5
^{124}Sb(n,γ)^{124}Sb	0.15	60	5

$$R = N_0 \sigma \Phi. \tag{3.1}$$

Here, N_0 is the number of target atoms, σ denotes the cross section for neutron absorption in barn (b), and Φ is the neutron flux given in $(cm^{-2} s^{-1})$. N_0 can be written as:

$$N_0 = \frac{m \, H}{M} N_A. \tag{3.2}$$

Here, m is the mass [g] of the element in sample, H is the abundance of the stable target isotope involved in the nuclear reaction,[14] M is the molar mass of the element (g mol^{-1}), and N_A is the AVOGADRO number. Already during the activation process, the number $\hat{N}$ of the radioactive nuclei formed decreases since they transform simultaneously according to the differential equation:

$$\frac{d\hat{N}}{dt} = R - \lambda \hat{N} \tag{3.3}$$

In this equation, the term $-\lambda\hat{N}$ describes the radioactive transformation of the nuclei. With

$$t^{1/2} = \frac{\ln 2}{\lambda} \tag{3.4}$$

integration of eq. (3.3) leads to the number of radioactive nuclides produced during an irradiation for the time period t^{irr}:

$$\hat{N}(t) = R \frac{1 - e^{-\lambda t}}{\lambda}. \tag{3.5}$$

For the activity $\hat{A}(t)$ of the radionuclide there is

---

14 For example, for neutron activation of a silver sample $H = 0.518$ for ^{107}Ag and $H = 0.482$ for ^{109}Ag (cf. Fig. 3.4).

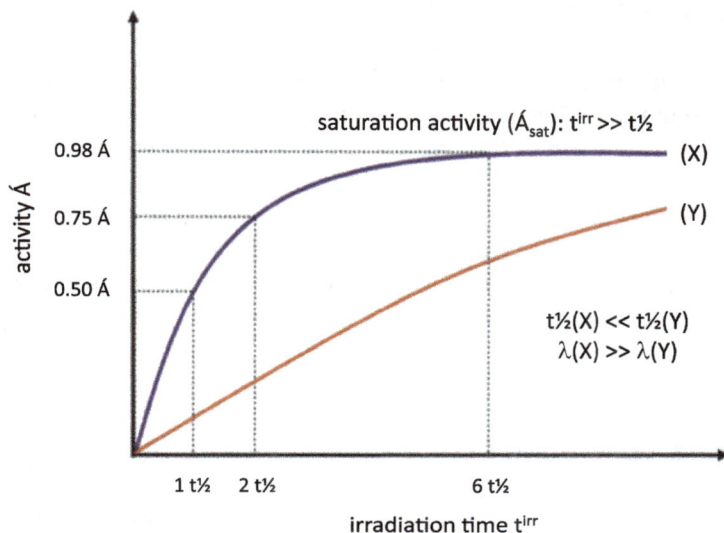

Fig. 3.8: Activation curves for two different nuclides x and y with significantly different half-lives.

$$\acute{A}(t) = \lambda \, \acute{N}(t) = R\left(1 - e^{-\lambda t}\right). \tag{3.6}$$

With eqs. (3.1) and (3.2) the activity $\acute{A}(t)$ becomes:

$$\acute{A}(t) = m\left(\frac{H}{M} N_A\right) \sigma \Phi \left(1 - e^{-\lambda t}\right). \tag{3.7}$$

For irradiation times that are considerably longer than the half-life of the produced nuclide ($t\frac{1}{2}$), the term $(1 - e^{-\lambda t})$ is close to unity, cf. Fig. 3.8, and the saturation activity $(\acute{A}_{sat})$ is defined as:

$$\acute{A}_{sat} = m\left(\frac{H}{M} N_A\right) \sigma \Phi. \tag{3.8}$$

However, saturation might not be reached and the absolute activity obtained after reasonable irradiation time limits the sensitivity of NAA for certain elements. In PGNAA the life-time of the nuclear levels involved is very short and saturation is reached almost immediately. Then the saturation activity becomes equivalent to the reaction rate as expressed in eq. (3.1).

In (delayed) NAA half-lives of product radionuclides can range from seconds (11 s for ^{20}F) up to several years (5.3 a for ^{60}Co). Figure 3.8 shows the activation curves for two nuclides with considerably different half-lives. Here, the half-life of nuclide x is much smaller compared to nuclide y and thus saturation is reached much faster. For long half-lives, the statistical accuracy as deduced from activity measurements can be improved by increasing the irradiation time and thus the produced activity.

3.3 Quantitative element determination

The determination of whether a chemical element is present in a sample or not as well as what its absolute mass (or its concentration) is, based on the activation equations where the activity produced during the irradiation is proportional to the mass of the analyte, see eq. (3.7). In reality, the sample is placed into a well-known neutron field for a certain period (irradiation time) to produce the unstable neutron capture nuclear reaction products. The activity of the produced isotope corresponds to its number of atoms $\hat{N}$ according to $\hat{A} = \lambda \hat{N}$. Two methods to determine the number of analyte atoms in a sample are commonly applied: the comparator method and the k_0-method.

3.3.1 The comparator method

For the determination of the concentration of an element in a sample by the comparator method, a standard sample with certified elemental composition, the so-called certified reference material, is irradiated and counted under the same conditions as the sample to be analyzed. This method is simple and precise since deviations in nuclear and reactor-based parameters like neutron flux density, flux density distribution within the sample, absorption cross sections, thermal-to-fast-neutron ratio, and transformation branching ratios that influence the activity to be measured are ruled out. The concentration of an element in the sample (c_{Sample}) can be determined using the following equation:[15]

$$c_{Sample} = c_{Standard} \frac{m_{Standard}}{m_{Sample}} \frac{A_{Sample}}{A_{Standard}}, \tag{3.9}$$

where $c_{Standard}$ is the concentration of the element in the comparator standard, $m_{Standard}$ and m_{Sample} denote the mass of the standard and the sample, respectively, and A denotes the corresponding activities as measured by γ-spectroscopy. In cases where the half-life of the nuclide to be determined (referred to as the indicator nuclide) is not considerably longer compared to the transformation time and the measurement time, respectively, the so-called transformation-corrected activity ($\hat{A}_{corrected}$) is calculated as:

---

15 In activity calculations according to eq. (3.9) a number of correction factors needs to be employed, e.g., if the isotopic abundance of indicator nuclide and sample nuclide are different or if sample and standard are irradiated at different positions in the reactor or if γ-ray counting occurs with different detectors or different sample geometries. For further reading see Zeisler et al. (2003) and Greenberg et al. (2011).

$$\acute{A}_{\text{corrected}} = \frac{N_{\text{indicator}} \, \lambda e^{\lambda t_{\text{transformation}}}}{1 - e^{-\lambda t_{\text{counting}}}}, \tag{3.10}$$

where $N_{\text{indicator}}$ is the number of counts in the indicator γ-ray peak, λ is the indicator nuclide transformation constant, $t_{\text{transformation}}$ denotes the time between the end of irradiation, and t_{counting} is the total time used for counting. The comparator method is simple and straightforward. However, for every element to be determined a separate standard needs to be prepared, irradiated, and then counted. Since the preparation of accurate standards for all elements of interest may be difficult and sample counting is time-consuming, an alternative approach was introduced employing a single-element standard, the so-called k_0-standardization method.

3.3.2 The k_0-method

This is a successful standardization method developed originally for NAA and later also applied for PGNAA (Revay and Kennedy 2012; Simonits et al. 1980; De Corte et al. 1989; De Corte and Simonits 1989). Here, activation with thermal and epithermal neutrons is treated separately taking also into account that the energy distribution of the epithermal neutron flux might differ from the ideal $1/\sqrt{E}$ behavior, cf. Fig. 3.4. The ratio of nuclear constants like thermal-to-epithermal flux, cross sections, γ-ray emission rates, and others for an analyte element and a standard are subsumed in a single constant, which can be measured with great precision using, e.g., gold as a neutron flux monitor.[16] The sample and a gold foil are irradiated together with and without a Cd shield and the corresponding k_0-factors for a certain nuclide are determined. The k_0-factors have been experimentally determined for many nuclides relative to a gold standard. It thus makes the use of certified reference materials unnecessary and is available for use in multi-element analysis. The drawback of the k_0-method is that it requires accurate absolute calibration of the detector efficiency with respect not only to energy but also to sample geometry and distance between sample and detector. If the irradiation position is changed, the whole set of k_0-constants for all nuclides to be determined has to be re-measured due to changes in the neutron flux distribution.[17] The k_0-method is widely applied and there are regular international workshops that cover all aspects of this method. The proceedings usually appear in a peer-reviewed journal and thus are easily available online; see Smodiš (2018) for a recent reference.

---

16 Gold provides ideal parameters as neutron flux monitor. The "monitor" reaction is ^{197}Au(n,γ)-^{198g}Au. ^{198g}Au has a half-life of 2.7 days and provides photons of 412 keV.
17 The IAEA provides an extensive compilation of k_0-constants and related nuclear data as well as software for data evaluation via its web-based "Neutron Activation Analysis Portal" (IAEA 2011).

3.3.3 Selectivity and sensitivity of neutron activation analysis

An outstanding feature of NAA is the selectivity and sensitivity of the method. Sensitivity is dependent on both the neutron flux density and the neutron absorption cross section (the probability for a nucleus to absorb a neutron of a certain energy) of the analyte nuclei. For a given neutron flux density a variety of experimental parameters like sample irradiation time, transformation times, and counting times, respectively, can be optimized in order to increase the sensitivity for a certain nuclide or a class of nuclides with comparable half-lives and thus overcome interferences with other nuclides of, e.g., the bulk sample material that might hinder the determination of an element present only in trace quantities (these interferences are referred to as matrix effects). For example, the sensitivity for a long-lived radionuclide that suffers from interferences with a short-lived nuclide can be improved by waiting for the latter to transform and measure the sample after a sufficiently long "cooling time".

DGNAA can be applied for the determination of about two thirds of the chemical elements in the periodic table with sensitivities in the µg/g-range [ppm]. In selected cases the ng/g-level can be reached. Figure 3.9 comprises estimated detection limits (LOD – Limit of Detection) for a number of elements assuming irradiation at a thermal neutron flux of $10^{12}\,\mathrm{cm}^{-2}\,\mathrm{s}^{-1}$ and γ-ray spectra free of interferences.[18] There are a number of elements where the LOD approaches a level of 0.1 ppb. One example is manganese with a thermal neutron capture cross section of 13.3 barn for ^{55}Mn and a half-life of the corresponding ^{56}Mn of (only) 2.6 h where the saturation activity is easily reached. There are selected cases where the LOD is even smaller and approaches the 0.01 ppb-level. For some light elements like hydrogen, boron, carbon, and nitrogen PGNAA can successfully be applied. However, DGNAA is not suitable for most of the light elements and also not for lead.

In reality, the sample is placed into a well-known neutron field for a certain period (irradiation time) to produce the unstable neutron capture nuclear reaction products. The radioactivity of the produced isotope corresponds to its number of atoms $\hat{N}$ according to $\hat{A} = \lambda \hat{N}$. Utilizing adequate methods to quantify the absolute radioactivity, both the number $\hat{N}$ and simultaneously the mass of this radionuclide can be obtained precisely (via the Avogadro constant: $6.22 \cdot 10^{23}$ atoms correspond to 1 mol).

Thus, sensitivity is dependent on both the neutron flux and the neutron absorption cross section (the probability for a nucleus to absorb a neutron of a certain energy) of the analyte nuclei. To obtain analyte element concentrations, a suitable standard that contains known amounts of the relevant trace elements is usually irradiated together with and counted under the same conditions as the unknown sample.

---

18 This neutron flux value typically corresponds to a low power reactor with a thermal power of less than 1 MW. For higher neutron fluxes the LOD values are even lower, in particular when nuclides with long half-lives are involved.

Detection limit [ppb]

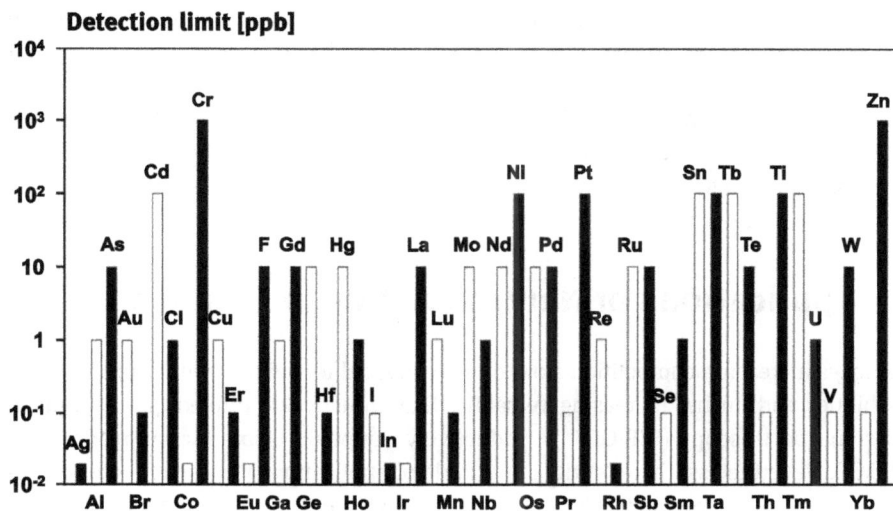

Fig. 3.9: Estimated detection limits [ppb] achieved by INAA for a number of chemical elements assuming a thermal neutron flux of 10^{12} cm^{-2} s^{-1} (Eberhardt and Trautmann 2007).

3.4 Fission reactors as neutron sources for neutron activation analysis

Although there are a number of nonreactor-based neutron sources available,[19] the most prolific sources for neutrons applicable for NAA are nuclear fission reactors. More than 220 research reactors are in operation worldwide spanning a range of thermal power from 0.1 W up to about 100 MW.[20] The vast majority uses ^{235}U as nuclear fuel and thus thermal neutrons to propagate the fission chain. Most widespread are swimming-pool reactors, where the reactor core is placed at the bottom of an open pool filled with demineralized (light) water for cooling and shielding purposes. Swimming-pool reactors allow easy access to in-core irradiation facilities and provide thermal neutron fluxes up to about 10^{15} cm^{-2} s^{-1}. In other cases the reactor core is enclosed in a pressurized tank with light or heavy water that can serve as moderator and coolant. In many reactors, the thermal flux has been optimized in distinct irradiation positions designated for NAA.

---

19 In addition to the already mentioned α-sources and sources available based on spontaneous fissioning nuclides like ^{252}Cf and ^{248}Cm.

20 A current overview is available at the website of the International Atomic Energy Agency, see IAEA (2016).

Neutrons that are not yet thermalized may be used for activation. Most of the high power and some of the low to medium power level reactors with a thermal power exceeding 1 MW have an epithermal flux greater than 10^{10} cm^{-2} s^{-1}, cf. Chapter 11. This can be advantageously employed for minimizing interferences from certain elements (e.g., Na, Cl, Mn, and V) that have high thermal neutron cross sections. Even faster neutrons employ (n,p), (n,d), (n,α), (n,n$'$), and (n,2 n) reactions.

3.5 Applications of Neutron Activation Analysis

NAA faces increasing competition from other analytical techniques with multi/element capabilities and further increasing versatility like Inductively Coupled Plasma Optical Emission Spectroscopy (ICP-OES) or Inductively Coupled Plasma Mass Spectrometry (ICP-MS). Nevertheless, there is still a multitude of applications for the different techniques of NAA that allows high selectivity and sensitivity for the determination of constituents present in very small concentrations (in the ppm to sub-ppm level). Very often no chemical pretreatment is necessary and the sample can be used "as it is". This is of special importance if the sample material needs to be conserved in its original shape or chemical form for further investigations, e.g., in the case of meteorites, gemstones, or archaeological artefacts. Thus, NAA is applied in many different fields including life science, agriculture, industry, food and nutrition studies, geology, environmental analysis, and forensic sciences.[21] NAA in its different forms can be used for elemental analysis of more than two thirds of the elements of the Periodic Table. Figure 3.10 shows which analytical technique is usually applied for determination of a certain element.

3.5.1 Prompt Gamma Neutron Activation Analysis

PGNAA is mainly used for the determination of light elements in a variety of sample matrices, among others geological samples (minerals and rocks), archaeological samples (glass, ceramics and coins), biological samples (blood samples and human tissue) as well as concrete and cement matrices. The determination of hydrogen and boron are prominent examples, where PGNAA offers unique opportunities, e.g., in the determination of moisture (as hydrogen) in concrete or boron in volcanic rocks as well as in human tissue. For an overview, see Greenberg et al. (2011) and references therein.

---

21 A recent and comprehensive overview of new techniques and applications of NAA is frequently presented in the proceedings of a series of conferences entitled "International Conference on Modern Trends in Activation Analysis (MTAA)". The proceedings usually appear in a peer-reviewed journal and thus are easily available online (see de Bruin and Bode (2012) for a recent reference).

3.5.2 Instrumental Neutron Activation Analysis

In INAA, sample preparation is limited to its physical treatment in order to maintain a suitable and representative portion for irradiation. In the ideal case the sample is weighed and without any further treatment placed into a proper irradiation container. Methods of treatment can involve dry freezing and pressing of pellets with a suitable geometry or dissolving the sample either in water or a diluted mineral acid.[22] Subsequent to irradiation the sample or a known portion of it should be transferred from the irradiation container into a new container used for γ-ray counting to minimize so-called blank correction. The blank values result from activation of, e.g., the container material or the solvent used that might also contain traces of the element to be determined in the sample only. Since in INAA no chemical pretreatment of the sample is carried out, the elemental composition is preserved and thus it is often used in the certification of reference materials. In principle, INAA can be carried out without perturbations of the traceability chain and thus INAA is regarded as a valuable tool in chemical metrology (Zeisler et al. 2005) (in metrology the so-called traceability chain can be defined as an unbroken chain of measurements and associated uncertainties, linked to certified reference materials or to SI standards). For the determination of uncertainties in nuclear measurements see IAEA (2004) and references therein. A very special example of activation analysis is the application of *in situ* INAA for elemental analysis during a future lunar mission. For this, a ^{252}Cf neutron source is discussed for sample irradiation on the moon (Li et al. 2011).

3.5.3 Radiochemical Neutron Activation Analysis

In RNAA, chemical treatment of the sample can be performed prior to or subsequent to irradiation to remove either the matrix elements or major contaminants. As a result of the chemical separation, interferences caused by other nuclides are avoided or minimized. In this way, detection limits can be lowered or the detection of additional elements present in low concentrations becomes possible. Solid samples are first dissolved using common techniques like, e.g., dry ashing, alkaline fusion, and the treatment with mineral acids applying microwave heating in closed vessels. Subsequent chemical separations might involve selective precipitation, ion exchange chromatography, and liquid-liquid extraction techniques. In applying these techniques, material losses are unavoidable and hence must be carefully controlled.

---

22 Here very often nitric acid (HNO_3) is used since it does not contain elements that form longer-lived isotopes during irradiation, like, e.g., chloric acid (HCl) that forms ^{38}Cl ($t^1/_2 = 37$ min), which might hinder the detection of shorter-lived isotopes.

For this – besides classical techniques – radioactive tracers are ideally suited for performing the necessary recovery measurements to determine the chemical yield for a certain chemical procedure applied to the sample.[23]

3.5.4 Delayed Neutron Activation Analysis

For the determination of fissile material (samples containing isotopes that undergo nuclear fission when exposed to a neutron beam like ^{235}U, ^{239}Pu, and ^{249}Cf, respectively), DNAA is most useful and in selected cases superior to more common techniques like α-particle spectroscopy or liquid scintillation counting (LSC). α-Particle spectroscopy usually requires intense chemical pretreatment and the production of a thin sample suited for α-counting. LSC is not applicable in cases where the sample is not readily soluble in the scintillator cocktail or when severe quench effects occur that hinder the detection of the scintillating light.

As an example, DNAA can be applied for the incorporation inspection of workers in the nuclear industry (e.g., in the production of nuclear fuel assemblies). For example, the uranium content in urine samples must be determined routinely. The analytical procedure involves the co-precipitation of uranium with Fe^{III}I-hydroxide to prepare a sample for DNAA for subsequent neutron irradiation. Knowing the

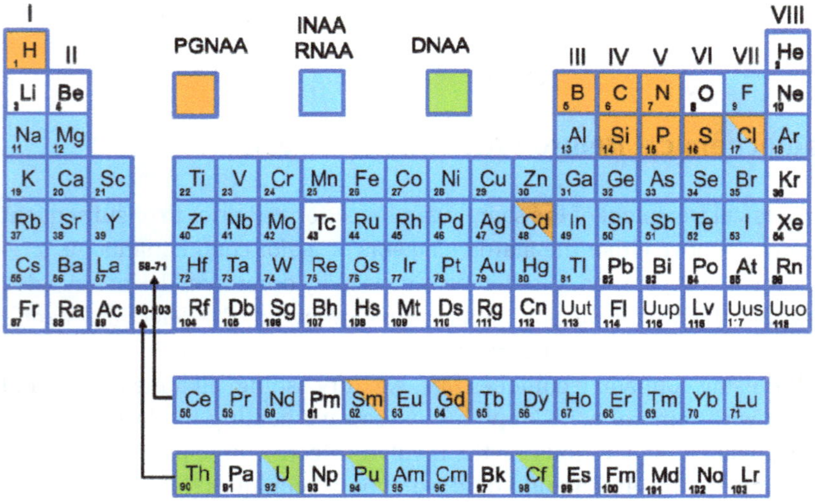

Fig. 3.10: Periodic Table of Elements indicating which neutron activation technique is commonly applied for the determination of a certain element.

---

23 For an overview of the different techniques applied in RNAA see Zeisler et al. (2003) and Greenberg et al. (2011) and references therein.

precise isotopic composition of the uranium used for fuel production, screening of a large number of persons can be carried out with the precision demanded by the authorities. Absolute detection limits of 10^{-11} g can be achieved for ^{235}U and ^{239}Pu, respectively (see Eberhardt and Trautmann (2007) and references therein). DNAA is faster than mass spectroscopic techniques and more sensitive than α-spectrometry; however, discrimination of fissile components in a mixture (e.g., ^{235}U besides ^{239}Pu) remains difficult. DNAA is ideally suited as a screening technique in nuclear forensics and in support of international nuclear safeguards by providing the amount of fissile material in swipe tests (Glasgow 2008).

3.6 Outlook

Besides NAA there are other so-called activation techniques. Among them are Charged Particle Activation Analysis (CPAA), Photon Activation Analysis (PAA), and Neutron Depth Profiling (NDP).

CPAA is based on the reaction of accelerated light charged particles like protons, deuterons, or α-particles. The heavier the particle is, the higher is the COULOMB barrier and thus protons are most often used for CPAA of light elements. The proton-induced nuclear reactions include (p,n) and (p,γ) reactions with protons up to an energy of about 10 MeV. With more energetic protons (p,α), (p,d), and (p,2 n) reactions can occur in the sample. CPAA is a method complementary to NAA, in particular for light elements with their typically low cross section for neutron capture. Since the particles lose energy when passing through the sample, the reaction cross sections change with depth. Consequently, CPAA is used for thin samples or as a surface technique (Erramli 2017).

Activation with photons can be performed at an electron accelerator using bremsstrahlung induced reactions. During photon activation, the target nucleus is activated by (γ,n) reactions, referred to as "photonuclear reaction". This type of reaction, which leads to neutron-deficient nuclei, is induced with photons exceeding 10 MeV. Typical irradiation energies in PAA are 30 MeV that can be achieved, e.g., with high-power linear electron accelerators. Like neutrons, photons have the advantage to penetrate the entire sample and thus methods and procedures applied in PAA are very similar to NAA. By means of detectors and connected data acquisition electronics suitable to record γ- or X-ray spectra, simultaneous multi-element analyses can be performed after bremsstrahlung exposure without chemical pretreatment of the sample. PAA has been applied in a variety of research areas including (but not limited to) geochemistry, forensic science, life science, and medicine and also for the certification of reference materials (Segebade and Berger 2008).

Location sensitive analysis of light elements is also possible using a special technique of neutron activation called NDP. NDP is a method that is very similar to neutron activation analysis. It is based on the fact that several isotopes of, e.g., lithium, boron, and nitrogen (and others) undergo charged particle emission as a result of the absorption of a low energy neutron. The corresponding nuclear reactions are ^{6}Li(n,α)^{3}H, ^{10}B(n,α)^{7}Li, and ^{14}N(n,p)^{14}C, respectively. The resulting charged particles (protons, α-particles as well as the corresponding recoil nuclei) are emitted with energies in the range of 40 keV for the ^{14}C recoil nucleus up to about 2.7 MeV for the tritium from the ^{6}Li(n,α) reaction. Protons and α-particles are emitted with intermediate energies of about 500 keV up to 2 MeV. As the charged particles travel through the material they lose energy due to interaction with the matrix atoms. When exiting the material the particles are detected with a surface barrier detector and their actual energy is determined. The detector is mounted in a vacuum chamber in order to prevent further energy loss when the particle travels the distance from the sample to the detector surface. From the energy loss one can conclude the depth of their origin from the characteristic stopping power of the material, given in keV per μm for a certain energy. For 1.4 MeV α-particles in silicon this value is about 280 keV per μm. Data is collected while the samples are under neutron irradiation. NDP can be applied to profile a few micrometer thin films and is particularly suitable for the semiconductor industry since devices are manufactured as thin films on Si wafers (Hossein 2001; Downing et al. 1993).

For many standard analytical techniques like Atomic Absorption Spectroscopy (AAS), Inductively Coupled Plasma Optical Emission Spectroscopy (ICP-OES), Inductively Coupled Plasma Mass Spectroscopy (ICP-MS), or X-ray Fluorescence Spectroscopy (XRF), respectively, standalone equipment with a high level of automization is available. Very often these devices have a table-top format and are easy to operate even by less experienced personnel. In contrast, neutron activation requires access to an intense neutron source and to special spectroscopy equipment. Furthermore, sample handling requires personnel trained in handling radioactive material. Consequently, NAA is less widely used and suffers partly from insufficient awareness within the research fields dealing with elemental analysis. However, it offers several almost unique advantages: (i) most of the "classical" techniques require the sample in the form of an aqueous solution, whereas NAA (in this respect INAA) is suitable for materials that are difficult to dissolve, (ii) NAA is largely independent of matrix effects, and (iii) NAA is a multi-element technique suitable for the majority of the elements in the periodic table, ranging from hydrogen up to the actinide elements. Due to these advantages, NAA still plays a special role in the certification of materials and thus is of utmost importance as a reference technique for other analytical methods that are more widely applied.

References

De Bruin M, Bode P. Proceedings of the 13th international conference on modern trends in activation analysis. J Radioanal Nucl Chem 2012, 291, 307–311.

De Corte F, Simonits A, De Wispelaere A, Elek A. k_0-Measurements and related nuclear data compilation for (n, γ) reactor neutron activation analysis IIIa: Experimental. J Radioanal Nucl Chem 1989, 133, 3–41.

De Corte F, Simonits A. k_0-Measurements and related nuclear data compilation for (n,γ) reactor neutron activation analysis IIIb: Tabulation. J Radioanal Nucl Chem 1989, 133, 43–130.

Downing RG, Lamaze GP, Langland JK, Hwang ST. Neutron depth profiling: Overview and description of NIST facilities. J Res Natl Inst Stand Technol 1993, 98, 109–126.

Eberhardt K, Trautmann N Neutron Activation Analysis at the TRIGA Mark II research reactor of the University of Mainz. In: International Atomic Energy Agency, ed. Utilization related design features of research reactors: A compendium, IAEA Technical Reports Series No. 455, Vienna, Austria, IAEA, 2007, 537–545.

Erramli H. Charged particle activation analysis. In: Meyers RA. ed. Encyclopedia of analytical chemistry. John Wiley and Sons, Hoboken, New Jersey, U.S. (published online, 2017. DOI: 10.1002/9780470027318.a6203.pub3. accessed February 04, 2022, at http://onlinelibrary.wiley.com/doi/10.1002/9780470027318.a6203.pub3).

Glasgow DC. Delayed neutron activation analysis for safeguards. J Radioanal Nucl Chem 2008, 276, 207–211.

Greenberg RR, Bode P, De Nadai-Fernandes E. Neutron activation analysis: A primary method of measurement. Spectrochim Acta B 2011, 3–4, 193–241.

Hevesy G, Levi H. Artificial radioactivity of dysprosium and other rare earth elements. Nature 1935, 135(3429), 103–106.

Hossain TZ Industrial applications of Neutron Activation Analysis. In: IAEA-TECDOC-1215, Use of research reactors for Neutron Activation Analysis, Vienna, IAEA, 2001.

International Atomic Energy Agency (IAEA). Research Reactors (database compiling all research reactors worldwide). Available at https://nucleus.iaea.org/rrdb/#/home, (accessed February 04, 2022) 2016.

International Atomic Energy Agency (IAEA). NGATLAS. Atlas of Neutron Capture Cross Sections. Available at https://www-nds.iaea.org/ngatlas2/, (accessed February 04, 2022) 2014.

International Atomic Energy Agency (IAEA). Nuclear Data Section Portal. Available at https://www-nds.iaea.org, (accessed February 04, 2022) 2011.

International Atomic Energy Agency (IAEA), ed. IAEA-TECDOC-1401: Quantifying uncertainty in nuclear analytical measurements. Vienna, Austria, International Atomic Energy Agency (IAEA), 2004.

Li X, Breitkreuz H, Burfeindt J, Bernhardt H-G, Trieloff M, Hopp J, Jessberger EK, Schwarz WH, Hofmann P, Hiesinger H. Evaluation of neutron sources for ISAGE – In-situ-INAA for a future lunar mission. Appl Rad Isotopes 2011, 69, 1625–1629.

Révay Z, Kennedy G. Application of the k_0 method in neutron activation analysis and prompt gamma activation analysis. Radiochim Acta 2012, 100, 687–698.

Smodiš B. Proceedings of the 7th international k_0-users workshop. J Radioanal Nucl Chem 2018, 315, 661–662. DOI: 10.1007/s10967-018-5707-6. J Radioanal Nucl Chem 2014, 300, 451–452.

Segebade C, Berger A. Photon activation analysis. In: Meyers RA. ed. Encyclopedia of analytical chemistry. John Wiley and Sons, Hoboken, New Jersey, U.S. (published online, 2008. DOI: 10.1002/9780470027318.a6211.pub2. accessed February 04, 2022, at http://onlinelibrary.wiley.com/doi/10.1002/9780470027318.a6211.pub2/abstract).

Simonits A, Moens L, De Corte F, De Wispelaere A, Elek A, Hoste J. k_0-Measurements and related nuclear data compilation for (n,γ) reactor neutron activation analysis. J Radioanal Chem 1980, 60, 461–516.

Tomlinson L. Delayed neutron precursors. At Data Nucl Data Tables 1973, 12, 179.

Zeisler R, Lindstrom M, Greenberg RR. Instrumental neutron activation analysis: A valuable link in chemical metrology. J Radioanal Nucl Chem 2005, 263, 315–319.

Zeisler R, Vajda N, Lamaze G, Molnár GL. Activation analysis. In: Vertes A, Nagy S, Klencsár Z. eds. Handbook of nuclear chemistry, Vol. 3. Dordrecht, NL, Kluwer Academic Publishers, 2003, 303–357.

Klaus Wendt

4 Radioisotope mass spectrometry

Aim: The sensitive determination of radionuclides, in no matter what kind of sample, could most easily utilize the detection of the emitted radiation. Nevertheless, there are quite a number of situations or special tasks in which this straightforward approach fails or is far too insensitive for the envisaged application. These primarily concern ultratrace analytics of lowest radioactivity levels, most often activities of radionuclides of extremely long half-lives. In these cases and in particular when dealing with low-energy β- or long-lived α-emitters, conclusive detection of radiation and sensitive quantification of the radiation sources often suffer from limited resolution, low efficiency, or unsuitability of the radiation detector systems in use. In addition, absorption and scattering processes in the sample itself, in the surrounding air or in the detector, as well as unavoidable interferences from radiation of other radionuclides could severely affect and obscure such analyses.

Consequently, the quantitative determination of the number of individual atoms of a specific radioisotope for quite a number of situations is much more sensitive and meaningful than the study of the transformation processes and the emitted radiation via radiometric techniques. In this approach, identification and suppression of disturbing influences and rejection of background by physical and/or chemical means are crucial for ensuring the significance of the results. The corresponding experimental approach is radioisotope mass spectrometry (RMS). It is applied frequently for routine analyses on a variety of samples and represents a lively field of ongoing research including steady optimization of specifications and expansion of application fields. Depending on the specific radionuclide, the sample matrix, and the specific technique in use, it allows to address and quantify sensitivities as low as 10^6 atoms per sample, which for long-lived species corresponds to activities in the low μBq range.

4.1 Introduction: Long-lived radioisotopes, origin, and relevance

4.1.1 Natural radionuclides

Today, more than 3000 radionuclides are known, belonging to all the chemical elements of the Periodic Table. The vast majority of them can only be produced artificially, e.g., by induced nuclear reactions in nuclear reactors or at particle accelerators and transform rapidly with relatively short half-lives towards stability; here, radiometric determination is obviously the detection method of choice and very powerful.

https://doi.org/10.1515/9783110742701-004

Only about 100 long-lived ones of these radioactive nuclides are found originally in nature. Most of these so-called *natural radionuclides* are of primordial origin from the supernova event that led to the formation of our solar system more than 10^9 years ago or belong to the corresponding transformation products.[1] Few other long-lived radionuclides are constantly produced cosmogenically via the various components of cosmic radiation. Such formation processes include direct cosmic ray-induced nuclear reactions or neutron capture, which occur typically in the atmosphere.[2] All these reactions have low production rates ranging from a few particles per square centimeter and second, e.g., for the production of radiocarbon ^{14}C, down to about 10^{-6} cm^{-2} s^{-1} for heavier radionuclides like the noble gas isotope ^{81}Kr. Correspondingly, overall contents and, in particular, relative isotopic abundances in respect to stable isotopes of the same element are extremely low. The latter, for example, span the range from 10^{-9} for ^{10}Be against stable ^{9}Be down to even less than 10^{-17} for the tritium abundance versus stable hydrogen. Due to this rare occurrence, natural radioactivity is usually very weak with values in the mBq range or even far below. For example, the most abundant radioisotope ^{40}K contributes only around 0.4 Bq per g of pure potassium in a sample. Comparable and very low radioactivity levels result from the various members of the three natural transformation series of the heaviest naturally occurring α-emitting isotopes ^{238}U, ^{232}Th, and ^{235}U, which are still active. The most long-lived in these series are ^{226}Ra, ^{238}U, and ^{232}Th; they represent the next abundant radioisotopes to ^{40}K but provide activity levels more than one order of magnitude lower. Their considerably long half-lives in the range of thousands of years or above account for lowest specific activities.[3] Consequently, for these cases, direct radiometric determination techniques using α-, β-, or γ-counting techniques are frequently insensitive or obscured by interferences with other shorter-lived radionuclides present in the sample, and alternative sensitive detection techniques are needed.

Figure 4.1 gives a plot of typical abundances for the portion of these long-lived natural radioisotopes, which also have stable isotopes, in the Earth's crust and atmosphere, additionally giving their individual origin. Showing prevailing isotopic abundances well below 10^{-9}, the plot indicates that standard analytical techniques including the majority of mass spectrometric approaches will not be able to quantify or even detect these species either.

Many of these radionuclides are of high relevance for fundamental research in geochemical and environmental studies[4] concerning, e.g.,

---

1 Cf. Chapter 5 of Vol. I.
2 Cf. Chapter 2 of this volume for details.
3 Specific activity is typically defined as the absolute radioactivity of a given radioisotope related to the sum of its own mass plus the mass(es) of its stable isotope(s); cf. Chapter 5 of Vol. I: eq. (5.12) and Fig. 5.12 for examples.
4 For applications in, e.g., dating based on radionuclides see Chapter 5 of this volume.

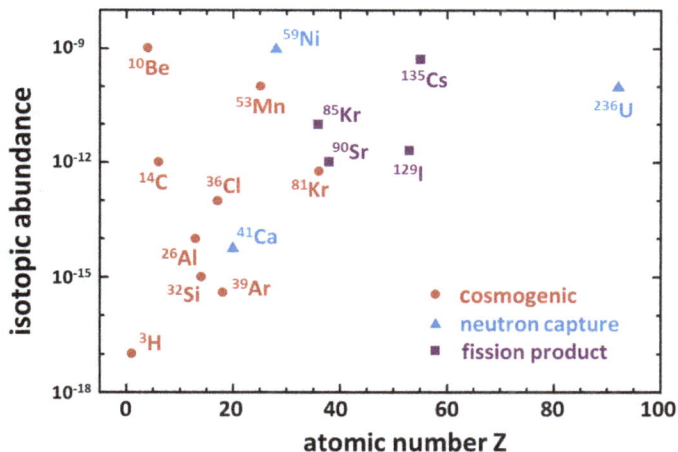

Fig. 4.1: Natural radioisotopes with abundances below 10^{-9} relative to the stable isotopes of the same element. Individual processes of origin are indicated by the symbol. Since in many cases, the distributions of these isotopes vary significantly among different environmental compartments, the abundances given should be seen as an order of magnitude estimate only.

- investigations on the flux and nature of cosmic rays in the atmosphere and the upper layers of the Earth crust,[5]
- origin and chemistry of trace gases in the atmosphere,
- circulation studies and dating of oceanic and groundwater in the hydrosphere,
- dating and variations of cosmic ray flux measured in the cryosphere,[6] and
- geomorphological and volcanism studies carried out predominantly in the lithosphere (Tykva et al. 1995).

Further interdisciplinary applications relate to cosmochemistry[7] and astrophysics with analyses of extra-terrestrial materials and characterization of meteorites to support nuclear and particle physics, e.g., by determination of the solar neutrino flux, or they contribute to archaeology, material sciences, and the vast fields of all kinds of biomedical studies and applications. For these tasks, highly sensitive and selective determination techniques have been developed, predominantly involving dedicated mass spectrometric ion counting (Lu et al. 2003).

---

5 Cf. also Chapter 2 of this volume.
6 Cryosphere subsumes the frozen water on the surface of the Earth (such as ice on oceans, seas, lakes and rivers, glaciers, and frozen ground, but also covers of snow).
7 Cosmochemistry or chemical cosmology is defined as the study of the chemical composition of matter in the universe and the processes that led to those compositions.

4.1.2 Artificial radionuclides

In addition to natural sources and obviously of much higher relevance regarding radioactive hazards, a number of anthropogenic, i.e., manmade radioisotopes nowadays are of highest concern too. They are distributed around the entire globe in a range from minuscule up to highest, strongly radiotoxic contamination levels, predominantly stemming from military or peaceful use of nuclear energy, medical applications of radioisotopes, and numerous miscellaneous releases of radioactivity, most of them fortunately on relatively low levels. Contaminations from nuclear industry predominantly comprise actinide isotopes, i.e., primarily uranium and plutonium, as well as the so-called minor actinides neptunium, americium and curium, on top of which a long list of fission products is produced. Longer-lived radionuclides originating from routine medical applications for diagnoses and treatment include ^{99g}Tc ($t_{1/2} = 2.1 \cdot 10^5$ a) and ^{60}Co ($t_{1/2} = 5.3$ a), which may similarly cause a long-term radiation risk. Another challenge is to control the nonproliferation treaty, ensure nuclear safety, and, in general, to protect humankind against any other kind of possible radiotoxic hazards. Furthermore, the chemical behavior of these artificial radionuclides, which determines speciation and migration characteristics, the transfer factors between individual compartments in nature and environment and the overall radioisotope circulation, must be investigated.

4.2 Radioisotope mass spectrometry

The choice of a most suitable determination technique for a particular radionuclide of interest or for a specific radioanalytical problem must be based on various aspects. These include the sample material and the overall amount of it being available for the study, the specific transformation characteristics, the expected overall radioactive content in the sample, and possible interferences with other radionuclides. A secondary aspect concerns the required measurement speed, the sample throughput as well as the required effort in experimental equipment and man-power. The first meaningful criterion is extracted from the half-life, when comparing the number of transformations during a given measuring time to the total number of atoms in the sample. This approach is demonstrated in Tab. 4.1 for some long-lived radionuclides of relevance, where an (unrealistic) 100% detection probability has been assumed for the radiometric technique. In absence of any background, a minimum value for a statistically significant result of, e.g., 10 counts for a 24 h measuring sequence has been adopted.

Correspondingly, if precise determination of radiation levels and individual exposure rates is required, conventional radiometric determination techniques can be applied only in case of sufficiently high specific activities in the sample. For all other cases, i.e., for determination of radionuclides of long half-lives at low concentration, in the presence of interferences and other disturbing species or for transformation

Tab. 4.1: Comparison of transformations per 24 h and atom numbers for some radioisotopes.

Radionuclide	Half-life $t\frac{1}{2}$	Transformations per 24 h ($\approx$0.01 mBq)	Atoms in the sample
^{3}H	12.3 a	1	$4.5 \cdot 10^3$
^{14}C	5730 a	1	$2.0 \cdot 10^6$
^{41}Ca	$1.03 \cdot 10^5$ a	1	$3.8 \cdot 10^7$
^{90}Sr	28.5 a	1	$1.0 \cdot 10^4$
^{99}Tc	$2.1 \cdot 10^5$ a	1	$7.7 \cdot 10^7$
^{129}I	$1.57 \cdot 10^7$ a	1	$5.7 \cdot 10^9$
^{239}Pu	$2.41 \cdot 10^4$ a	1	$8.8 \cdot 10^6$

spectra with strongly overlapping radiation, complementary sensitive techniques are required for unambiguous determination of ultratrace radioactivity levels. Generally, analyses with the highest validity already at low contents near the detection limits are needed. In order to compete with or even be superior to any radiometric technique, mass spectrometric determination must in addition be capable of dealing with sample sizes and atom numbers of the analyte that correspond to activities at the natural radiation levels found in the environment around the globe. Regarding reasonable sample sizes and measurement periods for analysis, the detectable contents, i.e., the level of detection (LOD) of radiometry lies in the range of down to well below µBq up to few mBq, depending on transformation characteristics and half-life. For many cases of relevance, these numbers are significantly above the concentrations found in nature and often inadequate.

In conclusion, above a certain threshold of half-life and for realistic conditions, counting the number of atoms is often much more sensitive than counting radioactive transformations. In the case of interfering radiation, the measurement of the atom number is even more important, while the analysis of the isotope pattern of an element provides additional meaningful information. Correspondingly, in many cases, RMS is the most suitable technique for unambiguous radionuclide determination. This is particularly valid for α- and β-emitters with considerable long half-lives, where radiometric techniques are severely limited in sensitivity or are impaired by background. The concept of RMS therefore encompasses a number of specialized inorganic mass spectrometry methods (Becker 2007), which are often specifically adapted and optimized for application to individual radioisotopes as discussed in detail in the next sections.

4.3 Components of radioisotope mass spectrometry

Mass spectrometry in general is one of the earliest developed, most widely used, and best known techniques in analytical chemistry. It combines a sequence of individual process steps, from which the five most prominent ones are pointed out schematically in Fig. 4.2. Of course, this layout applies to RMS too, and the individual steps will be discussed in detail in the successive sections below.

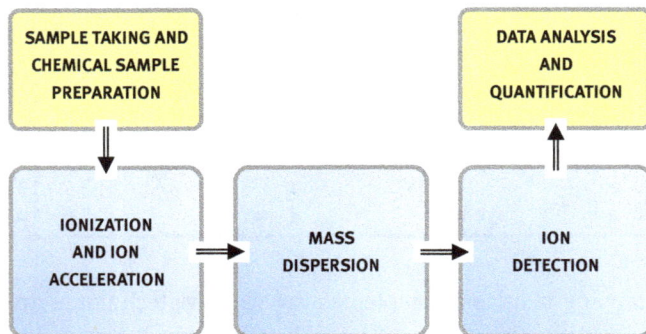

Fig. 4.2: Principal process steps of mass spectrometry in general as well as of RMS.

4.3.1 Sample taking and chemical sample preparation

Sampling location, the choice of sample material, and the procedure of sample taking itself are critical issues for RMS, as these steps might easily introduce artificial contaminations, interferences, or avoidable background. Correspondingly, in many cases, initial *chemical preselection* of the chemical element of interest – or at least a class of elements – is carried out prior to quantitative analysis by RMS. Conventional wet chemistry, primarily involving chromatographic separation methods, is applied frequently for conversion of the sample material into a suitable chemical form, i.e., a wet solution of the element of interest, to allow for the subsequent steps of efficient evaporation, atomization, and ionization.

Only in rare exceptional cases, direct sample analysis is possible. Here, for example, laser ablation or ion sputtering is used, but online coupling techniques of chromatography and mass spectrometry, in the form of a so-called hyphenation technique, have also been implemented. Much more common, however, is the discrete combination of the two individual steps of preseparation and mass spectrometry.

4.3.2 Ionization

After preparation, liquid sample material is introduced into the ion source of the mass spectrometer, usually employing electrolytic deposition and subsequent drying; alternatively, a solid or a powder sample can be introduced directly into the ion source cavity. A number of different techniques used for efficient *ionization* are discussed below; the individual method chosen not only distinguishes but even denominates the individual type of RMS. The ionization process also determines the required chemical sample form and preparation. After ionization, the formation of the ion beam includes the individual stages of ion acceleration and beam focusing. These are usually achieved by electrostatic fields shaped to achieve a suitable compromise between the optimum of the parameters of overall transmission (determining the efficiency), mass dispersion (determining the selectivity), and background suppression (limiting the sensitivity).

4.3.3 Mass dispersion

Mass dispersion in RMS, like in mass spectrometry, in general, is accomplished by selection of the mass-over-charge ratio of the accelerated ions. Typical for this purpose and most simple in construction and operation are static magnetic mass spectrometers, which use a single sector-field dipole magnet. We may discuss this geometry as a model case here somewhat in detail. These spectrometers apply the LORENTZ force F_{Lor} for spatial separation

$$F_{\text{Lor}} = ZevB \tag{4.1}$$

in a transversal homogenous magnetic field of field strength B of ions, which are characterized by their charge of Z times the elementary electric charge e and their velocity v. The LORENTZ force is always directed perpendicular to the ion motion and correspondingly acts as a centripetal force F_{centri}:

$$F_{\text{centri}} = \frac{mv^2}{r}, \tag{4.2}$$

which drives the ions onto a circular track with radius r inside the field. Here, the individual mass m of the ions enters, and from equalizing the two forces, the radius r of the circular path inside the magnet is easily computed as

$$r = \frac{mv}{ZeB}. \tag{4.3}$$

As a consequence, this radius in a homogeneous magnetic field is directly proportional to the momentum of the entering ion, $p = mv$. The ion velocity may be related to the electrostatic potential U_B of acceleration through $E_{\text{kin}} = \frac{1}{2}mv^2 = ZeU_B$, which leads to the fundamental equation of mass spectrometry:

$$\frac{m}{Z} = \frac{r^2eB^2}{2U_B}. \tag{4.4}$$

Correspondingly, the mass over charge ratio m/Z of an ion is unambiguously related to the radius r, the magnetic field strength B, and the acceleration potential U_B, ensuring proper mass resolution as long as different charge states can be separated. The bending of the ion beam in the magnetic field additionally implies a focussing of the angular distribution of the ion beam at a specific focal plane, in which a slit systems is installed and serves for the mass selection. By setting the width of this slit, the acceptance of the detector system is chosen to a fraction, an entire mass or a band of masses defining the resolution of the system, but also limiting transmission and thus overall sensitivity. Recording of a mass spectrum is finally accomplished by varying one of the easily accessible parameters, usually the magnetic field strength B, and measuring the transmitted ion current or alternatively counting the individual ions as function of this parameter at a suitable detector. Most common are sector field geometries with 60° or 90° deflection, which ensure favourable focussing conditions, in combination with a temporal scan of the magnetic field for point-wise registration of the mass spectrum. The use of photographic plates or spatially extended particle detectors for instantaneous recording of a full mass spectrum at the focal plane, as present in early mass spectrographs, has been abandoned due to numerous shortcomings regarding sensitivity, quantification, and resolution.

There are two prominent variables that characterize the quality and specific purpose of a particular mass spectrometry instrument: The mass resolution $\Re$ is given by

$$\Re = \frac{m}{\Delta m},$$

(4.5)

where Δm is the width of the mass peak at mass m, usually measured at 10% of the peak height. It defines the quality of the mass spectrum and is a measure for the accuracy with which a certain mass value may be determined. It is primarily depending on the emittance of the ion source, which is primarily given by the spatial, angular, and energetic distribution of the emitted ions and the value of the electrostatic potential U_B of acceleration. A high value of $\Re$ is of special relevance for investigation of complex molecules with high mass numbers well above 10,000 u, while for RMS, which is uniquely carried out on atoms and thus deals with maximum mass numbers well below 300, a mid-size mass resolution around $\Re = 500$ to 1000, as typical for sector field type spectrometers, is by far sufficient to completely separate neighboring masses. On the other hand, the mass resolution does not contain any information on the shape of and, in particular, about the outer wings of a mass peak. Accordingly, it is not a good indicator for the selectivity of a mass spectrometer, which is limited by the admixtures of contributions of either one of the two neighboring masses $m + 1$ and $m - 1$ on the mass peak at mass m. This contribution is described by the second characteristic quantity, the abundance sensitivity (AS), defined by

$$AS = \frac{H_{m\pm1}}{H_m}.$$

(4.6)

The abundance sensitivity may either be determined from the height H (as given) or from the area A of the neighboring mass peaks and, for asymmetrical peaks, may well be different for the left-hand side, corresponding to H_{m+1}, or the right-hand side, corresponding to H_{m-1}. The value of AS indicates the minimum abundance with which a specific trace isotope of mass m can be determined next to a significantly more abundant neighboring isotope of mass $m+1$ or $m-1$, respectively, as a quantity with negative exponent. The reciprocal value, $S = 1/AS$, given with positive exponent, is identified with the particular selectivity of a given mass spectrometric method.

Further developments and refinements of magnetic mass analyzers addressing the optimization of those characteristic quantities $\Re$ and AS, include the combination with an electrostatic energy analyzer for the ion beam. In this way, a double-focusing sector field mass spectrometer with considerably higher mass resolution in the order of up to $\Re = 100,000$ is formed.[8] It also allows for focussing both the angular and the momentum distribution of the initial ion beam from the ion source through the dispersive element and onto the detector, minimizing losses at the slit system.

Alternatively, dynamic mass spectrometers exploit the stability of ion trajectories in alternating electromagnetic fields in the radiofrequency (RF) range. Using a superposition of an electric alternating RF and an electrostatic (DC) potential, which are applied to precisely shaped electrodes, a two-dimensional quadrupole mass filter or a point-symmetrical three-dimensional ion trap is formed, in which ion oscillation and movement are determined by the mass-to-charge-ratio and can be analyzed accordingly. In both cases, specifically high abundance sensitivity in the order of 10^8 and beyond can be achieved. The general principle of these devices and their use for mass spectrometry were demonstrated for the first time by Wolfgang PAUL, who was honored with the Noble prize for this ground-breaking development, offering widespread applications in analytics as well as in fundamental research and high precision studies (Paul 1990). A combination of an electrostatic field with a parallel strong magnetic field also leads to a trapping geometry denoted Penning trap, which often is the central component of a Fourier-transform ion cyclotron resonance mass spectrometers, in which highest resolution of $\Re$ up to 2,000,000 is achieved by measuring the mass-dependent cyclotron frequency of the trapped ions. It is primarily used for applications in biochemistry and medicine, i.e., in proteomics or metabolism. Meanwhile, the fully electrostatic Orbitrap geometry, where the ions spiral multiple times around a central inner spindle-like electrode, has been developed, which also leads to an ion trap arrangement providing highest mass resolution with significantly reduced experimental effort.

---

8 Realized, e.g., as a MATTAUCH–HERZOG or NIER–JOHNSON-type mass spectrometer, they provide both spatial and energetic focusing and separate interfering components for most of the well-known "difficult" elements.

A completely different approach of mass spectrometry utilizes the precise recording of ion flight times along a given field-free drift length in a time-of-flight mass spectrometer (TOF-MS) apparatus. This device either involves a pulsed ion source, e.g., realized by a pulsed laser or gas inlet nozzle, while alternatively a pulsed acceleration field may be applied. In both cases, acceleration of the ions must be performed in an arrangement of a two-stage acceleration formed by electrodes with consecutive well-adjusted field gradients followed by the field-free drift length. In this way, a time focus with isochronic arrival of all ions of a given mass is achieved at a specific position, where the detector must be positioned. The total flight path length defines the achievable mass resolution. The latter can significantly be improved in a reflectron arrangement by using an electrostatic ion mirror about half the way along the path. In this way, influences resulting from the initial spatial spread and energetic distribution of the ions, caused by the ionization process, are widely compensated. In general, particularly fast and powerful data handling and acquisition for the multitude of full mass spectra registered in a short time period is required for TOF-MS.

In summary, the choice of the dispersive element in RMS depends on the specific analytical problem, which primarily dictates the required properties of the ionization process, but in most cases, is not really critical and may consider different arrangements. Highest experimental expenditure is caused by involving resonant laser radiation for the ionization process, by implementing a superconducting magnet, or by even using a high energy particle accelerator for providing highest specifications regarding the mass spectrometer performance. Here, a balance between the aspects of highest selectivity, corresponding to abundance sensitivity, highest efficiency, determined by the ionization probability and the spectrometer transmission and resolution in the spectrum must be made. As expected, those techniques with highest expenditure deliver ultimate specifications.

4.3.4 Ion detection

In the easiest way, quantitative *ion detection* on the selected mass is carried out by a direct and very sensitive ion beam current measurements. Using electrically well shielded collectors in the form of a so-called FARADAY cup together with high sensitivity amplifiers and amperemeters, currents in the 10^{-12} to 10^{-15} A range may well be recorded quantitatively, but influences from electronic noise and unspecific background must well be controlled. Individual counting of ions yields optimum sensitivity with detection of individual ions. Single ion counting is performed either in a discrete secondary electron multiplier (SEM), a single-channel continuous ion detector (channeltron), or in a channelplate with a multitude of identical channels. Also, multicollector arrangements with a set of 8 or more individual detectors are in use while the traditional setup of a highly sensitive photographic plates is outdated.

Nevertheless, all these highly sensitive devices are also affected by background, stemming from the electronic components but additionally from scattered ions, vagabonding electrons, and radioactive decay processes within the vacuum chamber of the spectrometer.

4.3.5 Data analysis and absolute quantification

The recording of mass spectra generally requires computerized data acquisition and process control. Analytical results are extracted from the spectra and deliver overall content and/or isotopic abundances of one or numerous elements of relevance in the sample by comparison of peak heights or areas. A particular challenge is the correct quantification of the absolute content of a species, i.e., one or more radionuclides of interest, in the initial sample from the measured count rates. For this purpose, different calibration routines have been developed. Very typical is the deliberate addition of a known amount of a tracer isotope of the element of interest, called a spike, which is used as an internal calibration standard for isotope dilution techniques. A further issue is the necessary deconvolution of molecular compounds, which might have similar mass and correspondingly mimic a peak of a radionuclide of interest in a spectrum. Their occurrence must either be prevented, e.g., by dissociation along the analytical determination process or distinguished by utilizing known isotope ratios. Correspondingly, a specific subset of dedicated techniques of inorganic mass spectrometry is found to be most suitable for radioisotope determination in RMS, which are used either in a very standard way or, for selected cases, in specifically adapted versions (Becker 2012). A somewhat quantitative comparison of the different experimental approaches discussed below will be attempted based upon the two relevant key factors: the analytical sensitivity and the selectivity.

Analytical sensitivity reflects the achievable *level of detection* (LOD) for a given species, i.e., a certain radioisotope of relevance in a sample. It is primarily determined by the *overall efficiency* of the method in respect to any kind of background, no matter if specific or unspecific. Quantitatively, the LOD is defined as the lowest quantity of a substance that can be distinguished from a blank (i.e., its absence) with a given statistical probability. The latter is determined by the count rates of the (wanted) signal and the (unwanted) background. Its numerical value is usually taken as the relative standard deviation (RSD) of the data, which is obtained from a statistically significant set of identical measurements according to mathematical statistics. The RSD also defines the significance of the results of this technique. A rough estimate on the overall efficiency may be made using assumptions on the performance of the individual process steps, while a quantitative measurement requires the use of a calibrated standard as initial sample and implies the quantitative assessment of the integrated ion counts at the detector.

In this context, special care must be taken to properly distinguish between the *precision* of an analytical technique, which specifies only the *reproducibility* of results and the *accuracy*, which represents the absolute *correctness* of the data, which may only be verified through an independent error-free method.

As already discussed above, in the general field of mass spectrometry, the *selectivity* is given by the reciprocal of the *isotopic abundance sensitivity* of the method. On the other hand, isobaric interferences from molecular or even atomic species of similar mass may play a major obstructive role. For some of the techniques mentioned below, they are completely unpredictable and may obscure results significantly, in particular close to the LOD. Correspondingly, these contaminants must either be removed beforehand, e.g., by proper sample preparation and preseparation using classical wet chemistry or chromatographic techniques or investigated in detail in independent measurements and properly taken into account during data analysis.

4.4 Individual mass spectrometric systems for RMS

Radioisotope mass spectrometry basically comprises the full range of inorganic mass spectrometer types. The selection of a particular system is governed by the sample material, the matrix, and the required specifications for sensitivity and selectivity. Most significant differences of the techniques concern the required chemical form of the sample, the sample supply, and the ionization process. As pointed out above, the specific choice of the dispersive element is a less important technical issue, which nevertheless affects the achievable sensitivity and selectivity. Disturbing background, caused by atoms, molecules, atomic clusters, which all might be present in highest surplus, or unspecific causes requires to configurate the specific mass spectrometric approach for highest selectivity in the suppression of these interferences.

A straight introduction of solid-state sample material preventing chemical sample preparation is possible in so-called direct mass spectrometric analysis using detachment by a primary particle beam or laser ablation, as carried out, e.g., in glow-discharge (GD), secondary-ion (SI), or laser ablation (LA) mass spectrometry (MS). For specific studies, these specific processes prevent the possible introduction of additional contaminations, which might occur during necessary preparations step for conventional MS. In addition, direct MS gives access to the analysis of smallest particles, which cannot be treated conventionally. On the other hand, pre-treatment is necessary for a number of dedicated applications, e.g., by removing isobars, and, in particular, for the most sensitive techniques, as there are thermal ionization (TI), resonance ionization (RI), or accelerator-based (A)MS techniques. Specifically, in aqueous solutions, isobaric interferences cannot be excluded completely, for instance, for inductively coupled plasma (ICP-)MS. However, proper quantification is much more difficult in direct solid-state mass spectrometry than for chemically treated liquid samples, as suitable calibrated reference materials are often not available.

Concluding these considerations, the most prominent RMS techniques in use are listed in Tab. 4.2.

Tab. 4.2: Important types of radioisotope mass spectrometers.

Abbreviation	Type	Applied usually to
TIMS	Thermal ionization mass spectrometry	Small volumes (≈10 µl) of liquid solution deposited on a filament
GDMS	Glow discharge mass spectrometry	Solid or powder samples
SIMS	Secondary ion mass spectrometry	Solids, particles, and surfaces
ICP-MS	Inductively coupled plasma mass spectrometry	Realized also in combination with laser ablation techniques (LA-ICP-MS)
RIMS	Resonance ionization mass spectrometry	Small volumes (≈10 µl) of liquid solutions deposited on a filament, also in combination with sputtering as direct technique secondary neutral mass spectrometry
AMS	accelerator mass spectrometry	Powder samples

The very specific and sophisticated techniques of RIMS and AMS presently are considered primarily for mono-elemental – or even mono-isotopic – ultratrace analysis of highest specifications regarding sensitivity, selectivity, and determination of precise isotope ratios spanning many orders of magnitude. The other four "conventional" methods of RMS in the table represent sensitive multi-element techniques, which permit for general determination of concentrations and no too extreme isotope ratios. In this way, isotopic compositions of trace and ultratrace elements become accessible. LODs for the overall content of a radioactive element or a certain radioisotope in a sample are generally in the concentration range of ng to sub-ng per g of sample material for solids and down to sub-pg/l for aqueous solutions. Precisions down to 10^{-4} RSD have been demonstrated for isotope ratio measurements in particular sample materials. Here, RSD is given by the standard deviation of results divided by the mean of the isotope ratio value. The achievable precision for AMS data is similar, while for RIMS, it is limited in the low percent range due to unavoidable fluctuations in the laser ionization process. Anyhow, LODs for the ultrasensitive techniques of RIMS and AMS are usually given as absolute atom numbers per sample with typical values of 10^6 atoms, corresponding to few fg, for RIMS down to 10^4 atoms, i.e., some atg, for AMS, respectively.

The abbreviations of the mass spectrometric methods, as given in Tab. 4.2, are chosen according to the evaporation, atomization, and ionization processes used, which also determine the chemical form of the sample accepted. For the analysis of compact solids and powder samples, evaporation and atomization are possible via various processes affecting the surface structure. These include thermal heating, laser ablation, or sputtering by interaction of a plasma, electron beam, or ion bombardment and may all already lead to some rate of ionization. Nevertheless, the efficient and deliberate ionization of the evaporated atomic species often require additional efforts: electron impact in a plasma or by an electron beam, surface ionization on a hot refractory metal filament, or optical multiphoton excitation with laser light are some possibilities. In more conventional ion sources, a separation of the evaporation, atomization, and ionization processes is not possible, e.g., for single-filament TIMS, GDMS, SIMS, ICP-MS, or AMS, where sputtering is used to produce negative ions. Distinct post-ionization on atoms is used in LA-ICP-MS, RIMS, and secondary neutral mass spectrometry (SNMS).

Specialities and application fields of the six specific techniques for RMS of Tab. 4.2, which are applied for the determination of particular long-lived radionuclides and isotope ratio measurements are described in the following.

4.4.1 Thermal ionization mass spectrometry (TIMS)

The release of neutral atoms or charged ions from hot surfaces in thermal equilibrium relies on energetic considerations as given by the BOLTZMANN constant k_B of thermodynamics.[9] According to the SAHA–LANGMUIR equation, the ratio of singly positively charged ions to the neutral species is

$$\frac{N_+}{N_0} = \frac{g_+}{g_0} \cdot e^{(W_e - I_p)/k_B T}, \tag{4.7}$$

where the statistical weights g_+ and g_0 are given by the multiplicity[10] of ionized and neutral ground state, I_p is the first ionization potential of the element under study, and W_e is the work function of the surface material.

For the formation of singly charged negative ions, the rather similar formula of (4.8) applies, primarily changing the sign of the exponent and replacing the ionization potential I_p by the electron affinity E_a.

$$\frac{N_-}{N_0} = \frac{g_-}{g_0} \cdot e^{(E_a - W_e)/k_B T}. \tag{4.8}$$

---

9 The BOLTZMANN constant bridges the energy of an individual particle level with temperature: $k_B = 8.617 \cdot 10^{-5}$ eV K^{-1}.

10 Cf. Chapter 1 in Vol. I, eq. (1.27).

In both cases, T is the absolute temperature of the filament used for ionization, while often the sample is evaporated from an independent atomizer filament on somewhat lower temperature used to decouple the processes of evaporation/atomization from ionization. The work function W_e determines the electron emission or absorption of the metal surface that initiates the ionization process. At filament temperatures of 1500 to 2300 K, all elements with ionization potentials below 7 eV, i.e., alkali, alkaline earth, lanthanide, and actinide elements,[11] are efficiently positively ionized from high temperature refractory metal surfaces of Ta, Re, or W. These materials have high work functions of 4.30 to 4.98 eV. For elements with higher first ionization potential, dedicated ion emitting reagents like silica gel with H_3PO_4 and/or H_3BO_3 are added for the formation of metal complex MeO_x^+ ions.[12]

Thermal ionization mass spectrometry, as schematically sketched in Fig. 4.3, is particularly well suited for precise isotope ratio measurements with accuracies up to 10^{-4}, no matter if concerning stable or radioactive isotopes. Careful sample preparation using ultrapure reagents minimizes isobaric contaminants. Quantitative determination of isotope ratios and absolute contents of radioisotopes may be hampered by isotope fractionation, which takes place due to a slight change of the ionization probability along a series of isotopes of one element. Attempts for correction use conventional isotope dilution techniques based on well-known isotope ratios by spiking with isotope tracers (Platzner 1997). Disturbing influences from fluctuations in the ion source operation are minimized by applying fast switching routines between isotopes or by using sophisticated multicollector detectors which enable parallel determination of ion currents or counts on several close lying isotopes.

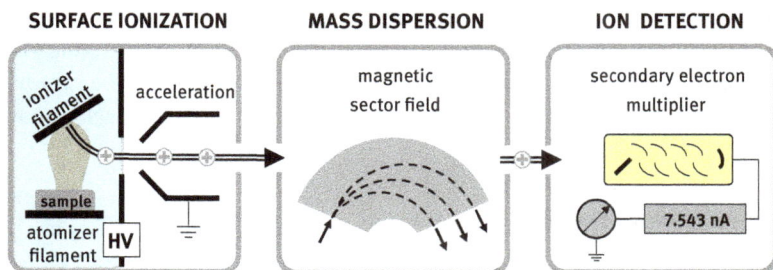

SURFACE IONIZATION   **MASS DISPERSION**   **ION DETECTION**

Fig. 4.3: Schematic view of a thermal ionization mass spectrometer with a double filament ion source, typical for alkaline earth, lanthanide, and actinide elements, for which the temperatures for evaporation and ionization are significantly different. HV = high voltage.

---

11 Cf. Chapter 1 in Vol. I, Fig. 1.11.
12 In contrast, negative ion production is not common in radioisotope mass spectrometry and will not be discussed here.

Applications of TIMS are manifold. Examples include the routine determination of U isotope abundances in natural U ore samples, carried out, e.g., by the Institute for Reference Measurements and Materials IRMM of the European Commission in Geel, Belgium as well as the investigation on unintentional lowest dose Pu intake by workers, which was measured in urine samples, e.g., at Los Alamos National Laboratory. Sensitivity in the order of 1 fg Pu per liter of sample material has been demonstrated, which ensures the determination of whole-body concentrations as low as 100 fg, corresponding to a minimum detectable total incorporation of about 10^8 Pu atoms in a human body. This is orders of magnitude below the LOD of radiometric techniques using α-spectrometry or whole body counters.

4.4.2 Glow discharge mass spectrometry (GDMS)

The principle of a glow discharge ion source, operating at a pressure of around 1 mbar (10 Pa) of argon as discharge gas, is based on Ar^+ ion formation and acceleration in the low-pressure plasma environment. By the impact of the Ar^+ ions onto the source container, which acts as cathode, sputtering processes on the sample deposited there take place and lead to formation of free atoms. These sputtered neutral species are ionized in the same plasma via electron impact. The broad energy distribution of the released ion ensemble implies the use of double-focusing mass spectrometers with selection of both the individual mass-to-charge and energy-to-charge ratio of each ion to give a meaningful mass spectrum.

A schematic drawing of a GDM spectrometer is shown in Fig. 4.4. Depending on the sample composition, either direct current (dc) or radiofrequency (rf) discharges may be used. Due to charging effects on the sample surface, the analysis of nonconducting materials is often obstructed, and admixture of graphite powder to the sample material, the use of a secondary cathode, or a dedicated radiofrequency discharge is required.

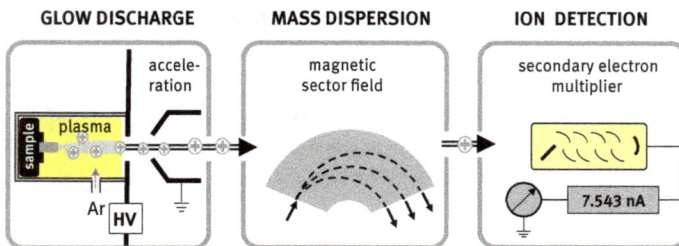

Fig. 4.4: Schematic picture of a glow discharge mass spectrometer, typically using a magnetic sector field mass analyzer and a Faraday cup or secondary electron multiplier for detection. HV = high voltage.

GDMS is very powerful for efficient and direct trace analysis of solid-state samples (Harrison 1988). In RMS, it has found widespread application for the comprehensive characterization of nuclear samples. It has been applied in the fields of nuclear technology and research for characterization of fuel elements, cladding materials, nuclear waste glasses, and smuggled undeclared nuclear materials, even addressing higher-level activities. In the latter case, it is necessary to fully enclose the GDMS, operate it in radiation protection areas, and control any waste gases for emitted radioactivity.

One main interest is the determination of the elemental and isotopic composition of all these materials, especially regarding concentrations of uranium, plutonium, and minor actinides, as an indicator of its origin and nuclear burnup. A comparison with TIMS and ICP-MS showed that GDMS is well competitive in terms of precision and accuracy for these elements. Nonconductive uranium and plutonium oxide compounds are investigated by adding a conducting host matrix, i.e., tantalum or titanium powder (Betti 2004).

GDMS has also been applied by the same authors for migration studies of radionuclides in the environment as a major aspect in safety assessments for nuclear waste repositories. An example is the determination of ^{237}Np or ^{236}U in Irish Sea sediment samples with detection limits in the pg/g range; in which the finding of both isotopes clearly indicates the presence of irradiated nuclear fuel.

4.4.3 Secondary ion mass spectrometry (SIMS)

Secondary ion mass spectrometry (SIMS) is based on generation of ions by the interaction of a well-focused primary ion beam, which impinges on the solid analytical sample with an energy in the range of about 1–25 keV. Depending on energy and species of the primary ions, which could be Ar^+, O_2^+, O_2^-, Cs^+, Ga^+, or few others, and the nature of the sample, the ions penetrate into the solid substance to depths between 1 and 10 nm. The impact of their kinetic energy on the atoms of the solid causes sputtering of both positively and negatively charged "secondary" ions as well as of neutral particles from the microscopic crater formed at the interaction spot. By a dedicated acceleration potential typically in the order of 10 kV and ion optical beam shaping, the released ions of either charge state may be extracted for transfer into and analysis in the subsequent mass separator.

For a specific element, the production yield of secondary ions depends on the ionization potential and, in addition, on particular properties of the sample material such as its chemical composition or structure. Different operational regimes of SIMS use either very low or alternatively rather high primary ion current density. The latter "invasive" technique is applied as "dynamic" SIMS for fast layer-by-layer erosion, depth profiling studies, and direct analysis of microparticles. It allows for three-dimensional characterization and imaging of the sample composition by scanning the well-focused primary ion beam over the surface in combination with depth profiling. Alternatively, quasi nondestructive "static" SIMS with low primary ion currents provides highest lateral resolution below 0.5 µm through dedicated ion optical focusing of the primary ion beam and well-adapted extraction of the secondary ion beam. A schematic presentation of the source region of a secondary ion mass spectrometer is given in Fig. 4.5.

SPUTTERING ION SOURCE

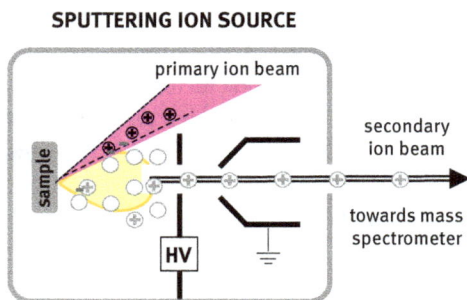

Fig. 4.5: Schematic sketch of the ion source region for secondary ion mass spectrometry. HV = high voltage.

Similar to GDMS, the broad energy distribution of the ions released by the sputtering process requires the use of either static double-focusing or dynamic TOF mass spectrometers. For the latter, a fast switching of the primary ion beam provides the required pulsed ionization. In both cases, very high mass resolution of up to $\Re = 10,000$ and detection limits in the ng/g to pg/g range have been demonstrated.

The quantification of analytical results obtained by SIMS requires the use of relative sensitivity coefficients (RSC), which depend on parameters like primary ion beam quality, physical and chemical properties of the element under study, composition of the sample matrix, and charge state distribution of the secondary ions. These have to be determined by the use of standard reference materials and limit the accuracy of the method somewhere in the percent region.

Due to the high lateral resolution, the application of SIMS to RMS focuses on the identification and analysis of particles including their radioactive inventory. Further examples for SIMS applications include the characterization of environmental samples including the distribution of interspersed natural radioisotopes, localization of contaminations originating from nuclear medical applications (^{99}Tc, ^{89}Sr, and ^{76}Br) or from the nuclear fuel cycle, investigations on surface contaminations, depth profile measurements on nuclear and shielding materials as well as general material science studies concerning on radioisotope content, e.g., for use in semiconductor industry.

The identification of individual hot particles of uranium and other actinides in respect to their isotopic composition is important for environmental monitoring, nonproliferation verification in nuclear safeguarding, and nuclear forensic studies. Hot particles with diameters of ≈10 µm can be transported as aerosols over long distances and are recovered in environmental and swipe samples. Isotopic compositions in those particles can be measured with accuracies and precision of 0.5% by SIMS. Application of the latest generation of SIMS instruments with lateral resolution in the range of down to ≈100 nm, presolar micrograins in meteorites were investigated on their $^{11}B/^{10}B$ and $^{9}Be/^{11}B$ compositions. These indicated intense incidents of γ-ray radiation in the very early solar system, which help understand the processes that led to its foundation.

4.4.4 Inductively coupled plasma mass spectrometry (ICP-MS)

Inductively coupled plasmas produce flame-like electrical discharges under atmospheric pressure, which provide highly efficient vaporization, atomization, and ionization of mainly all elements of the Periodic Table (Montaser 1998). Usually, the plasma is formed in a stream of 8 to 20 l/min of argon gas, flowing through an assembly of concentric quartz tubes denoted as plasma torch. Radiofrequency power in the kilowatt range at typical operation frequencies of 27 or 40 MHz ignites the plasma and ensures high-energy electron bombardment. Temperatures of the ions around 6000 K are reached, while the electrons in the plasma gain additional kinetic energy corresponding to even higher electron temperatures. Highest efficiency is obtained in ICP-MS through nearly complete ionization of all species from the liquid sample material. The latter is introduced by nebulization centrally into the torch within the central stream of the argon carrier gas. The introduction of the ions formed under atmospheric pressure conditions in the plasma torch into ultrahigh vacuum of the mass spectrometer chamber is carried out via very narrow gas inlet apertures and a differentially pumped interface including a sampler and skimmer unit as sketched in Fig. 4.6. Frequently, the ICP is combined with a compact quadrupole mass analyzer, which permits suitable operation also under not fully satisfying vacuum conditions.

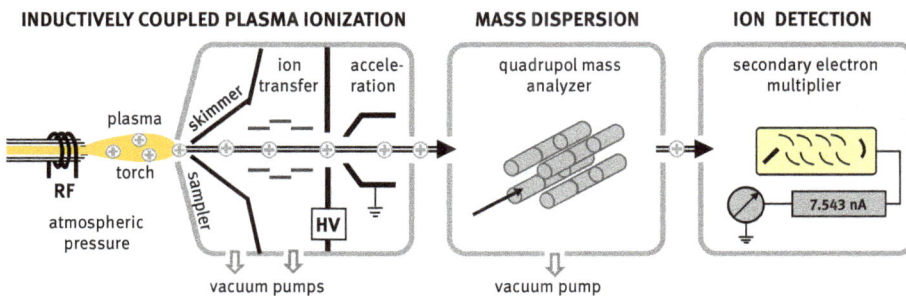

Fig. 4.6: Principal components of a quadrupole ICP-MS system. Ions are created in the plasma torch and transferred into the quadrupole mass spectrometer through the differentially pumped inlet system and interface.

The sampler forms the first differential pumping stage; an exit aperture of typically 1 mm diameter in combination with strong vacuum pumps results in a pressure reduction from atmosphere down to ≈5 mbar. The still smaller skimmer orifice is located somewhat later downstream for optimal extraction of the analyte ions into the ultrahigh vacuum of the mass spectrometer via transfer ion optics.

In ICP-MS, the introduction of the sample into the plasma is most essential. Liquids are dispersed into fine aerosols using either pneumatic or ultrasonic nebulizers. An analyte transport efficiency of typically 5% is obtained for conventional solution uptake rates of about 1 ml/min. For microsamples and well-adapted nebulizers with flow rates as low as only a few μl/min, transport efficiencies up to almost 100% can be achieved. In general, ICP-MS enables rather universal detection of a wide range of elements in various matrices at major, minor, trace, and ultratrace concentration level, also supported by preceding dissolution of solid samples into liquid form. Correspondingly, it has become the workhorse for inorganic mass spectrometry and is also omnipresent in RMS. As a further advantage, the liquid sample introduction for ICP-MS offers the very convenient opportunity of a direct combination of chromatographic preseparation steps leading to a hyphenated technique. Using ion or liquid chromatography or alternatively capillary electrophoresis, a detection limit in the low ppb range and precisions in the range of 1% are obtained, most often limited by unavoidable isobars or interferences from the plasma.

As an alternative sample preparation step, laser ablation (LA) is used for efficient evaporation of solid samples in ICP-MS. The plume of vapor and particulate matter from the pulsed laser laser-surface interaction is transported from a specific LA chamber into the plasma torch for atomization and ionization, as sketched in Fig. 4.7. LA-ICP-MS addresses a wide variety of materials including conducting and nonconducting inorganic and organic compounds both accessible as solid or powder. In addition to conventional bulk analysis, well-focused LA also permits for sampling of small areas in microanalysis or for spatially resolved studies (Gray 1985).

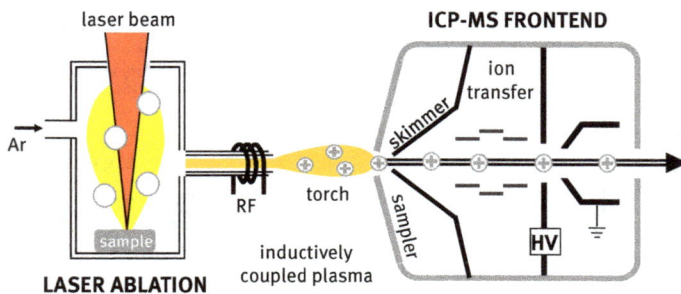

Fig. 4.7: Principle of LA-ICP-MS coupling. Interaction of a focused laser beam onto a solid surface ablates and evaporates species from the sample. The ablated material is transported using the argon carrier gas into the inductively coupled plasma torch for ionization. RF = radio frequency, HV = high voltage.

An effective approach of ICP-MS uses a combination of the ICP ion source with a high-resolution double-focusing sector field mass analyzer. This configuration allows for significant reduction of spectral interferences in the elemental analysis and thus permits the operation of the plasma torch in its optimum "hot" mode, in which interferences are abundant but also the strongly disturbing effect of oxide formation is strongly reduced in respect to a low background "cold" plasma. High-resolution ICP-MS thus is one of the preferred methods for multi-element investigations on low trace and ultratrace level in complex matrices as well as for precise measurements of isotope ratios.

A further upgrade of the ICP-MS technology implies the introduction of a collision cell into the mass spectrometer ion beam path. This device serves to dissociate disturbing molecular ions, to reduce the kinetic energy distribution of the ions, and to neutralize a major share of the disturbing argon ions of the plasma significantly reducing spectral interferences. It also results in higher ion transmission improving sensitivity and higher precision of isotope ratio measurements (Diez-Fernandez 2020). The most important technical refinement for isotope ratio measurements by sector-field ICP-MS was the implementation of a multiple-ion collector device, resulting in a precision of up to 0.002% in relative isotope abundances.

Already by using conventional low-resolution ICP-MS, the determination of ^{90}Sr, ^{137}Cs, lanthanides, and actinides is possible even in the presence of considerable isobaric interferences. In the determination of the radioisotopes of Cs at their nominal masses of 134, 135, and 137, for example, interferences by the natural isobaric isotopes of ^{134,145,137}Ba are efficiently eliminated. A quantitative detection limit of 16 pg/g for total Cs with a precision of 2.5% at concentration levels of 100 ppb was demonstrated. For the isotope ratio ^{134}Cs/^{137}Cs, a similar precision was achieved.

Soil samples from the former Semipalatinsk nuclear test site in Kazakhstan were investigated by ICP-MS in order to determine plutonium contamination levels together with isotope ratios at levels around 0.1 pg/g. ^{240}Pu/^{239}Pu ratios of 0.036(1) and 0.067(1) were measured in different samples and ^{242}Pu/^{239}Pu ratios of 0.000048(4) and 0.00029(5), respectively, showing the high precision of the technique. These values are significantly lower than values commonly accepted for global fallout from atmospheric nuclear weapons tests, which might indicate a specific composition of the nuclear explosive or additional sources of the Pu contamination.

The direct measuring capability of LA-ICP-MS was investigated on long-lived Th and U radionuclides in nonconducting concrete matrices as a very common material for nuclear waste packages. Thereby LODs in the low pg/g range were determined. This study also served to compare different calibration procedures, i.e., correction of analytical data by RSCs, the use of predetermined calibration curves, and calibration by coupling with an ultrasonic nebulizer. In high purity water, a detection limit for uranium in the lowest fg/ml range was achieved.

In comparison to the other conventional techniques of RMS, ICP-MS is one of the most sensitive techniques specifically regarding low-level long-lived radionuclide analysis in rather pure matrices, where interferences can be widely avoided. Correspondingly, it has also been frequently used for the pre-screening of materials to be employed in the framework of rare event searches, e.g., in detector material used in deep underground laboratories for investigations of properties of neutrinos or dark matter (Marchegiani 2021).

4.4.5 Resonance ionization mass spectrometry (RIMS)

For a variety of studies, e.g., on environmental or geological samples, the conventional methods of RMS discussed so far are limited in their performance and applicability. This is caused by the occurrence of isobaric interferences or by a surplus of strong neighboring masses due to the limited abundance sensitivity of the mass spectrometer. By applying a resonant optical excitation and ionization only to specific analyte atoms using laser light tuned to the unique resonance lines of this element, the ionization process in resonance ionization mass spectrometry (RIMS) removes these boundaries and introduces highest elemental and – for selected cases with dedicated experimental arrangements – even supplementary isotopic selectivity to the mass spectrometric process (Letokhov 1987). The resulting notable properties of resonance ionization in combination with subsequent mass spectrometry are:
- highest, almost complete suppression of any atomic and molecular isobaric interference in the sample;
- good overall sensitivity in the fg-range ($\approx 10^6$ atoms/sample);
- feasibility of ultra-high isotopic selectivity up to 10^{13} (by adding selection through optical isotope shifts in the atomic transitions to the abundance sensitivity of the mass spectrometer).

However, as a drawback and in contrast to other mass spectrometric methods, RIMS is not easily suitable for multi-element analysis but restricted to single-element determination. The principle of resonant ionization is shown in Fig. 4.8. The sample atoms are evaporated in ultra-high vacuum from a metallic filament into a beam or form a vapor inside a atomizer cavity. Starting from the atomic ground state or a thermally populated, still low-lying excited state, the free atoms are resonantly excited into a high-lying atomic state by step-wise absorption of photons from different lasers, which are well tuned to the individual optical transition. Typically, the first excitation step in the blue and the second one in the red spectral region are required. Ionization of the highly excited atoms can be performed in different ways by photons from another laser: either nonresonantly raising the overall excitation energy beyond the ionization potential or resonantly populating a so-

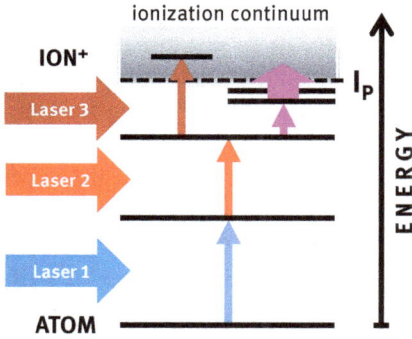

Fig. 4.8: Principle of three-step resonant excitation and ionization of an atom by laser light. Ionization into an auto-ionizing state situated above the ionization potential (I_p) is given on the left side, resonant population of a high-lying RYDBERG state, and subsequent ionization on the right side. The vertical error indicates the increase in excitation energy.

called auto-ionizing state situated above the first ionization potential of the atom. Alternatively, high-lying bound atomic, so-called RYDBERG states just below the ionization potential, may be populated and are subsequently ionized, e.g., via an electric field, black body radiation, far infrared photons, or by gas-kinetic collisions.

The resonant absorption of a photon by an atom exhibits a very high cross section of

$$\sigma_{\text{opt}} = \frac{\lambda^2}{2\pi} \approx 10^{-10}\text{cm}^2, \tag{4.9}$$

and the excitation probability per optical step may be estimated using the simple relation

$$W_{\text{opt}} = \sigma_{\text{opt}} \cdot n_{\text{Photon}} \cdot t_{\text{int}}. \tag{4.10}$$

State-of-the-art tunable laser systems deliver photon fluxes of $n_{\text{Photon}} \geq 10^{15}\text{–}10^{18}$ photons, either per second for continuous operation or alternatively within a short pulse (≈ 10 ns) for pulsed laser systems. Correspondingly, if interaction times in the order of $t_{\text{int}} \approx 1\,\mu\text{s}$ are arranged, the resonant optical excitation steps are "saturated" with optimum efficiencies near 100%. The bottleneck of the resonance ionization process is the final ionizing step. It exhibits much lower cross sections around $10^{-17}\text{–}10^{-19}$ cm^2 for the ionization into the continuum, while for auto-ionizing or RYDBERG states the overall ionization efficiency is increased significantly. Values up to 50% and beyond have been demonstrated on calibrated test samples.

In general, power, tunability, spectral stability, and bandwidth of the laser radiation are crucial factors of the RIMS technique and must be optimized in order to achieve highest performance. In particular, the spectral bandwidth in respect to the natural line width of the individual optical transitions influences both the optical selectivity and the overall efficiency.

4.4.5.1 Pulsed laser RIMS

For only moderate demands on isotopic selectivity, pulsed laser RIMS is used to ensure complete saturation of all optical excitation steps and the ionization for highest efficiency and analytical sensitivity. Pulsed laser ionization is ideally combined with mass separation by a time-of-flight mass spectrometer, for which the short laser pulse precisely defines the start signal and the ion arrival at the detector gives the stop signal for mass identification. Temporal and spatial overlaps of sample evaporation and laser pulses are critical and must be properly adjusted for well-resolved mass spectra.

Low repetition rate pulsed lasers operating in the 10 Hz regime are used together with well-synchronized pulsed laser ablation in the field of fundamental research on short-lived radionuclides. Unfortunately, this process is critically affected by insufficient pulse-to-pulse stability of power and spatial profile of the lasers preventing quantitative analytical application. In combination with continuous and smooth evaporation of an analytical sample by resistive heating from a filament or cavity, the use of high repetition pulsed lasers operating at around 10 kHz pulse rate is required for minimum losses. Well-controlled operation of such a laser system with up to three spectrally precisely tuned and temporally well-synchronized lasers is rather sophisticated and has so far restricted the use of RIMS to only a few research laboratories worldwide. Nevertheless, this approach, at a reflectron TOF-MS, as sketched in Fig. 4.9, is routinely used for ultratrace determination of plutonium and neptunium as well as of ^{99g}Tc in environmental and biological samples.

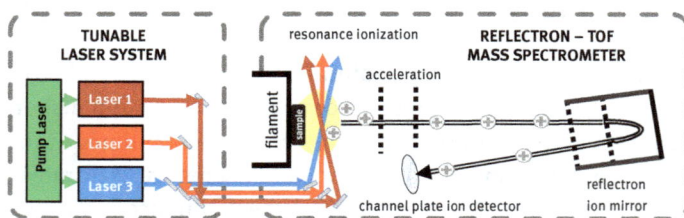

Fig. 4.9: Sketch of pulsed-laser RIMS with a reflectron time-of-flight mass spectrometer (TOF-MS) as used, e.g., for sensitive determination of plutonium and other actinides.

The necessary procedure for the particular case of Pu isotopes includes chromatographic preseparation from the sample for electrolytic deposition as hydroxide on a tantalum filament. For quantification, a known spike of a tracer isotope – usually ^{244}Pu ($t^{1/2} = 8.00 \cdot 10^7$ a) – is added to the sample prior to the chemical treatment while for monitoring of the chemical yield by α-spectrometry, a second spike of ^{236}Pu ($t^{1/2} = 2.858$ a) might be added, too. For efficient chemical reduction of Pu to its metallic form and for smooth atomization inside the vacuum recipient at around 1300 K, the filament is covered with a μm-thin titanium layer by sputtering. After selective resonant ionization, the ions are mass selected and detected in a reflectron TOF-MS system.

For highly selective direct and spatially resolved RMS, the very powerful hypernathed technique of secondary neutral mass spectrometry (SNMS) has been developed by the combination of sputtering with post-ionization via RIMS. The set-up is based upon a conventional SIMS device, which is upgraded by a suitable laser system. Secondary ions from the sputtering process are fully suppressed by pulsed electrostatic deflection, and remaining neutrals in the interaction region of the device are resonantly ionized for mass selection in a TOF-MS. The set-up permits for element selective analysis of smallest aerosol and hot particles as well as depth profiles or localized surface contaminations. The example of an 8-µm diameter hot particle, collected from the red forest of Chernobyl and shown in the inlay as an electron microscopy image, is given in Fig. 4.10. It shows the inventory and the ratios of the isotopes ^{239,240,241,242}Pu. The technique is also extended to minor actinide isotopes, e.g., for ^{242}Am determination (Bosco 2021).

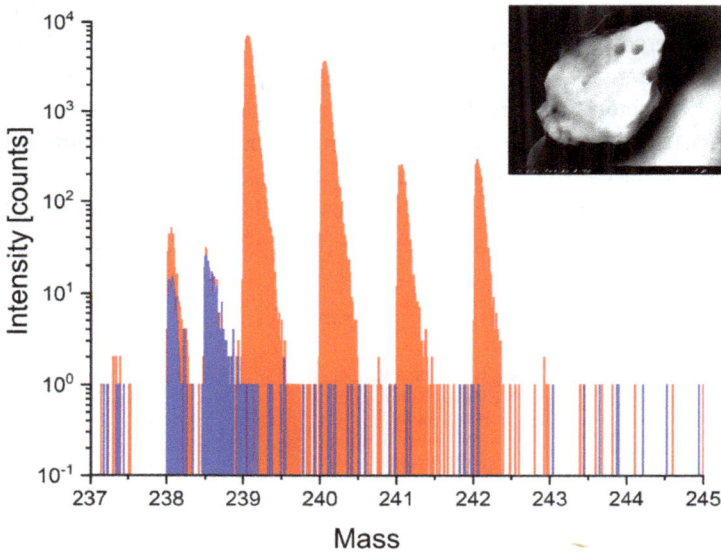

Fig. 4.10: Plutonium inventory of the 8 µm diameter hot particle shown in the electron microscopy inlay as obtained by SNMS. Lasers in resonance for plutonium (red), laser off-resonant showing uranium backgrounds (blue).

4.4.5.2 High-resolution RIMS

The method of high-resolution (HR)-RIMS offers ultimate isotopic selectivity well above 10^9 on top of the elemental isobar selection. Thus, it is used for cases requiring extremely high dynamic range in isotope ratios, e.g., for the determination of ultra-trace isotopes of an omnipresent element in the unavoidable presence of its stable isotopes. Using narrow bandwidth continuous wave lasers, high additional isotope selection is introduced into the optical excitation and ionization process by resolution of the optical isotope shift of atomic transitions. This small effect is due to the influence of the finite mass and size of the atomic nucleus of different isotopes onto the specific binding of the valence electron and can be exploited during optical excitation. In combination with a mass spectrometer which adds high abundance sensitivity, highest overall selectivity in respect to isotopes and isobars was demonstrated for a number of ultrarare radioisotopes, e.g., ^{90}Sr, ^{137}Cs, ^{210}Pb, and ^{236}U. For the long-lived radioisotope ^{41}Ca, a record value of up to S $\approx 3 \cdot 10^{12}$ has been achieved in the presence of a vast surplus of stable calcium, predominantly the neighboring isotope ^{40}Ca, with an LOD of 1 fg per sample, corresponding to about 10^6 atoms. This result serves as a benchmark for the achievable specifications of the technique and enables various applications, e.g., using ^{41}Ca as a biomedical tracer isotope for *in vivo* studies of human calcium kinetics and osteoporosis prevention studies. For these investigations, a minuscule µg amount (~10^{16}) atoms of the long-lived ^{41}Ca (half-life of 10^5 years) can be administered to volunteers. The resulting incorporation of ~3 kBq is well below the natural internal exposure of typically >10 kBq, stemming dominantly from the abundant radioisotope ^{40}K, and is not expected to cause any radiation hazard. As given in Fig. 4.11, the excretion curve for different test persons can be retraced, and long-term studies on the calcium metabolism become possible. In addition, fundamental research on extraterrestrial ^{41}Ca production cross sections prepare for cosmo-chemical investigations on the ^{41}Ca content and depth profile in meteorites, while the

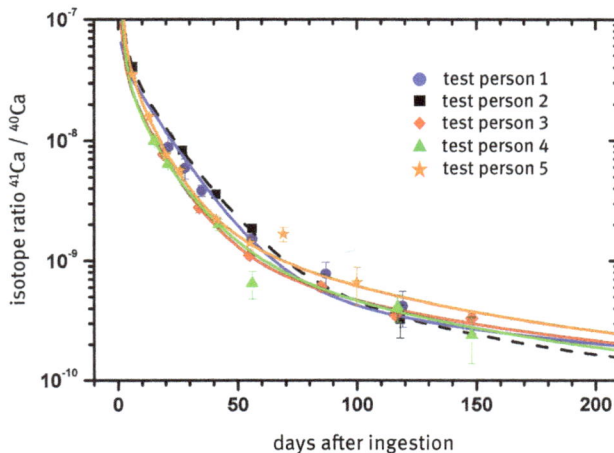

Fig. 4.11: Excretion curve of ^{41}Ca from five test persons, showing only small individual variability.

estimation of ^{41}Ca contamination from reactor concrete and other materials for inte-grated neutron dosimetry were rendered possible by this specification.

The setup for HR-RIMS applies a crossed-beam geometry for laser excitation on a well-collimated atomic beam, which effuses from a hot graphite furnace. In this way, spectral broadenings of the optical excitation lines are avoided, which would arise from the DOPPLER effect due to the different velocities of the individual atoms for a filament or in-cavity geometry and would prevent optical selectivity. For oper-ation with HR continuous laser ionization, the combination with a quadrupole mass spectrometer, for which abundance sensitivity in mass selection is optimum, is best-suited. Nevertheless, neighboring isotope selectivity is limited by various types of unspecific background to a range of 10^{13} and in addition require dedicated experimental arrangements.

For a fast and sensitive determination of the radioisotopes ^{89}Sr and ^{90}Sr in vari-ous environmental and medical samples, an alternative approach of HR-RIMS had to be customized. As isotope shifts in strontium are negligible, and resonance ioni-zation was applied on an accelerated fast atomic beam in collinear geometry. In this way, the isotope-dependent and very large DOPPLER shift of the optical reso-nance lines was used for selective ionization of individual isotopes. This technique was realized as an add-on to a large high-voltage mass spectrometer. The analytical sample is ionized unselectively in a conventional surface ion source, accelerated to a rather high ion beam energy of $\approx$30 keV and undergoes mass separation first. A fast atomic beam is formed by charge exchange processes induced through colli-sions in a cesium vapor cell with only little effect on the beam quality. Immediately afterwards, isotope selective laser excitation with subsequent ionization in a strong electric field is performed, followed by quantitative ion detection on a channeltron detector. An overall selectivity of S $\approx 10^{11}$ for the isotope ratio ^{90}Sr/^{88}Sr was demon-strated at an LOD for ^{90}Sr of $3 \cdot 10^6$ atoms ($\approx$1.5 mBq), and numerous low level sam-ples contaminated by the Chernobyl accident were analyzed.

4.4.6 Accelerator mass spectrometry

Accelerator mass spectrometry (AMS) is the most elaborated mass spectrometric technique, which is specifically tailored to provide highest selectivity with respect to any kind of isobaric and isotopic interferences for the determination of ultimate isotopic ratios. This feature involves high-energy ion acceleration in combination with a multitude of further process steps. The development of AMS was stimulated by the prevalent determination of the cosmogenic radioisotope ^{14}C with its out-standing applications for radiodating, atmospheric and oceanographic circulation studies, nuclear forensics, or biomedical studies, cf. Chapter 5. Many of these fields of ^{14}C determination just became possible by the development of AMS, which re-placed the formerly used low-level radiometric counting technique owing to its significantly higher sensitivity, smaller sample size requirement, and faster response. Mass spectrometric radiocarbon dating, carried out on ^{14}C isotopic levels as low as 10^{-15} in respect to stable carbon isotopes, represents the best known and widest field of high selectivity mass spectrometry with thousands of samples per year. On top of this, further applications of AMS on other radionuclides have contributed to the importance and acceptance of this highly sophisticated technique.

In the late 1970s, it was realized that isobaric contaminations in mass spectrometry may well be eliminated by production of negative ions in a sputter source. Moreover, after acceleration to high beam energy in the order of MeV, a variety of selection techniques can be applied and render possible quantitative, selective counting of the species of interest with unrivalled background suppression:

- Atomic isobars are eliminated already in the ion source by exploiting instability and corresponding inexistence of some negative ions, which would act as isobaric interferences like ^{14}N$^-$ in respect to ^{14}C$^-$.
- As pointed out by eq. (4.4) in Section 4.3.3, mass spectrometry selects according to the mass-to-charge ratio m/Z. Correspondingly, molecular ions like ^{28}Si^{2+}, ^{13}CH$^+$, or ^{12}CH$_2^+$ act as isobars and interfere with ^{14}C$^+$ ions in conventional RMS. In AMS, they are efficiently removed during acceleration: in the first half of the routinely used tandem accelerator, all negatively charged species, i.e., atomic and molecular ions are converted into positively charge states by stripping off outer electrons. In this process, molecular isobars are entirely removed by disintegration into their atomic fragments.
- On the high energetic ion, beam efficient and isobar discriminating nuclear physics detection techniques are used enabling element selective ion counting. The unambiguous determination of both mass and proton number for each ion removes any influence from remaining isobaric interferences even on the 10^{-15} level and beyond. The finally limiting background originates from effects of gas kinetics, wall scattering, or the charge-exchange collisions, which may be strongly, but never completely, suppressed by optimizing the experimental conditions (Tuniz et al. 1998).

A typical AMS system for radionuclides is based on a tandem accelerator with a terminal voltage around 2.5 to 5 MV. On the so-called low-energy side of the device, intense beams of negative atomic or molecular ions of the species of interest are produced within a conventional sputter ion source by Cs$^+$ ion bombardment of some keV beam energy focused onto the analyte. Typically, about 10 mg of sample material are needed for insertion into the sputter target, for which proper chemical pre-treatment is crucial. This includes removal of excessive material, dissolution, often addition of carrier, separation of the element of interest, and conversion into the form most suitable for generating negative ions in the sputtering process. These negative ions are accelerated towards the first low-energy magnetic sector field mass separator with typically about 50 keV ion beam energy for an initial mass selection. Afterwards, acceleration to the positive central terminal voltage of some MV of the accelerator takes place within the first half of the tandem accelerator. On the high positive potential of the accelerator terminal, the ions undergo the stripping process in a thin foil ($\approx 5\,\mu g/cm^2$) or alternatively in a gas target as mentioned before, which populates a spectrum of different positive charge states. Now positively charged, the ions are accelerated once more, this time from the positive terminal potential down to electrical ground. In general, charge states between 2+ and 7+ are populated depending on the ion velocity. For the 5+ ion as example, this leads to a total beam energy of six times the terminal potential of the tandem accelerator, i.e., 15 MeV for a typical 2.5 MV AMS machine. All molecular ions are dissociated during the stripping process, and their remaining fragments have significantly altered mass-to-charge ratios and are thus fully separated. For this purpose, a second magnetic sector field mass separator is installed together with an electrostatic deflector for energy filtering, this time on the high-energy side of the device. Species identification with respect to proton number Z and mass A is achieved by measuring energy and specific energy loss in a sensitive semiconductor detector telescope or in an energy-sensitive ionization chamber . The overall length of a typical AMS beam line ranges from a few meters for a 1 MeV up to several tens of meters for a typical 5 MeV machine. A sketch of the general layout of an AMS machine showing the individual components discussed above is given in Fig. 4.12.

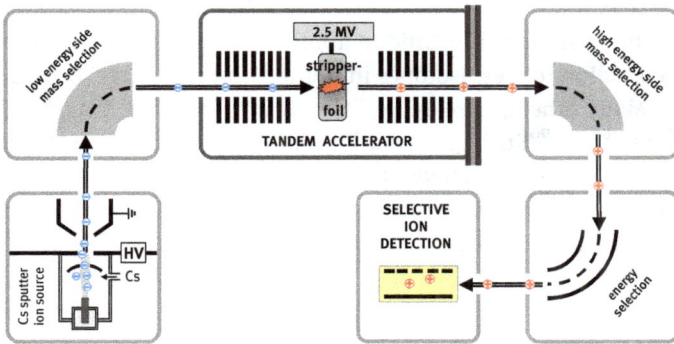

Fig. 4.12: Schematics of a typical AMS machine, highlighting the individual components of the Cs$^+$ sputter ion source, the first low-energy side mass selection, tandem accelerator with first acceleration stage, ion beam stripper, and second acceleration, second mass and energy selection on the high-energy side and finally selective ion detection.

Generally, AMS determines the concentration of the ultratrace radioisotope under study by comparing its count rate to the ion current of an abundant stable isotope of the same element as reference. Therefore, the measured ion count rates extend over many orders of magnitude and the achievement of reasonable accuracy in the order below 1% is a challenge. Experimentally, this is accomplished over a dynamic range of up to 15 orders of magnitude by operation in a fast cyclic mode, measuring the reference as electric current on a FARADAY cup detector and the ultratrace element as individual ion counts. Proper calibration and stability of both detectors as well as the whole apparatus during the measurement is mandatory. For this purpose, each analytical measurement is flanked by recordings on blind and calibration samples for background control and for the proper quantification of results.

For RMS studies of heavy radionuclides in the range of actinides, an additional time-of-flight measurement is usually implemented into the AMS high-energy ion beam path for further suppression of unselective background. In an alternative approach, the use of a gas-filled magnet in front of the detector provides spatial separation of ions with different proton number Z and further increases the suppression of those isobars that have not been filtered out·sufficiently during the stripping process. Recently, enormous efforts have been made to reduce sizes and costs of AMS systems. By incorporating optimized gas strippers and using state-of-the-art semiconductor detectors, compact low-voltage AMS machines of 200 kV to 1 MV have been realized, which provide comparable specifications to large machines but reduce expenditure and costs of analytical measurements significantly. These will be addressed in detail in Chapter 5.

4.4.7 Applications of AMS

AMS has nowadays become a standard technique applied for investigations on a vaste variety of long-lived radionuclides. Isotopes in routine operation are ^{10}Be, ^{14}C, ^{26}Al, ^{36}Cl, ^{41}Ca, and ^{129}I. Many others, namely ^{3}H, ^{3}He, ^{7}Be, ^{22,24}Na, ^{32}Si, ^{39}Ar, ^{44}Ti, ^{53}Mn, ^{55,60}Fe, ^{59,63}Ni, ^{79}Se, ^{81}Kr, ^{90}Sr, ^{93}Zr, ^{93}Mo, ^{99g}Tc, ^{107}Pd, ^{151}Sm, ^{205}Pb, ^{236}U, ^{237}Np, and $^{238-244}$Pu have also been investigated as research activities in analytics or to address fundamental problems. Most of them reveal natural abundances in the range of 10^{-9} down to 10^{-18}, cf. Fig. 4.1.

4.4.7.1 Nuclear dating

AMS is best known for its applications on the radioisotope ^{14}C with thousands of samples investigated per year worldwide. ^{14}C is produced by the neutron component of cosmic rays from ^{14}N in the upper atmosphere with a rate of about two atoms/(cm^2 s). Through the formation of CO_2, it is very homogeneously distributed throughout the global atmosphere and is easily assimilated by living organisms. With a half-life of $t^{1/2} = 5730$ a and an average abundance in living organisms of about 10^{-12}, the determination of the remaining ^{14}C abundance in fossils not only gives access to a dating range of about 60,000 years before present but, in addition, opens up a broad field of further investigations, e.g., circulation studies of the atmosphere or the oceans, authentication of minerals, plants, and even artificial pieces like artwork. This topic is discussed in detail in Chapter 5.

4.4.7.2 Pharmacological and medical applications

Studies involving radiotracer isotopes have become enormously relevant for various kinds of *in vivo* studies about influences of nutrition or deficiencies as well as on analyzing actuators and prevention of numerous diseases. In most cases, suitable enriched stable isotopes are not available, and radioisotopes must be chosen; thus, high donation doses would lead to a significant radiation dose for the test person. Furthermore, such high doses could affect the proper physiological response itself. Consequently, AMS with its high sensitivity and selectivity has become the most valuable system for analyzing biomedical samples. Apart from ^{14}C, that is also the key isotope for labeling numerous biomedically relevant compounds (cf. Chapter 12), radioisotopes used in this field are ^{10}Be, ^{26}Al, ^{41}Ca, ^{79}Se, and ^{129}I. Until recently, aluminum was considered an essential trace element, while only in the last decade has its toxicity been recognized. A number of AMS studies with ^{26}Al on gastrointestinal and intravenous absorption were carried out, which indicated pathways and established exchange rates for Al crossing the blood-brain barrier. ^{41}Ca, as another example, has become extremely versatile to analyze nutritional and other effects on *in vivo* calcium-kinematics and to work out means of prevention and treatments of Ca-related diseases like osteoporosis.

4.4.7.3 Nuclear energy

The determination of low level contaminations of radionuclides originating from the nuclear fuel cycle is of major importance for nuclear safeguarding, but it is often complicated by the necessity to distinguish these contributions from naturally existing radioactivity levels. A few AMS machines are specialized for measuring these radionuclides. Overall efficiencies of up to 10^{-4} enabling measurements with as few as 10^6 atoms of the radioisotope of interest have been demonstrated in favorable cases with typical selectivities above 10^{10}. AMS investigations in this field have been reported for ^{90}Sr, ^{99g}Tc, ^{206}Pb, ^{236}U, and several Pu isotopes. Applications concerning Pu also include biomedical studies on the uptake and retention of Pu in living organisms.

4.5 Outlook

The advantages and high performance of counting the individual atoms rather than the transformations by decay for specific long-lived radioisotopes in an analytical sample has been worked out in this chapter. Based on the well-established and widely applied techniques of inorganic mass spectrometry, approaches for radioisotope mass spectrometry (RMS) today are far developed and in many cases deliver more sensitive and more significant results than radiometric determination techniques based upon the emitted radiation.

Involving only rather minor adaptations, the conventional RMS methods are based on TIMS, GDMS, SIMS, and ICP-MS. Their performance competes with the specific sophisticated approaches of AMS and RIMS. The latter – on the other hand – delivers highest specifications regarding the key parameters of sensitivity and selectivity by implying advanced accelerator- and/or laser-based technology. A steady and steep progress is ongoing, regarding underlying aspects of mass spectrometry, upgrades of components, and combinations of methods. On one side, this includes the central mass-dispersive component involving novel computer-optimized magnet designs or new spectrometer geometries, like the orbital ion trap or a multi-reflectron approach. Similarly, continuous ion source development constantly increases the range of accessible atomic and molecular ion species in the beam as well as the stability and achievable ion currents. The latter is a prevalent factor for the study of ultratrace isotopes of lowest abundance at the required sensitivity as well as for investigations with highest dynamic range. The important add-on technologies of lasers in RIMS, accelerators in AMS, or a combination of both undergo a similar development, which enhances and simplifies the utilization of these techniques and ensures reliable and accurate results.

Most exciting is the ongoing miniaturization of spectrometers and the complete automatization of process control and most other aspects of RMS techniques, which is rapidly driven by the revolution in electronics and computer technology experienced in the last decades. Key words like "lab-on-a-chip" will also soon apply for mass spectrometric techniques in general and in particular for RMS applications on radionuclides. These new developments go well along with the quantitative and qualitative increase in the demand for radionuclide determination concerning analytics as well as fundamental problems. While on one side, the beneficial use of specific radioisotopes, e.g., in nuclear medicine, presently is steadily growing, the rising awareness of radioactive hazard to the population, and the environment fosters application and development in the field of radioanalytics. This also stimulates RMS as one of the most sensitive key techniques.

Appendix

Tab. A: Terminology in radioisotope mass spectroscopy.

Abundance sensitivity	Contribution of one peak in a mass spectrum to the neighboring peak, measured either in percentage of the height or area
Accelerator mass spectrometer	Mass spectrometric device involving a particle accelerator and numerous selection steps for ultimate isotope selectivity
Accuracy	Measure of the absolute correctness of analytical results
Auto-ionizing state	Energetic state of the atom, situated above the first ionization potential in the continuum of unbound energies
Channelplate	Single ion detector with high dynamical range, containing a multitude of individual single electron multiplier channels
Channeltron	Single ion detector based upon a single continuous electron multiplier channel
Chromatography	Physical method of species separation by distribution between two phases
Continuum	Energy range above the first ionization level of an atom
Cross section	Probability of an interaction, expressed as a corresponding interaction area
Dispersion	Ability of the mass spectrometric device to separate different masses
Efficiency	Ratio of detected particles in comparison to the initial sample content
Electron affinity	Binding energy of the extra electron of a negatively charged ion
Emittance	Quality parameter of a charged particle beam given by area times opening angle
Erosion	Process of slow excavation of material, e.g., by ion bombardment
Faraday cup	Well-shielded electrode arrangement for precise determination of delivered electric charge or current of an impinging ion beam
Filament	Heated metal carrier for introduction and evaporation of a sample at the ion source
Glow discharge mass spectrometer	Mass spectrometric device utilizing a glow discharge as ion source
Hot particle	Aerosol particle carrying significantly increased radioactivity level
Hyphenated technique	Direct coupling of chromatographic and further selection techniques like mass spectrometry

Tab. A (continued)

Inductively coupled plasma	Inductively heated argon plasma operating under ambient conditions
Inductively coupled plasma mass spectrometer	Mass spectrometric device using an ICP plasma torch and differential pumping stages as ion source
Interfering radiation	Disturbing α-, β-, γ-radiation with identical or close-lying energy from unwanted contaminations
Ionization potential	Binding energy of the valence electron of a neutral atom
Isobar	Atomic or molecular species with identical mass
Isochronous	Identical in time
Isotope dilution mass spectrometry	Intentional admixture of a quantified amount of isotopically enriched material to the sample enabling quantification
Isotope fractionation	Variation of detection efficiency as function of mass number in mass spectrometry
Isotope pattern	Significant series of isotopic abundances for an element
Isotope shift	Influence of the finite nuclear mass and size on the energetic levels of the valence electron and correspondingly the characteristic resonance lines
Isotope tracer	Quantified amount of a single isotope added to the sample enabling quantification
Laser ablation	Process of material evaporation by laser irradiation
Level of detection	Minimum detectable amount in a sample, determined by statistics
Multiplicity	Quantification of the maximum filling of degenerate atomic electron orbits
Orbitrap mass spectrometer	Ion trapping device based upon a central spindle on high potential around which the ion spiral
Photographic plate detector	Exposure to irradiation by particles gives sensitivity patterns on the plates, which show up after photographic development
Plasma torch	Radiofrequency discharge operating at atmospheric pressure as ion source in ICP-MS
Precision	Measure of the reproducibility of analytical results
Reflectron	Arrangement of a time-of-flight mass spectrometer (TOF-MS) involving an electrostatic ion reflector to compensate the initial energy distribution

Tab. A (continued)

Refractory metal	Class of metals that are extraordinarily resistant to heat and wear
Relative sensitivity coefficient	Experimental parameter characterizing the different influence of matrices in LA-ICP-MS
Relative standard deviation	Statistical quantity measuring the width of a normal distribution
Reproducibility	Measure of the ability of an apparatus to generate identical results in consecutive measurements
Resolution	The ability to distinguish mass peaks in a spectrum, determined by the mass number divided by the peak widths
Resonance ionization mass spectrometer	Mass spectrometric device using resonant optical excitation and subsequent ionization induced by tunable laser light as ion source
RYDBERG state	Highly excited bound atomic level with high principal quantum number of the valence electron, situated just below the first ionization potential
Sample	Specimen or small quantity of the material under investigation
Secondary electron multiplier	Electronic device based on discrete electrodes (dynodes) enabling amplification of single electron or ion impact for electronic recording
Sensitivity	Measure of the detection ability by an analytical method, quantified via the LOD
Sputtering	Process of ejection of atoms and ions from a solid target by impact of energetic particles
Statistical weight Thermal ionization mass spectrometer	Relative probability of a state Mass spectrometric device applying surface ionization on a single or double filament as ion source
TOF-MS	Mass spectrometric device using the flight time after pulsed ionization for mass selection
Work function	Required energy to remove an electron from a solid surface of a given material

References

Becker JS. Inorganic mass spectrometry: Principles and applications. Chicester, John Wiley & Sons, 2007.

Becker JS. Inorganic mass spectrometry of radionuclides. In: L'Annunziata M. ed. Handbook of radioactivity analysis, 3rd ed. Orlando, Academic Press, 2012, 833–871.

Betti M, Aldave de las Heras. L. Glow discharge spectrometry for the characterization of nuclear andradioactively contaminated environmental samples. Spectrochim Acta B 2004, 59, 1359–1376.

Bosco H, Hamann L, Kneip N, Raiwa M, Weiss M, Wendt K Walther C. New horizons in micro particle forensics: Imaging ^{238}Pu and ^{242m}Am in hot-particles. Science Adv 2021, 7(44).

Diez-Fernandez S, Isnard H, Nonell A, Bresson C, Chartier F. Radionuclide analysis using collision–reaction cell ICP-MS technology: A review. J Anal At Spectrom 2020, 35, 2793–2819.

Gray AL. Solid sample introduction by laser ablation for inductively coupled plasma source mass spectrometry. Analyst 1985, 110, 551–556.

Harrison, HW. Glow discharge mass spectrometry. In: Adams F, Gijbels R, van Griecken R. eds. Inorganic mass spectrometry, Chemical Analysis, Vol. 95. New York, John Wiley & Sons, 1988, 85–123.

Letokhov VS. Laser photoionization spectroscopy. Orlando, Academic Press, 1987.

Lu Z, Wendt KDA. Laser based methods for ultrasensitive trace-isotope analyses. Rev Sci Instr 2003, 74(3), 1169–1179.

Marchegiani F, Ferella F, Nisi, S. Material screening with mass spectrometry. Physics 2021, 3, 71–84.

Montaser A. Inductively coupled plasma mass spectrometry. New York, Wiley-VCH, 1998.

Paul W. Electromagnetic traps for charged and neutral particles, Angew. Chem Int Ed Engl 1990, 29, 739–748.

Platzner IT. Modern isotope ratio mass spectrometry. In: Chemical analysis, Vol. 145. Chicester, John Wiley & Sons, 1997, 1–530. ISBN 0471 974161.

Tuniz C, Bird JR, Fink D, Herzog GF. Accelerator mass spectrometry. Boca Raton, CRC Press LLC, 1998.

Tykva R, Sabol J. Low-level environmental radioactivity. Lancaster, Technomic Publishing Company, 1995.

Heinz Gäggeler and Sönke Szidat

5 Nuclear dating

Aim: To understand the past is instrumental for predicting the future. Therefore, dating of objects – be they from human activities or from nature – is of prime interest in science. Radionuclides are the most useful tool to tackle this problem. Their half-lives may be used as a clock to measure elapsed time. However, a number of conditions have to be met for proper application of such dating methods. Our environment is rich in radionuclides, mostly of natural origin but partly also from anthropogenic sources such as nuclear weapons testing or from nuclear accidents. This chapter summarizes applications of such radionuclides for dating purpose.

Special emphasis is given to radiocarbon, ^{14}C. For many applications this radioisotope of carbon is the most important dating radionuclide that enables to cover time horizons from the present until about 55,000 years ago. The use of parent/daughter systems is described as well.

Two other short sections address dating via stored signals in the samples from radioactive transformation. These are fission track and thermoluminescence dating.

5.1 Introduction

Knowledge on the age of important objects, e.g., of archeological or geological origin, environmental archives, or of our Earth and the solar system or even an object from space such as meteorites, is of great interest to mankind. This interest is based on the public understanding that present and future development can only be understood on the basis of historic information.

Until the discovery of radioactivity in 1896 by BECQUEREL it was essentially impossible to tackle this problem accurately except in cases where historic documentation was accessible (e.g., debris from documented volcanic eruptions stored in environmental archives such as lake sediments or glacier ice cores). Another example is seasonal information preserved, e.g., in accurately documented tree rings. Radioactive transformation of nuclides contained in our environment paved the way to solve many of the most challenging questions about the history and birth or death of many important samples or to study climate variability in the past.

The basis of nuclear dating lies in the fact that the number of atoms $\hat{N}$ of a given radionuclide is correlated with time according to the well-known exponential transformation law,

$$\hat{N}_1 = \hat{N}_0 \exp\left\{-\lambda(t_0 - t_1)\right\},\tag{5.1}$$

https://doi.org/10.1515/9783110742701-005

where $\hat{N}_1$ and $\hat{N}_0$ denote the number of atoms at time of measurement (t_1) and at time zero (t_0), respectively, and λ the transformation constant.

This equation may be transformed into

$$t_0 - t_1 = \Delta t = \frac{1}{\lambda}\ln\left(\hat{N}_0/\hat{N}_1\right), \qquad (5.2)$$

which yields the age Δt.

The transformation of a given radionuclide is essentially not altered by environmental influences such as chemical speciation, temperature, pressure, or other physical parameters. The transformation constant λ therefore acts as an absolute clock.[1] Radioactivity $\hat{A}$ may be quantified either by measuring the radioactive transformation rate (radiometric method, cf. Chapter 1) or by determining the mass of the nuclide of interest (mass spectrometry, cf. Chapter 4) to obtain the number of atoms. Both quantities are correlated via

$$\hat{A} = \lambda \hat{N}. \qquad (5.3)$$

As a rule of thumb, for short-lived radionuclides with half-lives lower than about 100 years it is advisable to perform a radiometric analysis and for nuclides with longer half-lives a mass spectrometric determination. The reason for this time limit is the achieved sensitivity: for short-lived radionuclides a large part of the ensemble of atoms transforms during the measuring time, say a few days at most. This yields an accurate determination of the activity because radiometric analysis may be performed with very high counting efficiency. Hence, for longer-lived nuclides the sensitivity of a mass spectrometric analysis becomes superior. Often, accelerator mass spectrometry (AMS) is used for the determination of the number of atoms.[2] This technique may reach overall efficiencies of a few percent and has the great advantage of being essentially background-free. Moreover, AMS enables counting of single separated ions with a counter placed in the focal plane, while conventional mass spectrometers determine separated ions via electric current measurement. This latter technique is not applicable to detection of only few ions. AMS has revolutionized dating with, e.g., the radionuclide ^{14}C ("radiocarbon", see below).

---

1 Except radionuclides transforming by electron capture, i.e., atoms that undergo a nuclear transformation via capture of an atomic electron. Such a process is slightly influenced by the chemical environment, cf. Chapters 7 and 11 of Vol. I. One example that has been studied in detail is the influence of chemical composition on the transformation of ^{7}Be, a radionuclide quite abundant in the atmosphere. This influence is even larger for nuclei traveling in space. Due to their high velocity such atoms are highly stripped, meaning that they exist only in form of highly positively charged ions with very few if any electrons. Clearly, this influences the probability to capture an atomic electron by the nucleus and changes the radioactive transformation probability significantly.

2 Cf. Chapter 4 on Radionuclide Mass Spectrometry.

In principle four situations exist that are commonly applied in nuclear dating, cf. Tab. 5.1.

Tab. 5.1: Principle situations commonly applied in nuclear dating.

Approach	Example radionuclide	Example application
Transformation from an equilibrium situation	^{14}C	Radiocarbon dating
Accumulation of a daughter nuclide	^{222}Rn	Dating applied in hydrology
Transformation from a single event injected into the environment	^{3}H	Dating of ice cores with maximum deposition in the years 1962–1963 from nuclear weapons testing
Exposure age dating	^{10}Be	Age determination of surface rocks

The following figures illustrate the four cases using baskets filled with sand where every sand corn represents one atom.

5.1.1 Equilibrium situation

In the equilibrium situation a basket is continuously filled with sand while, simultaneously, the basket loses sand corns (resembling atoms) via a leak at a rate that corresponds to the filling rate (see Fig. 5.1). Therefore, the sand level in the basket is constant (equilibrium level). If, at a given time, the supply of sand is stopped, meaning that the object is disconnected from the continuous supply, the level of sand starts to lower according to the leaking rate. This leaking rate resembles the transformation constant λ of the radionuclide. From knowing the equilibrium level at time zero (t_0) and the level at present time (t_1) the elapsed time Δt since the disconnection from supply can be determined using eq. (5.2). Here, $\hat{N}_0$ means the number of sand corns, i.e., the number of atoms of the radionuclide during the equilibrium situation and $\hat{N}_1$ the present number. A typical example of such a situation is radiocarbon dating.

5.1.2 Accumulation dating

The situation for accumulation dating is depicted in Fig. 5.2. Two baskets are connected via a tube. At time zero basket two is empty (or has a known level). As time elapses, basket one is leaking at a constant rate, resembling the transformation

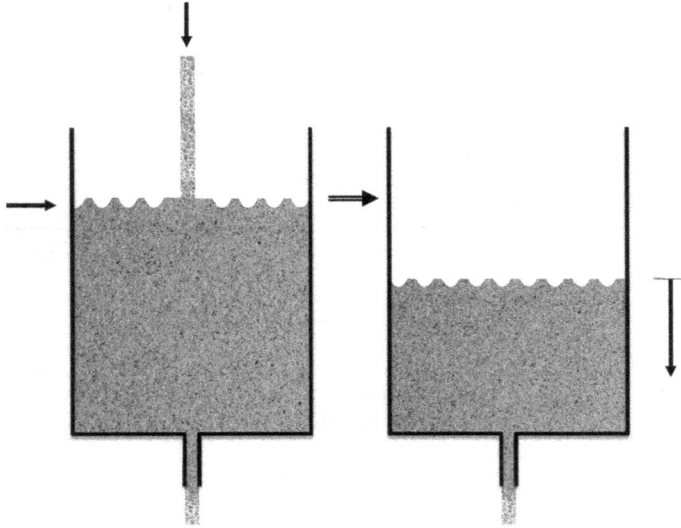

Fig. 5.1: Dating from an equilibrium state. Left side: equilibrium situation at time zero, t_0. The radionuclide is constantly produced at a rate which is equal to the transformation (i.e., loss) rate. Right side: situation at "present time". The radionuclide is no longer produced and transforms according by its half-life.

constant of the radionuclides contained in basket one. This leads to a gradual filling of basket two. Hence, with this method, the leaking time is determined from the measured accumulation in basket two at present – if needed corrected for the level at time zero – divided by the leak rate from basket one. This type of dating is frequently used in geology, e.g., using the $^{40}K \mid {}^{40}Ar$ method. In special cases, basket number 2 may also leak because the species accumulated in basket two is not stable but radioactive. Such a case is the ^{222}Rn dating applied in hydrology.

5.1.3 Decay from a single event

Decay from a single event means that due to an "injection" of a radioactive species into the environment, some compartments, e.g., the atmosphere, have been enriched with this radionuclide. By measuring the radioactivity at present and knowing the activity at time of injection, the elapsed time may be deduced. This situation is reflected in Fig. 5.3. A typical example for this situation is tritium dating using the well-known release ("injection") of this radionuclide into the stratosphere by nuclear weapons testing in the early 1960s.

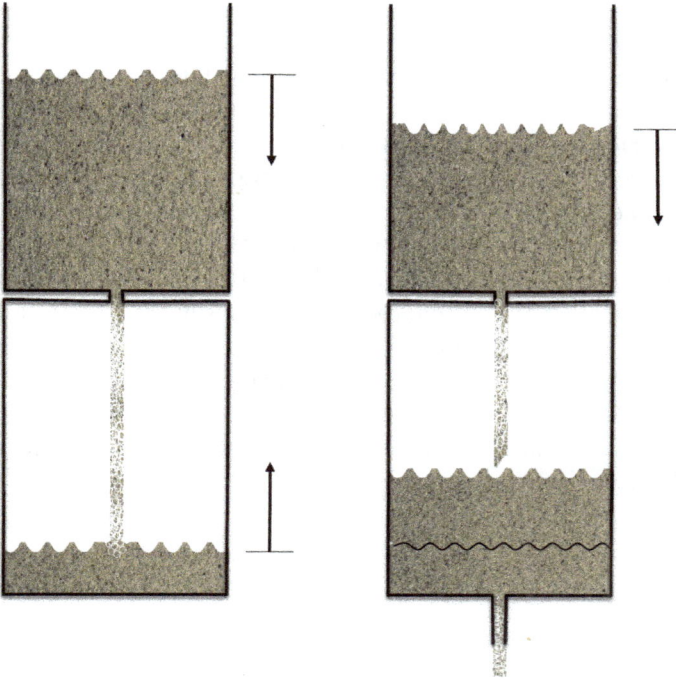

Fig. 5.2: Accumulation dating. Left side: at time zero (t_0) the lower basket has zero or a known filling. Right side: at "present time" the lower basket is partly filled due to transformation from the upper basket. It is possible that also the lower basket has a "leak", i.e., is radioactive.

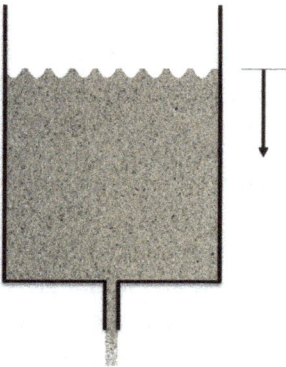

Fig. 5.3: Decay of a radionuclide injected into the environment.

5.1.4 Exposure age

Finally, exposure age dating uses the fact that radioactive nuclides are formed in samples from the Earth's surface because they are exposed to high-energetic primary and secondary cosmic ray particles.[3] Cosmic rays in space are predominantly protons that travel with very high velocities, i.e., high energies. Such cosmic rays and secondary particles (neutrons or myons) formed via interaction with the atmosphere – mainly nitrogen, oxygen, and argon – may interact at the Earth's surface with rocks or soil. As a consequence, radioactive nuclides are formed. This means that from the measured amount of such produced radionuclides in the sample and assuming that the amount was zero at time zero (i.e., at a time where the surface was not yet exposed to cosmic rays but covered by a layer of, e.g., soil or ice), the time interval since removal of the cover layer may be deduced. This situation is presented in Fig. 5.4. A typical example is exposure age determination of surface rocks by measuring ^{10}Be or ^{26}Al.

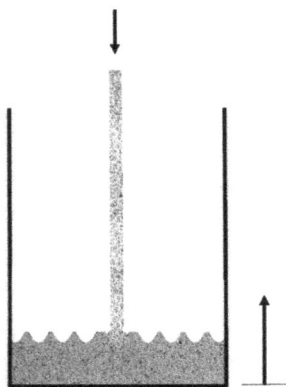

Fig. 5.4: Exposure age dating. At time zero (t_0) production of a radionuclide in the sample starts from zero.

5.1.5 Basic assumptions relevant in nuclear dating

All nuclear dating applications are based on the assumptions listed in Tab. 5.2.

Let us finally briefly touch on the reliability of a determined age with a given technique, being it radiometric or via mass spectrometry. Two terms are usually used to quantify the quality of the result: *accuracy* and *precision*. Accuracy describes the degree of agreement between measured and true age of a sample, while precision describes the statistical uncertainty of a measurement, usually defined by the number of detected events or atoms.

---

3 Cf. Chapter 2 for the origin and impact of cosmic rays.

Tab. 5.2: Basic assumptions relevant in nuclear dating.

1 The number of atoms at time zero, $Ñ_0$, must be known.	Quite often this prerequisite is not fully met. An example is ^{210}Pb dating. Usually, the equilibrium situation in the past is assumed to be reflected in constant activity concentration in precipitation or the deposition rate of ^{210}Pb per year at a given site to be constant (see text).
2 The system to be dated must be closed.	This means no physical or (geo)chemical processes have changed the sample in such a way that the radionuclide used for dating has been subjected to such processes.
3 The source of the sample is well defined.	Especially in hydrology, water samples quite often represent mixed sources. Such a situation may yield wrong average ages, especially if only one radionuclide is used for dating. Similarly, exposure age determinations with additional surface erosion of the samples may falsify the result.

5.2 Environmental radionuclides

Nuclear dating uses radionuclides from our environment. Figure 5.5 depicts all classes of "environmental radionuclides" plotted with their half-lives against the atomic number of the nuclides. Only those radionuclides are identified with symbols that are relevant for nuclear dating applications. They may be classified according to their origin, cf. Tab. 5.3.

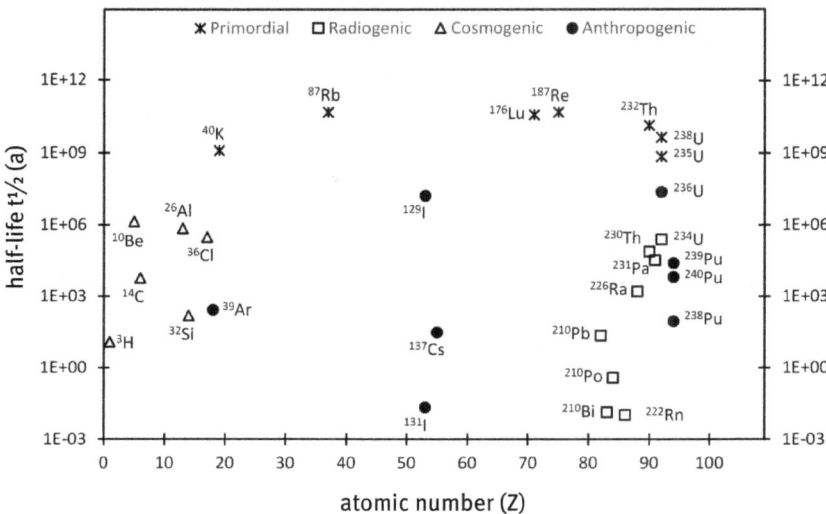

Fig. 5.5: Radionuclides in the environment of different origin as a function of their half-lives.

Tab. 5.3: Environmental radionuclides classified by origin.

Origin	Explanation	Examples
Primordial	Produced during element synthesis in our galaxy but having sufficiently long half-lives to still exist in part now	Actinides in nature: ^{235,238}U, ^{232}Th
Radiogenic	Formed in radioactive transformation of primordial radionuclides	^{226}Ra, ^{210}Pb
Cosmogenic	Produced in the interaction of high-energy cosmic ray particles (e.g., protons) with atmospheric constituents (e.g., nitrogen)	^{14}C, ^{10}Be
Fission	Formed in spontaneous fission of actinide nuclides, released by nuclear facilities, or in nuclear bombs	^{137}Cs, ^{131}I
Activation	Produced in nuclear reactions with particles (mostly neutrons) emitted in nuclear facilities or nuclear weapons testing	^{14}C
Breeding	Formed in nuclear reactions (e.g., with neutrons or heavy ion particles) with actinides	Plutonium isotopes

Of primordial origin are those radionuclides that were formed in space and partly survived since the birth of our solar system $4.55 \cdot 10^9$ years ago.

Radiogenic radionuclides are members of the natural transformation chains, hence rather short-lived radionuclides compared to the age of the Earth. They are constantly formed via radioactive transformation of the primordial radionuclides, ^{235,238}U and ^{232}Th, respectively.

Cosmogenic radionuclides are formed by interaction of primary (mostly protons) or secondary (mostly neutrons or myons) particles in the upper atmosphere or on the Earth's surface.

Mostly of anthropogenic origin are radionuclides formed in nuclear fission processes. This includes nuclear reactors as well as debris from nuclear weapons testing. In ultra-trace amounts, natural fission products also exist on Earth formed via interaction of neutrons with natural uranium or from spontaneous fission of ^{238}U.

Activation products are formed mainly in nuclear facilities or nuclear weapons testing via interaction of emitted particles such as neutrons with stable nuclides in close vicinity. An example is radiocarbon, formed from atmospheric nitrogen by interaction with neutrons emitted from nuclear reactors in the ^{14}N(n,p)^{14}C reaction.

Breeding means that radionuclides which are not abundant in nature, i.e., isotopes of all elements with atomic numbers higher than 92 (uranium) are formed in nuclear reactions. Artificial radioelements currently exist up to atomic number 118.[4]

---

4 Cf. Chapter 9.

Some of those radionuclides are also formed in nuclear explosions and have thus been distributed globally, such as the plutonium isotopes ^{239}Pu and ^{240}Pu, respectively. Another isotope of plutonium, ^{238}Pu is frequently used for electric supply in satellites. By accidental failure of a few satellites, ^{238}Pu has been deposited on the Earth's surface at the site of impact.

5.3 Dating with a single radionuclide

Most applications based on the detection of one radionuclide rely on the equilibrium situation, see Fig. 5.1. By some mechanism such as, e.g., via interaction of cosmic rays with the constituents of the atmosphere, the radionuclide of interest is formed at a constant rate or at least at a documented rate during the time interval to be dated. This enables a determination of the number of atoms or of the activity at time zero, i.e., $\hat{N}_0$ or $\hat{A}_0$, respectively. A special case is exposure age dating. Here, the radionuclide has zero concentration at time zero and is then formed when the object to be dated is exposed to cosmic rays. A summary of radionuclides discussed in this chapter is compiled in Tab. 5.4.

Tab. 5.4: Dating with a single radionuclide.

Radionuclide	Half-life in years	Typical application
^{14}C	5730	Dating of carbon containing objects up to 55,000 years
^{210}Pb	22.2	Accumulation rate determination of lake sediments, or ice cores over up to two centuries
^{32}Si	153	Accumulation rate determination of sediment cores or ice cores over about one millennium
^{10}Be	$1.39 \cdot 10^6$	Accumulation rates of rocks; indirect dating of archives based on chronology of solar activity
^{26}Al	$7.17 \cdot 10^5$	Accumulation rates of rocks
^{36}Cl	$3.01 \cdot 10^5$	Dating of archives over about one million years; dating of modern archives based on ^{36}Cl injection by nuclear weapons testing

5.3.1 ^{14}C (radiocarbon) ($t\,{}^{1}\!/_{2} = 5730$ years)

Radiocarbon is a cosmogenic radionuclide. As shown in Fig. 5.6, it is mainly produced in the stratosphere by interaction of thermal neutrons formed as secondary particles from primary cosmic rays (mostly protons) with nitrogen according to the nuclear reaction $^{14}N(n,p)^{14}C$. The ^{14}C atoms are instantaneously oxidized to ^{14}CO and in a slower process to $^{14}CO_2$. Due to long residence times of CO_2 in the stratosphere and a low exchange rate between stratosphere and troposphere, latitudinal differences of ^{14}C production are balanced. This results in a practically constant supply of $^{14}CO_2$ to the troposphere. Again, long residence times lead to a complete spatial mixing of $^{14}CO_2$ in the troposphere within each of both hemispheres, before it is transferred to the biosphere by photosynthesis or to the marine hydrosphere by bicarbonate dissolution. Consequently, living plants and higher organisms consuming such plants contain carbon on the contemporary radiocarbon level, cf. Fig. 5.1, left. After their death, "fresh" carbon ^{14}C is no longer taken up so that the radioactive clock "starts ticking" with the half-life of 5730 years, cf. Fig. 5.1, right. This allows dating of organic or carbonaceous material up to 55,000 years.

Although the stability of CO_2 in the atmosphere results in equilibration of spatial ^{14}C activity concentration, residence times are not long enough to eliminate temporal variability of ^{14}C production. Such variability occurs periodically on short to long timescales (i.e., 10 to 10,000 years) and may amount to 0.5% to more than 10%, respectively. As explained above, ^{14}C is produced practically exclusively from galactic cosmic rays.[5] Studies of meteoroids revealed that the energy spectrum and the amplitude of the interstellar galactic cosmic rays have been constant over the past million years. Actually, changes in the ^{14}C production rates are caused by two phenomena, the solar cosmic rays and the magnetic field of the Earth. Both components counteract the galactic cosmic ray intensity thus modulating the impact of the latter to the Earth's atmosphere.[6] This leads to decadal up to centennial wiggles, which stem from solar cycles of different frequency.[7] Additional wiggles on timescales of millennia to tens of millennia are caused by changes of the Earth's magnetic field inducing maximum differences of production rates in the order of >10%.

In early years of radiocarbon research (i.e., from the 1950s to the 1980s), gas proportional counters (and later liquid scintillator counters, LSC) were employed for measurement of the radioactivity of ^{14}C. Since 1977, accelerator mass spectrometry (AMS) has been increasingly used and has become the main technique for ^{14}C

---

5 The sun sporadically produces eruptive events leading to intense fluxes of solar energetic particles (SEPs). SEPs of extreme energies that occur very rarely may result in measurable signals of cosmogenic radionuclides such as ^{14}C on top of the background from the production by galactic cosmic rays.

6 A weak sun emitting cosmic rays of reduced fluence on the one hand or a weak magnetic field of the Earth on the other hand (both compared to average conditions) are less able to repel the high-energetic charged particles of intergalactic origin. This results in an enhanced ^{14}C production in the stratosphere.

7 The SCHWABE cycle with a periodicity of $\approx$ 11 years leads to an average difference of ^{14}C production between an active and a weak sun of about 5 ‰, whereas the GLEISSBERG and the DEVRIES cycles with a periodicity of 90 and 200 years, respectively, influence the variability of the radiocarbon production by about 25 ‰.

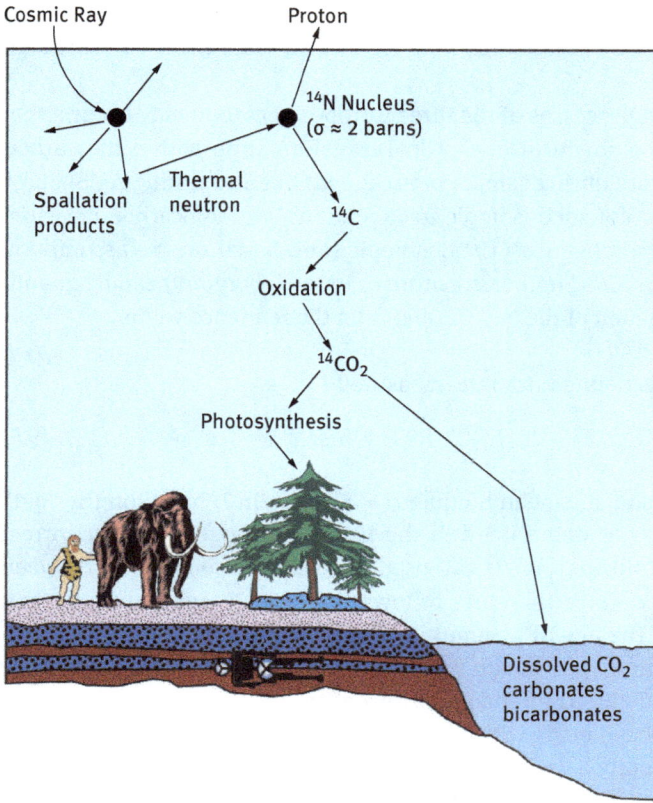

Fig. 5.6: Formation and distribution of natural ^{14}C; adapted from Currie (2002).

dating today. In AMS, ^{14}C is measured together with the stable carbon isotopes ^{12}C and ^{13}C quasi simultaneously in order to achieve high precision and accuracy of measurement reaching values of 3‰ or even better. Consequently, measurement results are given as $^{14}C/^{12}C$ ratios, which are related to the "present" level. The term $F\,^{14}C$ (fraction of modern radiocarbon) is defined as:

$$F^{14}C = \frac{(^{14}C/^{12}C)_{\text{sample}}}{(^{14}C/^{12}C)_{\text{modern}}} \tag{5.4}$$

Equation (5.4) includes several conventions such as:

1. The year 1950 is considered as the reference year of all radiocarbon measurements thus indicating the "present" or "modern" reference. Therefore, "before present" (often abbreviated as BP) denotes a time before AD 1950.

2. $(^{14}C/^{12}C)_{\text{modern}}$ is defined by convention from reference materials supplied by NIST (National Institute of Standards and Technology from the US Department of Commerce).

3. Isotopic fractionation processes of the three isotopes of carbon may occur either in nature during each chemical or physical transformation within the carbon cycle,[8] in the laboratory during sample pretreatment but also during AMS analysis. In order to correct for such isotopic fractionations, all radiocarbon measurements are related to a reference $^{13}C/^{12}C$ isotopic ratio based on the assumption that fractionations of $^{14}C/^{12}C$ (either in nature or in the laboratory) can be quantified as twice the deviation of the $^{13}C/^{12}C$ ratio from the reference value.

The so-called ^{14}C age Δt is defined from the measured $F^{14}C$ as

$$\Delta t = -\ln(F^{14}C) \cdot 8033\,\text{years}, \tag{5.5}$$

where 8033 years represents the mean lifetime $\tau (= \lambda^{-1} = t\frac{1}{2}/\ln 2)$ based on the "old" ^{14}C half-life of 5568 years as determined in the 1950s. Although the new, correct value is 5730 years, the "old" half-life is still used to guarantee consistency between old published and new data. It therefore follows that the ^{14}C age only reflects a first-order estimation of the age of a sample based on the wrong assumption of a constant ^{14}C production and applying an incorrect ^{14}C half-life. The correct calendar age of a sample is then deduced after comparison of the ^{14}C age with a calibration curve. A large effort has been made with radiocarbon analysis of environmental material that was independently dated with other techniques.

By comparison of the ^{14}C result with the independent (calendar) date, the radiocarbon formation at a given time has been quantified leading to a calibration of the ^{14}C age, Fig. 5.7. Tree rings have been used for the preparation of the calibration curve back to 11 950 BC. The independent dating was performed with dendrochronology. This method enables dating of trees by determination of their ring widths and by combination of overlapping tree-ring chronologies. For the Northern Hemisphere, this technique cannot be continued further back in time due to a lack of suitable trees at the conclusion of the last glacier maximum.

In order to continue calibration back to ≈55,000 years, other environmental archives have been used such as marine as well as lake sediments, marine corals, or speleothems.[9] The independent dating of these samples has been performed with the $^{230}Th\,|\,^{234}U$ method (see below) and other approaches, which, however, are less precise than dendrochronology.

---

8 Every physical or chemical process with carbon-containing compounds leads to a slight but significant change in the isotopic composition of its isotopes (i.e., ^{12}C, ^{13}C, ^{14}C). The reason is the mass dependence of equilibrium constants.

9 Speleothems are stalactites and stalagmites often found in caves.

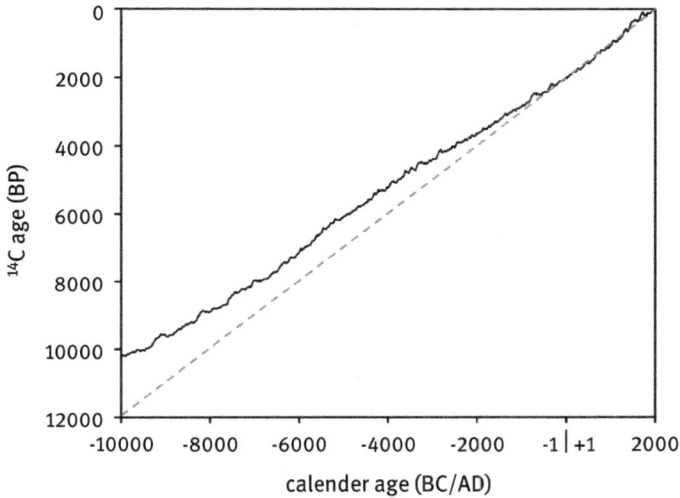

Fig. 5.7: The ^{14}C age in years before present (BP, i.e., before AD 1950) as determined from the radiocarbon measurement vs. the true calendar age.

Calibration of the ^{14}C age of a sample is achieved by convolution of the measurement result with the calibration curve. The calibration procedure increases the dating uncertainties substantially as a consequence of the measurement of the calibration curve itself and of ambiguities of the calibration curve from variable ^{14}C production. Therefore, probability distributions of the calibrated ^{14}C age are not Gaussian – as those of the ^{14}C age are. In some cases, it may be possible that two or even more probable periods for the calibrated date occur with unlikely periods between them. A famous example which yielded unexpected dating uncertainty is the ^{14}C age determination of Ötzi, the iceman found 1991 in the Italian Alps. The ^{14}C measurement resulted in ambiguous solutions for the calibrated date with three possible ages, covering roughly two centuries (see Fig. 5.8).

In many geochemical applications the absolute age of a sample is of minor importance, whereas the deviation of the ^{14}C level compared to the first-order model of constant ^{14}C production is very helpful, as this indicates processes such as the solar and geomagnetic impact on ^{14}C production or reservoir effects. An example is carbon storage in an environmental compartment. For such cases Δ^{14}C is defined as

$$\Delta^{14}C = \left(F^{14}C \exp\left[\frac{1950 - t_0}{8267}\right] - 1 \right) \cdot 1000‰, \qquad (5.6)$$

where t_0 is the year of origin of the sample and 8267 represents the mean lifetime in years based on the correct ^{14}C half-life of 5730 years. With the exponential term, a transformation correction is performed according to the age of the sample and based on the assumptions and conditions of eqs. (5.4) and (5.5), respectively. Therefore, all

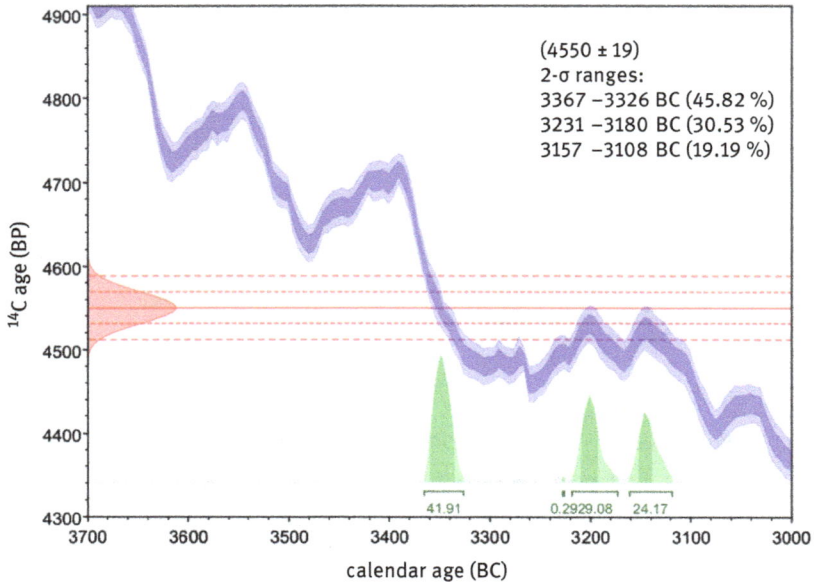

Fig. 5.8: ^{14}C age in years before present (BP) vs. the true calendar age for the Ötzi iceman. The measured value for the ^{14}C age was 4550 ± 19 BP (red distribution on y-axis). From the calibration curve (blue curve) it follows that within uncertainty several true ages are possible (green probability distributions on x-axis) between 3367 BC and 3108 BC; adapted from Bonani (1994).

samples meeting these requirements give $\Delta^{14}C = 0$ ‰ irrespective of their true age of origin. Positive or negative $\Delta^{14}C$ values indicate deviations of the reference conditions, cf. Fig. 5.9, which may be caused by (a) variability of ^{14}C production, (b) age offsets due to reservoir effects (these additional ages lead only to negative $\Delta^{14}C$ values), or (c) manmade emissions of ^{14}C from surface nuclear bomb explosions or of ^{14}C depleted material from combustion of fossil fuels (which are devoid of ^{14}C), respectively.

As already pointed out, ^{14}C can be determined by two fundamentally different techniques. Whereas radiometric techniques such as liquid scintillation or gas proportional counters measure radiocarbon via its radioactive transformation, i.e., $Á\,(^{14}C)$ by means of the emitting beta particle, accelerator mass spectrometry (AMS) detects the number $\hat{N}$ of ^{14}C atoms based on mass separation in magnetic and electric fields followed by an energy determination in gas ionization chambers (for details see Chapter 4). This fundamental difference has a substantial effect on the detection efficiencies. For a given number $\hat{N}\,(^{14}C)$, the efficiency of transformation counting is limited by the ratio of measurement time and half-life of the nuclide. Efficiencies in the order of 10^{-5} are reached for ^{14}C analysis if counting of the samples is restricted to not more than a few days because only very few atoms out of the entire ensemble transform during such a time period. Mass spectrometry of long-lived radionuclides does not depend on this restriction so that only its efficiencies of ionization,

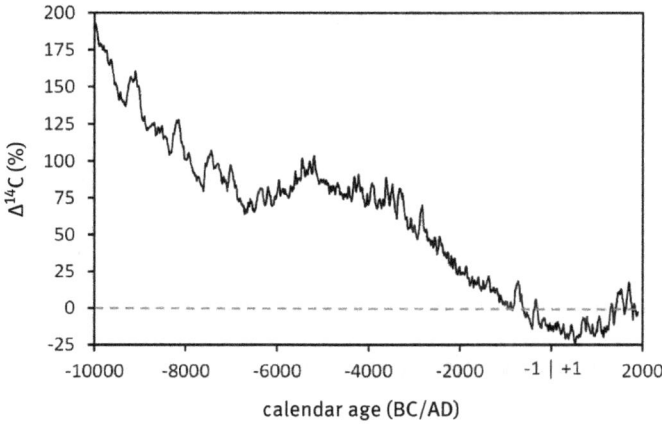

Fig. 5.9: $\Delta^{14}C$ during the last 10,000 years vs. the true calendar age; adapted from (Stuiver 1986).

transmission, and detection are relevant, which can be as high as 10% for ^{14}C-AMS. This results in a gain in detection efficiency between AMS and radiometric detection by a factor of about 10^4. Consequently, required sample quantities can be reduced from gram amounts for counting techniques to submilligram amounts for AMS, which entails a considerable benefit for dating of valuable samples.

5.3.2 ^{210}Pb ($t\frac{1}{2} = 22.2$ a)

^{210}Pb is a granddaughter-type member of the natural ^{238}U parent transformation series.[10] ^{210}Pb has a half-life of 22.2 years and transforms by β-emission to ^{210}Bi, which has a half-life of 5.01 days and again transforms via β-emission to ^{210}Po ($t\frac{1}{2} =$ 138 d), cf. Fig. 5.10. The β-transformation of ^{210}Pb populates an exited state of the β-emitter ^{210}Bi. This nuclear level de-excites via emission of a γ-ray of 46 keV energy to the ground state. However, this transition is highly converted which means that only in 4% of transformations is a photon with this energy emitted and in 96% of transformations an atomic electron. Decay continues then via α-emission with an energy of 5.3 MeV to the stable end product of the ^{238}U transformation series, ^{206}Pb.

The time interval accessible to ^{210}Pb dating is typically one to two centuries at most. It is therefore often applied in studies where environmental archives are investigated covering the industrial time period, often called the anthroposphere. Therefore, most applications of this radionuclide intend to study pollution records

---

10 Natural series of transformation are discussed in Chapter 3 of Vol. I.

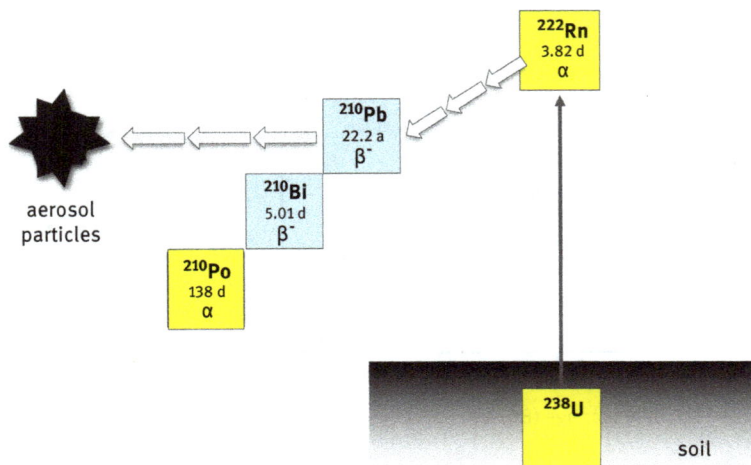

Fig. 5.10: Origin and fate of ^{210}Pb in the atmosphere. ^{222}Rn, formed in soil via transformation of ^{238}U, is emanating into the atmosphere. After further transformation, generated ^{210}Pb attaches to aerosol particles which are scavenged by precipitation and deposition onto to the Earth's surface. ^{210}Pb further transforms via short-lived ^{210}Bi to ^{210}Po and finally to stable ^{206}Pb.

caused by industrial emissions in appropriate archives (e.g., ice cores) or to determine accumulation rates during the last century in lake sediments.

Based on the rather short half-life of ^{210}Pb it is advisable to measure it via radiometric techniques and not via mass spectrometry. Three options exist to determine the ^{210}Pb activity concentration (i) β-counting of ^{210}Pb and/or ^{210}Bi, (ii) γ-counting of ^{210}Pb, and (iii) α-counting of the granddaughter ^{210}Po.

β-Counting has not found widespread application despite being very sensitive and of high precision. The reason is the high radiochemical purity of the sample needed prior to measurement. Since β-counting (e.g., using gas proportional or liquid scintillation counters) is not nuclide-specific, a sophisticated chemical purification of the sample is mandatory to suppress any contamination by the vast amount of environmental radionuclides that also transform via β-emission. Such a separation procedure is very time-consuming.

If the sample has a sufficiently high ^{210}Pb activity, γ-counting is the method of choice. This situation is often met when analyzing sediment samples. Unfortunately, for determination of ^{210}Pb in precipitation, the sensitivity achieved in γ-counting is not sufficient for reasonably sized volumes, say 1 liter. In this case, ^{210}Pb is usually determined via detection of ^{210}Po. This method, however, requires secular equilibrium[11] between the three radionuclides ^{210}Pb, ^{210}Bi, and ^{210}Po, respectively. This is

---

11 See Chapter 3 in Vol. I for mathematical aspects of radioactive equilibria.

reached after about one year. Advantage of this radiometric technique is the high sensitivity because α-counting yields energy-resolved data at a 100% counting efficiency of the used silicon detectors. The overall efficiency is then only determined by geometrical parameters, i.e., the size of the detector facing the sample.

^{210}Pb dating is frequently applied for lake sediments and glacier ice. Figure 5.11 depicts such a dating of a sediment core from Lake Zürich, Switzerland. Also inserted into the figure are measurements of ^{137}Cs, a radionuclide formed in nuclear fission. This radionuclide was globally distributed from nuclear weapons testing during Soviet Union and USA tests peaking in the years 1962–1963. The ^{210}Pb activity concentrations depicted in Fig. 5.11 are not the total but the so-called "unsupported" activities. Since sediment materials contain ^{238}U, ^{210}Pb is constantly formed via nuclear transformation. This "supported" ^{210}Pb has therefore to be subtracted from the measured total ^{210}Pb activity concentration. This is usually made by measuring the ^{210}Pb activity concentration in very deep samples, which are too old for ^{210}Pb dating. This value is then considered as the "supported" ^{210}Pb activity concentration and is subtracted from all measured values to yield the values for "unsupported" ^{210}Pb. From Fig. 5.11 it is evident that at the surface of the sediment the ^{210}Pb activity concentrations are nearly constant down to a depth of about 1.5 g/cm^2. From the ^{137}Cs measurement this depth can be dated to be roughly 1963. The explanation for this inconsistency between ^{137}Cs and ^{210}Pb data comes from a process often observed in lake sediments: biogenic perturbation (*bioturbation*). Microorganisms are actively mixing organic material until a certain depth, which influences the surface profile of ^{210}Pb. It is interesting to note that ^{137}Cs, which is predominantly attached to clay minerals, is not influenced by this process. The right axis of Fig. 5.11 depicts the combined age information from the ^{137}Cs and the ^{210}Pb measurements. From the slope of the age-depth relationship an average annual mass sedimentation rate of about 50 mg/cm^2 can be deduced.

It should be mentioned that analysis of the data depicted in Fig. 5.11 assume a constant initial concentration (CIC model) of unsupported ^{210}Pb. This means that the sediment core is assumed to have a constant sedimentation rate with always the same initial ^{210}Pb activity concentration. If, however, for the time period under study the sedimentation rate changed, then such an assumption is incorrect. It is then advisable to assume constant rate of supply (CRS model), meaning that the annual deposition rate of ^{210}Pb is constant even at variable sedimentation rate. As a consequence, for years with lower than average accumulation rates the activity concentrations are higher and vice versa. For a detailed description of ^{210}Pb in sediment dating using the two models and a combination of both, see Appleby (2008).

Figure 5.12 illustrates the ^{210}Pb activity concentration along an ice core drilled on Colle Gnifetti, Switzerland (saddle at 4450 m above sea level). Since ice contains negligible amounts of mineral dust as a source of "supported" ^{210}Pb, no corrections are needed. Hence, the data shown represent the measured values. From the radioactivity distribution along the ice core the age depicted on the lower horizontal axis may be deduced.

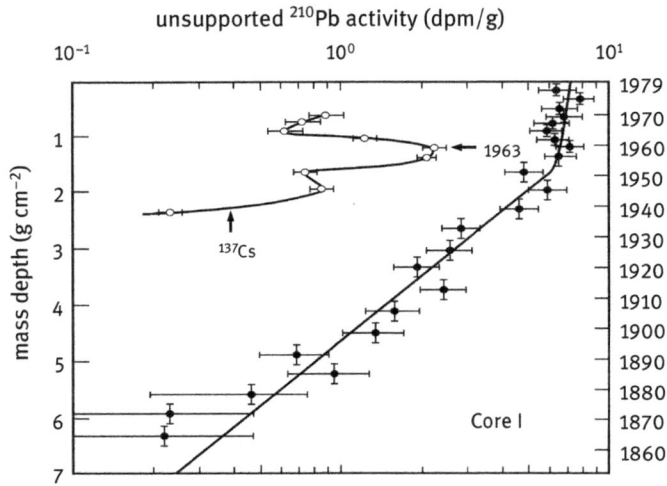

Fig. 5.11: ^{210}Pb activity concentrations as a function of depth for a sediment core from Lake Zürich. Also shown is the activity of ^{137}Cs from nuclear weapons testing (Erten 1985).

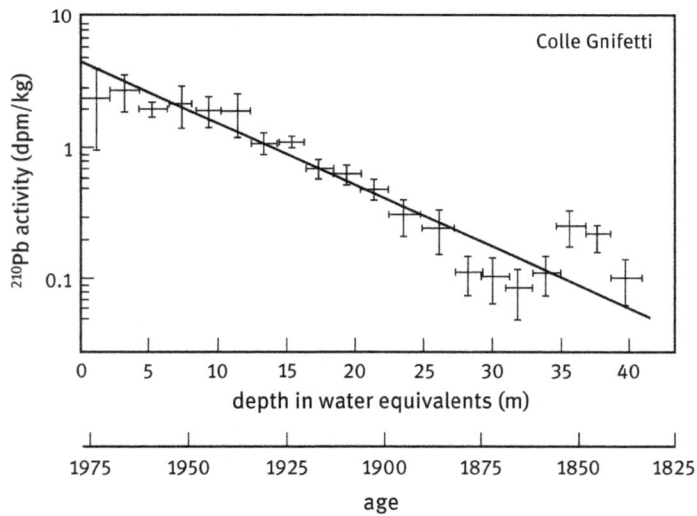

Fig. 5.12: ^{210}Pb activity concentrations along an ice core from the Colle Gnifetti at 4450 m above sea level, Swiss Alps (deduced from Gäggeler et al. 1983).

5.3.3 ^{32}Si ($t\frac{1}{2} \approx$ 153 a)

^{32}Si is produced in the upper atmosphere via interaction of cosmic rays with the argon isotope ^{40}Ar in the spallation reaction ^{40}Ar(p,2αp)^{32}Si. As a nonvolatile product it attaches to aerosol particles and is transported within about one week to the Earth's surface, where it is deposited mostly by wet precipitation.

^{32}Si has a half-life of 153 years, however with a large uncertainty.[12] ^{32}Si would be an ideal tool to date samples in the range between one century and one millennium, hence between the optimum dating ranges of ^{210}Pb and ^{14}C. Besides the poorly known half-life, there are several drawbacks that have prevented such applications so far. The activity concentration of ^{32}Si in air is very low because ^{40}Ar is a minor component of ambient air (0.9%). This means that a radiometric determination of ^{32}Si requires large sample volumes, e.g., about one ton of water. Typical values of ^{32}Si activity concentrations in one ton of precipitation (or snow) are a few mBq only. Therefore, in recent years attempts were made to measure ^{32}Si via accelerator mass spectrometry (AMS). But this approach is also limited due the fact that silicon is very abundant in nature. In environmental samples the ratio between the radioactive isotope ^{32}Si and the stable isotopes of silicon, mainly ^{28}Si with 92.2% abundance, is extremely low. Typical values are between 10^{-15} and 10^{-17}.[13] One of the very few examples where ^{32}Si dating was successfully applied is depicted in Fig. 5.13. It shows the ^{32}Si (and ^{210}Pb) depth profiles of a marine sediment core from Bangladesh. The measurements were performed using radiometric technique by measuring the nuclide ^{32}P with a half-life of 14.3 days in secular equilibrium, a high-energy β-emitter formed in the β-transformation of ^{32}Si. The interpretation of this study yielded a sedimentation rate for the uppermost part of the core (dated with ^{210}Pb) of 3.1 cm per year. Below this surface layer it was much lower, 0.7 cm per year. The proposed interpretation of this significant change in sedimentation rate is that it reflects increased sedimentation loads over the first 50 years due to human activities in this area.

Presently a great effort is being made to determine the half-life of ^{32}Si with high precision. In an international collaboration under the leadership of the Paul Scherrer Institute (D. Schumann) first a high activity of ^{32}Si was produced by bombarding a vanadium target with 590 MeV protons followed by a sophisticated radiochemical separation of silicon. A variety of techniques was then applied to determine the number of ^{32}Si atoms (including, e.g., AMS) and the activity (e.g., liquid scintillation counting (LSC)) of aliquots to accurately determine the transformation constant of ^{32}Si according to $\mathbf{A} = \lambda \cdot \mathbf{N}$.

---

12 The accuracy of current literature data is still surprisingly poor. The published values range from about 100 to 170 years.
13 Since AMS measures the ratio between the radioactive isotope and usually a stable isotope of the same element, a determination of ^{32}Si by this technique is at the cutting edge of AMS technology. Only in very pure samples, e.g., in ice samples from New Zealand glaciers or from Antarctica, has it so far been possible to measure ^{32}Si by AMS.

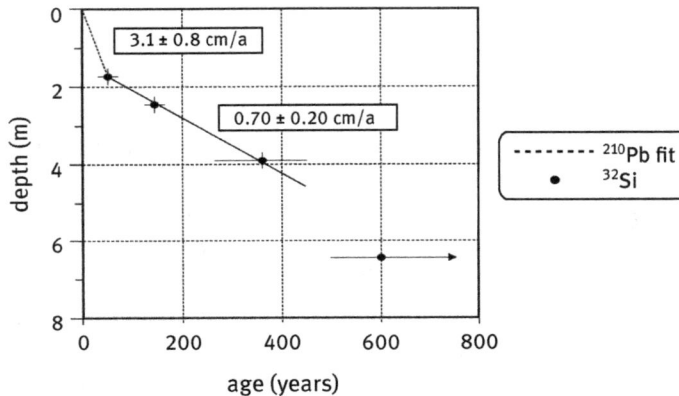

Fig. 5.13: Depth age relationship of a sediment core from Bangladesh deduced from measured ^{210}Pb and ^{32}Si activity concentration; adapted from Morgenstern (2001).

5.3.4 ^{129}I ($t_{1/2} = 1.57 \cdot 10^7$ a)

In nature two isotopes of iodine exist, ^{127}I which is stable, and the long-lived radioactive ^{129}I. ^{129}I has three different sources, (a) cosmogenic, (b) fissiogenic, and (c) anthropogenic. Cosmogenic ^{129}I is formed in the atmosphere via interaction of cosmic rays with xenon. Fissiogenic ^{129}I is produced via spontaneous fission of ^{238}U contained in the Earth's crust. Finally, nuclear weapons testing emitted ^{129}I in considerable amounts into the atmosphere. More recently, also reprocessing plants emitted ^{129}I into the atmosphere but also into the marine environment. These anthropogenic sources surpass the cosmogenic and fissiogenic sources by far. Usually ^{129}I is determined in environmental samples relative to the stable ^{127}I. Ratios are $1.5 \cdot 10^{-12}$ for pre-anthropogenic iodine. This value can be found in oceanic water that has not yet been influenced by anthropogenic input. Rain water in central Europe, however, is much influenced by emissions from the reprocessing plants in Sellafield (Great Britain) and La Hague (France). Typical isotopic ratios may be as high as 10^{-6}.

Application of ^{129}I for dating purpose is mostly in hydrology. Due to the long half-life of ^{129}I reliable dating starts at ages of several million years. This includes age determination of very old aquifers assuming the influence of fissiogenic and anthropogenic ^{129}I to be negligible. On the other hand, dating based on fissiogenic ^{129}I may be applied for geothermal fluids. In this case, the concentration of ^{238}U must be known as well as the transfer yield of formed ^{129}I from the ^{238}U source to the fluid. Finally, some applications focus on the detection of anthropogenic ^{129}I. In these cases the samples are very young.

In all these applications, ^{129}I is measured with highest sensitivity via AMS, relative to the stable isotope ^{127}I. Some applications applied neutron activation analysis as an analytic tool. Here, ^{129}I is determined via formation of ^{130}I ($t^{1/2} = 12.4$ h) after neutron capture at a research reactor.

5.3.5 ^{10}Be ($t^{1/2} = 1.39 \cdot 10^6$ a)

^{10}Be is a cosmogenic radionuclide formed in the upper atmosphere predominantly in the spallation reactions between neutrons (secondary cosmic rays) with nitrogen (^{14}N(n,3p2n)^{10}Be). This radionuclide has been used to trace variability of solar activity back in time because solar activity is anti-correlated with galactic cosmic ray intensity (see the section on radiocarbon dating). Ice cores from Antarctica have mostly been used as archives of this solar history. ^{10}Be is incorporated into such ice cores via precipitation. These investigations enabled dating of periods with low solar activity such as the MAUNDER minimum from 1645 to 1715, which is assumed to be responsible for a climatic depression ("Little Ice Age"). Hence, this dating is based on the "fingerprint" idea: fluctuations of ^{10}Be activity concentrations in the sample over time are compared to known fluctuations of solar variability in the past.

Another fast developing application of ^{10}Be is exposure age determination. This application is based on the fact that the Earth's surface is exposed to cosmic ray bombardment. Though the intensity of this bombardment is very small – due to a high absorption of cosmic rays in the atmosphere – it is still sufficiently intense to form some radionuclides via nuclear reactions, such as ^{10}Be in the reactions ^{16}O(α,2pn)^{10}Be or ^{16}O(μ^-,αpn)^{10}Be, respectively (see Fig. 5.14).

Therefore, based on known production rates of ^{10}Be in a given geological surface, the exposure age may be determined. However, one drawback under real conditions is the erosion rate. If the surface to be dated is significantly influenced by erosion, then the concentration of ^{10}Be levels off after a certain time (see Fig. 5.15). At saturation, production and erosion are in equilibrium and dating is no longer possible.

Measured production rates in surface rock samples from North America yielded about five ^{10}Be atoms produced per gram and year (Balco et al. 2008). To determine the erosion rate, a second radionuclide is required because the influence of erosion on the build-up of radioactive atoms depends on the half-life of the radionuclide. ^{26}Al is mostly used for this purpose (see below).

5.3.6 ^{26}Al ($t^{1/2} = 7.17 \cdot 10^5$ a)

^{26}Al is also a radionuclide formed on the Earth's surface via interaction with cosmic rays in different nuclear reactions, depending on the geological composition, e.g.,

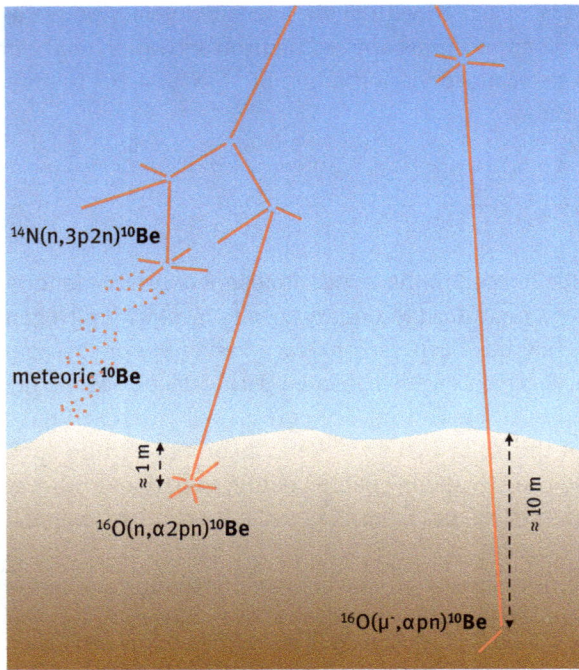

Fig. 5.14: ^{10}Be produced through interaction of secondary cosmic rays (neutrons or myons) with the Earth's surface.

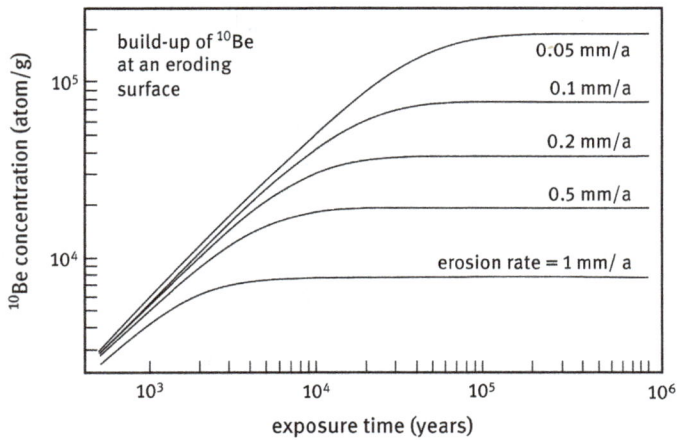

Fig. 5.15: Number of ^{10}Be atoms formed in an exposed surface as a function of time for different values of known surface erosion rates calculated via eq. (5.7).

in ^{28}Si(p,p2n)^{26}Al, or ^{28}Si(μ^-,3 n)^{26}Al, respectively. Measured production rates in surface rock samples from North America yielded about 30 ^{26}Al atoms produced per gram and year (Balco et al. 2008).

^{26}Al is often measured together with ^{10}Be using AMS in cases where erosion is assumed to be important. The reason is simply that the determination of two variables (exposure time and erosion rate) requires the measurement of two "clocks", e.g., both ^{10}Be and ^{26}Al. Measured exposure ages reach values of up to one million years (Balco et al. 2008). The exposure age can then be determined from the two radionuclides ^{10}Be and ^{26}Al via:

$$\hat{N}_x = P_{0,x}/\{\lambda_x + \varepsilon_{ero}/a_x\} \tag{5.7}$$

with

$\hat{N}_x$ number of measured atoms of x (^{10}Be or ^{26}Al) in the sample (atoms g^{-1})

$P_{0,x}$ production rate of species χ at the surface (atoms g^{-1}a^{-1}); e.g., about 5 for ^{10}Be (see Fig. 5.15) and 30 for ^{26}Al

λ_x transformation constant of species x (a^{-1})

ε_{ero} erosion rate (g cm^{-2}a^{-1})

a_x effective attenuation depth of spallation product formation for ^{10}Be or ^{26}Al, respectively, in the sample (assumed to be 160 g cm^{-2}).

Hence, eq. (5.7) has two variables, $\hat{N}_x/P_{0,x}$ (which is the exposure age in years) and ε_{ero}, the erosion rate.[14]

5.3.7 ^{36}Cl ($t_{1/2}$ = 3.01 · 10^5 a)

^{36}Cl is formed by cosmic rays, mostly in the atmosphere via interaction with argon (^{40}Ar(p,αn)^{36}Cl) but also in nuclear weapons testing. Cosmic ray produced ^{36}Cl has found application as a dating tool for archives up to a few million years. Another application is to use anthropogenic ^{36}Cl production via activation of sea-salt particles from the nuclear weapons testing in the Bikini Atoll in the late 1950s (see Fig. 5.21 from Section 5.6) as a dating horizon.

^{36}Cl is measured with highest sensitivity using AMS. Due to strong interference of ^{36}Cl with the isobaric stable isotope of sulfur (^{36}S), ^{36}Cl can only be measured with big AMS accelerators that have a high accelerator voltage of several MV. Under these conditions accelerated ions have high velocities (i.e., high kinetic energies), which enables efficient separation of the isobaric nuclides ^{36}Cl and ^{36}S by a gas-filled magnetic separator coupled to the accelerator. Recently ^{36}Cl measurements have also

---

14 The method for dating is available under http://hess.ess.washington.edu.

been successfully performed with smaller accelerators coupled to a gas cell to stop the ions and followed by a laser interaction mass spectrometer (Lachner et al. 2019).

5.4 Dating with a radioactive parent and an accumulating stable daughter

Most applications of this dating principle are in geochronology and cosmochronology because the half-lives of the mostly used systems are very long, reaching the age of the solar system. Knowing the actual activity and half-life of the transforming parent nuclide together with the concentration of the daughter nuclide (radioactive or stable) at time zero, the age can be inferred. Table 5.5 summarizes the examples discussed in this chapter.

Tab. 5.5: Parent-daughter dating systems for application in geochronology and cosmochronology as well as in hydrology (^{222}Rn | ^{226}Ra).

Mother	Half-life (a)	Daughter	Half-life (a)
^{238}U	$4.47 \cdot 10^9$	^{206}Pb	Stable
^{235}U	$7.04 \cdot 10^8$	^{207}Pb	Stable
^{232}Th	$1.40 \cdot 10^{10}$	^{208}Pb	Stable
^{87}Rb	$4.97 \cdot 10^{10}$	^{87}Sr	Stable
^{40}K	$1.25 \cdot 10^9$	^{40}Ar	Stable
^{234}U	$2.46 \cdot 10^5$	^{230}Th	$7.54 \cdot 10^4$
^{235}U	$7.04 \cdot 10^8$	^{231}Pa	$3.28 \cdot 10^4$
^{226}Ra	$1.60 \cdot 10^3$	^{222}Rn	3.82 days

The most important examples are the natural transformation series starting with ^{235,238}U and ^{232}Th and ending in the stable isotopes ^{206,207,208}Pb. Moreover, the ^{40}K | ^{40}Ar and ^{87}Rb | ^{87}Sr systems, where radioactive ^{40}K ($t\frac{1}{2} = 1.25 \cdot 10^9$ a) or ^{87}Rb ($t\frac{1}{2} = 4.97 \cdot 10^{10}$ a) transform into stable ^{40}Ar or ^{87}Sr, respectively, are also widely used. Some other systems of minor importance are ^{187}Re | ^{187}Os (^{187}Re: $t\frac{1}{2} = 4.33 \cdot 10^{10}$ a), ^{176}Lu | ^{176}Hf (^{176}Lu: $t\frac{1}{2} = 3.8 \cdot 10^{10}$ a), or ^{40}K | ^{40}Ca, which takes into account that 11% of ^{40}K transforms into ^{40}Ar (see above) while 89% transforms into ^{40}Ca.

Let us denote the number of radioactive parent nuclides with $\hat{N}^P$ and the number of stable daughter nuclei with $\hat{N}^P$. Then the time dependence of the number of parent nuclei at time t is:

$$\hat{N}^P(t) = \hat{N}^P(t_0)e^{-\lambda_P t}, \tag{5.8}$$

with t_0 referring to time zero. The number of daughter nuclei at time t is:

$$\hat{N}^{\mathrm{d}}(t) = \hat{N}^{\mathrm{d}}(t_0) - \left[\hat{N}^{\mathrm{P}}(t_0) - \hat{N}^{\mathrm{P}}(t)\right]. \tag{5.9}$$

Combining eqs. (5.8) and (5.9) leads to eq. (5.10). Finally, eq. (5.11) yields the required parameter: time.

$$\hat{N}^{\mathrm{d}}(t) = \hat{N}^{\mathrm{d}}(t_0) + \hat{N}^{\mathrm{d}}(t)\left(e^{\lambda p t} - 1\right) \tag{5.10}$$

$$t = \frac{1}{\lambda p}\ln\left\{1 + \left[\left(\hat{N}^{\mathrm{d}}(t) - \hat{N}^{\mathrm{d}}(t_0)\right)/\hat{N}^{\mathrm{P}}(t)\right]\right\}. \tag{5.11}$$

It requires the measurement of the concentrations of parent and daughter at time t, usually at present. Moreover, some assumptions are needed for the concentration of the daughter at time zero. It should be added that dating requires that the system remained "closed" during the time window to be dated, hence without any geochemical processes that might influence the parent and/or daughter nuclide.

5.4.1 Dating with the natural transformation series

The parent|daughter systems considered are ^{238}U | ^{206}Pb ($t\frac{1}{2}^{238}$U = 4.47 · 10^9 a), ^{235}U | ^{207}Pb ($t\frac{1}{2}^{235}$U = 7.04 · 10^8 a), and ^{232}Th | ^{208}Pb ($t\frac{1}{2}^{232}$Th = 1.40 · 10^{10} a). It is assumed that the intermediate transformation products between radioactive parent and stable daughter are too short-lived to be of importance for the dating application.[15] Dating with the natural transformation series requires an assumption about the concentration of the terminal daughters (i.e., ^{206,207,208}Pb) at time zero.[16] For determination of the concentration of lead isotopes mostly thermal ionization mass spectrometry (TIMS) is used. When needed, uranium and thorium are measured using high-resolution inductively coupled plasma mass spectrometry (HR-ICP-MS).

---

15 It is interesting to note that the very first application of dating with the natural transformation series was not based on the systems listed above but on the fact that in the transformation series many α-particles are emitted, namely 8 α-particles in the transformation chains of ^{238}U, 7 α-particles of ^{235}U, and 6 α-particles of ^{232}Th, respectively. After having captured two electrons from the surroundings these α-particles become stable helium atoms (^{4}He). This means that the number of helium atoms in a sample is correlated to the age. Based on this principle, RUTHERFORD estimated the age of our Earth from the concentration of helium in uranium ores to be about 0.5 · 10^9 years. This value turned out to be roughly one order of magnitude too short which was later explained as being caused by partial loss of helium during metamorphosis processes.

16 There is one stable isotope of lead, ^{204}Pb, which is not formed as end product of any natural transformation series. The natural isotopic composition of lead, not influenced by transformation from uranium or thorium (named nonradiogenic composition) is ^{204}Pb: 1.4%, ^{206}Pb: 24.1%, ^{207}Pb: 22.1%, and ^{208}Pb: 52.4%, respectively. The radiogenic contribution of ^{206}Pb, ^{207}Pb, or ^{208}Pb in an ore can then be determined by measuring the ^{204}Pb concentration and subtracting the nonradiogenic contribution.

In practical applications, usually two of the three transformation chains are combined. This enables to check whether the system under study was indeed closed, hence, has not been subjected to any geological processes that might have led to a loss of lead, which is a volatile element. Such curves are called *concordia*.

Let us consider the two uranium transformation series and define R_6 and R_7 as follows:

$$R_6 = \frac{\hat{N}\left(^{206}Pb/^{204}Pb\right)(t) - \hat{N}\left(^{206}Pb/^{204}Pb\right)(t_0)}{\hat{N}\left(^{238}U/^{204}Pb\right)(t)} \tag{5.12}$$

and

$$R_7 = \frac{\hat{N}\left(^{207}Pb/^{204}Pb\right)(t) - \hat{N}\left(^{207}Pb/^{204}Pb\right)(t_0)}{\hat{N}\left(^{235}U/^{204}Pb\right)(t)}. \tag{5.13}$$

Figure 5.16 shows the concordia as a plot of R_6 against R_7. If experimental results lie on the line, then the system was closed. If the values are below the line then this indicates loss of lead or a gain of uranium, and vice versa for data above the line. The numbers listed along the lines represent the age in 10^9 years.

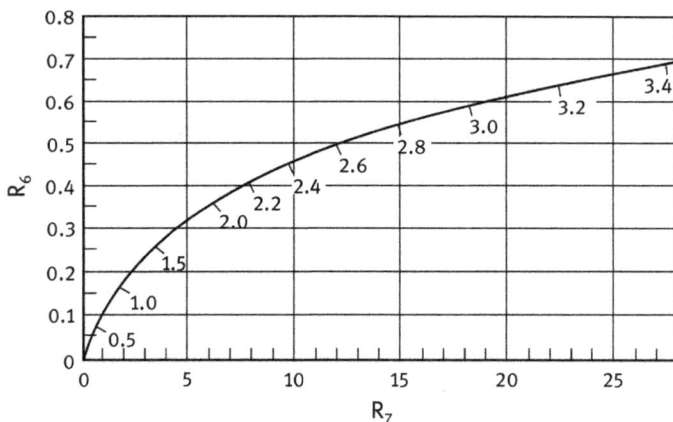

Fig. 5.16: Concordia for R_6 (eq. (5.12)) against R_7 (eq. (5.13)) (Friedländer et al. 1981). The numbers listed along the lines represent the age in 10^9 years.

It is also possible to date solely via the concentration ratio between ^{206}Pb and ^{207}Pb, if the sample contains no natural (nonradiogenic) lead, meaning that the concentration of ^{204}Pb is zero. Taking eq. (5.10) and setting $\hat{N}^d(t_0) = 0$ leads at time t to

$$\hat{N}\left(^{206}\text{Pb}/^{207}\text{Pb}\right) = \frac{\hat{N}\left(^{238}\text{U}\right)\left(e^{\lambda(^{238}\text{U})t} - 1\right)}{\hat{N}\left(^{235}\text{U}\right)\left(e^{\lambda(^{235}\text{U})t} - 1\right)}. \tag{5.14}$$

With the natural isotopic composition of uranium ($^{235}\text{U} = 0.72\%$ and $^{238}\text{U} = 99.2745\%$, i.e., $\hat{N}\left(^{238}\text{U}\right)/\hat{N}\left(^{235}\text{U}\right) = 137.8$), eq. (5.14) results in:

$$\hat{N}\left(^{206}\text{Pb}/^{207}\text{Pb}\right) = 137.8\,\frac{e^{\lambda(^{238}\text{U})t} - 1}{e^{\lambda(^{235}\text{U})t} - 1}. \tag{5.15}$$

This correlation is depicted in Fig. 5.17.

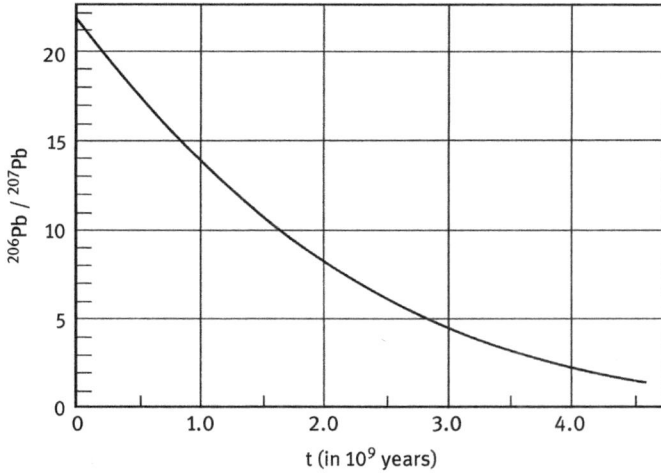

Fig. 5.17: Dating based solely on $^{206}\text{Pb}/^{207}\text{Pb}$ for cases where the sample does not contain nonradiogenic lead (Friedländer et al. 1981).

If in a class of samples of identical age, e.g., meteorites of a given type, the concentration of lead isotopes ^{206}Pb and ^{207}Pb are measured relative to the nonradiogenic isotope ^{204}Pb for different subsamples, then a correlation is obtained named an *isochron*.

$$\frac{\hat{N}\left(^{207}\text{Pb}/^{204}\text{Pb}\right)(t) - \hat{N}\left(^{207}\text{Pb}/^{204}\text{Pb}\right)(t_0)}{\hat{N}\left(^{206}\text{Pb}/^{204}\text{Pb}\right)(t) - \hat{N}\left(^{206}\text{Pb}/^{204}\text{Pb}\right)(t_0)} = \frac{\hat{N}\left(^{235}\text{U}\right)\left(e^{\lambda(^{235}\text{U})t} - 1\right)}{\hat{N}\left(^{238}\text{U}\right)\left(e^{\lambda(^{238}\text{U})t} - 1\right)}$$

$$= \frac{1}{137.8}\frac{e^{\lambda(^{235}\text{U})t} - 1}{e^{\lambda(^{238}\text{U})t} - 1}. \tag{5.16}$$

An example of an isochronous dating for L-chondrites, a class of meteorites, is depicted in Fig. 5.18. All data on the isochronous correlation yield an age of $4.55 \cdot 10^9$ years, which is the age of our solar system.

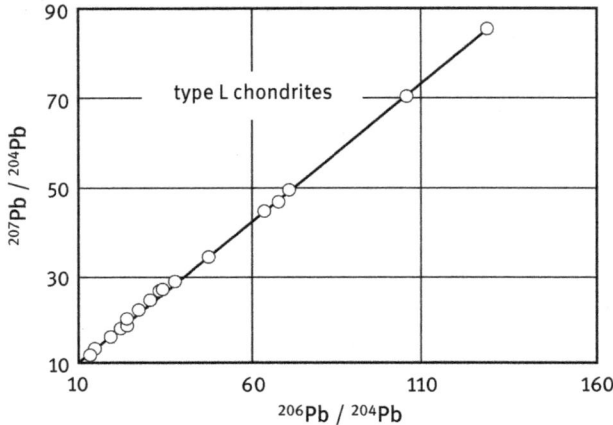

Fig. 5.18: Isochronous dating of a meteorite. Applying eq. (5.16), the slope defines an age of $4.55 \cdot 10^9$ years (Manhes 1982).

5.4.2 ^{87}Rb | ^{87}Sr system

^{87}Rb transforms with a half-life of $4.97 \cdot 10^{10}$ years into stable ^{87}Sr. The measurement of the two nuclides is usually made via mass spectrometry of ^{87}Rb and of ^{87}Sr/^{86}Sr, respectively. This requires a quantitative chemical separation of both elements from aliquots of the same sample. The age then follows from:

$$\left(^{87}\mathrm{Sr}/^{86}\mathrm{Sr}\right)_t = \left(^{87}\mathrm{Sr}/^{86}\mathrm{Sr}\right)_0 + \left(^{87}\mathrm{Rb}/^{86}\mathrm{Sr}\right)_t \left(e^{\lambda\left(^{87}\mathrm{Rb}\right)t} - 1\right), \qquad (5.17)$$

While the concentration of stable ^{86}Sr is of course constant at all times, the concentration of stable ^{87}Sr at time zero is not known. It may be deduced from an isochronous plot, which includes a set of data from subsamples of the same object.

Figure 5.19 depicts the experimental data of ^{87}Sr/^{86}Sr plotted against ^{87}Rb/^{85}Sr for a gneiss sample from Greenland. The value for $\left(^{87}\mathrm{Sr}/^{86}\mathrm{Sr}\right)_0$ results from the intercept of the isochronous line with the ordinate, yielding for the example depicted in Fig. 5.19 the value 0.702. Applying eq. (5.17), the slope of the isochronous line represents one age, in this case $3.66 \cdot 10^9$ years.

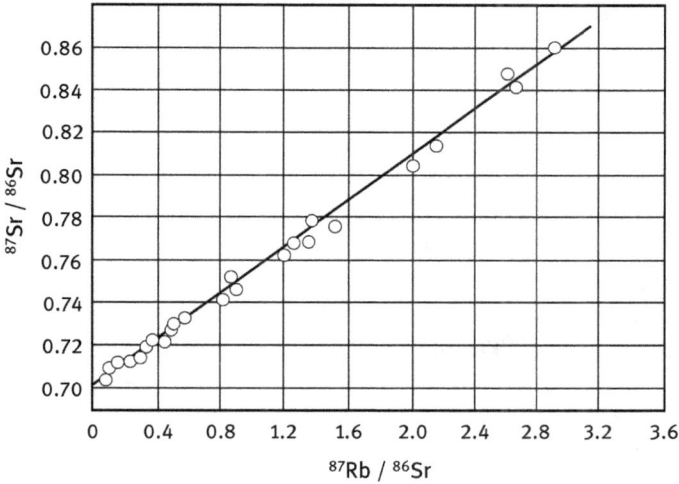

Fig. 5.19: Example of isochronous dating of several subsamples from a piece of gneiss from Greenland with the ^{87}Rb|^{87}Sr system. The slope of the line represents an age of $(3.66 \pm 0.09) \cdot 10^9$ a (Moorbath 1972).

5.4.3 ^{40}K | ^{40}Ar system

The dating is based on the 10.7% transformation branch of ^{40}K into stable ^{40}Ar.[17] The long half-life of ^{40}K of $1.25 \cdot 10^9$ a makes this system ideal for dating igneous rocks, lunar samples, or meteorites, hence, very old objects of our solar system. Argon is analyzed using a gas mass spectrometer after heating the sample to high temperatures close to the melting point. ^{40}K is usually analyzed from an aliquot of the sample using atomic mass spectrometry (AAS) or ICP-MS. As for the ^{87}Sr | ^{87}Rb system, it is very important to know the yield of the chemical separation of both elements with high accuracy. Analogous to eq. (5.17), the equation is:

$$\left(^{40}\text{Ar}/^{36}\text{Ar}\right)_t = \left(^{40}\text{Ar}/^{36}\text{Ar}\right)_0 + f\left(^{40}\text{K}/^{36}\text{Ar}\right)_t \left(e^{\lambda\left(^{40}\text{K}\right)t} - 1\right), \tag{5.18}$$

where f represents the branching ratio of 10.7% in the transformation of ^{40}K. The ^{36}Ar concentration is constant as a function of time and therefore used as reference. The value for $(^{40}\text{Ar}/^{36}\text{Ar})_0$ can be determined from an isochronous plot at intercept with the ordinate at ^{40}K/^{36}Ar value at zero.

---

17 The more abundant branch, namely the 89.3% of ^{40}K transforming to ^{40}Ca is of little practical use, since calcium is very abundant and the isotope ^{40}Ca has a natural abundance of 96.4%. Hence, the very little additional ^{40}Ca formed via transformation of ^{40}K can hardly be distinguished from the vast amount of natural ^{40}Ca.

5.4.4 ^{39}Ar | ^{40}Ar system

As for the ^{87}Sr | ^{87}Rb method, the drawback of the system ^{40}K | ^{40}Ar lies in the fact that measurements have to be made on different aliquots of the sample since two different elements have to be analyzed. This can result in erroneous ages, e.g., due to heterogeneity of the sample. In recent years, a modification of the ^{40}K | ^{40}Ar method has found widespread application: the ^{39}Ar | ^{40}Ar method. The dating range is identical to the ^{40}K | ^{40}Ar method described above.

Radiogenic ^{39}Ar is a short-lived nuclide and has a half-life of 269 years. In this approach stable ^{39}K is measured indirectly via ^{39}Ar formed after neutron irradiation in a research reactor in the reaction ^{39}K(n,p)^{39}Ar. The rock samples are then heated to temperatures close to the melting point and released ^{39,40}Ar is collected and is measured via gas mass spectrometry. Typical applications include, besides Earth sciences, also age determinations of hominid fossils in Africa. In Ethiopia *Herto hominids* were dated with this method at between 160,000 and 154,000 years, making them the oldest record of modern *Homo sapiens*.

5.5 Dating with a radioactive parent and an accumulating radioactive daughter

This situation refers to Fig. 5.2 (right part) and is met in natural transformation series. Parent and daughter nuclides are from the same transformation series, but are not the stable end products such as ^{206,207,208}Pb. Two situations exist in reality. First, at time zero the activity concentration of the daughter is zero – or negligible – (called *daughter deficient situation*, DD), but increases with time. Second, the activity concentration of the parent nuclide is negligibly small at all times, but the activity concentration of the daughter increases with time (*daughter excess situation*, DE).

5.5.1 ^{234}U | ^{230}Th system

These two nuclides from the ^{238}U transformation series resemble the most important example for DD-dating. The half-lives of ^{234}U and ^{230}Th are $2.46 \cdot 10^5$ years and $7.54 \cdot 10^4$ years, respectively. This limits the dating horizon to about $0.4 \cdot 10^6$ years.

The applications include solid objects formed in marine or terrestrial fluorapatites, carbonates such as corals or mollusks, etc. Uranium is very soluble in water.[18]

---

18 Uranium exists in marine environment in form of $[UO_2^{2+}(CO_3)_3]^{4-}$ ionic complexes which are very soluble.

Thorium on the other hand is fully hydrolyzed at typical marine pH conditions, hence very insoluble and mostly adsorbed to solid particles contained in the water. These particles sediment to the ocean floor. Therefore, the concentration of dissolved thorium in sea water is negligibly small. If objects such as stalactites in speleothems or organisms that build a shell via carbonate formation (e.g., corals) in a marine environment are formed, they will have a high concentration of uranium but are depleted in thorium. Only via radioactive transformation of uranium is thorium then formed in the solid objects. From the measurement of the activity concentration of ^{230}Th the time elapsed since formation may be deduced.

The situation refers to Fig. 5.4, being described specifically by

$$\hat{N}_{^{230}\text{Th}} = \hat{N}_{^{234}\text{U}}\left(1 - e^{\lambda\left(^{230}\text{Th}\right)t}\right). \tag{5.19}$$

$\hat{N}_{^{230}\text{Th}}$ is the concentration at time t and $\hat{N}_{^{234}\text{U}}$ is assumed to be independent of time for a given sample.[19] The reason is that ^{234}U is permanently formed via its parent ^{238}U.

5.5.2 ^{235}U | ^{231}Pa system

This couple of nuclides acts very similar to the system ^{234}U | ^{230}Th. As mentioned above, uranium is very water soluble. Protactinium, however, has a very low solubility. In its predominantly pentavalent state it is similar to the tetravalent state of thorium and very hygroscopic. The half-lives of the nuclides are for ^{231}Pa $3.28 \cdot 10^4$ years and for ^{235}U $7.04 \cdot 10^8$ years. This limits the accessible dating interval to less than about 150,000 years.

5.5.3 ^{222}Rn | ^{226}Ra system

This application represents the DD situation. In glaciofluvial[20] deposits surface water from rivers infiltrate into nearby groundwater. The infiltration velocity of such

---

19 For objects to be dated the condition of a closed system is not always met. Many samples such as bones or mollusks may experience some uranium uptake after formation. This influences the age deduced from eq. (5.12). Another possible source of error may come from the fact that activity concentration of the daughter ^{230}Th at time zero is not negligible. If, e.g., the sample has some intrusions from geological materials, then this contamination contains the natural transformation series, especially ^{234,238}U. The measured ^{230}Th activity concentration then needs to be corrected for this contribution.
20 Retreating glaciers from the last ice age left many deposits in the Swiss midlands of several hundred meter thickness.

surface water into the groundwater is an important parameter for hydrological modeling. This process may be traced using the noble gas isotope ^{222}Rn ($t\frac{1}{2} = 3.82$ d) being generated from ^{226}Ra ($t\frac{1}{2} = 1600$ a), a member of the ^{238}U transformation series. The activity concentration of ^{222}Rn is very low in surface water due to outgassing into the atmosphere, hence there is no equilibrium between its precursor ^{226}Ra and ^{222}Rn. This reflects the daughter-deficient situation at time zero. This is not the case in the adjacent soil acting as bed of the groundwater. Here, secular equilibrium exists between the members of the natural transformation series. By measuring the increase of the ^{222}Rn activity concentration as a function of the distance from the river into the groundwater body, infiltration velocities may be deduced. Such an example is shown in Fig. 5.20, which yielded a mean groundwater flow velocity of 5 m per day.

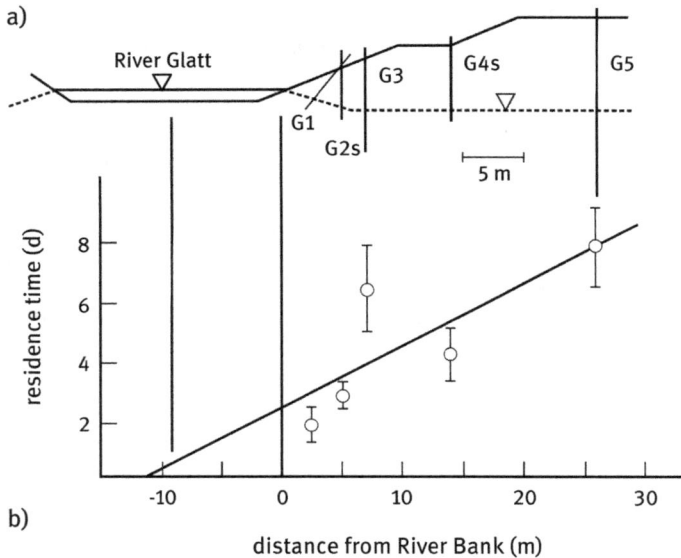

Fig. 5.20: Example of ^{222}Rn dating in hydrology. Upper part: cross section of the river (Glatt, Switzerland) with the locations of the boreholes (G1 to G5) to sample groundwater. Lower part: ^{222}Rn activity measured in water samples extracted from the boreholes (von Gunten 1995).

5.6 Dating with nuclides emitted during nuclear weapons testing and nuclear accidents

On several occasions, manmade radionuclides have been emitted into the environment. These events may be used as dating horizons if stored in environmental archives. One of the most important emission sources were the aboveground[21] nuclear weapons testing series conducted by USA and Soviet Union. Many radionuclides such as fission products, bomb material, or secondary products from interaction with neutrons were emitted into the atmosphere. In the late 1950s such tests were performed by USA in the Bikini Atoll that led to activation of sea-salt particles contained in the plume of the explosion. Figure 5.21 depicts some typical debris isolated from the ice of a glacier in Switzerland. It includes material from the bomb itself such as ^{239}Pu ($t^{1/2} = 2.41 \cdot 10^5$ a), products formed in nuclear fission such as ^{137}Cs ($t^{1/2} = 30.1$ a), activated (by the enormous intensity of neutrons) sea-salt aerosol particles

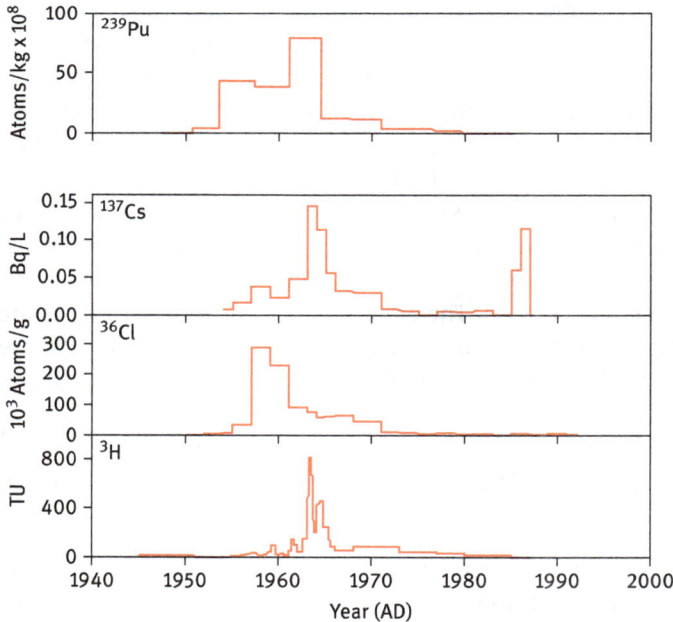

Fig. 5.21: Radionuclides detected in ice cores from glaciers in Siberia (Belukha, Russia, ^{239}Pu) and from the Alps (Grenzgletscher, Switzerland, ^{137}Cs, ^{36}Cl, ^{3}H) that can be assigned to nuclear weapons testing in the late 1950s and early 1960s and to the Chernobyl accident in 1986 (activities in units of Bq/l or TU (1 TU = 0.18 mBq/l)) (Olivier 2004 (upper part); Eichler 2000 (lower part)).

---

21 "Aboveground" nuclear tests in contrast to "underground" tests.

such as ^{36}Cl ($t\frac{1}{2} = 3.01 \cdot 10^5$ a) from the early tests over the Bikini Atoll, and tritium ($t\frac{1}{2} = 12.3$ a), which was formed in the hydrogen bombs also via interaction with neutrons. These nuclides were largely injected into the stratosphere from where they returned into the troposphere within several years. For the fission product ^{137}Cs, the emission into the European atmosphere caused by the Chernobyl accident in April 1986 is also observed. In this case, however, the transport occurred within the troposphere. These stratigraphic radioactive horizons are also often found in other archives such as lake sediments (see ^{137}Cs in Fig. 5.11 in Section 5.3.2).

5.7 Fission tracks

^{238}U is ubiquitous in nature. The average concentration in the Earth's crust is 5 ppm. ^{238}U mostly transforms via α-emission and to a much lesser extent by spontaneous fission (sf).[22] Primary fission fragments have very high kinetic energies and therefore damage the surrounding matrix leading to tracks. These so-called latent fission tracks have a length of typically 10 to 20 µm. These tracks become visible by chemical etching, cf. Chapter 2, and can be counted using a microscope. From the number of observed fission tracks and knowing the concentration of ^{238}U, the age may be inferred based on the known half-life of ^{238}U. Best suited for determination of the ^{238}U concentration is to measure ^{235}U via thermal neutron induced fission. ^{235}U has an abundance in natural uranium of 0.72%. The procedure then goes as follows: first, the sample is etched and the number of fission tracks from ^{238}U is determined. Then the sample is irradiated at a research reactor with thermal neutrons that initiate induced fission (if) of ^{235}U. The additionally formed fission tracks are again etched and counted. From the added number of new fission tracks and the natural abundance of ^{235}U the concentration of ^{238}U can be inferred. This leads to the age Δt of a sample according to:

$$\Delta t = \frac{1}{\lambda(^{238}\text{U})} \ln\left\{1 + \left[1.61 \cdot 10^4 \frac{D(\text{sf})}{D(\text{n, if})} \delta_{(\text{n, if})} \Phi_n t_{\text{irr}}\right]\right\} \tag{5.20}$$

with $D(\text{sf})$ and $D(\text{n,if})$ being the fission track densities from ^{238}U and from ^{235}U in the sample, respectively, $\delta_{(\text{n,if})}$(n,if) the cross section for thermal neutron induced fission of ^{235}U (= $5.86 \cdot 10^{-22}$ cm^2), Φ_n the flux of thermal neutrons in n cm^{-2} s^{-1}, and t_{irr} the irradiation time.

One major drawback in fission track dating is annealing. In the course of time some tracks may fade away due to thermal processes or due to ion diffusion. This may lead to narrower and shorter tracks. As a consequence, underestimated ages

---

22 Cf. Chapter 10 in Vol. I. The probability for sf in ^{238}U is only $4.5 \cdot 10^{-5}$%.

may result. Annealing depends greatly on the material. Zircon or titanite is influenced very little by annealing while, e.g., apatite is much less stable.

Fission track dating has been successfully applied in dating of tephra and obsidian from volcanic eruptions for up to several hundred millennia. It was, e.g., possible to date tephra in the lee of the Andes (Rio Grande Valley) to have an age of 225,000 ± 25,000 years (Espizua 2002).

5.8 Thermoluminescence dating

Soil or rock usually contain radionuclides such as isotopes of uranium, thorium, and potassium. The emitted radiation in the material forms ion-electron pairs. Often, these electrons are trapped in structural defects of crystal lattices and can then be released in the laboratory via controlled heating. As a consequence, light signals are observed. Loosely bound electrons lead to light pulses at heating temperature of about 100 °C, while highly bound electrons need heating temperatures in excess of 500 °C to ignite light emission. Photon detection is usually plotted against the heating temperature (glow curve), in which peaks reflect the various electron trap configurations in the material. It turned out that high temperature thermoluminescence is proportional to the irradiation dose of the material. The age of a sample can then be determined from the known natural dose per year in the sample. However, the energy deposited in the material is strongly dependent on the nature of radioactivity: α-particles have a high LET (linear energy transfer) value, but on a short distance of typically some 20 to 50 μm, while β-particles are stopped with a much lower LET value within a range of several mm. Gamma rays, finally, have very few interactions and as a consequence very long stopping ranges. This means that for each type of irradiation a calibration is required for the dose to light emission relationship. In addition, the linearity in this relationship is limited: e.g., quartz, saturation of the light emission is observed at 100–500 Gray (Gy), while calcite saturates at a much higher dose of about 3 kGy. To conclude, thermoluminescence dating should be applied with a certain degree of caution and, if possible, verified with some other dating options. In real applications, thermoluminescence dating turned out to be suited for the time window between ^{14}C and ^{40}K | ^{40}Ar dating.

5.9 Outlook

Nuclear dating has found widespread application in many fields of science. The time range that may be analyzed is huge: it covers hours to billions of years. However, nuclear dating very much depends on the known accuracy of the half-life of the radionuclides involved. Some environmental radionuclides still have poorly known half-lives that need further improvement. One example mentioned in this chapter is ^{32}Si. Published values range from about 100 to 170 years with an average value of 153 years. This situation prevents application of this nuclide for accurate dating purposes. Another example is ^{60}Fe (not mentioned in this chapter). Its half-life was published to be $1.5 \cdot 10^6$ a. Recently, it was re-measured and the new value is now $2.62 \cdot 10^6$ a (Rugel 2009).

One drawback in nuclear dating is the analytical challenge: usually the samples to be dated contain only tiny amounts of the radionuclide that serves as a "clock" to determine the age. Earlier, the determination of such radionuclides occurred mostly via radiometric techniques. Recently, thanks to technological progress, mass spectrometry has gained increased importance due to considerably improved sensitivity. This made it possible to measure, e.g., radionuclides with half-lives greater than about 100 years in environmental samples with accelerator mass spectrometry (AMS) down to a level of less than 10^6 atoms (cf. Chapter 4).

The "workhorse" in nuclear dating is still radiocarbon, ^{14}C. This situation will probably remain so in the future, simply because many objects to be analyzed contain carbon. Hence, many laboratories have recently ordered miniaturized AMS devices, now available on the market, that are fully dedicated to measure radiocarbon with supreme sensitivity and accuracy.

References

Appleby PG. Three decades of dating recent sediments by fallout radionuclides: A review. Holocene 2008, 18(1), 83–93.

Balco G, Stone JO, Lifton NA, Dunai TJ. A complete and easily accessible means of calculating surface exposure ages or erosion rates from ^{10}Be and ^{26}Al measurements. Quat Geochronol 2008, 3, 174–195.

Bonani G, Ivy SD, Hajdas I, Niklaus TR, Suter M. AMS ^{14}C age determinations of tissue, bone and grass samples from the Ötztal Ice Man. Radiocarbon 1994, 36, 247–250.

Eichler A, Schwikowski M, Gäggeler HW, Furrer V, Synal HA, Beer J, Saurer M, Funk M. Glaciochemical dating of an ice core from upper Grenzgletscher (4200 m asl). J Glaciol 2000, 46, 507–515.

Espizua LE, Bigazzi G, Junes PJ, Hadler JC, Osorio AM. Fission-track dating of a tephra layer related to Poti-Malal and Seguro drifts in the Rio Grande basin, Mendoza, Argentina. J Quat Sci 2002, 17, 781–788.

Gäggeler HW, von Gunten HR, Rössler E, Oeschger H, Schotterer U. ^{210}Pb-dating of cold Alpine firn/ice cores from Colle Gnifetti, Switzerland. J Glaciol 1983, 29(101), 165–177.

von Gunten HR. Radioactivity: A tool to explore the past. Radiochim Acta 1995, 70/71, 305–316.

Friedländer G, Kennedy JW, Macias ES, Miller JM. Nuclear and radiochemistry, 3rd ed. Wiley, 1981.

Lachner J, Marek Ch., Martschini M, Priller A, Steier P, Golser R. ^{36}Cl in a new light: AMS measurements assisted by ion-laser interaction. Nucl Instr Meth Phys Res 2019, B 456, 163–168.

Thèse MG. Univ. de Paris VII. cited in von Gunten, 1995, 1982.

Moorbath S, O'Nions RK, Pankhurst RJ, Gale NH, McGregor VR. Further rubidium-strontium age determinations on the very early Precambrian rocks of the Godthaab District, West Greenland. Nature Phys Sci 1972, 240, 78–82.

Morgenstern U, Geyh MA, Kudrass HR, Ditchburn RG, Graham IJ. ^{32}Si dating of marine sediments from Bangladesh. Radiocarbon 2001, 43, 909–916.

Olivier S, Bajo S, Fifield KL, Gäggeler HW, Papina T, Santschi PH, Schotterer U, Schwikowski M, Wacker L. Plutonium from global fallout recorded in an ice core from Belukha Glacier, Siberian Altai. Env Sci Techn 2004, 38, 6507–6512.

Rugel G, Faestermann T, Knie K, Korschinek G, Poutivtsev M, Schumann D, Kivel N, Günther-Leopold I, Weinreich R, Wohlmuther M. New measurement of the ^{60}Fe half-life. Phys Rev Lett 2009, 103, 072502/1–072502/4.

Stuiver M, Earson GW, Braziunas T. Radiocarbon age calibration of marine samples back to 9,000 CAL yr BP. Radiocarbon 1986, 28, 980–1021.

Clemens Walther and Michael Steppert

6 Chemical speciation of radionuclides in solution

Aim: The term *speciation* subsumes the quantification of the chemical species with regard to oxidation states, complexation, formation of large molecules, and polymerization. It comprises physicochemical properties, molecular mass, charge, valence or oxidation state, complexing ability, structure, morphology, and density. Furthermore, the term speciation is used for the distribution of a given element amongst defined chemical species in a system.

Many radionuclides have a very rich chemistry in solution. Solutions may simultaneously contain many different ions, complexes, molecules colloidal, or solid phases. Different species might vary considerably in their molecular growth mechanisms, complexation behavior, dissolution properties, and their sorption behavior onto and incorporation behavior into solid phases. To understand each individual species and the behavior of the system as a whole the speciation needs to employ a variety of different analytical techniques. No single analytical technique is capable of measuring a complete speciation but typically a multitude of complementary characterization techniques are required.

The present chapter introduces some speciation methods ordered by the size range of species the methods can be applied to. None of the methods is used for radionuclides only, rather are they frequently applied techniques and for each of the methods one representative example of radionuclide speciation is discussed.

6.1 Introduction

The solution chemistry of some of the radionuclides is rather "complicated" – this is particularly true for some of the actinides. Under certain chemical conditions as many as four different oxidation states of plutonium are present in solution simultaneously, each of them having unique properties.[1] Consequently, quantification of the chemical species with regard to oxidation states, complexation, formation of large molecules, polymerization, and also sorption onto or incorporation into solids needs to be investigated. This in short is called *speciation*. Speciation comprises physicochemical properties, molecular mass, charge, valence or oxidation state, complexing

---

[1] To cite Glen SEABORG, "In a real sense, by virtue of its complexity, plutonium exhibits nearly all of the solution behavior shown by any of the other elements."

https://doi.org/10.1515/9783110742701-006

ability, structure, morphology, and density. The term *speciation* is also used for the *distribution* of a given element amongst defined chemical species in a system.

Chemical species can vary over a wide size range, from small ions or molecules with sizes of less than 1 nm via small molecule with simple ligands to macromolecules, organometallic compounds, polymeric species, nanoparticles, and colloids ranging from 1 nm all the way to macroscopic mineral particles and other solid phases with sizes of more than 1 μm. The species of different size ranges are related to each other, often in terms of chemical equilibria. One simple example is the equilibrium between two oxidation states determined by the redox potential. An example from the upper size range is the solubility of a defined solid phase in contact with an (aqueous) solution, which determines the amount of atomic/ionic species in solution for defined chemical conditions of the liquid phase.

Solutions may contain many different ions, complexes, and molecules with a multitude of molecular growth mechanisms, complexation and dissolution processes taking place simultaneously. It is rather obvious that in an ensemble of many different and interacting species reactivities, mobilities, sorption behavior, and so on can be difficult to obtain. Furthermore, it is not necessarily the most abundant species which dominates the chemical properties of the system but rather some minor but highly reactive species – for instance due to high charge. To understand the behavior of the system as a whole the speciation needs to be obtained including those minor species. No single analytical technique is capable of measuring a complete speciation but typically a multitude of complementary characterization techniques are required. Table 6.1 lists the many techniques used in speciation, which will be mentioned in this chapter.

The smallest solution species are single ions, and the chapter starts with techniques for probing oxidation states of radionuclides in solution. Next, detection and quantification of complexes and small molecules including structural probes are discussed. The separate subsections are structured according the corresponding most representative analytical techniques. The chapter continues with methods applied for polymers and colloids and concludes with a short outlook on surface-sensitive techniques. Bulk analysis is not in the scope of this chapter.

6.2 Absorption spectroscopy (UV-VIS, LPAS)

The chemical environment of any atom ranging from formation of covalent electronic bonds to changes in oxidation state influences its electronic structures. Since the atom's electrons also absorb photons, it is straightforward to probe oxidation states by absorption spectroscopy, which is indeed applied for many elements as a routine method. The principle is very simple (Fig. 6.1). A beam of light enters a sample with a well-defined path length. The intensity of the light in front of the sample is

Tab. 6.1: Analytical techniques used in speciation.

Method	Acronym	Specific feature/physicochemical basis	Main application
Absorption spectroscopy	UV-VIS	The absorption coefficient of light (wavelength dependent) is indicative of oxidation state and concentration of solution species. Caution: UV-VIS measures extinction = absorption plus scattering	Oxidation state distribution, complexation, concentration
Laser photoacoustic spectroscopy	LPAS	Same as UV-VIS, but more sensitive and no interference by light scattering	Same as UV-VIS
X-ray absorption near edge structure	XANES	Absorption of X-rays with sufficient energy causes the emission of photoelectrons from samples. The absorption coefficient is element and oxidation state specific.	Determination of oxidation states and in some cases the geometry of the first coordination shell.
Capillary electrophoresis	CE	A fractionation technique that uses the different mobilities of ions of different charge and different sizes.	Pure fractionation technique. For speciation a suitable detection system needs to be coupled.
Electrospray ionization mass spectrometry	ESI-MS	Soft ionization technique: Solution is placed in capillary at high voltage. This causes formation of some tens of nm-sized charged droplets which further shrink during introduction into vacuum for subsequent mass spectrometric identification.	Mass over charge ratio of solution species is determined. Most sensitive for large organic molecules but also well suited for inorganic ions and polymers in aqueous solution
Extended X-ray absorption fine structure	EXAFS	Photoelectrons generated by the absorption of X-rays scatter with the electrons of neighboring atoms. This causes oscillations in the x-ray absorption spectra that contain information about the number and type of the neighbor atoms of the photoelectron emitting atom.	Determination of the nearest neighbors of the probed atom as well as the numbers of the nearest neighbors.

(continued)

Tab. 6.1 (continued)

Method	Acronym	Specific feature/physicochemical basis	Main application
Infrared spectroscopy	IR	Molecules can show absorption of light in the infrared region that induces vibrations in the molecules. Different coordinating ligands can alter the electronic density in the moiety and shift its vibration frequency.	Chemical information on complexation of the respective IR-active moiety.
Raman spectroscopy		Inelastic light scattering at molecules that induces vibrations (and rotations) of the molecules. As in the case of IR spectroscopy, changes in electronic densities cause shifts of the frequencies.	Chemical information on complexation of the respective Raman-active moiety.
Nuclear magnetic resonance spectroscopy	NMR	Nuclei that have a spin different from zero and that are brought into an external magnetic field will take different possible orientations in that magnetic field. Transitions between these orientations can be induced by electromagnetic radiation. The resonance frequencies are sensitive to the chemical environment of the excited nucleus.	Information on the chemical environment of the probed nucleus. Due to spin-spin coupling, structural information can be deduced.
Time-resolved laser-fluorescence spectroscopy	TRLFS	Ions are excited by the light of a pulsed laser causing subsequent luminescence which is recorded (time and wavelength resolved). Transformation time and spectral shifts depend on chemical surrounding of the luminescing ion. Works in liquid and solid states and also at low temperature.	Chemical information on complexation, number of ligands (e.g. water molecules), and geometry of neighboring ions (ligand field), down to trace concentrations
Laser-induced breakdown detection	LIBD	A plasma is generated selectively on nanoparticles (colloids) which diffuse into the tightly focused beam of a pulsed laser. Efficiency of plasma formation depends on size of colloids, frequency of plasma formation on the concentration.	Ultratrace detection and determination of size distribution of (inorganic) aquatic colloids

Field flow fractionation	FFF	In a small channel liquids travel faster in the center than at the edges. FFF applies a force perpendicular to the flow driving molecules to the edges. However, diffusion counteracts this directed motion. Since diffusion is faster for small particles they remain in the fast central flow and are eluted prior to the large particles	Size separation of particles and molecules from some 0.5–100 nm. Sensitivity strongly depends on detector used (optical detection, mass spectrometry, etc.)
X-ray photoelectron spectroscopy	XPS	By measuring the kinetic energies of photoelectrons emitted from solid samples, the binding energies of the electrons can be determined. These are element and oxidation state specific. By XPS very thin surface layers are accessible.	Probing of the oxidation state and the chemical environment of thin surface layers.

compared to the one behind, the difference between both intensities being defined as *extinction*.[2] It is important to keep in mind that extinction does not only comprise the desired absorption signal (the interaction with the electrons), but also scattering of light which often is unwanted, e.g., if the sample is opaque or contains a large number of small particles. A wavelength-dependent spectrum is obtained by either using a tenable monochromatic light source or by introducing "white" light and adding a wavelength dispersive element such as a grating or a prism behind the sample. Precision is increased by so-called double path instruments where the light beam is split into two parts of equal intensity and guided via identical beam paths to two independent detectors. One beam penetrates the sample, the second a reference solution or a blank. The difference in extinction is measured accurately to typically 1 part in 10 000 or better. For dilute solutions (low extinction) BEER's law is linearly approximated relating the absorption A_{abs}

$$A_{abs} = \varepsilon_{mol} c d, \tag{6.1}$$

with the molar absorptivity ε_{mol} (in $M^{-1}\,cm^{-1}$), thickness of the sample d (in cm), and concentration c of an absorbant (in M = mol/l). Typical absorptivities are $100\,M^{-1}\,cm^{-1}$ leading to lowest measurable concentrations in the µM range. However, organic molecules, e.g., dyes, absorb light up to 1000 times stronger, metal ions with so-called forbidden transitions may absorb light weaker by the same factor. The limit of detection (LOD) is lowered by increasing the path length.

A very convenient method is the use of a so-called "liquid wave guide capillary". The sample is filled into a narrow capillary of a material with an index of refraction[3] n_{refr} smaller than the one of water ($n_{refr}\,H_2O = 1.33$). In analogy to the principle of a fiber light guide, the light cannot leave the capillary due to total internal reflection. While the standard absorption cell has a path length of only 1 cm, waveguides of up to 10 m length are available. The extinction is hence 1000 times higher.

A beam of light enters a sample with a well-defined path length. The intensity of the light in front of the sample is compared to the one behind, the difference between both intensities being defined as extinction. The interaction of the light with the sample can be detected by different methods: the absorption can be monitored, fluorescence can be detected (this will be discussed in more detail in Section 6.11) or the deposition of energy in the sample can be detected by a photoacoustic method.

Figure 6.2 illustrates the application of the technique for the case of plutonium in solution. Under acidic, slightly oxidizing conditions Pu^{III}, Pu^{IV}_{aq}, Pu^{V}, Pu^{VI}, and

---

2 Extinction is the loss of light which passes through a medium. Extinction is the sum of absorption and scattering.
3 Index of refraction of a substance (optical medium) is a dimensionless number that describes how light propagates through that medium.

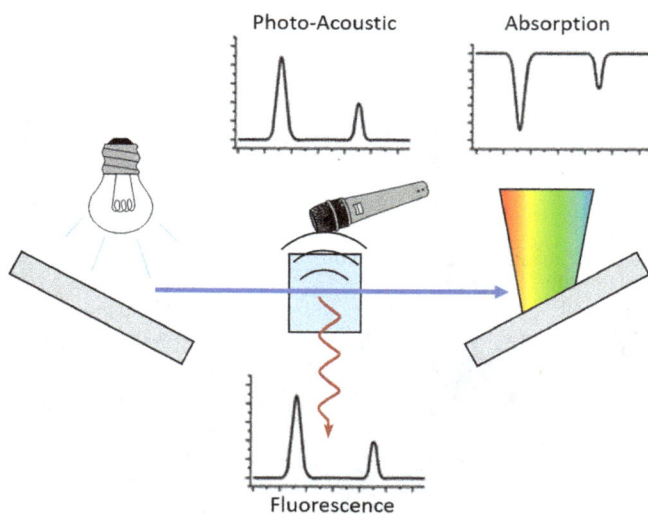

Fig. 6.1: The principle of absorption spectroscopy.

Pu^{IV}-polymer species[4] (colloids) are stable in solution making the exact speciation particularly challenging. Fortunately, all Pu oxidation states exhibit quite narrow and characteristic absorption peaks, due to simple LAPORTE forbidden intraband transitions[5] of the 5f electrons, as shown in the upper left part of Fig. 6.2. The corresponding colors of plutonium solutions in the respective oxidation states are shown in the lower part of Fig. 6.2. If a mixture of different oxidation states is present in solution, the absolute concentration of all oxidation states is obtained by mathematical deconvolution of the sum spectra. The achievable precision strongly depends on the set of reference spectra available taking in account the influence of the matrix due to complexation of in particular the tetra- and hexavalent ions with, e.g., nitrate or chloride ions. Thus, by this method, complexation of actinides by organic compounds can be studied in order to gain thermodynamical complexation data, as shown for Np(V) by Maiwald et al. (2020).

In samples with strong light scattering, the absorption signal is often considerably disturbed as illustrated in Fig. 6.2 (upper right, blue line), for the case of a solution containing trivalent praseodymium and 100 nm polystyrene nanoparticles. Effects of absorption and scattering should be considered. The light scattered elastically by particles leaves the sample and does not deposit energy. Absorption, in contrast, deposits energy, leads to local heating and, consequently initiates volume

---

4 The subscript in Pu^{IV}_{aq} indicates summation over all aqueous species containing tetravalent Pu. Without subscript, only the ion Pu^{4+} was considered.

5 LAPORTE forbidden intraband transitions are electronic transitions including a spin change, which are strongly suppressed.

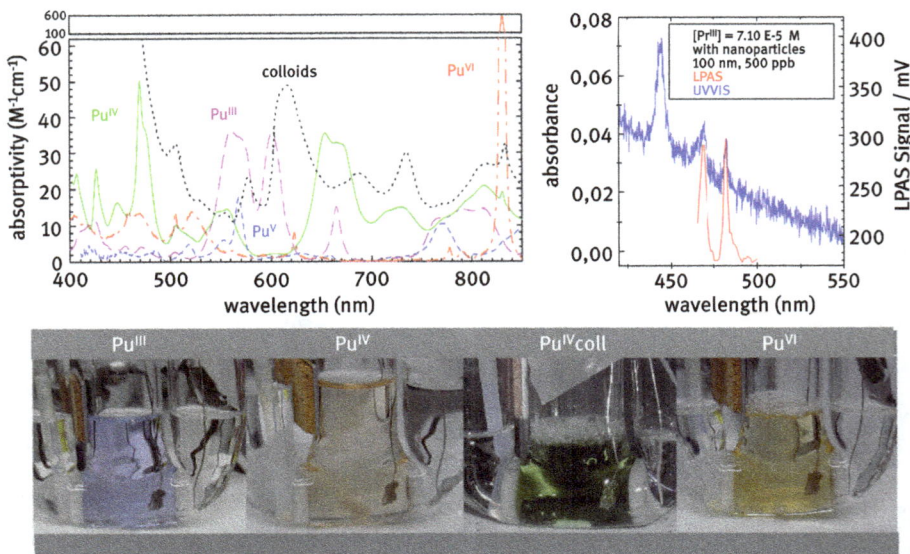

Fig. 6.2: Application of the UV-VIS method to determine oxidation states of Plutonium solution samples. Left: Absorption spectra of four reference solutions are given in the molar absorptivity scale. The solutions were prepared at different concentrations and in different media: Pu^{III} at $2.6 \cdot 10^{-3}$ M in 1 M $HClO_4$, $Pu^{IV}{}_{aq}$ at $3.7 \cdot 10^{-4}$ M in 0.5 M HCl, Pu^{V} at $3.0 \cdot 10^{-43}$ M in 0.001 M $HClO_4$, Pu^{VI} at $2.3 \cdot 10^{-3}$ M. Bottom: Plutonium solutions with the respective pure oxidation states show characteristic colors. Right: In samples with strong light scattering, the absorption signal is often considerably disturbed as illustrated in the right figure (blue line), for the case of a solution containing trivalent praseodymium and 100 nm polystyrene nanoparticles. The red curve shows the same sample as before, this time measured by LPAS.

expansion of the illuminated fraction of the sample. The higher the absorptivity at a given wavelength is, the stronger the shock wave caused by the expansion. This effect is exploited by LPAS. A narrowband tunable laser is scanned over the spectral range to be measured and a piezo detector[6] transforms the shock wave into an electric signal, which is strictly proportional to the absorptivity of the absorbent. The red curve in Fig. 6.2 shows the same sample as before, this time measured by LPAS. The slope due to RAYLEIGH scattering[7] is clearly suppressed. Furthermore, LPAS is up to three orders of magnitude more sensitive compared to conventional UV-VIS. However, LPAS is technically more demanding and more time consuming.

---

6 A piezo detector is a very sensitive microphone.
7 RAYLEIGH scattering is scattering of light by particles much smaller than the wavelength of the light.

6.3 X-ray absorption near edge structure spectroscopy (XANES)

A method closely related to absorption techniques in the visible range, is an X-ray absorption based technique called XANES (X-ray absorption near edge structure). The technique is comparable to the absorption spectroscopy methods described in Section 6.2 but differs in the energy region of the incident radiation, ranging from ≈500 eV to 100 keV, namely the X-ray region in the electromagnetic spectrum.

Being an absorption technique, it follows BEER's law,

$$I = I_0 e^{-\mu d}, \tag{6.2}$$

with the intensity of the incident beam I_0, the intensity I after absorbed by a sample of the thickness d, as shown in Fig. 6.3. The absorption coefficient μ is element specific and depending on the sample density ρ, the atomic number Z, the atomic mass A, and the X-ray energy:

$$\mu \approx \frac{\rho Z^4}{A E^3}. \tag{6.3}$$

This technique is based on the photoelectric effect:[8] By use of a synchrotron, X-ray photons of defined (and variable) energy can be produced. These photons ionize atoms by the photoelectric effect, possible for all electrons with a binding energy below the one of the photon entering.

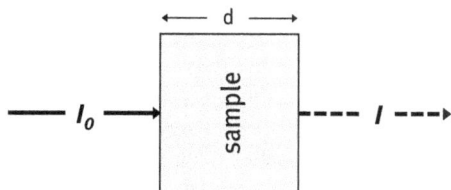

Fig. 6.3: X-ray absorption near edge structure is an absorption technique and thus follows Beer's law. The intensity of the incident beam I_0 is reduced to the intensity I by passing the sample of thickness d. The absorption coefficient μ is dependent on the atomic number and the energy. Thus the spectra contain information about the element composition.

To understand the effects of the emission of electrons, we have to recall what we learned about the electron shell, cf. Sections 1.3 and 1.4 in Vol. I. Electrons are

---

8 The photoelectric effect was introduced in Chapter 11 of Vol. I in the context of secondary transitions of unstable nuclides, related to excited nuclear states. Here, it is induced by external excitation on stable atoms.

present in different shells n of discrete energies ($n = 1, 2, 3, \ldots, 7$). These shells are often symbolized by Latin capital letters (K, L, M, ..., Q), as shown in Fig. 6.4. The electrons in the outer shell are called valence electrons and are responsible for the chemical behavior of an element. All the other electrons present in the shells of lower n than the valence electrons are referred to as "core electrons" in X-ray absorption spectroscopy methods. Let us revisit the calcium atom from Sections 1.3 and 1.4 in Vol. I: it contains 20 electrons in its shells from $n = 1, \ldots, 4$ (see Fig. 6.4). The valence electrons are situated in the shell with $n = 4$ (the N-shell), the other electrons distributed over the K-, L-, and M-shells are the core electrons.

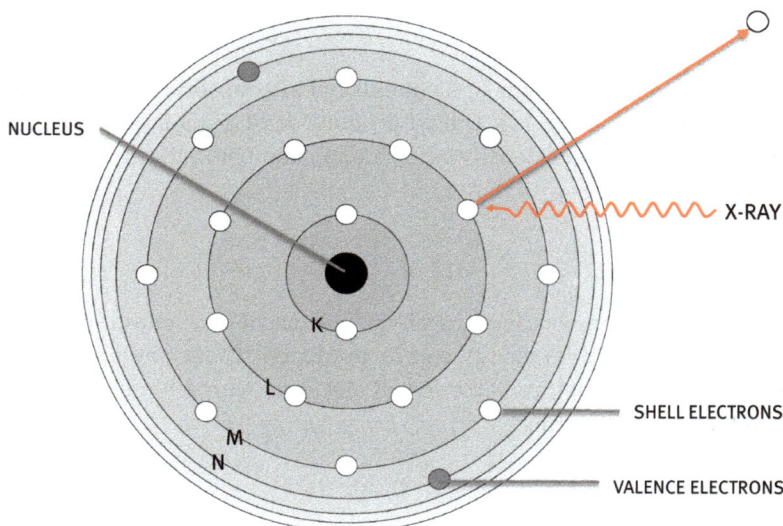

Fig. 6.4: X-ray absorption near edge structure spectroscopy is based on the photoelectric effect. X-rays of sufficient energy can be absorbed by core-level electrons of the probed element (here Ca) in the sample. These electrons are excited to unbound states, the so-called continuum. K, L, and M denote the electron shells of the respective probed element.

If we now bring a calcium sample (or any other element) into an X-ray beam with tunable energy at a synchrotron, the X-ray photons can interact with the electron shells of the atoms. By increasing the X-ray energy, corresponding to eq. (6.3), the absorption coefficient μ will decrease. Thus, with increasing X-ray energy every probed sample would become more and more transparent. However, there are pronounced resonances in those cases where a core-level electron is excited to so-called unbound states (see Fig. 6.4). One observes a sudden increase as shown in Fig. 6.5 in the X-ray absorption cross section called "absorption edge" when the X-ray energy $E^{\text{X-ray}}$ corresponds to the binding energy of the respective core-level electrons. These binding energies are not only element specific, they also depend on

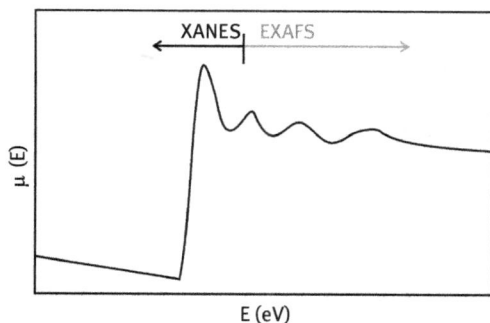

Fig. 6.5: Schematic sketch of an X-ray absorption spectrum. The energy of the X-ray photons is plotted on the x-axis, the absorption coefficient μ on the y-axis. With increasing energy, the absorption coefficient decreases until the X-ray energy resonantly excites a core electron into an unbound state. This leads to a sudden increase in the absorption, a so-called absorption edge. The position of the absorption edge is both element and oxidation state specific. The X-ray absorption near edge structure (XANES) spectroscopy focuses on the area close (typically within 30 eV of the main absorption edge) to the absorption edge. The EXAFS region in the plot will be discussed in Section 6.8.

the effective charge at the probed atoms and thus can be applied to determine the oxidation state of the investigated element.

An example for the identification and quantification of oxidation states of radionuclides is shown in Fig. 6.6. It shows typical absorption spectra of the different oxidation states of ^{237}Np in solution recorded at the Np L3-edge.[9] The spectral features close to the absorption edge are the X-ray absorption near edge structure (XANES), being element and oxidation state specific. The part of the spectrum at higher energies, called the EXAFS (extended X-ray absorption fine structure) will be discussed later in this chapter (Section 6.7), when the interactions with the nearest neighbor atoms of the radionuclide are of interest.

The different oxidation states of neptunium were prepared by electrolysis and the XANES measurements were performed *in situ* in an electrolysis cell. The versatility to combine X-ray spectroscopy with *in situ* preparative methods in combination with the high selectivity of excitation is one of the strong advantages of X-ray spectroscopy. The shift in the absorption peak for the respective oxidation states follows the expected trends of the effective charge on the metal center. The discontinuity from the Np^{IV} to Np^{V} results from the formation of transdioxo neptunyl ions NpO$_2^+$ and NpO$_2^{2+}$, respectively. The doubly bound O atoms donate electron density back to

---

9 Np L3-edge means that the excitation energy of the X-ray corresponds to the electrons in the Np-L-shell. As introduced in Section 1.4.1 of Vol. I, all shells except the K-shell split into subshells. The L-shell is divided into three subshells: L1, L2, and L3. L1 corresponds to the 2s energy level, L2 to $2p_{1/2}$, and L3 to $2p_{3/2}$ (see also Section 1.6.2 in Vol. I).

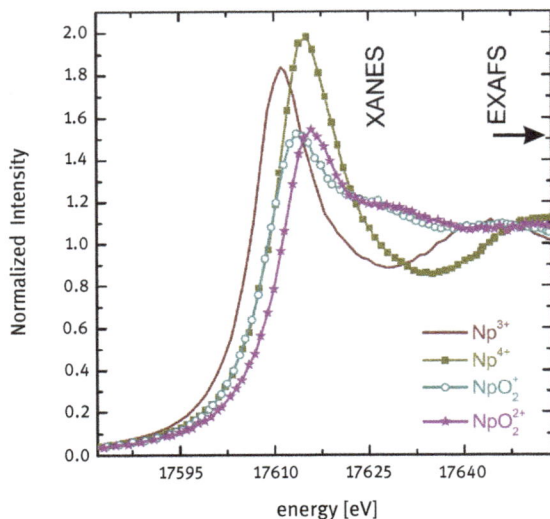

Fig. 6.6: The dependency of the peak positions of a probed element on the oxidation state is shown in this example for ^{237}Np in its different oxidation states prepared by electrolysis. The spectra of the pure oxidation states clearly show distinctive shifts. Moreover, the peak forms differ due to the different ionic species characteristic for the Np oxidation states: while in the lower oxidation states Np exist as Np^{3+} and Np^{4+}, the higher oxidation states form dioxo-ions $NpO_2^{+/2+}$. The different coordination geometry causes different peak forms. Figure reproduced from Gaona et al. (2012) with permission.

the Np atom causing a shift of the absorption edges. From the spectra of the pure oxidation states, one can determine the abundance of the different oxidation states in solution of unknown composition by fitting these spectra.

Recently a high energy resolution fluorescence detection (HERFD) mode was developed and applied to the investigation on nanometer sized plutonium nanoparticles (see also Section 6.12). Gerber et al. (2020) showed that the plutonium oxidation state in PuO_2 nanoparticles is exclusively Pu(IV). Before, a mixture of Pu(IV) and higher oxidation states was proposed. The technique can be applied on solid samples, solutions, and gaseous samples.

6.4 Liquid-liquid extraction

Alternative approaches to measure a radionuclide's oxidation state in solution are based on chemical separation applicable to even minute concentrations and hence particularly suited for radiochemistry.[10] All of these techniques make use of oxidation state selective complexation by ligands, some kind of extraction or separation of the noncomplexed residues followed by (e.g., radiochemical) detection of the radionuclide of interest. One separation method is the liquid-liquid extraction technique. An organic phase with selective extraction ligands[11] is added to the aqueous phase containing radionuclides in different oxidations states (see Fig. 6.7). The organic phase needs to solvate both the extraction ligand and the M-L-complexes formed by the ligands with the respective radionuclide. Furthermore, the organic phase should be resistant against radiolysis and should not mix with the aqueous phase. Ligands are selected according to oxidation state selective complexation of radionuclides, allowing for extraction of one characteristic oxidation state only. The two-phase mixture is shaken for a sufficient time, to enlarge the phase boundary during the extraction process. Afterwards, the mixtures are allowed to phase separate again. Consequently, one oxidation state is enriched in the organic phase, while the other oxidation states

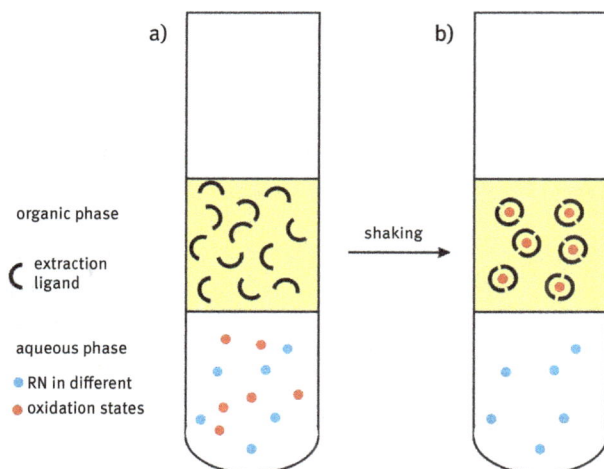

Fig. 6.7: Liquid-liquid extraction: a) An organic phase containing selective extraction ligands is added to an aqueous solution of radionuclide in different oxidation states. If the organic phase is nonmiscible with the aqueous phase, two phases form. On the phase boundary, the extraction ligands complex the metal ions from the aqueous solution and stabilize them in the organic phase. The mixture is shaken vigorously to enlarge the phase boundary. b) After subsequent phase separation one oxidation state is enriched in the organic phase if ligands were applied that selectively stabilize one oxidation state in the organic phase. The other oxidation states remain in the aqueous phase.

---

10 Radiochemical separation methods are discussed in Chapter 7.
11 "Extraction" ligands L are chelating agents which form M-L-complexes, which change the lipophilicity of the initial hydrophilic cationic M species to a lipophilic complex.

remain in the aqueous solution (see Fig. 6.7). The quantification can be performed by first separating the two liquid phases or by taking aliquots from each phase; followed by analyzing the activities of the two phases by standard radioanalytical techniques like liquid scintillation counting, γ-spectrometry, etc.

6.5 Capillary electrophoresis (CE)

Another chemical separation technique often applied in radiochemical speciation is capillary electrophoresis[12] (CE). Here, a conductive solution within a thin capillary is subjected to an electric field. A schematic figure of the principle of the technique is depicted in Fig. 6.8. Two mechanisms play a role: First, charged species will experience a force that is proportional to their charge and the electrical field. Thus, species with different charge will be separated. This is called electrophoretic mobility[13] (μ_{EF}):

$$\mu_{EF} = \frac{v}{E} = \frac{ze_0}{6\pi\eta r_{hd}},$$ (6.4)

with z being the effective charge of the ion, e_0 the elementary charge of the electron, η the dynamic viscosity in $N\,s\,m^{-1}$, and r_{hd} the hydrodynamic ionic radius of the hydrated ion. It connects the migration velocity v of the ions with the electrical field strength E.

Anions will move towards the anode, cations towards the cathode; neutral species are only separated when they are associated to charged molecules. When an ion moves through a solution, it is surrounded by a number of solvent molecules. This increases its effective radius. This effect is accounted for with the so-called hydrodynamic radius which describes this effective radius. Larger ions have lower mobilities and are separated from the smaller ones.

The second mechanism is called electroosmotic flow[14] (μ_{EOF}):

$$\mu_{EOF} = \frac{\varepsilon_0\varepsilon_r\zeta}{4\pi\eta},$$ (6.5)

---

12 The principles of electromigration are introduced in Chapter 7. Many of the parameters defined there also apply to CE.

13 Electrophoretic mobility is defined as the rate of migration of the substance measured in cm/s under the influence of a potential gradient of 1 V/cm and given in units of $cm\,s^{-1}\,V^{-1}$, cf. also Chapter 7.

14 Electroosmotic flow describes the motion of a liquid in a capillary induced by a voltage applied to the two capillary ends.

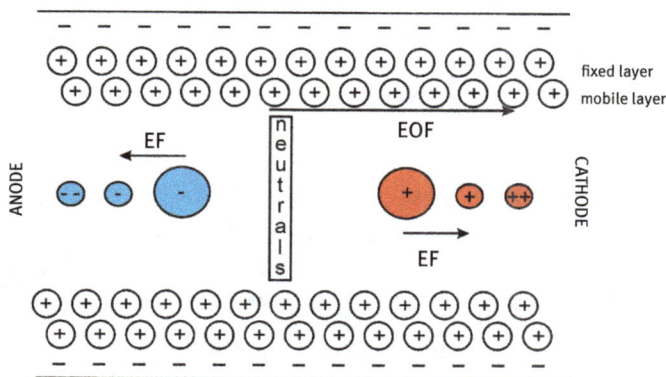

Fig. 6.8: Principle of capillary electrophoresis: A glass capillary filled with a solution of radionuclide containing electrolytes is subjected to a strong electric field. This leads to a motion of the ions present in solution towards the electrode of opposite charge. Ion electrophoretic mobility μ_{EF} depends on size and charge causing the intended fractionation. Moreover, the solution is dragged through the capillary towards the cathode by the electroosmotic flow μ_{EOF}, induced by a double layer of electrolyte cations that forms at the capillary walls due to its negative charge.

with ε_0 being the electric field constant in $A\,s\,V^{-1}\,m^{-1}$, ε_r the dielectricity constant, and ζ the zeta potential.[15] Electroosmotic flow is induced by the capillary walls coated with silanol groups. These are deprotonated at pH > 2, leading to a negative charge at the capillary walls that is compensated by cationic electrolyte ions from the buffer solution. The electrolyte cations form a double layer: the layer closer to the wall is fixed, the second layer is mobile. Due to the voltage applied, this positive layer will move towards the cathode and drag the whole solution in that direction. The total mobility (μ_{tot}) results from the sum $\mu_{EF} + \mu_{EOF}$.

After separation, the different fractions have to be detected by a suitable detection method, for instance by a mass spectrometer, a UV-VIS spectrometer, an NMR spectrometer or an activity counter. An example for the separation of ^{237}Np in different oxidation states applying an ICP-MS[16] as detection method is shown on the left side of Fig. 6.9. The different oxidation states pass the capillary after different times and are thus fractionated. Due to the size dependency of the electrophoretic mobility, molecules can also be separated by this method. The right side of Fig. 6.9 shows the fractionation of the first three mononuclear hydrolysis species of ^{242}Pu in the oxidation state Pu^{IV}, i.e., $[Pu(OH)_n]^{4-n}$. The three different species were separated clearly from each other and quantified by ICP-MS afterwards.

---

15 The zeta potential ζ describes the potential gradient at the cationic double layer formed at the capillary walls. Its value is crucial for the macroscopic behavior of an electrolyte in a capillary.
16 ICP-MS is an inductively coupled plasma mass spectrometer. It is a widely applied tool for elemental analysis and concentration determination of metals in solutions.

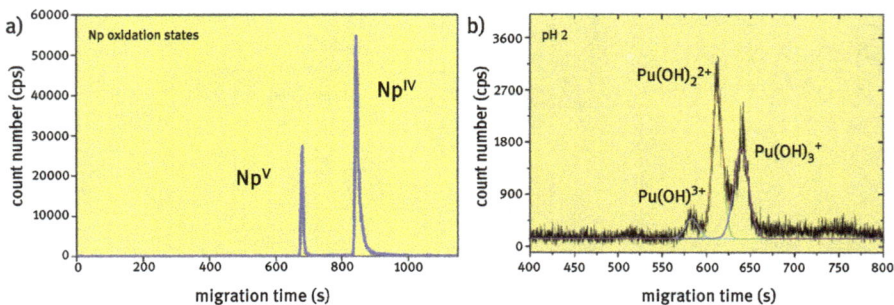

Fig. 6.9: Data obtained by capillary electrophoresis from Banik et al. Left: Different oxidation states of ^{237}Np are separated by CE and the fractions are quantified by ICP-MS. Here Np^{IV} shows a higher electrophoretic mobility due to its higher charge than Np^V, which forms NpO$_2^+$ ions in aqueous solution (Banik et al. 2010). Right: Fractionation and quantification of ^{242}Pu^{IV} hydrolysis products (Banik et al. 2009).

6.6 Electrospray ionization mass spectrometry (ESI)

A direct and at the same time low invasive[17] technique for speciation of molecules in solution is electrospray ionization (ESI). A liquid is introduced into a small capillary with a fine tip. By applying a high voltage to the tip the electric field generates charged droplets, as shown in Fig. 6.10. In an electrospray ion source, coupled to mass spectrometer, these primary droplets are guided into the vacuum where they shrink due to solvent evaporation and coulomb explosion[18] until only one molecule with a small number of solvent molecules remains. These charged species are characterized by a mass analyzer, such as a time-of-flight drift tube, magnetic sector field analyzer, quadrupole (HF) analyzer, or ion trap. Developed for the low invasive detection of fragile, large biomolecules, as shown in Hoyer et al. (2019), this technique also proved useful for detection of inorganic complexes and polymers.[19]

During the droplet shrinking process the ratio of volatile to nonvolatile species may be changed, so-called up-concentration. This effect is strongly reduced by generating smaller primary droplets in the case of so-called nano-ESI, a particularly

---

[17] "Noninvasive" means that the species are not changed during transfer to the detection system (in this case into the mass analyzer).

[18] "Coulomb explosion" means that the charge density on the outside of a droplet increases during shrinkage (constant charge but decreasing surface area). This causes instability and leads to coulomb explosion.

[19] In contrast to detection of large organic molecules that are charged in the source by attachment of protons, inorganic complexes need to be present as charged species already in solution. Neutral molecules cannot be detected.

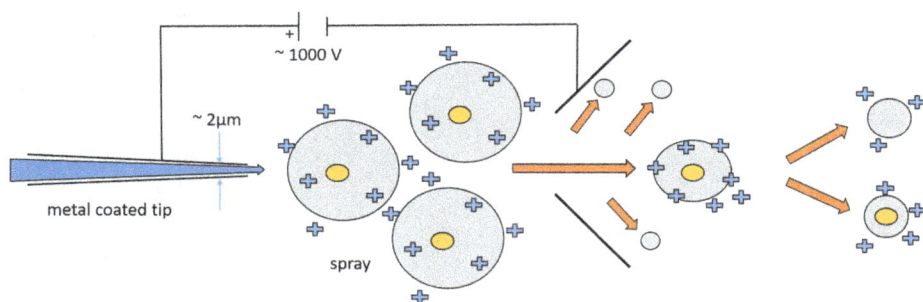

Fig. 6.10: A spray is generated by applying high voltage to a metal-coated quartz capillary. The droplets shrink and Coulomb explode as they enter vacuum, forming single charged molecules with a small shell of water molecules.

soft variation of electrospray. By substituting the metal capillary by a glass capillary of only approximately 2 μm inner tip diameter, droplets of less than 50 nm radius are generated. These droplets contain only a small number of nonvolatile compounds, making polymerization during droplet shrinkage less likely. Second, evaporation is much faster than for conventional ESI. Operative parameters of the source can be chosen to be mild such that a solvent shell is maintained surrounding the molecules during all stages of detection, minimizing also the risk of fragmentation or dehydration.

Typical samples may contain analyte concentrations of < 10 mM and ionic strength $I < 100$ mM, since the amount of electrolyte present in solution is limited by the ESI technique: Salt deposition clogs the tip of the needle or the small orifice where the spray enters the vacuum chamber.

The method is illustrated for the analysis of polymers of (tetravalent) thorium[20] ions in solution, as shown in Fig. 6.11. It is well known that thorium hydrolyzes rather fast, which in turn leads to formation of polynuclear hydroxide complexes and colloids. These species play a crucial role in understanding thorium solution chemistry, since their presence causes apparent solubilities several orders of magnitude higher than in the absence of polymerization. Reactivity and extractability of polymeric species differ strongly from monomers due to their high charge. It is hence necessary to understand formation of polymers quantitatively on a molecular level. While at high acidity (ca. pH 1) monomeric species account for the majority of thorium in solution $[Th^{IV}]_{tot}$, the fraction of polynuclear hydroxide complexes increases with increasing pH. Close to the limit of solubility for amorphous thorium hydroxide $Th(OH)_4$, polymers often dominate the solutions. One especially stable and pronounced species detected by ESI is the pentamer shown in Fig. 6.11.

---

20 Polymers comprise as many as 20 metal ions bridged by hydroxide groups or oxygen.

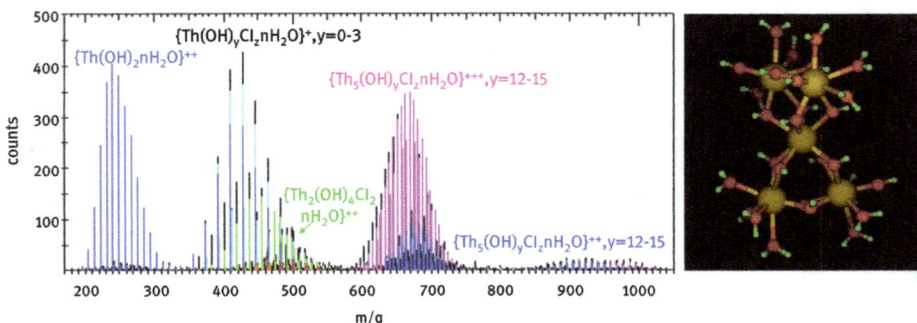

Fig. 6.11: Time-of-flight spectrum of a solution at pH_C 3.17 containing $[Th^{IV}]_{tot} = 2.89 \cdot 10^{-2}$ M. The two peak groups around $m/q = 250$ amu/e[21] and $m/q = 400$ amu/e comprise mononuclear hydrolysis complexes still inside tiny water droplets (15–30 water molecules). Doubly charged dimers are superimposed on the latter group. The peaks around $m/q = 650$ amu/e comprise triply charged pentamers. Right: Structure of the Th-pentamer according to molecular modeling.

Another strength of the method is that it can determine the presence of ternary species in solution: Recently Lösch et al. studied[21] the complexation of hexavalent uranium with aqueous silicates by a combination of ESI-MS and fluorescence spectroscopy (see Section 6.11). Uranium(VI) tends to hydrolyze in the investigated pH-range, leading to the formation of different uranium hydroxo species. The results in Lösch et al. (2020) clearly show the presence of ternary uranium-hydroxo-silicate species in solution. These ternary species are not accessible by means of spectroscopic measurements alone; ESI-MS results show the sum formulas of the species present in solution and can give the crucial hints to the presence of such species in solution.

The method is suited for detecting the distribution of all charged species in solution, including minor species. Using mass analyzers of high resolving power, species contributing only a few ppm to the total metal concentration in solution are efficiently detected.

6.7 Extended X-ray absorption fine structure spectroscopy (EXAFS)

After having determined characteristic oxidation states of a radionuclide and quantified the fraction of individual complexes formed, the next step in speciation is to gain information on the structure of the compounds. This asks for information on

---

21 amu/e represents atomic mass units divided by the charge.

the close environment of the (radio)metal and the kinds of infractions between the "central" metal Me and atoms representing the "environment". To identify the nearest neighbor atoms of the radionuclide, the extended X-ray absorption fine structure (EXAFS) technique, mentioned before in Section 6.3 on XANES, can be utilized.

When a core-electron (see Fig. 6.4) absorbs an X-ray photon of sufficient energy and wave length λ, it becomes a "photoelectron", an electron induced by interaction with electromagnetic radiation, with a specific energy ejected from the absorbing atom. Its kinetic energy can be expressed as wave number $k = 2\pi/\lambda$.

On its path, the photoelectron is backscattered from the electrons of neighboring atoms to the absorbing atom. Presence of the backscattering electron modulates the amplitude of the photoelectron wave function at the absorbing atom. A simplified scheme is depicted in Fig. 6.12.

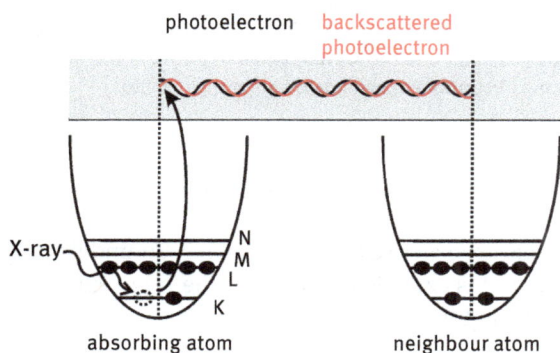

Fig. 6.12: An incident X-ray beam excites an electron from a core level to the continuum. The photoelectron can be backscattered by electrons from neighboring atoms. The backscattered photoelectron wave interferes with the outgoing parts of the initial photoelectron wavefunction.

The waves of the ejected electron and the backscattered electron interfere. Depending on the wavelength of the electrons and the distance between the emitting and backscattering atom, interference can be constructive or destructive. This interference gives rise to the oscillations in the absorption function called the extended X-ray absorption fine structure (EXAFS), depicted in Fig. 6.13. It plots the absorption coefficient (E) as a function of the X-ray energy $E^{\text{X-ray}}$ (in eV).

The oscillations can be described by the EXAFS fine structure function (E):

$$\chi(E) = \frac{\mu(E) - \mu_0(E)}{\Delta\mu_0(E)}, \tag{6.6}$$

where (E) is the measured absorption coefficient (black line in Fig. 6.13), $\mu_0(E)$ is a background function (plotted in gray) that represents the absorption of an isolated

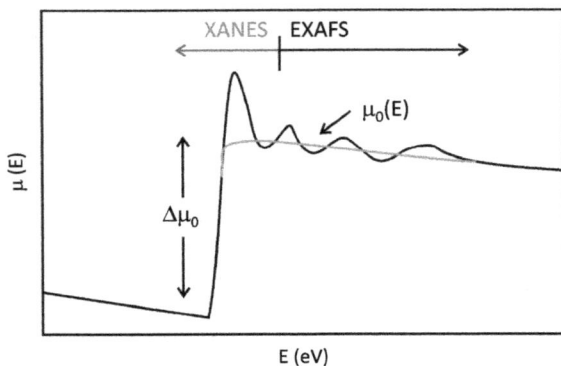

Fig. 6.13: The absorption coefficient (E) is plotted against the kinetic energy of the incident X-ray photons in eV. The interfering photoelectron waves cause modulations in the absorption coefficient (E) called the extended X-ray absorption fine structure (EXAFS). This oscillation can be seen in the EXAFS region at higher energies compared to the XANES region depicted in the spectrum. The gray line $\mu_0(E)$ represents the absorption of an isolated atom that has no backscattering neighbors.

atom without backscattering neighbors, and $\Delta\mu_0(E)$ is the jump in the absorption at the energy E_0 at the absorption edge.

As mentioned above, the photoelectron shows a wave behavior. To understand the fine structure in the EXAFS region, it is expedient to convert the X-ray energy to k, being the wavenumber of the electron. It is defined as follows:

$$k = \sqrt{\frac{2m_e(E - E_0)}{\hbar^2}}, \qquad (6.7)$$

with E_0 being the absorption edge energy, m_e the electron mass, and $\hbar$ the reduced PLANCK constant.[22]

Now the oscillations in the fine structure (k) can be described as a function of the photoelectron wave number k. The different frequencies in the oscillations in (k) result from different near-neighbor coordination shells and can be modeled by the EXAFS equation:

$$\chi(k) = \sum_j \frac{N_j f_j(k) e^{-2k^2\sigma_j^2}}{k R_j^2} \sin\left[2kR_j + \delta(k)\right], \qquad (6.8)$$

where $f(k)$ and $\delta(k)$ are scattering parameters of the atoms neighboring the excited atom (both of these depend on Z, the atomic number of the neighbor atoms and consequently are element specific), N is the number of neighboring atoms, R is the

---

22 $\hbar = h/(2\pi)$.

distance to the neighbors, and σ^2 a factor that describes the disorder in the neighbor distance. From this somewhat complicated equation we can see that the oscillations contain information about the atomic number Z of the neighboring atoms, and the distance R between absorbing atom and neighbor atom up to 5 Å.

As the EXAFS (k) oscillations decay rather quickly, the signal is weighted with respect to the wavenumber k by multiplying it with a power of k (k^2 or k^3), to amplify the oscillations at high k-values as shown in Fig. 6.14. The EXAFS data basically consist of sums of damped sine waves. FOURIER transformation[23] of the k-space spectrum to the position space delivers a direct relation of the distances of the neighbor atoms and the oscillation magnitude: the oscillation magnitude is then plotted as a function of the bond distances in Å. In this representation the peaks correspond to groups of atoms at different distances.

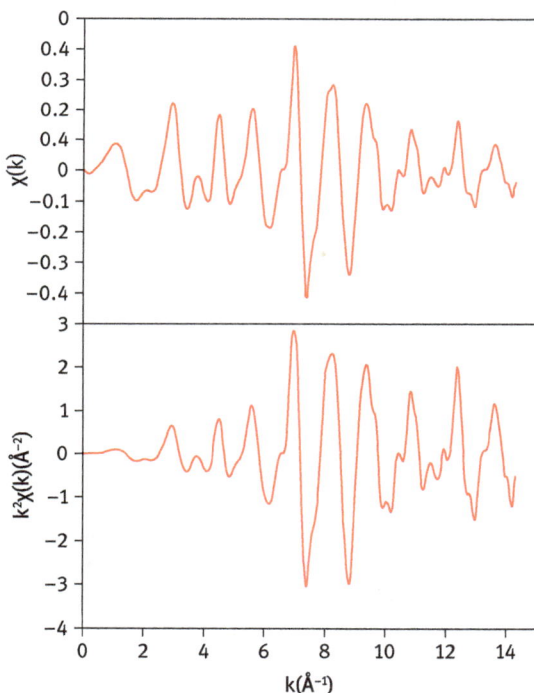

Fig. 6.14: Isolated (k) oscillations of a molybdenum sample. The top shows the oscillations extracted from the X-ray absorption spectrum by background subtraction, normalization, and conversion to the k-space without weighting. The unit of k is 1/Å. On the bottom the same data is k-weighted ($k^2(k)$) in order to emphasize the oscillations that transform quickly with increasing k-values.

---

23 The FOURIER transformation is a mathematical method that describes a signal as a sum of sine and cosine waves of different frequencies and amplitudes.

An example for an EXAFS spectrum is given in Fig. 6.15. It shows the EXAFS spectra of $^{242}Pu^{VI}$ and $^{242}Pu^{VII}$ in solution. The k^3 weighted EXAFS oscillations are plotted against the X-ray energy. A FOURIER transformation results in a spectrum showing the EXAFS oscillation magnitude at varying Pu–O bond distances to the neighboring atoms. Here the distances to the next-neighbor atoms are discernible. EXAFS confirmed that Pu^{VII} does not form dioxo-ions analog to PuO$_2^{2+}$, but a tetraoxo-ion PuO$_4^-$ with four oxygen atoms at a distance of 1.88 Å corresponding to four doubly bound oxygen atoms, and two further oxygen atoms at a distance of 2.3 Å corresponding to two hydroxide moieties in the first coordination shell. The respective structures are depicted.

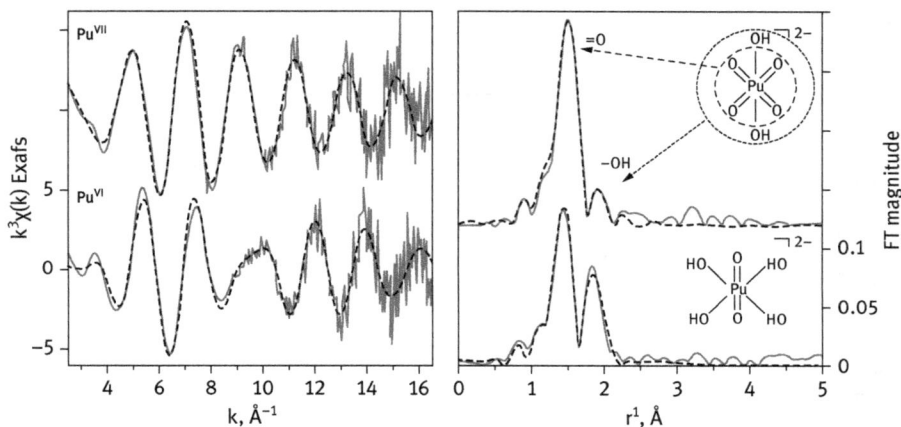

Fig. 6.15: EXAFS data of $^{242}Pu^{VII}$ and $^{242}Pu^{VI}$ solutions by Antonio et al. (2012) Left: By background subtraction, normalization and conversion to the k-space, see eqs. (6.7) and (6.8), the EXAFS oscillations $x(k)$ can be isolated from the raw data. The oscillations are weighted with respect to the wavenumber k (multiplied by a power of k) in order to enhance the quick oscillations for high k values. This is depicted in the two k^3 weighted spectra for Pu^{VI} (bottom) and Pu^{VII} (top). These oscillations contain information about the types, number, and distances of neighboring atoms. Right: After Fourier transformation into the r-space (Å) the spectrum visualizes the EXAFS oscillations depending on the bond distance r in Å. The different coordination spheres around the absorbing atom and their respective bond distances are visible. The two spectra for Pu^{VI} and Pu^{VII} show two different oxygen shells for the oxygen atoms bound to the plutonium in two different ways. The plutonium oxygen double bond has a smaller bond length than the plutonium hydroxide bonds, as shown in the upper right part of the figure. If one compares the spectra of the two plutonium species, one can clearly see that the EXAFS magnitude contains information about the number of atoms in the respective shell: For Pu^{VII} in the upper spectrum the EXAFS magnitude for the Pu =O shell is much higher, as four oxygen atoms are bound to the plutonium center in this fashion. For Pu^{VI} in the lower spectrum the peak corresponding to the Pu–OH bonds becomes more pronounced as the number of hydroxide moieties is larger in this species. Figure reprinted and adapted with permission from Antonio et al. (2012).

As many organic ligands ubiquitous in the geosphere, might enhance the mobility of actinides in the far field of a nuclear waste repository due to their relatively high complexation constants, understanding of actinide complexation by organic ligands is crucial. Taube et al. (2019) studied the complexation of americium by malic acid by a combination of EXAFS and other techniques. The EXAFS spectra give insight into the first coordination sphere of a coordinated metal, the method revealed that americium is coordinated in a tridentate ring structure by malate.

6.8 Infrared (IR) vibration spectroscopy

The absorption of electromagnetic radiation in the much lower energy range, i.e., the infrared region $(1.2–1.7\,\text{eV})$[24] is not capable of exciting shell electrons to non-binding or unbound states, as seen in the UV-VIS or X-ray region. However, molecules can absorb light in this region. This absorption induces vibrations of the molecules, as shown on the left side in Fig. 6.16 and can be applied for infrared (IR) spectroscopy.

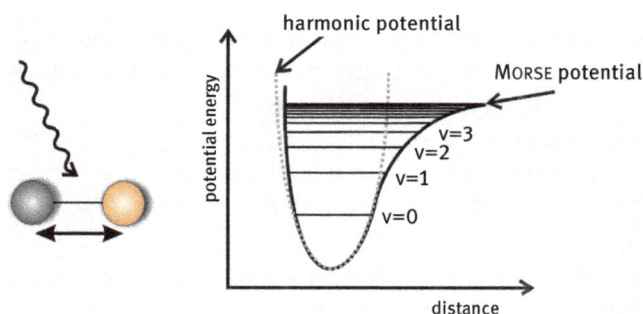

Fig. 6.16: Left: Simplified scheme of infrared (IR) spectroscopy: Electromagnetic radiation in the infrared region can induce vibrations of molecules. Right: Potential curves for the harmonic (gray dots) and anharmonic oscillator. The real system is approached by the Morse potential depicted as a black curve. The energy levels calculated by the Schrödinger equation are shown for the Morse potential. The higher the v-values, the closer the energy levels get to each other until they form a continuum. At a certain energy value the molecule dissociates. In IR spectroscopy the energy transitions between the different vibrational levels are measured.

The energy levels of the different vibrational states are quantized: each vibration that can be induced has its own discrete energy. In the simplest approach the

---

24 For the classification of electromagnetic radiation cf. Chapter 3 in Vol. I.

energy levels of a vibrating diatomic molecule can be described by the harmonic oscillator.[25]

It uses the analogy of a vibrating diatomic molecule to the vibration of two masses linked by a spring, dependent on the masses of the atoms in the respective molecule and the bond strength. The energy levels are described by the following equation:

$$E_{vib} = \left(v + \frac{1}{2} \right) \hbar \omega, \tag{6.9}$$

with ω being the angular frequency and v the quantum number of the vibration levels. In this simple approach the potential energy curve would have a parabola shape (as depicted in Fig. 6.16 in gray) and equidistant energy levels. The special selection rule for vibrational transitions is $\Delta v \pm 1$. This means that only transitions from and into neighboring energy levels are allowed.

The real system however, is better described by the MORSE potential,[26] depicted in Fig. 6.16 in black. In contrast to the harmonic approach, it allows the description of the dissociation of molecules when they have absorbed a certain threshold value of energy. Here also other transitions with $\Delta v \pm 1, 2, 3, \ldots$ are allowed.

Neither do all molecules show absorption bands in the infrared region, nor does every possible vibration. There is a general selection rule[27] demanding that the dipole moment has to change upon the vibration movement. Consequently homonuclear diatomic molecules (like H_2) do not show absorption bands in the IR region, as their dipole moment does not change upon vibration.

A variation of the dipole moment is, for example, given for asymmetric stretching modes of linear triatomic molecules, like the actinyl ions AnO_2^{2+}, shown in Fig. 6.17. The respective vibration frequency of the free uranyl ion ($961 \, cm^{-1}$) can be altered by coordination of hydroxide moieties to the uranium metal center upon hydrolysis. This coordination leads to a variation in the electronic density of the uranium atom, and a subsequent weakening of the U=O bonds.

An example for the changing IR transitions due to the formation of dimeric and trimeric hydrolysis species is shown in Fig. 6.17. Here, the different hydrolysis species can be assigned by their characteristic vibration frequencies. The abundance of the respective species in solution can then be determined by peak deconvolution.

---

25 The harmonic oscillator was introduced in Section 1.5.2 in Vol. I.

26 The MORSE potential describes the electronic potential of a diatomic molecule depending on the distance between the two nuclei.

27 A vibrating dipole emits electromagnetic waves. Looking at it the other way round, the induction of a vibration upon absorption of an electromagnetic wave has to induce a dipole moment during the vibration. Whenever light interacts with matter, a dipole moment is induced. It is described by $\vec{\mu}_{ind} = \alpha E_0 \cos 2\pi v_0 t$, where E_0 is the amplitude of the electromagnetic field, v_0 is the frequency of the field, and α is the polarizability, which describes how effective an electromagnetic field induces a dipole moment.

(a)

(b)

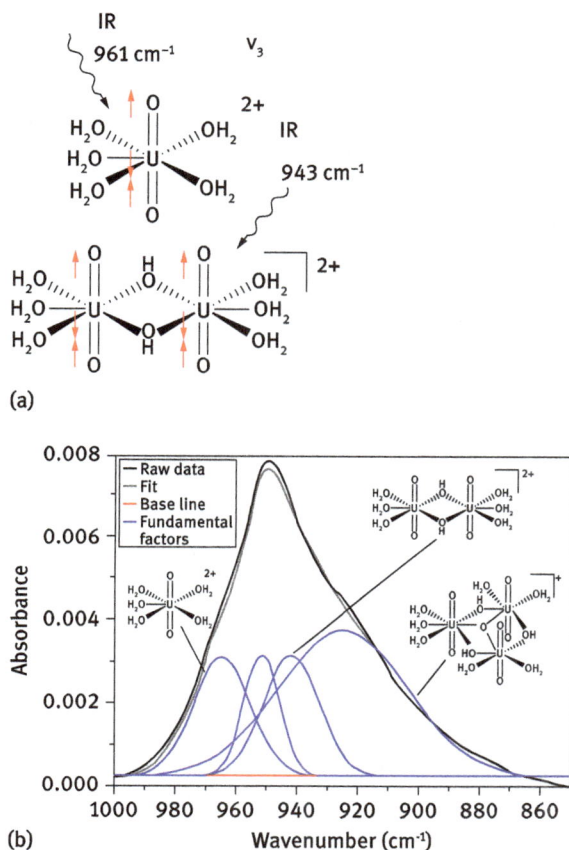

Fig. 6.17: Changing IR transitions due to the formation of dimeric and trimeric hydrolysis uranium species. Left: The v_3 asymmetric stretching mode of UO_2^{2+}-moieties is IR active and causes absorption bands in the infrared region. For uranyl aquo ions (shown above) this absorption band has its maximum at 961 cm^{-1}. The coordination of strong OH$^-$ ligands to the uranyl cation varies the bond strength of the U=O bonds, causing shifts in the absorption bands, as shown for the case of the $UO_2(OH)_2^{2+}$ species on the lower part. Right: IR spectrum of a U^{VI} solution containing different uranium species. Due to the different absorption bands of the respective species the integral absorption plotted in black can be deconvoluted and the abundances of the three different species $UO_2(OH)^+$, $(UO_2)_2(OH)_2^{2+}$, and $(UO_2)_3(OH)_5^+$ can be obtained.

6.9 Raman spectroscopy

In addition to IR spectroscopy, there is another type of vibrational spectroscopy (though not based on *absorption* of electromagnetic radiation, but on *scattering*), called RAMAN spectroscopy. If a beam of monochromatic light passes through a sample, some of the photons collide with the molecules. A number of the photons are scattered elastically. These keep their initial energy. The corresponding process is called RAYLEIGH scattering.

A small fraction of the photons change their direction and either lose part of their energy (so-called STOKES scattering), or gain energy from the molecules (anti-STOKES scattering; for both see Fig. 6.18). The RAMAN process is described as an excitation of the molecules to virtual states.[28] This excitation occurs from the ground state in the case for STOKES scattering, shown in the right part of Fig. 6.18 with the red arrows. From the virtual excited state the molecule relaxes into a vibrational excited mode of the electronic ground state. Due to the energy loss of the photon in the scattering process, vibrations are induced in the molecule.

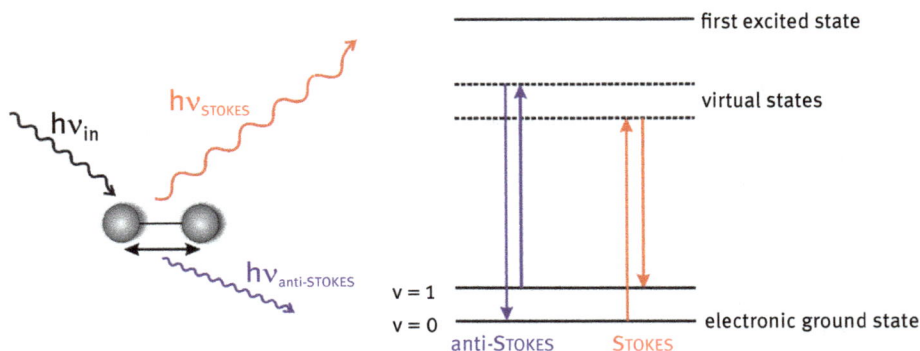

Fig. 6.18: Schematic sketch of the Raman effect. Left: Incident photons can be scattered on the molecules. If they scatter inelastically the photons can either lose energy in this process (Stokes scattering, red) or gain energy (anti-Stokes scattering, blue). Right: Raman scattering is described as an excitation of the molecule into virtual states and a relaxation into different vibration states. The virtual states are below the first excited electronic state. Transitions into and out of these states cannot be observed separately; they only occur in scattering processes.

The anti-STOKES scattering occurs in molecules that are already in an excited vibrational state. It is depicted on the right side of Fig. 6.18 with the blue arrows. Excitation into the (electronic) virtual state is followed by relaxation into the ground state: the photon gains energy from the molecule. Only one photon out of 10^8 shows RAMAN scattering. Thus for the detection of RAMAN photons, the analyzer is placed perpendicularly to the beam of monochromatic light in order to reduce the background of

---

28 "Virtual" states: When a photon whose energy is too low to excite the electrons in a molecule from the electronic ground state to the first excited state is absorbed, the molecule can end up in a so-called "virtual" state which does not correspond to a real energy level of the molecule. The RAMAN scattering process consists of a dipole transition from the electronic ground state into these virtual states and from there back to the ground state. As neither of these two transitions can be observed separately, these states are called virtual states.

the elastically scattered light, which is strongly enhanced in a forward and backward direction. By analyzing STOKES and anti-STOKES scattering, these energy levels reflect vibrational states of the molecules. The STOKES and anti-STOKES signals are measured as shifts in wavenumbers from the RAYLEIGH peak.

But when is a vibration RAMAN active and which transitions are allowed? The key property is the polarizability α of the molecule. It was briefly introduced in Section 6.8 on infrared spectroscopy:

$$\vec{\mu}_{ind} = \alpha E. \tag{6.10}$$

It is a value for the ability of an electric field E to induce a dipole moment in a molecule (or atom). Polarizabilities in molecules might change with changing positions of the atoms during vibrations (or rotations):

$$\alpha = \alpha_0 + \left(\frac{\partial\alpha}{\partial r}\right)\Delta r, \tag{6.11}$$

with Δr being the distance from the molecule's equilibrium geometry. When we write the change in the distance between the atoms (during the vibration) Δr as a cosine based function[29] in terms of the frequency of the vibration v and time t and express the inducing electric field E also in terms of frequency and time,[30] we obtain an expression that describes the scattering processes of light at a molecule:

$$\mu_{ind} = \alpha_0 + E_{max}\cos(2\pi v_{in}t)$$
$$+ \frac{E_{max}r_{max}}{2}\left(\frac{\partial\alpha}{\partial r}\right)(\cos(2\pi r(v_{in}+v)) + \cos(2\pi r(v_{in}-v))). \tag{6.12}$$

v_{in} is the frequency of the incoming light. The first term describes the RAYLEIGH scattering: the scattered photon has the same frequency, as the incoming one. The other two terms describe the shifts in the frequency of the STOKES scattering ($v_{in}-v$) and the anti-STOKES scattering ($v_{in}+v$). The second term is multiplied with the derivative that describes the change in the polarizability with the change in bond-length. From eq. (6.12) we can see the selection rule for RAMAN scattering: the polarizability α has to change during the vibration, otherwise the derivative becomes zero and the second term becomes zero and there will be no RAMAN scattering.

IR and RAMAN spectroscopy complement one another. Some vibrations that do not show IR activity are accessible by this light-scattering technique. An example for RAMAN active vibrations in radionuclide species are the v_1 stretching modes[31] of actinyl-ions some actinides form in the oxidation states +V and +VI. This is shown for the case of Pu^{VI} in Fig. 6.19 for the PuO_2^{2+}-ion.

---

29 As $\Delta r = r_{max}\cos(2\pi vt)$.
30 As $E = E_{max}\cos(2\pi v_{in}t)$.
31 "v_1 Stretching" is the symmetric stretching vibration of the actinyl ion AnO_2^{2+}. It is shown for the case of PuO_2^{2+} in the upper spectrum of Fig. 6.20.

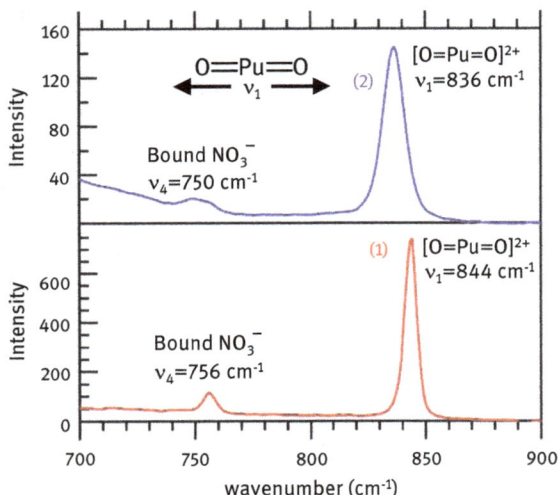

Fig. 6.19: Raman spectra of $^{242}PuO_2(NO_3)_2$ in the solid state (lower part, red) and in solution (upper spectrum, blue). The symmetric stretching mode of the plutonyl ion (O =Pu =O) changes the polarizability of the molecule. Thus this vibration mode shows Raman absorption bands at 836 and 844 cm^{-1}, respectively. The second absorption band in the spectra results from a vibration of coordinated nitrate anions. This band is also present in the upper spectrum of the solution species, showing that nitrate anions are also coordinated in the solution species. Figure reprinted and adapted with permission from Gaunt et al. (2011).

6.10 Nuclear magnetic resonance (NMR)

Another versatile tool providing insight into the configuration of radionuclide solution species, being capable also of delivering dynamic information on molecules, is nuclear magnetic resonance (NMR) spectroscopy. A nucleus that has an overall spin I ($I \neq 0$) also has a magnetic moment. If such a spin is brought into an external magnetic field, it can obtain $2I + 1$ possible orientations, which differ in the magnetic quantum number[32] m_I. For a spin 1/2 nucleus this is shown in Fig. 6.20.

A nucleus with the spin I in a magnetic field B can only have the following energy values:

$$E_{ml} = -g_I \mu_N m_I B. \tag{6.13}$$

---

[32] The magnetic quantum number was introduced for electrons in Section 1.4.1 in Vol. I. Here m_l is used for a nucleus with a spin I. It describes the projection of the angular momentum of the nucleus on one axis of the coordination system.

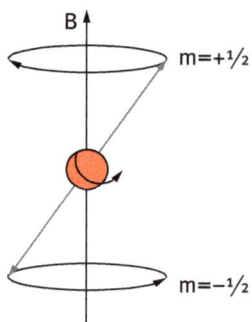

Fig. 6.20: A nucleus with a nuclear spin I can occupy $2I + 1$ orientations in a magnetic field. For a spin 1/2 nucleus, such as the proton, there are two different orientations $m = +1/2$ and $m = -1/2$. The respective spin axes show a precession around the magnetic field axis with a specific frequency ω_0. It is dependent on the magnetic field applied to the system.

g_I is the nucleon g-factor,[33] μ_N is the nuclear magnetron.[34] The allowed values for m_I are also given in eq. (6.13). There is an energy gap between the two possible orientations given by:

$$\Delta E = g_I \mu_N B. \tag{6.14}$$

The varying energies of the two orientations with a varying magnetic field B are shown schematically in Fig. 6.21. Slightly more nuclei of the system are present in the state where the spin is aligned with the magnetic field. Thus in the whole system, there is a small net magnetization along the axis of the B-field.

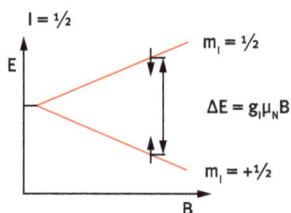

Fig. 6.21: Nuclear spin levels of a spin 1/2 in a magnetic field. With increasing magnetic field strength B, the energy gap between the energetically favored $m_I = +1/2$ and the $m_I = -1/2$ state increases. Radio frequency pulses can induce transitions from favored to nonfavored spin states when the energy of the photons has the same value as the energy gap. It corresponds to the value of the precession frequency around the axis of the magnetic field ω_0.

---

33 The nucleon g-factor characterizes the magnetic moment and the gyromagnetic ratio of a nucleus. It is a dimensionless quantity.
34 The nuclear magnetron μ_N is a physical constant that describes the magnetic moment. It is defined by: $\mu_N = e\hbar/(2m_p)$, with m_p being the mass of the proton and e the elementary charge.

Radio frequency pulses induce transitions from favored to nonfavored[35] spin states. More spins are aligned with the magnetic field. As the population of the two energy states is nearly equal for weak magnetic fields, very strong magnetic fields are applied. Not every atom of the same element shows the resonances at the same energies. The resonance is influenced by the local magnetic field at the nucleus. It does not have the exact same value as the field applied to the system with the magnet. The applied field induces an orbital angular momentum in the electrons surrounding the nucleus, which causes a small magnetic field δB. It is proportional to the applied magnetic field. The induced field δB causes a shielding of the nucleus, where the nucleus seems to "see" a smaller magnetic field. The induced field can be expressed as $\delta B = -\sigma B$, where σ is the shielding constant. The value of this shielding is dependent on the electron density around the nucleus. Thus protons bound in different chemical groups have different shielding constants. The local magnetic fields influence the resonance condition to:

$$h\nu = g_I\mu_N B_{local} = (1-\sigma)B. \qquad (6.15)$$

Chemically equivalent nuclei show the same resonance frequency, the so-called "chemical shift" σB, describing the differences in magnetic shielding with respect to a standard compound. This chemical shift is plotted against the absorption intensities in the spectra. Changes in the electron density on the probed nucleus will lead to differently pronounced shielding of the nucleus. Additionally the presence of paramagnetic atoms[36] leads to a shift of the resonances.

In modern NMR techniques the resonance frequencies of the different nuclei is no longer measured by scanning the applied radiofrequency field or the magnetic field strength. Very short, high intensity pulses of radiofrequency radiation are applied to the nuclei in the magnetic field. This short radiofrequency pulse excites all frequencies within a spectral width. It can be compared to hitting a bell with a hammer: all frequencies of the bell will start to ring due to this action and will decay over time. The radiofrequency pulse is introduced into the system along the x-axis, as shown in Fig. 6.22. This forces the spin showing a precession around the z-axis into the x-y-plane, as shown in the middle of Fig. 6.22. If the system is undisturbed after the pulse, it will show a relaxation of the spins back into the ground state. This relaxation into the ground states is monitored in NMR spectroscopy as so-called "free induction decay" (FID), monitored along the x-axis of Fig. 6.22. It contains all resonance frequencies of the system. A subsequent

---

35 "Favored" or "nonfavored" nuclear states can be explained by Fig. 6.21. With increasing magnetic field B the $m_I = +1/2$ state has lower energies. It is energetically favored with respect to the $m_I = 1/2$ state whose energy increases with increasing magnetic field strength.

36 Paramagnetic atoms have unpaired electrons in their electron shell. When these are brought into a magnetic field, they are attracted by it.

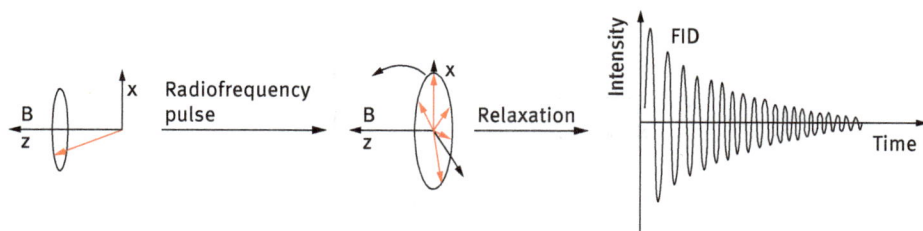

Fig. 6.22: In modern Fourier transformation NMR systems very intense directed radiofrequency pulses with a specific spectral width are used to excite all spins of the system at once. After the pulse, these relax into the ground state. As every nucleus in a different chemical environment has different local magnetic fields, the precession frequencies ω_0 of these differ during the relaxation process. The relaxation of all the different frequencies is monitored as the free induction decay (FID) over time shown on the right side. This FID contains the whole spectral information. Subsequent Fourier transformation of the signal transforms the signal from the time domain to the frequency domain. Afterwards it shows the resonance frequencies of all excited nuclei in the system. An example for this is shown in Fig. 6.23.

FOURIER transformation[37] of the signal leads to the NMR spectrum containing information about the resonance frequencies of the excited spins.

The integral of the peaks is proportional to the number of chemically equivalent nuclei. Figure 6.23 shows this for the so-called n-propyl-BTP extraction ligand, designed for the selective separation of trivalent actinides from trivalent lanthanides. Magnetically equivalent nuclei, like the protons in position 3 and 5, give rise to one signal in the spectrum. The protons, bound in different chemical environments, result in multiplets shifted by different characteristic ppm values from the standard compound tetramethylsilane (TMS) at 0 ppm.

The spins of the nuclei are not independent from each other: neighboring atoms show an arrangement of their spins due to magnetic interactions. This spin-spin coupling is mediated via covalent bonds by the electrons, which are also spin 1/2 particles.

It leads to fine structures in the spectra and can furthermore be applied for correlation spectroscopy where nuclei linked by bonds can be correlated to each other, as performed in Fig. 6.24 for ^{15}N and ^{1}H nuclei. Thus by NMR spectroscopy one can obtain structural information of molecules, containing NMR active nuclei, in solution.

NMR spectroscopy can be applied for a wide variety of questions: Structural characterization of organic molecules is the most broadly applied one. Furthermore, monitoring the magnetic relaxation of the excited nuclei can follow dynamic processes reaction kinetics.

---

37 The FOURIER transformation was introduced in Section 6.8 of this chapter. Here it is applied in order to transform the signal from the time domain into the frequency domain.

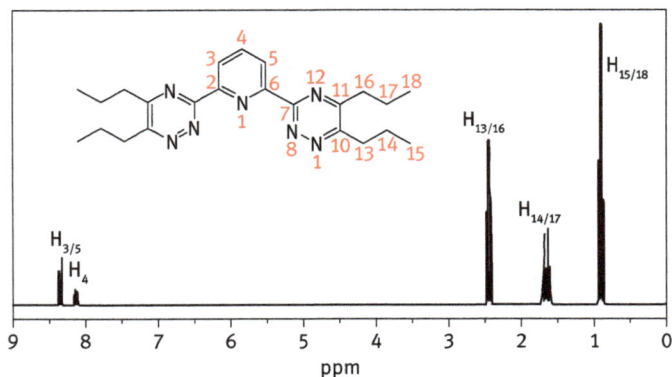

Fig. 6.23: Proton-NMR spectrum of an alkylated bis-triazinyl-pyridine (BTP). After Fourier transformation the resonance frequencies relative to an internal standard are obtained. These are plotted as a chemical shift with respect to the internal standard in ppm. The integrals of the respective peaks contain information about the relative abundances of the nuclei with the respective resonance frequency in the measured compound. Five groups of chemically equivalent protons are in the BTP molecule. Electron mediated spin-spin interaction between the different protons via covalent bonds lead to the splitting of the signals into multiplets (Adam et al. 2013).

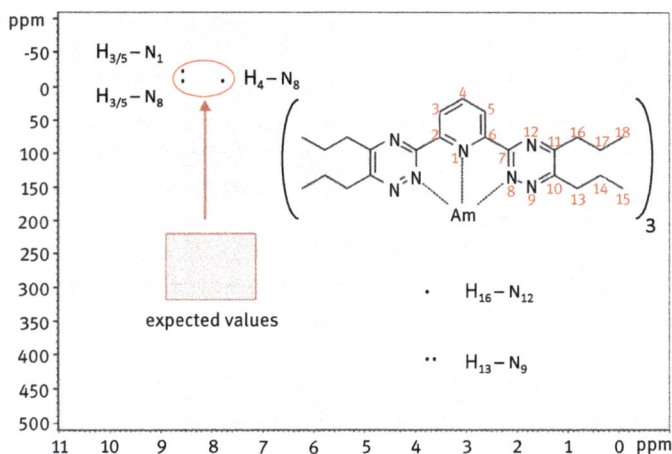

Fig. 6.24: ^{15}N-^{1}H correlation NMR spectrum of an Am-BTP complex. The x-axis shows the chemical shift of the protons (^{1}H) and the y-axis the chemical shift of the nitrogen nuclei (^{15}N). The overall spins of different nuclei can be correlated to each other in correlation NMR-experiments. The spins of these are coupled by the binding electrons when the ^{15}N-^{1}H are relatively close to each other. These experiments deliver structural information about which proton is bound to which nitrogen in the compound. The spectrum of an Am(BTP)$_3$ complex furthermore shows that due to the coordination of BTP to the Am cation, the electron density of the metal ion is transferred to the coordinated nitrogen atom in the central pyridine rings. As the Am^{3+} metal ion is paramagnetic, this causes characteristic peak shifts in the NMR spectra. Reproduced and adapted from Adam et al. (2013) with permission of the Royal Society of Chemistry.

As mentioned above, not only the differences in electron density around the excited nucleus, causing different local magnetic fields, can cause shifts in the resonance frequencies. Also the presence of an atom with unpaired electrons in its shell, thus being paramagnetic, causes characteristic shifts in the resonance frequencies of the nuclei close to the paramagnetic center. An example for the application of NMR spectroscopy to radionuclides is the measurement of complexes of f-elements with organic and inorganic ligands. The f-elements show exceptionally large paramagnetic properties, as they might have several (up to seven) unpaired electrons in their f-shell. These paramagnetic properties influence the chemical shifts of the ligands coordinated, causing characteristic shifts in the NMR spectra.

Their so-called uncoupled spin density[38] can be transferred via covalent bonds or through space due to dipolar interaction. This means that nuclei directly coordinated to f-elements show a distinct shift of their resonance positions in the spectra.

The following example focuses on the aforementioned n-propyl-BTP ligand shown in Fig. 6.23. As mentioned above, it can be applied to partition the trivalent actinides (Am^{3+} and Cm^{3+}, being 5-f-elements) from the also mostly trivalent lanthanides (being 4-f-elements) during reprocessing steps of spent nuclear fuel (for the *Partitioning and Transmutation* concept, cf. Chapter 10). During liquid-liquid-extraction[39] steps in the reprocessing of spent fuel n-propyl-BTP ligand show a selectivity to extract Am^{3+} and Cm^{3+} some orders of magnitude better than the chemically similar trivalent lanthanides. In order to understand this selectivity better, NMR measurements of the complexes n-propyl-BTP forms with Am^{3+} and different trivalent lanthanides were performed. In Fig. 6.24 a correlation spectrum of an Am-BTP-complex is shown.

The paramagnetic shift due to the uncoupled electrons of Am^{3+} is clearly visible in the correlation spectrum. The ligand binds to the americium ion with the nitrogen atoms number 1 and 8, as shown in the upper right part of the figure. These atoms, directly coordinated to the actinide ion, show an exceptional paramagnetic shift. The values for these nitrogen resonances in the respective lanthanide complexes are located in the red field in Fig. 6.24. These results indicate a different behavior of the trivalent actinides compared to the chemically very similar trivalent lanthanides. The Am^{3+}-ion seems to form stronger bonds to the extraction ligand. The orbitals of the Am^{3+} seem to have a larger overlap with the ones of the nitrogen atoms of the ligand than the ones of the lanthanide-ions show. Consequently, a larger paramagnetic shift for the coordinating nitrogen atoms is detected with the NMR technique.

As mentioned before organic ligands present in nature can form complexes with actinide ions and change their behavior in the environment. Citric acid is ubiquitous in nature and present in all living organisms. Complex formation with uranium was

---

38 Spin density is the difference in electron density of electrons with different spin alignments.
39 See Chapter 7, Section 7.4.

studied by Kretzschmar et al. (2021). Temperature dependent NMR studies revealed, that the dimeric and trimeric complex species form in aqueous solution. The ^{17}O-NMR results give insight to the structures of these species. Species distribution calculations showed, that the mobility of U(VI) in the environment is not enhanced by these complexes, being an important information for the assessment of the environmental fate of uranium.

6.11 Fluorescence spectroscopy

When the atoms of a sample are excited by photons, light of equal or lower energy can be emitted. This is called fluorescence or phosphorescence depending on the dipole selection rules of the transition, determined by the angular momenta[40] of initial and final state. Figure 6.25 shows the absorption and emission processes for the case of fluorescence.

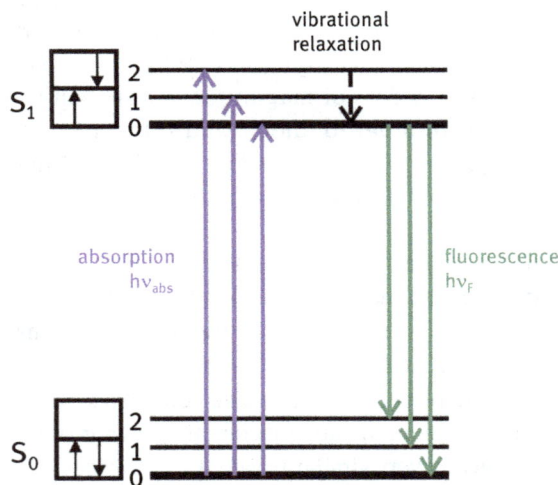

Fig. 6.25: Schematic view on fluorescence processes: Electrons in atoms can be excited from the electronic ground state s_0 to the first electronic excited state s_1, as shown on the left side. After vibrational relaxation to the vibrational ground state of the s_1 state (black arrow) the excitation energy can be dissipated by the emission of light. This effect is called fluorescence when the spin of the excited electron is kept during the whole process, as shown in the boxes to the left. Due to the vibrational relaxation step, the fluorescence light (hv_F, green arrows) shows a shift to lower energies compared to the stimulating light (hv_{Abs}, blue arrows) or the same energy value.

---

40 The angular momentum of an atom is a good quantum number. The difference between the angular momenta of the initial and final atomic state must be carried by the photon.

The experimental setup is similar to the one of absorption spectroscopy and hence included in Fig. 6.26. The emitted light is usually analyzed by a spectrometer. This method is suited to probe the chemical environment of an atom, hence doing speciation at the nanoscale.

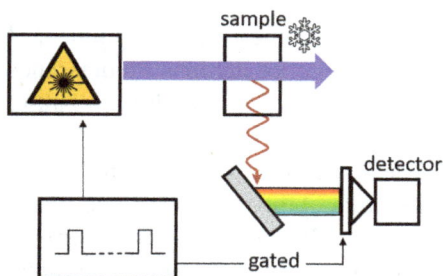

Fig. 6.26: Schematic view of a TRLFS experiment. The probed element in the sample is excited by laser-light of a suitable wavelength. As TRLFS is a fluorescence technique, the fluorescence signal of the sample is detected perpendicular to the incident laser beam. The signal is then analyzed by a spectrograph and recorded by a camera. A trigger system regulates both the laser and the camera. The sample can be cooled by a cryostat in order to obtain very sharp fluorescence lines.

As mentioned in the chapter on absorption spectroscopy, the energetic levels of an atom's electrons are influenced by neighboring atoms or ions, an effect called ligand field shift. Consequently, the energy of fluorescence photons is also shifted, since it reflects the difference between excited state and final (e.g., ground) state provided the levels are shifted by different amounts. In the case of actinides, complexation frequently leads to a shift to lower energy (longer wavelength) called bathochromic shift. It is very pronounced for trivalent Cm^{III} (shown in Fig. 6.27 on the left), hexavalent U^{VI}, tetravalent U^{IV} less so for Eu^{III} and Am^{III}. Pa^{IV} is an exception because complexation causes a shift to shorter wavelengths.

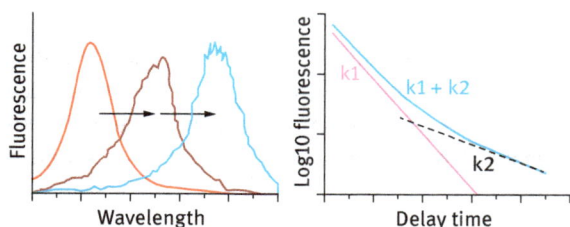

Fig. 6.27: Left: Spectral shift indicates complexation of the fluorescing ion. Right: a single species shows monoexponential transformation, a mixture of species with different numbers of quenchers show biexponential transformation.

Additional information is obtained from the lifetime of the emission.[41] If there is only one species present in the system, the decrease in the fluorescence intensity is described by the first order relation:

$$(t) = N_0 e^{t/\tau}, \tag{6.16}$$

where $N(t)$ is the number of atoms in the excited state at time t, N_0 is the number of atoms in the excited state at $t = 0$, and τ is the fluorescence lifetime. Fluorescence lifetime describes the time it takes for the fluorescence intensity to drop to half of its initial value.

The fluorescence rate (inverse lifetime) of the undisturbed ion (e.g., Cm^{3+} in the gas phase) is determined by level spacing dipole order of the transition (JUDD–OFELT theory).[42] Close contact of the luminescing ion Cm^{3+} to ligands can cause an energy transfer from the photo-excited metal ion to the ligand via a competing nonradiative de-excitation process. The most prominent example is the so-called fluorescence quenching by water or hydroxide ions. The electronic energy of the excited states of Cm^{3+} and Eu^{3+} equals five times the energy of the O–H stretching vibration. Energy transfer from the metal ion excites the fifth overtone[43] of the water molecule, vibration frequency being a very efficient energy transfer process. This is shown in Fig. 6.28. The more water molecules there are in direct neighborhood of the metal center (coordination shell), the more efficient the process gets. If there is only one species present in solution, it shows a monoexponential fluorescence decay. If there are two species, with different amounts of quenchers surrounding them, the fluorescence decay shows a biexponential behavior, as shown in Fig. 6.27 on the right side.

To good approximation the rate of this energy transfer correlates linearly with the number of O–H oscillators in the first coordination shell of the ion and hence the apparent fluorescence rate (k_F), which is the sum of inherent rate (k_I) and quenching rate (k_Q), increases

$$k_F = k_Q + k_I, \tag{6.17}$$

with $k_I \approx n(H_2O)$, n is the number of quenching ligand molecules, for instance water. For Eu^{3+} this relation is called HORROCK's equation for Cm^{3+}: KIMURA equation. Vice versa, one can deduce the number of water or OH ligands from the observed fluorescence decay rate, which is measured by use of a pulsed light source and a

---

41 In this context emission means de-excitation by emission of photons.
42 The JUDD–OFELT theory describes the intensities of lanthanide and actinide transitions in solids and solutions, providing a theoretical expression for the line strength.
43 Fifth overtone of the water molecule frequency is five times the energy of the lowest excitation mode.

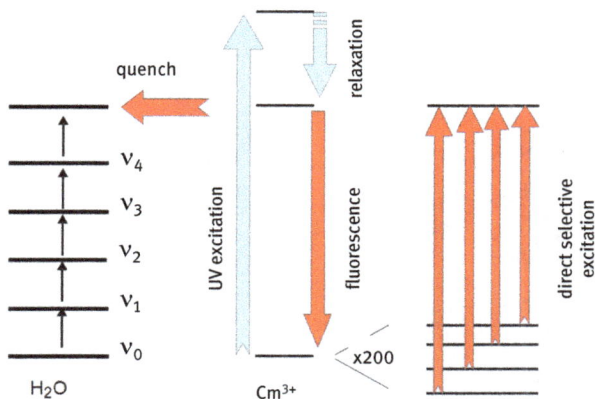

Fig. 6.28: TRLFS: Level scheme of excitation, fluorescence and quenching for Cm^{3+}. The energy levels of the fifth overtone of the O–H- vibrations of water molecules lie in a comparable energy range to the excited state, from which the Cm^{3+} ion would normally relax under emission of fluorescence light. This is why a part of the excitation energy will be transferred to the water molecules surrounding the metal center. Thus the fluorescence lifetime decreases with increasing number of water molecules surrounding the Cm^{3+} ion.

detector which measures the light after a variable time span (so-called *time-resolved laser-induced fluorescence spectroscopy*, TRLFS).

This is particularly useful when surface sorption or incorporation is investigated in aqueous environment, as shown in Fig. 6.29. Due to sterical reasons each of these processes requires at least partial loss of the coordinated water shell causing a decrease in quenching rate and an increase in observed fluorescence lifetime.

Huittinen et al. investigated the complexation of Cm(III) with aqueous phosphates. Inorganic phosphates are widely present in the environment due to the natural decomposition of phosphate-containing rocks or minerals and to anthropogenic sources such as fertilizers and phosphate-based detergents. When Cm(III) aquo ions are complexed by phosphate species, water molecules in the first coordination shell are replaced by phosphate species, leading to a change in the fluorescence lifetimes. Furthermore, the complexation has effects on the fluorescence emission wavelength and intensity. Taking into account these parameters, Huittinen et al. established the presence of $Cm(H_2PO_4)^{2+}$ and $Cm(H_2PO_4)_2{}^+$ species in solution.

Additional information is obtained when the fluorescing states are nondegenerated.[44] A strongly asymmetric ligand field reflecting the geometry of the binding site causes a splitting of these levels and in turn a splitting of the fluorescence emission peaks. These splittings[45] of 2 to 30 cm^{-1} are resolved by high-resolution spectrometers.

---

44 Nondegenerate means the levels have different energies.
45 Inverse centimeters are used as a measure of energy (inverse wavelength: $E \approx 1/\lambda$).

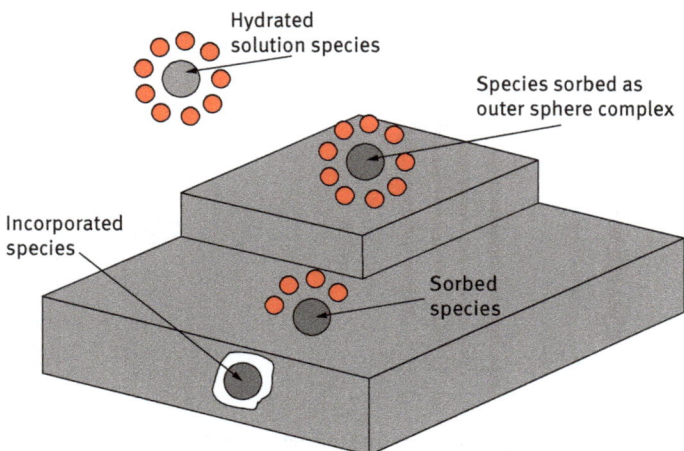

Fig. 6.29: The interaction of solution species with solid material can lead to the sorption on the surface or even the incorporation of the solution species in the solid material structure. The solution species (represented in gray), has a water shell that surrounds it (the water molecules are represented in orange). If the species now sorbs to the surface, it loses a part of its water shell. If the species gets incorporated, it loses the whole water shell. This loss of the water shell (of Cm^{3+} for instance) can be followed by TRLFS measurements. Thus from the spectra, one can deduce which mechanisms play a role when the solution species is in contact with a solid phase.

Often experiments are performed at low temperature to have narrow peaks. When several compounds with different geometry around the Cm^{3+} ion are present simultaneously, each shows a unique peak shift, making it possible to selectively excite a single site by narrow band laser light. In combination with the time-resolved measurements, a speciation of single incorporation sites in, for instance a crystal lattice with regard to geometry and number of water adducts, is possible down to trace concentration of as little as ppt (or 10^{10} atoms in absolute number).

The fact that TRLFS can be applied to solid and liquid samples is used to gain information on coordination numbers. The radius of actinide ions (as is the case for lanthanide ions) decreases with increasing atomic number (so-called actinide contraction). As a consequence, the number of water molecules in the first coordination sphere of Ln^{3+}_{aq} decreases from 9 in the case of the early light actinides to 8 for the later ones. The transition point is close to Cm, which has a half-filled f-shell. At room temperature, the distribution of 9 coordinated to 8 coordinated Cm^{3+} ions is 9 : 1. The TRLFS signal (lifetime, shift and spectral splitting) is calibrated with hydrated solids of well-defined coordination number, where $Cm^{3+}(H_2O)_n$ is part of, e.g., Ln- or Y-based crystals, as shown for example in Fig. 6.30. Increasing the temperature in the liquid phase stepwise reveals a decrease in the coordination number ratio 9-fold: 8-fold = 4 : 6 at 200 °C. This effect was also shown in Huittinen et al. (2021) on aqueous Cm-phosphate species in the temperature range from 25 °C to 90 °C.

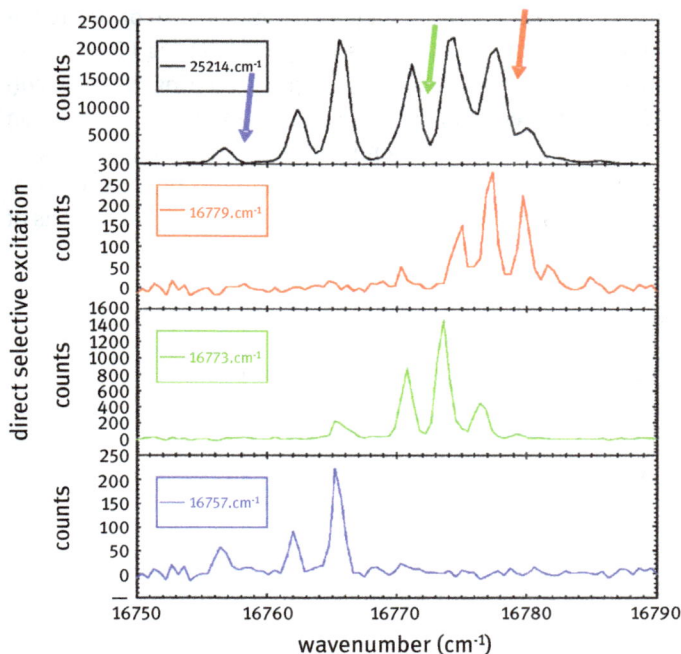

Fig. 6.30: TRLFS: Example of three different species identified in a Cm doped YBr3 compound. Figure reprinted and adapted with permission from Lindqvist-Reis et al. (2006). © 2012 American Chemical Society.

6.12 Colloid detection

Nanoparticles in liquids, so-called colloids,[46] are omnipresent in natural waters as well as in chemical solutions and also play an important role in the chemistry of radionuclides. Colloid size distribution spans a range from some nanometers to micrometers and natural distributions often follow PARETO's law,[47] i.e., the number concentration N_{conc} of colloids (differential number of colloids per unit volume and unit diameter) scales with radius r as:

$$N_{conc} = Ar^{-\beta}; \quad 3 < \beta < 5 \tag{6.18}$$

---

46 "Colloids" by definition are nanoparticles in aqueous systems which do not settle in solution due to their small size and/or density close to 1 g/cm^3.
47 Size distribution of colloids in natural waters.

as shown in Fig. 6.31. If these colloids are rather large (micrometer size) or present at very high concentration, they are detected by means of light scattering (TYNDALL effect[48]). Very sensitive instruments are able to detect the light scattering of single colloids larger than about 50 nm. Smaller colloids are measured by photon correlation spectroscopy, which basically measures the interference of light before and after scattering by the particle, an effect due to the shift of the scattered light. The latter method ranges down to particles as small as 2 nm but requires ppm concentrations of colloids.

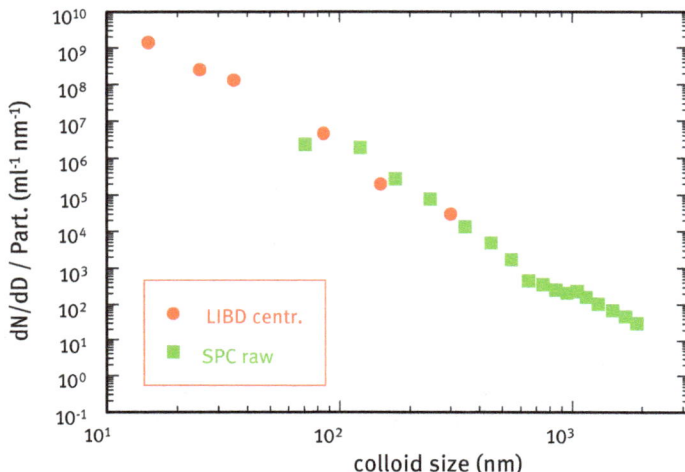

Fig. 6.31: Colloid size distribution as typical for natural waters following Pareto's law; eq. (6.18).

The most relevant colloids, however, are those below 50 nm, and characterized by their high mobility and a large surface-to-volume ratio. In actinide chemistry a detection method combining ppt sensitivity with a lower size limit of about 5 nm was developed, "Laser-Induced Breakdown Detection" (LIBD). This technique utilizes plasma formation when a colloid moves into the high electric field of a focused beam of a pulsed laser (some 10^{10} W/cm^2). The bright light emission of the plasma and the pronounced sound following its rapid thermal expansion are easily detected, allowing single colloid counting. The experimental setup is depicted in Fig. 6.32.

By varying the laser energy, size distributions of colloids between 5 nm and 1000 nm of mass concentrations down to 10^{-9} g dm^{-3} can be measured, often without sample pre-treatment, both in the laboratory and in experiments in the field. Although the method is "noninvasive", the detected colloid itself is destroyed in the plasma of the microprobe because only a marginal fraction of the total colloid

---

48 The TYNDALL effect refers to light scattering in solutions containing colloids.

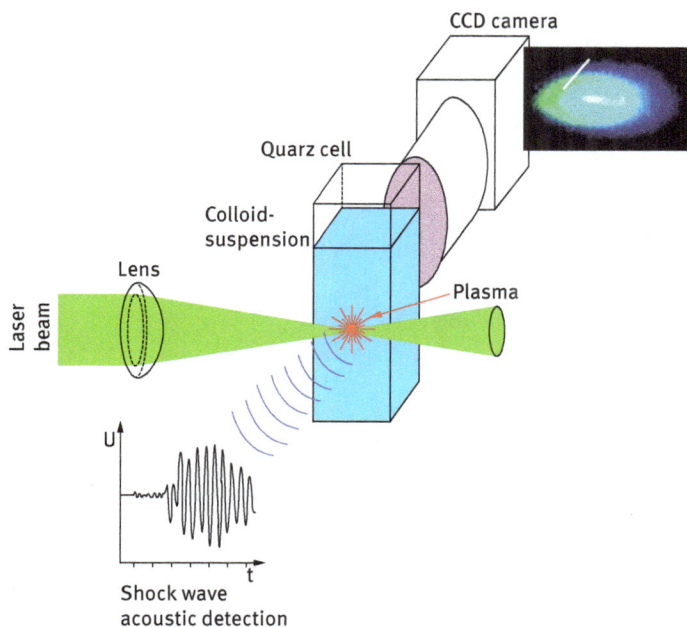

Fig. 6.32: Schematic illustration of the LIBD process. A pulsed laser initiates plasma on single colloids that is detected either optically or via emission of an acoustic shock wave.[49]

number is used for characterization, i.e., the sample is virtually unaffected. This allows repeated measurements of a single sample, for instance when studying colloid stability as a function of time (colloids tend to aggregate when their surface charge is neutralized).

Applications in actinide sciences are twofold: First, naturally occurring colloids effectively act as a carrier for radionuclides: Due to their small size colloids remain suspended in water for considerable time. Especially highly charged ions such as Pu^{IV} or Am^{III} sorb to the colloids' surface and migrate wherever the colloids are going, as shown in the upper part of Fig. 6.33. This so-called pseudo-colloid transport needs to be considered for radionuclides, which would otherwise be insoluble.

The relevance of colloid-mediated transport of radionuclides became obvious after reports on unexpected fast migration of Pu close to former nuclear weapons test sites in Nevada, USA, and in the vicinity of a reprocessing plant close to Mayak, Russian Federation. In experiments during the last decade, radionuclide transport has been investigated in fractured granite rock, e.g., in the Swedish Äspö Laboratory.

The second application of LIBD is measuring the solubilities of actinides. Especially tetravalent and to some extent hexavalent actinides have a high tendency for

---

49 An acoustic shock wave is a (super)sonic density fluctuation (similar to a MACH cone).

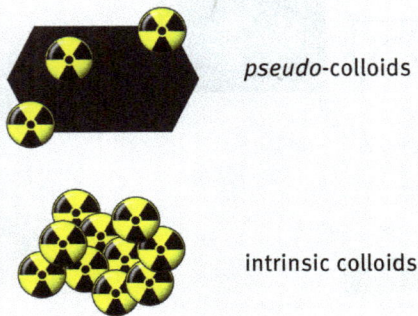

Fig. 6.33: Many radionuclides form intrinsic colloids, i.e., colloids consisting only of radionuclides typically bridged by OH or O groups, which are mobile in aqueous solution. In addition natural colloids act as efficient carriers for radionuclides, so called pseudo-colloids.

hydrolysis. The formation of polynuclear complexes continues by the formation of intrinsic colloids, as shown in the lower part of Fig. 6.33. In a classical sense the solubility is determined via the concentration of atomic/ionic or molecular species found in the liquid phase after the solid is brought into contact with the liquid. Vice versa, when the concentration of solvated ions is increased in the solution, at some point precipitation will occur. LIBD allows detecting very small (nonsettling) particles of precipitates – namely the colloids. Solubility data of Pu^{IV} (see Fig. 6.34), U^{IV}, and Th^{IV} were obtained by this method.

6.13 Field flow fractionation (FFF)

A drawback of LIBD is its restriction to probes of rather high density, particularly exceeding that of water. This is well fulfilled for inorganic colloids, such as intrinsic actinide colloids, alumina, iron oxides, and clays, but is not the case for organic compounds, e.g., humic substances. However, these organic compounds can complex actinide ions very effectively and cause enhanced mobility. Single humic acid molecules have mass distributions centered around 500 amu but depending on chemical conditions they form aggregates up to tens of nanometers in size. Humic acids readily complex metal ions in solution. This leads to charge compensation and formation of very mobile species – even if the uncomplexed ion might have been strongly sorbing or of low solubility.

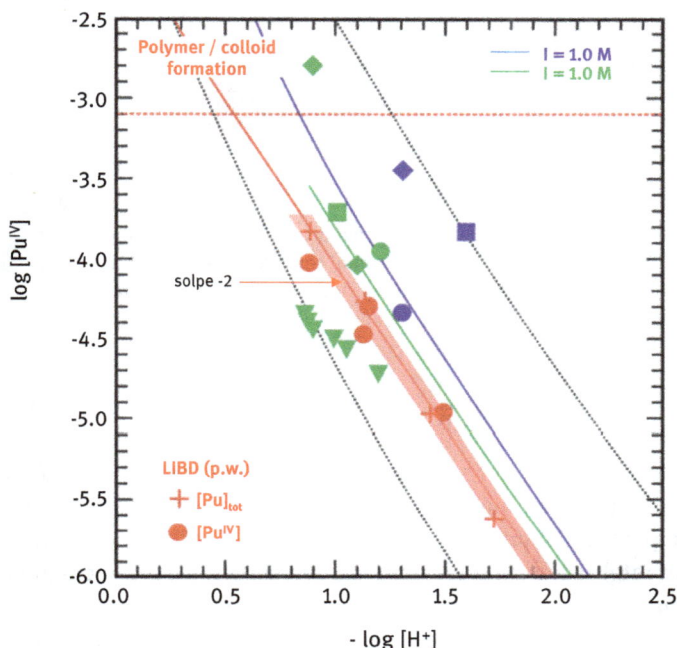

Fig. 6.34: Pu^{IV} solubility is measured by LIBD by probing the formation of colloids.

A size fractionation method covering this full size range is the field flow fractionation (FFF). The sample (diluted by a carrier liquid where required) is pumped through a channel of some 100 μm thickness. For low REYNOLDS number[50] a laminar Newtonian flow[51] is established which exhibits a parabolic flow profile as in a capillary tube, i.e., the velocity distribution varies from zero at the wall to its maximum value in the center of the channel, Fig. 6.35.

A force field perpendicular to the flow direction acts as a net force on the particles towards the opposite wall and causes an accumulation of particles in a cloud of a few μm thickness. The perpendicular force can be established by various means such as temperature gradients, fluid crossflow, electric field or gravitational/centrifugal force. Due to FICK's law this concentration enhancement causes a back diffusion towards the center of the capillary (similar to osmosis), which is proportional to the diffusion coefficient of the particle: The diffusion counteracts the directed flow. After a short equilibration time an exponential concentration distribution is established,

---

50 The REYNOLDS number is a measure of turbulence in a flow tube or channel.
51 Newtonian flow is inelastic and laminar.

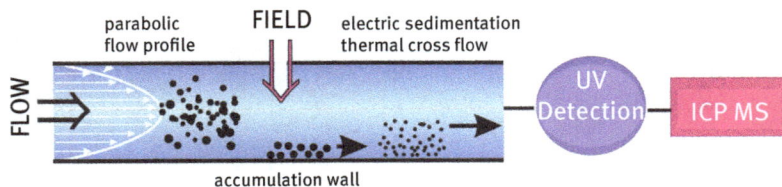

Fig. 6.35: Schematic representation of FFF. A force acting perpendicular to the flow of the carrier liquid counteracts the isotropic diffusion. Large particles remain close to the wall, whereas the distribution of small ones extends into the center. The higher flow velocity at the center causes small particles to be eluted first.

$$J = D\frac{d\Phi}{dx}, \qquad (6.19)$$

where J is the "diffusion flux" amount of substance per unit area per unit time, D is the diffusion coefficient, Φ (for ideal mixtures) is the concentration in dimensions of amount of substance per unit volume, x is the position. The distribution of small particles (high diffusion rate) reaches far into the high velocity flow of the channel whereas larger ones remain close to the wall and experience only a reduced flow field. Consequently, the small particles travel faster and are eluted first out of the channel – a fractionation has taken place.

An example for the application of the FFF method coupled to an ICP-MS and an UV-VIS detector is shown in Fig. 6.36. Here the competing reactions of the metal ions of europium and thorium with mixtures of bentonite[52] colloids and humic acid of small size fraction were studied. The combination of FFF with UV-VIS (for the detection of the humic acid) and ICP-MS for the detection of the clay mineral, Th and Eu, revealed that the Eu^{3+} ions are bound to the humic acid whereas the Th^{4+} ions show sorption on the bentonite particles. The bentonite particles were detected via the aluminum content in the clay particles.

Detailed discussions on achievable retention factor, sensitivity and resolution as well as technical details of channel layout and fluid velocities are given in Cölfen and Antonietti (2000). The effluent is subsequently characterized by sensitive colloid detection methods such as fluorescence or absorption techniques, or coupled to element analysis (ICP-MS, GFAAS,[53] . . .).

---

52 Bentonite is a mixture of different clay minerals.
53 Graphite furnace atomic absorption spectrometry.

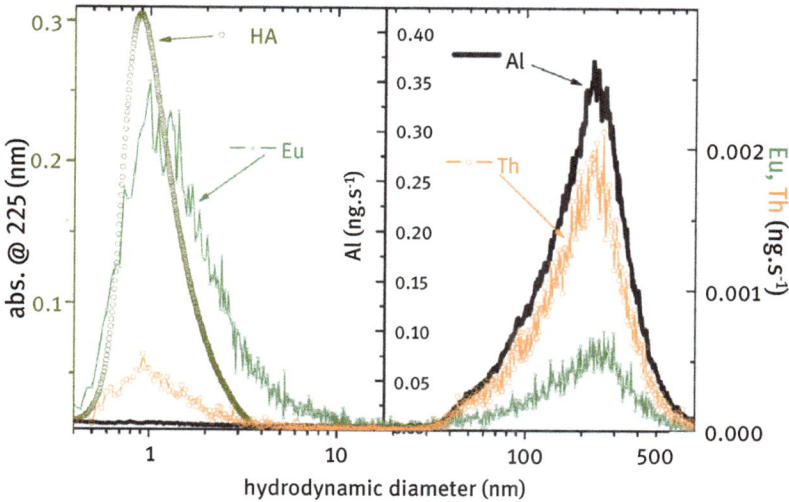

Fig. 6.36: Application of FFF: Complexation of metal ions by humic acid (small size fraction) and bentonite (clay) colloids (larger size fraction, detected via the aluminum Al content) is observed by coupling FFF with UV detection and ICP-MS. (The scope of this work was the quantification of competing reactions of the metal ions with humic acids and colloids).

6.14 X-ray photoelectron spectroscopy (XPS)

The liquid-solid interface can have a strong influence on the behavior of solution species. This interaction is analyzed with a surface sensitive technique, the X-ray photoelectron spectroscopy (XPS, see Fig. 6.37). The chemical composition within the first nanometers of a solid compound, i.e., the first surface layer, can be probed by making use of the photoelectric effect, more precisely by measuring the energy of the core-level electrons emitted due to the absorption of X-ray radiation. Equation (6.20) shows that if the energy of the incident X-ray beam is known, by measuring the kinetic energy of the emitted electron (E_{kin}), one can determine its binding energy (E_{bind}).

$$E_{kin} = h\nu - E_{bind} - \phi_s. \tag{6.20}$$

Here $h\nu$ is the energy of the incident X-ray and Φ_s is the work function of the sample.[54]

---

[54] The work function Φ_s is the minimum energy needed to remove an electron from a solid to a point in the vacuum outside the solid surface. The work function is a property of the surface of the material.

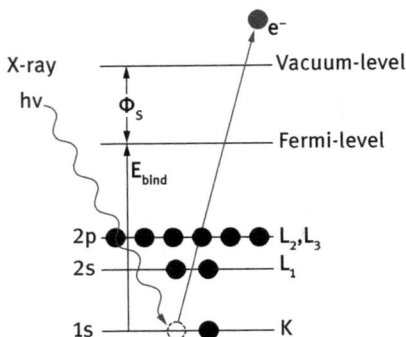

Fig. 6.37: Principle of X-ray photoelectron spectroscopy (XPS): Incident X-ray radiation of sufficient energy can excite core-level electrons in matter. Here it is shown for an electron of the K-shell. If the energy is sufficiently high, the excited electron will leave the matter. Its kinetic energy depends on the binding energy in the core level (E_{bind}), which is element and oxidation state specific, the energy of the incident radiation $h\nu$ and the work function of the sample Φ_s.

As the emitted electrons are emitted from only the upper layers of solid material, this technique is particularly surface sensitive. The energy needed to emit the photoelectron is element specific and depends on the element's electron shell structure.[55] Thus the elemental composition and the respective oxidation states of a sample can be probed by XPS.

The photoelectrons emitted from p-, d-, or f-electron levels show splitting due to the spin-orbit coupling[56] into doublets: $p_{1/2}$, $p_{3/2}$, $d_{3/2}$, $d_{5/2}$, $f_{5/2}$, and $f_{7/2}$ bands form in the XPS spectra. Changes in the oxidation states or electronic density in the atom probed cause chemical shifts that can be used to quantify the different oxidation states present. Moreover, the formation of a core hole[57] and following relaxation of the valence electrons can leave the formed ion in a specific excited state a few eV above the ground state. This leads to a reduction of the kinetic energy of the photoelectron by this excitation energy, causing satellite peaks, the so-called "shake-up" peaks in the spectra shifted to higher energies up to an additional 15 eV.

An example for the application of XPS is depicted in Fig. 6.38. A plutonium solution was brought into contact with colloidal humic substances. The initial oxidation state of the plutonium was trivalent. The XPS spectrum shows that with increasing

---

55 Cf. Chapter 9 in Vol. I for the correlation of X-ray energies and photoelectron energies in the context of secondary transformation processes.

56 Cf. Section 1.6.2 in Vol. I for the spin-orbit coupling and the fine structure of p, d, and f-levels.

57 After the photoelectron leaves one of the inner electron shells, the atom is in an excited state. There is a "hole" in the shell the electron was emitted from. To relax from this excited state, electrons from the higher shells will move to the lower shell. Their additional energy can be dissipated by two different mechanisms: the emission of X-rays (X-ray fluorescence) or the emission of electrons (so-called AUGER electrons).

time, this plutonium is oxidized to Pu^{IV}, due to the contact with the colloidal material. This can be followed by the chemical shift in Fig. 6.38. The right part of Fig. 6.38 also shows the spectrum of the solution after a long time span. The different expected peak positions of Pu in the respective oxidation states caused by the chemical shift are marked in the spectrum. Besides the $4f_{7/2}$ main peak, a "shake-up" satellite peak is observed. This can be assigned to the formation of polymeric Pu^{IV} species upon oxidation by the colloidal humic substances.

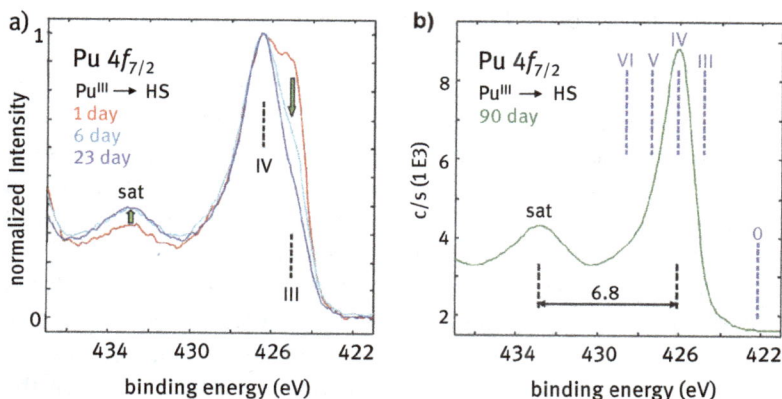

Fig. 6.38: XPS spectra: a) XPS of a colloidal humic substance which was brought into contact with a Pu^{III} solution. The binding energy of the $4f_{7/2}$ electrons is plotted against the intensity. The humic substance oxidizes the Pu^{III} to Pu^{IV}. This leads to a change in the peak position and a formation of a satellite peak caused by electron emission from an excited state of the plutonium. b) XPS of the suspension after longer contact times of Pu with the colloidal humic substance. Plutonium is totally converted to its tetravalent oxidation state. The peak positions of the other oxidation states are marked (Schild et al. 2006).

6.15 Outlook

Taking a close look at current scientific papers on the speciation of radionuclides in solution it becomes clear that very often only a combination of different suitable techniques can help to understand a system comprehensively. To this end many of the techniques presented in this chapter are combined by different research groups working on the very often challenging task of understanding the solution behavior of radionuclides.

Of course the development of the techniques introduced does not stand still. There is always an interest in making the existing methods more sensitive or to applying new methods to the investigated solution systems. Sensitivity is definitely an issue when one is interested in the speciation of radionuclides in environmental samples where the radionuclides are usually present in very low concentrations. Some of the current developments are presented here.

For instance the LPAS method was recently developed to reach limits of detection for Pu^{IV} of less than 10^{-8} mol/l corresponding to about 2.4 microgram per liter of Pu. Even though capillary electrophoresis coupled to ICP-MS experienced its breakthrough for ultratrace speciation (for metal concentrations of less than 10^{-10} mol/l), the limit of detection is expected to be lowered by two more orders of magnitude in the near future. The force field fractionation technique (FFF) with various transversal flow fields is coupled to detectors tailored to each respective measured compound in order to enhance sensitivity.

The electrospray ionization mass spectrometry (ESI; especially nano-ESI) finds increasing application for speciation of inorganic compounds. Coupling of ESI sources to even more advanced mass spectrometric techniques leads to higher sensitivity and higher accessible masses, which can improve the strengths of the method even more. Moreover, by the application of ion trap techniques successive induced fragmentation reactions can give information on the stability of the species.

Another approach is to implement spatial resolution to some of the introduced techniques. This is true for the X-ray absorption and fluorescence techniques like EXAFS and XANES that become increasingly more sensitive and also can obtain high spatial resolution of some tens of nm. This is important for studies on the water-solid interface where sorption and incorporation processes take place. Furthermore, temporal resolution for the X-ray absorption spectroscopy methods is also being developed, allowing to study process dynamics on the picosecond scale. For chemical speciation on submicrometer levels the coupling of the time-resolved laser-fluorescence spectroscopy (TRLFS) method with near field microscopes is promising.

Another approach for the development of speciation techniques is the implementation of the possibility to make measurements in the field. This has been reached for the quantification and determination of colloidal species present in solution via the laser-induced breakdown detection (LIBD) technique: the method has recently become commercially available in a small, robust package for experiments in the field under harsh conditions, for instance for the detection of colloid migration in water conducting shears in granite studied in rock laboratories.

With the consistent further development of existing methods, more and more complicated solution systems and lower concentration ranges can be studied. The processes that influence the oxidation states of the radionuclides and their chemical behavior, like complexation with organic and inorganic ligands, have an influence on the mobility of radionuclides and their sorption behavior. Only if all these processes are understood on a molecular level, can sound predictions for the behavior of radionuclides in the geosphere and biosphere be made.

References

Adam C, Kaden P, Beele BB, Mullich U, Trumm S, Geist A, Panak PJ, Denecke MA. Evidence for covalence in a N-donor complex of americium(III). Dalton Trans 2013, 42, 14068–14074.

Antonio MR, Williams CW, Sullivan JA, Skanthakumar S, Hu YJ, Soderholm L. Preparation, stability, and structural characterization of plutonium(VII) in alkaline aqueous solution. Inorg Chem 2012, 51, 5274–5281.

Banik NL, Büchner S, Fuss M, Geyer F, Lagos M, Marquardt CM, Rothe J, Sasaki T, Steppert M, Walther C. Mass spectromety methods. KIT-SR 7600, Karlsruhe, Karlsruhe Institute of Technology, 2010, 74–77.

Banik NL, Büchner S, Burakham R, Fuss M, Geyer F, Geckeis H, Lagos M, Marquardt CM, Walther C, Kuczewski B, Aupiais J, Baglan N, Topin S, Maslennikov AG. Mass spectromety methods. KIT-SR 7559, Karlsruhe, Karlsruhe Institute of Technology, 2009, 81–84.

Cölfen H, Antonietti M. Field-flow fractionation techniques for polymer and colloid analysis. Adv Polym Sci 2000, 150, 67–187.

Gaona X, Tits J, Dardenne K, Liu X, Rothe J, Denecke MA, Wieland E, Altmaier M. Spectroscopic investigations of Np(V/VI) redox speciation in hyperalkaline TMA-(OH, Cl) solutions. Radiochimica Acta 2012, 100, 759–770.

Gerber E, Romanchuk AY, Pidchenko I, Amidani L, Rossberg A, Hennig C, Vaughan GBM, Trigub A, Egorova T, Bauters S, Plakhova T, Hunault MOJY, Weiss S, Butorin SM, Scheinost AC, Kalmykov SN, Kvashnina KO. The missing pieces of the PuO2 nanoparticle puzzle. Nanoscale 2020, 12(35), 18039–18048.

Gaunt A, May I, Neu MP, Reilly SD, Scott BL. Structural and spectroscopic characterization of plutonyl(VI) nitrate under acidic conditions. Inorg Chem 2011, 50, 4244–4246.

Hoyer E, Knöppel J, Liebmann M, Steppert M, Raiwa M, Herczynski O, Hanspach E, Zehner S, Göttfert M, Tsushima S, Fahmy K, Oertel J. Calcium binding to a disordered domain of a type III-secreted protein from a coral pathogen promotes secondary structure formation and catalytic activity. Sci Rep 2019, 9, 7115.

Huittinen N, Jessat I, Réal F, Vallet V, Starke S, Eibl M, Jordan N. Revisiting the complexation of Cm (III) with aquoeous phosphates: What can we learn from the complex structures using luminescence spectroscopy and ab initio simulations?. Inorg Chem 2021, 60, 10656–10673.

Kretzschmar J, Tsushima S, Lucks C, Jäckel E, Meyer R, Steudtner R, Müller K, Rossberg A, Schmeide K, Brendler V. Dimeric and trimeric uranyl(VI)–citrate complexes in aqueous solution. Inorg Chem 2021, 60, 7998–8010.

Lindqvist-Reis P, Walther C, Klenze R, Eichhöfer A, Fanghänel T. Large ground-state and excited-state crystal field splitting of 8-fold-coordinate Cm^{3+} in $[Y(H_2O)_8]Cl_3 \cdot 15$-crown-5. J Phys Chem B 2006, 110, 5279–5285.

Lösch H, Raiwa M, Jordan N, Steppert M, Steudtner R, Stumpf T, Huittinen N. Temperature-dependent luminescence spectroscopic and mass spectrometric investigations of U(VI) complexation with aqueous silicates in the acidic pH-range. Environ Int 2020, 136, 105425.

Maiwald MM, Dardenne K, Rothe J, Skerencak-Frech A, Panak PP. Thermodynamics and structure of neptunium(V) complexes with formate. Spectroscopic and theoretical study. Inorg Chem 2020, 59, 6067–6077.

Schild D, Marquardt CM, Seibert A, Fanghänel T. Plutonium-humate complexation in natural groundwater by XPS. In: Alvares R, Bryan ND, May I. eds. Recent advances in actinide science. The Royal Society of Chemistry, 2006, 107–109. https://pubs.rsc.org/en/content/ebook/978-0-85404-678-2

Taube F, Drobot B, Rossberg A, Foerstendorf H, Acker M, Patzschke M, Trumm M, Taut S, Stumpf T. Thermodynamic and Structural Studies on the Ln(III)/An(III) Malate Complexation. Inorg Chem 2019, 58, 368–381.

Norbert Trautmann, Steffen Happel and Frank Roesch

7 Radiochemical separations

Aim: As in chemical separation in general, radiochemical separation separates the radionuclide from another stable or unstable element or from groups of other elements so that maximum yield and purity are obtained in a time that is short compared to the half-life of the radioisotope. One should be more precise concerning the radionuclide: the separation is about a defined radioisotope of a given element present as a defined chemical species.

There is a broad spectrum of individual radiochemical separation methods. Most of them, such as precipitation, liquid-liquid extraction, ion exchange, volatilization, or electrochemical methods are techniques common to conventional chemistry. Nevertheless, if applied to the separation of radionuclides, the radiochemical version of these methods needs special modifications. Short half-lives of radionuclides ask for extremely fast separation procedures, low concentrations of radionuclides ask for sophisticated separation concepts.

In addition to modified but still common methods, there is a separation technique absolutely unique to radiochemistry. It is based on the "hot atom" chemistry of recoil nuclei as generated in nuclear reactions (SZILARD–CHALMERS reactions) and on the "post effect" of transforming unstable nuclei. These effects allow the separation of a radioisotope from a stable isotope of the same chemical element.

In this chapter the emphasis is put on characteristic features, rather than repeating the basic principles of well-established separation procedures. It covers both the preparative and the analytical separations. Characteristic examples are presented, highlighting the unique options of radiochemical separations. These examples include some prominent applications for each procedure, used among others for the identification of isotopes of new elements, for the separation of radionuclides produced for medical applications and for techniques to design medically relevant radionuclide generators.

7.1 Introduction

7.1.1 Historical aspects

Since the discovery of radioactivity by H. BECQUEREL in 1896, chemical separation procedures have been applied for the identification and investigation of radioactive isotopes. For decades, chemical methods provided one of the most important tools for unravelling mixtures of radioactive species and assigning atomic numbers. Among others, the use of chemical separations has led to major discoveries in nuclear chemistry: the

https://doi.org/10.1515/9783110742701-007

identification of the first radioactive elements in nature, polonium and radium (P. CURIE and M. CURIE, 1898); the discovery of artificial radioactivity by F. JOLIOT and I. CURIE (1934); the identification of the first synthetic element, technetium (PERRIER and SEGRÈ, 1937), the discovery of nuclear fission by HAHN and STRASSMANN (1939); and discovery of the first transuranium element, neptunium (McMILLAN and ABELSON, 1940).

A radiochemical separation defines the process to isolate the radionuclide (a defined radioisotope of a given element present as a defined chemical species) from those of groups of other elements so that maximum yield and purity are obtained in a time that is short compared to the half-life of the radioisotope. With respect to the unique features of radiochemistry (short half-life and low amount of the radionuclide, yet still emitting radiation which requires protection), radiochemical separations are special. In many cases, these addresses both a sophisticated separation chemistry and an adequate separation technology.

7.1.2 Basic considerations

This chapter will address the most important radiochemical procedures.[1] There is a huge variety in individual radiochemical separation procedures originating from conventional chemical separations – both inorganic and organic. Early radiochemistry, i.e., the era of the discovery of radioactivity until the middle of the twentieth century, was mainly "inorganic" and utilized the broad experiences obtained over decades in analytical chemistry. Over the last 50 years, new and more automated methods are being applied in radiochemistry. This is either initiated by the increasing interest in studying short-lived radionuclides and isotopes of the heaviest elements produced at the atom-at-a-time scale and – conversely – in the routine utilization of very high batches of radioactivity, which cannot be handled by traditional manual procedures due to radiation safety concerns.

There are two main principal approaches to radiochemical separations: the *batchwise*, discontinuous and the *continuous* operation. Whereas in the discontinuous procedure a nuclide is produced, chemically separated, and measured sequentially, in the continuous mode a target is permanently irradiated, and the release of the reaction products from the target, their chemical separation, and the counting are performed online.[2] In parallel, there are two main principal intentions: the *analytical* separation and the *preparative* separation. For the analytic separation, it is sufficient to obtain a pure fraction of the desired species for, e.g., nuclear spectroscopy and

---

1 This chapter does not refer to physical separation methods, which are nevertheless relevant for several purposes, such as mass separations of radioactive isotopes or species, or to other methods such as ICPMS, etc. These are introduced in Chapter 6.
2 The latter version is relevant to the production and identification of superheavy elements; see also Chapter 9 for a systematic description.

nuclide identification, the identification of the chemical form of the species, and the quantification of the relative amount of an individual species within a mixture of co-existing ones, etc. A typical example is the research in superheavy element production and identification. Radiochemical separations are also very widely finding use in the analysis of environmental, biological, and waste samples, especially for the determination of natural and anthropogenic alpha and beta emitters. The preparative separations aim at isolating a radioactive element or species to make use of it for a variety of individual purposes: in industry, in medicine, in research, etc. A typical example is the design of radionuclide generators, where the radioactive daughter is isolated from the radioactive parent nuclide and subsequently used for radiopharmaceutical applications in medical units.

As mentioned earlier, most radiochemical separations rely on traditional separations in general chemistry. However, there are some specific aspects for radiochemistry:

1. The chemical behavior of species at ultralow concentration, a typical situation in radiochemistry, does not always follow the characteristics of the same element at macroscopic concentrations. For example, formation of di-atomic species (or multimeric ones) can be excluded in single atom chemistry and thus, the chemical behavior of the radiospecies is different to that one may expect from the "normal" species.

2. Relativistic effects can influence the chemistry of radioelements: The investigation of the chemical properties of the transactinide elements is an example and also a great challenge because strong relativistic effects were predicted for those elements, leading to deviations from the periodicity of the chemical properties as expected by linear extrapolations from the lighter homologues in the Periodic Table of Elements.

3. Contrary to conventional chemical separations, very often very fast separations are required for the isolation of radionuclides due to the short half-life of the isotopes (in the range of seconds to minutes), requiring the development of dedicated separation chemistry and its technical realization. Already in the year 1940 HAHN and STRASSMANN used fast chemical separations for the discovery of radioisotopes with half-lives down to less than one minute. Since then, experimental techniques employing computer-controlled automated separation systems have been developed for studies on short-lived isotopes. They are a typical application for radiochemical separations of transactinide elements. An overview on fast radiochemical separations and their application to nuclear research is presented in review articles: Herrmann and Trautmann (1982), Rengan and Meyer (1993), Kratz (2011), and Gäggeler (2011).

4. The demands on radiochemical separations depend on the particular case and are also influenced by instrumental developments like high-resolution γ- and X-ray spectroscopy enabling *simultaneous* detection of many unstable nuclei.

7.1.3 Procedures

In the following sections various radiochemical separation procedures are outlined organized according to the principal chemical steps and not according to selected elements:[3]
- precipitation and co-precipitation
- liquid-liquid extraction
- solid phase-based ion exchange
- thin layer chromatography/high pressure liquid chromatography
- electrochemical procedures (electrodeposition, electrophoresis, electromigration)
- dry methods (volatilization, sublimation, evaporation)
- hot atom chemistry/SZILARD–CHALMERS-type separations

7.2 Precipitation and co-precipitation

Precipitation of a compound in an aqueous solution with subsequent filtration of the precipitate is one of the traditional techniques in radiochemical analysis. Precipitation occurs when the ionic product is greater than the solubility product of the compounds under consideration. The purity of a precipitate is not only influenced by the kinetics of formation and the size of the insoluble particles formed but also by factors like adsorption, formation of mixed crystals, and trapping processes known as co-precipitation. In general, contamination of a precipitate can happen even when the product of the contaminant does not exceed its solubility product. Whereas often the absorbed contaminants can be removed by proper washing of the precipitate, this is not the case for (a) the formation of mixed crystals as a consequence of chemically similar electrolytes, (b) the so-called occlusion resulting from trapping of adsorbed impurities inside the precipitate during its growth, and (c) mechanical entrapment. The various factors affecting the formation of a precipitate are described in detail by Walton (1967).

One milestone in the application of precipitation procedures was the discovery of fission in December 1938. HAHN and STRASSMANN (1939) used fractional crystallization of barium salts like bromide and chromate to investigate the behavior of the artificial "radium" isotopes obtained by neutron-induced fission of uranium. It was well known since the days of M. CURIE that radium concentrates in the head fractions. In the indicator experiment on December 17, 1938, they performed the fractionation of 86 min "radium III" in the presence of natural 5.8 a ^{228}Ra with BaBr$_2$. From the transformation curves of the various fractions one can see the enrichment

---

of natural radium in the order fraction I → II → III, as expected, whereas the 86 min nuclide is distributed equally in the three fractions, i.e., it must be Ba-139 resulting from a break of uranium during bombardment with neutrons into barium, namely the fission process as interpreted by L. MEITNER und O.R. FRISCH (1939). These precipitation experiments rank among the most careful and unambiguous ever carried out in radiochemistry.

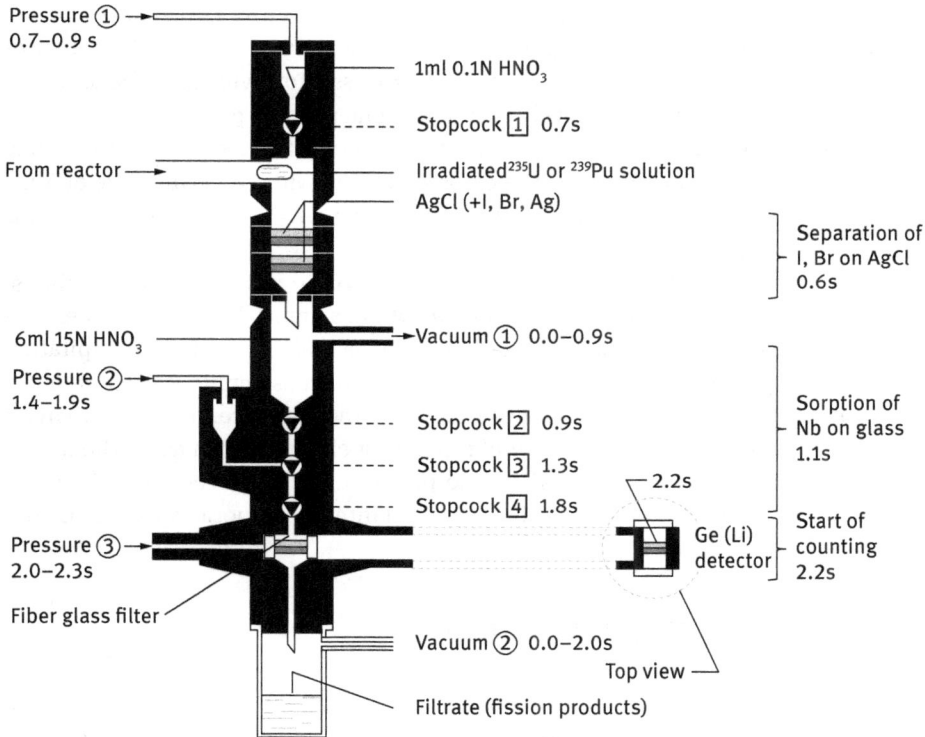

Fig. 7.1: Apparatus and time schedule for the rapid separation of niobium from fission products. The time for each separation step is given on the right-hand side (Ahrens et al. 1976).

For rapid separations precipitation and filtration may occur too slowly. Therefore, another technique has found application, the exchange reaction with a preformed precipitate: the solid is prepared in advance and the solution with the radioactive species to be separated is filtered through a thin layer of the reactive precipitate. Exchange reactions with a preformed precipitate were used for the separation of short-lived iodine nuclides from fission products by means of a programmer-controlled discontinuous procedure. For a rapid separation procedure for niobium from fission products, for example, the procedure consists of five steps: (1) A solution of ^{235}U or ^{239}Pu sealed in a capsule is irradiated in the tube system of the TRIGA Mainz reactor. (2) The capsule is transferred pneumatically into the separation apparatus and smashed by impact on its walls. (3) The solution is filtered through a layer of silver chloride to remove bromine and iodine isotopes. (4) Niobium is separated from 10 N HNO_3 by filtration through fiberglass filters. (5) The filters with the niobium fraction are projected to the detectors. Apparatus and time schedule are presented in Fig. 7.1. The different steps are controlled by an electronic programmer, which opens and closes stopcocks operated pneumatically. Counting is started 2.2 s after the end of irradiation. Niobium isotopes with half-lives down to 0.8 s such as ^{104}Nb were studied in this way.

The contamination of a precipitate by co-precipitation of elements forming isomorphous crystals is normally an undesired process. However, this phenomenon can also be used to carry trace levels of certain elements on a precipitate. For example, insoluble rhenium compounds are used to co-precipitate technetium traces like tetraphenylarsonium perchlorate. The co-precipitation with Fe^{III} hydroxide or Mn^{IV} dioxide is often applied to scavenge unwanted elements from aqueous solutions (Townshend and Jackwerth 1989).

Co-precipitation also frequently finds application in radioanalytical methods, the determination of radionuclides in environmental samples being a prominent example. A separation method may even include several types of co-precipitation mechanisms. Adsorption of selected radioelements such as actinides calcium phosphate or ferrous hydroxide precipitates is used for the preconcentration of analytes from water samples, and the removal of interfering elements such as K. The formation of mixed crystals can be used for the preparation of sources for alpha spectrometry (e.g., co-precipitation of actinides with lanthanide fluorides such as CeF_3).

7.3 Liquid-liquid extraction

Solvent extraction or liquid-liquid extraction is based on the distribution of a chemical species between usually an aqueous solution and a liquid organic phase. However, liquid-liquid extraction can also be performed in nonaqueous systems; e.g., in a system consisting of a molten metal in contact with molten salts the metal can be transferred from one phase to the other. In the following, only aqueous and organic phases will be considered as liquid phases. The extraction method can be applied over a wide concentration range, from rather highly concentrated solutions, such as in nuclear fuel reprocessing, to one-atom-at-a-time separations of the heaviest elements. Depending on the timescale, discontinuous, or continuous techniques are applied.

In solvent extraction very often the distribution ratio is given as a measure for the quality of the separation. This ratio D^{SE} is equal to the concentration of a solute in the organic phase divided by its concentration in the aqueous phase:

$$D^{SE} = c_{\mathrm{org}}/c_{\mathrm{aq}}. \tag{7.1}$$

The separation factor expresses the ability of a system to separate two solutes (A and B) and is defined as:

$$\mathrm{SF}_{AB} = D_A/D_B. \tag{7.2}$$

There are different types of extraction mechanism. An example for a very simple procedure is the extraction of iodine from an aqueous phase into carbon tetrachloride or the extraction of ruthenium tetroxide or osmium tetroxide into an organic solvent. The interaction of a metal ion with an organic anion can form a neutral complex. Often the anion is a bidentate and the mechanism is called chelate extraction. Cupferron is a very important chelating agent. The use of a solvating agent is another way to extract a metal ion into the organic phase. The solvating reagent can be the organic phase itself or it can be dissolved in a diluent. Examples for this type of mechanism are the extractions of uranium and plutonium by tri-n-butyl phosphate. Other important reagents are trioctylphosphine oxide and bi-2-ethylhexyl phosphoric acid. In ion pair extraction an inner sphere complex is formed which is extracted with an organic cation. Here primary, secondary, tertiary amines or a quaternary ammonium salt are used often referred to as liquid ion exchangers due to the similarity between the ion-pair formation and an ion-exchange process. Ionic liquids are another class of extractants that are of interest for the separation of, e.g., uranium, lanthanides, palladium, rhodium, and ruthenium. Bis(trifluoromethanesulfonyl)imide is well suited for the separation of the uranium oxides UO_3, UO_2, and U_3O_8 from each other. Crown ethers with a ring structure that fits an atom with a certain ionic radius forming hydrophobic and extractable species are well suited for the isolation of cesium and strontium.

The batchwise solvent extraction can be done in separation funnels or test tubes whereas for continuous extraction mainly mixer-settlers, columns, centrifuges, or centrifugal extractors are used.

Solvent extractions play an important role in reprocessing spent nuclear fuel with the separation of uranium and plutonium from high active waste (HAW). The plutonium uranium redox extraction (PUREX) process is the most common one for this purpose. In this process the fuel elements are first cut in pieces with a length of 3–5 cm and then dissolved in 6–11 N HNO_3 leaving the cladding hulls undestroyed. The solution is then diluted to 3–4 N HNO_3, nitride is added as a reducing agent to assure that plutonium is present as Pu^{IV} and uranium as U^{IV}. Uranium and plutonium are selectively extracted into tri-n-butyl phosphate (TBP) in aliphatic kerosene, Fig. 7.2. Most of the fission products and the trivalent actinides remain in the aqueous phase. In the next step, plutonium is back-extracted into the aqueous phase by reduction to Pu^{III} using, e.g., Fe^{II} sulfamate. Uranium is not reduced under these conditions and remains in the organic phase. Its back extraction occurs

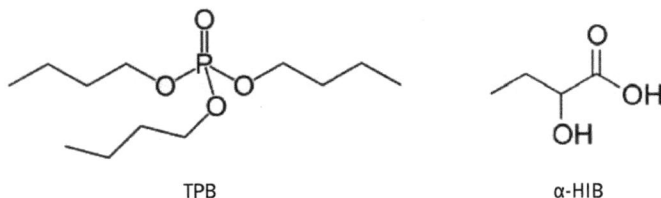

Fig. 7.2: Left: chemical structure of TBP: n-tributyl phosphate (OP(OC$_4$H$_9$)$_3$; right: chemical structure of α-HIB (or 2-hydroxyisobutyric acid or 2-hydroxy-2-methylpropanoic acid), see Section 7.4.

with 0.01 N nitric acid. The plutonium and uranium fractions are further purified by several extraction cycles and finally converted to the oxides.[4]

In liquid-liquid extraction the time-determining step is in general the separation of the two phases. For fast separations, fixing the organic solvent on a fine-grained carrier and filtering quickly the aqueous solution through a layer of such a quasi-solid solvent can accelerate this. The rapid discontinuous separation of technetium from fission products is an illustration of this technique. In a first step the halogens are removed by exchange with silver chloride. Then, technetium in form of the pertechnetate ion is extracted into a solution of tetraphenylarsonium chloride in chloroform adsorbed on a fine-grained carrier. Next, the pertechnetate is re-extracted from this solution with 2 N HNO$_3$ plus NH$_4$ReO$_4$ and co-precipitated with the homologous tetraphenylarsonium perrhenate. After filtration and projection of this precipitate to the detection system the start of counting is 7.5 s after the start of the separation procedure. Technetium isotopes with half-lives down to 5 s [108]Tc could be studied. Similar to the illustration in Fig. 7.1, the different operations are controlled by an electronic programmer delivering signals at predetermined times to open valves or to move stopcocks pneumatically.

The introduction of high-speed centrifuges for rapid phase separation enabled the performance of continuous rapid extraction procedures. With various versions, nuclides with half-lives down to ≈1 s could be separated from complex reaction product mixtures and investigated. Figure 7.3 shows as an example the on-line isolation of technetium from fission products.

A [249]Cf target was irradiated at the Mainz TRIGA reactor with neutrons and the recoil fission products were transported from the target position by a nitrogen/KCl gas jet to the centrifuge separation system (SISAK 3). The KCl clusters with the attached fission products were dissolved in sulfuric acid plus an oxidizing agent in a static mixer. Afterwards nitrogen and the fission noble gases (Kr, Xe) were removed from the liquid in a degassing centrifuge. Then, technetium was extracted into tetraphenylarsonium chloride dissolved in chloroform. After the extraction the organic

---

4 For details relevant to the treatment and recycling of used fuel, see Chapter 10, Section 10.6.5.

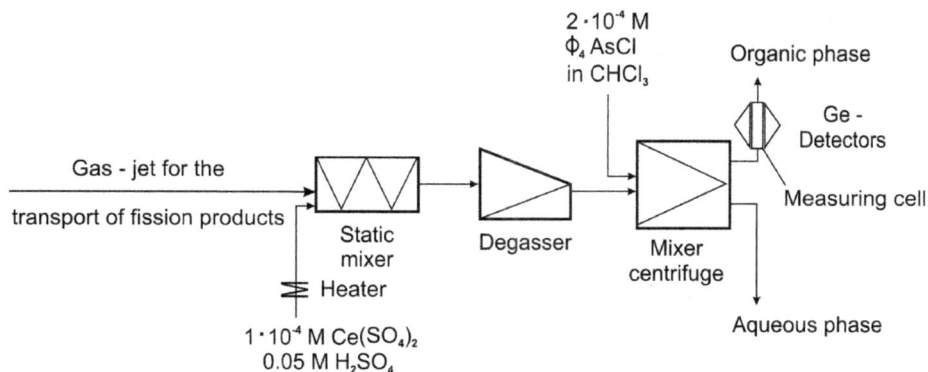

Fig. 7.3: Flow diagram for the continuous separation of technetium from fission products by solvent extraction using the high-speed centrifuge system SISAK (Altzitzoglou et al. 1990).

phase was pumped to a measuring cell placed in front of two Ge detectors. The overall hold-up time from the production (neutron pulse) to the counting unit was 2.5 s. This procedure allowed detailed studies on the transformation of 1.0 s ^{110}Tc and investigation on the β^--transformation of 0.9 s ^{109}Tc.

7.4 Solid phase-based ion exchange

7.4.1 Ion exchanger columns

Ion exchange is a method that allows, among others, separations within groups of very similar elements such as the lanthanides or the trivalent actinides. It is based on small differences in the distribution of the elements between a solid phase consisting of a polymer with reactive groups, the ion exchanger, and a solution in which the elements to be separated are present in form of cations or anions.

For a good separation the distribution step must be repeated very often which can be achieved by chromatography: the mixture is placed on top of a column filled with the solid phase through which the liquid phase is passed. Ion exchangers are either cation exchangers that exchange cations or anion exchangers that exchange anions. The cation exchangers are classified into strong, moderate, and weak acid types while the anion exchangers are typically classified into strong and weak base exchangers. There are also other types of ion exchangers like amphoteric exchangers that can exchange cations or anions simultaneously, chelating resins and bifunctional resins.

An ion exchanger consists of functional groups attached to a polymer matrix. Sulfonic groups bound to cross-linked polystyrene matrices are typical for strong acid type cation exchangers, while moderate cation exchangers contain the group –$PO(OH)_2$ and weak cation exchangers the group –COOH. Strong base anion exchangers have the group –$(CH_2NR_3)Cl$ and weak base anion exchangers contain either –$(CH_2NHR_2)Cl$ or –$(CH_2NH_2R)Cl$ with R representing an alkyl chain.

The exchange reaction with, e.g., a strong cation exchanger can be written as follows:

$$R-SO_3^- H^+ + Me^+L^- \leftrightharpoons R-SO_3^- Me^+ + H^+L^-, \tag{7.3}$$

where Me^+ is a metal ion and L^- an anionic ligand.

For the ion exchanger similar to solvent extraction a distribution ratio can be defined:

$$D^{IE} = c_{res}/c_{aq}, \tag{7.4}$$

where c_{res} is the equilibrium concentration of the exchanged ion in the resin (mol/kg) and c_{aq} the concentration in the aqueous solution (mol/m^3).

Ion exchangers are mostly operated as columns. However, membranes or surfaces covered with ion exchanger material have also found application. Depending on the separation problem, the size of the columns can vary from a few millimeters to several meters and flow rates from a few milliliters per minute to several cubic meters per hour. For ion exchangers used as cartridges cf. Section 7.4.2. In recent years, the speed of complex separations could be improved by using high-performance solid phases and an operation of the column at higher pressure. With this method a separation of the whole series of lanthanide elements can be achieved within about 20 min.

7.4.2 Ion exchange cartridges

The ion exchange separation methods described so far used chromatographic columns to separate two or more species. However, for many of the separation problems the basic intention is to isolate the main (radioactive) species for further analytical purposes (e.g., spectroscopy) or applied purposes (e.g., synthesis of radiopharmaceuticals). In these cases, the most practical version is to transform a chromatographic column into a cartridge. Cartridges (commercially available in versatile versions) are small plastic containers filled with the chromatographic material – the ion exchange resin. Their performance is simple: following conditioning of the cartridge, the solution containing the desired radioactive species passes through the cartridge in one step, while all the other species remain absorbed in the cartridge. The cartridge thus filters off all unwanted material. An alternative version consists in trapping the radioactive species selectively in the cartridge while the contaminants are not absorbed. This protocol, nevertheless, requires subsequent steps: washing the cartridge with a solution to remove traces of unwanted species and finally desorbing the pure radioactive species with an adequate solution.

7.4.3 Examples

Ion exchange techniques have been applied for separations of chemically similar elements, such as trivalent lanthanides and trivalent actinides. Figure 7.4 illustrates the high-resolution separation of trivalent actinides with α-hydroxy isobutyric acid (α-HIB, cf. Fig. 7.2 for chemical structure). The actinides are separated according to their complex formation with α-HIB; those with the highest complex formation constant being eluted first.

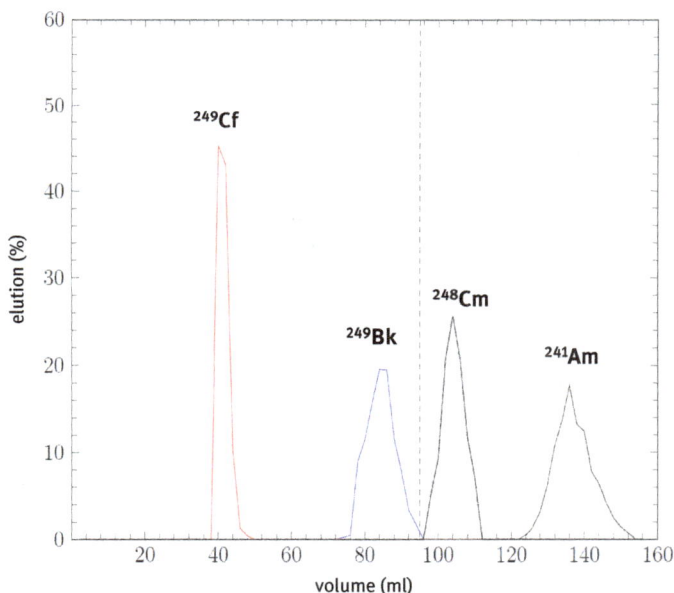

Fig. 7.4: Separation of trivalent actinides on a cation exchange column CK 10 Y (Mitsubishi Corp.) of 150 mm × 4 mm obtained with α-hydroxyisobutyric acid of 0.14 M (0–96 ml) and 0.17 M (97–150 ml) at pH of 4.8.

Further examples for the use of the ion exchange technique are separations of the transactinide elements. One example is the isolation of the first transactinide element, rutherfordium. From the extrapolations of the lighter homologues, zirconium, and hafnium, it was expected that rutherfordium should pass through a cation exchange column in a weakly acid solution of α-HIB acid whereas the heavier trivalent actinides should be strongly absorbed. This was demonstrated by means of α-spectroscopic measurements for the rather long-lived isotope 78 s ^{261}Rf. Already at that time a large part of the chemical system was automated to obtain samples within about a minute. Hundreds of irradiations had to be performed, each followed by a chemical separation in order to measure few events because only ≈10% of the atoms produced were observed after chemistry due to transformation, chemical losses, and counting geometry (Silva et al. 1970).

For aqueous chemistry studies of the heaviest elements mostly automated discontinuous batchwise procedures have been applied. In order to get statistically significant results, the same experiment must be repeated several hundred or thousand times with a predetermined cycle time. Recently two arrangements have mainly been used for such applications, namely the Automated Rapid Chemistry Apparatus (ARCA) and the Automated Ion exchange separation apparatus coupled with the Detection System for Alpha Spectroscopy (AIDA). Both systems are computer-controlled and have been applied for repetitive high-performance ion exchange chromatography in combination with a gas jet for the transportation of the reaction products. With the AIDA apparatus detailed investigations on the fluoride complexation of rutherfordium by cation exchange chromatography and the adsorption of dubnium on an anion exchanger in HF solution in comparison with its homologues niobium and tantalum have been performed. For this 34 s ^{262}Db was produced in the ^{248}Cm(^{19}F,5 n) reaction at the JAEA tandem accelerator in Tokai with a production rate of about 0.4 atoms per min. The reaction products were transported by means of a He/KCl gas jet to AIDA and deposited there for 75 s. Then, the products were dissolved in 13.9 N HF and fed into the anion exchanger column (1.0 mm i.d. × 3.5 mm) operated at a flow rate of 1.2 ml/min. The effluent was collected on tantalum disks and evaporated to dryness with hot helium gas and a halogen heat lamp. The sorbed products on the column were then stripped with a mixture of HNO$_3$ plus diluted HF, collected on tantalum disks and evaporated to dryness. The disks were transferred automatically to an α-spectrometry station consisting of eight passivated ion-implanted planar silicon (PIPS) detectors. The chromatographic separation was accomplished within ≈30 s and the α-particle measurement was started after ≈50 s for the first fraction. All events were registered event by event together with the time information.

Figure 7.5 shows the adsorption probability of Db and the homologous elements niobium and tantalum on an anion-exchange resin as a function of the initial HF concentration. The adsorption of Db from 13.9 N HF is smaller than those of Nb and Ta. Taking all the experiments into account including the behavior of the pseudo-homologue protactinium, the experiments showed that the adsorption of Db on the anion-exchange resin is different from that of Nb and Ta and that the adsorption of anionic fluoro complexes decreases in the sequence of Ta ≈ Nb > Db ≥ Pa. In order to compare the experimental adsorption sequence with that from predictions, the element species of these elements must be determined.

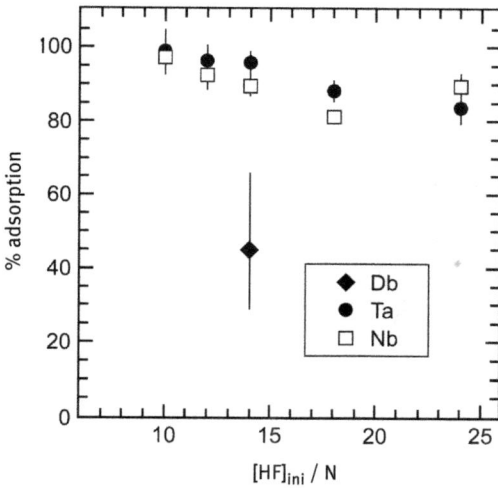

Fig. 7.5: Adsorption of Db, Ta, and Nb (in percent) on the anion-exchange resin as a function of the initial HF concentration, $[HF]_{ini}$, as obtained from on-line column experiments (Tsukada et al. 2009).

7.5 Extraction chromatography

Extraction chromatographic resins can be described as a combination of solvent / liquid-liquid extraction and solid phase extraction. In some publications they are also referred to as solvent impregnated resins or supported solvent extraction. A very good overview over the history of extraction chromatography is given in Braun and Ghersini (1975). While extraction chromatographic resins are used in a similar way than ion exchange or solid phase extraction resins (e.g., use in columns or cartridges) it is based on very different extraction mechanism. Figure 7.6 illustrates the principal composition and the main 3 components.

Extraction chromatographic resins consist of an inert support that is impregnated with a selective stationary phase, also referred to as extractant. It is typically comprised of a nonwater miscible organic compound having distinctive extraction properties. The separation is taking place on the surface of the impregnated inert support via distribution of the respectively present elements between the aqueous phase and the stationary phase, similar to liquid-liquid extraction. Accordingly, the diffusion into the resin bead is generally of lower importance than, e.g., in case of ion exchange resins.

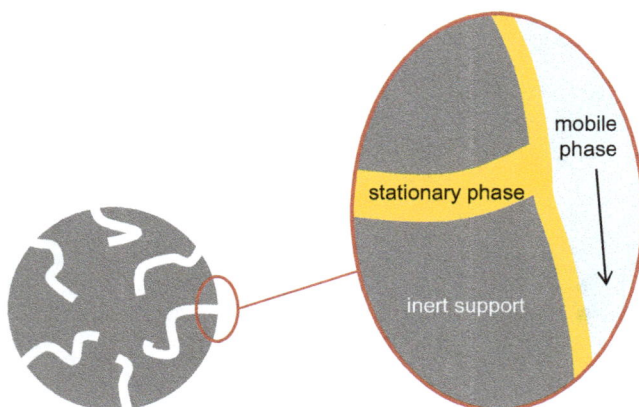

Fig. 7.6: Schematic representation of an extraction chromatographic resin bead. Extraction chromatographic resins are comprised of an inert support and a stationary phase, typically a nonwater-miscible organic compound. Modified with permission from Triskem (2021).

7.5.1 Support materials

A variety of different support materials are described in literature, from kieselguhr to silica and a variety of polymers. Nowadays spherical polymer particles of well-defined sphericity, particle and pore size distribution are used preferably. Inert supports need to meet several criteria. They have to be chemically stable with respect to the extractant to be impregnated and the volatile solvent used during the impregnation process. They further need to be inert with respect to the aqueous phases that will be employed during the separation and the analyte to be separated, e.g., to avoid nonspecific retention of cations of higher valence on a silica surface. Further, the inert support should be mechanically stable to allow facile column packing and working at elevated flow rates without mechanical degradation of the resin. Ideally, it should have high capacity for the extractant to allow for a stable physisorption of the latter. The degree of hydrophobicity of the inert support will have an impact on the wettability of the resin.[5]

---

5 Often slightly less hydrophobic resins are preferred (such as aliphatic (acrylic ester) polymer-based polymers) as they allow for more facile wetting of the resin, thus simplifying column packing. It could be shown on the other hand that PS-DVB (or other polymers containing aromatic groups) based polymers, which are more hydrophobic and thus more difficult to wet, can be of interest as they offer higher resistance against radiolytic degradation. In such cases wetting the resin with EtOH: water mixtures containing up to 30% (V/V) EtOH can still allow for facile column packing.

It is important to use a well-defined particle size distribution and that these supports contain pores with suitable, defined pores size and distribution, cf. Fig. 7.7. The choice of an inert support may also be driven by external factors, especially when working in very high radiation environments, such as the separation of radionuclides from irradiated targets. The decision which inert support to employ should in any case be based on the foreseen use of the resin.[6]

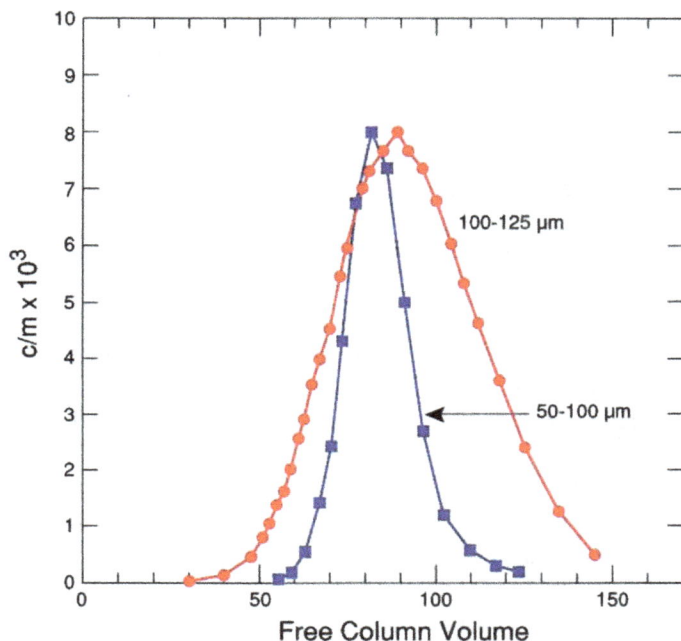

Fig. 7.7: The impact of the particle size on the peak shape and thus also peak resolution shown in the case of the elution of Sr from the commercially available SR Resin. At similar flow rates using particles of smaller size will result in narrower elution peaks. Sr elution curves obtained with 3.2 M HNO_3, 23–24 °C, on 50–100 μm SR Resin and 100–125 μm SR Resin (Horwitz et al. 1992a).

---

6 Using polymers that are containing scintillating compounds, also called Plastic Scintillation microspheres (PSm), as inert support allows the preparation of extraction chromatographic resins that can also be used as detection medium, as described, i.e., by Barrera et al. (2016). In such a case, as e.g. shown for Aliquat 336 impregnated PSm, the analyte is retained on the resins, preferably prepacked in a cartridge of suitable size, which can then be placed in a liquid scintillation counter and measured directly without elution of the analyte or addition of a liquid scintillation cocktail.

7.5.2 Stationary phase

The second component of an extraction chromatographic resin is the stationary phase impregnated onto the inert support. The compounds used as stationary phase again need to meet a certain number of specifications: not soluble in water, not degrade the inert support, stable with respect to the aqueous phase (e.g., stability against nitric acid of elevated concentration), not volatile, but be soluble in a volatile solvent (i.e., methanol, ethanol, or acetone). Generally, extractants employed in extraction chromatography are of amphoteric character, containing a hydrophilic part that is complexing the ions and a hydrophobic part limiting the water solubility of the compounds. This hydrophobic part is often comprised of long chained alkyl groups. Very hydrophobic compounds tend to show slower extraction kinetics accordingly they will require better control of the flow rate during the separation. On the other hand, they will show less bleeding (loss of the stationary phase due to a transfer into the aqueous phase); it will thus usually be possible to re-use such extraction chromatographic resins more often.

The fact that the extractant is only impregnated (*physisorbed*) onto the inert support and the potential bleeding resulting from this, might need to be addressed in a suitable manner for some applications. Possible approaches include mineraliz-ing the aqueous phase (e.g., wet-ashing with nitric acid) or using a reverse phase resin to extract the organic material from the aqueous phase. It could further be shown by Horwitz et al. (1992b) that pre-rinsing the resin with large volumes of water or acid can remove loosely bound parts of the extractant and thus lower the amount of extractant removed during the actual separation. Further the extractant can be stabilized via cross-linking as described, i.e., by Trochimczuk et al. (2004) although the latter will impact the kinetics of the resins.

Most of the extractants employed in extraction chromatography are liquid, however it is also possible to use solid compounds if they are soluble in nonvolatile and nonwater miscible solvents, the so-called diluent. In such cases the diluent needs to be chosen carefully as it can impact the selectivity of the organic phase significantly.[7] Diluents can further be used to avoid so called third-phase formation,

---

7 Diluents may even be chosen specifically to modify or adjust the selectivity of an extractant sys-tem. Examples include the improved retention of selected elements in presence of ionic liquids or modifying the pH at which a certain element is retained. Extraction chromatographic resins based on crown ethers show differences in their selectivity depending on the diluent used as shown for example by the commercially available SR, PB, TK100, and TK101 Resins. All four resins are based on the same crown-ether (4,4'(5')-di-t-butylcyclohexano 18-crown-6) however by using diluents such as octanol (SR Resin), decanol (PB Resin), HDEHP (TK100) or a short chained ionic liquid (TK101) the selectivity of the resin can be modified to match a particular purpose. Using an alcohol with a longer chain such as decanol (PB Resin) instead of octanol (SR Resin) for example will de-crease the retention mainly of Pb to facilitate its elution as described by Horwitz et al. (1994). Using a short chained ionic liquid or HDEHP on the other hand allows extracting elements such as Sr and Pb also under neutral or slightly acidic conditions as shown, i.e., by Surman et al. (2014).

a well-known example being the TBP/CMPO system, or to exploit synergetic effects between different extractants. As the resins are comprised of impregnated beads it is not possible to employ pure organic solvents in a separation, the only exception being the intentional use of an organic solvent, i.e., an alcohol to elute the extractant together with the analyte as demonstrated by Eikenberg et al. (2004). Adding EtOH to an aqueous phase (up to 10%) can improve chromatographic separations as discussed, e.g., in the technical information of the commercially available TK211/2/3 Resins (TrisKem 2020). As the phase transfer is taking place on the surface of the inert support, separation kinetics are generally quite fast, often faster than, e.g., for classical ion exchange resins (the diffusion of the elements to be separated to the functional groups is of lower importance). This frequently allows for performing separations at higher flow rates compared to ion exchange resin-based methods.

Using smaller amounts of extractants can lead to a better peak resolution due to even less diffusion within the stationary phase which can be desirable for example in case of the separation of trace amounts of lanthanides. One of the main advantages of this type of resins is that many different extractants with a wide range of selectivity maybe employed. Extractants typically used in the fabrication of these resins include acidic extractants (e.g., HDEHP), amines, and ammonium salts (e.g., Aliquat 336), solvating and chelating extractants, especially compounds containing $P = O$ groups (e.g. CMPO, TBP), crown-ethers and other macrocycles. Figure 7.8 gives an overview over some of these typically used extractants. Selected examples of the application of resins based on some of these extractants will be discussed later.

Generally, the extractants employed in extraction chromatography, like in liquid-liquid extraction, can be divided in three groups:

Acidic extractants following, e.g., the equilibria:

$$M^{3+} + 3\,HY \rightleftharpoons MY_3 + 3\,H^+ \tag{7.5}$$

$$M^{3+} + 3\,(HY)_2 \rightleftharpoons M(HY_2)_3 + 3\,H^+ \tag{7.6}$$

The driving force is the proton concentration, and it follows a cation exchange mechanism. Accordingly, the pH of the solution will be decisive for the resin's selectivity, although steric effects of the formed complex can have a strong impact, too. Generally, these extracting systems will show rather limited differences in terms of selectivity between different acids, given that the counter anion is not complexing the respective cation.

Neutral extractants following, e.g., the equilibrium:

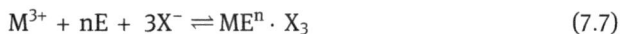

$$M^{3+} + nE + 3X^- \rightleftharpoons ME^n \cdot X_3 \tag{7.7}$$

where X is typically Cl^- or NO_3^-

Tributylphosphate
(TBP)

2-Ethylhexylphosphonic acid mono-2-ethylhexyl ester (HEH[EHP])

Aliquat 336 (R = 8, 10)   Tertiary amine

Hydroxamate

Phosphine sulfide

Dipentyl pentyl phosphonate (DPPP)

Bis(2-ethylhexyl) hydrogen phosphate (HDEHP)

4,4'(5')-di-t-butylcyclohexano-18-crown-6

Trioctylphosphine oxide (TOPO)

Octyl (phenyl)-N-N-diisobutyl-carbamoylmethylphosphine oxide (CMPO)

N,N,N',N'-tetra-n-octyl-diglycolamide (TODGA)

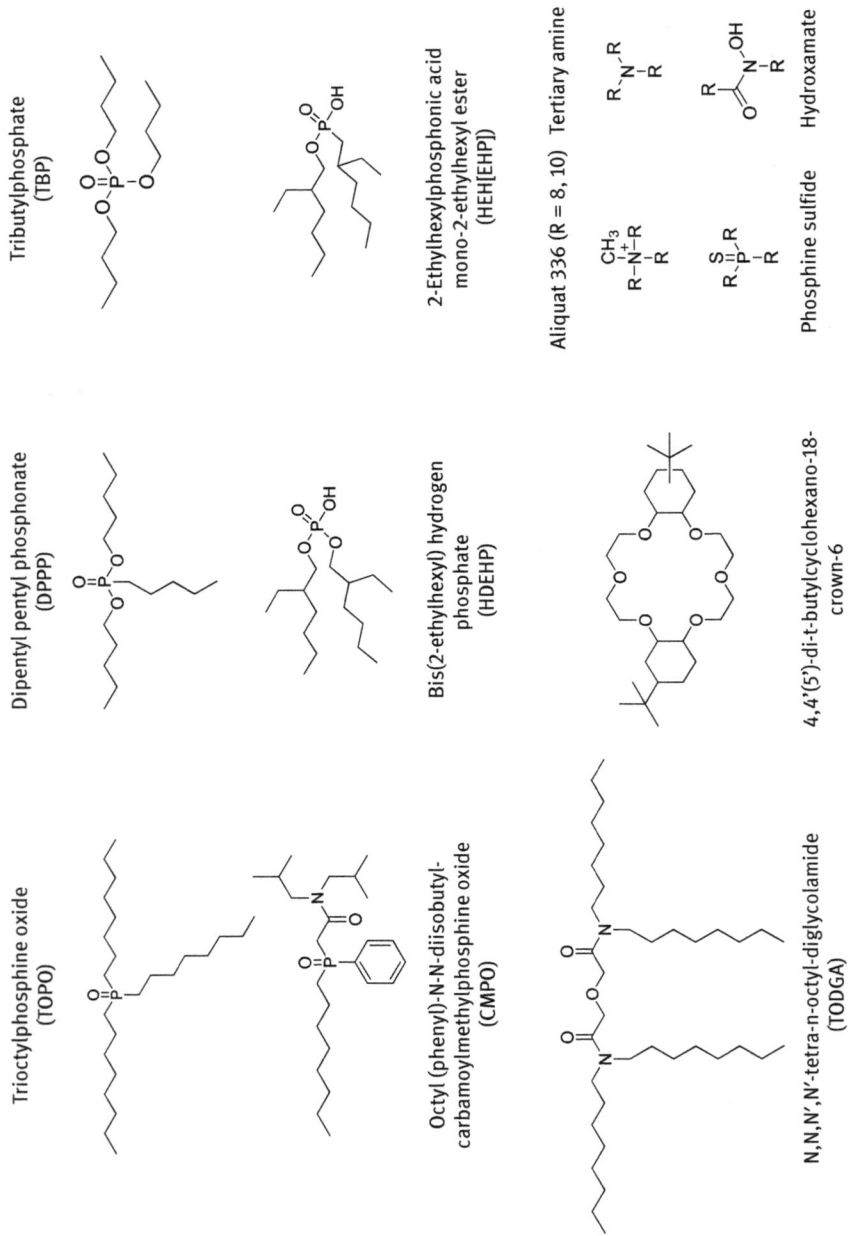

Fig. 7.8: Selected examples of extractants typically used in extraction chromatography.

In this case the driving force is the anion concentration. Typically, the transfer of the cation into the organic phase will be favored in presence of a large amount of suitable anions. A useful, although simplified, explanation is that the cation and the anions form a 'neutral species' in the aqueous phase that is transferred into the organic phase where it is forming a neutral complex involving the cation, the extractant and the anions. The selectivity can differ very strongly from one anion to another, a fact that is very often used to facilitate separations. Generally, the anion concentration is adjusted through the addition of the corresponding acid, however other means may be employed, e.g., adding Li^+ or NH_4^+ salts of the respective anion.

Basic extractants, e.g., following the equilibria:

For amines: $$R_3N + HX \rightleftharpoons R_3NH^+ X^- \tag{7.8}$$

$$R_3NH^+ X^- + MX_3 \rightleftharpoons R_3NH^+ MX_4^- \tag{7.9}$$

For ammonium salts: $$R_4N^+ X^- + MX_3 \rightleftharpoons R_4N^+ MX_4^- \tag{7.10}$$

where X is typically Cl^- or NO_3^-

These extractions follow an anion exchange type of mechanism, accordingly in this case the driving force is also the concentration of anions in the system. Like for neutral extractants, the nature of the anion, mainly its ability to form anionic complexes with a cation, will have a strong impact on the selectivity of the system.

As described above, there are generally two axes of selectivity that can be exploited to optimize a chemical separation, one is the choice of the resin, and the other is, especially in the case of resins based on neutral or basic extractants, the choice of the anion (typically nitrate or chloride). Especially in analytical separations, but also in some methods for the purification of radionuclides for use in radiopharmaceutical chemistry and nuclear medicine, several different resins maybe used in sequence to facilitate a separation or to allow separating several elements from the same sample.

To optimize a separation method, it is thus important to first characterize the resins with respect to their selectivity. This is generally done by performing a series of batch experiments where a defined amount of the resin (w) to be characterized is contacted with a well-defined volume (v) of a solution of a given acid of defined concentration containing a known concentration of analyte(s) (A_0). After contacting the resin and the solution for a suitable amount of time (typically 30–60 min) the test sample is centrifuged or filtered, and an aliquot of the solution is withdrawn and analyzed (A_S). Based on the results of this analysis the so-called weight distribution ratio (D_w) can be calculated:

$$D_w[mL/g] = \frac{A_0 - A_s}{w(g)} / \frac{A_s}{v(mL)} \tag{7.11}$$

The weight distribution ratio (D_w) is a measure for the retention of an element on the resin: the higher the D_w value the higher the retention of the resin for the respective element under the given conditions. An even more useful measure for the retention of an element is the so-called capacity factor k'. The k' value is difficult to determine experimentally but it can be calculated as it is proportional to the D_w value. Unfortunately, the proportionality factor depends on several parameters including the volume of extractant impregnated onto the inert support. This information is not always readily available; however a value of 0.5 g/mL often is a suitable approximation. Accordingly, for most extraction chromatographic resins it may be assumed that

$$k' = D_w \cdot 0.5 \tag{7.12}$$

Plotting these D_w or k' values against tested acid concentrations allows visualizing a resins selectivity for a range of different conditions and elements. Figure 7.9 shows an example of such k' value plots.

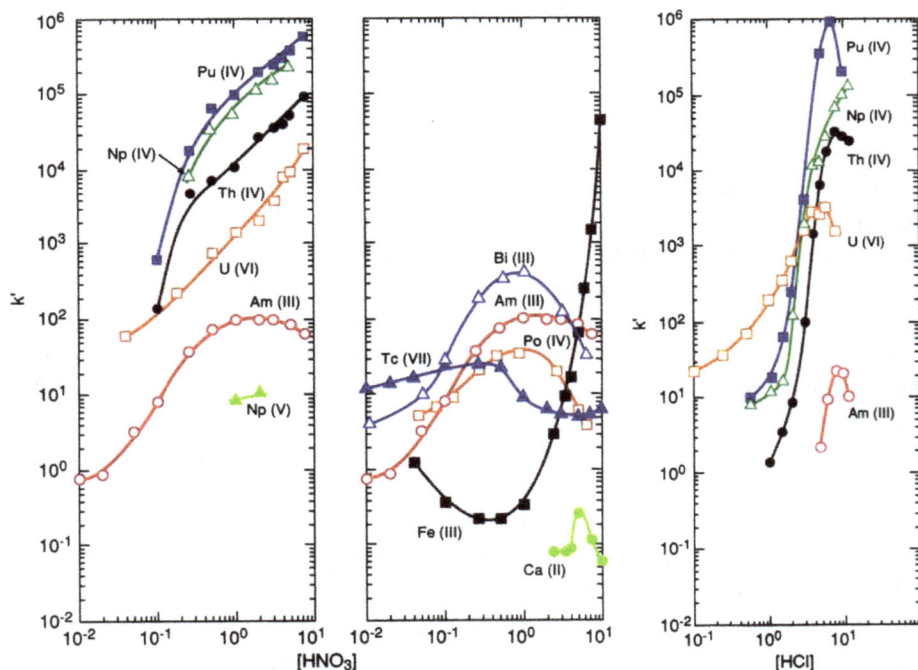

Fig. 7.9: k' values for actinides and other selected elements on TRU Resin in HNO₃ and HCl of varying concentration. These k' graphs visually summarize the selectivity of the presented resin (Horwitz 1993).

The same resin (in this example, the TRU Resin) may have very different selectivities in different acids (generally HCl and HNO₃), a fact that is frequently used in complex separations. It should be noted that historically k' values and acid concentrations are very frequently plotted double logarithmically in these kind of graphs.

This visual representation of the respective selectivity can facilitate the identification of experimental conditions suitable for the extraction and elution of analytes and interferents. A rather simple example is the separation of Am^{III} from matrix elements and other actinides), cf. Fig. 7.9.[8] An interesting aspect of the capacity factor k' is that it is proportional to the volume necessary to elute an analyte, or to be more precise to the volume necessary to get to the maximum of the elution peak (V_p). Accordingly, knowing the k' value allows to estimate the volume necessary to elute an analyte, and estimating what volume can be loaded onto the column before the analyte breaks through, respectively. The proportionality factor between this volume V_p and the capacity factor is the so-called "free column volume" (fcv), the interstitial space or void volume in a chromatographic column, with $k' = V_p/fcv$. The fcv of extraction chromatographic columns is frequently about 65% of the column volume. Based on this is possible to estimate that at a k' of 5 and an fcv of 1.3 mL (typical value for a 2 mL column of an extraction chromatographic resin), the V_p is approximately 6.5 mL, meaning that the peak maximum is reached, under ideal conditions, after ~6.5 mL. To assure quantitative elution, an elution volume of at least 15 mL should be chosen. On the other hand, at a k' of 1000 and a fcv of 1.3 mL the V_p is approximately 1.3 L under ideal conditions, it is thus safe to assume that even under nonideal conditions loading, e.g., 100 mL onto the column should not lead to a breakthrough of the analyte. A rather widely applied rule of thumb in this context is that k' values ≥100 are suitable for retention of an element (the higher the better) and k' values of ≤10 suitable for elution (the lower the better).

Interferents, cations as well as anions, can have an impact on the retention of elements on the resin, this influence too can be represented in k' value plots, cf. Fig. 7.10. These kind of graphs help identifying whether a resin is suitable for a certain separation, or if a particular cation or anion would need to be removed from the sample before performing the separation. In some cases identifying a problem also allows finding a solution, e.g., in case a large amount of phosphate is known to be present in a sample (e.g., after a calcium phosphate co-precipitation) and interfering, e.g., the extraction of tetravalent actinides, it is often sufficient to add Al^{III} to the solution. The added Al^{III} is then preferably complexed by the phosphate, thus lowering the amount of free phosphate in the sample. Other means of removing interferences are the complexation of interfering cations with complexants like oxalate or fluoride, or changing the oxidation state of an interferent, e.g., Fe^{II}/Fe^{III}.

---

8 In 3 M HNO_3, for example, Am^{III}, as well as other actinide elements such U, Th, Np, and Pu, show high k' values accordingly, they are all well retained, while typical matrix elements such as Fe and Ca are not or only very weakly retained. Am^{III} may then be selectively eluted from the resin using, e.g., 4–5 M HCl, as it shows, other than the remaining actinides, very low k' values under these conditions. The remaining actinides can then be separated using other means like different acids/ acid concentrations, the use of complexing agents such as oxalate or the use of redox agents, e.g., reduction of Pu^{IV} to Pu^{III}.

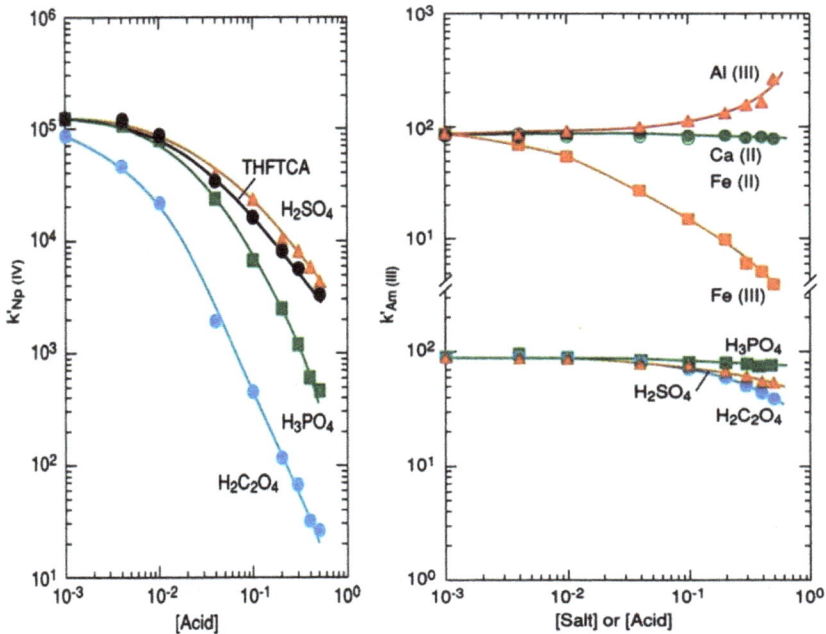

Fig. 7.10: left: k' values for Np^{IV} on TRU Resin in 2 M HNO_3 and in presence of varying concentrations of different acids; right: k' values for Am^{III} on TRU Resin in 2 M HNO_3 and in presence of varying concentrations of different acids, cations respectively (Horwitz 1993). These k' graphs allow identifying the impact a compound will have on the retention of the analyte on the resin. For tetravalent actinides like Np^{IV} oxalic acid and phosphoric acid interfere strongly with the retention and should be removed – through mineralization (oxalic acid) or addition of Al^{III} (phosphoric acid). On the same resin these acids interfere very little with Am^{III} extraction. Al^{III} has, due to a salting out effect, a retention enhancing effect while Fe^{III} interferes strongly with the Am^{III} extraction. Reduction to Fe^{II} eliminates this interference.

7.5.3 Applications

As described before, extraction chromatographic resins are finding use in a wide range of applications, some examples of such applications are discussed in the following paragraphs.

Applications in radioanalytical chemistry: One of the first applications of extraction chromatographic resins was the analysis of radioactive elements in a variety of samples. This is mainly due to the fact that many of the extractants used originated in nuclear research performed at the Argonne National Laboratory (ANL). The CMPO/TBP system employed in the TRU Resin for example originated in the TRUEX process, the crown ether used in the SR Resin in the SREX process. An early example of the use of extraction chromatography for the separation of radioactive elements dates to

Horwitz (1969) when the group used a HDEHP-based resin for the separation of a number of actinides. Many applications for the separation and quantification of radioactive elements in environmental, bioassay, waste, and decommissioning samples have been developed using extraction chromatographic resins. These methods mainly cover anthropogenic radionuclides like actinides, activation and fission products such as ^{90}Sr, ^{63}Ni, and ^{99}Tc. However naturally occurring radionuclides like U isotopes, Th isotopes, Ra isotopes, ^{210}Pb, and ^{210}Po are very frequently separated and analyzed, too.

Extraction chromatographic resins are generally employed to separate the radionuclide to be quantified from spectral interferences (e.g., other alpha or beta emitters) and from matrix elements - this is of particular importance in case of alpha spectrometric determinations to allow obtaining alpha sources with very low thickness and thus high peak resolution. To allow obtaining as much information as possible from one sample several resins with different selectivity are employed sequentially.

In such applications the resins are frequently used as cartridges as this allows stacking them in a predefined order. A typical separation is described, i.e., by Maxwell (2011), cf. Fig. 7.11A and B. After suitable sample preparation and oxidation

Acid digestion
1) Redissolve in 6 ml 6 M HNO$_3$ and 6 ml 2 M Al(NO$_3$)$_3$
2) Add 0.5 ml 1.5 M sulfamic acid + 1.25 ml 1.5 M ascorbic acid
3) Add 1.25 ml 3.5 M sodium nitrate

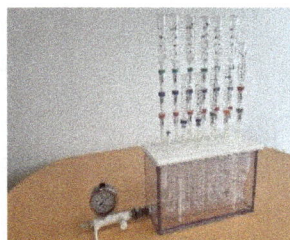

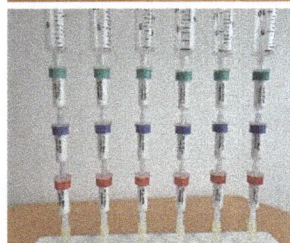

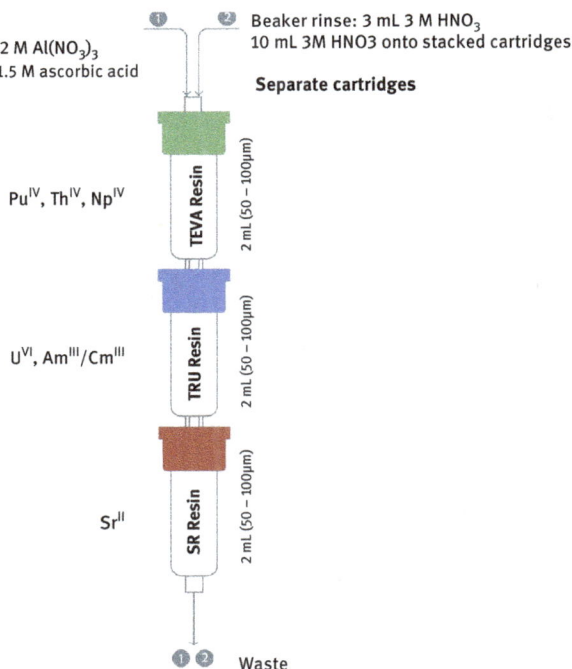

❶ ❷ Beaker rinse: 3 mL 3 M HNO$_3$
10 mL 3M HNO3 onto stacked cartridges

Separate cartridges

Pu^{IV}, Th^{IV}, Np^{IV} — TEVA Resin — 2 mL (50 – 100μm)

U^{VI}, Am^{III}/Cm^{III} — TRU Resin — 2 mL (50 – 100μm)

Sr^{II} — SR Resin — 2 mL (50 – 100μm)

❶ ❷ Waste

Fig. 7.11A: Scheme of a stacked cartridge-based separation of U, Th, Pu, Am/Cm, and Sr from air filters (from Maxwell 2011). Load and rinse on stacked cartridges. The photos show: left – vacuum box setup for vacuum assisted separation methods, right – stacked cartridges (TEVA, TRU, SR).

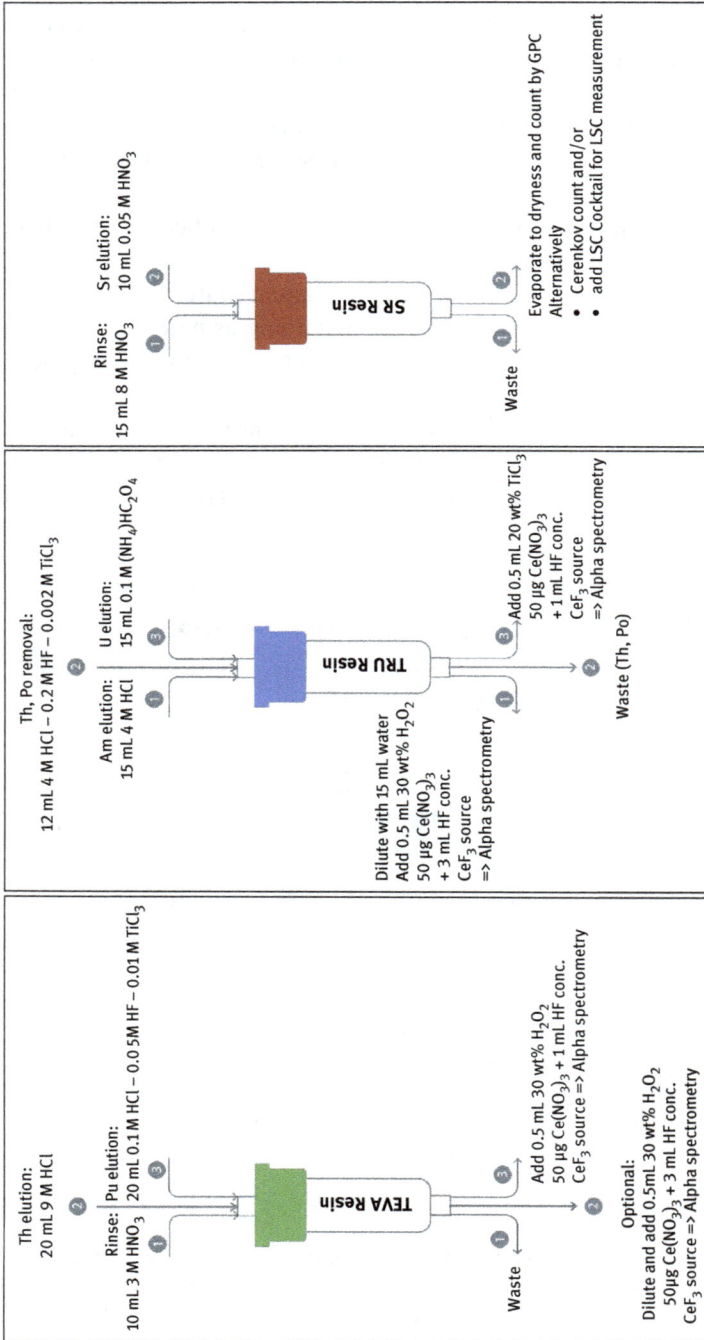

Fig. 7.11B: Scheme of a stacked cartridge-based separation of U, Th, Pu, Am/Cm, and Sr from air filters (from Maxwell 2011). Separation steps after cartridge splitting. Left: Pu and Th separation on TEVA Resin. Middle: U and Am/Cm separation on TRU Resin. Right: Sr separation on SR Resin.

state adjustment to assure presence of Pu as Pu^{IV} and Np as Np^{IV} the test solution is loaded through three stacked cartridges (TEVA, TRU, and SR Resin) from 3M HNO_3 in presence of Al^{III} to remove potential phosphate interference. Under these conditions Th^{IV} and Pu^{IV}/Np^{IV} are retained on TEVA Resin, while other elements such as U^{VI}, Am^{III}/Cm^{III}, and Sr^{II} pass through onto the TRU Resin cartridge. This resin retains U^{VI} and Am^{III}/Cm^{III} while Sr^{II} passes through onto the following SR Resin cartridge where it is finally retained, cf. Fig. 7.11A. The three cartridges can then be separated and treated separately to allow obtaining clean U, Th, Pu/Np, Am/Cm, and Sr fractions, cf. Fig. 7.11B.

A particular example of an extraction chromatographic resin is used for the separation of ^{63}Ni, e.g., from decommissioning samples. The commercially available NI Resin is based on the dimethylglyoxime (DMG) complexing agent impregnated onto the inert support as a solid. DMG is known to form, under suitable conditions, bright pink nickel-bis(dimethylglyoximate) complex precipitates Fig. 7. 12. In case of the NI Resin this precipitate is formed on a column, it is thus based on a precipitation mechanism. ^{63}Ni can be separated employing the NI Resin (following precipitation and matrix removal via anion exchange chromatography) from nuclear waste samples with high chemical yield and purity, allowing its quantification by liquid scintillation counting after destruction of the eluted Ni-DMG complex.

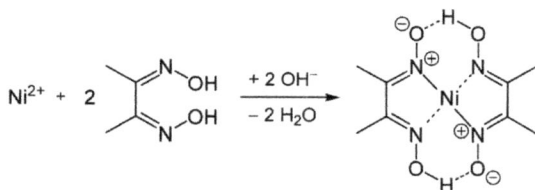

Fig. 7.12: Dimethylglyoxime (DMG) and Ni-DMG complex nickel-bis(dimethylglyoximate) Eichrom (2021).

The ability of extraction chromatographic resins to strongly retain cations can also be used to perform separations that are otherwise difficult to achieve, like the separation of Cl and I from each other, and from matrix elements, under alkaline, neutral or acidic conditions. While co-precipitation with silver is frequently used for the separation of Cl and I, such co-precipitation methods are difficult to automize. An alternative consists in loading Ag^+ onto a resin that can retain it over a very wide pH range while keeping it accessible for interaction with analytes. In such cases, as the Ag^+ is retained on a resin, it is possible to perform the separation chromatographically using columns or cartridges.

An example for such a resin is the phosphine sulfide-based CL Resin, which retains silver very strongly under highly alkaline and acidic conditions. It further has the advantage of having no, or only very little, selectivity for other radionuclides that might interfere with the respective quantification. Once the resin is loaded with Ag⁺ it will show a selectivity very similar to Ag precipitations, accordingly Cl^- and I^- are strongly retained on the resin. By choosing suitable elution agents it is then possible to sequentially elute Cl^- and I^- as shown in Figure 7.13. As expected from AgCl and AgI complex stability data, NH_4SCN elutes Cl^- while keeping I^- retained on the resin as AgI. It may then be eluted using a stronger complexing agent for Ag⁺ such as sulfide for example, cf. Fig. 7.13.

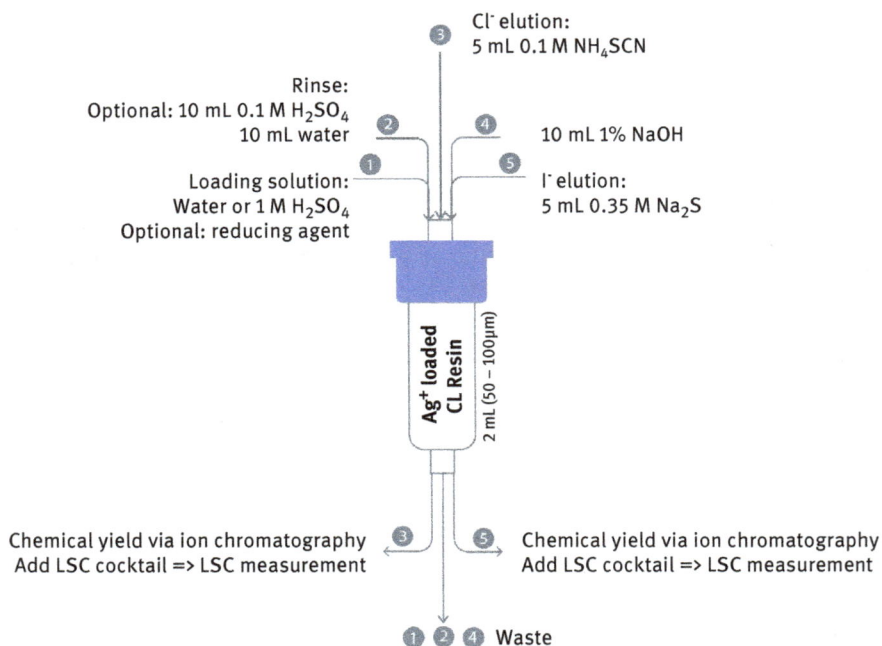

Fig. 7.13: Scheme of a ³⁶Cl/¹²⁹I separation method described by Zulauf et al. (2010) based on Ag⁺ loaded CL Resin. The Ag⁺ loaded CL Resin retains chloride and iodide over a wide pH range. Using a complexing agent that has higher affinity for Ag⁺ than Cl^- but lower than I^-, here SCN^-, allows selectively eluting chloride from the column. Iodide is then eluted with an even stronger complex ligand (here S^{2-}).

Applications in radiopharmacy and nuclear medicine: The purification of radionuclides, particularly radiometals, for use in diagnostics (PET or SPECT) or therapy (alpha, beta or Auger emitters) typically requires the separation of ultra-trace amounts of the desired radionuclide from macroscopic amounts of irradiated target materials and trace amounts of co-produced radionuclidic impurities. In most cases irradiated targets are subsequently dissolved in acid of elevated concentration. The final product on the other hand should ideally be a highly pure solution of the radionuclide in a small volume of a dilute acid or a buffer solution. As the element(s) comprising the target are present in a vast excess compared to the radionuclide to be separated it is generally advisable to choose an extraction chromatographic resin for the separation that has no, or as little as possible, selectivity for the target material, while very strongly retaining the radionuclide to be separated. In case radionuclidic impurities are known to be present their removal should be taken into account when optimizing the separation scheme (choice of resin, acid concentrations, . . .). The aim is keeping the amount of resin employed during the separation as small as possible to allow performing the separation in a short time – particularly important when separating short-lived radionuclides – and eluting the targeted radionuclide in a small volume.

In some cases adding another resin upfront to the resin used for the actual separation can simplify the purification of a radionuclide. In the case of separating ^{61}Cu or ^{64}Cu from solid Ni targets a TBP Resin cartridge removes Fe and Ga impurities potentially present in the dissolved irradiated target, while letting Ni, Co, Cu, and Zn pass. Removing these two elements facilitates the subsequent separation of the remaining elements on the amine-based TK201 Resin significantly. This pre-

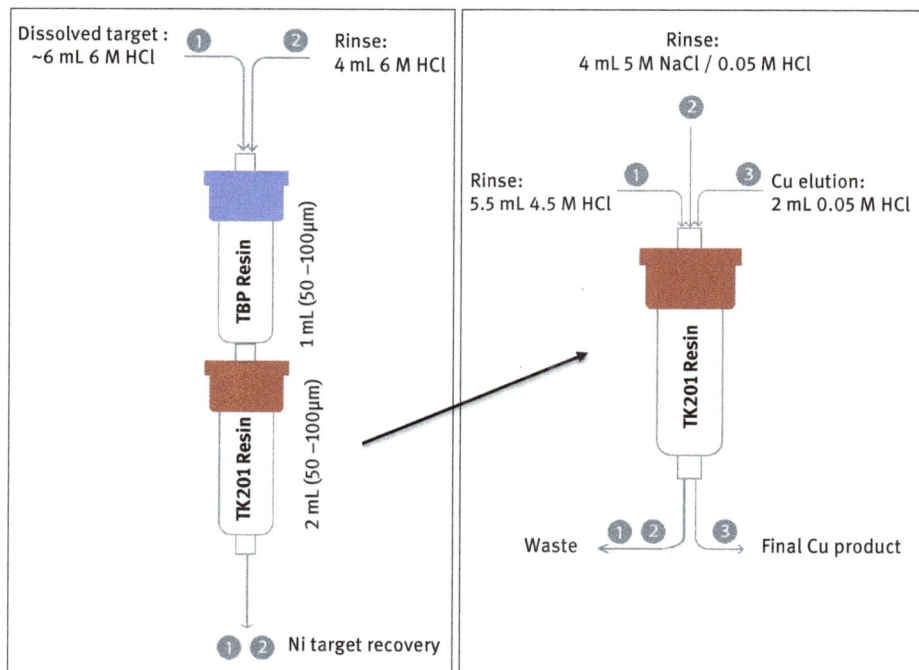

Fig. 7.14: Scheme of a cassette-based method for the separation of ^{61}Cu or ^{64}Cu from irradiated solid Ni targets from Svedjehed et al. (2020). Left: The dissolved target is passed through TBP and TK201 Resin cartridges. The TBP Resin retains Fe and Ga impurities, it is subsequently discarded. Cu radioisotopes, as well as potentially present Co and Zn impurities, are rinsed onto the amine-based TK201 Resin. The Ni target material passes through both cartridges and is recovered for recycling. Right: ^{61}Cu/^{64}Cu purification on TK201 Resin and elution in dilute HCl.

purification, combined with a NaCl/HCl-based rinsing step, allows eluting the final Cu fraction in a small volume of 0.05 M HCl of defined concentration with high purity and yield, cf. Fig. 7.14.

Another example is the separation of ^{68}Ga from irradiated liquid or solid ^{68}Zn targets using a hydroxamate resin (e.g., ZR Resin). After the purification of the ^{68}Ga on the ZR Resin it is eluted using 1.5–2 M HCl, conditions too acidic for direct use in injection or labelling. Using a TOPO Resin-based cartridge (e.g., TK200 Resin) it is possible to extract ^{68}Ga from this solution, and subsequently elute it water or in dilute HCl, cf. Fig. 7.15.

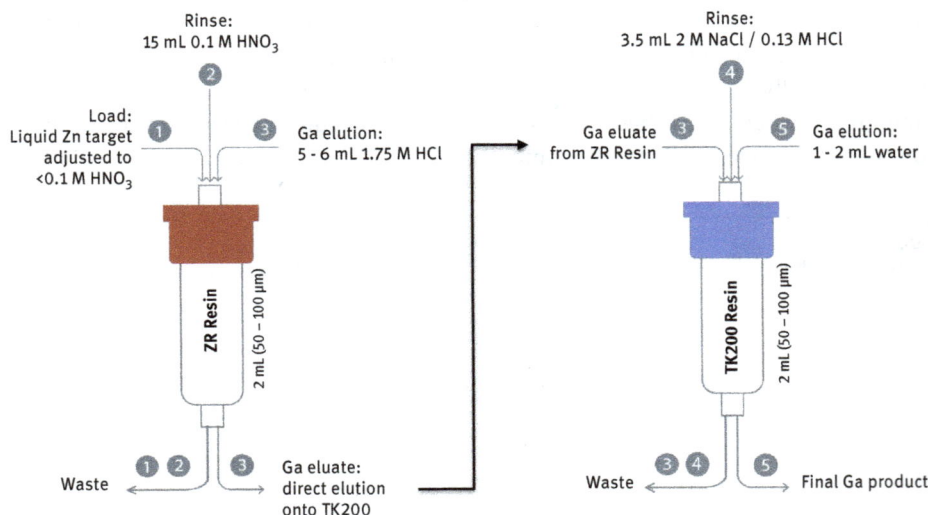

Fig. 7.15: Scheme of a cassette-based method for the separation of ^{68}Ga from liquid ^{68}Zn targets, Rodnick et al. (2020). Left: Initial ^{68}Ga separation from the target material and other radionuclidic impurities on a hydroxamate resin (ZR Resin). The obtained purified ^{68}Ga solution is too acidic for direct use (1.75 M HCl). Right: using a TOPO-based extraction chromatographic resin (TK200 Resin) converts the ^{68}Ga to dilute HCl. Performing an additional rinse with 3.5 ml of 2 M NaCl in 0.13 M HCl allows for obtaining ^{68}Ga in a well-defined HCl concentration.

A similar approach, using one resin to convert a solution from one acid concentration to another, is also used in the separation of nca ^{177}Lu or nca ^{161}Tb from their respective macroscopic target materials ^{176}Yb and ^{160}Gd. Extraction chromatographic resins based on organo-phosphoric (e.g., HDEHP), organo-phosphonic or organo-phosphinic acids are used for the separation of lanthanides from another in a wide range of applications as they often show elevated selectivity even for neighboring lanthanide pairs. For example the HEH[EHP] (see Fig. 7.8) based LN2 Resin, is used in combination with the TO-DGA-based DGA, N Resin, for the separation of nca ^{177}Lu from 300 mg Yb (Horwitz 2005) or even larger targets.

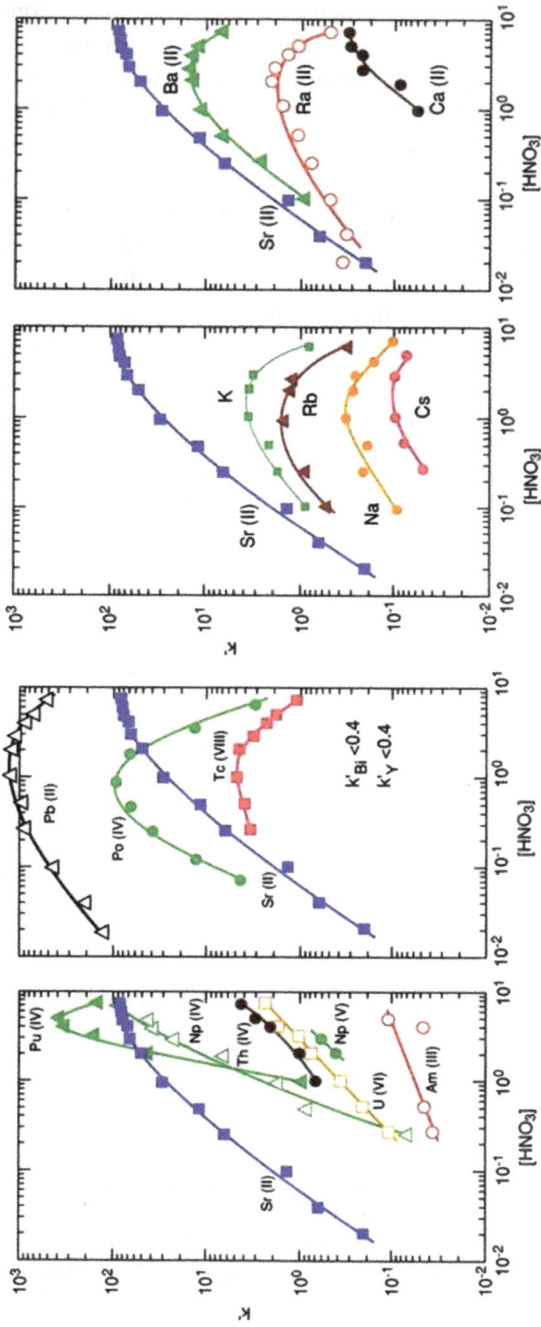

Fig. 7.16: k' values for selected elements on the crown-ether-based SR Resin in HNO_3. While Sr is very well retained, i.e., at 3 M HNO_3 Y is not retained – thus allowing for their facile separation. Horwitz et al. (1992a).

Another important application of extraction chromatography is the quality control of a produced radionuclide, particularly the determination of its radionuclidic purity with respect to the presence of non-gamma-emitting radionuclides. A typical example is the determination of long-lived ^{90}Sr in ^{90}Y obtained, e.g., from a ^{90}Sr/^{90}Y generator. This control is especially important as ^{90}Sr is a β^- emitter with a long half-life of 28.8 years. It further is, due to its chemical similarity with Ca, easily incorporated into bones, making it particularly radiotoxic. To separate and detect any potentially present trace amounts of ^{90}Sr the most frequently used resin in this context is the crown ether-based SR Resin. Figure 7.16 shows the selectivity of the SR Resin for a number of elements, including Sr and Y. After adjustment of the sample to 3 M HNO$_3$ and loading onto the resin, ^{90}Sr is retained while ^{90}Y passes through. ^{90}Sr may then be eluted with water or dilute HNO$_3$ and quantified, e.g., via LSC counting.

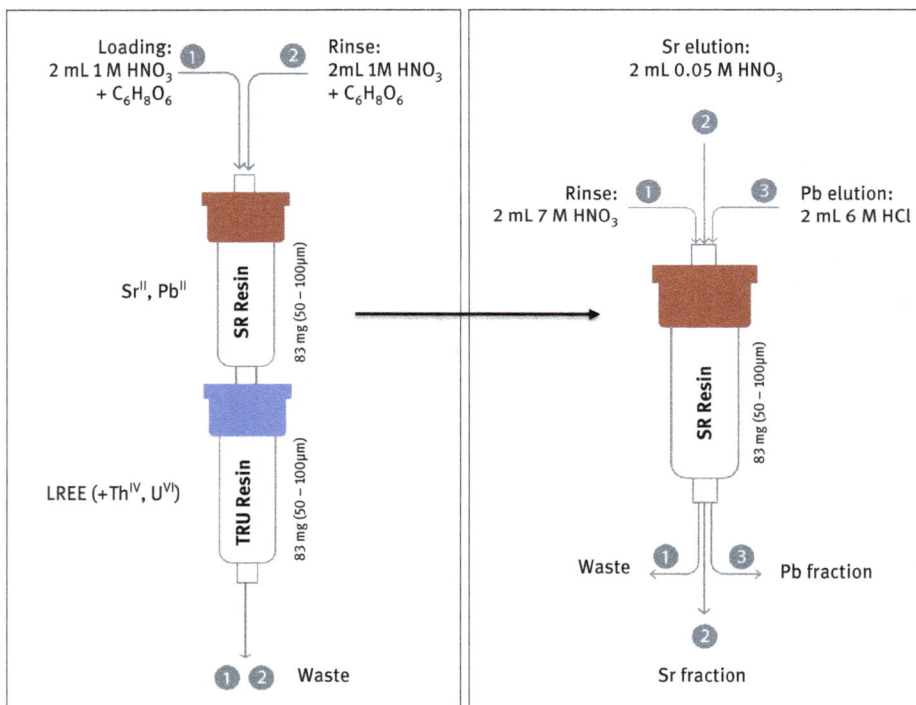

Fig. 7.17A: Schematic separation of Nd, Sm, Sr, Pb, Ba from actinides, other lanthanides and matrix elements from Pin et al. (2014). Left: Loading of the dissolved sample onto stacked SR Resin and TRU Resin. Sr and Pb are retained on SR Resin. Light rare earth elements (LREE), Th^{IV} and U^{VI} are retained on TRU Resin. Matrix elements pass through both resins. Cartridges are split after this step. Right: Ba, Sr, and Pb separation on the SR Resin cartridge.

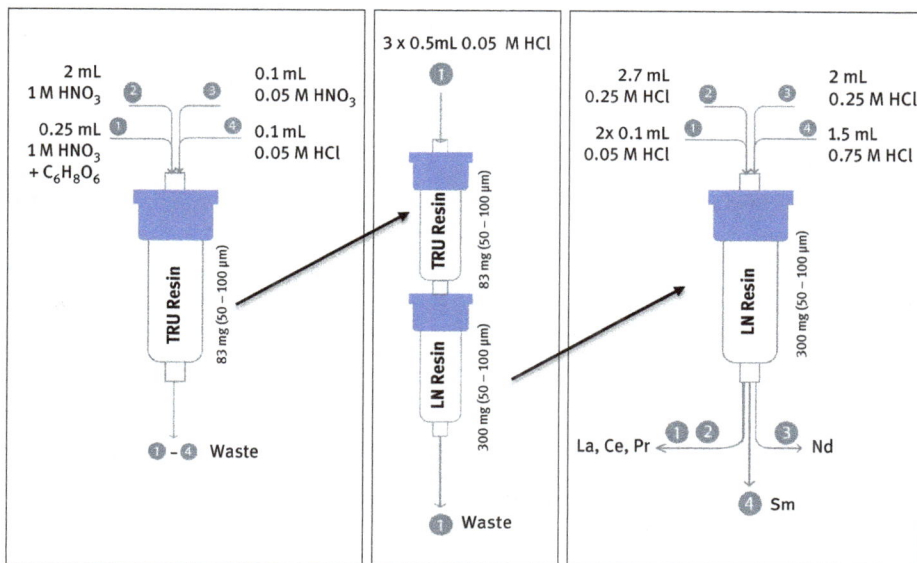

Fig. 7.17B: Schematic separation of Nd, Sm, Sr, Pb, Ba from actinides, other lanthanides and matrix elements from Pin et al. (2014). Left: Further matrix removal from the TRU Resin, LREE remain retained. Middle: TRU resin is stacked above LN Resin. LREE are rinsed from the TRU Resin onto the LN Resin using 0.05 M HCl. Right: Chromatographic separation of the LREE on LN Resin. La, Ce, and Pr are discarded, Nd and Sm are recovered for mass spectrometric measurement.

Applications in geochemistry: The mass spectrometric determination of isotope ratios of stable elements is finding very wide use in a number of fields such as dating of samples (such as rocks or zircons), nuclear forensics, analysis of biomedical samples, metallomics or food provenancing (the confirmation of the origin of food and beverages such as wine, cheese, and olive oil). To obtain precise isotope ratios, isobaric interferences (elements or polyatomic compounds of the same mass) need to be removed prior to the mass spectrometric measurement. As the amount of the analyte or sample mass available is usually very small it is also very important to remove the sample matrix thoroughly.

A typical example is the dating of rock samples. It is of high interest to obtain information on a variety of different elements, e.g., to allow applying two different radiometric dating techniques such as U/Pb dating and Sm/Nd dating, cf. Chapter 5. After dissolution the sample is separated using a suitable separation chemistry. The method shown in Figs. 7.17A and 7.17B, e.g., allows obtaining clean fraction of Ba, Sr, Pb, Nd, Sm, and if required U and Th. In this example the dissolved sample is passed through stacked columns starting with the SR Resin which preferably retains divalent elements such as Ba, Sr, and Pb. After passing through this column the solution passes through the CMPO/TBP-based resin described earlier (TRU Resin) which retains U, Th, and rare earth elements (REE) while matrix elements pass through. U and Th may be recovered at this point for further purification, however in the given example only the light rare earth elements are recovered. The latter are further separated on the organo-phosphoric acid (HDEHP)-based LN Resin, resulting in clean Nd and Sm fractions.

Other applications: Extraction chromatography is not limited to use as resin, it is also possible to impregnate supports of other geometries such as TLC paper, membrane filters or hollow fiber membranes, to give just a few examples. Extraction chromatographic resins may further be used in hydrometallurgical applications such as the recycling of critical metals or the recovery of valuable metals from waste streams, and the decontamination of radioactively contaminated effluents.

7.6 Thin layer chromatography/high-performance liquid chromatography

7.6.1 Radiochemical separation

Both thin layer chromatography (TLC) and high-performance liquid chromatography (HPLC) are separation methods common in conventional chemistry and well adapted to radiochemistry. The separation of the species in liquid phase (either aqueous or nonaqueous or mixed) is due to its sometimes quite complex interaction with a solid phase it passes through. For TLC the solid phase is paper or a chromatographic material (such as silica gel, for example) organized on a thin backing of, e.g., a thin aluminum foil, and the transfer of the liquid phase is due to diffusion. In HPLC a column is used containing the chromatographic material, and the transport of the liquid phase carrying the species is organized by pressure. In TLC, once the liquid phase has almost reached the end of the chromatographic strip, the distribution of the radioactive species is detected by measuring the radioactivity distribution along the strip, yielding retention fraction r_f, which is the distance migrated relative to the length of the solvent front, i.e., with r_f ranging between 0 and 1. In HPLC, the radioactive species are registered when leaving the end of the column through a capillary tube by passing a radioactivity detector. Performances are expressed as retention times r_t, which is the time (in minutes) the species needed between entering the column and leaving it.

TLC and HPLC have found wide application in radiochemical analytics. Important specific cases refer to the determination of the chemical or radionuclidic purity of a given radioactive species prior to further application. Another specific feature, mainly for HPLC, is the preparative separation of the radioactive species of interest from its chemical and radiochemical impurities.

7.6.2 Radiochemical purification

Particularly for the preparation of radiopharmaceuticals which are neurotransmitter analogs applied to imaging the living human brain, the amount of radiopharmaceutical must be extremely low to prevent toxic effects. In this case, the final radiopharmaceutical must be isolated from the macroscopic amount of its nonradioactive labeling precursor, which is often a pharmacologically active component negatively influencing the imaging and eventually causing toxic effects. The radiosynthesis thus must consider specific activity issues, and must in all cases separate the radiopharmaceutical from it precursor by means of HPLC.

7.6.3 Radiochemical quality control

The field of quality control for radiopharmaceuticals demands effective radiochemical separations. These separations should be relatively fast, since many radiopharmaceuticals – despite the short-lived radionuclides – need to undergo quality control prior to applications in patient diagnosis and therapy according to the requirements of drug syntheses.

Figure 7.18 shows the quality control for radiochemical purity of the most relevant radiodiagnostic for PET, 2-deoxy-2-[^{18}F]fluoroglucose (FDG) as synthesized by reacting the mannose triflate labeling precursor with ^{18}F-fluoride.[9] The requirement is to quantify the percentage of the product FDG relative to the fraction of [^{18}F]fluoride unreacted, which accumulates *in vivo* in bone structures. According to national and international pharmacopeia, this parameter should be >95%.

In other cases, TLC is accepted for quality control of radiopharmaceuticals, at least in cases when the lower accuracy of TLC compared to HPLC conforms to legal requirements. Nevertheless, both HPLC and TLC must yield identical results. Figure 7.19 shows HPLC and TLC methods to separate the ^{68}Ga-labeled peptide [^{68}Ga]DOTA-TOC[10] used for molecular imaging of neuroendocrine tumors by means of PET in the context of quality control of that radiopharmaceutical.[11]

---

9 See Chapter 12 on its synthesis and application.

10 [^{68}Ga]Ga-DOTA-TOC ((DOTA = 1,4,7,10-tetraazacyclododecane-1,4,7,10-tetraacetic acid; TOC = D-Phe-cyclo[Cys-Tyr-D-Trp-Lys-Thr-Cys]-Thr(ol)) with TOC representing an octapeptide with analogy to somatostatin, cf. Chapter 12.

11 See Chapter 12 this volume on its synthesis and application.

species	r_t (min)	fraction (%)
[^{18}F]Fluoride	05.38	0.06
[^{18}F]FDM	07.51	0.03
[^{18}F]FDG	09.07	99.91

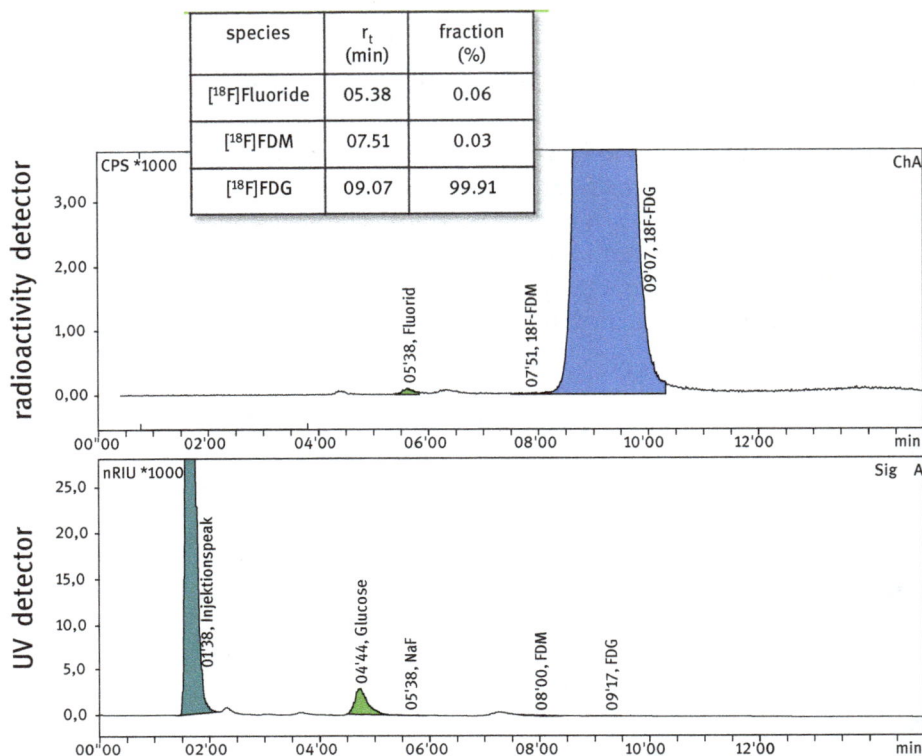

Fig. 7.18: HPLC to separate species relevant for [^{18}F]FDG quality control: Upper chromatogram: Radioactivity detection identifying retention times of [^{18}F]FDG and "free" [^{18}F]fluoride and [^{18}F]FDM (fluorodeoxymannose). Percent fractions for a specific batch of [^{18}F]FDG are summarized in the table. Lower chromatogram: IR detection of nonradioactive amounts of glucose, NaF, FDM, and FDG. (Dionex CarboPac PA10 250*4 (guard column CarboPac PA10 50*4), eluent: 0.1 N NaOH, flow rate: 1 ml/min, first peak at $r_t = 1.38$ min is HPLC injection signal.).

7.7 Electrodeposition/electrophoresis/ electromigration

In addition to the various kinds of physicochemical interaction between a (radioactive) species with the surface of a solid material such as paper, oxides, salts, metals, glass etc., the charge of the species can influence this behavior. Its behavior is then composed of the interaction with the so-called supporting material (a thin layer strip or any kind of column packing material as in HPLC or GC, or a surface of a container) and with the migration of the species in an electric field. Focusing on the effect of the electric charge of the species, electrophoretic separation of species

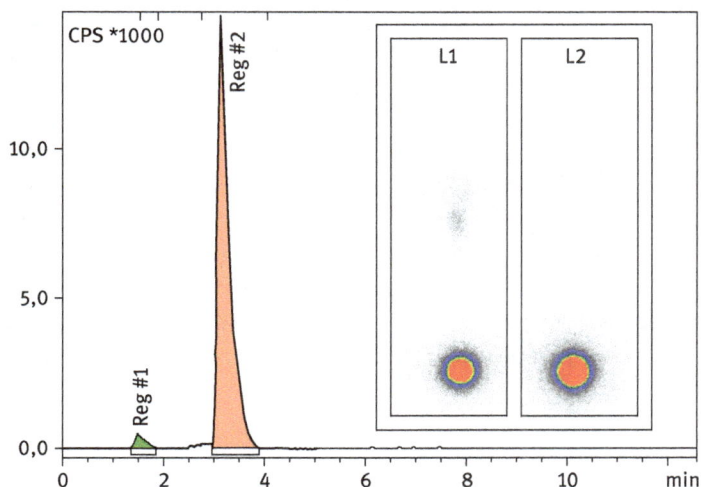

Fig. 7.19: Quality control of a synthesis of [^{68}Ga]Ga-DOTATOC for medical application. HPLC shows the radiopharmaceutical and the nonlabeled ^{68}Ga. HPLC: Nucleosil 100–5 C$_{18}$ AB with 25% acetonitrile/75% 0.1% TFA in water at 0.75 ml flow. Retention and content of ^{68}Ga and ^{68}Ga-DOTATOC is 1.5 min, 2.7% and 3.2 min, 97.3%, respectively. The TLC result is inserted showing on lanes L1 and L2 the syntheses batch prior and after purification: TLC (inserted right): Merck Silica SG60 plates developed with Na-citrate buffer pH = 4, L2: before purification over Strata-X; 93.5%, L1: after purification; 99.0%.

migrating in an external electric field is well known in conventional analytical chemistry, and also finds application in radiochemistry. Here, it offers a straightforward possibility to detect and to quantify the electrophoresis-chromatograms via radiation measurements of the radioactive species on-line or off-line. Another advantage is the use of radionuclides at very low concentrations, which facilitates the chromatographic separation capacities.

7.7.1 Electrodeposition

Similar to the approach of precipitating species A and thereby separating it from the nonprecipitated species B remaining in solution (cf. Section 7.2), radiochemical separations use the transfer of a species A to the surface of a solid material driven by electro-deposition (ED). In the case of a radioactive cation of type Me^{3+} to be isolated, for example, a cathode (usually made of platinum wire or plate) is placed into the solution and an electric current is applied. The radio-cation is deposited on the cathode surface in the form of an Me(OH)$_3$ layer.

ED is mainly applied to separate a carrier-free nuclear reaction product from macroscopic stable target material. A proof-of-principle separation was demonstrated for ^{86}Y, a positron emitter used in medical imaging, from ^{86}SrO targets following a ^{86}Sr(p,n)^{86}Y nuclear reaction. Figure 7.20 illustrates the concept. The irradiated target (in amounts of 100–200 mg of ^{86}SrO) is dissolved in 3 ml of 0.6 N HNO$_3$ in a quartz vessel, acting as the anode. A platinum cathode is positioned in the middle of the vessel. At a constant electric field of 450 mA (corresponding to ca. 20 mA/cm^2), the ^{86}Y^{3+} cation approaches the cathode and adsorbs (electro-deposits) on its surface, basically forming ^{86}Y(OH)$_3$. After the platinum cathode is removed (with the electric current not switched off), it is washed and finally ^{86}Y is desorbed in diluted hydrochloric acid. Another option is to transfer the cathode into a second vessel with fresh electrolyte and to alternate the type of electrode: The initial cathode is turned into the anode, and ^{86}Y is again electrodeposited onto the new Pt cathode, thereby acting as a second purification cycle.

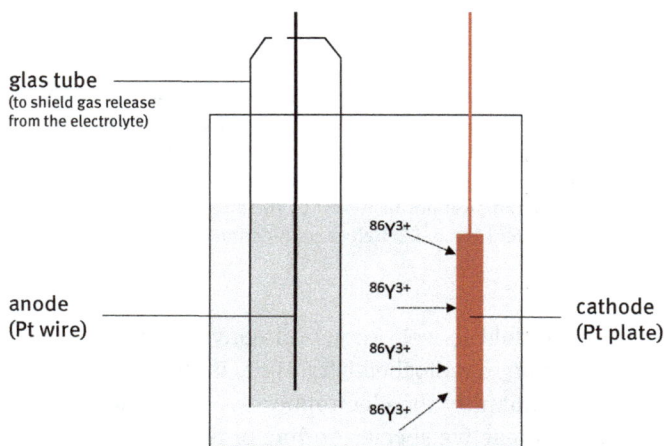

Fig. 7.20: Radiochemical separation of ^{86}Y^{3+} from ^{nat}SrCO$_3$ by means of electrodeposition. The ^{86}Y adsorbs on a Pt cathode, the strontium salt remains in solution (Reischl et al. 2002).

The process only takes a few minutes, and the recovery of ^{86}Y is >90%. The level of ^{86}Sr transferred into the final ^{86}Y fraction is < ppm levels. A similar approach was suggested for the separation of ^{90}Y from the ^{90}Sr/^{90}Y generator, where the no-carrier-added radionuclide daughter ^{90}Y is separated from the no-carrier-added radionuclide parent ^{90}Sr.

This approach to separate either generator-derived trivalent radiometals from di- or four-valent radionuclide generator parents or stable target materials used for cyclotron or nuclear reactor based radionuclide production processes may be extended to a number of currently relevant examples such as ^{68}Ga and ^{44}Sc. Other examples are the isolation of ^{188}Re from aqueous solutions of the ^{188}W/^{188}Re generator and ^{99m}Tc from aqueous solution of the ^{99}Mo/^{99m}Tc generator, cf. Roesch and Knapp (2011) for a review.

Instead of electrochemically adsorbing a trivalent radioactive nuclear reaction product onto a surface of an inert cathode, another process is known to use redox reactions with a corresponding absorber material. This is particularly relevant for two chemically almost identical trivalent metals, namely, the nuclear reaction target material and the radioactive nuclear reaction product, which would behave very similarly in conventional separation strategies such as ion exchange, liquid-liquid extraction, most kinds of chromatography etc., but also ED. This situation is typical for two neighbored trivalent lanthanides. In a few cases, one of the lanthanides may form oxidation states different from III. This is the case for the separation of ^{177}Lu as produced in the ^{176}Yb$(n,\gamma)^{177}$Yb $\to \beta^-$ (t½ = 1.9 h) $\to$ ^{177}Lu (t½ = 6.71 d) reaction. From a mixture of carrier-free ^{177}Lu and macroscopic ^{176}Yb ions in aqueous solution, the macroscopic ^{176}Yb^{3+} (following dissolution of the, e.g., 100 mg of target material ^{176}Yb$_2$O$_3$) is transferred into Hg(Na) amalgam driven by reduction of ^{176}Yb^{III} to ^{176}Yb^{II}, which forms an amalgam with the mercury cathode. ^{177}Lu remains in solution, which is easily separated from the solid Hg(Na)^{176}Yb amalgam, cf. Fig. 7.21.

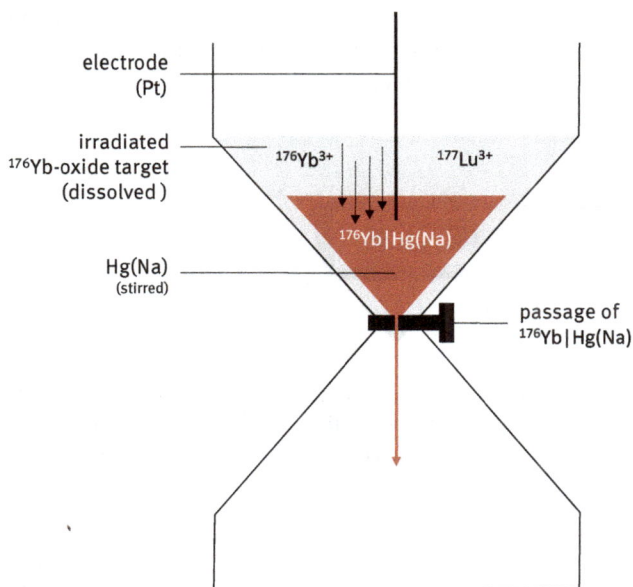

Fig. 7.21: Radiochemical separation of ^{177}Lu^{3+} from neutron-irradiated ^{176}Yb$_2$O$_3$ by means of electrodeposition. The nonradioactive, macroscopic ^{176}Yb forms solid ^{176}Yb | Hg(Na) amalgam, whereas the carrier-free nuclear reaction product remains in solution. This procedure is repeated several times with fresh Hg(Na) in order to remove the macroamount of ^{176}Yb (Lebedev et al. 2000).

7.7.2 Electrophoresis

According to the species to be separated and the kind of interaction between species and supporting material, there are many technological types of electrophoresis (EP), such as paper EP, capillary EP, gel EP, but also affinity EP, etc. The most conventional version is paper EP. The mixture of radioactive species is added in a few μl to the middle of a paper strip, soaked with an adequate (most often aqueous) electrolyte. The two ends of the paper strip are immersed in the electrolyte medium. At the two ends of the strip, two electrodes induce a constant electric field, and the species start to move (to migrate) towards the corresponding electrode. After a certain time, the electric field is switched off; the paper strip is removed and dried. The position of the radioactive species is detected by either scanning the whole paper or by cutting the strip into segments and measuring each segment separately.

It is obvious that the separation depends on whether the species has a negative or a positive charge or is electrically neutral. It is difficult to separate two cationic or two anionic species. However, by forming metal-ligand complexes ML_x of, e.g., two cations, different electrical charges of the complexes can be induced enabling a separation with this method. Figure 7.22 demonstrates this for the separation of radiostrontium and radioyttrium.[12] Here, the complexing ligand L is citrate.

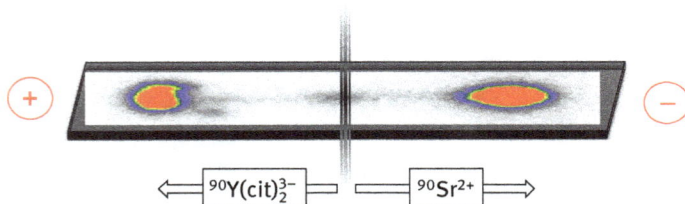

Fig. 7.22: Radiochemical separation of $^{90}Y^{3+}$ and $^{90}Sr^{2+}$ by means of paper electrophoresis. The electrolyte contains citrate anions. Its concentration is high enough to form a $[^{90}Y(cit)_2]^{3-}$ complex, moving towards the anode, whereas $^{90}Sr^{2+}$ does not form a citrate complex and moves towards the cathode.

7.7.3 Electromigration

In contrast to electrophoresis, electromigration is a method which tries to exclude the physicochemical interaction between a (radioactive) species and the material of

---

12 This is of practical interest, because ^{90}Y is commercially distributed for medical applications, and is obtained as the daughter in a $^{90}Sr/^{90}Y$ generator. The separated ^{90}Y should not contain traces of the parent ^{90}Sr. However, both nuclides are β^--emitters and the content of ^{90}Sr within an excess of ^{90}Y cannot be analyzed spectrometrically directly from that mixture. Thus, both nuclides must be separated first.

a surface of a solid medium such as paper, oxides, salts, metals, glass, etc. Electromigration (EM) thus focuses on the "pure" impact of the charge of the species on their (electro)migration velocity. Capillary EP would belong to this category. EM basically utilizes on-line detection of the electromigration of the radioactive species. The simple concept would be a glass tube, filled with a pure aqueous electrolyte, to locally inject a small volume (such as 1 to 10 µl) of the radioactive species and to monitor the electromigration by scanning the tube with a radioactivity detector device. The electromigration velocity u (in m/s V) constitutes a parameter, representing the migration ion, the type and the ionic strength of the electrolyte, and the temperature. This parameter is as unique and characteristic for the species as the diffusion coefficient is.

While EM is most often used for chemical speciation of a single radiospecies, for determining its electomigration velocity at infinite dilute aqueous solution, to investigate hydrolyses and Me-L-complex formation equilibria of radiometals – thus representing an analytical method – it can also be used to separate two species, even of the same electrical charge. For reviews on and examples of electromigration studies, see Roesch and Khalkin (1990) and Mauerhofer et al. (2002).

7.8 Volatilization/sublimation/evaporation

The radiochemical separations discussed in this section are generally referred to as "volatilization",[13] which basically means that the radiospecies to be isolated is in a gaseous state. Sublimation, evaporation, and distillation are synonymously used for the volatilization processes referred to in this section. In almost all cases, the procedures start from solid rather than liquid phases and the separation is "dry", which substantially differentiates them from the other procedures described so far.[14] Thus, "distillation techniques" were readily adopted for radiochemical separations. Already in 1962 a compilation of separations applied to radioisotopes of almost the entire range of chemical elements appeared (DeVoe 1962).

The most obvious version is the release (i.e., separation) of a radionuclide belonging to a gaseous element. The most prominent example in radiochemistry is the isolation of isotopes of radon as formed in the naturally occurring transformation chains that radon isotopes naturally emanate from, for example, uranium ores.

---

13 Also generally referred to as "distillation". However, distillation in conventional chemistry describes separations starting from liquid samples. In radiochemistry, in contrast, most of the samples to be treated are solid ones (irradiated metals or alloys, oxides, salts).
14 For many separations, chemical purity is essential. This also refers to contaminants of traces of stable species present in the initial sample, but also those introduced by the chemical solvents. Here, dry separations avoid the latter contaminations.

If (at room temperature) the partial gas pressure of the species is zero or low, their separation can be induced by simply heating a sample to separate and isolate the most volatile species. The process starts from solid samples and yields gaseous species without a notable appearance of a liquid intermediate. Supposing the differences in thermal volatility are significant, the process is thus similar to *sublimation* known in conventional chemistry. It is the most effective way to isolate radiohalogens in their element state: ^{123}I or ^{131}I from irradiated tellurium oxides, cf. Fig. 7.23, and ^{211}At from irradiated bismuth metal. Alternatively, the pressure may be lowered down to vacuum, where even radiometals volatilize – in particular if temperatures are increased.

In a more advanced version, the increasing temperature applied to the solid mixture is accompanied by letting a gas stream continuously pass the surface of the solid to allow the volatile component to be "gas-transported" away from the mixture and deposited later at a different place and on the surface of a cold material. The nonvolatile species remains at the origin.

A rather sophisticated version is to separate two (or more) volatile species. In this case the radiochemical method resembles conventional gas chromatography but modified to a preparative rather than analytical version. Here, the different physicochemical kinds of interaction between the volatile species and the absorber material define the separation quality. However, the ultimate approach is to avoid column material packings and to exclusively rely on the interactions of the volatile species with the surface of the chromatographic column. Ideally, the two (or more) volatilized radiospecies deposit on the inner surface of quartz or metal columns at different positions according to a temperature gradient of that column. The desorption is driven by the adsorption enthalpy of the species on the surface material selected. The process is defined as "thermochromatography".

Finally, an adequate volatility of a radioactive species naturally existing in a nonvolatile chemical species in the sample can be induced. It in can be reacted chemically to form a new molecule, which is volatile – and which sublimates (or is gas-transported) away from the original sample, returning to a nonvolatile status.

7.8.1 Volatilization/sublimation/evaporation

One of the most conclusive "dry" radiochemical separations is the isolation of isotopes of iodine and of astatine following their production by particle-induced nuclear reactions. Medically relevant ^{123}I (for SPECT), ^{124}I (for PET), or ^{131}I (for scintigraphy and therapy) can be produced by (p,xn) or (d,xn) nuclear reactions on macroscopic tellurium oxide TeO_2 targets.[15] The irradiated TeO_2 is transferred into a horizontal quartz apparatus as, for example, shown in Fig. 7.23. Either inert gases or just air

---

15 Cf. Chapters 12 and 13 for the production of the medically relevant nuclides.

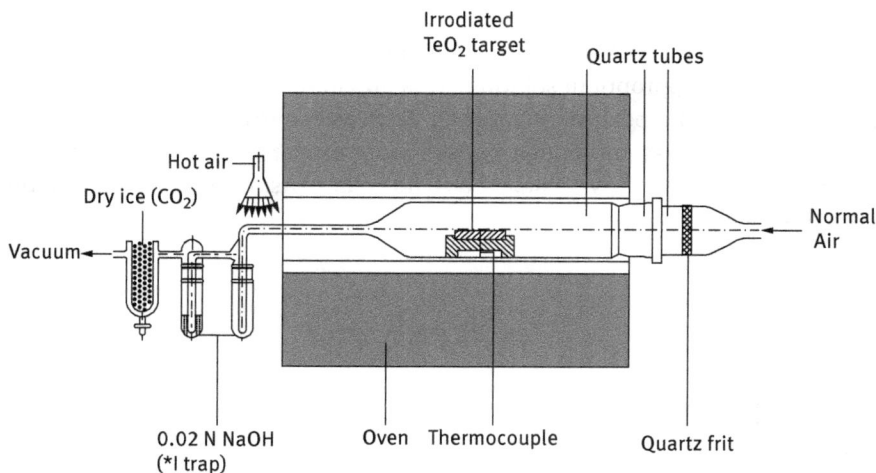

Fig. 7.23: Apparatus for separating radioiodine from irradiated TeO$_2$ targets (after Qaim et al. 2003).

flows over the irradiated target. Once the temperature at the position of the irradiated target is increased by heating the quartz apparatus close to the melting point of 732 ° C of the tellurium oxide, all the radioiodide inside the crystalline tellurium oxide diffuses to the surface. From this surface, iodine is volatilized and transported within the stream of the gas into an aqueous solution of NaOH to trap the isolated radioiodine as iodide. The tellurium oxide, typically made of isotopically enriched and expensive Te isotopes, remains intact and is used for repeated irradiations.

Alternatively, specific absorbers may be inserted into the gas stream to absorb the volatile radioalogen isotope. This is typically the case for separating ^{211}At from irradiated bismuth targets following a Bi(α,2n) nuclear reaction. Typical absorbers are short (1–2 cm) pieces of silver or platinum wire, positioned within the glass tube exactly at a position where the temperature is low enough for the deposition of the halogen radioisotopes.

Another version is also of interest in which the irradiated target is heated in the reactant gas (O$_2$, Cl$_2$, and others) with formation of a nonvolatile residue while the nuclear reaction product is transferred further by the gas flow. These separations were usually carried out in a *horizontal apparatus* similar to that displayed in Fig. 7.23, but recently the vertical arrangement has come to be viewed as more rational. This *vertical arrangement* greatly decreases the carry-away of gaseous products against the direction of incoming reactant gas flow and allows the process to be easily automated. The target is loaded at the bottom of the open apparatus, the apparatus is closed using inserted smaller quartz tubes, the reactant gas is passed through the quartz tube system, and then the assembly is placed in the vertical oven with the required temperature distribution. After the termination of the separation process the assembly (or its inner segment only) is taken out of the oven and cooled. The adsorbed product is washed off from the inner quartz tube surface with a minimum volume of an appropriate solution. This apparatus cf. Figure 7.24, first developed to separate the positron emitter ^{94m}Tc from the irradiated molybdenum oxide in 25 minutes, was subsequently used more universally for separations of the systems ^{110}Sn/^{110m}In, ^{186}W/^{186}Re, ^{188}W/^{188}Re, ^{72}Se/^{72}As; cf. Roesch and Knapp (2011) for a review.

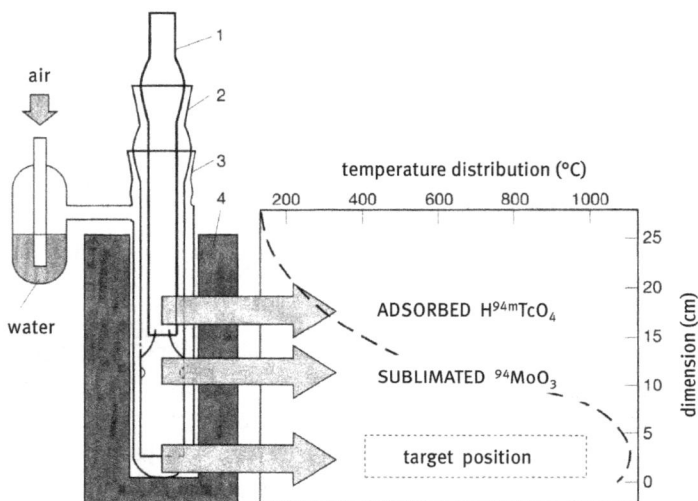

Fig. 7.24: Vertical-type thermochromatographic apparatus for routine separation of ^{94m}Tc from irradiated ^{94}Mo targets. The irradiated target is transferred into a vertical quartz container (3) and a flow of air is passed over it. An oven (4) induces a temperature gradient (right) with the hottest temperature at the bottom of the quartz apparatus (ca. 1100 °C). At elevated temperature, both the target element (Mo) and the nuclear reaction product (Tc) oxidize and form the corresponding oxides. The less volatile MoO$_3$ sublimates within the middle quartz tube (2). The H^{94m}TcO$_4$ adsorbs at an inner quartz tube (2), which is removed after the separation process. H^{94m}TcO$_4$ is easily dissolved from the quartz surface (Roesch et al. 1994).

Furthermore, when separation of nuclear reaction products does not entail changes in their chemical composition, multiple use of the target material holds the greatest promise for large-scale production of, e.g., ^{67}Ga, ^{111}In, and ^{201}Tl. The separation procedure consists in merely heating massive targets at controllable pressure of the reactant gas. The target material loss does not exceed 0.1% in one cycle of radiochemical separation, but it is easily restored after 20–30 target irradiation-radionuclide separation cycles. In addition, multiple use of the target material gradually leads to a high specific activity of the produced radioactive samples due to periodic separation of impurities. Furthermore, it should be noted that similar protocols might be used for constructing generator systems. The concept is to thermally adsorb the generator parent nuclide, and to thermally volatilize the generator daughter – *and* to be able to repeat this separation protocol many times. By now the parent-daughter radionuclide pairs ^{188}W/^{188}Re and ^{72}Se/^{72}As have been studied.

7.8.2 Thermochromatography

Similar to gas chromatography (GC),[16] thermochromatography (TC) is a stationary nonisothermal gas-solid chromatography based on thermal variation of the interaction between the volatile radioactive species (atoms or molecules) and the adsorbent. In this method a negative temperature gradient is produced in the direction of the carrier gas flow or, alternatively, in vacuum.[17] It is, consequently, this temperature gradient that retards the transport of species in the column. If the slowing down is different for different species, they are separated chromatographically.

It allows collecting components from a complex mixture at a certain position, which is at a certain temperature of the inner surface of the TC column. This can be achieved either off-line, i.e., injecting a mixture once (as in common GC), or on-line, i.e., continuously passing a mixture of on-line produced radionuclides through the TC column (in contrast to common GC). In radiochemistry, the most popular version is TC with a stationary temperature gradient, in which the concentration of a substance and its local distribution depend on the distance x (or temperature, which depends on the distance x according to the temperature gradient), i.e., $c = (x$ or $T)$. While this is similar to conventional GC, in TC the volatile compounds typically remain in the column.[18] For a comprehensive description of the physicochemical

---

16 Historically, "thermochromatography" refers to the gas-chromatographic method for separation of mainly volatile organic compounds in which the temperature field with a negative gradient moves along the column, so that the outlet temperature gradually increases ousting the separated substances from the column. In this case the *gas-chromatogram* is the time t (or temperature T) dependence of the outlet concentration, i.e., $c = (t)$ or $f(T)$.

17 Note that the ultralow concentrations of the radioactive species undergo various interactions with the many chemical species that are part of aqueous solvent systems or in condensed matter – also with those that are typically disregarded as "impurities" at ppm levels in conventional chemistry. Here, gases or ultimately, vacuum, represent much "cleaner" system.

18 It is particularly noteworthy that Marie and Pierre CURIE used the method in question as far back as the end of the 19th century when they discovered Ra and Po (Curie and Curie 1898). Polonium sulfide was distilled in a carbon dioxide flow into a tube with a negative temperature gradient, where it was precipitated in a narrow zone. Unfortunately, this experience and method was long forgotten. What may also be considered the first publication on the use of TC in radiochemistry is the paper by Merinis and Bouissieres (1961) on separation of no-carrier-added (n.c.a.) spallogenic products: radioactive Hg, Pt, Ir, Os, and Re isotopes have been separated from macroscopic gold targets bombarded with medium-energy protons. Another important work that stimulated the further development of TC was an attempt to separate and identify the artificial element 104 (now named rutherfordium), an analogue of Hf, which clearly showed the advantages of the method for

and theoretical background see the review by Novgorodov et al. (2011) and the references therein.

7.8.2.1 Thermochromatographic setup

Figure 7.25 displays a typical thermochromatographic apparatus. The main part is a *gradient oven* providing the required temperature field along the column. The *thermochromatographic columns* (TCC) are usually made of glass, quartz, porcelain, or metallic tubes, chemically resistant to reactant gases. They may be open (hollow) or filled with various substances (metals, oxides, halides, etc.). The gradient oven may be made of a large number of short sections or by winding a heat-releasing wire (nichrome, superkanthal, platinum, etc.) on an insulating tube (the TCC, see below) etc. in a varying mode. The simplest design is an oven heating at one end of a long heat-conducting tube surrounded with a vacuum enclosure. This design ensures almost linear temperature distribution along the apparatus. The "cold" end of the tube is usually cooled with water or even liquid nitrogen.

The apparatus also comprises devices to organize and monitor the *gas system*. It consists of a system for the chemically inert carrier gas, to which traces of reactants (which form volatile species of the radionuclides under investigation) may optionally be added, or which may be completely substituted by the reactant gas. This system involves supply of the gas, purification, and flow rate control, but optionally also the addition of aerosol particles as inert carriers. Alternatively, the gas system is completely avoided and an appropriate vacuum system is applied instead. The apparatus needs a system for injection/inlet of the radionuclide mixture to be separated (periodic or continuous). Probes containing the radionuclides of interest are inserted into the TCC at places of maximum temperature (starting temperature T_s) in a batch mode before the experiments (off-line) or continuously (on-line) using gas jets. In this latter version, microdisperse aerosol particles (KCl, MoO_3, etc.) are added to the gas stream in order to support the transfer of the radioactive species with the radionuclides of interest adsorbed on it. Figure 7.25 illustrates this situation. Finally, there is also a system for trapping outgoing reactant gases and possible radioactive gases or aerosols.

The distribution of the radionuclides adsorbed on the inner wall of the TCC must be registered according to their radioactive emission characteristics. This is organized with an array of radioactive detectors on-line, or by scanning the TCC with one detector, again either on-line or off-line. A method for determining the profile

---

studying short-lived radionuclides (Zvara et al. 1971). The simple equipment required and the high efficiency of the method stimulated its rapid theoretical and practical development in the 1960s, marked with some outstanding results, for example, the characterization of previously unknown isotopes of many elements from the Periodic Table and even more significantly, the discovery of new transactinide elements.

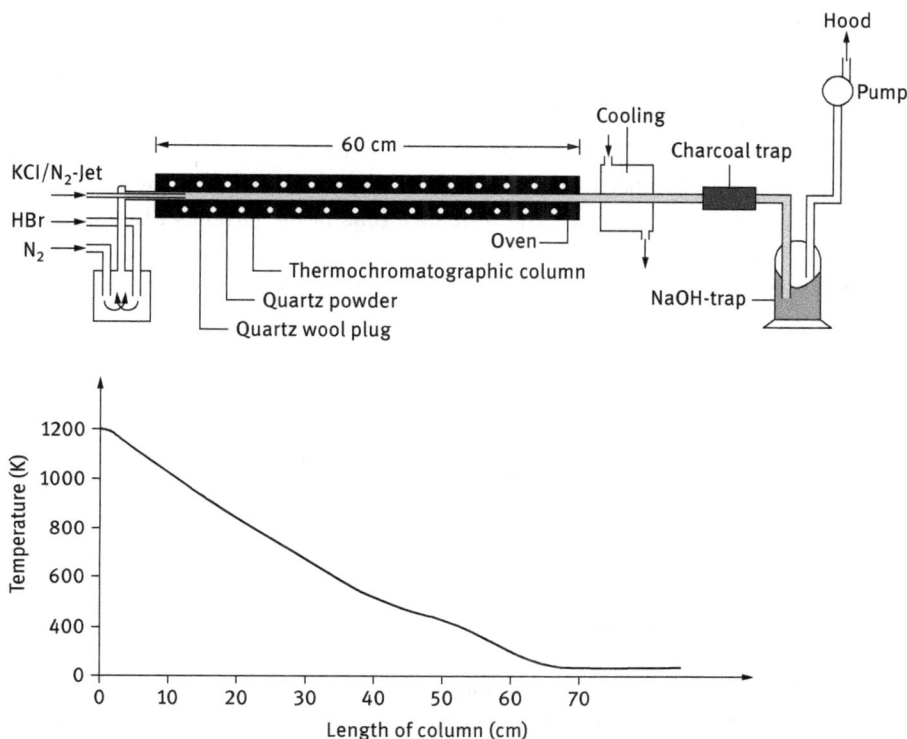

Fig. 7.25: Horizontal-type thermochromatographic apparatus and typical profile of temperature along the column (Hickmann et al. 1993).

of the distribution of separated substances in the column is of major importance in thermochromatographic experiments. Most of them are carried out with radionuclides emitting radiation that penetrates the columns walls (relatively "hard" β- and γ-rays), which allows the radioactivity within the column sections to be measured off-line either by cutting the column into small parts or by screening the column using collimated detectors. If sufficient activities of radionuclides transforming with emission of high-energy γ-rays are used, in principle it is possible to monitor the substance distribution in the column on-line during the thermochromatographic experiment by using one or several well-collimated movable detectors. Mixtures of radionuclides usually require the use of HPGe detectors because of the high-energy resolution.[19]

---

19 The situation becomes much more complicated when the radionuclides used undergo α decay or spontaneous fission, i.e., if nonpenetrating radiation needs to be measured. If these radionuclides are long-lived, the column is cut into small sections, the substances adsorbed on its inner surface are washed off with appropriate solutions, and the resulting solutions are analyzed. In the

7.8.2.2 Transport of volatile species along the column

Theories on the thermochromatographic process (Merinis and Bouissieres 1969; Eichler and Zvara 1975; Rudolf and Bachmann 1978; Novgorodov and Kolatchkowski 1979), beginning with works by Merinis and Bouissieres (1969),[20] are based on the fact that for a given species (atom, molecule), the *transfer time* within the TCC breaks into two components, i.e., the *residence time in the gaseous phase* and the *residence time on the surface of the column*. The latter is governed by the average lifetime of the species in the adsorbed state $\tau_a(s)$ related to the temperature by the FRENKEL equation:

$$\tau_a = \tau_0 \exp\left(\frac{Q}{RT}\right), \tag{7.5}$$

where τ_0 is the period of oscillation of the species in the adsorbed state perpendicularly to the surface (s); Q is the *adsorption heat* (J mol^{-1}); R is the molar gas constant (≈ 8.314 J mol^{-1} K^{-1}); and T is the absolute temperature (K).

7.8.2.3 The thermochromatogram

The experiment results in a *thermochromatogram*. An example is given in Fig. 7.26, which displays the distribution of bromides of uranium fission products continuously supplied to the column for 15 minutes. The thermochromatogram for each chemical compound (or element) allows the determination of the basic parameter T_a, the adsorption temperature of the substance (center of gravity of its distribution in the column).

7.8.2.4 Thermochromatography applied to the investigation of new transactinide elements

Discovery and investigation of new elements is an extreme challenge, because their half-life is a few seconds or fractions of seconds and the rate of their production at modern accelerators is very low (one atom per a few hours or even days). On the assumption of analogy between transactinide elements and group 4–18 elements, the

---

case of short-lived radionuclides, e.g., transactinide elements, semiconductor, or track detectors (mica, quartz) placed inside the column are used. They allow single decays to be determined at each point of the column. In principle, other methods may also be used to determine the distribution of substances (for microamounts) in the column, e.g., photospectrometry, X-ray fluorescence, etc.

20 For details and the references mentioned, see the review by Novgorodov et al. (2011).

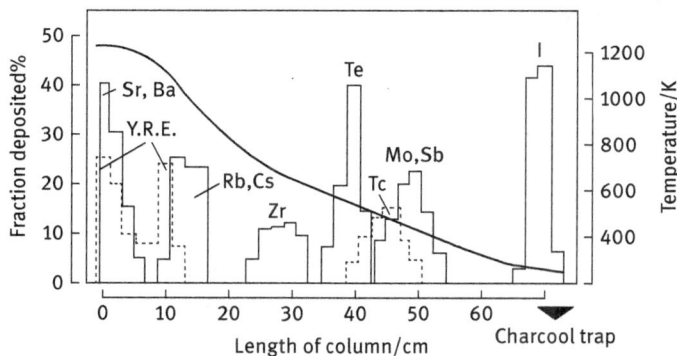

Fig. 7.26: Element distribution in a thermochromatographic column with 9 vol. % hydrogen bromide as reactive additive to the carrier gas nitrogen, together with the temperature profile along the column. Quartz-powder column, 15 min exposure time (Hickmann et al. 1993).

halides, oxohalides, and oxides of the first transactinide elements were expected to have increased volatility, which might allow good selectivity of their separation and concentration by gas chemistry methods (IC and TC). The first identifications of Rf (element 104) were carried out by isothermal chromatography (IC) of its most volatile compound, the chloride. It was shown that the behavior of Rf chloride was analogous to the behavior of Hf tetrachloride. Trends among volatilities in the groups of the Periodic Table have systematically been studied, cf. Tab. 7.1, and contributed to the chemical identification of the heaviest homolog of several groups. For a detailed discussion cf. Chapter 9.

Tab. 7.1: Relative volatilities of some analog compounds of groups of the periodic table of elements indicating the heaviest homolog of each group.

$HfCl_4$	$<RfCl_4$	$\approx ZrCl_4$
$HfBr_4$	$<RfBr_4$	$\approx ZrBr_4$
$DbCl_5$	$<NbCl_5$	
$TaOCl_3$	$<DbOCl_3$	$<NbOCl_3$
WO_2Cl_2	$\approx SbO_2Cl_2$	$<MoO_2Cl_2$
BhO_3Cl	$<ReO_3Cl$	$<TcO_3Cl$
HsO_4	$<OsO_4$	

7.9 Hot atom chemistry/SZILARD–CHALMERS-type separations

7.9.1 What is a "hot" atom?

Let us start with the "cold" atom? Temperatures of atoms are cross-linked to their energies and to their average speed of movement. The energy for a certain state of a particle is $E_{kin} = (m/2)v^2$. This is quantitatively expressed by the MAXWELL–BOLTZMANN correlation bridging speed, mass and temperature with probability density functions (v), which quantifies the probability of the particles speed per unit of speed, where the translational energy corresponds to a temperature:

$$f(v) = \sqrt{\left(\frac{m}{2\pi k_B T}\right)^3} 4\pi v^2 \exp\left(-\frac{mv^2}{2 k_B T}\right). \tag{7.6}$$

The most probable speed $\hat{v}$ is

$$\hat{v} = \sqrt{\frac{2 k_B T}{m_M}} = \sqrt{\frac{2RT}{M}}, \tag{7.7}$$

where R is the gas constant, k_B the BOLTZMANN constant, and m_M is the molecular mass of the atom or molecule.

At room temperature, atoms or molecules move or "vibrate" within their bounds according to "thermal" energy. For the nitrogen gas N_2, "most probable" velocities $\hat{v}$ are 422 m/s at "room" temperature, i.e., 26.85 °C = 300 K, and 688 m/s at 800 K (526.85 °C). Doubling the absolute temperature (Kelvin scale) increases this velocity by a factor of $\sqrt{2} \approx 1.4142$.[21]

Historically, a "hot atom" connoted an atom "freshly" formed by a nuclear process – which could be a spontaneous transformation of an unstable, radioactive nuclide, but also the product of a nuclear reaction induced on a stable target nucleus.[22] The term "hot atom chemistry" now is synonym for "chemical effects of nuclear

---

21 Note that these speed distributions, by definition, apply to ideal gases, where interactions between the atoms or molecules are zero. Real gases, of course, differ, and so do their speed distributions. In condensed matter, where many of the "hot" atom chemical reactions proceed, the situation is much more complex.

22 For isolated neutrons, once they were slowed down by numerous interactions with atoms and molecules existing at room temperature until they had left all their initial kinetic energy and acquired the terminal energy equivalent to that of their neighboring atoms and molecules, this "thermal" equilibrium energy amounted to 0.0025 eV. (Note that this energy level is much lower than most of the *recoil energies* of "hot" atoms by many orders of magnitude.)

transformations". According to the IUPAC,[23] it is defined as "The rupture of the chemical bond between an atom and the molecule of which the atom is part, as a result of a nuclear reaction of that atom".

In an abstract sense, "hot" means "higher temperature", and higher temperature means "higher energy". For a particle, "higher energy" means "higher velocity". Since chemical effects are concerned, those "hot" atoms should have a translational or electronic energy above conventional chemical bond energies.[24] As a consequence of primary nuclear transformations of unstable nuclei or of nuclear reactions, the recoil atom with its corresponding impulse and kinetic energy can be attributed with an incredible quasi-temperature of the "hot" atom – yielding an estimation of the chemical "reactivity" of such an atom. Both aspects (translational or electronic energy) are the reason for a "rupture" of the chemical bonds the atom is/was involved in. However, referring to either "translational" or "electronic" energy indicates that the chemical reasons for how "hot atoms" rupture chemical bonds may differ. The first aspect is related to the pure "speed" and/or energy of the atom. Chemically speaking, this refers to the primary effect of the *recoil energy of the nucleus* of an atom formed. The second aspect addresses the rapid and dramatic changes of the *electron shell occupancies and structures* within the shell of the atom, being inherent consequences of primary and secondary radioactive transformation processes (i.e., mainly induced by electron vacancies generated by positron emission or electron capture). The sciences of hot atom chemistry were studied systematically in the 1960s–1990s, and comprehensive books collect a variety of experimental data and theoretical concepts, such as Matsuura (1984) and Adloff et al. (1992).

7.9.2 Recoil energies in nuclear transformations

Recoil energies are defined in the corresponding chapters of Vol. I for α-, β-, IC-, and γ-emissions. Basically, nuclear recoil follows the conservation of impulses, and impulses $p = mv$ and kinetic energies $E_{kin} = \frac{1}{2}mv^2$ refer to the two species formed in the primary or secondary nuclear transition. The larger the recoiling nucleus is, the lower is its kinetic energy. The kinetic energy of recoiling nuclei of identical mass is greater if the mass and the energy of the second component are larger, i.e., either α- or β- particle or IC electron or photon, cf. e.g., Tab. 9.3 in Vol. I. Accordingly, recoil

---

23 IUPAC (International Union of Pure and Applied Chemistry) Compendium of Chem. Terminology, 2nd edition, 1997.

24 There is some overlap between *hot atom chemistry* and *radiation chemistry*: Radiation chemistry also studies chemical effects following the delivery of excitation energy to a single atom larger than chemical bond and ionization energies. However, in the latter case this energy is caused by *external* sources, i.e., not from inside the molecule or atom itself induced by nuclear reactions or nuclear transformations.

energies are large for α-emissions (reaching the 0.1 MeV level), lower for β-emissions, and lowest for photon-emission (keV scale).

7.9.2.1 Primary α-emissions

It was already realized in 1904 that the α-transformation process gives the new nucleus a significant recoil energy – an effect measured in terms of its (chemical) separation from the parent radionuclide. The product nucleus appeared to be "volatile" relative to the parent nuclide fixed on the surface of a solid. For example, ^{214}Pb (t½ = 26.8 min), the product of ^{218}Po (t½ = 3.05 min) undergoing α-transformation, was radiochemically isolated "without" chemistry in 1904.[25] Using the same concept and further along the ^{226}Ra natural transformation chain,[26] ^{210}Tl (t½ = 1.3 min) was isolated from this ^{214}Pb (and because of this: discovered!) by HAHN and MEITNER in 1909.

7.9.2.2 Primary β-emissions

The β-electrons are of relatively high maximum kinetic energy, reaching MeV scales. Due to their low mass, however, and because most of the β-particles emitted are of much lower kinetic energy than the E_{max} value (continuous β spectrum), the average recoil energy of the formed nucleus is small. Nevertheless, the nuclei formed in β-transformation processes are always chemically different from the initial element (by ΔZ = + 1 for β⁻ and ΔZ = – 1 for β⁺ and EC). Thus, in all cases the new element is a different one – and except for example in transitions between trivalent lanthanides – has chemical properties significantly different from those of the progenitor. For example, a radioactive halogen atom may transform into a noble gas (β⁻) or a group VI element (β⁺ and EC); a divalent radioactive metal may transform into a trivalent (β⁻) or monovalent metal (β⁺ and EC). In all cases, either covalent or coordination bounds are broken and changed into different ones. This is referred to as "post-effects" of nuclear transformations and has a significant impact on speciation chemistry and, in particular, separation chemistry.

---

25 "Since radium A breaks up with an expulsion of an α particle, some of the residual atoms constituting radium B, may acquire sufficient velocity to escape into the gas and are then transferred by diffusion to the walls of the vessel." (RUTHERFORD1904.) Radium A = ^{218}Po, Radium B = ^{214}Pb.
26 Cf. Chapter 4 in Vol. I; ^{214}Pb → α → ^{210}Hg → β⁻ (t½ > 0.3 μs) → ^{210}Tl.

7.9.2.3 Secondary γ-emissions

In the case of de-excitation of excited nuclear states (such as in transition processes of metastable nuclear isomers) the recoil energy is small (ca. 5 keV). This energy is generally too low to induce molecular rupture directly. This relates to most of the isomeric transitions.

7.9.3 Orbital electron capture and cascades of releasing electrons and filling electron vacancies

AUGER cascades are synonymous with extreme ionization of the residual atom. Electrons initially involved in chemical bonds are lost. Moreover, the high positive charge of the residual atom is immediately spread to neighbor atoms of the same initial molecule, inducing a "COULOMB explosion". Atomic and molecular fragments are formed, each with individual secondary recoil energy (depending on their mass). Finally, the high number of AUGER electrons emitted, though of relatively low kinetic energy, creates intense local ionization in condensed matter (cf. the LET values of the various particles). Local ionization is radiolysis of, e.g., water, again causing rupture of chemical bonds. The electrons themselves are, at least in principle, able to oxidize surrounding atoms. A similar process occurs when refilling a vacancy in an inner electron shell caused by internal conversion or by positron emission.[27]

7.9.4 Recoil energies in nuclear reactions

When in 1934 SZILARD and CHALMERS bombarded C_2H_5I with neutrons causing a ^{127}I $(n,\gamma)^{128}I$ nuclear reaction,[28] they observed experimentally that surprisingly ca. 60% of the ^{128}I activity detected was "inorganic" ^{128}I iodine, while the remaining ca. 40% was still present as $C_2H_5{}^{128}I$.[29] In terms of radiochemical separation, the effect

---

27 One may expect that these processes occur relatively seldom. However, most of the recoil nuclei formed in primary or secondary transformation processes do not necessarily exist at ground-state levels. Instead, excited nuclear states are populated and de-excite via photon emission, IC or pair formation, and thus simultaneously induce hot atom effects. Nucleus recoil and electron shell effects overlay, and sometimes it is the latter effect that dominates.

28 . . . actually, the "real" intention of SZILARD and CHALMERS was to create new chemical elements according to $_zE(n,\gamma) \rightarrow \beta^- \rightarrow _{z+1}E$ . . . .

29 In nuclear reactions, recoil energies depend on the type and energy of the projectile, the type of the ejectile and the mass of the product nucleus. For example, recoil energies are in the order of 0.1 keV for (n,γ) reactions and 0.1 MeV for (n,p) reactions. For example, a "hot" ^{18}F nucleus is formed in, e.g., $^{19}F(\gamma,n)^{18}F$ reactions. The ^{18}F released from the compound nucleus forms new

is obvious: The fraction of "inorganic" ^{128}I (precipitated as Ag^{128}I) is easily separated from the "organic" C$_2$H$_5$^{128}I and from the irradiated, nonradioactive target material C$_2$H$_5$^{127}I. This is a clear evidence of a hot atom chemistry way to isolate radioactive and nonradioactive isotopes of the same element.[30] Consequently, an additional term was introduced for the option of using hot atoms for chemical separations: the SZILARD–CHALMERS reactions (Szilard and Chalmers 1934).

An early assumption was that the momentum imparted on the iodine atom by the impinging neutron would rupture the C–I bond. However, this rupture can also be observed also by low-energetic, thermal neutrons of ca. 0.02 eV – energy much too low to simply knock out the iodine atom. So, in the present context, it is not the direct effect of nucleus A displacement induced by the incoming projectile a of the nuclear reaction process of type (a, b), cf. Chapter 13 in Vol. I that matters: It is the recoil of the radioactive product nucleus B caused by the ejectile b of the nuclear reaction – in the present case the photon of the (n,γ) process.

7.9.5 Hot atom effects applied to radiochemical separations I: Radioisotope separation and radioisotope enrichment

One of the unique features of radiochemistry is its ability to separate two isotopes of one and the same chemical element in the course of radioactive transformation processes.[31] This is exclusively due to hot atom effects, and is mostly related and applied to the production of relevant radionuclides in the course of neutron-capture nuclear reactions, mostly with thermal neutrons, i.e., A(n$_{th}$,γ)B reactions with A being a stable isotope A of a chemical element E ($_Z$E$_N$) and B being $_Z$E$_{N+1}$. In this case, the radioactive product B is isotopic with the target element A, and it cannot be separated chemically. Consequently, the product nucleus B co-exists with the target atom. Its specific activity is its absolute radioactivity related to the cumulative mass of B and A, cf. Chapter 4 in Vol. I. Depending on the mass of A used, the neutron flux and the neutron capture cross section, specific activities of B are relatively low, in particular at low flux reactors. Hot atom chemistry, however, allows isolating a certain fraction of the product radioisotope B from the target element.

There are different options for the fate of a radioisotope produced in a nuclear reaction illustrated for (n,γ) processes: A stable target nucleus chemically bound

---

fluorinated species with surrounding molecules, which was studied systematically with fluorinated hydrocarbons, where ^{18}F/H and ^{18}F/^{19}F replacement reactions occur.

30 SZILARD himself wondered that to " . . . separate the radioactive iodine from the bulk of bombarded iodine . . . sounds 'blasphemic' to the chemist . . . " (Lanouette 1992).

31 Different isotopes of one and the same chemical element can also be separated by other technologies, however, these are more tedious and expensive: including gaseous diffusion (cf. the enrichment of ^{235}U from ^{238}U as hexafluoride compounds), electromagnetic (cf. mass separation), etc.

within a molecular or crystalline matrix (the "cage" as illustrated schematically in Fig. 7.27 as a three-dimensional box with dotted lines indicating chemical bonds) is irradiated. Fate 1 is thermalization of the hot atom nucleus and a terminal chemical stabilization as a new species. Fate 2 is a process in which the hot atom "returns" to a vacancy created within the cage and finally remains there. Fate 3 is a concept in which the hot atom hits a stable atom to eject it from the cage to finally take over its position within the cage. Both fates 2 and 3 are called "retention" or "annealing reactions". In both cases, the radioisotope B finally finds itself in the "cage", i.e., as the chemical species of the irradiated target material. Experimentally, it depends on the physicochemical properties of this material (including metal-ligand complexes, oxides, salts etc.).[32]

The other option is B thermalizing as it is. Now it terminates in a new chemical form that is outside the "cage". While these processes are continuously ongoing with the irradiation, the distribution of the different chemical species of B is fixed at the end of bombardment. Yet, the mechanistic explanation of the individual processes is quite complex and difficult.[33] Anyhow, the formation of a new chemical species of the radioisotope produced is the essence of the separation of a radioisotope from a stable isotope of one and the same chemical element.

The lower the amount of stable A (2 and 3 in Fig. 7.27) is in the isolated fraction of B (1 in Fig. 7.27), the higher its isotopic "enrichment"[34] is and the larger the increase in the final specific activity of that fraction of B that is isolated.

The key factor to achieving maximum enrichment lies in the stabilization of the hot atom B in a defined chemical environment and to blocking the backward transfer of B into the cage. A conclusive example is the enrichment of a radiolanthanide isotope produced by thermal neutron capture as produced from an isotopic, macroscopic

---

32 In condensed matter, i.e., solids such as metal-ligand crystals, or liquids, both organic and aqueous, the fate of the recoil or hot atom is difficult to quantify. Reforming of the *initial* chemical bonds often follows the primary ejection of the hot atom. Crystal structures tend to "hold" the recoil atom to a significant degree due to reforming initial bonds and the fraction of newly formed species is rather small. In ligands, these "cage effects" are weaker and the recoiling nucleus diffuses into the liquid phase. However, mechanistic studies are best performed in the gas phase, where the isolated recoil nucleus cannot undergo rapid electron-exchange processes with surrounding species.

33 Clearly, there is a relevant recoil of ^{128}I determined by the ejected photon, which plays a major role. However, in addition, de-excitation of nuclear states in ^{128}I accompanied by internal conversion and AUGER electron emissions contribute to coulomb and charge effects and ionizations. Nevertheless, the basic aspects are initial bond rupture, energy transfer, charge transfer, ionization, radical formation, and ultimate chemical speciation. This altogether influences the ^{128}I product distribution pattern.

34 In laboratory experiments, enrichment factors of 1000 and more have been reported. In large-scale isotope production, however, this is somewhat lower for many reasons. The main factor is the simultaneous decomposition of the chemical bonds of the target material itself due to the high ionization density caused by the irradiation, yielding a spectrum of species fragments. Consequently, enrichment factors are higher, the lower the neutron flux is and the shorter the irradiation period.

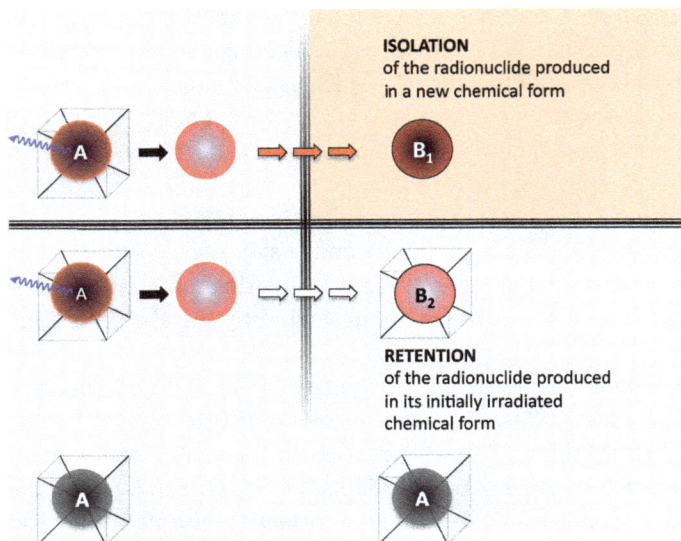

Fig. 7.27: Concept of radiochemical separation of stable and radioactive isotopes of the same element in Szilard–Chalmers type processes: The fate of a radioisotope produced in a nuclear reaction (stable A to radioactive B) is either to terminate as a new, chemically different chemical species (B_1) or to return into the initial species (B_2) where it co-exists with an excess of the initial target compound containing the stable target nucleus isotope (A) that did not undergo any nuclear reaction. Radiochemical separation isolates the fraction of B_1 from both the inactive target material and the fraction B_2 of the radioisotope that returned back into the cage.

lanthanide target. Both A and B are isotopes of one and the same chemical element, and both A and B are trivalent lanthanides of exactly identical chemistry.

Let us take a proof-of-principle nuclear reaction: ^{165}Ho(n,γ)^{166}Ho. The "trick" is to use a target material of type metal-ligand complex A–L, with L representing a macrocyclic chelate such as DOTA. Complexes of type Ln^{III}-DOTA are extremely stable, both in terms of thermodynamic and kinetic stability, cf. Fig. 7.28.

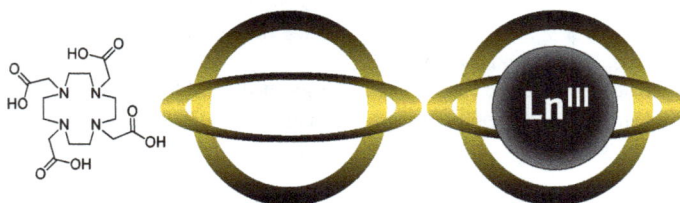

Fig. 7.28: The macrocyclic ligand DOTA as chemical structure (left) and pictogram (middle). Right: A trivalent lanthanide Ln-DOTA complex.

These complexes are formed only at elevated temperature (close to 100 °C). So if the "hot" radiolanthanide created in the course of the nuclear reaction is released from the irradiated complex (and this is typically at room temperature), it will not again form a DOTA complex: The "hot" ^{166}Ho thermalizes as a ^{166}Ho^{3+} cation. Consequently, it is easily and quantitatively separated from the intact ^{165}Ho-DOTA target species, cf. Fig. 7.29. This combines unique *thermodynamic* and *kinetic* parameters of Ln^{III}-DOTA complexes with the hot atom chemistry of radionuclides *following nuclear reactions*. Within this experimental design, enrichment factors of 1000 have been observed at the TRIGA research reactor Mainz.

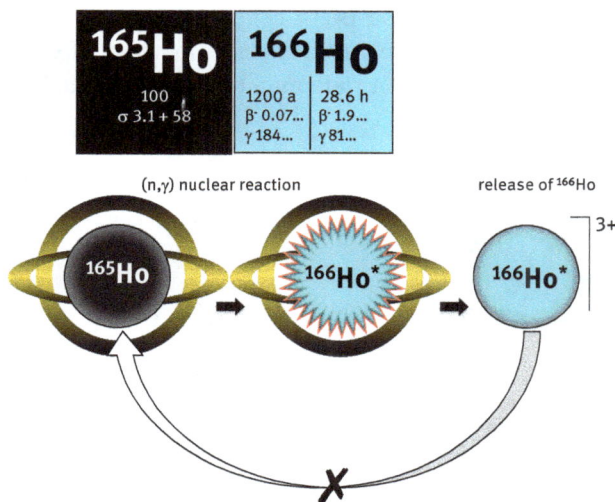

Fig. 7.29: Radiochemical separation of the "hot" ^{166}Ho^{3+} cation from neutron-irradiated ^{165}Ho-DOTA targets.

7.9.6 Hot atom effects applied to radiochemical separations II: Radioelement separation

The physicochemical processes occurring after the primary radioactive transformation, such as electron capture (EC) and/or AUGER electron emission or X-ray emission, cause changes in the chemical state of the generated daughter nucleus. In the case of electron capture or internal conversion, these "post-effects" are known to provide the possibility of separation of different chemical forms of the parent and daughter radionuclides.[35] This particular kind of hot atom chemistry-based radiochemical

---

[35] Segrè et al. (1939). Stenström and Jung (1965), Mirzadeh et al. (1993), Beyer et al. (1969).

separation utilizes the effects of post-processes subsequent to a radioactive transformation process of an unstable nuclide *K1 → *K2. Thus, it is not (mainly) based on recoil energies of the radionuclide *K2 formed within the transformation, but on the processes within the electron shell of *K2.

Attempts to separate neighboring lanthanides have been reported for several cases as well. A prototype of continuous separation (a radionuclide generator system) of type $_Z^A\text{Ln}/_{Z-1}^A\text{Ln}$ was reported for the ^{140}Nd (t½ = 3.37 d)/^{140}Pr (t½ = 3.4 min) pair. The daughter ^{140}Pr is formed via electron capture transformation of ^{140}Nd. This ^{140}Nd ($_Z^A\text{Ln}$) is chemically almost identical to ^{140}Pr ($_{Z-1}^A\text{Ln}$) and is released from its chemical microenvironment. Thus, the fraction of "free" $^{140}\text{Pr}^{3+}$, i.e., non ^{140}Pr-DOTA, can be effectively separated from the ^{140}Nd-DOTA complex, cf. Fig. 7.30 for the concept. This combines unique *thermodynamic* and *kinetic* parameters of Ln^{III}-DOTA complexes with the hot atom chemistry of radionuclides *following nuclear transformations.*

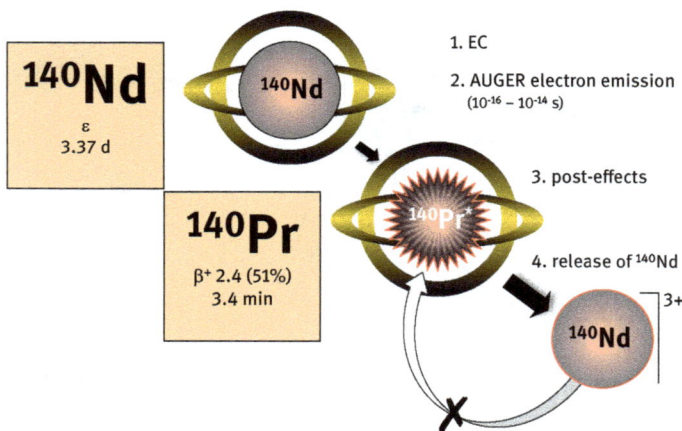

Fig. 7.30: Concept of radiochemical separation of two radioactive isotopes of chemically almost identical elements in Szilard–Chalmers type processes: The $^{140}\text{Pr}^{3+}$ cation generated via the primary β⁺-process from neutron-deficient ^{140}Nd co-ordinatively bound in a DOTA complex, and the fate of the hot ^{140}Pr atom (Zhernosekov et al. 2007).

How to realize the $^{140}\text{Nd}/^{140}\text{Pr}$ separations? The different options start with the synthesis of a ^{140}Nd-DOTA complex or a ^{140}Nd-DOTA-conjugated compound. In a first version of a radiochemical separation, ^{140}Pr is isolated from ^{140}Nd-DOTA by means of cation exchange chromatography. The mixture of ^{140}Nd-DOTA + ^{140}Pr is eluted through a strong cation exchanger resin, which selectively adsorbs the $^{140}\text{Pr}^{3+}$ cation. The ^{140}Nd-DOTA complex, in contrast, passes through the cartridge.

This approach is, however, not an applicable radionuclide generator design, as the generator parent nuclide is mobilized rather than immobilized. So, how to realize a "real" ^{140}Nd/^{140}Pr radionuclide generator system? This is achieved, for example, by chemically stably adsorbing the parent radionuclide ^{140}Nd in form of a ^{140}Nd-DOTA complex or a ^{140}Nd-DOTA-conjugated compound with particular adsorption characteristics for a particular surface on a solid phase (such as on selected resins/cartridges), and to remove (elute) the generated transformation product ^{140}Pr from this solid phase. While the ^{140}Nd remains chemically fixed, the effective isolation of ^{140}Pr depends on the composition of the aqueous medium. The challenge is to provide a chemical environment that rapidly stabilizes the hot ^{140}Pr created in a cationic or complex form easily to separate.[36]

7.10 Outlook

Radiochemical separations have found widespread applications in different areas of nuclear chemistry. This is true for investigations on the chemistry of the radioelements with the emphasis on the heaviest elements, for radiopharmaceutical chemistry, for studies on the behavior of radionuclides in the environment and in connection with nuclear power.

A fundamental application is the radiochemical separation of carrier-free radionuclides produced in nuclear reactions from the corresponding macroscopic target material. One of the main challenges of radiochemical separations was and still is to isolate the radioisotope of interest in extreme radionuclidic purity to allow for precise nuclear spectroscopy. This refers to basic parameters such as half-life and quality and quantity of the radiations emitted. In the larger picture, the nuclear data obtained, however, open access to fundamental parameters of the nuclear reaction itself or on the fundamental properties of matter or on astrophysics and the understanding of the genesis of the universe. Consequently: No or imperfect radiochemical separation = no progress in basic scientific research.

Half-life regions down to a few tenths of a second can be covered with fast chemical separation procedures. A further exploration of yet unknown regions of the chart of nuclides may require chemical separations on a time scale so far inaccessible. However, in the future the existing limits mainly determined by the velocity of the diffusion process for chemical species through boundary layers and by the velocity of phase separations can possibly be overcome by a combination of

---

36 The elution yield of ^{140}Pr in DTPA, citrate and NTA solutions is a function of the ligand's concentration. About 20% of ^{140}Pr^{III} generated could be eluted with 1 ml of pure water. In contrast, >93% of the ^{140}Pr activity could be obtained in 1 ml of 10^{-3} M DTPA eluate. The elution yield decreased with decreasing ligand concentration and was around 20% at a concentration $\leq 10^{-4}$ M DTPA. The elution capacity of citrate and NTA was evidently poorer. About 90% of ^{140}Pr could be eluted only at 0.1 M concentration of citrate and NTA solutions. This is due to lower complex stability of the trivalent Pr^{III} lanthanide with citrate and NTA (nitrilo triacetic acid) ligands.

specific nuclear effects with chemical selective steps. Here, new techniques have to be explored.

The second principal direction is the radiochemical separation of carrier-free radionuclides produced in nuclear reactions from the corresponding macroscopic target material for practical application. This is the case today for many commercially distributed radionuclides used in radiochemistry, but most important in radiopharmacy and nuclear medicine: ^{11}C from N_2 targets (gas chemistry), ^{18}F from water targets (solid phase adsorption), ^{89}Zr from Y targets (ion exchange), ^{124}I from TeO_2 targets etc. for PET, ^{67}Ga from Zn targets (ion exchange), ^{111}In from Ag targets (ion exchange), ^{123}I from TeO_2 targets (volatilization), ^{131}I from TeO_2 targets (volatilization), ^{201}Tl from Tl targets (ion exchange), etc. for SPECT, or ^{177}Lu from Yb targets (ion exchange) and ^{211}At from Bi targets (volatilization) etc. for therapy. The same is true for radionuclides obtained in nuclear reactions functioning as parent nuclides for radionuclide generators: ^{99}Mo and ^{90}Sr as isolated from n-induced fission on ^{235}U targets, ^{68}Ge as isolated from, e.g., $^{69}Ga(p,2n)$ reactions, etc.

Radionuclide generators are constructed to proceed with the latter primary nuclear reaction products once they have been isolated from macroscopic targets. Here, the radiochemical separation means separating two carrier-free radionuclides. The additional challenge is to guarantee for fast, reproducible separations of these systems of long shelf-life: $^{99}Mo/^{99m}Tc$ (solid phase adsorption), $^{68}Ge/^{68}Ga$ (solid phase adsorption or solid phase extraction), $^{90}Sr/^{90}Y$ (liquid-liquid extraction). Consequently: No or imperfect radiochemical separation = no means for the practical application of radionuclides in, e.g., industry and medicine.

References

Adloff J-P, Gaspar PP, Imamura M, Maddock AG, Matsuura T, Sano H, Yoshihara K. eds. Handbook of hot atom chemistry. Tokyo, Kodansha Ltd., and VCH, Weinheim|New York|Cambridge|Basel, 1992.

Ahrens H, Kaffrell N, Trautmann N, Herrmann G. Decay properties of neutron-rich niobium isotopes. Phys Rev C 1976, 14, 211–217.

Altzitzoglou T, Rogowski J, Skålberg M, Alstad J, Herrmann G, Kaffrell N, Skarnemark G, Talbert W, Trautmann N. Fast chemical separation of technetium from fission products and decay studies of ^{109}Tc and ^{110}Tc. Radiochim Acta 1990, 51, 145–150.

Barrera J, Tarancón A, Bagán H and García JF. A new plastic scintillation resin for single-step separation, concentration and measurement of technetium-99. Anal Chim Acta 2016, 936, 259–266. DOI: 10.1016/j.aca.2016.07.008.

Beyer G-J, Grosse-Ruyken H, Khalkin WA. Eine Methode zur Trennung genetisch verknüpfter isobarer und isomerer Nuklidpaare der Seltenen Erden. J Inorg Nucl Chem 1969, 31, 1865–1869.

Braun T and Ghersini G. Extraction chromatography, 1st ed., Vol. 2. Journal of Chromatography Library, 1975, iii–viii, 1–566.

DeVoe JR. Application of distillation techniques to radiochemical separations. Subcommittee on Radiochemistry, National Academy of Sciences – National Research Council, USA, 1962. Available from the Office of Technical Services, Department of Commerce, Washington 25, DC.

Ebihara H, KE Collins. Production of radionuclide using hot atom chemistry. In: Matsuura H. ed. Hot atom chemistry: Recent trends and applications in the physical and life sciences and technology. Tokyo, Kodansha Ltd., 1984, 378–382.

Eichler B, Zvara I. Dubna report, JINR-r12-8943, 1975.

Eichrom. NI Resin product sheet. https://www.eichrom.com/eichrom/products/nickel-resin/, accessed 14/12/2021

Eikenberg J, Bajo S, Beer H, Hitz J, Ruethi M, Zumsteg I and Letessier P. Fast methods for determination of antropogenic actinides and U/Th-series isotopes in aqueous samples. Appl Rad Isot 2004, 61, 1–106. DOI: 10.1016/j.apradiso.2004.03.020.

Gäggeler HW. Gas chemical properties of heaviest elements. Radiochim Acta 2011, 99, 503–513.

Hahn O, Strassmann F. Nachweis der Entstehung aktiver Bariumisotope aus Uran und Thorium durch Neutronenbestrahlung: Nachweis weiterer aktiver Bruchstücke bei der Uranspaltung. Naturwissenschaften 1939, 27, 89–95.

Hahn O, Strassmann F. Verwendung der "Emanierfähigkeit" von Uranverbindungen zur Gewinnung von Spaltprodukten des Urans: Zwei kurzlebige Alkalimetalle. Naturwissenschaften 1940, 28, 54–61.

Herrmann G, Trautmann N. Rapid chemical methods for identification and study of short-lived nuclides. Ann Rev Nucl Part Sci 1982, 32, 117–147.

Hickmann U, Greulich N, Trautmann N, Herrmann G. On-line separation of volatile fission products by thermochromatography: Comparison of halide systems. Radiochim Acta 1993, 60, 127–132.

Horwitz EP, Bloomquist CAA, Henderson DJ and Nelson DE. The extraction chromatography of americium, curium, berkelium and californium with di(2-ethylhexyl)orthophosphoric acid. J Inorg Nucl Chem 1969, 31(10), 3255–3271. DOI: 10.1016/0022-1902(69)80113-2.

Horwitz EP, Chiarizia R and Dietz ML. A novel strontium selective extraction chromatographic resin. Solv Extr Ion Exch 1992a, 10(2), 313–336. DOI: 10.1080/07366299208918107.

Horwitz EP, Chiarizia R, Dietz ML and Diamond H. Separation and preconcentration of actinides from acidic media by extraction chromatography. Anal Chim Acta 1993, 281, 361–372. DOI: 10.1016/0003-2670(93)85194-O.

Horwitz EP, Dietz ML, Chiarizia R, Diamond H, Essling AM and Graczyk D. Separation and preconcentration of uranium from acidic media by extraction chromatography. Anal Chim Acta 1992b, 266, 25–37.

Horwitz EP., Dietz ML, Rhoads S, Felinto C, Gale NH, Houghton J. A lead selective extraction chromatographic resin and its application to the isoloation of lead from geological samples. Anal Chim Acta 1994, 292, 263–273.

Kratz, J. V. (2011). Aqueous chemistry of the transactinides. Radiochim. Acta 99, 477–502.

Lanouette W, Szilard B. Genius in the shadows: A biography of Leo Szilard. New York|Oxford| Singapore|Sydney, Maxwell MacMillan International, 1992.

Lebedev NA, AF Novgorodov AF, Misiak R, Brockmann J, Roesch F. Radiochemical separation of no-carrier-added ^{177}Lu as produced via the ^{176}Lu(n,γ)^{177}Yb → ^{177}Lu process. Appl Radiat Isot 2000, 53, 421–425.

Matsuura H. ed. Hot Atom Chemistry: Recent trends and applications in the physical and life sciences and technology. Tokyo, Kodansha Ltd., 1984.

Mauerhofer E, Kling O, Rösch F. Dependence of the mobility of tracer ions in aqueous perchlorate solutions on the hydrogen ion concentration. Phys Chem Chem Phys 2002, 5, 117–126.

Maxwell SL and Culligan BK. Rapid radiochemical analyses in support of Fukushima. Presented at the Eichrom Workshop as part of the 57. RRMC (Destin, FL), Nov 2, 2011.

Merinis H, Bouissieres G. Fractional sublimation of the radioisotopes of mercury, platinum, iridium, osmium, and rhenium formed by bombardment in a gold target. Anal Chim Acta 1961, 25, 498–504.

Merinis M, Bouissières G. Migration of radioelements in a tube with a temperature gradient. Radiochim Acta 1969, 12, 140–152.

Mirzadeh S, Kumar K, Gansow OA. The chemical fate of ^{212}Bi-DOTA formed by β-transformation of ^{212}Pb(DOTA). Radiochim Acta 1993, 60, 1–10.

Novgorodov AF, Kolatchkowski A. Dubna report, JINR-p6-12457, 1979.

Novgorodov AF, Rösch F, Korolev NA. Radiochemical separations by thermochromatography. In: Vértes A, Nagy S, Klencsár Z, Lovas RG, Rösch F. eds. Handbook of nuclear chemistry, 2nd ed., Vol. 5. Berlin-Heidelberg, Springer, 2011, 2429–2458.

Pin C, Gannoun A and Dupont A. Rapid, simultaneous separation of Sr, Pb, and Nd by extraction chromatography prior to isotope ratios determination by TIMS and MC-ICP-MS. J Anal At Spectrom 2014, 29, 1858–1870. DOI: 10.1039/C4JA00169A.

Qaim SM, Hohn A, Bstian T, El-Azoney KM, Blessing G, Spellerberg S, Scholten B, Coenen HH. Some optimisation studies relevant to the production of high-purity ^{124}I and ^{120g}I at a small-sized cyclotron. Appl Radiat Isot 2003, 58, 69–78.

Rengan K, Meyer RA. Ultrafast chemical separations. Nuclear Science Series. NAS-NS-3118. Washington, DC, National Academy Press, 1993.

Reischl G, Roesch F, Machulla H-J. Electrochemical separation and purification of yttrium-86. Radiochim Acta 2002, 90, 225–228.

Rodnick, M.E., Sollert, C., Stark, D., et al. Cyclotron-based production of 68Ga, [68Ga]GaCl3, and [68Ga]Ga-PSMA-11 from a liquid target. EJNMMI Radiopharm Chem 2020, 5, 25. DOI: 10.1186/s41181-020-00106-9.

Roesch F, Khalkin VA. Ion mobilities of trivalent f-elements in aqueous electrolytes. Radiochim Acta 1990, 51(3), 101–106.

Roesch F, Knapp FF. Radionuclide generators. In: Vértes A, Nagy S, Klencsár Z, Lovas RG, Rösch F. eds. Handbook of nuclear chemistry, 2nd ed., Vol. 4. Berlin-Heidelberg, Springer, 2011, 1935–1976.

Roesch F, Novgorodov AF, Qaim SM. Thermochromatographic separation of ^{94m}Tc from enriched molybdenum targets and its large scale production for nuclear medical application. Radiochim Acta 1994, 64, 113–120.

Rudolf J, Bachmann K. The use of radionuclides for the determination of adsoroption isotherms of volatile chlorides. Radioanal Chem 1978, 43, 113–120.

Rutherford E. Radioactivity. Cambridge, Cambridge University Press, 1904.

Segrè E, Halford RS, Seaborg GT. Chemical separation of nuclear isomers. Phys Rev 1939, 55, 321–322.

Silva RJ, Harris J, Nurmia M, Eskola K, Ghiorso A. Chemical separation of rutherfordium. Inorg Nucl Chem Lett 1970, 6, 861–877.

Stenström T, Jung B. Rapid chemical separation of isomeric states of lanthanides. Radiochim Acta 1965, 4, 3–5.

Svedjehed, J., Kutyreff, C.J., Engle, J.W., et al. Automated, cassette-based isolation and formulation of high-purity [61Cu]CuCl2 from solid Ni targets. EJNMMI Radiopharm Chem 2020, 5, 21. DOI: 10.1186/s41181-020-00108-7.

Szilard L, Chalmers TA. Chemical separation of the radioactive element from its bombarded isotope in the Fermi effect. Nature 1934, 134, 462.

Surman JJ, Pates JM, Zhang H, Happel S. Development and characterisation of a new Sr selective resin for the rapid determination of ^{90}Sr in environmental water samples, Vol. 129. Talanta, 2014, 623–628. DOI: 10.1016/j.talanta.2014.06.041.

Townshend A, Jackwerth E. Precipitation of major constituents for trace preconcentration: Potential and problems. Pure & Appl Chem 1989, 61, 1643–1656.

TrisKem International. Technical Documentation "Extraction Chromatography – All Resins", 2021, V_10_HT, https://www.triskem-international.com/technical-documents.php, accessed on 14/12/2021

Triskem International. Product sheet 'TK211/2/3 Resins', EN_200803, 2020, https://www.triskem-international.com/catalog/products/resins-and-accessories/tk211-resin/bl,product,5318,0, accessed 14/12/2021

Trochimczuk, AW, Kabay N, Arda M and Streat M. Stabilization of solvent impregnated resins (SIRs) by coating with water soluble polymers and chemical crosslinking. React Funct Polym 2004, 59(1), 1–7. DOI: 10.1016/j.reactfunctpolym.2003.12.011.

Tsukada K, Haba H, Asai M, Toyoshima A, Akiyama K, Kasamatsu Y, Nishinaka I, Ichikawa S, Yasuda K, Miyamoto Y, Hashimoto K, Nagame Y, Goto S, Kudo H, Sato W, Shinohara A, Oura Y, Sueki K, Kikunaga H, Kinoshita N, Yokoyama A, Schädel M, Brüchle W, Kratz JV. Adsorption of Db and its homologues Nb and Ta, and the pseudo-homologue Pa on anion-exchange resin in HF solution. Radiochim Acta 2009, 97, 83–89.

Walton AG. The formation and properties of precipitates. Interscience 1967.

Zhernosekov KP, Filosofov DV, Qaim SM, Rösch F. A 140Nd/[140]Pr radionuclide generator based on physico-chemical transitions in [140]Pr complexes after electron capture transformation of 140Nd-DOTA. Radiochim Acta 2007, 95, 319–327.

Zhernosekov KP, Filosofov DV, Rösch F. The Szilard-Chalmers effect in macrocyclic ligands to increase the specific activity of reactor-produced radiolanthanides: Experiments and explanations. Radiochim Acta 2012, 100, 669–674.

Zulauf A, Happel S, Mokili M.B., et al. Characterization of an extraction chromatographic resin for the separation and determination of 36Cl and 129I. J Radioanal Nucl Chem 2010, 286, 539–546. DOI: 10.1007/s10967-010-0772-5.

Zvara I, Belov VZ, Domanov VP, Korotkin Yu S, Chelnokov IP, Shalajevskii MR, Shegolev VA, Hussonois M, Dubna report, JINR-d12-5845, 1971.

Norman M. Edelstein and Lester R. Morss

8 Radioelements: Actinides

Aim: The actinide elements constitute a second f element series and are found in the periodic table under the lanthanide elements. However the chemical properties of the naturally occurring actinide elements, thorium, protactinium, and uranium are quite different from their lanthanide counterparts and in some respects, behave like their homologs in the 4d and 5d transition metal series. With the discovery of the transuranium elements and investigations of their properties, it became clear that these heavier actinides became more "lanthanide-like" in their chemical properties as the atomic number increased. Spectroscopic studies also showed that the 5f shell was being filled in this series as a function of atomic number. The aim of this chapter is to give a brief introduction to the chemical and physical properties of the elements in this series.

8.1 Introduction

The actinide series, starting with actinium, atomic number 89, and ending with lawrencium, atomic number 103, is an analogy of the 4f lanthanide transition series which starts with lanthanum, element 57, and ends with lutetium, element 71.[1] The elements Th, Pa, and U were considered part of the 6d series in the periodic table before 1940. However, with the discovery and characterization of the electronic and chemical properties of the transuranium elements, it became clear that the elements with atomic numbers 89–103 were members of another inner f transition series. The actinide concept was first formulated by Glenn T. Seaborg in 1944. Figure 8.1 gives the historic and the modern versions of the arrangements of the f-elements within the periodic table.

8.2 General properties of the actinides

8.2.1 Electronic configurations

The electronic configurations of the actinide atoms and some ions are shown in Tab. 8.1. For the early actinides, the configurations with 6d electrons form the lowest energy (ground) configuration. As the atomic number increases, configurations with

---

[1] The lanthanides are subsumed as "Ln", the actinides as "An".

Note: Lester R. Morss – April 6, 1940 | June 14, 2014. This chapter is dedicated to his memory.

https://doi.org/10.1515/9783110742701-008

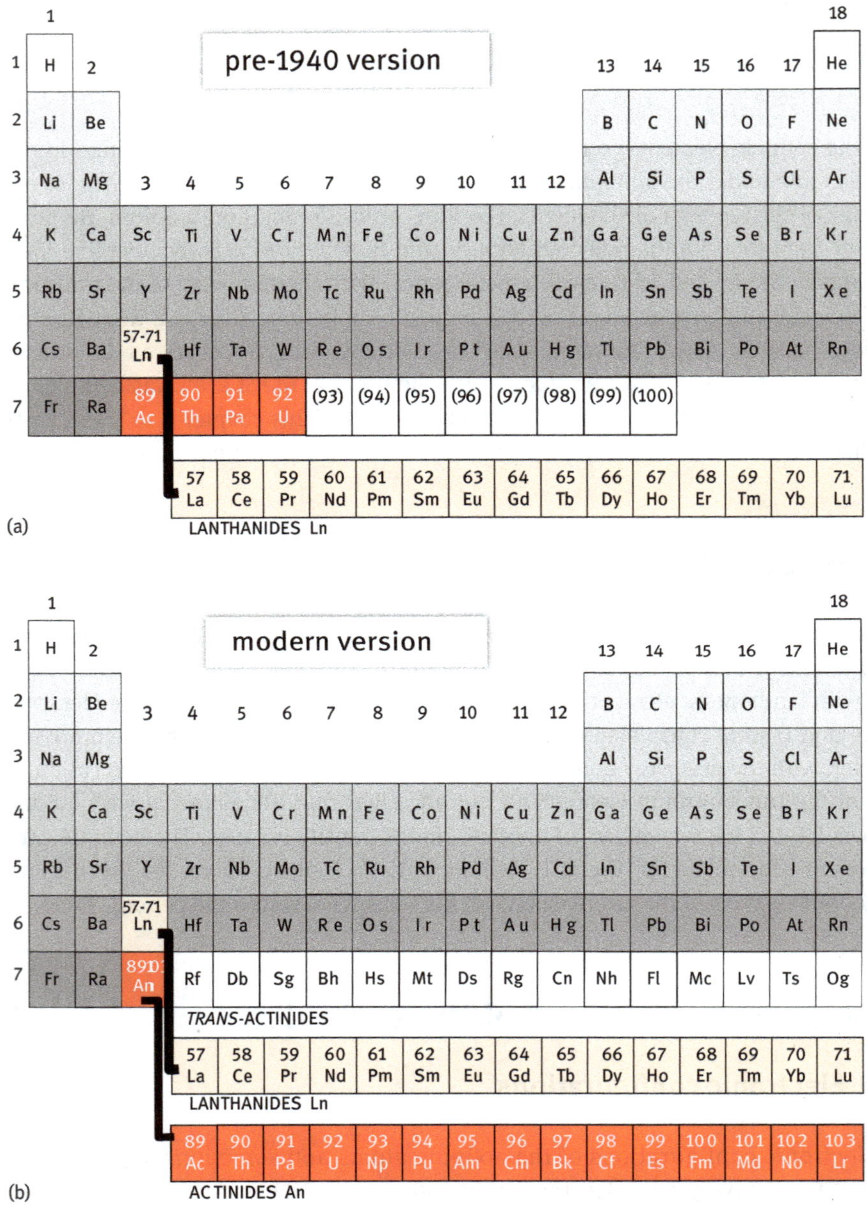

Fig. 8.1: Pre-1940 version and modern versions of the periodic table. Note that the only actinide elements known at the time (shown in red) are listed as 6d transition elements (under the 5d transition elements). The unknown higher atomic number elements (at the time) are also listed as 6d elements. In modern periodic tables, the actinide elements (atomic numbers 89–103) are listed separately in a second row under the lanthanides, cf. Fig. 9.1 in the following chapter on the *trans*actinide elements.

5f electrons become more stabilized, as shown in Fig. 8.2. As the oxidation state of an element increases, again the configurations with 5f electrons increasingly become more stable. This trend is shown in Fig. 8.3 for one-electron ions of the actinide series. Triply ionized gaseous Th^{3+} has one 5f electron as the ground state with the 6d electronic configuration about $\approx 1.2\,eV$ (or $\approx 9100\,cm^{-1}$) higher in energy. In the few chemical compounds of Th^{3+} characterized to date, the 6d configuration is stabilized by the crystal field from surrounding ligands relative to the 5f configuration and is found to be the ground configuration in these compounds.

Tab. 8.1: Electronic configurations for the actinide series.

Element	Ground gaseous atom	Ground M$^+$ (gaseous)	Ground M^{2+} (gaseous)	Ground M^{3+} (gaseous)	Ground M^{3+} (aq)	Ground M^{4+} (gaseous)
Ac	$6d7s^2$	$7s^2$	$7s$			
Th	$6d^27s^2$	$6d7s^2$	$5f6d$	$5f$	$6d^*$	
Pa	$5f^26d7s^2$	$5f^27s^2$	$5f^26d$	$5f^2$		$(5f)$
U	$5f^36d7s^2$	$5f^37s^2$	$5f^37s^2$	$5f^3$	$5f^3$	$5f^2$
Np	$5f^46d7s^2$	$5f^46d7s$	$5f^46d7s$	$5f^4$	$5f^4$	$(5f^3)$
Pu	$5f^67s^2$	$5f^67s$	$5f^67s^2$	$5f^6$	$5f^5$	$(5f^4)$
Am	$5f^77s^2$	$5f^77s$	$5f^77s^2$	$5f^7$	$5f^6$	$(5f^5)$
Cm	$5f^76d7s^2$	$5f^77s2$	$5f^77s2$	$5f^7$	$5f^7$	$(5f^6)$
Bk	$5f^97s^2$	$5f^97s$	$5f^97s$	$5f^8$	$5f^8$	$(5f^7)$
Cf	$5f^{10}7s^2$	$5f^{10}7s$	$5f^{10}7s$	$5f^9$	$5f^9$	$(5f^8)$
Es	$5f^{11}7s^2$	$5f^{11}7s$	$5f^{11}7s$	$5f^{10}$	$5f^{10}$	$(5f^9)$
Fm	$5f^{12}7s^2$			$5f^{11}$	$5f^{11}$	$(5f^{10})$
Md	$5f^{13}7s^2$			$5f^{12}$	$5f^{12}$	$(5f^{11})$
No	$5f^{14}7s^2$		$(5f^{14})$	$5f^{13}$	$5f^{13}$	$(5f^{12})$
Lr	$(5f^{14}6d7s^2$ or $5f^{14}7s^27p)$			$5f^{14}$	$5f^{14}$	$(5f^{13})$

*In known solid organometallic Th^{3+} complexes.
Note: All configurations are built upon the Rn core $[1s^22s^22p^63s^23p^64s^23d^{10}4p^65s^24d^{10}5p^66s^24f^{14}5d^{10}6p^6]$ which is not included in the table. For example, the ground gaseous ion Ac^+ configuration is $[Rn]6d7s^2$ and the Ac^{2+} ion is $[Rn]7s^2$. Predicted configurations are given in parentheses.

8.2.2 Oxidation states

The lanthanide elements all have stable trivalent oxidation states; Ce, Pr, and Tb also form relatively stable tetravalent states. Eu and Yb form relatively stable dipositive compounds and Sm forms less stable dipositive compounds. The early actinides, from Th to Am, have quite different chemical properties than the later actinides and all the lanthanides. Table 8.2 lists the stable oxidation states of the actinide ions. As the atomic number increases in the actinide series, the heavier actinide ions become more "rare-earth like" or "lanthanide like".

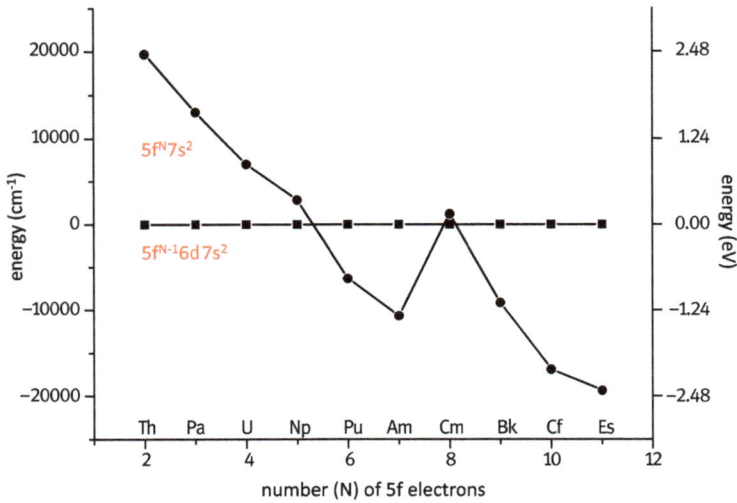

Fig. 8.2: Relative energies of the $5f^N 7s^2$ and $5f^{N-1}6d7s^2$ electron configurations for gaseous actinide atoms. This graph shows the stabilization of the $5f^N 7s^2$ with respect to the $5f^{N-1}6d7s^2$ as the atomic number Z ($Z = 88 + N$) increases.

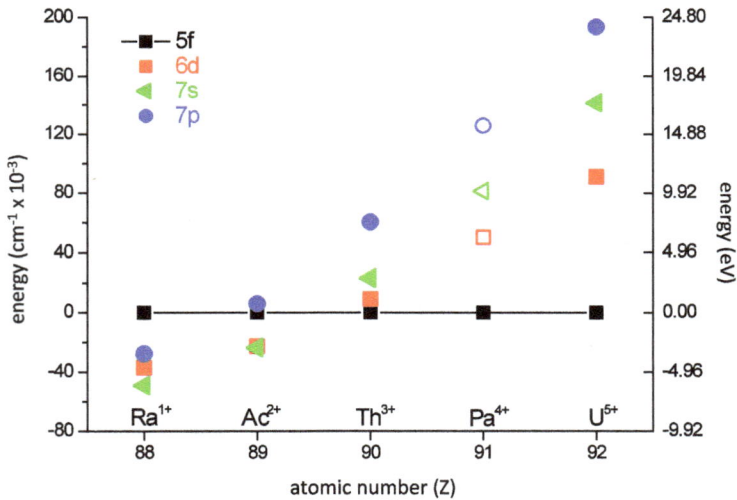

Fig. 8.3: Energies of the lowest level of various configurations of the one-electron ions near the beginning of the actinide series, relative to the 5f level. The 5f configuration is stabilized as the atomic number increases. The data for Pa^{4+} are interpolated.

Tab. 8.2: Oxidation states of the actinide elements.

Atomic no. (Z)	89	90	91	92	93	94	95	96	97	98	99	100	101	102	103
Element	Ac	Th	Pa	U	Np	Pu	Am	Cm	Bk	Cf	Es	Fm	Md	No	Lr
Oxidation states															
I															
II							(II)			(II)	(II)	II	II	II	
II	III	(III)		III	III	III	III	III	III	III	III	III	III	III	III
IV		IV	IV	IV	IV	IV	IV	IV	IV	IV					
V			V	V	V	V	V								
VI				VI	VI	VI	VI								
VII					VII	(VII)									

Bold type: most stable; (. . .): unstable.

In some ways, the chemical properties of the actinides Th, Pa, and U resemble the 5d transition metals, Hf, Ta, and W, cf. Fig. 8.1. Th^{4+} is the most stable oxidation state of Th, similar to Hf, and Th^{3+} compounds can only be prepared with great difficulty. Pa^{5+} is the most stable oxidation state of Pa, similar to Ta, although stable oxides and halides of Pa^{4+} can be prepared. Both W and U form the volatile compounds MF_6, as do Np and Pu, although the stability of the MF_6 compounds decreases from U to Pu. The actinides U through Am form compounds with the actinyl configuration, that is with strong axial bonds (O = An = O) where An can be pentavalent AnO_2^{1+} or hexavalent AnO_2^{2+}. Actinyl ions contain the nearest neighbor atoms from complexing ligands in an equatorial plane perpendicular to this axis, Fig. 8.4. For uranyl in compounds and aqueous solution, the stable oxidation state is hexavalent; for neptunyl compounds and aqueous solution, the stable oxidation is pentavalent. Pu^{3+} and Am^{3+} trivalent ions in aqueous solution are stable to oxidation by air but can be oxidized by stronger agents to higher oxidation states including the actinyl ions. The next element, Cm, is found only in the trivalent state under normal conditions in aqueous solutions. Cm^{4+} compounds are known but there is no evidence for any higher

$UO_2(H_2O)_5^{2+}$

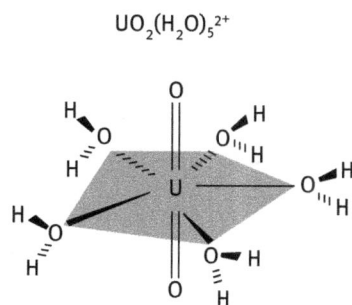

Fig. 8.4: Structural motif for the uranyl ion in acidic aqueous solution. Most actinyl ions follow this structural motif (Courtesy of Los Alamos National Laboratory).

oxidation states. The chemical behavior of the transamericium ions of the actinides is much closer to that of the lanthanide series than the early actinides.

8.2.3 Occurrence and production

Note that in comparing the lanthanide and actinide transition series, the electronic configuration of any given element may be significantly different in the gaseous atomic state, in the metallic state, or for trivalent ions in solution. For example, in the trivalent lanthanide series, the 4f shell is filled sequentially beginning with Ce^{3+} ($4f^1$ in gaseous atomic state as well as in solution and compounds). In the actinide series, Th^{3+} in the gaseous atomic state is $5f^1$; although very few Th^{3+} compounds have been synthesized, they have been shown to have a 6d electron in the ground state. Pa^{3+} compounds are presently unknown; however, U^{3+} has been shown to have a $5f^3$ configuration similar to that of its lanthanide analog Nd^{3+}, and heavier An^{3+} ions add 5f electrons sequentially, again similar to the lanthanide series. The principal differences between the two f transition series arise from the lower binding energies and less effective shielding of the 5f electrons in the actinides as compared to the 4f electrons in the lanthanides.

Actinium through uranium, occur naturally on earth; only thorium and uranium are present in large enough quantity to extract by conventional mining processes. Thorium and uranium are found associated with various minerals in the earth's crust and are present in large enough quantity to extract by conventional mining processes. Ores rich in thorium include monazite (a rare-earth thorium phosphate), thorianite ($ThO_2 + UO_2$), and thorite ($ThSiO_4$). Pitchblende (impure UO_2), and carnotite (a potassium uranium vanadate) are some of the most common uranium ores.

Natural thorium, almost entirely ^{232}Th with traces of ^{228}Th, is more abundant than natural uranium, whose terrestrial isotopic composition is 99.27% ^{238}U, 0.72% ^{235}U, and 0.0005% ^{234}U. Thorium is estimated to occur in the earth's crust at ≈6 ppm compared to the estimated abundance of uranium at ≈1.8 ppm. Protactinium, as ^{231}Pa, is formed terrestrially from the alpha transformation of ^{235}U. The isotope ^{227}Ac is formed from the alpha transformation of ^{231}Pa. These nuclei are part of the uranium-actinium ($4n + 3$) series shown in Fig. 8.5. The second naturally occurring isotope of actinium is ^{228}Ac, also known as *meso*-thorium, comes from the alpha transformation of ^{232}Th to ^{228}Ra which then beta transforms to ^{228}Ac. These reactions are part of the 4 n or thorium series, which is also shown in Fig. 8.5.[2]

---

2 Cf. also Chapter 5 in Vol. I on the natural transformation chains.

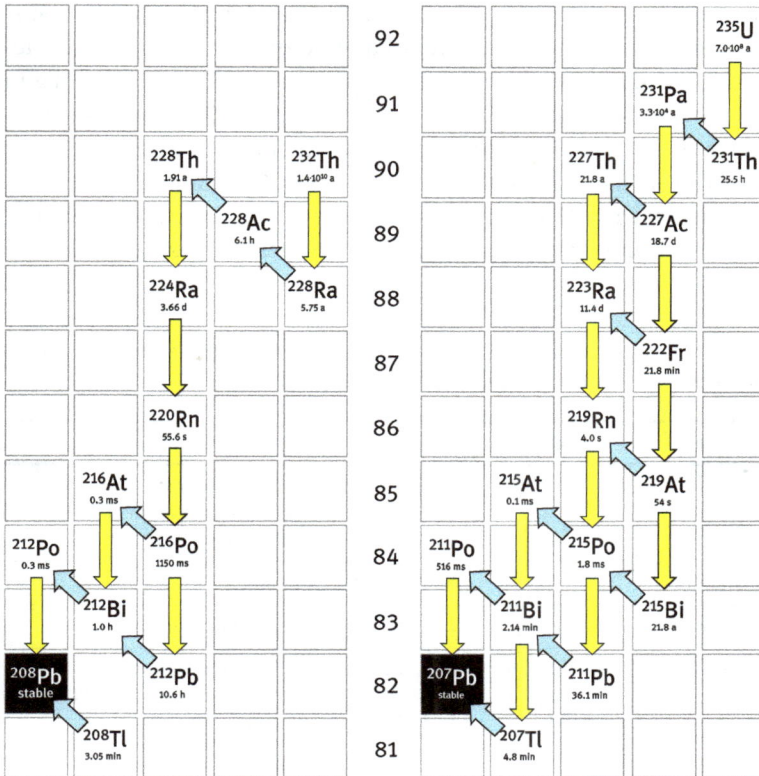

Fig. 8.5: Right: The 4n + 3 radioactive transformation series starting from ^{238}U. Left: The 4n radioactive transformation series (starting from ^{232}Th).

Pure Pa products have been produced by processing the waste residues from Ra ores (pre-1940) or from uranium refinery waste material. Sludges from U production were processed in the United Kingdom in the late 1950s and resulted in 127 g of pure ^{231}Pa. This material has been utilized in the subsequent years to establish the chemical properties of Pa and its compounds.

The transuranium elements Np and Pu are produced in large quantities through neutron capture and beta transformation in nuclear reactors. The principal reactions for their production are:

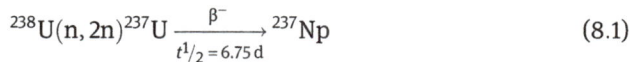

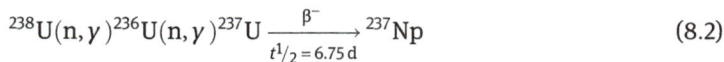

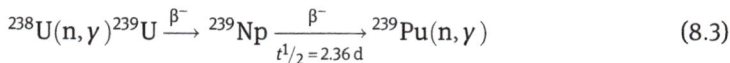

$$^{238}\text{U}(n, 2n)^{237}\text{U} \xrightarrow[t^{1}/_{2}=6.75\,\text{d}]{\beta^{-}} {}^{237}\text{Np} \tag{8.1}$$

$$^{238}\text{U}(n, \gamma)^{236}\text{U}(n, \gamma)^{237}\text{U} \xrightarrow[t^{1}/_{2}=6.75\,\text{d}]{\beta^{-}} {}^{237}\text{Np} \tag{8.2}$$

$$^{238}\text{U}(n, \gamma)^{239}\text{U} \xrightarrow{\beta^{-}} {}^{239}\text{Np} \xrightarrow[t^{1}/_{2}=2.36\,\text{d}]{\beta^{-}} {}^{239}\text{Pu}(n, \gamma) \tag{8.3}$$

Irradiation of large amounts of Pu in a high neutron flux reactor produces the trans-plutonium elements Am and Cm in relatively large quantities. The actinide elements Bk, Cf, and Es have been produced only in relatively small quantities (mg to µg) because of small nuclear cross sections and short half-lives. Fm isotopes (transformation products of Es) have been produced only in submicrogram quantities and have relatively short half-lives (see Tab. 8.3). Transfermium isotopes can be produced only by nuclear reactions at accelerators with very specialized particle beams and targets and only "one-atom-at-a-time" studies can be performed on these elements.

Tab. 8.3: Long-lived actinide nuclei found in nature or synthesized by neutron irradiation and/or subsequent nuclear transformation and suitable for chemical studies.

Atomic no. (Z)	Element	Isotope	Half-life	Production
89	Actinium	^{225}Ac	10.0 d	Separation from ^{229}Th
		^{227}Ac	21.772 a	Neutron irradiation of ^{226}Ra
		^{228}Ac	6.15 h	Separation from ^{228}Ra
90	Thorium	^{232}Th	$1.405 \cdot 10^{10}$ a	Mining
91	Protactinium	^{231}Pa	$3.28 \cdot 10^{4}$ a	Separation from uranium wastes
92	Uranium	^{233}U	$1.592 \cdot 10^{4}$ a	^{232}Th neutron irradiation
		^{235}U	$7.038 \cdot 10^{8}$ a	Separation from natural U
		^{238}U	$4.468 \cdot 10^{9}$ a	Mining
93	Neptunium	^{237}Np	$2.144 \cdot 10^{6}$ a	Production reactor
94	Plutonium	^{238}Pu	87.7 a	Production reactor
		^{239}Pu	$2.411 \cdot 10^{4}$ a	Production reactor
		^{242}Pu	$3.75 \cdot 10^{5}$ a	Production reactor
		^{244}Pu	$8.08 \cdot 10^{7}$ a	Production reactor
95	Americium	^{241}Am	432.7 a	Production reactor
		^{243}Am	$7.38 \cdot 10^{3}$ a	Production reactor
96	Curium	^{244}Cm	18.10 a	Production reactor
		^{248}Cm	$3.48 \cdot 10^{5}$ a	Separation from ^{252}Cf
97	Berkelium	^{249}Bk	320 d	High neutron flux reactor
98	Californium	^{249}Cf	351 a	High neutron flux reactor
		^{252}Cf	2.645 a	High neutron flux reactor
99	Einsteinium	^{253}Es	20.47 d	High neutron flux reactor
		^{254}Es	275.7 d	High neutron flux reactor
		^{255}Es	39.8 d	High neutron flux reactor
100	Fermium	^{257}Fm	100.5 d	High neutron flux reactor

8.2.4 Physicochemical properties

8.2.4.1 The metallic state

The actinide metals are unique in the periodic table. Many actinide compounds be-have in a predictable way (except for the fact that they are radioactive). For example, the trivalent actinide ions in compounds and solution are very similar to the trivalent lanthanide ions in solution and the solid state. Tetravalent Th behaves similarly to tet-ravalent Zr and Hf and the chemical properties of pentavalent Pa are similar to penta-valent Ta. However, the larger sizes of the actinide ions relative to the d transition elements must be taken into account. This is not true for the actinide metals or inter-metallic compounds. They have properties that are not found in conventional metals and new theories have to be formulated in order to explain many of their properties.

Actinium metal and the transplutonium actinide metals, americium through einsteinium, have electronic structures with an empty or partially filled 5f subshell and physical properties similar to those of the corresponding lanthanides with an empty or partially filled 4f subshell. These elements have "localized" 5f electrons as characterized by their magnetic properties. The 5f electrons of the intermediate ele-ments, thorium through neptunium, due to the greater radial distribution of their 5f electrons as compared to their 4f homologs, can interact with other valence elec-trons to form a delocalized electronic band structure; that is, there is metal-metal bonding (itinerant behavior) through the 5f orbitals, similar to the d transition met-als. Figure 8.6 shows the effects of forming bands (metal-metal bonds) on the atomic volume of the metals. Plutonium metal is unique in the periodic table, hav-ing seven allomorphs between room temperature and its melting point. Table 8.4 lists some properties of actinide metals. The metals, protactinium through curium, have been studied at high pressures because high pressure compresses the inter-atomic distances, changes the band structures (phase transitions) of the metals and forces localized metals to exhibit delocalized band behavior.

8.2.4.2 Optical and magnetic properties

The 5f electrons of the actinide series can be regarded as an inner electron shell similar to the 4f shell of the lanthanide series. This means, especially for the lower oxidation states, the optical spectra are somewhat insensitive to changes in the surroundings. Since the main features of the spectra are due to intra-electronic repulsion (coulomb repulsion) and spin-orbit coupling, crystalline compounds and aqueous solutions of the same ion with different surroundings have in general similar spectra. Actinyl spectra are dominated by a strong axial crystal field arising from the short An-O bond found in the AnO_2^{x+} $(x = 1, 2)$ with smaller interactions coming from the intra-electronic repulsion, spin-orbit coupling, and longer An-equatorial ligand bonds. In

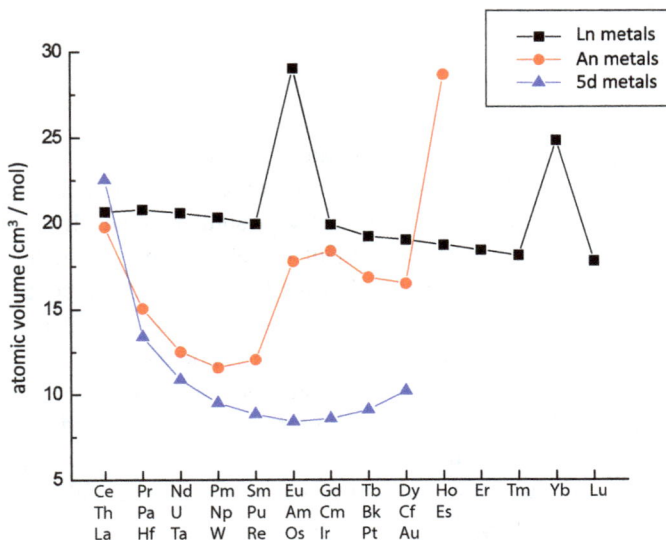

Fig. 8.6: The atomic volumes for the lanthanide (Ln) metals, actinide (An) metals, and 5d metal transition series plotted as a function of atomic number for each series at room temperature and 1 atmosphere pressure. The 5d metals form electronic bands (metal-metal bonding) that result in the volume being reduced as the atomic number increases. By contrast, the lanthanide metals exhibit localized trivalent ion behavior (no band structure) and the atomic volume remains almost constant across the series. The early actinide metals also form bands as shown by the decrease in volume across the series (similar to the 5d metals) up to plutonium. At americium, the actinide series reverts to localized trivalent behavior with a large increase in volume, and the atomic volume remains almost constant for the heavier actinide metals. Eu, Yb, and Es form divalent metals with localized electrons, which accounts for their large increase in volume compared to the trivalent localized lanthanide and actinide metals.

Tab. 8.4: Some properties of actinide metals.

Element	Melting point (K)	Structure at 298 K	Density (g cm⁻³)	Metallic radius (nm)
Actinium	1323 ± 50	fcc	10.01	0.1878
Thorium	2023 ± 10	α fcc	11.724	0.1798
Protactinium	1845 ± 20	α bc tetrag.	15.37	0.1642
Uranium	1408 ± 2	α orthorh.	19.04	0.1542
Neptunium	912 ± 3	α orthorh.	20.48	0.1503
Plutonium	913 ± 2	α monocl.	19.85	0.1523
Americium	1449 ± 5	α dhcp	13.67	0.1730
Curium	1619 ± 50	α dhcp	13.5	0.1743
Berkelium	1323 ± 50	α dhcp	14.79	0.1704
Californium	1173 ± 30	α dhcp	15.1	0.1691
Einsteinium	1133	fcc	8.84	0.203

fcc: face-centered cubic;   bc tetrag.: body-centered tetragonal;   orthorh.: orthorhombic;
dhcp: double hexagonal close packed.

this case, the main features of the spectra arise from the strong axial field. Figure 8.7 shows the near infrared and visible absorption spectrum of Np^V (NpO_2^+) complex, which is due to transitions within the $5f^2$ configuration.

The magnetic properties of actinide compounds with a $5f^N$ configuration are determined by the population of the low-lying energy levels. For an f^N ion where N is even,

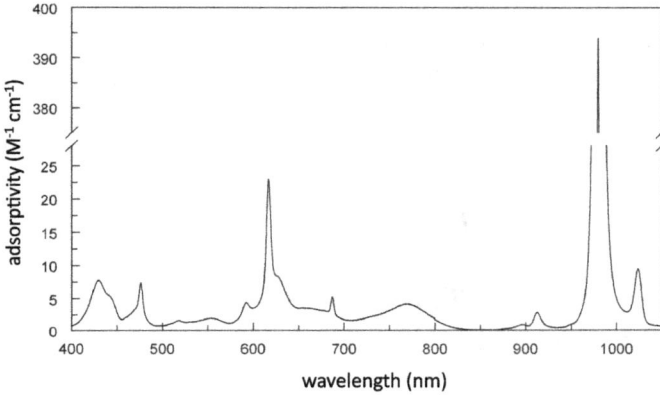

Fig. 8.7: Absorption spectrum of 2.6 mM NpO_2ClO_4 (NpO_2^+, $5f^2$ configuration) in aqueous solution at 25 °C and pH ≈ 6 showing transitions within the $5f^2$ configuration. The peak at 980 nm is an f–f transition that is anomalously intense.

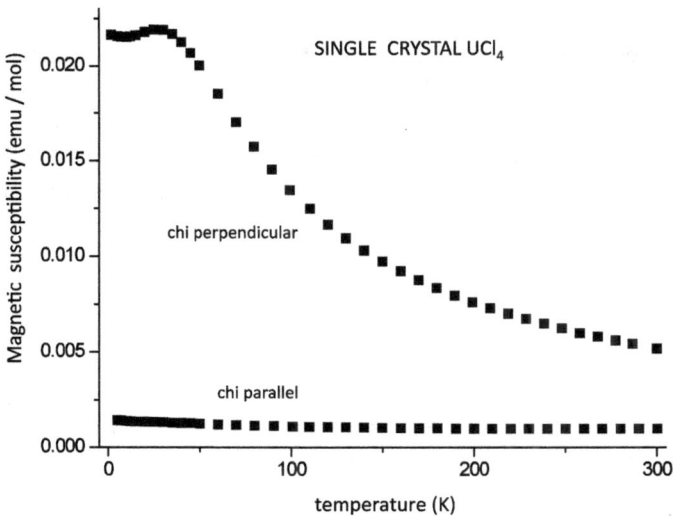

Fig. 8.8: The magnetic susceptibility, chi, of a single crystal of UCl_4 showing the anisotropic response to a magnetic field as a function of temperature. The crystal has one fourfold (S_4) axis; measurements were performed with the magnetic field parallel and perpendicular to this axis.

it is possible to have a singlet state lowest, and depending on the temperatures of the measurements and the splitting of the crystal field states, temperature-independent magnetism may be observed. If N is odd then all levels are at least two-fold degenerate and temperature-dependent susceptibility should be observed. In cases when the $5f^N$ wavefunctions are known from analysis of optical data, the magnetic properties can be calculated and compared with experimental data such as magnetic susceptibility and EPR (electron paramagnetic resonance). Measurements of magnetic susceptibility of single crystals of f^N compounds can be highly anisotropic, i.e. the response to a magnetic field can vary with respect to the orientation of the crystal to the direction of the field. Figure 8.8 shows the magnetic susceptibility of a single crystal of UCl_4 (U^{4+}, a $5f^2$ configuration) with respect to the orientation of the magnetic field with respect to the single crystal axes.

8.3 The elements: Discovery, availability, properties, and usage

8.3.1 Actinium

8.3.1.1 Discovery

Actinium was first claimed to be isolated by Deberine, in the laboratory of the Curies in Paris, in 1899. Although Deberine named this new element, it appears now that its chemical properties (titanium or thorium-like properties) were not consistent with the presently known nuclear and chemical properties of actinium. In 1902, Giesel reported and separated a new "emanating-producing" material found in the radium samples he had obtained from pitchblende residues. Giesel explored many of its chemical properties and found that it followed the lanthanide elements in its chemistry. Nevertheless, the name given by Deberine has been accepted.

8.3.1.2 Availability

Actinium can be separated from uranium ores, but it is difficult to obtain it in high purity due to the large percentage of rare earths found in natural uranium ores. However, by irradiating ^{226}Ra with thermal neutrons, macroscopic amounts of pure ^{227}Ac have been produced. Cation exchange chromatography is utilized for separating Ac from its transformation products. Macroscopic amounts of ^{227}Ac have been produced at the Belgian Nuclear Research Centre (SCK-CEN, Mol) by irradiating $RaCO_3$. The production capacity was about 20 g per year. This production facility is closed.

8.3.1.3 Chemical properties

Ac^{3+} is more basic than the lanthanide ions due to its larger ionic radius. Ac^{3+} exhibits the same general aqueous chemistry as the trivalent lanthanide ions. Ion exchange methods are used to obtain pure samples.

Multi-curie amounts of ^{227}Ac have been placed in capsules and used as radio-isotope thermoelectric generators. ^{225}Ac, with a 10-day half-life, transforms by the emission of alpha particles. Its short-lived daughters also emit only alpha and beta particles with no high-energy gamma rays. Thus, this isotope and its daughters can deliver high-energy radiation to a tumor without greatly affecting the surrounding tissue. Compounds labeled with ^{225}Ac are being evaluated for their use in nuclear medicine. ^{225}Ac simultaneously can be used as a generator of isotopically pure ^{213}Bi, which has a 45.6 min half-life and again emits α-particles plus a 440 keV gamma ray. Thus, ^{213}Bi, eluted from the generator, can be used for therapy and imaging.[3]

8.3.2 Thorium

8.3.2.1 Discovery

The Swedish chemist Berzelius, in 1815, analyzed a rare mineral, which he erroneously thought contained a new element that he named thorium after the Norse god of thunder and weather. He later discovered his mistake and in 1828 when he actually found a new element in another mineral (which he called thorite) used the same name. Thorite is $ThSiO_4$ and usually contains not only Th but also a significant amount of U.

8.3.2.2 Chemical properties

Thorium chemistry is almost exclusively the study of the tetrapositive ion except for the metal and its alloys and compounds that exhibit metal-metal bonding. Binary compounds of tetravalent Th have been prepared with all the halides and ternary compounds with alkali and alkaline earth metals are known. In some cases, where the stoichiometry of Th to halide (or hydride) is less than $1:4$, it is likely that metal-to-metal bonds are involved in the materials. ThO_2 is the stable oxide and ternary compounds of this oxide have been prepared. Hydrides, chalcogenides (S, Se, and Te compounds), nitrides, carbonates, phosphates, vanadates, and molybdates of Th are known and have been characterized.

---

3 Cf. Chapter 13 on nuclear medicine therapy.

Th^{4+} is the largest tetravalent actinide ion (ionic radius 0.105 nm for eight-fold coordination) and has the smallest charge to radius ratio. Consequently, it is less hydrolyzable than most ions and can be studied up to a pH range of ≈ 4. Th^{4+} in aqueous solution undergoes polynuclear reactions, colloid formation, and readily forms carbonate complexes. Furthermore, the hydrous oxide/hydroxide is quite insoluble. Thus, solution measurements are quite difficult. There have been comprehensive studies of the coordination chemistry of the Th^{4+} ion in nitric acid and with various complexing agents that are useful for separations.

Since Th^{4+} is a very large metal ion, it can attain a very high coordination number with small ligands. Th(BH$_4$)$_4$ is a polymeric material and has a coordination number of 14 with two (BH$_4$)$_4$ groups attached through a triple hydrogen bridge from the B atom (and one terminal H atom) and four (BH$_4$)$_4^-$ groups attached through a double hydrogen bridge from the B atom (with the other two H atoms bridging to another Th atom). However, in the molecular compound [Th(H$_3$BNMe$_2$BH$_3$)$_4$], the Th ion attains a coordination number of 15 by having seven bridges from each of two hydrogen atoms from each boron bond to the Th ion and the last bridge from the eighth boron bond with only one hydrogen bridge. The structure is shown in Fig. 8.9.

A large number of thorium organometallic compounds have been prepared and characterized. They have been investigated as stoichiometric or catalytic compounds to promote organic transformations and compared with early or late d transition metals and with uranium compounds for similar purposes. In some cases, thorium compounds exhibit totally different reactivities. A few very reactive Th^{3+} organometallic compounds have been prepared which all exhibit magnetic properties characteristic of a 6d^1 configuration.

8.3.2.3 Uses

Past uses of thorium include ceramic production, carbon arc lamps, alloys, and mantles for lanterns. Thorium metal burns with a brilliant white flame and thorium dioxide has an extremely high melting point, 3050 °C. However, in recent years, the use of thorium for nonnuclear purposes has decreased markedly due to environmental concerns about its radioactivity.

The relative ratios of various isotopes of U and Th in natural samples have proven useful in the dating of geologic and coral formations. However, the most promising use of ^{232}Th is as a nuclear fuel. ^{232}Th is nonfertile but can be converted to fertile (that is, fissionable) ^{233}U by the following nuclear reactions:

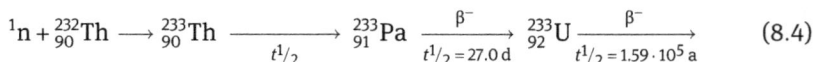

$$^1\text{n} + {}^{232}_{90}\text{Th} \longrightarrow {}^{233}_{90}\text{Th} \xrightarrow[t^{1/2}]{} {}^{233}_{91}\text{Pa} \xrightarrow[t^{1/2}=27.0\,\text{d}]{\beta^-} {}^{233}_{92}\text{U} \xrightarrow[t^{1/2}=1.59\cdot10^5\,\text{a}]{\beta^-} \tag{8.4}$$

^{232}Th is about three times as abundant as U and is nonfissionable. ^{233}U is fissionable but would have to be separated from the irradiated nuclear fuel to be utilized as a

weapon. The buildup of transuranium actinides (including plutonium) is much reduced compared to that from the uranium fuel cycle. The fission products are similar
to those from the uranium fuel cycle. Since ^{232}Th cannot undergo fission, initially
other isotopes such as ^{233}U, ^{235}U, or ^{239}Pu must be added to get to criticality. As shown
above, the production of ^{233}U occurs from the beta transformation of ^{233}Pa so an appreciable amount of this isotope is produced in the nuclear fuel, cf. Chapter 13, Vol. I.
Significant amounts of ^{231}Pa ($t\frac{1}{2} = 3.23 \cdot 10^4$ a) are also produced and which creates a
long-term radioactive hazard.[4]

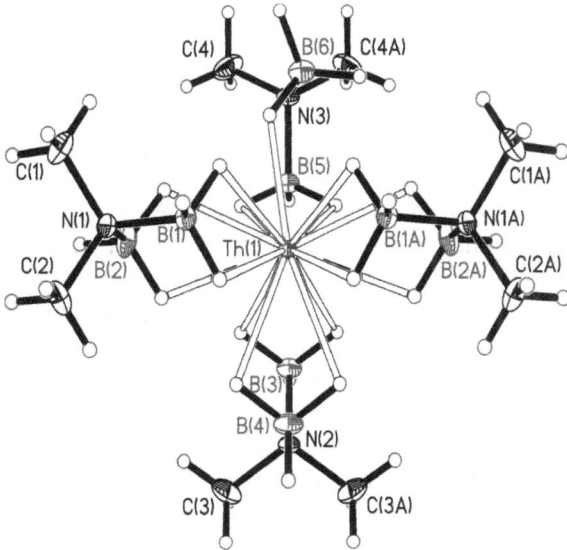

Fig. 8.9: Molecular structure of Th(H$_3$BNMe$_2$BH$_3$)$_4$ from neutron diffraction data. The thorium atom
is coordinated to 15 hydrogen atoms (the hydrogen atoms are represented by circles) illustrating
the large size of the Th atom.

8.3.3 Protactinium

8.3.3.1 Discovery

Mendeleev, in his periodic table of 1871, noted that an element was missing between Th (placed under Hf) and U (placed under W); see Fig. 8.1. From the placement of these two known elements, he predicted that the missing element would

---

4 The THOREX process for treating the spent fuel from a Th-based nuclear reactor is in the development stage.

have properties similar to Ta. In 1913, Fajans and Göhring found a new element, which was a high-energy beta-particle emitter with chemical properties similar to those of Ta and a half-life of 1.15 min. They named this element *brevium* because of its short half-life. In 1918, Soddy and Cranston and, independently, Hahn and Meitner, found a long-lived alpha-particle activity with Ta-like properties and a half-life between 1200 and 180,000 years. Hahn and Meitner proposed the name protactinium for element 91 because the name *brevium* clearly was inappropriate for a long-lived element. The name itself comes from the Greek word *protos* (meaning first, before) as ^{227}Ac is created by the α-transformation of ^{231}Pa.

8.3.3.2 Chemical properties

The long-lived protactinium isotope, ^{231}Pa, is found terrestrially; it can be obtained by exhaustive treatment of the sludge left after processing of uranium ores. Therefore, it is a scarce material. Pure ^{231}Pa (after immediate separation from daughter activities) can be utilized with the usual procedures for handling alpha-active materials. However, growth of new daughter activities occurs rapidly ($\approx$six months), which again limits the amount of Pa and its ease of handling. Protactinium metal has been prepared and melts at $\approx$1560 °C. It is superconducting at 1.4 K. Protactinium forms pentavalent and tetravalent compounds. Binary compounds of pentavalent and tetravalent Pa have been prepared with all the halides. Ternary halide compounds with alkali and alkaline earth metals are known. Pa_2O_5 and PaO_2 are the stable oxides in each of the known oxidation states. Some ternary Pa oxide compounds have been prepared. Some protactinium pnictides[5] have been prepared as well as a few other miscellaneous compounds.

The aqueous solution chemistry of protactinium is highly complex. The oxidation states Pa^V and Pa^{IV} are known in aqueous solutions. Both oxidation states hydrolyze extensively so most work has been done at the tracer level. With macroamounts of Pa, the aqueous chemistry is dominated by the hydrolyzed Pa^V ion, which precipitates as hydrated Pa_2O_5. Macroscopic quantities of protactinium may be dissolved in hydrofluoric and sulfuric acids. With other mineral acids, Pa solutions are unstable, forming precipitates or colloidal suspensions over time. Recent studies have shown that Pa^V in concentrated sulfuric acid solution forms a single short Pa–O bond (similar to the vanadyl ion) and does not form a *trans*-dioxo bond as found for U through Am for the pentavalent and hexavalent states. This short Pa–O bond is not found for Pa^V in HF solutions because Pa–F complexes are stronger than Pa–O complexes.

---

5 *Pnictides* is the common name of compounds with the chemical elements in group 15 of the periodic table (also known as the *nitrogen family*): nitrogen (N), phosphorus (P), arsenic (As), antimony (Sb), and bismuth (Bi).

The relatively rapid growth of daughter products necessitates their separation rather frequently.

Only a few organometallic compounds have been prepared because of the difficulties of working with Pa. Earlier work produced $Pa(C_5H_5)_4$, $Pa(C_8H_8)_2$, and $Pa[(CH_3)_4C_8H_4]_2$. A recent study has shown that $PaCl_5$ is soluble in hydrocarbons such as chlorobenzene, toluene, or benzene. The reaction of dissolved $PaCl_5$ in toluene with excess $(CH_3C_5H_4)Na$ readily produced $Pa(CH_3C_5H_4)_4$. Thus, if the problems of handling Pa can be overcome, a straightforward route to new Pa organometallic compounds is available.

8.3.3.3 Uses

There are no industrial or commercial uses of protactinium. ^{231}Pa in conjunction with other naturally occurring Th and U isotopes has been utilized for radiometric dating[6] and paleoceanography.

8.3.4 Uranium

8.3.4.1 Discovery

Although uranium has been found in mosaic glass produced near Naples, Italy almost 2000 years ago, its discovery is credited to German chemist Klaproth, who dissolved the mineral pitchblende (impure UO_{2+x}) in nitric acid and then obtained a yellow precipitate (probably potassium uranate) upon neutralizing the acid solution with KOH(aq). Klaproth concluded that the mineral contained a new element, naming it uranium after the recently discovered planet Uranus in 1789.

8.3.4.2 Chemical properties

As indicated in Tab. 8.2, uranium can exist in oxidation states +3 through +6. In aqueous solution, the yellow uranyl ion UO_2^{2+} is the most common and most stable. It was the species observed by Klaproth and has the linear actinyl geometry as described in the Introduction (Fig. 8.3). It cannot be oxidized and is stable in air with respect to reduction. Uranyl ions can be reduced to the green U^{4+} ion at low pH by

---

6 Cf. Chapter 3.

electrolysis or with $H_2(g)$ in the presence of a platinum black catalyst. The $U^{4+}(aq)$ ion is stable in oxygen-free acidic solution; it hydrolyzes readily, e.g.

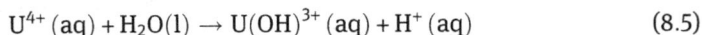

$$U^{4+}(aq) + H_2O(l) \rightarrow U(OH)^{3+}(aq) + H^+(aq) \tag{8.5}$$

Further reduction to the red $U^{3+}(aq)$ ion is possible with zinc amalgam or careful electrolysis in the absence of oxygen. The $U^{3+}(aq)$ ion is a strong reductant, unstable even to oxidation by water. Its oxidation is sufficiently slow that solutions are metastable for minutes or hours.

The pentavalent uranyl ion UO_2^+ is also linear, but it is rarely seen in aqueous solution because it is hard to prepare. It is unstable with respect to oxidation, reduction, and disproportionation:

$$4H^+(aq) + 2UO_2^+(aq) \rightarrow U^{4+}(aq) + UO_2^{2+}(aq) + 2H_2O \tag{8.6}$$

Under ideal conditions, very dilute solutions of UO_2^+ can be maintained for several hours. Recently a variety of new complexes of UO_2^+ have been prepared in nonaqueous, oxygen-free solutions.

The relative stability of the uranium aqueous ions is $UO_2^{2+} > U^{4+} > UO_2^+ \approx U^{3+}$. Although the U^{4+} and U^{3+} ions exist in acidic solution, U^{4+} complexes readily react with many simple anions such as carbonate and many complexants. All uranium ions exhibit hydrolysis, that is, the reaction of a cation with water to produce OH^- complexes. An example of hydrolytic behavior even at pH 1 is

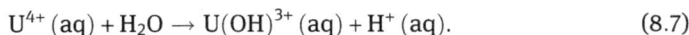

$$U^{4+}(aq) + H_2O \rightarrow U(OH)^{3+}(aq) + H^+(aq). \tag{8.7}$$

The uranyl ion UO_2^{2+} hydrolyzes in solution significantly when pH is above 3. The degree of hydrolysis of uranium ions increases in the sequence $UO_2^+ < UO_2^{2+} < U^{3+} \ll U^{4+}$.

8.3.4.3 Ionic compounds

Uranium halides have been prepared in all known oxidation states +3 through +6. The preparation of uranium halides must exclude the presence of water (typically water vapor) because of hydrolysis, e.g. reactions such as

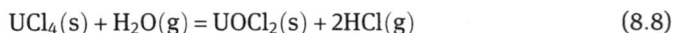

$$UCl_4(s) + H_2O(g) = UOCl_2(s) + 2HCl(g) \tag{8.8}$$

Thus uranium halides are not found in nature. All binary trihalides (UF_3, UCl_3, UBr_3, UI_3) and tetrahalides are known. Among pentahalides, UI_5 is unknown. The only binary uranium hexahalides are UF_6 and UCl_6. By far, the most important uranium halide is UF_6. All metal hexafluorides are volatile solids. UF_6 corrodes many solid surfaces and hydrolyzes rapidly when exposed to water or water vapor. Because of its volatility and despite its corrosive properties, UF_6(g) is the best compound to use to separate the fissile ^{235}U isotope from the more abundant ^{238}U by gaseous diffusion or centrifugation.[7] Another volatile compound is the borohydride $U(BH_4)$.

Uranium oxides are known with uranium oxidation states from +4 through +6. Uranium dioxide (UO_2) is the most important uranium compound because it is the main component of most nuclear fuels. In nature, it exists as uraninite (hyperstoichiometric UO_{2+x}), a common mineral because the U^{4+} ion is quite insoluble. UO_2 possesses the highly symmetric fluorite (CaF_2) structure, isostructural with actinide dioxides ThO_2 through CfO_2. In the laboratory, as in nature, it is typically hyperstoichiometric UO_{2+x}; at temperatures between about 300 and 1100 °C, the fluorite structure persists in the range $UO_{2.0}$–$UO_{2.2}$. At high temperatures under reducing conditions, hypostoichiometric UO_{2-x} can also exist. Unlike the lanthanides and transuranic elements, there is no sesquioxide U_2O_3 because UO_2 is so stable that U_2O_3 would disproportionate following the reaction

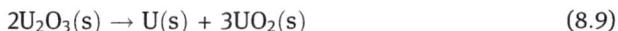

$$2U_2O_3(s) \rightarrow U(s) + 3UO_2(s) \tag{8.9}$$

The other main uranium oxides are U_3O_8 and UO_3. The mineral name "pitchblende" refers to impure, somewhat amorphous, oxidized uranium oxide with a composition between UO_2 and U_3O_8. Fully oxidized UO_3 is quite difficult to prepare; it has seven crystalline modifications. There are a vast number of "complex" uranium oxides, referred to as uranates and also as ternary or polynary uranium oxides. These are stoichiometric oxides prepared in the laboratory with U^{IV} such as $BaUO_3$ or with U^{VI}, for example Na_2UO_4, or found as uranium minerals in nature.

An enormous number of complex halides and oxides of uranium are known, mostly with U^{VI}. Some complex halides have had importance in nuclear technology as reactor fuels or as molten salts for electrochemical recycling of spent (used) nuclear fuel, so that the ^{235}U can be reused after removing fission products. Most uranium minerals are complex U^{6+} oxides with O–U–O networks, typically sheets with edge- and corner-sharing polyhedra. Uranium minerals also include silicates and phosphates.

Another important group of uranium compounds is the hydrides. When finely divided uranium metal is heated in hydrogen to 250 °C or higher, β-UH_3 forms. At high temperature, β-UH_{3-x} has a wide hypostoichiometric range. Another phase, α-UH_3, can be prepared below −80 °C but is unstable. UH_3 decomposes at high temperature in a vacuum, leaving finely divided and pyrophoric uranium metal. An interesting related compound is the borohydride $U(BH_4)_4$, a volatile compound that was once considered as an alternative to UF_6 for gaseous-diffusion separation of uranium isotopes. Its chemical reactivity, in particular pyrophoricity, made it unsuitable for this application.

---

7 Cf. Chapter 11 on nuclear energy.

Compounds of uranium with most elements of the periodic table are known. Although almost all compounds are to some degree ionic, many binary uranium compounds are metallic or semiconductors. Uranium borides, group 13 compounds (carbides, silicides, germanides), pnictides (U–N, U–P, U–As, U–Sb, and U–Bi systems), chalcogenides (U–S, U–Se, and U–Te systems) are important classes of actinide binary compounds. Most of these are refractory and some are important in nuclear technology.

In most solid compounds, U^{4+} is the most stable oxidation state. For example, the tetravalent halides are all known. However, stability must always be related to reaction conditions: all uranium halides are unstable with respect to hydrolysis. For example, $UCl_4(s)$ reacts with water vapor with reactions such as

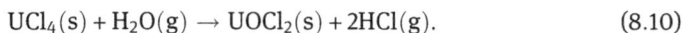

$$UCl_4(s) + H_2O(g) \rightarrow UOCl_2(s) + 2HCl(g). \tag{8.10}$$

There are exceptions to the rule that U^{IV} compounds are the most stable. The dioxide UO_2 oxidizes to U_3O_8 upon heating in air above 600 °C. UO_3 reduces to U_3O_8 upon heating in air to about 800 °C. Therefore, U_3O_8 is the most stable binary uranium oxide in the presence of air. Uranium forms many ternary oxides, the most stable of which contains U in the +6 oxidation state.

8.3.4.4 Coordination chemistry

In aqueous solution, each uranium ion displays its own coordination behavior. In uncomplexed aqueous solution, X-ray absorption spectra and other methods show U^{IV} to be coordinated by 8 to 10 water molecules. These methods show UO_2^+ and UO_2^{2+} typically to have $5\,H_2O$ in the equatorial plane (Fig. 8.3). For U^{III}, quantum chemical calculations and analogies with Pu^{III} and Ln^{III} studies indicate 8 to 10 coordinated water molecules.

Uranium solid coordination compounds are mostly known for U^{VI}. The linear uranyl ion dominates most such complexes as it does in solution. The $O = U = O$ "axis" has oxygens at both "poles" with four to six ligand atoms bonded to the uranium on or near the equatorial plane that is perpendicular to the axis. The $UO_2(OH)_4^{2-}$ complex found in solution and in some compounds has a square bipyramidal structure with four oxygens in the equatorial plane. More common is the pentagonal bipyramid formed by monodentate ligands (binding at one oxygen) and bidentate ligands (chelating, binding at two oxygens); an example is the $UO_2(SO_4)_4^{6-}$ ion in $Na_{10}[UO_2(SO_4)_4]$ $(SO_4)_2 \cdot 2H_2O$ in which three sulfates are monodentate and one is bidentate. Hexagonal bipyramids are found in carbonate complexes such as $K_4UO_2(CO_3)_3$ with three bidentate carbonates in the equatorial plane. The $UO_2(CO_3)_3^{4-}$ also exists in solution, typically above pH 8.

Because U^V is unstable with respect to oxidation as well as reduction, few U^V complexes are known in aqueous solution. In recent years, a large number of new

U^V compounds have been synthesized and characterized in nonaqueous solutions under an inert atmosphere.

Uranium^{IV} forms normal salts with inorganic ligands that have eight to ten oxygens coordinated with U^{IV}. Because U^{IV} is acidic, it forms oxo-salts, silicates, phosphates, molybdates, etc., many of which occur in minerals. For example, $USiO_4$ has 8-coordinated uranium with oxygens at apexes of a triangular dodecahedron (two interpenetrating tetrahedral with U–O distances of 0.232 and 0.252 nm). The mineral coffinite is $USiO_4$ (possibly with some OH^- substituting for the SiO_4 ion).

The few known U^{III} complexes resemble other An^{III} and Ln^{III} complexes, with high coordination numbers (eight or nine).

There are a vast number of uranium (mostly UO_2^{2+} uranyl) complexes with "organic" ligands (ligands that have one or more carbon-carbon bonds). The simplest such ligand, oxalate ($C_2O_4^{2-}$), shows experimental evidence for three bidentate oxalates (six oxygen atoms) coordinated in the equatorial plane.

The coordination chemistry of uranium (UO_2^{2+} uranyl) with ligands of biological relevance has been extensively studied. Amino acids, for example, coordinate as oxygen donors at biological pH, where the amino acid is typically a zwitterion ^+H_3N (R)$COOH^-$. At higher pH values, the amino end is deprotonated and the amino acid can be a nitrogen donor as well as an oxygen donor. Thus, in proteins, bonding takes place at carboxylate and amino sites. In DNA/RNA, bonding occurs via phosphate sites, and in polysaccharides at deprotonated OH sites.

Within the past decade, unique uranium clusters with peroxide and hydroxide links have been found to self-assemble as nanospheres or spheroids from aqueous solution, Fig. 8.10. Most of these uranium clusters have symmetry resembling that of fullerene C_{60}. Uranium clusters have also been prepared as tubes or crowns, some forming very quickly and subsequently rearranging into more stable solids. (Related clusters have been prepared with ions of the transuranium elements neptunium and plutonium.)

8.3.4.5 Organouranium chemistry

Organoranium chemistry is the study of molecules or ions that have uranium-organic ligands bonded via uranium-carbon bonds. In traditional organouranium complexes, the organic ligand is the five-membered ring cyclopentadienyl anion $(C_5H_5)^-$, which typically forms σ U–C bonds in U^{III} compounds, e.g. $(\eta^5\text{-}C_5H_5)_3U$, and in U^{IV} compounds, e.g. $(\eta^5\text{-}C_5H_5)_3UCl$. (The notation η^5 refers to five bonds between each ring carbon and the uranium.) Many substituted cyclopentadienyl complexes form organouranium complexes with unusual properties. When each hydrogen of cyclopentadienyl is replaced by a methyl group, the bulky $\eta^5\text{-}C_5(CH_3)_5$ ligand forms reactive complexes such as $[\eta^5\text{-}C_5(CH_3)_5]_2UCl_2$ that have homogeneous and heterogeneous catalytic properties similar to those of some transition-metal organometallic complexes.

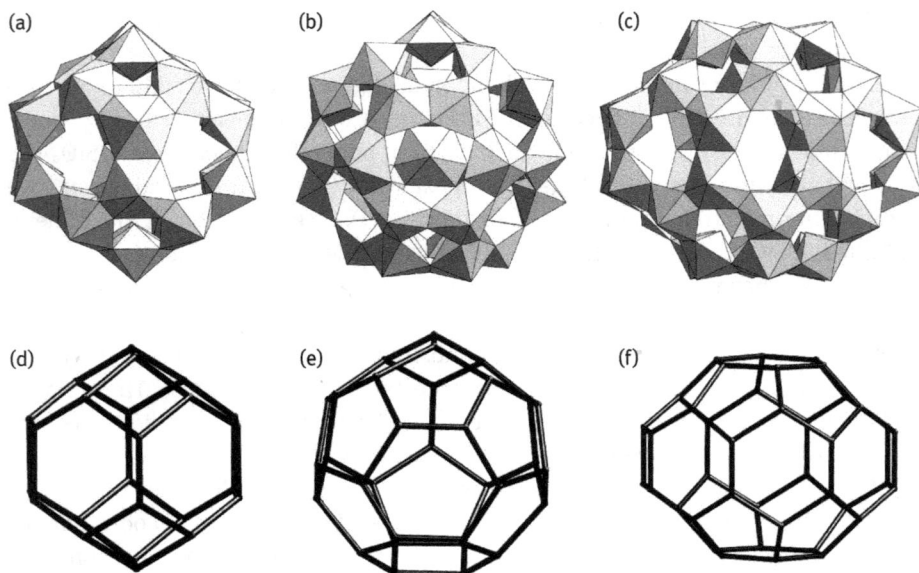

Fig. 8.10: Clusters of uranyl peroxide hexagonal bipyramids containing topological squares. Shown are the polyhedral representations of the clusters and topological graphs representing U sites: (a), (d) U24; (b), (e) U32; (c), (f) U40.

Uranocene, η^8-(C_8H_8)U is the archetype of a class of organometallic compounds and was first synthesized in 1968. In this symmetric "sandwich" compound, the U–C bonds are π bonds rather than σ bonds; uranocene is a 5f-orbital homolog of the "sandwich" ferrocene η^5-$(C_5H_5)_2$Fe. In the absence of oxygen and water, uranocene is remarkably stable: it can be sublimed and numerous substituted analogs have been synthesized. With this discovery, it became clear that the actinides (primarily U and Th due to the higher radioactivity of the other actinides) can react differently with organic ligands than d transition elements and initiated a new era of research in organoactinide chemistry.

8.3.4.6 Uses

Before 1940, uranium had very few uses. In small amounts, it made attractive glass and ceramic objects, often with green fluorescence (from the uranyl ion (UO_2^{2+})). The thermal expansion coefficient of uranium glass is similar to that of tungsten metal, so uranium glass has been used in "graded seals" that allow tungsten electrodes to pass through glass. By far, its most important use has resulted from the radioactivity of its natural isotopes. In 1896, Becquerel discovered that uranium salts fogged photographic plates even when the plates were protected from light.

We know now that he was observing the penetrating gamma rays from the predominant natural uranium isotopes ^{235}U (0.72%) and ^{238}U (99.27%) and from the isotopes in their transformation chains, almost all of which are radioactive. Subsequent discovery of more useful radioactive isotopes that are daughters of natural uranium isotopes, particularly radium-226 in 1898 by M. and P. Curie, resulted in uranium mining in order to extract the tiny amount of radium-226 in secular equilibrium.

After the discovery of nuclear fission in 1938 by Hahn and Strassmann, many scientists became aware that the neutrons released during the fission process could lead to chain nuclear reactions. Therefore, the dominant use of uranium since 1940 has been for nuclear fission, controlled in nuclear reactors[8] or uncontrolled in nuclear weapons.

8.3.5 Neptunium

8.3.5.1 Discovery

Neptunium was the first transuranium element to be discovered. During the 1930s, many elements, including uranium, were bombarded with neutrons in Fermi's research group in Rome and elsewhere. Although scientists hypothesized that transuranium elements were produced, almost all the radioactive particles produced were shown to be fission products by Hahn, Meitner, and Strassmann in 1939. At Berkeley, California, Abelson and McMillan bombarded uranium samples (actually thin films of U_3O_8) with intense neutron sources produced in a cyclotron. In 1939, they found two beta-emitting radioactive isotopes remaining in these films, whereas the fission products were ejected. The half-lives of the two beta-emitters were 23 min and 2.3 days. By chemical separation,[9] Abelson and McMillan identified the nuclear reaction sequence as

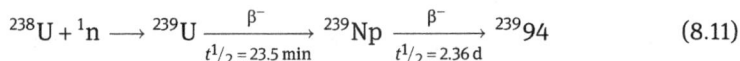

$$^{238}\text{U} + {}^1\text{n} \longrightarrow {}^{239}\text{U} \xrightarrow[t^{1}/_2 = 23.5\,\text{min}]{\beta^-} {}^{239}\text{Np} \xrightarrow[t^{1}/_2 = 2.36\,\text{d}]{\beta^-} {}^{239}94 \qquad (8.11)$$

although they could not detect any radiation emitted by the $^{239}94$. They hypothesized that $^{239}94$ was an alpha-emitting or spontaneous-fission isotope with very long half-life. McMillan suggested that element 93 be named neptunium, because Neptune is the planet beyond Uranus.

---

8 Nuclear power plants are discussed in Chapter 11.
9 Cf. Chapter 7.

8.3.5.2 Chemical properties

As indicated in Tab. 8.2 neptunium can exist in oxidation states +3 through +7. In aqueous solution, Np^V exists as the linear neptunyl ion NpO_2^+ $(O = Np = O)^+$, the most common and most stable neptunium ion. It was the species used by Abelson to separate Np from U in 1940. It has the same linear configuration as UO_2^+. Hexavalent NpO_2^{2+} is also linear, and is produced in aqueous solution by oxidation of Np in lower oxidation states with strong oxidants such as Ce^{IV}, BrO_3^-, and Cl_2. In aqueous solution, Np in lower oxidation states exists as monatomic ions Np^{3+} and Np^{4+}. NpO_2^+ can be reduced to Np^{4+} with weak reductants such as Fe^{2+} or I^-. Further reduction to Np^{3+} requires stronger reductants such as amalgamated Zn.

The deep green heptavalent ion was first produced in 1967 by ozone oxidation of Np^{VI} in strongly basic solution. In basic solution, Np^{VII} has been shown by EXAFS and density-functional theory calculations to exist either as $Np(O_4)(OH)_2^{3-}$ or as $Np(O_4)(H_2O)^-$ ions, with four short nearly coplanar (equatorial) oxygens at 1.87 Å and two axial hydroxides with Np-O distances of 2.24 Å. This species also is found in solid compounds of Np^{VII}.

8.3.5.3 Ionic compounds

Neptunium fluorides have been prepared in all known oxidation states +3 through +6. Like UF_6 and other metal hexafluorides, NpF_6 is a volatile solid that is highly corrosive. Among other halides, only $NpCl_3$, $NpCl_4$, $NpBr_3$, $NpBr_4$, and NpI_3 are known.

Neptunium oxides are known with oxidation states from +4 through +7. The only common binary oxide is NpO_2, easily prepared by decomposition of neptunium oxalate and many other compounds. Np_2O_5 can be prepared by careful decomposition of $NpO_3 \cdot H_2O$ or NpO_2OH, by ozonation of some molten salt mixtures, or by hydrothermal reaction of NpO_2^+ solution at 200 °C in the presence of calcite. Although there are no known binary oxides of Np^{VI} or Np^{VII}, neptunium ions in these high oxidation states have been prepared in many ternary oxides. For example, finely ground stoichiometric mixtures of Na_2O or Na_2O_2 and NpO_2 react in oxygen at 400 °C to oxidize Np from Np^{IV} to Np^{VI}, producing Na_2NpO_4 that can exist up to 1000 °C. Finely ground mixtures of excess Li_2O or Li_2O_2 with NpO_2 react in oxygen at 400 °C to oxidize Np from Np^{IV} to Np^{VII}, producing Li_5NpO_6.

An unusually effective tool to identify oxidation states of neptunium compounds is Mössbauer spectroscopy. The 60 keV Mössbauer resonance of the ^{237}Np isotope has a linear correlation with Np formal charge (oxidation state); this correlation has been used to identify or confirm oxidation state of many neptunium compounds.

8.3.5.4 Coordination and organometallic chemistry

In aqueous solution, neptunium coordination chemistry follows that of uranium. Solid coordination compounds are mostly known for Np^V. Linear NpO_2^+ dominates most such complexes as it does in solution.

Organoneptunium chemistry has followed the parallel organometallic chemistry of uranium, although with far fewer studies. Trivalent organoneptunium complexes are known, e.g. $Np(C_5H_5)_3 \cdot 3THF$, although synthesis of $Np(C_5H_5)_3$ has not been successful. Tetravalent organoneptunium species, such as $Np(C_5H_5)_4$ and complexes derived from it, have been prepared and characterized. Neptunocene, η^8-$(C_8H_8)Np$, is isostructural with uranocene.

8.3.5.5 Uses

Several nuclear processes produce the chemically useful isotope ^{237}Np during the irradiation of uranium in reactors, for example:

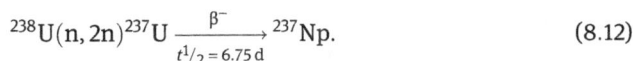

$$^{238}U(n, 2n)^{237}U \xrightarrow[t^1/_2 = 6.75\,d]{\beta^-} {}^{237}Np. \tag{8.12}$$

The rate of production of ^{237}Np in reactors is approximately 0.1% that of ^{239}Pu. Because of its long half-life, $2.144 \cdot 10^6$ a, this isotope accumulates in the reactor fuel and can be separated after irradiation. Its long half-life makes it convenient for chemical studies. Although ^{237}Np does not have other uses, targets of separated ^{237}Np can be irradiated to produce ^{238}Pu in kilogram quantities:

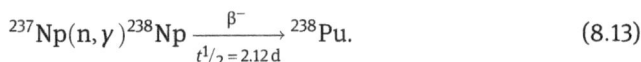

$$^{237}Np(n, \gamma)^{238}Np \xrightarrow[t^1/_2 = 2.12\,d]{\beta^-} {}^{238}Pu. \tag{8.13}$$

The moderately long alpha-emitting transformation (87.7 a half-life) makes ^{238}Pu the best radioisotope to produce electrical energy for deep space probes by thermoelectric conversion of the heat generated by the alpha transformation energy.

8.3.6 Plutonium

8.3.6.1 Discovery

^{239}Np was discovered in 1939 and 1940 and transforms by β^--particle emission. McMillan and Abelson immediately realized that the daughter should be element 94. However the scale, of the early experiments did not permit its identification. In 1941, Seaborg, McMillan, Kennedey, and Wahl produced the first identifiable isotope of Pu by bombarding uranium with deuterons to produce the isotope ^{238}Pu by the following reactions:

$$^{238}U(d, 2n)^{238}Np \xrightarrow[t^{1}/_{2}=2.12\,d]{\beta^{-}} {}^{238}Pu, \tag{8.14}$$

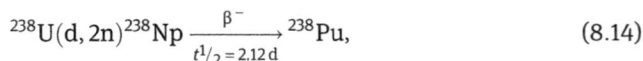

where the initial product ^{238}Np transforms by β^{-}-particle emission with $t^{1}/_{2}$ of 2.1 days to ^{238}Pu, which has a $t^{1}/_{2}$ of 87.7 years and transforms by α-particle emission. The short half-life of this isotope allowed Seaborg and his colleagues to evaluate the chemistry of this new element and to develop procedures for the separation of this element from the reactants and other products. With these new procedures, the Seaborg group produced and separated the most important isotope of Pu, ^{239}Pu in 1941 by the following reactions:

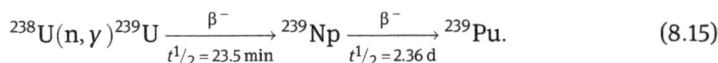

$$^{238}U(n, \gamma)^{239}U \xrightarrow[t^{1}/_{2}=23.5\,min]{\beta^{-}} {}^{239}Np \xrightarrow[t^{1}/_{2}=2.36\,d]{\beta^{-}} {}^{239}Pu. \tag{8.15}$$

The element with atomic number 94 was named by Seaborg and Wahl in 1942 as plutonium (Pu) for the planet Pluto (the second planet beyond Uranus) following McMillan who had named element 93 as neptunium (Np) for the planet Neptune (the first planet beyond Uranus).

The half-life for the β^{-}-particle transformation of ^{239}U is 23.5 min followed by the β^{-}-particle transformation of ^{239}Np with a half-life of 2.36 days. ^{239}Pu itself has a half-life of 24 110 years. The fission properties of ^{239}Pu were established in 1941 by Kennedey, Seaborg, Segrè, and Wahl who showed that this nucleus was indeed fissionable and a likely nuclear energy source.

8.3.6.2 Availability

Following the discovery of ^{239}Pu in 1941 and its possible use as a nuclear energy source, it became urgent to determine its chemical properties in order to be able to separate the newly formed plutonium from the uranium, neptunium, and fission products in the irradiated material. Initially, one microgram of ^{239}Pu was produced in 5 kg of uranyl nitrate hexahydrate at the Berkeley 60-inch cyclotron. A team at Berkeley performed initial processing and the processed material was then sent to the Manhattan Project laboratories at the University of Chicago. Cunningham and Werner further purified the ^{239}Pu and on August 18, 1942 isolated approximately 1 microgram sample of PuF_4 in the tip of a microcone, which was clearly visible with the aid of a $30 \times$ microscope.[10] A larger sample from 90 kg of uranyl nitrate hexahydrate containing 20 micrograms of ^{239}Pu had been irradiated while the experiments described above were performed. From this batch of material, a sample of 2.77 micrograms of PuO_2 was weighed on September 10, 1942 in a unique quartz fiber microbalance. The alpha counting of

---

10 This material was the first prepared from a man-made element that could be isolated and viewed. See Cunningham and Werner, J Amer Chem Soc 1949, 71, 1521–1528.

an aliquot of this material obtained after dissolution permitted the first determination of the specific activity of ^{239}Pu from which its half-life could be calculated. From these types of microscale experiments, large-scale separations processes were developed which were utilized at the wartime separations plant at Hanford to process the irradiated uranium fuel rods produced at the Hanford reactors.

8.3.6.3 Separations

Almost all of the hundreds of tons of plutonium that have been extracted so far have utilized the PUREX (**p**lutonium–**ur**anium **ex**traction) solvent extraction process.[11] The following equilibrium equations represent the major separation steps of this process

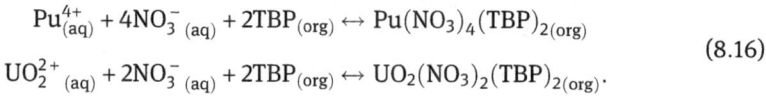

$$Pu^{4+}_{(aq)} + 4NO_3^-{}_{(aq)} + 2TBP_{(org)} \leftrightarrow Pu(NO_3)_4(TBP)_{2(org)}$$
$$UO_2^{2+}{}_{(aq)} + 2NO_3^-{}_{(aq)} + 2TBP_{(org)} \leftrightarrow UO_2(NO_3)_2(TBP)_{2(org)}.$$

(8.16)

In these equations, the subscripts (aq) and (org) refer to the aqueous and organic phases, respectively. The organic solvent consists of 30% tributylphosphate (TBP) in kerosene and is used as shown in the first two equations to extract the plutonium and uranium into the organic phase. The fission products and the higher trivalent actinides (Am^{3+} and Cm^{3+}) stay in the aqueous phase. The plutonium in the organic phase is then selectively reduced to Pu^{3+} and goes into the aqueous phase. The uranium can then be removed from the organic phase by back extraction into nitric acid. These steps are shown in the generalized PUREX flow sheet in Fig. 8.11.

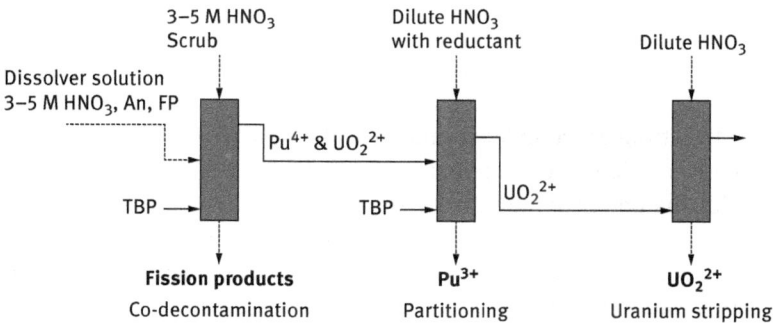

Fig. 8.11: Schematic representation of the extraction and stripping steps for separating Pu in the processing of spent fuel rods (PUREX process) by solvent extraction. The organic fractions (dark continuous lines) consist of 30 wt% tributylphosphate (TBP) in the organic solvent kerosene. The aqueous streams are shown as dashed lines.

---

11 Cf. also Section 7.3 on "Liquid-liquid extractions" and Section 10.6 on "The fuel cycle".

8.3.6.4 Chemical properties

Plutonium compounds are known with the Pu ion in oxidation states 3+ to 7+ (Pu^{III} to Pu^{VII}). In aqueous solution, Pu ions can exist in any of these states depending on the complexing ions and oxidizing (or reducing) power of other reagents in solution. In oxidation states 5+ or 6+, Pu forms the plutonyl ion where the Pu^V or Pu^{VI} atom forms a linear array between two O atoms, O = Pu = O, with the atoms from other ligands in the first coordination shell surrounding the Pu atom forming a plane perpendicular to the linear O = Pu = O array. Usually, there are four or five atoms in this first coordination shell. For example, Pu^V or Pu^{VI} in aqueous acidic solution forms the complex ion PuO$_2$(H$_2$O)$_5^{q+}$ with q = 2 or 1 and the O atoms of the five water molecules forming an equatorial plane around the linear O = Pu = O array. This geometry is similar to that of the uranyl ion shown in Fig. 8.3. Under acidic conditions with noncomplexing acids, Pu^{III} or Pu^{IV} ions form complex ions of the composition Pu(OH$_2$)$_n^{q+}$ with n = 8 or 9 and q = 3 or 4. The water molecules can be displaced by stronger coordinating ligands, and many coordination complexes of Pu^{III} and Pu^{IV} are known. In general, most Pu^{III} and Pu^{IV} compounds are relatively insoluble, but Pu^V and Pu^{VI} compounds are more soluble. Thus, the oxidation state of a compound of Pu affects its properties and is important to determine. This knowledge is essential in deciding what remediation efforts to utilize in cases where Pu contamination has occurred. Pu^{IV} hydrolyzes strongly to form small particles of Pu hydroxides or Pu oxides (colloidal solids) that behave similarly to solutions. These Pu colloids can attach themselves to natural colloids and can cause migration of Pu under conditions where no migration by normal processes is expected.

The relative tendency of Pu ions for hydrolysis or for complex formation is as follows:

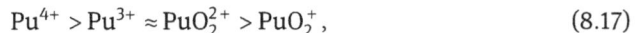

$$Pu^{4+} > Pu^{3+} \approx PuO_2^{2+} > PuO_2^+, \tag{8.17}$$

where Pu^{4+} forms the strongest complexes and PuO$_2^+$ the weakest. In acidic solution (low pH), the energy needed to change oxidation states is rather small so small changes in the oxidizing power of the solution (E_h) can result in changes in the oxidation state of the Pu species. Thus, it is possible for multiple oxidation states of Pu to exist in the same solution. Figure 8.12 shows the range of oxidation states that can exist in aqueous solution containing various complexing anions.

Many coordination compounds of Pu in all oxidation states can be precipitated from solution. However, structural information from single crystal X-ray analysis is sparse due to the problems of handling radioactive compounds. Recently, a number of these problems have been overcome and more structural information is now becoming available.

8.3.6.5 Ionic compounds

The known binary trihalides of Pu are PuF_3, $PuCl_3$, $PuBr_3$, and PuI_3. PuF_4 and $PuCl_4$ are the only tetravalent tetrahalides known. $PuCl_4$ is formed at high temperatures as a gas by the reaction

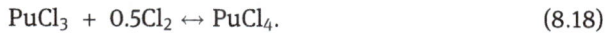

$$PuCl_3 + 0.5Cl_2 \leftrightarrow PuCl_4. \qquad (8.18)$$

When the gaseous $PuCl_4$ is condensed, $PuCl_3 + Cl_2$ are formed again. No characterized pentavalent halides are known and the only Pu hexahalide is PuF_6. The volatile and corrosive solid PuF_6 may be formed from the reaction of PuF_4 with F_2 at high temperatures. This volatile solid is of technological importance and many studies have been reported. Although solid Pu fluoride compounds with stoichiometry between PuF_4 and PuF_6 have been reported such as Pu_4F_{17}, no pure solid compounds of PuF_5 are known. Its existence has been proposed in gas-phase studies of the decomposition of PuF_6.

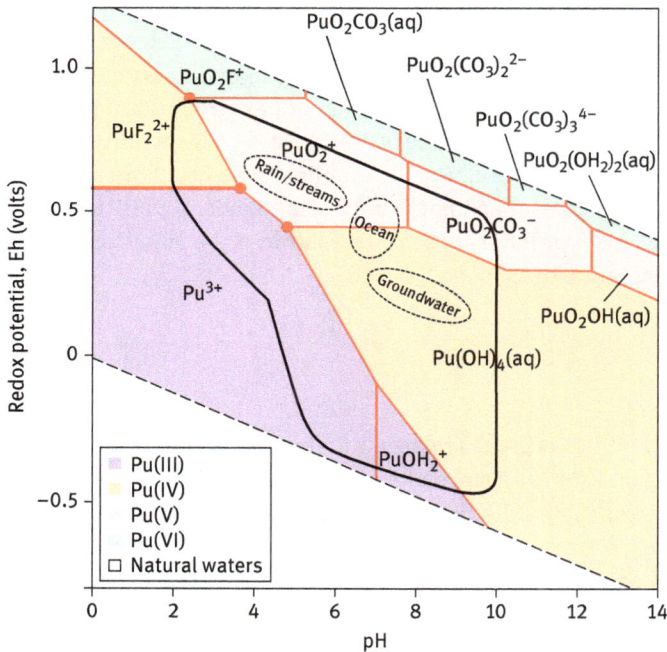

Fig. 8.12: The distribution of various Pu oxidation states in aqueous solution as a function of pH and oxidizing power (courtesy of Los Alamos National Laboratory).

The binary plutonium oxides are of great technological importance and quite complex. Stoichiometric materials such as Pu_2O_3 ($PuO_{1.5}$) and $PuO_{2.00}$ can be prepared but there are also other compounds with different ratios of Pu/O_x where x is close to (but not exactly) 1.5 or 2.00. The most important compound is PuO_2 which is used as a nuclear fuel, as a long-term storage material for spent nuclear fuel and for surplus weapons material, and in power generators. Thermodynamic predictions and early results indicated that PuO_2 was the highest stable binary phase in the Pu–O system. However, recent results suggest that there may be a solid PuO_{2+x} phase. This is now an area of active investigation.

Plutonium forms oxyhalides with valence states Pu^{III} and Pu^{VI} but not with Pu in the tetra- or pentavalent states. $PuOX$, where X = F, Cl, Br, and I, are known as are PuO_2F_2, $PuOF_4$, and $PuO_2Cl_2 \cdot 6H_2O$.

Plutonium hydrides are formed by the reaction of Pu metal with carefully controlled amounts of gaseous H_2 at elevated temperatures because the stoichiometry of the product PuH_x ($1.9 \leq x \leq 3$) can vary. Plutonium hydride can be decomposed under vacuum at temperatures higher than 400 °C to form finely divided, highly reactive plutonium metal. Disassembly of metallic plutonium from nuclear weapons can be followed by a hydride-dehydride process for conversion of the plutonium to a less reactive form such as PuO_2.

Compounds of plutonium with most elements of the periodic table are known. Although almost all compounds are to some degree ionic, many binary plutonium compounds are metallic or semiconductors. Plutonium borides, group 13 compounds (carbides, silicides, and germanides), pnictides (Pu–N, Pu–P, Pu–As, Pu–Sb, and Pu–Bi systems), and chalcogenides (Pu–S, Pu–Se, and Pu–Te systems) are some important, primarily refractive classes of actinide binary compounds. A particularly interesting ternary compound of Pu, $PuCoGa_5$ has been shown to be superconducting at the relatively high temperature of 18.5 K (or –255 °C).

8.3.6.6 Uses

The isotope ^{239}Pu has a half-life of 24,110 years so that large amounts can be prepared. Its high cross section for slow neutrons makes it suitable for nuclear fuel in both weapons and in power plants. Its major use has been in nuclear weapons[12] and the "trigger" for most thermonuclear bombs. The processing of spent nuclear fuel results in the separation of the uranium and plutonium fractions. These fractions can be combined to produce a mixed oxide fuel of uranium and plutonium $(U,Pu)O_2$ (abbreviated MOX), which can be reutilized in nuclear power reactors.

At present, nuclear power plants are producing on the order of 70–90 megatons of plutonium per year. The total global inventory is over 1900 megatons. Most of this material is in spent fuel rods; however, there is also a significant inventory in

---

12 The Trinity test in New Mexico, July, 1945, the bomb dropped over Nagasaki the following month.

nuclear weapons, from disassembled nuclear weapons, and from recycled reactor fuel. Making sure that this plutonium material is safe and secure is a very significant problem.

The isotope ^{238}Pu has a half-life of 87.7 years, a specific activity of 17.2 Curies/gram, and a power density of approximately 7 watts/gram. These characteristics make it useful in nuclear power systems that get their energy from the heat generated by radioactive alpha transformation, which is converted into electrical energy, rather than from nuclear fission. ^{238}Pu was initially used in cardiac pacemakers in the 1970s but improved batteries and electronic advances made these devices obsolete. Electrical power sources fueled by ^{238}Pu have been widely used in the United States for space exploration purposes. For example, Space Nuclear Auxiliary Power (SNAP) units have been utilized for instrument packages on the five Apollo missions to the moon, the Viking Mars Lander, and the Pioneer and Voyager probes to the outer planets. Other types of ^{238}Pu heat and power sources have been utilized on other space missions.

8.3.6.7 Plutonium metal

Plutonium metal and its alloys and intermetallic compounds are important technologically because the nuclear properties of ^{239}Pu enable this material to be utilized for nuclear weapons, for advanced designs of nuclear fuel rods, and for safe storage and disposition of excess weapons plutonium. There are many ways to produce Pu metal; they usually involve the reduction of a compound of Pu (such as PuO_2 or PuF_4) with a metal such as Ca at elevated temperatures. Pure Pu metal (α phase) at room temperature has a dense, monoclinic structure that is brittle and reactive. The Pu phase diagram is quite complex as shown by the thermal expansion curve and densities in Fig. 8.13. The δ phase is less dense and workable and is formed and stable in pure Pu at higher temperatures. This phase can be stabilized down to room temperature by the addition of small amounts of aluminum or gallium. Pu metal is one of very few substances that increases in density when heated. The complex behavior of Pu metal is due to the presence of 5f electrons in the valence band. These 5f electrons in the earlier actinide metals (protactinium through neptunium) form bands with metal-metal bonding (itinerant behavior) through the 5f orbitals. This behavior is in contrast to the lanthanide metals (with 4f electrons) that exhibit localized 4f electron behavior and where the metals have magnetic characteristics similar to trivalent lanthanide compounds. In the actinide series at plutonium, there is a transition between itinerant behavior and localized behavior, thus small changes such as alloying or temperature or pressure can cause drastic changes in the properties of the metal. This can be clearly seen from Fig. 8.6 where the atomic volumes of the actinide metals are compared with those of the 5d and 4f transition metals.

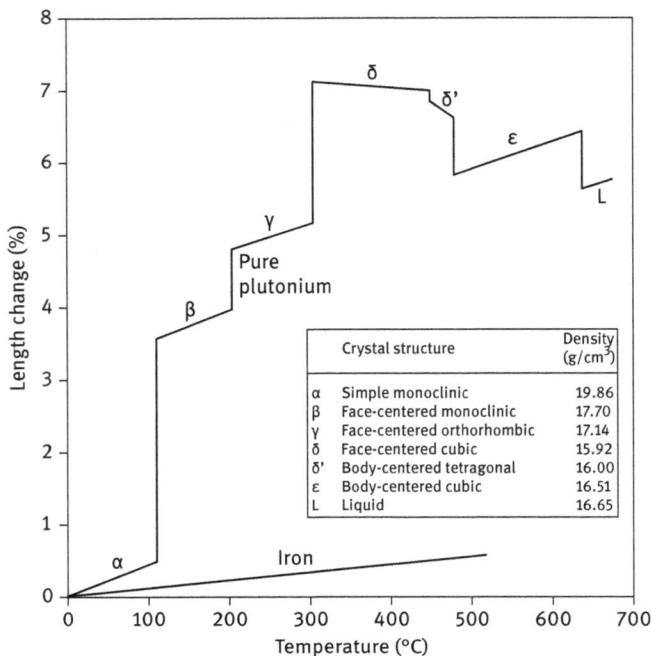

Fig. 8.13: Thermal expansion of pure plutonium at one atmosphere showing six different solid phases and the liquid phase. Iron is shown for comparative purposes. The crystal structures of the various phases are given as well as their density (courtesy of Los Alamos National Laboratory).

Experimentally, plutonium metal and its alloys and intermetallic compounds require special handling because of their radioactivity and criticality issues (the amount of material that can be handled or stored safely without causing a spontaneous nuclear reaction or excess radiation hazards). Nevertheless, these materials present some outstanding challenges to our fundamental understanding of complex materials.

8.3.6.8 Organoplutonium chemistry

There have been very few molecules synthesized that contain plutonium-organic ligands bonded via plutonium-carbon bonds. The major reason is because of safety concerns in the handling of plutonium materials and organic compounds and solvents in oxygen-free conditions. Triscyclopentadienyl plutonium ($\eta^5C_5H_5)_3$Pu and Pu($C_8H_8)_2$ have been synthesized and characterized and are structurally similar to the analogous uranium compounds. Compounds of stoichiometry ($\eta^5C_5H_5)_3$PuX compounds are known where X = Cl or NCS as well as some mono-cyclopentadienyl Pu^{IV} halide-Lewis base complexes.

8.3.7 Americium

8.3.7.1 Discovery

The element americium was first produced in 1944 by the beta transformation of ^{241}Pu, which was produced by an (α,n) reaction on ^{238}U

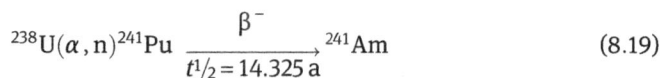

$$^{238}U(\alpha,n)^{241}Pu \xrightarrow[t^{1/2}=14.325\,a]{\beta^-} {}^{241}Am \tag{8.19}$$

by Seaborg, James, and Morgan. The estimated half-life of the ^{241}Pu was approximately 10 years (present value 14.4 a). Several samples of ^{241}Pu were carefully purified to remove all traces of rare-earth materials and then were checked again after some time. When the rare earth fractions were removed again it was found that the α-activity previously removed had grown in again. From these types of experiments the following observations were made:
1. The rate of formation of α activity was constant over a period of several years.
2. The yield from a given sample was a linear function of the time of growth.
3. The amount of growth of the α-activity in similar periods of time was directly dependent on the amount of ^{241}Pu in the sample.

It was concluded that the α activity was due to the isotope ^{241}Am, a new element. The new element was named americium, after the Americas, on the basis of its position as the sixth member of the actinide series, analogous to europium in the lanthanide series.

8.3.7.2 Uses

The two major isotopes of americium are ^{241}Am and ^{243}Am with half-lives of 433 and 7380 years, respectively. The major industrial uses of ^{241}Am are as a low energy gamma source in smoke detector alarms and to produce neutrons by the (α,n) reaction with beryllium. These ^{241}Am(Be) neutron sources are used in soil moisture measurements, well-logging applications, and neutron radiography.

8.3.7.3 Chemical properties

As indicated in Tab. 8.2, americium can exist in oxidation states +3 through +7. In strong aqueous basic carbonate solution, americium can exist in the 3+, 4+, 5+, and 6+ oxidation states. A few divalent Am compounds have also been prepared. The most stable oxidation state in aqueous solution is Am^{3+}, which can be prepared by the dissolution of AmO_2 in hot HCl. Am^{4+} in aqueous solution can only be stabilized with complexing agents such as concentrated NH_4F solutions. AmO_2^{2+} can be prepared from Am^{3+} with strong oxidizing agents in dilute acidic solution. However, AmO_2^+ can only be oxidized by ozone or hypochlorite from Am^{3+} in near-neutral or basic solution. Aqueous solution of Am^V and Am^{VI} exist as the linear americyl ion; AmO_2^+ $(O = Am = O)^+$ or AmO_2^{2+} $(O = Am = O)^{2+}$ with the same linear configuration found in the earlier actinides in these oxidation states. Am^{VII} can be prepared by the oxidation of $\approx 0.001\,M$ AmO_2^{2+} in 3–4 M NaOH at 0 °C to form the green Am^{VII} solution. In dilute acidic aqueous solution, Am^{3+} exhibits similar solution behavior as the trivalent lanthanide ions including precipitation by hydroxide, fluoride, oxalate, and phosphate.

8.3.7.4 Metal

Americium metal can be formed by the reaction of AmF_3 with barium or lithium. Several intermetallic compounds are known. Am metal melts at 1149 K and is the first actinide metal to exhibit localized 5f behavior at room temperature and pressure, similar to the lanthanide metals.

8.3.7.5 Solid compounds

The only binary Am fluorides known are AmF_3 and AmF_4. $AmBr_3$ and AmI_3 are also known binary compounds. Interestingly, the divalent compounds $AmCl_2$, $AmBr_2$, and AmI_2 also have been prepared, the first divalent compounds in the actinide series. The divalent chemistry of Am is somewhat similar the divalent chemistry of its lanthanide analog Eu, although it is much more difficult to prepare the divalent Am compounds. The binary compound AmO_2F_2 is known as are ternary compounds such as Rb_2AmF_6, $RbAmO_2F_2$, and $Cs_2AmO_2Cl_4$.

8.3.7.6 Oxides

The binary oxides of Am known with certainty are Am_2O_3 and AmO_2. AmO_2 is the most stable oxide and is prepared by heating Am precipitates such as hydroxides, carbonates, or oxalates in air or oxygen at temperatures of 600–800 °C. Am_2O_3 is obtained by heating AmO_2 in hydrogen at temperatures of 600 °C and above. Am_2O_3 oxidizes easily in air even at room temperature. The intermediate phases between $AmO_{1.5}$ and AmO_2 have been determined.

8.3.7.7 Other compounds

Americium forms compounds with hydrogen, chalcogenides, pnictides, carbon, and a number of mineral acids. The identification of these compounds is primarily from powder X-ray diffraction data. Only a few organometallic compounds have been prepared because of the difficulties of working with Am which include $Am(C_5H_5)_3$ and $KAm(C_8H_8)_2$.

8.3.8 Curium

8.3.8.1 Discovery

The element curium was first discovered in 1944 by Seaborg, James, and Ghiorso utilizing the nuclear reaction

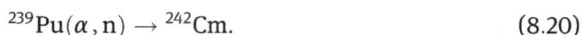

$$^{239}Pu(\alpha, n) \rightarrow {}^{242}Cm. \tag{8.20}$$

The half-life of the ^{242}Cm is 163 days. Many curium isotopes are known, ^{244}Cm is the most abundant isotope ($t^{1}/_2 = 18.1$ a) and is available on the multigram scale. The isotope ^{248}Cm has a very long half-life ($t^{1}/_2 = 340,000$ a) but is of only limited availability. ^{248}Cm has an 8% spontaneous fission yield, which limits the amount of material that can be handled in a glove box. If this isotope is available, it is the isotope of choice for chemical experiments. Curium was named after Pierre and Marie Curie in honor of their seminal contributions to radiochemistry. This name was selected by analogy to the corresponding lanthanide element, gadolinium, named after Gadolin, who is known for his contributions to lanthanide chemistry.

8.3.8.2 Uses

^{242}Cm and ^{244}Cm have been used for power sources (thermal and electrical) for space and medical applications. However, their uses in these applications have been supplanted by the availability and use of ^{238}Pu. The strong luminescence of Cm^{3+} in compounds and solution due to its favorable electronic energy levels has made it the trivalent actinide element of choice for analytical applications.

8.3.8.3 Chemical properties

As indicated in Tab. 8.2, curium can exist only in oxidation states +3 and +4. For Cm^{4+} only compounds of oxides and fluorides are known. Almost all aqueous chemistry of Cm is for trivalent Cm. Cm^{4+} complexes have been reported in concentrated alkali metal solutions. Cm^{3+} exhibits similar solution behavior as the trivalent lanthanide ions including precipitation by hydroxide, fluoride, oxalate, and phosphate.

8.3.8.4 Metal

Curium metal can be formed by the reaction of CmF_3 with barium or lithium. A number of intermetallic compounds are known. Cm metal melts at 1618 K and exhibits localized 5f behavior at room temperature and pressure with a magnetic moment similar to its lanthanide counterpart, gadolinium, and confirming the half-filled shell configuration, radon core, $5f^7$.

8.3.8.5 Solid compounds

The only binary Cm fluorides known are CmF_3 and CmF_4. $CmCl_3$, $CmBr_3$, and CmI_3 are also known binary compounds. CmO_2 is prepared by heating Cm precipitates such as hydroxides, carbonates, nitrates, or oxalates in air or oxygen at temperatures below 400 °C. Cm_2O_3 is obtained by heating thermally unstable CmO_2 to 900 °C. Intermediate phases between $CmO_{1.5}$ and CmO_2 are known. Curium forms compounds with hydrogen, chalcogenides, pnictides, carbon, and a number of mineral acids. The identification of these compounds is primarily from X-ray powder diffraction data. The only organometallic compound that has been prepared because of the difficulties of working with Cm is $Cm(C_5H_5)_3$. This compound fluoresces bright red under ultraviolet irradiation.

8.3.9 Transcurium actinides

8.3.9.1 Discoveries

Elements 97 and 98: Berkelium was first produced and separated in December 1949 in Berkeley, California, by cyclotron irradiation of ^{241}Am with α-particles. Chemical separation and identification of Bk^{3+} were achieved by cation exchange with citric acid eluant. Two months later, cyclotron irradiation of ^{242}Cm with α-particles achieved a parallel discovery of the first californium isotope ^{244}Cf, again separated with cation exchange and citric acid. Similar to other transuranium elements, the first-discovered isotope is typically not that of longest half-life or most readily available for chemical studies. ^{249}Bk has been produced and separated from high-neutron-flux reactor targets in multi-milligram quantities at Oak Ridge National Laboratory; this isotope has a

320 day β^--transformation to ^{249}Cf, a 351 year half-life α-emitting isotope that is also suitable for chemical studies.

Elements 99 and 100: Einsteinium and fermium were produced in neutron-irradiated uranium at the "Mike" thermonuclear explosion after its detonation in November 1952. The isotopes ^{253}Es and ^{255}Fm were subsequently discovered at US national laboratories by promptly separating transcurium isotopes from samples of uranium, using cation exchange of their +3 ions with citric acid eluant.

Element 101: Mendelevium was synthesized in 1955 by bombarding a target of $\approx 10^9$ ^{253}Es atoms with α-particles. Separation was done by cation exchange with the new eluant α-hydroxyisobutyric acid (α-HIB),[13] which had recently been shown to have better separation of adjacent +3 f-element ions than citric acid. Existence of the isotope ^{256}Md, which transforms mostly by electron capture to ^{256}Fm, was inferred by observation of spontaneous fission from the daughter isotope.

Element 102 (nobelium) was first claimed in 1957 by an international group carrying out bombardments of ^{244}Cm with ^{13}C in Sweden. Again separation was carried out with α-HIB. This result was never confirmed. Later experiments in California and Russia achieved reproducible syntheses of ^{254}No using novel nuclear physics, in particular the "double recoil technique" that captured the product atoms on an electrically charged moving belt to separate it and to estimate its half-life. As explained below, the most likely oxidation state is No^{2+} in solution, so the α-HIB separation as No^{3+} was improbable. Nevertheless, by 1968 all competing research groups agreed to accept "nobelium" as the name for element 102.

Element 103 Several isotopes of lawrencium were discovered as the result of many bombardment experiments in California and Russia between 1961 and 1971. Chemical properties have been elucidated by use of the relatively long-lived isotopes ^{256}Lr ($t^{1/2}$ = 26 s) and ^{260}Lr ($t^{1/2}$ = 3.0 min).

8.3.9.2 Uses

The only transcurium isotope with practical uses is ^{252}Cf. This isotope has a half-life of 2.645 years. Almost all of the transformations (97%) are by α-emission, but a significant fraction (3.1%) of the transformations are spontaneous fission. As a result, ^{252}Cf samples can be fabricated as very small intense neutron sources (3 · 10^6 n/s/μg)

---

13 Cf. Section 7.4.

that have been utilized for neutron radiography (e.g., to prospect for underground spontaneous fission petroleum sources) and medical brachytherapy[14] (to deliver intense neutron doses to tumors).

8.3.9.3 Chemical properties

As indicated in Tab. 8.2, the most common and most stable oxidation state of all transcurium elements is +3, with the notable exception that No^{2+} is the most stable oxidation state of nobelium. A few tetrapositive species of Bk and Cf are known. For berkelium, Bk^{4+} has the half-filled $5f^7$ subshell so that it is possible to remove one electron from Bk^{3+} ($5f^8$). The resulting Bk^{4+} is a strong oxidant (similar to Ce^{4+}, $4f^0$) in aqueous solution; tetravalent stoichiometric compounds such as BkO_2 and BkF_4 are known. The next element, californium has an accessible Cf^{4+} ($5f^8$) state in solids, but Cf^{4+} is too strong an oxidant to exist in aqueous solution except when complexed with phosphotungstate. CfF_4 is known but CfO_2 exists as hyperstoichiometric CfO_{2-x}. There is only fleeting evidence for EsF_4 as a metastable species in hot flowing $F_2(g)$.

In stark contrast with the 4f elements, transcurium elements display increasingly stable +2 states from Bk through No. Bk^{2+} has been observed in a pulse radiolysis study, in which the strong reductant $e^-(aq)$ is produced by bombarding aqueous solutions with a high-energy electron beam. $Cf^{2+}(aq)$ and $Es^{2+}(aq)$ are slightly too strong reducing agents to be even metastable; their presence has been implied by radiopolarography (reduction of an ion at a mercury-drop electrode with identification by its radioactivity). The existence of Fm^{2+} has been inferred from reduction of $Fm^{3+}(aq)$ by radiopolarography and in molten salts. $Md^{2+}(aq)$ is stable in the absence of oxygen. $No^{2+}(aq)$ is the most stable oxidation state of No in solution; $No^{3+}(aq)$ has been shown to be an oxidant as strong as $Ce^{4+}(aq)$.

8.3.9.4 Ionic compounds

As expected from trends in solution, trivalent compounds (oxides and halides) are known for Bk, Cf, and Es. These compounds, and the tetravalent compounds mentioned above, have properties consistent with other ionic compounds. Divalent halides of californium and einsteinium, $CfCl_2$, $CfBr_2$, and CfI_2 parallel the properties of alkaline earth halides and so are relatively ionic. CfF_2 and EsF_2 are unknown. The half-lives of all isotopes of all elements beyond einsteinium are so short that sample sizes would be orders of magnitude too small for even nanoscale synthesis and characterization.

---

14 Cf. Chapter 13.

8.3.9.5 Coordination and organometallic chemistry

In aqueous solution, Bk^{3+} and Cf^{3+} have between 8 and 9 water molecules (CN between 8 and 9) in the inner coordination sphere, paralleling the behavior of Sm^{3+} and Eu^{3+}. The lighter actinide +3 ions have CN 9 in aqueous solution and the heavier actinide +3 ions are expected to have CN 8. The relatively few known coordination complexes of these ions parallel the coordination complexes of rare earth ions with similar ionic radii.

Due to their short half-lives and limited availability, only a few organometallic compounds of these elements are known. $Bk(C_5H_5)_3$ and $Cf(C_5H_5)_3$ have been synthesized and characterized structurally. $Bk(C_5H_5)_2Cl$ is also known, probably as a dimer.

Recommended reading

Morss LR, Edelstein NM, Fuger J (eds.). The chemistry of the actinide and transactinide elements, 4th ed., 6 volumes. Dordrecht, Springer, 2010.

Kaltsoyannis N, Scott P. The f elements. Oxford, Oxford University Press, 1999.

Cotton S. Lanthanide and actinide chemistry. Chichester, Wiley, 2006.

Seaborg GT, Loveland WD. The elements beyond uranium. New York, Wiley Interscience, 1990.

Edelstein NM, Navratil JD, Schulz WW (eds.) Americium and Curium Chemistry and Technology: Papers from a Symposium given at the 1984 International Chemical Congress of Pacific Basin Societies, Honolulu, HI, December 16–27, 1984, 2012, Springer Netherlands, 978-94-009-5444-1.

Christoph E. Düllmann

9 Radioelements: Transactinides

Aim: The heaviest chemical elements in the periodic table are all artificial, man-made elements, produced one atom at a time at accelerator facilities by fusing nuclei of lighter elements. Following the actinide series, they are called the transactinides and are all radioactive and transform rather quickly – with a few exceptions with half-lives of the order of minutes to seconds or less. The first 14 transactinides up to the element with atomic number $Z = 118$ are known. They fill the 7th period and are homologs of the elements Hf–Rn.

Despite their short half-lives and minute production rates, chemical properties of elements 104–108 and 112–114 have been studied and compared to those of their lighter homologs. While the gross trends established within the groups by the lighter, well-known, stable members are mostly followed, detailed studies revealed significant deviations from simple extrapolations within the respective groups. In many cases, these can only be understood if the influence of the very high nuclear charge on the electrons in the atomic shell is fully taken into account. In the transactinides, electrons are accelerated to a substantial fraction of the speed of light. A correct theoretical treatment requires fully relativistic quantum chemical calculations. These elements are thus an ideal laboratory to test the influence of relativity on chemical properties and experimental data help refine theoretical models.

In this chapter, the discovery of the transactinides, their nuclear aspects, as well as the methods to study chemical properties of single short-lived atoms are treated, and the known chemical properties of the transactinides are summarized.

9.1 Introduction

Chemical elements beyond the last actinide element, lawrencium, cf. Fig. 9.1, are called *trans*actinide elements.[1] Thus, the first transactinide element is the one with atomic number $Z = 104$, rutherfordium. Transactinide elements with $Z = 104$ to $Z = 118$ have been discovered and named, most recently – in 2016 – the elements with $Z = 113, 115, 117$, and 118. All currently known transactinides are members of the 7th period of the periodic table. Search experiments for the elements $Z = 119$ and 120, which would start the 8th period, are ongoing, but so far these elements have not been

---

1 The terms "superheavy elements" and "transactinide elements" are frequently used almost interchangeably. As the former expression emphasizes the nuclear aspect while the latter the chemical aspect, superheavy elements is frequently preferred when nuclear phenomena are central to the discussion. When chemical aspects are in the center of attention, transactinides is more frequently used.

https://doi.org/10.1515/9783110742701-009

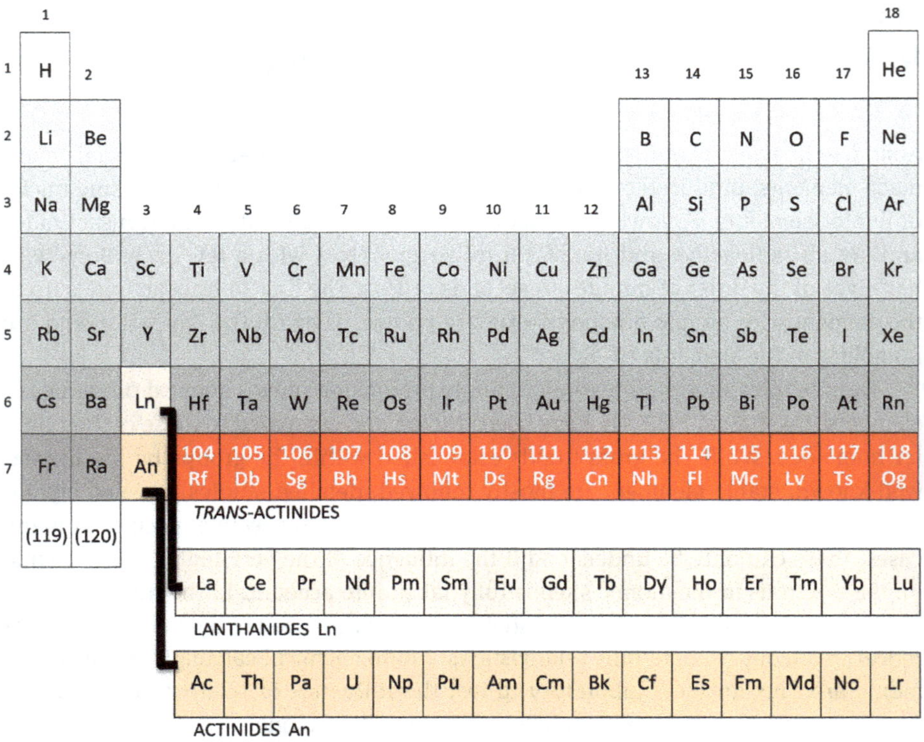

Fig. 9.1: Positioning of the transactinide elements within the periodic table. Transactinides of the 7th period are indicated by the red box. The 8th period will start with elements Z = 119 and 120, which are yet undiscovered and will be the heavier homologs of francium and radium.

discovered. All transactinides are radioactive elements; with few exceptions, isotopes with half-lives of at most one minute are known.

9.2 The concept of the "Island of Stability of Superheavy Elements"

The *liquid-drop model* (LDM) of the atomic nucleus[2] predicts that nuclei with $Z \gtrsim 104$ will transform by spontaneous fission (sf) within less than 10^{-14} s due to the strong COULOMB repulsion of the many protons contained in a single nucleus. As the formation of an electron shell takes about 10^{-14} s, this is equivalent to the prediction that the periodic table of the elements ends at $Z \approx 104$. However, based on the *nuclear*

---

2 Cf. Chapter 3 in Vol. I for the two models of the nucleus of the atom.

shell model (NSM), which takes into account that the liquid drop model significantly underestimates the stability of nuclei with certain "magic" numbers of protons and neutrons, the existence of long-lived nuclei with atomic numbers far larger than those of any element found on earth was predicted in the 1960s. The basic underlying idea was that the filling of the next proton and neutron shells beyond those filled at $Z = 82$ and $N = 126$ (which give rise to the heaviest stable doubly-magic nucleus, $^{208}_{82}Pb_{126}$) should lead to a next doubly magic nucleus. The extra stability associated with its doubly magic nature was predicted to be large enough to largely compensate the strong COULOMB repulsion, leading to significant half-lives of nuclei at and around the next shell closures. Owing their existence to nuclear shell effects, such nuclei were termed superheavy nuclei. From a chemist's point of view, elements where all isotopes with half-lives $\geq 10^{-14}$ s exist exclusively thanks to these nuclear shell effects are referred to as superheavy elements. Coincidentally, these are all elements with $Z \geq 104$, and hence the terms transactinides and superheavy elements can be used almost synonymously.

Different models led to different predictions for the next magic proton number. Even today, it is not clear where the next proton shell is filled. Most models favor $Z = 114$, $Z = 120$, or even $Z = 126$. The next magic neutron number is most often predicted to occur at $N = 184$. Recent experiments found no effect at $Z = 114$, albeit at neutron numbers that were still many units away from $N = 184$. Potentially, there

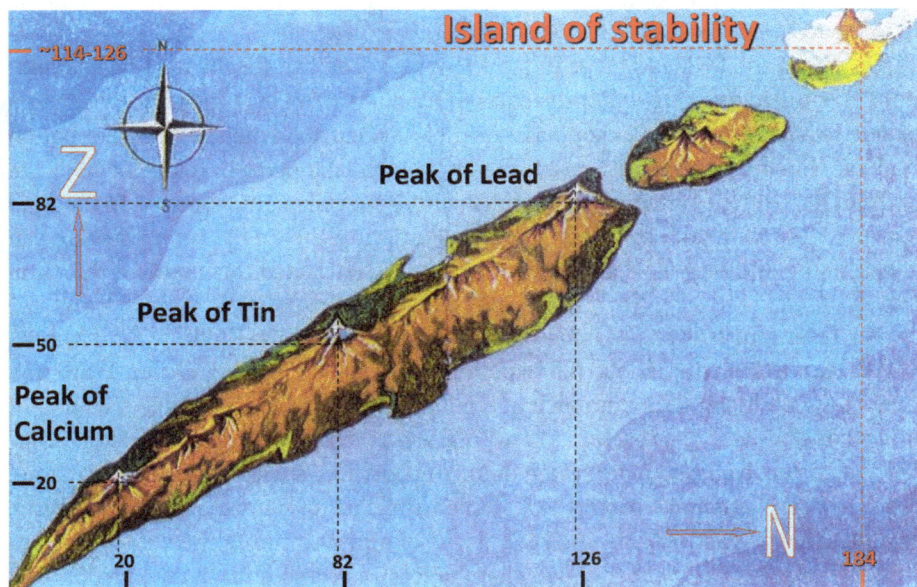

Fig. 9.2: The landscape of the nuclear chart. The third dimension, i.e., the elevation, is a measure for nuclear stability. The *island of stability* as predicted by the nuclear shell model is located in the midst of a *sea of instability* (after Flerov and Ilyinov 1986).

will no longer be a single magic number, but rather a whole region of enhanced nuclear stability, the "Island of Stability" (see Fig. 9.2).

9.3 The discovery of the transactinide elements

After the discovery of the last actinide[3] element, Lr (Z = 103), in 1961, competing teams from the United States of America and from the Soviet Union continued their pursuit of ever heavier elements. The discovery of the first element beyond the actinides, **element 104**, was heavily contested, with claims coming from the Lawrence Berkeley National Laboratory (LBNL) in Berkeley in the USA and from the Joint Institute for Nuclear Research (JINR) in Dubna in the USSR. Conclusive proof was obtained in Berkeley from the systematic study of the nuclear reactions ^{15}N + ^{249}Bk, ^{16}O + ^{248}Cm, and ^{18}O + ^{249}Cf, which yielded a ≈20 ms sf transforming isotope, ^{260}104. In the first two cases, this was the 4n evaporation channel product,[4] in the last case the α3n evaporation channel product. In the following few years, longer-lived isotopes with mass numbers 257 (t½ ≈ 4 s) and 261 (t½ ≈ 1 min) were discovered, which allowed chemical experiments to be performed. These established the transition metal character of the new element, which was in agreement with its position as the first transactinide element in the periodic table. Its name, rutherfordium (Rf), was only agreed on by all involved parties in 1997, when the names of all elements up to Z = 109 were fixed.

Groups in the USA as well as in the USSR also searched for the next heavier **element 105**, again by using light projectiles to irradiate actinide targets. By observing a genetic relationship of the isotope assigned to element 105 with its α-transformation product, which was identified as an already known isotope of Lr (element 103), the group in Berkeley unambiguously established this new element, obtained in the ^{249}Cf (^{15}N,4n)^{260}105 reaction. To honor the contribution of the Soviet group working in Dubna to the discovery of the heaviest elements, IUPAC fixed this name as dubnium (Db).

Also **element 106** was searched for in the USA and in the USSR, without the involved parties knowing about the experiments of the competing group. While the Americans continued their approach of bombarding heavy actinide targets with light projectiles and used the reaction ^{18}O + ^{249}Cf, the Dubna group adopted a new strategy, which involved bombarding the doubly-magic, i.e., shell-stabilized ^{208}Pb target with the complementary projectile – ^{54}Cr in the case of element 106. Using more strongly bound reaction partners such as ^{208}Pb allows forming the compound nucleus at lower excitation energy, which enhances its chances for survival in the exit

---

3 Cf. Chapter 8 on the actinides.
4 Cf. Chapter 13, Vol. I on nuclear reactions.

channel. This new "cold fusion"[5] concept was very successful in producing element 102 in the reaction of two doubly-magic nuclei, ^{48}Ca + ^{208}Pb, where the cross sections are extraordinarily large. Both groups simultaneously announced discovery of element 106, through the observation of different isotopes, though. Berkeley claimed ^{263}106 formed in the 4n-evaporation channel, and Dubna a more neutron-deficient isotope, likely ^{259}106. To avoid an unsatisfactory situation with the element simultaneously having different names in different parts of the world, the two groups agreed to withhold naming of the element until either (or both) of the reported experiments were confirmed in independent studies. This finally happened in 1994, about 20 years after the initial reports, when researchers at LBNL repeated the ^{18}O + ^{249}Cf experiment and were able to reproduce the data of the original experiment. In honor of G. T. SEA-BORG, who (besides other fine work) invented the actinide concept and is the codiscoverer of 10 elements, element 106 is named seaborgium (Sg).

Element 107 (bohrium, Bh), **element 108** (hassium, Hs), **element 109** (meitnerium, Mt), **element 110** (darmstadtium, Ds), **element 111** (roentgenium, Rg), **element 112** (copernicium, Cn), and **element 113** (nihonium, Nh) were discovered using the cold fusion concept. At the Gesellschaft für Schwerionenforschung (GSI) in Darmstadt, Germany, the universal linear accelerator (UNILAC), capable of accelerating ions of all elements up to U to energies above the COULOMB barrier, was ideally suited to accelerate the medium heavy ^{54}Cr, ^{58}Fe, ^{62}Ni, and ^{70}Zn ions, which were used to bombard ^{208}Pb and ^{209}Bi targets. A new technique was introduced to separate the evaporation residues from unreacted beam and unwanted reaction products, e.g., from *multinucleon-transfer reactions*, where some nucleons are exchanged between the projectile and target nuclei. In previous approaches, the evaporation residues were thermalized in gas and transported with the flowing gas to a detector station, or they were implanted in foils, which were transported to such a station. The setup at GSI, on the other hand, used the well-defined velocity of the fusion products, which differs significantly from that of all other particles – and which can easily be calculated by considering momentum conservation – to provide the separation. This was achieved in the *Separator for Heavy Ion reaction Products* (SHIP), an ion-optical device comprising crossed electric and magnetic fields in an arrangement that transmits only ions of a specific velocity and deflects all others. The SHIP provided very fast separation – on the order of microseconds – basically given by the flight-time through the 11-meter-long separator. The names of elements 107–112 derive from eminent scientists (BOHR, MEITNER, RÖNTGEN, and COPERNICUS) or the location of the GSI in Darmstadt, in the German state of Hessen (Latin: *hassia*). Using a similar technique, element 113 was discovered at the RIKEN Nishina Center for Accelerator-Based Science in Wako-shi, Japan; its name comes from Nihon, which is

---

5 In the Fermi gas model, the nucleus can be characterized by usual thermodynamic quantities, including the potential energy E and the nuclear temperature t. For significantly large excitation energies (given in MeV), E is proportional to t^2.

one of the two ways to say "Japan" in Japanese, and literally means "the Land of Rising Sun". Despite an exceedingly small cross section, the researchers succeeded in the observation of three atoms of nihonium in the course of nine years.

In the GSI experiments, which were carried out between 1984 and 1996, the cross section was found to steadily decrease by a factor of about 10 per two protons that were added. The production rate of Cn is about two atoms per week with current technology. The elements beyond Nh were discovered by collaborators working in Dubna, who again adopted a change in concept. Instead of irradiating magic ^{208}Pb/^{209}Bi targets, the projectile was fixed at ^{48}Ca, a doubly magic and very neutron-rich projectile, which was used to bombard long-lived actinide isotopes. In experiments at the JINR's Flerov Laboratory of Nuclear Reactions (FLNR) in Dubna, using the Dubna Gas-Filled Recoil Separator – a recoil separator based on separating different ions emerging from the target in a magnetic field – **element 114** (flerovium, Fl)**, element 115** (moscovium, Mc), **element 116** (livermorium (Lv), **element 117** (tennessine, Ts), and **element 118** (oganesson, Og) were produced. The first names honor the contributions of Dubna's FLNR, named after the eminent Russian chemist G.N. FLEROV, who discovered, e.g., the sf transformation mode, and was the founder of the heavy ion laboratory in Dubna, as well as the Moscow region, where Dubna is located. Livermorium was chosen after the Lawrence Livermore National Laboratory, a collaborating research center based in Livermore in California, USA, and tennessine after the American state of Tennessee, where Oak Ridge, home to the Oak Ridge National Laboratory (ORNL) that operates a high-flux research reactor suitable to breed[6] heavy actinides including ^{249}Bk that was used as a target material in the synthesis of tennessine, is located. Oganesson, finally, is named after Yu.Ts. OGANESSIAN, who discovered the cold fusion concept and is the leader of the superheavy element program in Dubna (Oganessian 2015). The endings of the names of tennessine and oganesson were fixed as "-ine" and "-on" to conform to the endings of the lighter members of the halogen and noble gas groups in the periodic table.

The era of ^{48}Ca-induced reactions comes to an end at element 118, which is produced in the irradiation of ^{249}Cf, a target isotope with $Z = 98$. No isotopes of elements beyond Cf are available in quantities sufficient for making a target for an accelerator experiment. Hence, to proceed beyond element 118, the successful ^{48}Ca route has to be left and reactions with heavier projectiles will have to be used.

The main transformation modes of the heaviest nuclei are α-transformation and sf, two processes that are covered in more detail in Chapters 9 and 10 of Vol. I.

Table 9.1 summarizes the currently known transactinides and Fig. 9.3 displays the discovery timeline.

Figure 9.4 displays a cutout of the chart of nuclides showing experimentally observed nuclei of the transactinide elements.

---

6 Cf. Chapter 13.7.2 in Vol. 1 on the slow neutron capture process (s-process)

Tab. 9.1: Currently known transactinides.

Element	Symbol	Z	Year of discovery	Discoverers
Rutherfordium	Rf	104	1969	Ghiorso et al. (Berkeley)
Dubnium	Db	105	1970	Ghiorso et al. (Berkeley)
Seaborgium	Sg	106	1974	Ghiorso et al. (Berkeley)
Bohrium	Bh	107	1981	Münzenberg et al. (Darmstadt)
Hassium	Hs	108	1984	Münzenberg et al. (Darmstadt)
Meitnerium	Mt	109	1982	Münzenberg et al. (Darmstadt)
Darmstadtium	Ds	110	1995	Hofmann et al. (Darmstadt)
Roentgenium	Rg	111	1995	Hofmann et al. (Darmstadt)
Copernicium	Cn	112	1996	Hofmann et al. (Darmstadt)
Nihonium	Nh	113	2004	Morita et al. (Wako-shi)
Flerovium	Fl	114	2004	Oganessian et al. (Dubna)
Moscovium	Mc	115	2010	Oganessian et al. (Dubna)
Livermorium	Lv	116	2004	Oganessian et al. (Dubna)
Tennessine	Ts	117	2010	Oganessian et al. (Dubna)
Oganesson	Og	118	2006	Oganessian et al. (Dubna)

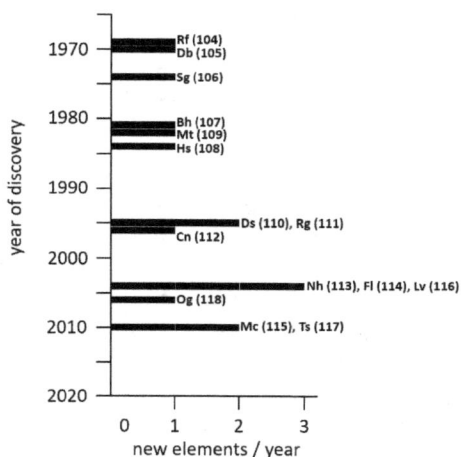

Fig. 9.3: Timeline of the discovery of the transactinide elements.

9.4 Naming of new elements

An intriguing aspect of superheavy element research is the naming of new elements. Traditionally, the discovery of a new element implied the right to name the new substance. With the body of the periodic table up to elements substantially heavier than uranium complete, and the rate at which new elements are discovered being small (cf. Fig. 9.3), the naming of a new element is a rare occasion, which consistently

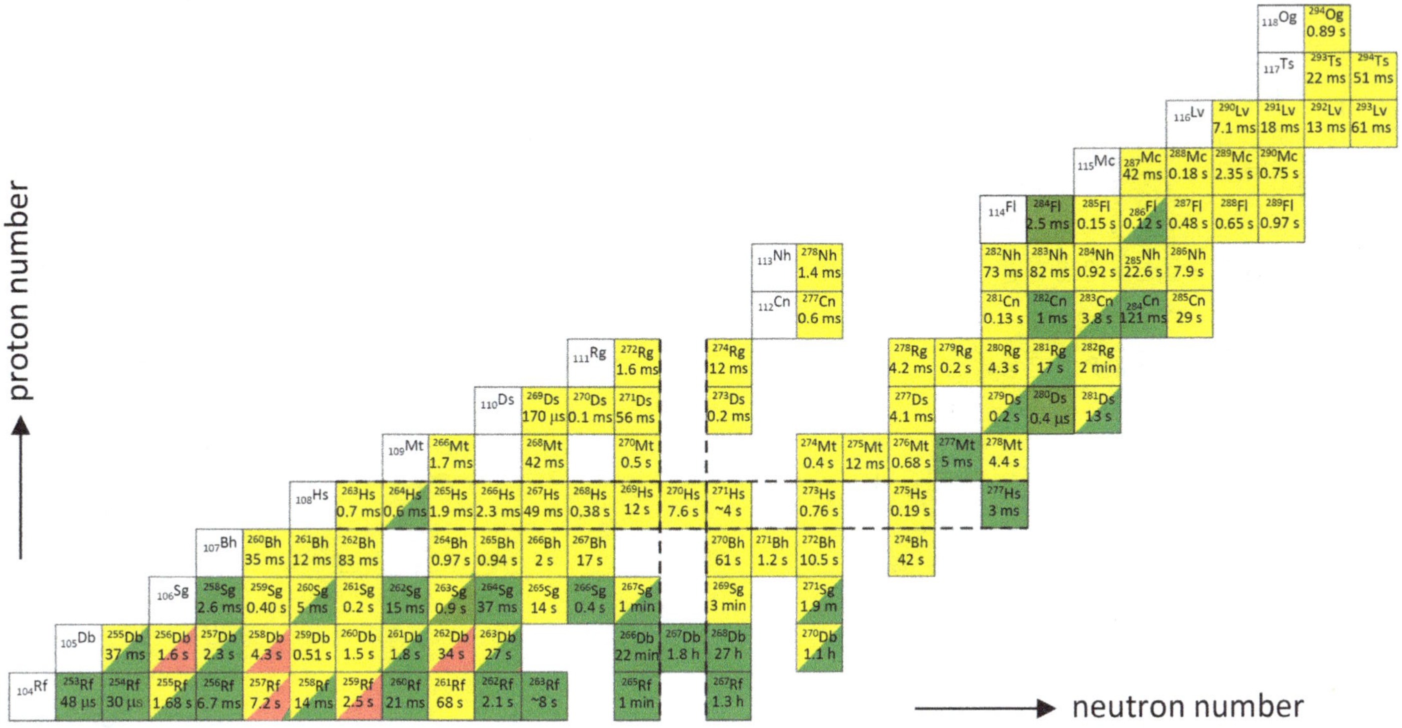

Fig. 9.4: Cut out of the chart of nuclides showing experimentally observed isotopes of the transactinide elements. Color coding follows that of the Karlsruhe Chart of Nuclides. Only the ground state (or most long-lived state if the assignment of multiple known states is unknown) and the one or two dominant decay modes are indicated.

raises large interest also among nonscientists. The history of the naming of transactinides is rich in interesting stories of cases that were far from straightforward, probably with elements 104 and 105 having been the most heavily contested ones. For quite some time, different names were used for these elements in different parts of the world. For example, element 104 was referred to as *rutherfordium* in the United States while it was called *kurchatovium* in the Soviet Union. Now what is the formal process – indeed, it is a very formal process – to name a new element? The official body in charge of chemical nomenclature is the International Union of Pure and Applied Chemistry (IUPAC), and names of elements definitely are one aspect of chemical nomenclature. The IUPAC, along with its sister organization, the International Union of Pure and Applied Physics (IUPAP) has installed a Joint Working Party (JWP) to consider claims for the discovery of new elements. Regularly, the JWP invites persons and/or teams that claim the observation of a new element to submit their material along with a detailed analysis describing how the observation was made, and why the submitting party is certain to have observed a new element. Submitted claims are then scrutinized with respect to a defined set of criteria that need to be fulfilled for the discovery of a new chemical element to be recognized (Hofmann 2020). In general, the requirements for official recognition involve an independent verification of the discovery, ideally by a different team using a different technique. This should avoid what happened for several transactinide elements in the past – controversies due to a priority dispute as to who first found conclusive evidence for the existence of an element, or as to what evidence was in fact conclusive. The JWP then arrives at a conclusion whether the claim is substantiated – in which case the discovery of the new element is officially recognized – or not, in which case the element remains officially undiscovered. In case of a discovery, the discoverers are credited with the right to propose a name for the new element. Most recently, the elements 113, 115, 117, and 118 were acknowledged and added to the periodic table. Certain rules that new names have to adhere to exist, but are comparatively loose. The naming right therefore still is indeed a true, free right. Once a name is proposed, this is announced publicly along with the invitation for the submission of comments on the proposed name. After a certain time, comments – if any – are analyzed, and in general, the new name is then officially approved by the IUPAC General Assembly and published in the IUPAC journal "Pure and Applied Chemistry".

9.5 Synthesis and nuclear properties of superheavy elements

9.5.1 Facilities

No transactinide elements have been found in nature, despite recurring announcements of such findings. Confirmation experiments have so far always been negative, cf. Korschinek and Kutschera (2015).

Therefore, these elements have to be produced artificially, atom by atom. To date, the fusion of lighter nuclei is the only viable method. Such experiments are carried out at specialized accelerator research centers, e.g., in Berkeley, CA (USA), Dubna (Russian Federation), Wako (Japan), or Darmstadt (Germany). A prerequisite is the availability of intense (10^{11}–10^{13} particles s^{-1}) beams of heavy ions that can be accelerated to energies at around the Coulomb barrier, i.e., $\approx$5–6 MeV/u, corresponding to about 10% of the speed of light.

These beams are directed at a target made from the complementary sort of nuclei. The proper choice of the projectile and target nuclei is a crucial aspect for the synthesis of the superheavy elements. Not all combinations hold the promise to have large enough cross sections to be suitable. In fact, there are only very few combinations that are attractive, and obtaining a full understanding of the heavy ion fusion process is still a topic of intense research.

9.5.2 Nuclear reaction strategies

Historically, two pathways have proven most advantageous for the synthesis of transactinide elements. Both involve acceleration of projectiles to energies high enough that the Coulomb repulsion encountered upon approaching the target nuclei can be largely overcome. After two nuclei have reached a distance smaller than the range of the strong attractive nuclear force, the nuclei fuse, thus forming a single nucleus containing all protons and neutrons of the two initial nuclei, the compound nucleus.[7] The two pathways are usually referred to as "hot" vs. "cold" fusion reactions, depending on how highly excited the compound nucleus formed after fusion of projectile and target is. This depends critically on the beam energy, which is transformed into internal excitation energy. A lowering of the beam energy to values that ensure forming an ideally "cold", i.e., nonexcited compound nucleus in its ground state, is impossible, as such beam energies are too small to allow overcoming the Coulomb barrier, which however is a prerequisite for fusion. Typical reactions already used early to produce and study transactinides involved relatively

---

7 Cf. Chapter 13 of Vol. I on nuclear reaction mechanisms.

light projectiles in the $A = 10-20$ range, which were used to bombard actinide targets. Due to the curvature of the β-stability line[8] and the increasing neutron excess with increasing proton number, neutron-rich isotopes were used to allow coming as close to this stability line as possible. The compound nuclei of such reactions (when using beam energies at the COULOMB barrier) are usually formed at some 40–50 MeV excitation energy, i.e., relatively "hot".

In the 1970s, it was realized that the nuclear structure of projectile and target plays an important role: the use of more strongly bound shell-stabilized reaction partners allows forming the compound nucleus in a relatively cold state, i.e., at rather low excitation energy. Therefore, nuclei around the $Z = 82$ and $N = 126$ shell closures like ^{208}Pb and ^{209}Bi were used as targets and allowed obtaining compound nuclei at some 10–20 MeV excitation energy.

Why is this temperature of the compound nucleus so critical? This has to do with the possible options of the compound nucleus to release its excitation energy. One option is prompt fission, i.e., disintegration into two lighter fragments, caused by the large COULOMB repulsion of the many protons. Other options include the evaporation of nucleons. Neutrons are evaporated most easily. In contrast to charged particles like protons, the neutrons are not trapped inside the nucleus "behind" a COULOMB barrier. The evaporation of a single neutron lowers the excitation energy by the sum of the neutron binding energy ($\bar{E}_B \approx 8$ MeV per nucleon)[9] and the neutron's kinetic energy whose spectrum is BOLTZMANN-distributed with typical values being around 2 MeV. In total, the excitation energy decreases by about 10 MeV through the evaporation of a single neutron.

The question is now which process is the dominant one: fission or neutron evaporation? This depends mainly on the timescale. The faster process will be the more frequent one. Here, prompt fission generally wins by a large margin. As a rule of thumb – obtaining more accurate numbers is another topic of intense current research – in 99% of all cases, a compound nucleus will fission rather than evaporate a single neutron. In the lucky case of neutron evaporation, the product is most often still excited, just less highly by about 10 MeV – the competition between prompt fission and neutron evaporation starts again. Only after several neutron evaporation steps – typically four or five – the excitation energy is so low that the remaining energy is carried away by electromagnetic radiation (γ-rays), whose emission is a rather slow process and thus not competitive as long as fission or neutron evaporation are energetically possible. The final product at the end of the neutron evaporation cascade is a nucleus in its ground (or sometimes *meta*-stable) state, which is called the evaporation residue (EVR). More detailed, newer research seems to indicate that the ratio of neutron evaporation to prompt fission changes

---

8 Cf. Chapter 6 of Vol. I on primary β-transformations.
9 Cf. Figure 3.4 in Vol. I showing mean nucleon binding energies as $\bar{E}_B = f(A)$.

along the evaporation cascade, with the first steps leaning less overwhelmingly towards fission than later ones.

The just described competition explains the success of cold fusion reactions: the competition has to be won fewer times than in hot fusion reactions. This leads to the expectation that cold fusion reactions should have by far larger cross sections than hot fusion ones. However, experimental data suggest both types to have similar cross sections, say within a factor of 10, if nuclei of the same element are produced. Why is this? Here, the answer lies in the first step of the nuclear reaction, the entrance channel, which is dominated by the mass asymmetry $(m_a - m_A)/(m_a + m_A)$ of the reaction where m_a and m_A are the masses of the projectile and target, respectively. The larger this mass difference, i.e., the more asymmetric the reaction, the easier the fusion of the two nuclei. Graphically speaking, a small nucleus will simply be sucked up by the large one. In the LDM picture of the nucleus, the surface tension would offer a good explanation. In contrast, if the projectile has a significantly larger mass, leading to a more symmetric reaction, fusion upon contact is no longer guaranteed. Rather, the two nuclei are captured inside their common potential, forming a di-nuclear system, which often re-separates before reaching the fused, i.e., compound nucleus configuration. This is referred to as the *quasi-fission* process. Fusion and hence the formation of a compound nucleus are then hindered.

A cold fusion and a hot fusion example of reactions leading to Hs nuclei are given in Fig. 9.5.

As follows from Fig. 9.5, the neutron number of the EVRs is a further difference between cold and hot fusion reactions, besides the different masses of the projectiles and targets and the excitation energy of the compound nucleus. In the hot fusion reactions, more neutron-rich isotopes are obtained (e.g., ^{269}Hs vs. ^{265}Hs, cf. Fig. 9.5). The more neutron-rich nuclei are generally more long-lived, which is why hot fusion reactions are still mostly used in chemical studies of transactinides, which require longer half-lives than physics experiments.

Current theoretical models describing EVR cross sections σ_{EVR} for the formation of transactinides are usually based on the following approach (eq. (9.1)),

$$\sigma_{EVR} = \sigma_{cap}\, p_C^*\, W_{sur}, \tag{9.1}$$

where σ_{cap} is the capture cross section, p_C^* is the fusion probability with which two captured nuclei transform into a compound nucleus C^*, and W_{sur} is the probability for the system to survive the evaporation cascade yielding an EVR. In the simple picture of a fusion-evaporation reaction consisting of two independent steps, as suggested by BOHR, the product of the first two terms, $\sigma_{cap}p_C^*$ equals the fusion cross section σ_{fus}, and σ_{EVR} is then $\sigma_{fus}W_{sur}$.

a)
"cold fusion"

b)
"hot fusion"

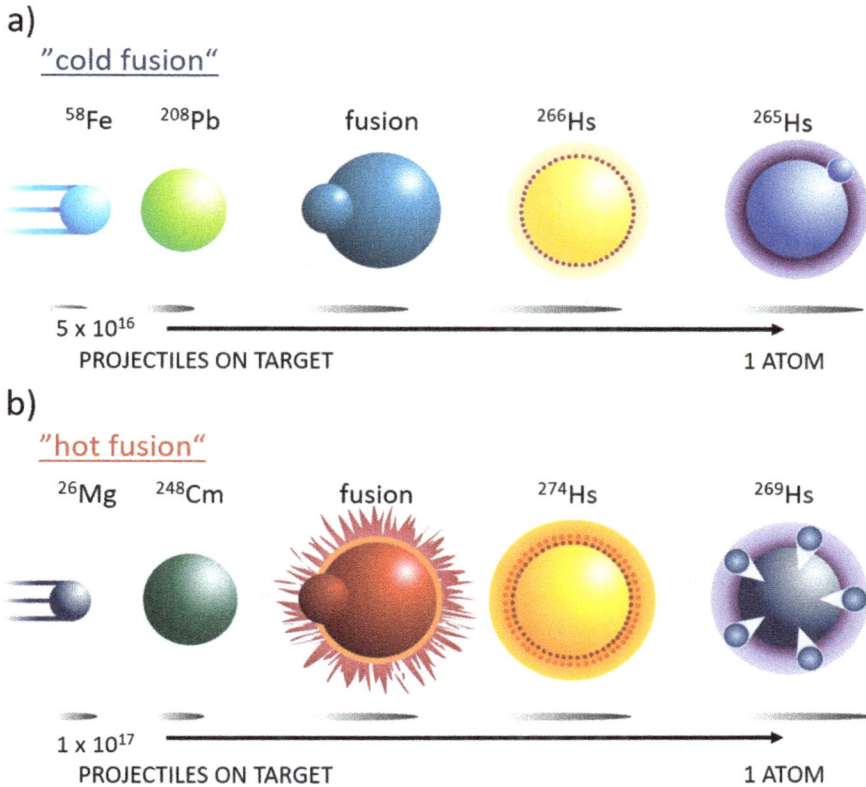

Fig. 9.5: Schematic representation of the synthesis of hassium nuclei (a) in a cold fusion reaction: $^{208}Pb(^{58}Fe,1n)^{265}Hs$, and (b) in a hot fusion reaction: $^{248}Cm(^{26}Mg,5n)^{269}Hs$ (after Schädel 2006).

As described earlier, the cross sections to reach a given element with $Z \approx$ 102–108 are quite similar for hot and cold fusion reactions. As shown in Fig. 9.6, the cross sections decrease exponentially when going to heavier elements. The reason for this drastic decrease, however, is quite different for the two reaction types. In case of the hot fusion reactions, it is mainly due to the fissility of the compound nucleus, i.e., connected with exit channel effects, where the probability for fission compared to neutron evaporation becomes increasingly more disadvantageous for the formation of an EVR. In the language of eq. (9.1), it is the decrease of W_{sur} that leads to ever smaller cross sections. In contrast, in cold fusion reactions, the fusion of the two nuclei in the entrance channel is the limiting process, and the ever smaller values for p_c^*, which diminish the fusion cross section, are responsible for the drop in EVR cross sections.

Figure 9.7 shows basic features of a hot and a cold fusion reaction, both leading to the $^{266}Hs^*$ compound nucleus.

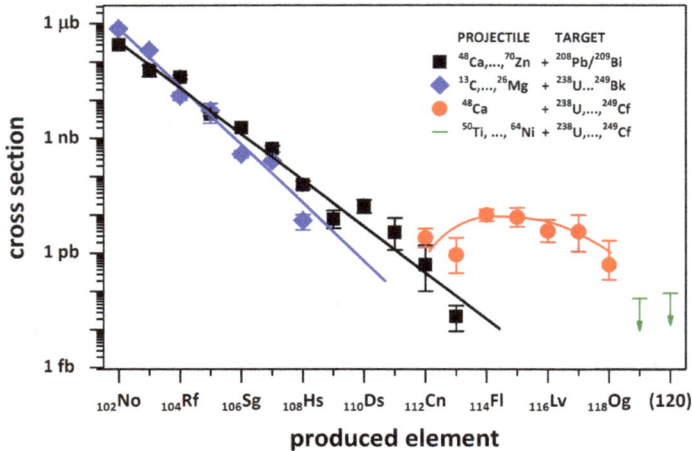

Fig. 9.6: Highest measured cross section leading to a given element in cold fusion – 1*n* – reactions (black squares), hot fusion – 5*n* – reactions (blue diamonds), and ^{48}Ca-induced reactions on actinide targets (red circles). The lines are drawn to guide the eye. The most sensitive reported upper limits for search experiments for the yet undiscovered elements 119 and 120 are indicated in green.

An obvious feature in Fig. 9.6 is that the cross sections of ^{48}Ca-induced reactions leading to elements beyond $Z = 113$ deviate significantly from the exponentially decreasing trend observed for the other reaction types. The cross sections are surprisingly constant at values around 1–10 pb over a range of six elements ($Z = 112$–118). In lighter elements, the cross section loss over a similar range amounts to several orders of magnitude!

A full understanding of the underlying mechanism for this deviation is still being accumulated. Theoretical approaches that reproduce experimental cross sections invoke differing explanations for this behavior. Some models ascribe it to the approach to the next spherical shell closure, which gives rise to large fission barriers. Another approach predicts the synthesized nuclei to be oblately deformed instead of spherical. It then suggests that fission along the de-excitation cascade from the compound nucleus to the EVR, which passes over a prolate saddle point, occurs with reduced probability due to the necessary large-scale rearrangement of the nucleus along this path. All models, though, agree that exit channel phenomena are likely responsible for the surprisingly large cross sections to produce elements up to $Z = 118$.

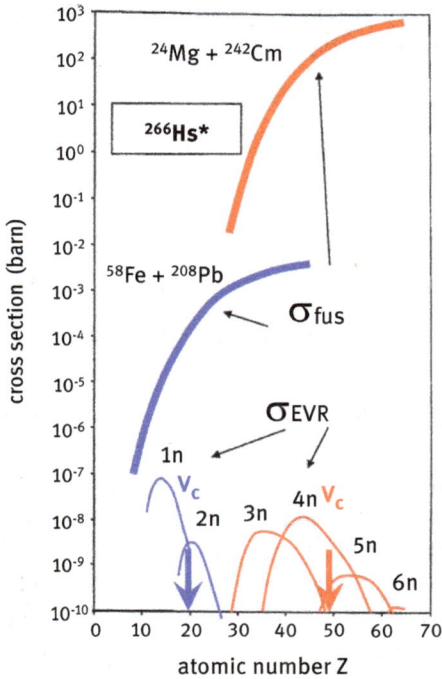

Fig. 9.7: Excitation functions for the hot fusion reaction ^{24}Mg + ^{242}Cm and the cold fusion reaction ^{58}Fe + ^{208}Pb, both leading to the same compound nucleus. The fusion cross sections $\sigma_{fus} = \sigma_{cap}p_c^*$ and the σ_{EVR} are shown. The arrows indicate the position of the Coulomb barrier, below which the σ_{fus} is rapidly diminishing (from Türler, PSI).

9.6 Relativistic effects and the periodic table of the elements

The Bohr model of the hydrogen-like atom describes the motion of an electron around an atomic nucleus. A circular motion occurs if the attractive COULOMB force is balanced by the centrifugal force. The basic equation that holds is

$$\frac{Ze^2}{4\pi\varepsilon_0 r^2} = \frac{m_e v^2}{r}, \tag{9.2}$$

where e is the elementary charge, ε_0 the dielectric constant, r the orbital radius, m_e the mass of the electron, and v its velocity. Inserting the quantum condition $m_e v r = n\hbar$ and solving for r yields

$$r_n = n^2 \frac{4\pi\varepsilon_0 \hbar^2}{Z m_e e^2}, \tag{9.3}$$

from which follows that the radius of the nth electron shell, r_n, is a function of m_e. If the speed of light were infinite, the above equation would be correct. However, the speed of light is finite, and EINSTEIN's theory of relativity tells that the mass of an object depends on its velocity and increases as the velocity approaches the speed of light:

$$m_{\text{rel}} = \frac{m_0}{\sqrt{1-(v/c)^2}}. \tag{9.4}$$

Using the simple relationship which holds for 1s electrons,

$$v/c = Z\alpha, \tag{9.5}$$

where c is the speed of light and α the fine structure constant, eq. (9.3) modifies to

$$r_n = n^2 \frac{4\pi\varepsilon_0 \hbar^2}{Z m_e e^2} \sqrt{1-(Z\alpha)^2}, \tag{9.6}$$

The last term in this equation is always <1. Orbitals with a significant electron probability density near the atomic nucleus – most pronouncedly s- and $p_{1/2}$-orbitals, see below – are thus spatially contracted and energetically stabilized by the influence of relativity. This is called the *direct relativistic effect.* As a consequence, orbitals without electron density at the nucleus experience a better shielding from the attractive positively charged nucleus, and are thus less bound. Consequently, they expand in space and are energetically destabilized – the *indirect relativistic effect.* Finally, the orbital quantum number l and the spin quantum number s are no longer good quantum numbers in the high-Z regime, but their vector sum, the total angular momentum number j, still is. The coupling scheme changes from RUSSEL–SAUNDERS (or L-S) coupling[10] to jj coupling. This leads to two energetically distinct states for each l > 0 (p, d, f, . . .) orbital into j = l ± 1/2, thus giving rise to the $p_{1/2}$ and $p_{3/2}$, $d_{3/2}$, and $d_{5/2}$, . . . orbitals.

In the transactinides, the *spin-orbit splitting* becomes comparable to (or even larger than) binding energies in compounds. A special case is the $p_{1/2}$ orbital: in the relativistic treatment, where shape is determined by j and no longer by l, this adopts a spherical shape, just like the $s_{1/2}$ orbital. Consequently, it is affected by the direct relativistic effect, in contrast to the $p_{3/2}$ orbital, which is influenced by the indirect effect.

Why is it exactly the transactinide series of elements that is most pronouncedly affected by relativistic effects? At the heart of this is eq. (9.4). Figure 9.8 shows the relativistic mass increase, m_e^{rel}/m_e^0 of an electron in hydrogen-like systems.

This figure immediately suggests that the heaviest elements are most pronouncedly affected by relativistic effects. Their influence scales roughly $\approx Z^2$, where, however, the individual effects may scale somewhat differently. The figure also

---

10 Cf. Figures 1.21 and 1.22 in Chapter 1 of Vol. I on electron spin–orbit coupling.

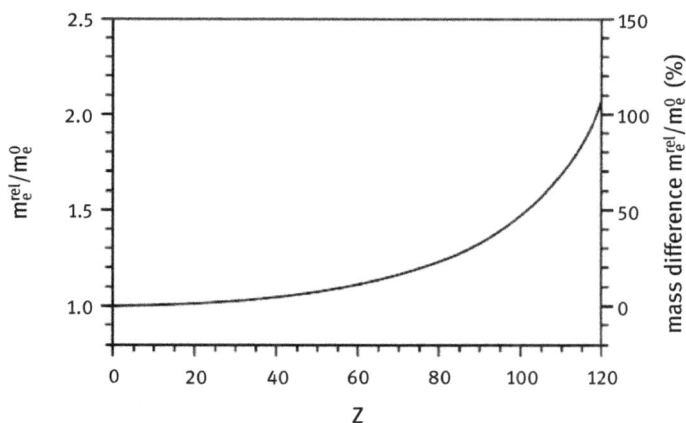

Fig. 9.8: Magnitude of the relativistic mass increase for a 1s electron in hydrogen-like atoms.

shows that lighter elements are influenced by relativity, too.[11] Theoretical chemical studies of the transactinides are a rich field of research, which provides guidance in the preparation of experimental studies and facilitates a correct interpretation of obtained experimental results. In this chapter, though, the focus will be on experimental studies.

A strong motivation for the experimental chemical investigation of the transactinides is the study of the influence of relativistic effects on their chemical properties (Pyykkö 1988). In light of the most obvious difficulties – on average, at most one single, short-lived atom is available for study at any given time – investigations are often limited to addressing a few fundamental properties. Nevertheless, technical advances have allowed studies of all elements up to Hs and more recent work addressed the elements Cn, Nh, and Fl, with Mc coming into reach.

Figure 9.9 shows the current periodic table, with elements that have never been studied experimentally slightly offset. The expectation that elements 119 and 120 will be members of the alkaline and earth alkaline series agrees with fully relativistic quantum chemical calculations. These then suggest element 121 to have an 8p electron in its ground state electronic configuration, unlike its lighter homologs. Element 122 has a 7d electron in addition, rendering the structure to be different from that of its lighter homolog Th.

---

11 Indeed, several phenomena well known from daily life can only be understood in terms of the influence of relativistic effects on the electronic structure of involved elements. These include the color of gold, the liquid standard state of mercury, or the instability of the tetravalent oxidation state of lead, a prerequisite for the lead acid battery that starts our cars.

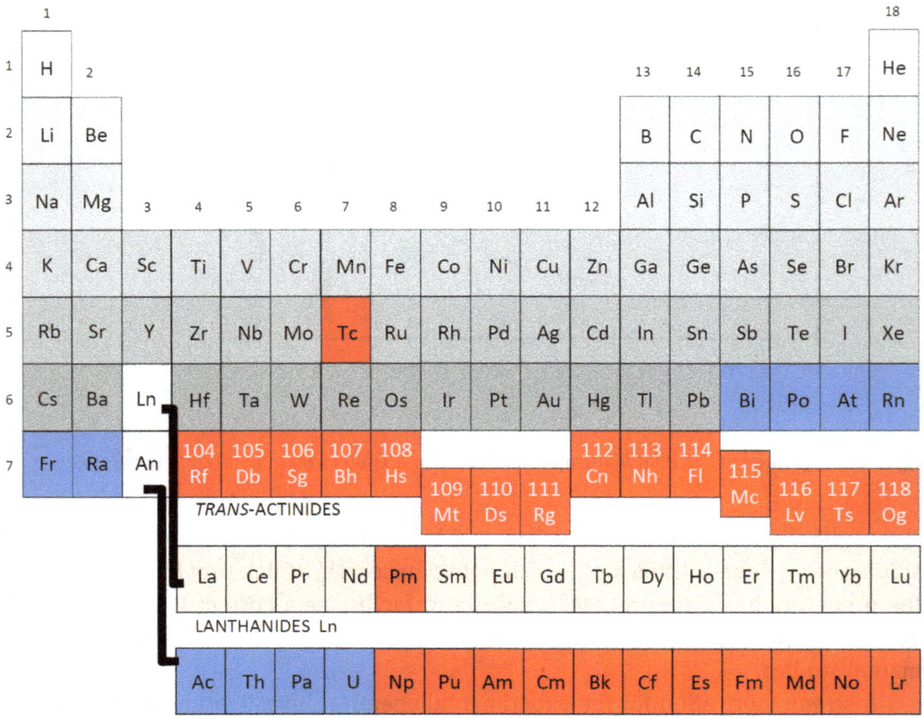

Fig. 9.9: The periodic table as of 2021. Naturally occurring radioactive elements are shown in blue and artificial ones are shown in red. The elements with Z = 109–111 and 116–118 are offset because they have never been studied chemically to date. For element 115, first chemical data is currently being accumulated.

9.7 Methods for experimental chemical studies of transactinide elements

9.7.1 Overview

Any experimental study of chemical properties of transactinide elements involves the following basic steps:

(1) synthesis in a heavy-ion induced fusion reaction;
(2) fast transport from the place of synthesis to a chemistry setup;
(3) chemistry experiment;
(4) preparation of a sample suitable for registering the nuclear transformation of the studied isotope.

Tab. 9.2: Isotopes frequently used in chemical studies of transactinides.

Isotope	Half-life t½	Production reaction	Production rate*
^{257}Rf	≈4 s	^{208}Pb(^{50}Ti,n)	8 min^{-1}
^{261}Rf	68 s	^{248}Cm(^{18}O,5n)	6 min^{-1}
^{262}Db	34 s	^{249}Bk(^{18}O,5n)	2 min^{-1}
^{268}Db**	28 h	^{243}Am(^{48}Ca,3n)	6 d^{-1}
^{265}Sg	≈10 s	^{248}Cm(^{22}Ne,5n)	10 h^{-1}
^{267}Bh	17 s	^{249}Bk(^{22}Ne,4n)	3 h^{-1}
^{269}Hs	12 s	^{248}Cm(^{26}Mg,5n)	5 d^{-1}
^{283}Cn	3.8 s	^{238}U(^{48}Ca,3n)	2 d^{-1}
		^{242}Pu(^{48}Ca,3n)***	2 d^{-1}
^{284}Nh**	1 s	^{243}Am(^{48}Ca,3n)	6 d^{-1}
^{289}Fl	2.1 s	^{244}Pu(^{48}Ca,3n)	7 d^{-1}
^{288}Mc	0.18 s	^{243}Am(^{48}Ca,3n)	6 d^{-1}

*under typical conditions
**via α-transforming ^{288}Mc precursor
***via α-transforming ^{287}Fl precursor

With few exceptions, hot fusion reactions with neutron-rich actinide isotopes are used for the synthesis of transactinide isotopes to be studied chemically. Some reactions and key properties of typical isotopes are given in Tab. 9.2. No isotopes of elements 109–111 and ≥115 are included in Tab. 9.2, as no chemistry experiments with these elements have been performed to date.

9.7.1.1 Gas jet

After production, the isotopes have to be transported to the chemistry setup. A very successful technique for this is the *gas jet* technique. In this approach, a gas-filled (≈1 bar) volume is installed behind the target, the so-called recoil chamber, in which EVRs recoiling from the target are stopped. The gas is seeded with ≈ 5 · 10^6 particles[12] with diameters of about 100 nm. These particles can be transported through small plastic tubes (inner diameter of a few mm) with high yields (up to 90%) over distances of several meters, as they are so light that they are not affected

---

12 Typically, carbon or KCl clusters are employed.

by gravity. On the other hand, they are large enough for their diffusion to be very slow, which prevents them from touching the plastic tube.

Nonvolatile EVRs stopped in the recoil chamber attach to any surface they encounter. With very high probability, this will be the surface of an aerosol particle. These particles, with the adsorbed EVRs, are flushed with the rapidly flowing carrier gas to the chemistry setup within a few seconds. Gas jets are very efficient for the transport of nonvolatile atoms, but there is no selection among the different nuclear reaction products. Atoms of many elements are produced in the nuclear synthesis of a transactinide, e.g., in collisions where the projectile and target atom do not fuse but only exchange a few nucleons. Some of these by-products have nuclear transformation properties similar to the transactinide atom of interest, and interfere with its unambiguous detection.

In cases where volatile elements are studied – e.g., noble gas atoms – or when compounds volatile at room temperature can easily be formed – a prominent example is the case of the group 8 tetroxides – no aerosol particles are needed and transport is possible with pure gas.

9.7.1.2 Liquid- and gas-phase experiments

Chemical studies of the transactinides were conducted in the gas phase and in the liquid phase. Because gas-phase chemical reactions are generally faster and – maybe even more importantly – the preparation of samples suitable for nuclear spectroscopy is easier, many studies used such techniques and yielded chemical data, sometimes based on the observation of only a handful of atoms or even less. On the other hand, liquid-phase studies give access to other properties and may yield a more detailed insight into the chemical behavior of an element, especially if conducted with many atoms. Their main drawback is the long time needed to evaporate all the liquid to obtain a sample suitable for nuclear spectroscopy. This is reflected in the fact that all transactinides up to Hs as well as Cn to Fl have been studied in the gas phase. In the liquid phase, the first three transactinides Rf, Db, and Sg, have been studied.

9.7.1.3 Chemical data from the study of single atoms

As only a very small number of transactinide atoms is available for study, let us clarify how the obtained results can be related to those obtained in studies of macroamounts ($\approx 10^{20}$ atoms) of lighter elements. Let us take as an example the law of mass action, in which the equilibrium constant is expressed as the ratio of two concentrations. This works well as long as sufficiently many atoms are present that the concept of a concentration is valid. In case of one single atom being present in the whole experiment, one of the concentrations entering the expression will naturally be one atom per volume unit while the other zero. The ratio is thus zero or infinity, certainly different compared to a study where more atoms would have been available. Fortunately, however, there is a way out: it was shown that we can simply replace the concentration of the species under study in a certain phase by the *probability* to find one single atom in this phase at a given time. The way forward towards statistically meaningful results is now clear: we simply have to give the atom enough chances to change between the different states that the probability that we measure is a true, statistically significant expression of this probability. This is best fulfilled in chromatographic setups where an atom will change from one state to the other many times.

9.7.2 Gas phase experiments

In gas-phase experiments, two chromatographic techniques have been widely applied (see Fig. 9.10):
- isothermal chromatography (IC) and
- thermochromatography[13] (TC).

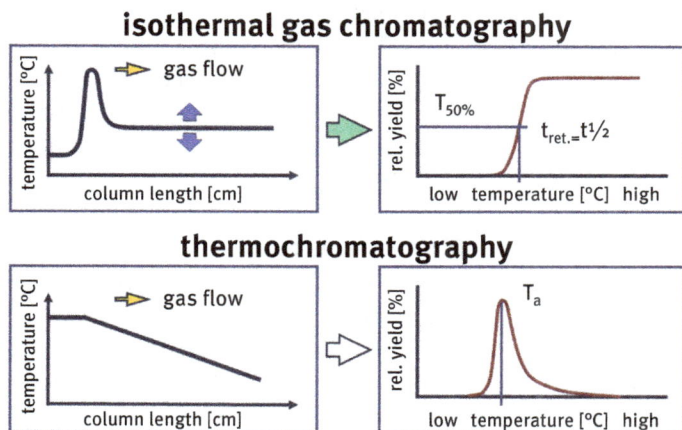

Fig. 9.10: Temperature profile applied in isothermal chromatography (top left) and obtained isothermal chromatogram, often called "breakthrough" curve (top right). Temperature profile applied in thermochromatography (bottom left) and observed deposition peak (bottom right). The variables are defined in the text (after Türler and Pershina 2013).

---

13 See also Chapter 7 on radiochemical separation techniques for details.

In both techniques, a gas flow through an open chromatography column is established. In IC, the goal of the experiments is the measurement of the retention time of a volatile species on the column material. In a first step, the species has to be synthesized. For this, reactive gas (e.g., Cl_2 or HCl) is added to the gas stream before it enters the column. The beginning of the column is usually heated to temperatures above 1000 °C, where decomposition of the aerosol particles and the formation of the volatile species take place. The main part of the column is then kept at a fixed temperature (Fig. 9.10, top left panel), and the rate at which the transactinide element under study is detected behind the column is measured. Experiments at different temperatures are conducted and the measured yields normalized to the maximum yield are plotted as a function of the column temperature. At high temperatures, species hitting the wall desorb instantaneously and there is no retention on the column surface. The retention time increases with decreasing temperature. Therefore, the yield decreases at low temperatures, as an increasing fraction of the transactinides transforms inside the column before reaching its exit. This is reflected in the shape of the yield curve (Fig. 9.10, top right panel). To evaluate the time the species spent adsorbed to the surface (i.e., the retention time), the radioactive transformation of the transactinide atom is used as an internal clock: at the temperature where the yield is 50% of the maximum yield 50% transform before reaching the exit, and the retention time equals the nuclear half-life. From this known retention time at a known temperature, the interaction strength of the atom or molecule with the surface material can be evaluated and is usually expressed as adsorption enthalpy $-\Delta H_a$. In thermochromatography, a negative temperature gradient is applied along the chromatography column (Fig. 9.10, bottom left panel) and the adsorption temperature, T_a, at which the species is adsorbed for so long that nuclear transformation sets in, is measured (Fig. 9.10, bottom right). From this temperature, again the $-\Delta H_a$ can be evaluated.

The on-line-gas chemistry apparatus (OLGA) was successfully used for chemical studies of (oxy)halides of Rf, Db, Sg, and Bh. It is schematically depicted in Fig. 9.11.

For highly volatile species, thermochromatography systems were developed where the chromatography column itself consists of detectors for registering the transformation of species adsorbed on the detector surface. Temperature gradients from ambient temperature to about −180 °C can be applied.

Empirically, many chemical systems were found to yield significant correlations if experimentally determined $-\Delta H_a$ values are plotted against experimental sublimation enthalpies, $-\Delta H_s$. Two examples of such correlations are shown in Fig. 9.12.

Implying that the nature of the chemical processes in the interaction of the transactinide element compound with the surface is identical to those present in the lighter homologs, this approach allows extracting a macroscopic property, the sublimation enthalpy, of an element or a compound based on experiments with a few single atoms.

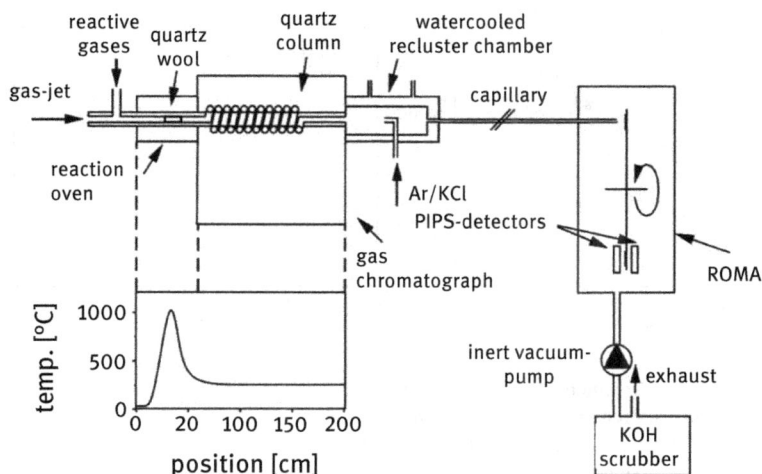

Fig. 9.11: Schematic of the OLGA apparatus. The transactinides are transported to OLGA by a gas jet. Reactive gases (depending on the chemical system to be studied; e.g., HCl for the study of chlorides) are added. Aerosol particles are collected and destroyed by a chemical reaction with the reactive gas on quartz wool kept at 1000 °C in the reaction oven. There, formation of the volatile transactinide compound takes place. This enters the chromatography column and reaches the exit, if the retention time is short enough. Behind the exit, a second gas jet (typically KCl particles in Ar gas) is used to transport the transactinides to the detection system for registering the nuclear transformation. For this, the aerosol particles (with the attached transactinide) are deposited on a thin foil mounted on a rotating wheel. The foil is placed between PIPS[14] detectors, suitable for registering the transformation of the transactinide (reprinted with permission from Elsevier from Gäggeler 1998).

Note that theoretical/computational chemistry provides data that do not rely on similarities of chemical processes among groups of elements. In many cases, results based on trends in the periodic table and results from fully relativistic quantum chemical calculations agree at least qualitatively, while examples where the results differ significantly are also known. The motivation for performing ever more sophisticated experiments is therefore very strong.

9.7.3 Liquid-phase experiments

Standard liquid-phase separation techniques like liquid-liquid phase separations or column-based separations have also been adapted for application in transactinide element chemical studies. We focus here on the latter approach, which was by far more widely applied. To allow studies of single atoms with short lifetimes, a natural

---

14 PIPS: passivated implanted planar silicon.

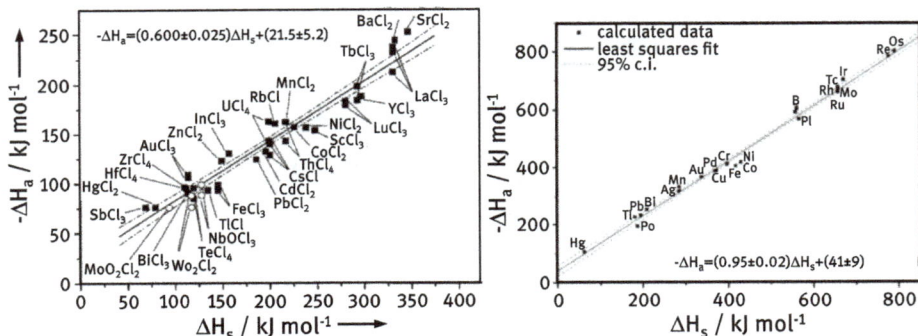

Fig. 9.12: Correlation between the microscopic property "adsorption enthalpy" and the macroscopic property "sublimation enthalpy". Left: experimentally determined adsorption enthalpies for chlorides and oxychlorides deposited on quartz surfaces plotted versus experimental sublimation enthalpies. Right: the same but for volatile elements deposited on Au surfaces (left: from Türler and Pershina 2013, right: from Eichler 2005).

modification compared to classical setups was the miniaturization of the chromatography columns. A successful example of the implementation of such a system is the automated rapid chemistry apparatus (ARCA), see Fig. 9.13.

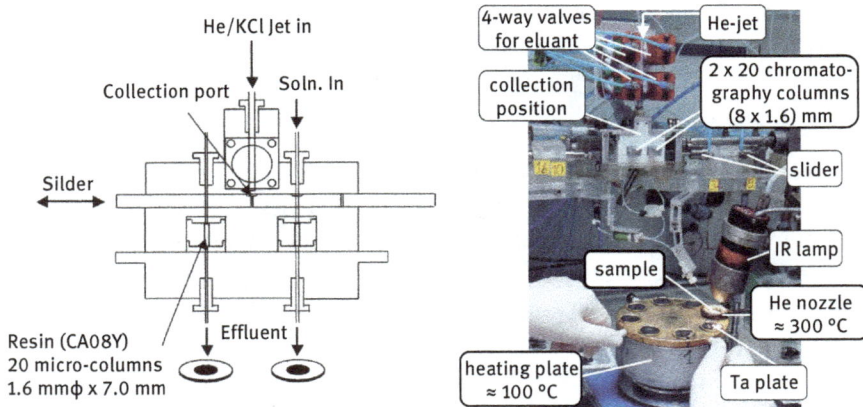

Fig. 9.13: Schematic (left) and photograph (right) of ARCA (left: reprinted with permission from Elsevier from Nagame et al. 2004; right: photo: Zschau, GSI).

Isotopes of interest were transported by a salt aerosol (e.g., KCl) gas jet to ARCA, where the aerosol particles are collected on a frit mounted on a slider. The frit is automatically moved to a position on top of a miniaturized (7 mm length, 1.6 mm i.d.) chromatography column. The liquid phase is fed through the frit, thus dissolving the aerosol particles, releasing the isotope under study, and bringing this into the column, where the chromatography process takes place. The liquid leaving the column is collected drop-wise on Ta plates and then rapidly (within a few seconds) evaporated to dryness with hot He gas, a heat lamp, and a heating plate. The plate is subjected to α-spectroscopy, thus yielding the fraction of atoms that passed the chromatography column in a certain amount of time. From this, the distribution ratio, K_d, between the mobile and the stationary phase can be obtained. ARCA was further developed to the automated ion-exchange separation apparatus coupled with the detection system for alpha-spectroscopy (AIDA), which includes a robotic system for fully automated sample transfer and α-spectroscopy. Depending on the chemical nature of the resin, anion exchange chromatography (AIX) as well as cation exchange chromatography (CIX) experiments can be conducted with ARCA and AIDA.

9.8 Chemical properties of the transactinide elements Rf–Rg

In Sections 9.8 and 9.9, the current knowledge of the chemical properties of the transactinide elements is presented. Naturally, many more experiments have been performed in the past decades than are described here. The presented selection focuses on highlighting the trends present in the respective groups of the Periodic Table. In some cases, we show how chemical separation systems were exploited as a "tool" for nuclear physics experiments, e.g., to measure excitation functions or transformation properties. The Section 9.8 focuses on the transition metal transactinides Rf – Rg, in which the 6d shell is being filled. Due to the refractory nature of the early transactinides, gas-phase chemical studies focus on volatile compounds rather than the elemental state. For the first three transactinides, liquid-phase chemical studies have been performed. The chemical properties of the elements beyond Rg is treated in Section 9.9.

9.8.1 Rutherfordium (Rf, Z = 104)

First attempts to study chemical properties of Rf were intimately connected with the goal to discover this element in the first place and then to confirm it to be a "trans"-actinide element, with chemical properties resembling those of the group 4 elements rather than the actinides. Zr and Hf, the lighter homologs in group 4, are known to form volatile tetrachlorides, MCl_4 (M = Zr, Hf). Therefore, these studies, performed in the 1960s, aimed at converting Rf into the tetrachloride and separating it from the actinides, which do not form volatile chlorides. The setups used were only sensitive to sf transformation, which is, however, unspecific and does not yield any information on the transforming isotope.[15]

Nevertheless, the obtained results were interpreted to point at the existence of a volatile Rf chloride. The results were, however, discussed quite controversially in the field. Later experiments with OLGA-type setups, capable of registering correlated α-transformation chains, confirmed that Rf forms a volatile tetrachloride. The volatility trend of the MCl_4, expressed as $-\Delta H_a$, indicates a decrease in volatility when going from $ZrCl_4$ ($-\Delta H_a = 79 \pm 5$ kJ/mol) to $HfCl_4$ ($-\Delta H_a = 103 \pm 5$ kJ/mol). A straightforward extrapolation to Rf might thus suggest $RfCl_4$ to be yet less volatile. However, the $-\Delta H_a$ of $RfCl_4$ was measured to be 82 ± 5 kJ/mol, similar to that of $ZrCl_4$ and indicating a reversal of the trend. This higher volatility of $RfCl_4$ was attributed to relativistic effects, which render the Rf–Cl bond to be more covalent, leading the $RfCl_4$ molecule to be less reactive towards the chromatography column surface material. In experiments where the reactive gas was enriched with O_2, a less volatile species was observed, likely $RfOCl_2$. Finally, the tetrabromide, $RfBr_4$, was studied as well.

Rf was also frequently studied in liquid phase experiments. The AIDA setup allows obtaining very detailed information, and typical results are nowadays based on the observation of several hundred atoms. We will consider four classes of experiments. In the first, reversed phase extraction chromatography with AIDA was performed. Rf was extracted from a hydrochloric acid phase into tributylphosphate (TBP), which was bound to a polymer-based support material with $\approx$30 μm particle size. The chemical process taking place in the HCl solution can be described as $[(H_2O)_8]^{4+} + 4\,Cl^- \rightleftharpoons MCl_4\,(aq) + 8\,H_2O$. The formed tetrachloride is then extracted by the TBP into the organic phase according to $MCl_4\,(aq) + 2\,TBP\,(aq) \rightleftharpoons MCl_4(TBP)_2\,(org)$. The H_2O in the coordination sphere is increasingly replaced by chloride ligands Cl^- with increasing acid concentration, which leads to higher extraction yields with increasing concentration of hydrochloric acid. When comparing the extraction behavior of the three homologs Zr, Hf, and Rf, the sequence Zr > Rf > Hf was observed, which is well reproduced by relativistic calculations[16] that consider both, the chloro-complex formation as well as hydrolysis of these complexes. In different experiments, the chloride complex formation was studied in AIX experiments using pure HCl as the aqueous phase. Rf was found to bind more strongly to the ion exchanger than its two homologs, which behaved almost identically. The results of the TBP and the chloride complex experiments were interpreted using the "hard/soft acid/base" concept, which ascribes certain hardness to all cations

---

15 Fission fragments are emitted with a wide energy spectrum and cannot be traced back to the nuclide that fissioned. In contrast, α-particle energies are characteristic for their original nuclide, and in many cases the daughter product after α-transformation is again a radioactive nuclide whose transformation can be measured, thus adding proof to the identity of the initial nucleus.
16 Calculations that include the effects described in Section 9.6.

and anions.[17] The relatively large, readily polarizable and only singly charged Cl^- is a rather soft anion. Consequently, its complexes with Rf^{4+}, which is larger and hence softer than Zr^{4+} and Hf^{4+}, should be stronger than those with Zr and Hf.

In a third class of experiments, the fluoride complexation of Rf was investigated. CIX experiments in mixed HNO_3/HF solutions confirmed again the formation of neutral and then anionic species – e.g., $(Rf^{IV}F_6)^{2-}$ – with increasing fluoride concentration, see Fig. 9.14. Rf exhibited a behavior in between those of the Zr/Hf pair and the pseudo-homolog Th.

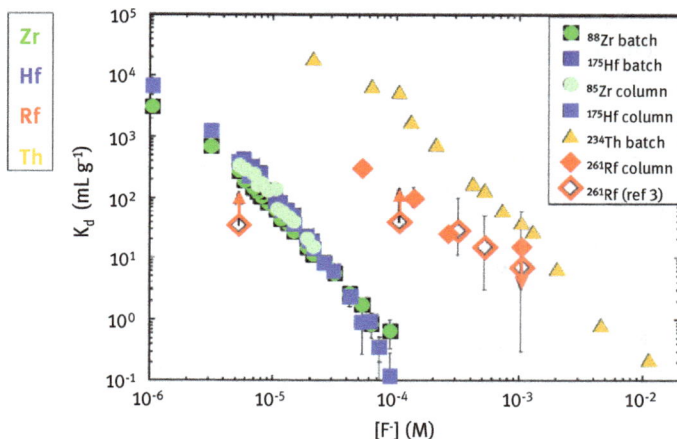

Fig. 9.14: Distribution coefficients (K_d) of Zr, Hf, Th, and Rf on a cation-exchange resin in 0.1 M HNO_3 depending on the concentration of the fluoride ion [F^-] (from Ishii et al. 2008).

The valence state of the chemical species undergoing the chromatographic process can be extracted from the slope, if the measured K_d values are plotted against the ion concentration in a doubly logarithmic diagram as shown in Fig. 9.14. For all investigated elements, a slope of −2 was determined, suggesting that the chemical species of Rf was the same as for Zr and Hf under the present experimental conditions.

In separate studies, it was investigated whether the nitrate ion, also present in these studies, participated in the chromatography process. The results showed that indeed, the nitrate concentration had a large effect on the measured K_d values at fixed fluoride concentration. Therefore, experiments with pure HF solution were carried out. Somewhat surprisingly, the measured slope for Rf was −2, different

---

17 The smaller and more highly charged a species is, the harder it is. The concept states that pairs of similar ions form the strongest complexes. In contrast, e.g. the complex formed by a hard cation and soft anions is expected to be less stable.

from those of Zr and Hf, which were −3. This may point to Zr/HfF_7^{3-} and RfF_6^{2-} as the involved species.

9.8.2 Dubnium (Db, Z = 105)

Gas-phase experiments focused on the study of volatile halides and oxyhalides in SiO_2 columns, very much in the spirit of the studies described in Section 9.8.1 for Rf. Using OLGA, the lighter homologs Nb and V were extensively studied. The O_2 concentration in the reactive gas turned out to be a critical parameter in this system, as is highlighted in Fig. 9.15.

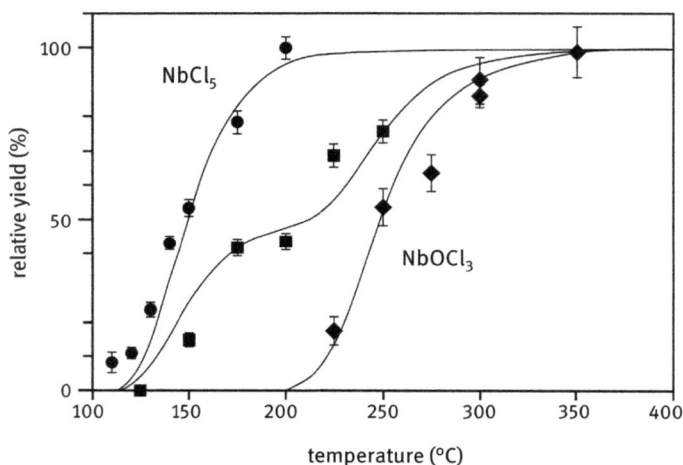

Fig. 9.15: Relative yields of ^{99}Nb pentachloride and oxytrichloride as a function of isothermal temperature and O_2 concentration. Circles: <1 ppmv O_2: $NbCl_5$. Diamonds: ≥80 ppmv O_2: $NbOCl_3$. Squares: 1–80 ppmv O_2: 1 : 1 mixture of the two compounds (from Türler et al. 1996).

At low O_2 concentration (≤1 ppm by volume), the highly volatile pentachloride forms. At high concentration (≥80 ppm by volume), the less volatile oxytrichloride is observed. At intermediate oxygen concentrations (1–80 ppm by volume), the oxytrichloride seems to be present at higher temperatures, and the pentachloride at lower temperatures. This rendered early studies of Db (oxy)chlorides difficult to interpret, as both chemical species of Db seemed to be present, depending on the temperature of the IC column. The unambiguous determination of the a pure $DbOCl_3$ was reported in 2021 by using a setup similar to OLGA; a volatility sequence $NbOCl_3 > TaOCl_3 \approx DbOCl_3$ resulted.

The results were interpreted to hint at increased covalency of the transactinide compound, accompanied by a decrease in the ionic character, due to the influence of relativistic effects on the molecular structure. The study of the pure dubnium pentachloride is still an open task. Besides the chlorides, the bromides were also studied, including DbBr$_5$. The derived adsorption enthalpies are given in Tab. 9.3.

Tab. 9.3: Adsorption enthalpies of group 5 (oxy)halides on SiO$_2$ surface.

Compound	$-\Delta H_a$ (kJ/mol)
NbCl$_5$	80 ± 1
NbOCl$_3$	99 ± 1
NbBr$_5$	89 ± 5
NbOBr$_3$	155 ± 5
TaCl$_5$	–
TaOCl$_3$	157 ± 12
TaBr$_5$	103 ± 5
TaOBr$_3$	–
DbCl$_5$	≤ 97
DbOCl$_3$	130 ± 6
DbBr$_5$	71 ± 5
DbOBr$_3$	121 ± 11

Db was also extensively studied in the liquid phase, though not in as much detail as its lighter neighbor in the periodic table, Rf. This is mainly due to the smaller cross sections – which lead to a lower production rate – and the shorter half-lives of the isotopes most frequently used for chemical studies of these two elements: ≈ 30 s for ^{262,263}Db compared to 68 s for ^{261}Rf. Initial studies were aimed at studying the chloride complexation of Db in hydrochloric acid. Based on experience with the lighter group 5 elements, minute amounts of hydrofluoric acid were added to the solution. This had been found to inhibit the sorption of Ta at walls, e.g., of beakers, which causes significant loss of Ta, which is unacceptable in experiments with transactinides. The extraction sequence among Nb, Ta, Db, and the "pseudo"-homolog[18] Pa was found to be Ta > Nb > Db ≥ Pa, i.e., Db behaved differently from Ta. The observed extraction sequence was not understood in light of the theoretically predicted one, which was Pa > Nb ≥ Db > Ta. More detailed calculations were performed and revealed that,

---

18 While elements within the same group of the periodic table are called homologs, elements with chemical similarity due to similar electronic structure but positions in different groups of the periodic table are referred to as *pseudo*-homologs.

most likely, Db was not present in the form of chloride complexes, but rather mixed chloride/fluoride complexes or even pure fluoride complexes, despite the fluoride ion concentration being about four orders of magnitude lower compared to the chloride ion concentration. This was expected to be due to the much higher stability of fluoride complexes of Db compared to chloride complexes. A subsequent experiment series was performed in pure hydrochloric acid, which revealed an extraction sequence Pa > Nb ≥ Db > Ta, exactly as theoretically predicted.

Experiments in pure hydrofluoric acid were also performed and resulted in the observation of four Db transformations, which was sufficient to point to a behavior of Db more similar to that of Nb and V than that of Pa.

After the observation of Mc transformation chains, which terminate in a long-lived isotope assigned to Db (^{268}Db, $t^{1/2} \approx 1$ d), this was chemically studied with the goal to prove that it was indeed a Db isotope. This, in turn, would support the assignment of the transformation cascade as originating from Mc. While currently still being inconclusive, the chance to perform chemical studies of Db with an isotope with such a long half-life is intriguing, despite the cross section and hence the production rate being orders of magnitude smaller than in the case of ^{262,263}Db. As ^{268}Db transforms by unspecific sf, the chemical separation from all other elements that yield long-lived sf-transforming isotopes is a special challenge in these experiments.

9.8.3 Seaborgium (Sg, Z = 106)

Seaborgium is the heaviest element that has been studied both in gas phase as well as in liquid-phase experiments. First gas-phase experiments focused on the volatility of the oxychlorides. The MO_2Cl_2 species (M = Mo, W, Sg) were studied. A first experiment using the thermochromatography method with SiO_2 columns, which were sensitive for sf-transforming nuclei, yielded indications on the formation of a volatile Sg compound. This is in line with the results of two isothermal chromatography experiments with the OLGA apparatus, where the breakthrough curves of ^{104}MoO$_2$Cl$_2$, ^{168}WO$_2$Cl$_2$, and ^{265}SgO$_2$Cl$_2$ were recorded. In total, 13 transformation chains of ^{265}Sg were observed in the OLGA experiments. The evaluated $-\Delta H_a$ values (see Fig. 9.16) yield a volatility sequence of Mo > W ≈ Sg.

A further gas-phase study focused on the oxide-hydroxide system. The obtained results were interpreted to point to a chemical reaction occurring upon each adsorption and desorption step, with the species participating in the transport along the hot (≈1000 °C) SiO_2 chromatography column being MO_3 (ads) and $MO_2(OH)_2$ (g). The corresponding reaction is MO_3 (ads) $\Leftrightarrow$ $MO_2(OH)_2$ (g) + H_2O (g). The behavior of Sg in humid oxygen was typical for both U^{VI} and the group 6 elements.

Until 2013, gas-phase chemical studies of transition metal transactinides focused on halogen and/or oxygen-containing compounds. In these compounds, the metal is present in its highest formal oxidation state, with all its valence electrons being

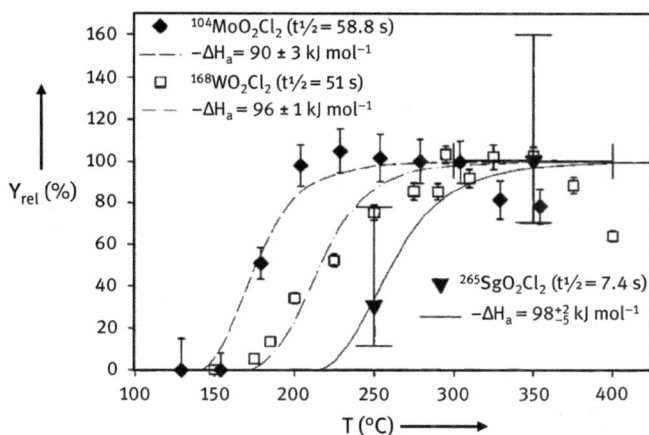

Fig. 9.16: Cumulative fraction of MO_2Cl_2 (M = Mo, W, Sg) that passes the TC column as a function of isothermal temperature (from Türler et al. 1999).

occupied in covalent bonds. It has been a long-time dream of transactinide element chemists to also get access to compounds in which the metal is present in a reduced oxidation state. Potentially, the influence of relativity might be manifested in a more easily detectable manner than in the compounds studied to date. In 2014, finally, a first such compound was synthesized, namely seaborgium hexacarbonyl (Even et al. 2014). In this compound, seaborgium is bound to six carbon monoxide molecules through metal-carbon bonds, in a way typical of organometallic compounds. To avoid destruction of the only moderately stable hexacarbonyls by the intense heavy-ion beam, the group 6 elements were isolated in a gas-filled recoil separator. Behind the separator, the atoms were thermalized in a gas volume flushed with CO-containing carrier gas, in which the complexes formed. These were then transported to a cryothermochromatography detector, where they adsorbed on silicon dioxide surfaces at similar temperatures as the lighter homologous compounds of Mo and W. This approach holds the promise to give access to further new compound classes, including chemical systems in which relativistic effects are potentially much more clearly visible than in systems available so far. In the liquid phase, two experiments on the fluoride complexation with ARCA were performed. In a first CIX experiment, the liquid phase consisted of 0.1 M nitric acid and $5 \cdot 10^{-4}$ M HF. Eluted species included $[M^{VI}O_2F_3(H_2O)]^-$, $[M^{VI}O_2F_2]$, and $[M^{VI}O_4]^{2-}$. The group 4 elements were retained on the column. This is of crucial importance, because all three detected transformation chains consisted of the pair [261]Rf/[257]No. The transformation of [265]Sg was missed; as 38 s elapsed between the elution of the solution from the chromatography column to the beginning of the α-measurement, this is not unexpected in light of the considerably shorter half-life of [265]Sg (see Tab. 9.2).

To differentiate among the different possible chemical species present in the chromatography column, a second experiment was conducted in which no hydrofluoric acid was present in the liquid phase. If a nonfluoride containing species had been responsible for the results obtained in the first experiment, a total of about five transformation chains from ^{265}Sg were expected. However, none were observed that could unambiguously be attributed to ^{265}Sg, indicating that Sg was present in a positively charged chemical form that adsorbed on the CIX column. The lighter homolog, W, was eluted from the column and thus present in neutral or anionic form, in contrast to Sg. Likely, the hydrolysis of Sg is weaker than that of W, resulting in the species $Sg(OH)_5(H_2O)^+$, but $WO_2(OH)_2$ being present in dilute nitric acid.

9.8.4 Bohrium (Bh, Z = 107)

Knowledge on the chemical behavior of bohrium derives from one single experiment. Using isothermal chromatography, the volatility of a bohrium oxychloride compound, most likely BhO_3Cl, was measured. The obtained breakthrough curves for the three homolog elements Tc, Re, and Bh are shown in Fig. 9.17. As follows from the deduced $-\Delta H_a$ values, the volatility trend is decreasing when going down group 7, with the Bh compound being the least volatile one.

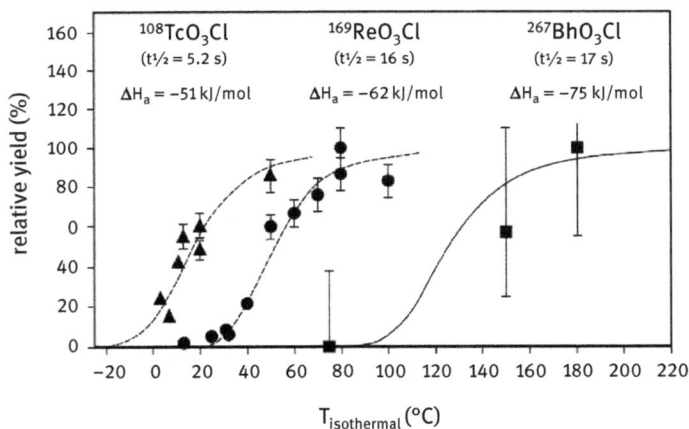

Fig. 9.17: Breakthrough curves of Tc, Re, and Bh oxochlorides (from Eichler et al. 2000).

9.8.5 Hassium (Hs, Z = 108)

The group 8 elements Ru and Os are known for their highly volatile tetroxides with boiling points around only 130 °C.[19] The stability of the tetroxide increases from RuO_4 to OsO_4. No other metal is known to form an oxide with similar properties. For a long time, this has rendered the hypothetical HsO_4 appearing to be an outstanding compound for the highly selective and efficient gas-phase chemical separation and investigation of Hs. With the expected volatility being much higher than that of any of the previously studied Rf, Db, Sg, or Bh compounds, a new experimental method was developed for the chemical study of Hs. Instead of transporting the EVRs with a particle gas jet, the formation of the tetroxide should occur directly behind the target from which the Hs EVRs recoiled. The gas in the recoil chamber therefore contained about 10% of O_2. Os was found to be rapidly converted into the OsO_4, which could be transported with the gas stream over several meters and then deposited on a cold surface, where it adsorbed by physisorption. Other elements were retained quantitatively, which provided very high separation factors for Os from elements with isotopes that potentially interfere with the unambiguous identification of the nuclear transformation of single Hs atoms. A schematic of the setup that was used for the first chemical study of Hs in the form of the HsO_4 is shown in Fig. 9.18.

The obtained thermochromatogram (inset in Fig. 9.18) indicates that HsO_4 is indeed formed and transported in the gas phase at room temperature. This is similar to the behavior of Os. However, on the Si_3N_4 surface of the chromatography column, HsO_4 deposited at slightly higher temperatures than OsO_4. The slightly lower volatility of HsO_4 compared to OsO_4 has since been confirmed in further experiments, also on SiO_2, Al_2O_3, and Au surfaces.

In a next suite of experiments, the highly selective separation of Hs in the form of the tetroxide was exploited to study the nuclear synthesis and the transformation properties of isotopes around ^{270}Hs. The experiment was conducted at different beam energies to measure the excitation function of the ^{26}Mg + ^{248}Cm reaction. Three distinct different types of transformation chains were observed, which were attributed to ^{269}Hs (5n evaporation channel), ^{270}Hs (4n evaporation channel), and ^{271}Hs (3n evaporation channel), with the latter two isotopes and their daughters being discovered in these studies. The average transformation properties and the excitation function are displayed in Fig. 9.19.

In a subsequent experiment, focusing again on chemical aspects, HsO_4 was deposited on NaOH covered detectors in a moist atmosphere through chemisorptions

---

19 In case of Os, where the tetroxide is continuously forming at the surface of the metal when exposed to air, the smell of this compound even gave the element its name (*osme*, Greek: smell).

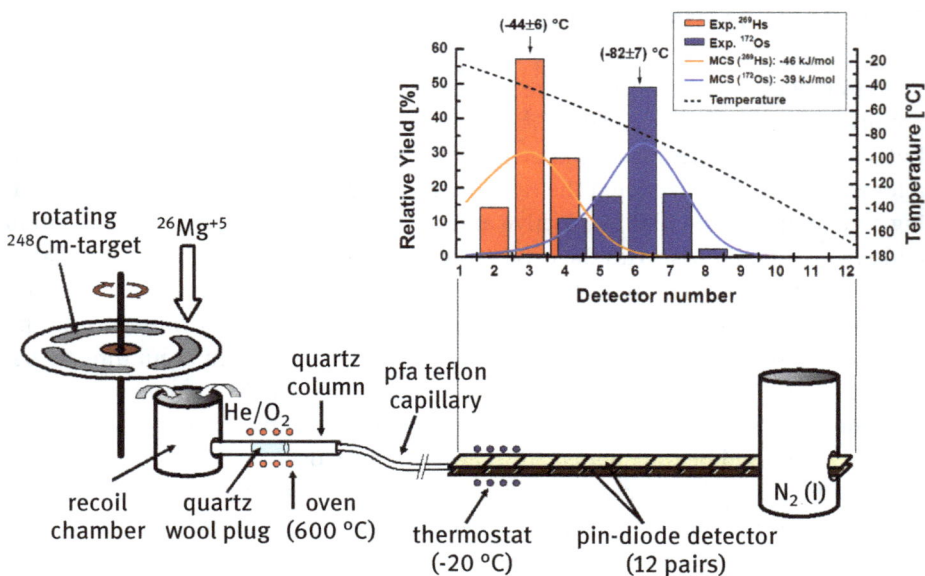

Fig. 9.18: Setup used for the synthesis, separation, and detection of volatile HsO_4. ^{269}Hs was produced in the $^{26}Mg + ^{248}Cm$ reaction and stopped in the He/O_2 flushed recoil chamber, where HsO_4 formed. This was transported with the gas through a perfluoroalkoxy (pfa) Teflon capillary to a series of detectors suitable for detecting the nuclear transformation of ^{269}Hs. The insert shows the temperature profile established along the detector array and the observed deposition of ^{269}Hs and ^{172}Os, from which the volatility was deduced. Nonvolatile species were retained on heated quartz wool (from Düllmann et al. 2002).

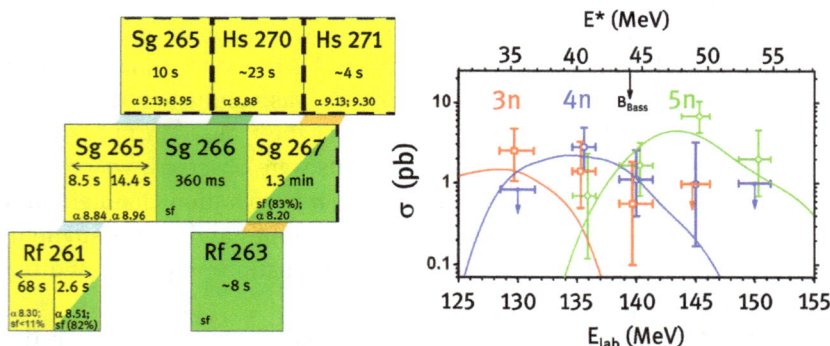

Fig. 9.19: Decay properties of $^{269-271}Hs$ and their daughters and excitation function for their production in the $^{26}Mg + ^{248}Cm$ reaction (right: after Dvorak et al. 2008).

via the formation of the Hs^{VII}-hassate, $Na_2[HsO_4(OH)_2]$. Therefore, Hs behaved chemically similar to Os also in this experiment.

9.8.6 Meitnerium (Mt, Z = 109), darmstadtium (Ds, Z = 110), and roentgenium (Rg, Z = 111)

To date, no chemical experiments have been performed with these elements; they appear to remain somewhat neglected. This is connected to the facts that (i) no sufficiently long-lived isotopes can be produced directly in a nuclear reaction, and (ii) the cross sections for producing these elements are very small, cf. Fig. 9.6 (where it should be kept in mind that the isotopes accessible in cold fusion reactions with cross sections as depicted in that figure are very short-lived). In addition, the choice of a suitable chemical system is far from trivial in the respective groups 9–11. In all of these three groups, the lighter homologs in the 4d and 5d series of the periodic table behave quite differently in many chemical systems, rendering a classification of the transactinide member based on (dis)similarities with its lighter homologs difficult.

9.9 Chemical properties of elements Cn–Og

The superheavy elements from Cn onwards have long been predicted to be volatile in the elemental state, which makes them suitable for chemical studies in the gas phase using similar experimental setups as for the characterization of the highly volatile HsO_4. Cn and Fl are particularly interesting due to their closed or quasi-closed-shell electronic configurations in the ground state, which is $7s^2 6d^{10}$ and $7s^2 7p_{1/2}^2$, respectively. A strong relativistic stabilization and contraction of the 7s and $7p_{1/2}$ atomic orbitals (see Fig. 9.20), renders these less accessible for chemical bonding. Fully relativistic calculations of their $-\Delta H_a$ on, e.g., gold and quartz surfaces – the materials used in experimental studies – predicted a much weaker interaction than those exhibited by their lighter homologs. Fl is expected to interact more strongly with the gold surface due to its active 7p-atomic orbitals, whereas Cn has closed-shell structure and is expected to be more volatile and inert. On the other hand, Nh and Mc are expected to be chemically reactive due to the presence of one and three valence p-electrons in the ground states respectively. This is also confirmed by relativistic calculations of their $-\Delta H_a$ on gold and quartz.

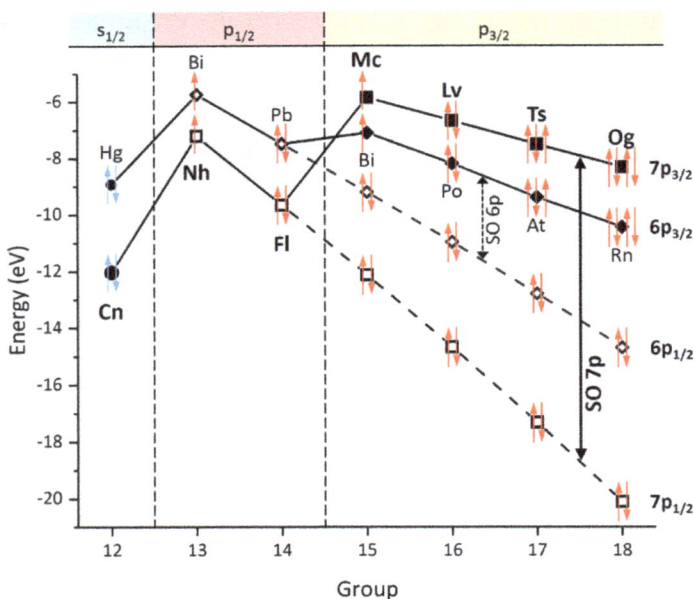

Fig. 9.20: Binding energies of the valence atomic orbitals of the elements from Cn to Og with the values for their lighter homologs, calculated in a fully relativistic framework (Pershina et al. 2010). Red arrows indicate electrons, successively filling the orbitals; these can accommodate two ($s_{1/2}$ and $p_{1/2}$) and four ($p_{3/2}$) electrons. The (sub)orbital that is occupied by the last electron is denoted above the figure. SO denotes spin-orbit splitting.

9.9.1 Copernicium (Cn, Z = 112)

Copernicium as a member of group 12 of the Periodic Table is a homolog of Zn, Cd, and Hg. The valence electronic configuration of these elements is s^2d^{10}, i.e., both these electronic shells are fully occupied. The valence s orbitals are increasingly influenced by the direct relativistic effect. Consequently, the elements become more and more inert when going down along group 12. Already for a long time, Cn was predicted to potentially rather resemble a noble gas than a noble metal. In 2007, this element was for the first time successfully studied in a thermochromatography experiment, in its elemental state. Being a homolog of Hg, which forms amalgams with Au, i.e., is known to bind strongly with Au, Cn can also be expected to form metallic bonds with Au. Hence, the goal of first experiments was to discriminate between an Hg-like and an Rn-like behavior of Cn.

To this end, a setup was built which was able to also observe both Hg and Rn. The thermochromatography method was applied, where the detectors forming the chromatography column (cf. Fig. 9.18) were covered with an Au layer thick enough to represent an Au surface, but thin enough for α-particles to penetrate without

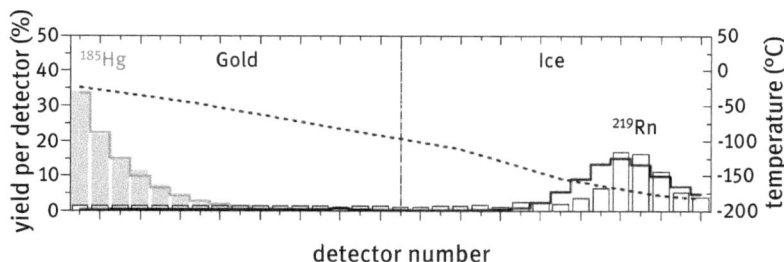

Fig. 9.21: Thermochromatogram for Hg and Rn. Bars indicate the experimental distribution of Hg (gray) and Rn (white). The dash-dotted vertical line indicates the beginning of the ice coverage of the gold surface at temperatures below −95 °C. Solid lines are results from simulations of the migration of Hg and Rn along the chromatography channel, assuming adsorption enthalpies $-\Delta H_a$ > 47 kJ/mol (Hg on gold) and 20 kJ/mol (Rn on ice) (from Eichler et al. 2007).

significant energy loss. A thermochromatogram obtained for Hg and Rn is shown in Fig. 9.21.

Due to the very low temperatures, the trace amounts of water still present in the He carrier gas despite massive efforts for drying the gas, led to a dew point of about −95 °C. Correspondingly, detectors at higher temperatures were covered by a reactive Au surface, and those at colder temperatures by a thin ice layer, as indicated in Fig. 9.21. As follows from the figure, Hg deposits at the beginning of the chromatography column, and Rn – as expected for an element with a very low chemical reactivity – at very low temperatures near the exit. In a series of experiments, a total of five transformation chains from ^{283}Cn were registered. From the measured adsorption temperatures, the $-\Delta H_a$ for Cn on Au was evaluated to be 52^{+4}_{-3} kJ/mol. Figure 9.22 shows the trend along the group 12 elements, which is nicely followed by Cn.

Using a correlation (in the spirit of correlations shown in Fig. 9.12) between the adsorption enthalpy and the boiling point, the latter was derived to be 357^{+112}_{-108} K for Cn. It is thus the most volatile member of group 12.

9.9.2 Nihonium (Nh, Z = 113)

Nihonium, a member of group 13 and the heavier homolog of In and Tl, has a valence electronic configuration of s^2p, i.e., it has an unpaired $p_{1/2}$ electron and is thus expected to be significantly more reactive than the transactinide neighbors Cn and Fl. Relativistic effects stabilizing the $p_{1/2}$ orbital are predicted to render it more inert than In and Tl. Several attempts to study its chemical behavior in a setup similar to those used for studies of Cn have been reported. The isotopes ^{284,286}Nh were produced as daughter and granddaughter of the initial nuclei in the ^{48}Ca(^{243}Am,3n)-^{288}Mc and ^{48}Ca(^{249}Bk,3n)^{294}Ts reactions, respectively. Under the conditions of the experiment, the formation of metallic Nh or of the hydroxide NhOH was expected.

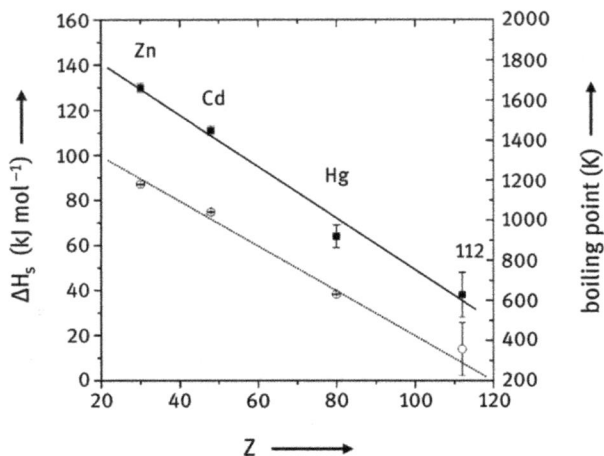

Fig. 9.22: Sublimation enthalpy on Au surface (left-hand *y*-axis) and boiling point (right-hand *y*-axis) for the group 12 elements (from Eichler et al. 2008).

Based on known cross sections of the used nuclear reactions and all partial efficiencies of the setup, observation of 10 to 20 atoms would be expected among the ^{243}Am-target based experiments, however, only five decay chains were observed, and in most of the cases the observed decay properties of at least one chain member are in poor agreement with known data based on over 100 decay chains measured in physics experiments. In the ^{249}Bk-target based study, one decay chain was reported, which is again substantially fewer than expected. Nuclear reaction products were thermalized directly behind the target and directly transported to the chemistry setup in these studies. Volatile nuclear reaction byproducts like Rn also reached this system, and overall, substantial amounts of unwanted species contributed to a large background in the nuclear spectra. This casts significant doubt on whether indeed Nh was observed in these experiments. To obtain cleaner spectra, further experiments were performed, in which the heavy-ion beam as well as unwanted nuclear reaction products were deflected in a gas-filled recoil separator. Two experiments devoted to studying Nh did not observe any decay chains. This might be due to a higher reactivity of Nh compared to that of Cn and Fl, preventing Nh from reaching the detection system. Improved setups will be needed, and are already under development (Yakushev et al. 2021).

9.9.3 Flerovium (Fl, Z = 114)

Flerovium is currently the heaviest element that has been studied by chemical means. As a member of group 14, its nearest homolog is the heavy metal Pb. The valence electronic configuration is $s^2p_{1/2}^2$. Similarly to the case of the group 12 elements, the valence electrons occupy orbitals that are relativistically stabilized. Due to the large spin-orbit splitting between the $7p_{1/2}$ and the $7p_{3/2}$ orbitals, Fl was predicted to be significantly different from Pb, potentially resembling a noble gas. More recent quantum chemical calculations still predict Fl to be less reactive than Pb. However, upon contact with, e.g., an Au surface, the formation of metallic bonds is expected.

Quite surprisingly, a first experiment using the same setup as applied in the studies of Cn yielded indications for a more volatile behavior of Fl, with deposition observed at significantly lower temperature, corresponding to $-\Delta H_a = 34^{+20}_{-3}$ kJ/mol (68% confidence limit), which was interpreted to point at weak physisorption processes governing the interaction of Fl with Au. However, the experiment suffered from some unwanted background and is thus not universally accepted. A follow-up study conducted by a different team, which first separated Fl from other nuclear reaction products in a gas-filled recoil separator, found adsorption of Fl on Au surfaces at room temperature. As only two transformations were detected, only a lower limit for the $-\Delta H_a$ could be obtained from these studies, which is 48 kJ/mol. Such high values hint at the formation of a metallic bond between Fl and Au. The studies of the chemical properties of Fl are one of the most interesting topics in current transactinide chemical research and further experiments are under way.

9.9.4 Elements with Z ≥ 115

To date, these elements are still "unknown" to chemistry: no chemistry experiment has been reported for any of these elements, which means that none of even their most basic properties have been determined experimentally. Mc is the element within closest experimental reach. Fast gas-phase chromatography setups, as developed for the study of Nh and to be used behind a recoil separator might offer access to 0.18-s ^{288}Mc (cf. Tab. 9.2). Right now, though, our knowledge for all of these elements is based on theoretical calculations. There is still a lot of exciting work waiting for the next generation of experimental nuclear chemists: proving a "new" element to rightfully belong to a certain group of the periodic table, based on measured similarities to the lighter homologs within this group.

9.10 Outlook

Though only single, short-lived atoms of the transactinide elements are available, all elements up to Hs as well as Cn to Fl have been studied chemically. Most of these studies so far rely on MENDELEEV's idea of homology within the groups of the

Periodic Table. As the cross sections for the nuclear synthesis generally become smaller with increasing atomic number, most data are naturally available for the early transactinides.

Topics of highest current interest include the clarification of the chemical behavior of Fl, cf. Section 9.8.8. New experiments will provide higher statistics than just the two or three transformation chains of the two experiments conducted so far. The goal is to clearly establish the volatility and reactivity sequence among the group 12 and 14 elements Hg, Cn, Pb, Fl, and the noble gas Rn. The characterization of Nh has started and reproducible data is expected to become available shortly. Shedding light on the evolution of reactivity in the early 7p elements, cf. Fig. 9.21, comes into reach. Beyond Fl, half-lives drop to 180 ms for the most readily available Mc isotope and to the level of tens of ms for the yet heavier elements. Development of techniques that are fast enough to access such isotopes has already started. High selectivity will be achieved by using physical preseparators and the necessary speed can be achieved, e.g., by applying electrical fields to rapidly guide the separated ions into a chemistry setup. Alternatively, chemical studies in vacuum appear promising, but do not yet exhibit the needed high efficiencies. Building up on the pioneering work on $Sg(CO)_6$, carbonyl complexes of heavier elements come into focus. Preparatory work with, e.g., Re and Ir, have been successful, paving a way to the first chemical study of Mt in this chemical system. Also liquid-phase chemical studies, beyond Sg, will remain a topic of interest, The necessary gain in speed can be achieved by coating α-detectors with functional groups acting as ion exchangers, which avoids the time-consuming step of evaporating a sample to dryness for nuclear counting. The importance of theoretical calculations will remain high in transactinide element chemistry and may in many cases remain the only way for their study. Experimentally accessible systems, however, still may continue to be the most exciting way to gain more insight into the chemical behavior of the heaviest elements.

References

Chiera NM., Sato TK, Eichler R, et al. Chemical characterization of a volatile dubnium compound, $DbOCl_3$. Angew Chem Int Ed 2021, 60, 17871–17874.

Düllmann ChE, Brüchle W, Dressler R, et al. Chemical investigation of hassium (element 108). Nature 2002, 418, 859–862.

Dvorak J, Brüchle W, Chelnokov M, et al. Observation of the 3n evaporation channel in the complete hot-fusion reaction $^{26}Mg + {}^{248}Cm$ leading to the new superheavy nuclide ^{271}Hs. Phys Rev Lett 2008, 100, 132503.

Eichler R. Empirical relation between the adsorption properties of elements on gold surfaces and their volatility. Radiochim Acta 2005, 93, 245–248.

Eichler R, Brüchle W, Dressler R, et al. Chemical characterization of bohrium (element 107). Nature 2000, 407, 63–65.

Eichler R, Aksenov NV, Belozerov AV, et al. Chemical characterization of element 112. Nature 2007, 447, 72–75.

Eichler R, Aksenov NV, Belozerov AV, et al. Thermochemical and physical properties of element 112. Angew Chem Int Ed 2008, 47, 3262–3266.

Even J, Yakushev A, Düllmann ChE, et al. Synthesis and detection of a seaborgium carbonyl complex. Science 2014, 345, 1491–1493.

Flerov GN, Ilyinov AS. On the way to the super-heavy elements. Moscow, MIR Publishers, 1986.

Gäggeler H, Chemistry gains a new element: Z=106. J. Alloys Comp. 1998, 271–273, 277–282.

Hofmann S, Dmitriev SN, Fahlander C, et al. On the discovery of new elements (IUPAC/IUPAP Report). Pure Appl Chem 2020, 92, 1387–1446.

Ishii Y, Toyoshima A, Tsukada K, et al. Fluoride complexation of element 104, rutherfordium (Rf), investigated by cation-exchange chromatography. Chem Lett 2008, 37, 288–289.

Korschinek G, Kutschera W. Mass spectrometric searches for superheavy elements in terrestrial matter. Nucl Phys A 2015, 944, 190–203.

Nagame Y, Haba H, Tsukada K, et al. Chemical studies of the heaviest elements. Nucl. Phys. A 2004, 734, 124–135.

Oganessian Yu Ts, Utyonkov VK. Superheavy nuclei from ^{48}Ca-induced reactions. Nucl Phys A 2015, 944, 62–98.

Pershina V, Borschevsky A, Anton J, Jacob T. Theoretical predictions of trends in spectroscopic properties of homonuclear dimers and volatility of the 7p elements. J Chem Phys 2010, 132, 194314.

Pyykkö P. Relativistic effects in structural chemistry. Chem Rev 1988, 88, 563–594.

Schädel M. Chemistry of superheavy elements. Ange Chem Int Ed 2006, 45, 368–401.

Türler A, Eichler B, Jost DT, et al. On-line gas phase chromatography with chlorides of niobium and hahnium (element 105). Radiochim Acta 1996, 73, 55–66.

Türler A, Brüchle W, Dressler R, et al. First measurement of a thermochemical property of a seaborgium compound. Angew Chem Int Ed 1999, 38, 2212–2213.

Türler A, Pershina V. Advances in the production and chemistry of the heaviest elements. Chem Rev 2013, 113, 1237–1312.

Yakushev A, Lens L, Düllmann ChE, et al. First study on nihonium (Nh, element 113) chemistry at TASCA. Front Chem 2021, 9, 753738.

Gert Langrock and Peter Zeh
10 Nuclear energy

Aim: The discovery of induced nuclear fission led to many civil and noncivil applications utilizing the enormous amount of either the energy and/or the neutrons released by nuclear fission. While military applications, such as the bombing of Hiroshima and Nagasaki and countless nuclear weapons tests, demonstrated the devastating potential of nuclear fission if used for noncivil purposes, the civil use of nuclear fission created remarkable progress in scientific research and energy production.

Nuclear reactors are nowadays mainly used for electricity supply. Some special reactor designs are also used for fundamental research, for the production of radionuclides, mainly for medical applications, or for the production of plutonium or tritium for nuclear weapons.

All reactor types are based on the use of a nuclear chain reaction where fissile material such as ^{235}U or ^{239}Pu undergoes neutron capture reactions. This chapter continues topics described in Vol. I on spontaneous fission and nuclear reactions, and illustrates special scientific and technological features of nuclear reactor technology.

During the last decades, nuclear energy has become an important and reliable electrical power source in many countries. The civil nuclear industry is now a multibillion business with hundred thousands of employees. Each year, new reactors are being sold, erected, and connected with the grid – a good reason to have a look at some basics of nuclear energy production.

10.1 Introduction

This chapter gives a more general overview of nuclear reactors, with focus on nuclear power reactors, and some interesting aspects of the radiochemistry in light-water reactors during normal operation. It is not intended and not possible in the frame of this introductory book to cope with the whole field of nuclear reactor technology. Therefore, some important topics are not subjects of this chapter, e.g., detailed neutron physics and neutron balance, safety design of nuclear power plants, and incidents and accidents. This chapter is an invitation to the reader to further step into the wide area of nuclear technology.

10.1.1 Historic notes

Nuclear technology and research and especially that of nuclear reactors are all based on the discovery of induced nuclear fission by HAHN, MEITNER, and STRASSMANN. The

https://doi.org/10.1515/9783110742701-010

first manmade nuclear reactor worldwide became critical on December 2, 1942 in Chicago,[1] USA, beneath a football stadium, in the framework of the Manhattan project.[2] Most of the knowledge necessary for the development of nuclear power reactors had already been worked out in these early years of research on nuclear fission and nuclear reactors. Many reactor concepts were developed, but only few of them were brought beyond conceptual or experimental stage into operation as reactors for power generation. Some of them bore an obvious "dual use" concept as they were constructed to both generate power for civil purposes and produce plutonium for military purposes.

10.1.2 Nuclear reactors

All common reactor types are based on the use of a nuclear chain reaction[3] where fissile material such as ^{235}U or ^{239}Pu undergoes neutron capture reactions. These neutrons can – depending on their energy – initiate a nuclear fission of the fissile isotopes. The released fission neutrons have high energies ("fast" neutrons). To initiate further fission processes, they have to be slowed down ("moderated") to lower, "thermal" energies. If the neutrons have the right energy to be absorbed by the heavy nuclei, again fission takes place. If this process becomes repetitive, a nuclear chain reaction establishes and the reactor becomes critical. The chain reaction is stopped when the fissile material (often called "fuel") is consumed ("burned up"). It is also stopped when no more neutrons are available, e.g., by loss of the moderating material or by use of neutron absorbing material ("neutron poisons") to intentionally stop the chain reaction. **Power reactors** convert the released energy into heat and subsequently into electrical energy. They take advantage of the high energy content and the relatively low cost of nuclear fuel; 1 kg of uranium-235 releases by fission an energy of about $2 \cdot 10^7$ kWh. This energy is equivalent to the combustion of 2700 t coal. **Research reactors** serve as neutron sources. These reactors are typically used for scientific, engineering, or medical purposes.

---

1 This air-cooled reactor, called CP-1 (Chicago Pile-1), was built with uranium metal und uranium oxide as fuel and graphite as moderator and achieved the first self-sustaining chain reaction. More reactors, sometimes also using D_2O as moderator, were put into operation in the years after, aiming at the production of plutonium for nuclear weapons. The heat produced in those reactors was then not yet utilized.

2 The Manhattan project was a large research project of USA, United Kingdom, and Canada during World War II, devoted to the development of atomic weapons. More than 100,000 people worked on fundamental research as well as the production of fissile material and the construction of the bombs. However, the results of the project also laid the foundation for the considerable growth of nuclear sciences and nuclear energy after the war.

3 Table 10.1 gives some short explanations of frequently used terms in nuclear technology.

Tab. 10.1: Terminology in nuclear reactor technology.

(Neutron) absorber	Material that readily catches neutrons.
AGR	Advanced gas-cooled reactor, second generation of British gas-cooled reactors with CO_2 as coolant and graphite as moderator.
Burn-up	Quantity for a kind of "fuel consumption" during operation, typically given as thermal energy released by fuel in a time unit, divided by the mass of this fuel (GW d/t); a typical burn-up for PWR fuel assemblies is in the region of about 40–60 GW d/t.
BWR	Boiling water reactor. Type of nuclear reactor. Light water (H_2O) is coolant and moderator. The water is heated in the reactor core, turned into steam in the reactor pressure vessel, and led to the turbine(s) to drive them.
CANDU	Canadian deuterium uranium reactor. Nuclear reactor type, heavy-water moderated and cooled. Instead of one large reactor pressure vessel, many small pressure tubes containing the fuel are used.
Chain reaction	Fission reaction, in which the neutrons, produced by fission, induce new fissions and are thereby continuously regenerated, so that the fission becomes repetitive (self-sustaining).
Control rod	Instrument in a reactor core. Contains or is made of a neutron absorber and is used to control the nuclear chain reaction by inserting (slowdown/termination of the reaction) or withdrawing (acceleration of the reaction) the rod.
Conversion	Purification of uranium concentrate after mining/leaching and transformation (conversion) into the chemical species required. For LWR, the required species is UF_6 (to be enriched thereafter), followed by another conversion to UO_2. For PHWR, the conversion is from purified uranium to uranium oxide (UO_2 or UO_3).
Coolant	Material (liquid or gas) to which the heat from the reactor core is transferred.
Criticality	Criticality is reached when a nuclear chain reaction becomes self-sustaining, i.e., when the number of neutrons produced by fission is equal to the number of neutrons consumed.
Critical mass	Minimum mass of an object consisting of fissile material to sustain a chain reaction; depends on the material (concentration of the fissile isotopes, density, size, and shape) and the presence and nature of neutrons.
Enrichment	Increase of the ^{235}U content in a unit quantity of uranium.
Fertile	Material that is able to be transformed by absorption of neutrons into a fissile material.
Fissile	Material that is able to undergo fission after absorption of neutrons.
Fuel	Fissile material that is able to undergo fission reactions and release energy by fission.

Tab. 10.1 (continued)

Fuel assembly	Unit in a nuclear reactor, holding fuel rods or fuel elements in a defined geometry.
Grace period	Time period after an incident in which a reactor is safe without human intervention.
Heavy water	If the light hydrogen isotope ^1H in water (H_2O) is replaced by the heavier hydrogen isotope $^2H = D$, "heavy" water is formed: D_2O.
Moderator	Material used to slowdown ("moderate") neutrons to lower kinetic energies so that they can be absorbed by fissile nuclei. Typically, light elements (H/D/B/C) are used for moderators, since the energy transfer upon collision between the neutron and the light nucleus is then very efficient.
MOX fuel	Mixed oxide fuel. Contains not only uranium oxide as fuel, but also plutonium oxide recovered from spent nuclear fuel or weapons.
Reactivity	Reactivity is a term describing the neutron economy in a reactor. Reactivity is zero (reactor is critical) when neutron production is equal to neutron consumption (neutron absorption and leakage). Reactivity is positive (reactor is supercritical) when more neutrons are formed than are consumed. Reactivity is negative if more neutrons are consumed than produced.
Thermal	The term "thermal" refers to neutrons with low energy ($E_n < 1$ eV).
PHWR	Pressurized heavy-water reactor. Similar to PWR, uses heavy water (D_2O) as coolant and moderator (instead of light water (H_2O) as for PWR). This enables the use of both natural and enriched uranium as fuel.
Poison	Neutron absorbing material. Often fission products that accumulate in the fuel and absorb neutrons are meant by this term (e.g., ^{135}Xe, ^{149}Sm, ^{155}Gd, ^{157}Gd). Sometimes, poison is also used to generally designate neutron absorbing material. This can be mixed with fresh fuel to compensate for an excess initial reactivity.
PWR	Pressurized water reactor. Most common nuclear reactor type. Water is circulated at high pressure in a primary circuit and transfers the heat to a secondary water/steam circuit in a steam generator. The produced steam drives the turbine.
Re-processing	Process to recover fissile material (remaining ^{235}U and fissile isotopes produced during reactor operation, e.g., ^{239}Pu) from spent nuclear fuel. Comprises physical and chemical steps to separate the required material so that it can be reused in the reactor (typically in MOX fuel).
RBMK	Russian designation (reaktor bolshoi moshnosti kanalny: high power channel reactor) of a specific design of a graphite-moderated light-water cooled reactor
(Power) uprate	Increase in the power level of a nuclear power plant by increasing the thermal power level in the reactor (e.g., by increasing fissile material) and/or by optimization of thermal efficiency by various constructional and engineering measures (e.g., replacement or refurbishment of turbines).

Tab. 10.1 (continued)

SMR	Small modular reactors are a recent development comprising a wide variety of reactor concepts with lower power (typically 10–300 MW$_{el}$) output than common PWR's, PHWR's, and BWR's.
VVER	Russian designation (Vodo-Vodianoy Energeticheski Reaktor: water (cooled)-water (moderated) power reactor) for a specific PWR design.

10.2 Relevant basic physics

10.2.1 Nuclear reactions

Although the physical processes in a nuclear reactor are very complex, some basics shall be introduced for a deeper understanding. For details, see Vol. I Chapters 10 on "Spontaneous fission" and 13 on "Nuclear reactions". Four of the most important neutron-induced nuclear reactions in a nuclear reactor are:

(i) the absorption of neutrons and subsequent fission with production of fission fragments and further neutrons, e.g., $A(n,\gamma)B,C$ (B and C designate the primary fission fragments)
(ii) absorption of neutrons with subsequent particle emission, e.g., $A(n,\gamma)B$
(iii) elastic collision of neutrons, e.g., $A(n,\gamma')A'$
(iv) inelastic collision of neutrons, e.g., $A(n,n')A'$

Whereas in the first two reactions, the absorption of neutrons leads to a transformation of the original material, the collision of neutrons only transfers parts of the neutron energy to the colliding nucleus.

10.2.2 Prompt and delayed neutrons

The released fission neutrons evaporate from primary fission fragments, which have high neutron excess. About 99% of the "fission neutrons" are released within 10^{-14} s after fission and are called "prompt" neutrons. The remaining 1% of the neutrons released within milliseconds to minutes after the fission are called "delayed" neutrons. Primary fission fragments are neutron-rich. Consequently, they undergo β^--transformation,[4] leading in many cases to a highly excited state of the product nucleus.[5] If the energy of this *excited state* is higher than the partial binding energy δE_B

---

4 See Chapter 8 of Vol. I on "β-transformations II: β⁻-process, β⁺-process, and electron capture".
5 See Chapter 6 of Vol. I on "Processes of transformations: overview".

of the last neutron, this particular nucleus may transform by primary neutron emission[6] instead of secondary photon emission. A common example is ^{87}Br ($t\frac{1}{2}$ = 56 s). If ^{87}Br transforms to a highly excited state of ^{87}Kr, ^{87}Kr can transform to ^{86}Kr and release a "delayed" neutron, cf. Fig. 10.1. More examples are listed in Tab. 10.2.

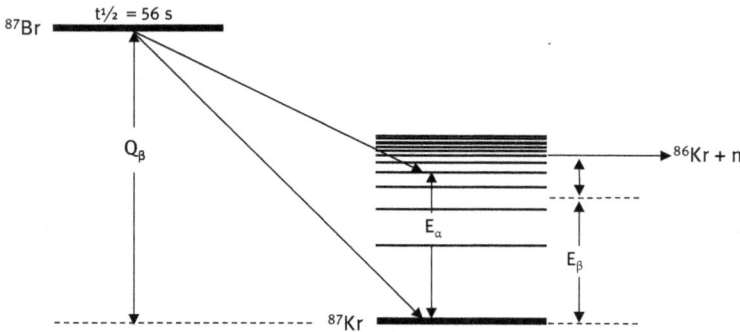

Fig. 10.1: Generation of delayed neutrons from excited nuclear states of nuclei formed in β⁻-transformation processes.

Tab. 10.2: Fraction v_d of delayed neutrons after the induced fission of fissile U and Pu isotopes.

Fissile nucleus	Type of fission	Total fraction v_d of delayed neutrons after thermal or fast induced fission
		v_d (neutrons/100 fissions)
^{233}U	Thermal	0.67 ± 0.003
^{235}U	Thermal	1.62 ± 0.05
^{238}U	Fast	4.39 ± 0.10
^{239}Pu	Thermal	0.63 ± 0.04
^{240}Pu	Fast	0.95 ± 0.08
^{241}Pu	Thermal	1.52 ± 0.11
^{242}Pu	Fast	2.21 ± 0.26

The total fraction of delayed neutrons per 100 fissions is 1.62 for fission of ^{235}U by thermal neutrons, 0.63 for ^{239}Pu fission by thermal neutrons and 4.4 for ^{238}U fission by fast neutrons (Tab. 10.2). The sum of prompt neutrons v_p and delayed neutrons v_d is $v = v_p + v_d$, and thus, the relative fraction of delayed neutrons can be defined with $\beta = v_d/v$.

---

6 See Chapter 9 of Vol. I on "Other 'particle' emissions".

Delayed neutrons are divided into six effective groups representing the half-life of their precursor nuclei. An example is given in Tab. 10.3, where the fraction of delayed neutrons is given for each group j as the relative yield β_j/β of all delayed neutrons.

Tab. 10.3: Groups of delayed neutrons originating from induced fission of ^{235}U and ^{239}Pu.

Group (j)	Half-life $(t\frac{1}{2})$ (s)	Relative yield β_j/β	Group (j)	Half-life $(t\frac{1}{2})$ (s)	Relative yield β_j/β	Precursor nuclei (examples)
^{235}U 1	55.9	0.033	^{239}Pu 1	54.2	0.035	^{87}Br, ^{142}Cs
2	22.7	0.219	2	23.0	0.298	^{88}Br, ^{137}I, ^{136}Te, ^{141}Cs
3	6.2	0.196	3	5.6	0.211	^{87}Se, ^{89}Br, ^{93}Rb, ^{138}I
4	2.3	0.395	4	2.1	0.326	^{85}As, ^{88}Se, ^{90}Br, ^{93}Kr, ^{94}Kr, ^{139}I, ^{143}Xe
5	0.6	0.115	5	0.6	0.086	^{86}As, ^{140}I, ^{145}Cs
6	0.2	0.042	6	0.3	0.044	various: Br, Rb, As, etc.

Without delayed neutrons, it would be impossible to technically control the reactor power. How can this be explained? A nuclear chain reaction can only be sustained if neutron production is equal to or larger than the inevitable neutron absorption in the reactor or their escape ("neutron leakage") from the reactor. If they are equal, the so-called effective neutron multiplication factor k_{eff} is 1. This factor is the ratio of the number n_i of neutrons in one neutron generation (i) over their number n_{i-1} in the previous generation ($i - 1$). Neutron multiplication is only possible if k_{eff} is greater than 1.[7]

$$k_{eff} = n_i/n_{i-1}. \tag{10.1}$$

Another quantity, the reactivity ρ[8] is then defined by

$$\rho = \frac{k_{eff} - 1}{k_{eff}}. \tag{10.2}$$

If $k_{eff} = 1$, reactivity $\rho = 0$, then a reactor is said to be "critical". This is the case during normal operation, i.e., at stationary conditions. $k_{eff} > 1$ and $\rho > 0$ is fulfilled, e.g., during start-up of the reactor or during transients. This state ($\rho > 0$) is called *supercritical*. If $k_{eff} < 1$ and $\rho < 0$, a reactor is subcritical, as is the case for shutdown.

---

7 For nuclear reactors, k_{eff} is only in the range between 1 and 1.01, whereas for atomic bombs k_{eff} is much greater than 1.

8 It should not be forgotten in the light of this very simple definition and simplified presentation that reactivity actually is an important characteristic number of a reactor and a function of many status parameters such as reactor power, temperature, pressure, density of the coolant, fuel temperature, burn-up, and boron concentration.

Taking into account the number of prompt neutrons at time $t = 0$ and the time τ as the time of each neutron generation, the multiplication of neutrons after m cycles $t = m\tau$ is

$$n = n_0 k_{\text{eff}}^m = n_0 e^{(m \ln k_{\text{eff}})}. \tag{10.3}$$

If k_{eff} is close to 1, the term $\ln k_{\text{eff}}$ can be simplified, eq. (10.4), and the multiplication of the neutrons is described by eq. (10.5):

$$\ln k_{\text{eff}} = -\ln(1 - \rho) \approx \rho \tag{10.4}$$

$$n \approx n_0 e^{t\rho/\tau}. \tag{10.5}$$

Assuming a multiplication factor $k_{\text{eff}} = 1.003$ and hence a reactivity $\rho = 0.003$ and also a τ for the neutron of $\tau = 3 \cdot 10^{-5}$ s (this is a characteristic slowdown time for neutrons in the moderator H_2O), then the number of neutrons and also the power of the reactor would increase within 1 s by a factor of e^{100}. The reactor period T is

$$T = \tau/\rho \tag{10.6}$$

in which the neutron number increases by a factor e (e = 2.717 ...) and is then 10 ms. It is obvious that no such reactor could be controlled – but thanks to the delayed neutrons, it is possible: The neutron multiplication equation must be corrected for the fraction of delayed neutrons. Their average lifetime can be calculated to 0.0146 s for ^{235}U (cf. Tab. 10.2). So, the average neutron lifetime τ is

$$\tau = \left(1 - \sum_{j=1}^{6} \beta_j\right)\tau_{\text{prompt}} + \sum_{j=1}^{6} \beta_j \tau_j \tag{10.7}$$
$$= (1 - 0.0162) \cdot 3 \cdot 10^{-5}\text{s} + 0.1\text{s} \approx 0.1\text{s}.$$

The reactor period then becomes $T = \tau/\rho = 34$ s. A reactor period of this magnitude is technically well controllable.

10.2.3 Energy of neutrons

The fission neutron energy is between 0.1 and 10 MeV, and the resulting energy spectrum $\chi_0(E)$ can be approximated by a MAXWELL–BOLTZMANN function with the (fictive) temperature T and the BOLTZMANN constant $k_B = 1.387 \cdot 10^{-23}$ J/K:

$$\chi_0(E) = \frac{2}{\sqrt{\pi}}(k_B T)^{-3/2}\sqrt{E}e^{-E/(k_B T)}. \tag{10.8}$$

The mean kinetic energy $\bar{E}_{\text{kin}}$ of a prompt fission neutron is calculated by integration according to:

$$\bar{E}_{kin} = \int_0^\infty E\chi_0 \, dE = \frac{3}{2} k_B T. \tag{10.9}$$

$\bar{E}_{kin}$ of fission neutrons from fission by thermal neutrons is typically about 2 MeV. Taking the kinetic energy $E = \frac{1}{2}mv^2$, the resulting speed v of such neutrons with 2 MeV can be calculated to $v = 1.96 \cdot 10^7$ m/s. Neutrons with energies between 0.1 and 10 MeV are also called "fast" neutrons. Neutrons with only 0.025 eV have a speed of $2.2 \cdot 10^3$ m/s and a fictive temperature of 290 K (17 °C), neutrons at temperature equilibrium have around 0.04 eV. Those neutrons and the ones with temperatures of up to 770 K (0.1 eV) are called "thermal" neutrons. The neutrons in-between the energy/velocity of the "thermal" and "fast" ones are called "epithermal

10.3 Evolution of nuclear power reactors

10.3.1 Generation I

The first commercially used civil nuclear power plant with an electrical power output of 5 MW$_{el}$[9] was put into operation in 1954 in Obninsk (Soviet Union). The reactor was a boiling water reactor using graphite as moderator and light water (H_2O) as coolant (LWGR). In 1956, the gas-cooled reactor (GCR) in Calder Hall (United Kingdom) was connected to the grid. This reactor used metallic uranium as fuel, graphite as moderator, and carbon dioxide (CO_2) as coolant and had an electrical output of 50 MW. Based on the experience with US naval propulsion reactors, another reactor type, a pressurized reactor, cooled with and moderated by light water (H_2O), was developed. The first commercial pressurized water reactor (PWR) with 60 MW$_{el}$ started its operation in 1957 in Shippingport (USA). The first commercial light-water cooled and moderated boiling water reactor (BWR), Dresden-1 in Morris (USA), started operation in 1960. (The basic principles of PWR and BWR are described below and their similarity to conventional coal-fired plants is illustrated in Fig. 10.2.) Canada developed its own reactor type, a heavy-water moderated and cooled pressurized tube reactor. The first 22 MW$_{el}$ prototype of this CANDU (Canada deuterium uranium) reactor became critical in 1962 in Rolphton (Canada).

---

9 It is common to distinguish thermal and electrical power output by using the subscripts "th" and "el" after the unit (or sometimes to append an "e" to the unit of the electrical power, i.e., "MWe"). Knowing both values of a specific NPP, one can simply calculate its thermal efficiency by dividing both. For example: an NPP with a PWR with an electrical power of 1100 MW$_{el}$ and a thermal power of 3200 MW$_{th}$ would have a thermal efficiency of 1100/3200 = 34%.

All these first prototypes of commercial nuclear power reactors developed until about 1960 are often called the first generation of nuclear power reactors. They demonstrated successfully the possibility to produce electrical power by nuclear reactors, although the economic competitiveness of nuclear energy was not clear at that time. They also demonstrated that civil nuclear technology had loosened the tight connection to military applications and nuclear technology can be responsibly practiced.

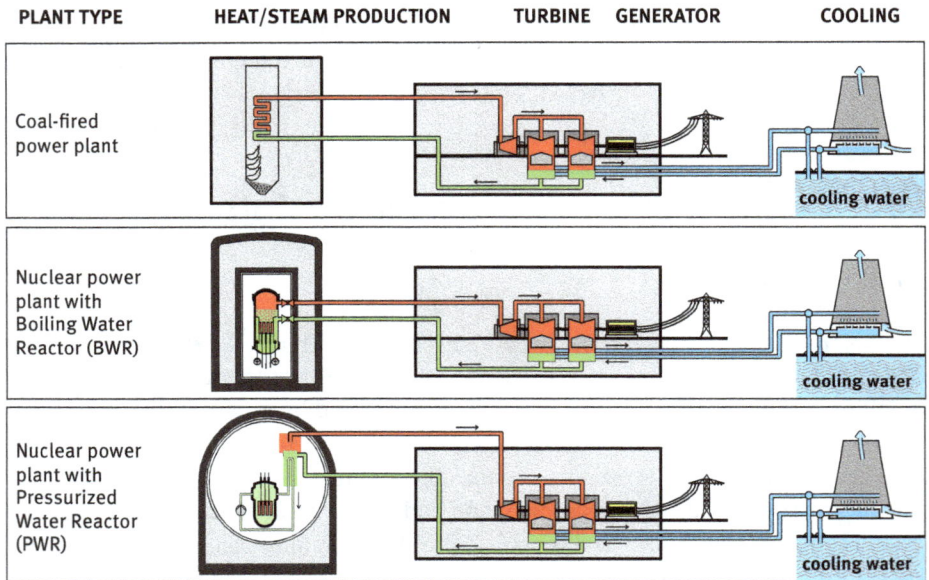

Fig. 10.2: Scheme of conventional coal-fired and light-water nuclear power plants to demonstrate their basic similarities. All types are used to generate steam that drives a turbine and a subsequent generator for electricity production. The difference is the methods to produce heat.

10.3.2 Generation II

Nevertheless, the result of these promising experiences was a boom in orders and construction of advanced and uprated nuclear power reactors in the 1960s and early 1970s. This so-called second generation of commercial nuclear power plants is characterized by a fast and continuous development of their technology to date and a continuous increase of the installed nuclear power capacity since then, cf. Fig. 10.3. It resulted in reactors with increasing dimensions and power output and numerous new technological solutions for the safe and reliable operation of nuclear power

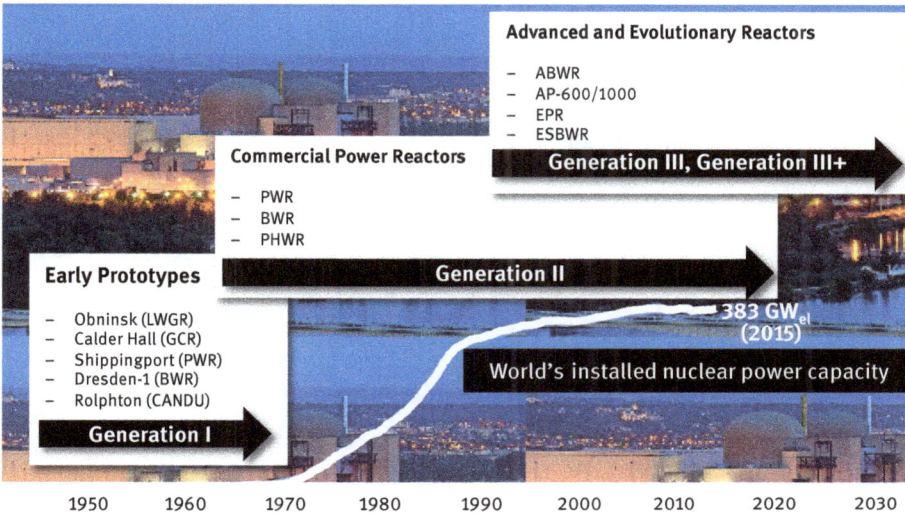

Fig. 10.3: Evolution of reactor technology and installed nuclear power capacity in the world.
(Ich habe die Datenreihe ergänzt, aber mir fehlt die Datei mit dieser Abbildung).

plants.[11] The classification "generation II reactors" comprises nearly all commercial reactors constructed until about 2000. Many of them are still operational.

As of 2022, more than 400 reactors were operational worldwide in 34 countries (cf. Tab. 10.4) and more than 50 were under construction. Most of these reactors are PWRs, BWRs and pressurized heavy-water reactors (PHWR, mostly of the CANDU type) with an electrical capacity of typically 200–1400 MW$_{el}$ for each reactor.

The reactor accidents of Three Mile Island (1979), Chernobyl (1986), and Fukushima (2011) sparked ongoing public and scientific discussion about the use and role of nuclear energy in many countries. The results of this discussion were different: some countries decided to phase out nuclear energy, in some countries no new reactors were licensed, and other countries maintained their nuclear programs but with significantly tightened technical and regulatory demands for safe operation of the plants. This led to a number of substantial improvements and backfittings of existing reactor installations, new safety design features, and new procedures for operation of plants.

---

11 The designed lifetime of these reactors is typically 30–40 years. Many of them were or are being licensed to operate for 50–60 years in total, depending on the energy policy in the countries where they are located to ensure energy supply.

Tab. 10.4: Reactors in operation and under construction (2022), grouped by reactor type, and some typical parameters.[10] Source: IAEA PRIS, http://www.iaea.org/pris.

Reactor Type	Operational		Under construction		Fuel	Moderator	Coolant	Typical core power density	Maximum coolant temperature	Coolant pressure	Thermal efficiency
	No. of reactors	Total net electrical capacity	No. of reactors	Total net electrical capacity							
		(MW$_{el}$)		(MW$_{el}$)				(MW/m^3)	(°C)	(bar)	(%)
PWR	305 (69.3%)	289,796 (74.0%)	44	49,289	Enriched UO$_2$ (PuO$_2$)	Light water	Light water	100	325	160	33
BWR	61 (13.9%)	61,849 (15.8%)	2	2653	Enriched UO$_2$ (PuO$_2$)	Light water	Light water	50–60	285	70	33
PHWR	48 (10.9%)	24,463 (6.2%)	3	1890	Natural or enriched UO$_2$	Heavy water (D$_2$O)	Heavy water (D$_2$O)	10–15	310	110	33
LWGR	12 (2.7%)	8358 (2.1%)			Enriched UO$_2$	Graphite (C)	Light water	5	285	70	33
GCR	10 (2.3%)	5650 (1.4%)			Natural U metal	Graphite (C)	CO$_2$	2	650	40	40

10 All values are approximate and can vary between different reactors of the same reactor type.

FBR	3 (0.7%)	1400 (0.4%)	3	1412	UO_2 and PuO_2	None	Liquid Na	400	550 (750)	10	40
HTGR	1 (0.2%)	200 (0.05%)			UO_2	Pyrocarbon/silicon carbide (C/SiC)	He	3–6	750 (950)	50	40–48
Total	440	391,716	52	55,244							

PWR Pressurized light-water moderated and cooled reactor; **BWR** Boiling light-water cooled and moderated reactor;
PHWR Pressurized heavy-water moderated and cooled reactor; **LWGR** Light-water cooled, graphite-moderated reactor;
GCR Gas-cooled, graphite-moderated reactor; **FBR** Fast breeder reactor; **HTGR** High-temperature gas-cooled reactor.

10.3.3 Generation III

Beginning in the 1980s, several manufacturers started to develop a new class of reactors, which is nowadays called generation III[12] of nuclear power reactors. Based on the experiences with generation II reactors and the results of extensive safety research, the existing designs were overhauled. New design features were implemented resulting in evolutionary reactor designs of the most common reactor types. These include more passive safety features. These are features that rely on automated, inherent actions, foreseen in the design to keep the reactor controlled or to safely shut it down. Passive safety features do not require any human or electronic intervention; they automatically work, only based on laws of physics (e.g., gravity or pressure differences) in case of malfunction of other systems. Passive safety functions aim to increase the grace period, after an accident when no operator action is required and the plant remains in a safe state with no risk for the environment, to typically more than 72 h.

Other envisaged characteristics of generation III reactors are a longer lifetime of 60 years and more, standardized reactor designs, better fuel utilization based on improved fuel technology, lower probability of core damage by an accident, and higher thermal efficiency, making reactors economically more competitive.

10.3.4 New developments

Common reactor designs are mature, not at least with the development of the Generation III reactors, and have reached a status where not much more can be done to make them more efficient and economical. The construction of such large-power water-cooled reactors is usually accompanied by long erection times, high investments (including frequent delays and cost overruns of new build projects), and requires a suitable electrical grid and grid capacity in a country.

More recent developments aim at providing smaller-sized reactors, capable of supplementing installed capacity and renewable energy sources (e.g., wind, solar). They are often referred to as small modular reactors (SMR), pointing both at their size and the simplified and standardized design as "modules", allowing easy scale-up by addition of more modules. As of 2021, more than 70 SMR designs are under investigation, research and development, or even construction, from small, inherently safe water-cooled reactors to alternative concepts with other moderators and heat transfer media. While most of these designs were in a conceptual or design stage, some were already realized.

---

12 There is sometimes also a differentiation between generation III and generation III+ reactors, pointing at the fact that early generation III reactors were mainly advanced versions of already existing reactors (e.g., the CANDU 6 PHWR or AP 600 PWR) while later generation III + reactors (e.g., ABWR, AP1000, ACR-1000, EPR) have more evolutionary and safety characteristics.

The Russian "Akademik Lomonosov" is a floating power unit, based on PWR technology already used for nuclear-driven vessels, with an electrical power output of 35 MW of each of both reactors. It started power operation in 2019. In Argentina, a 27 MW_{el} prototype of a modified PWR design which integrates major PWR components in one single unit, is under construction (CAREM project).

China has developed a High Temperature Gas-Cooled Reactor (HTGR) which came into operation in 2021. It is based on already existing experience with helium-cooled reactors with a special fuel of spherical balls (pebbles of 6 cm diameter) enveloping the fuel with pyrolytic carbon (moderator). These pebbles are arranged in the reactor pressure vessel in a pebble bed and this reactor is therefore also known as high-temperature reactor pebble-bed module (HTR-PM). The use of helium as main coolant allows to operate at higher temperature, giving a higher efficiency of the plant (211 MW_{el} at 500 $MW_{th} = 42\%$).

10.4 Main components of a nuclear power reactor

Independent of their specific design, most currently operated power reactors have several components in common. This section describes the typical components and their functions.

10.4.1 Fuel

Uranium, typically in the form of uranium dioxide (UO_2) pellets, is the basic fuel in most power reactors. UO_2 is suitable as it is dense, has a high melting point of 2865 °C, and good thermal conductivity. The uranium used is often enriched which means that the content of the fissile isotope ^{235}U is increased from the natural content of 0.7% to higher values.[13] This is necessary especially in light-water reactors where light water (H_2O) absorbs many neutrons (physically speaking, it has a high neutron capture cross section) and is therefore a less efficient moderator than, e.g., heavy water (D_2O) or graphite.

Plutonium can serve as fuel, too. It is usually obtained from the reprocessing of spent uranium nuclear fuel. ^{239}Pu is "bred" during reactor operation from the fertile ^{238}U according to the reaction sequence of eq. (10.10) and can later be extracted from the spent uranium fuel by reprocessing[14]

---

13 The isotope composition of natural uranium is 0.72% ^{235}U, 99.2745% ^{238}U, and 0.0055% ^{234}U.
14 This reaction sequence is also used for the production of plutonium for military purposes.

$$^{238}U(n,\gamma)\ ^{239}U\ \xrightarrow[t^{1/2}=23.45min]{\beta^-}\ ^{239}Np\ \xrightarrow[t^{1/2}=2.36d]{\beta^-}\ ^{239}Pu \tag{10.10}$$

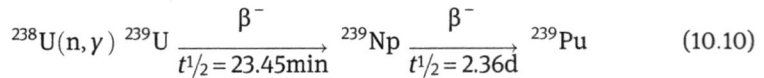

The reprocessed PuO_2 is blended with UO_2 to a mixed oxide (MOX) fuel. The use of PuO_2 containing MOX fuel allows the use of lower enriched or even natural uranium as such a mixed oxide reacts similarly to low enriched uranium, depending on the detailed composition.

Thorium is another possible fuel to be used in MOX fuel, containing ^{232}Th and ^{233}U or ^{235}U. In a thorium-based reactor, the fissile nuclide ^{233}U is produced from the fertile ^{232}Th via eq. (10.11). To date, industrial experience with thorium reactors is by far not as high as with U and Pu fuel, although thorium high-temperature reactors (THTR) are already operating and promising experimental experiences with thorium-based heavy-water reactors have been reported

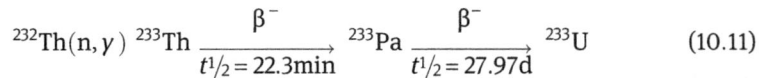

$$^{232}Th(n,\gamma)\ ^{233}Th\ \xrightarrow[t^{1/2}=22.3min]{\beta^-}\ ^{233}Pa\ \xrightarrow[t^{1/2}=27.97d]{\beta^-}\ ^{233}U \tag{10.11}$$

^{233}U, ^{235}U, and ^{239}Pu are certainly not the only radionuclides undergoing neutron-induced fission in a reactor. The situation is much more complex. The composition of the fuel changes during operation due to a number of transmutations by neutron capture and subsequent radioactive transformations and leads to the buildup of a number of higher actinide isotopes. Some of them are also fissile (e.g., ^{241}Pu). The main pathways of transuranium elements buildup in fuel are illustrated in Fig. 10.26.

The fuel should meet some specific requirements during operation concerning material properties, e.g., good thermal conductivity, mechanical strength, no reaction with cladding or coolant, operation temperature below the melting point, and retention of fission products in the matrix, cf. Tab. 10.5.

10.4.2 Fuel rods

The fuel is mostly arranged in cladding tubes which form a fuel rod. The fuel cladding is made of a specific metal alloy, mostly zirconium alloy (zircaloy), sometimes steel. It has got to be both corrosion and radiation resistant and have properties like good thermal conductivity, low neutron absorption, small coefficient of thermal expansion, mechanical strength, little or no reaction with water under operating conditions, and it should also not react with the fuel inside. The fuel rod for an LWR contains a stack of cylindrical UO_2 pellets, cf. Fig. 10.4, isolated by Al_2O_3 pellets at the end positions, cf. Fig. 10.5. The pellets can also be made of UO_2/PuO_2 MOX fuel.

Tab. 10.5: Some properties of some fuel and cladding materials under normal conditions (298 K, 0.1 MPa).

Unit	U	Pu	Th	UN	UC	UO$_2$	PuO$_2$	MOX*	Zircaloy-2	Zircaloy-4	SS316**
Density ρ in g/cm	19.05	19.84	11.72	14.3	13.63	10.96	11.46	11.07	6.55	6.58	7.95
Melting point T_{melt}											
in K	1405	913	2023	3123	2638–2780	3120 ± 30	2663 ± 20	3023	≈ 2110	2118	1710
(in °C)	(1132)	(640)	(1750)	(2850)	(2365–2507)	(2850)	(2390)	(2750)	(≈ 1840)	(1845)	(1440)
Boiling point T_{boil}											
in K	4018–4400	3500	5063		4691–4866	3815	3600	3811	≈ 4600	≈ 4600	3090
(in °C)	(3745–4127)	(3230)	(4790)		(4418–4593)	(3542)	(3327)	(3538)	(≈ 4330)	(≈ 4330)	(2820)
Heat capacity C_p in J/(kg K)	116.3	130	26.23	190	200	235	240	240	290	293	500
Thermal conductivity λ in W/(m K)	22.5	150	37–54	13.0	25.3	8.68	6.3	7.82	17	14.1	16
Linear expansion coefficient α in 10^{-6}/K	13.9		11.2	7.52	10.1	9.75	7.8	9.4	5.8	5.8	16

*U$_{0.8}$Pu$_{0.2}$O$_2$; **Stainless steel.

Also burnable poisons such as gadolinia[15] (Gd_2O_3) are often mixed in the fuel to compensate for excess initial reactivity[16] of the fuel. At the upper end of the fuel rod (often called upper plenum) is a free space[17] with a pressure spring keeping the pellet stack under tension. The fuel rod is pressurized with helium to about 20 bar primary pressure and sealed.

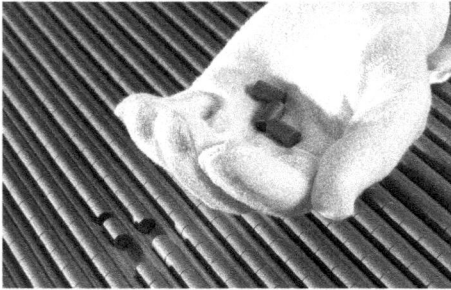

Fig. 10.4: Fuel pellets, FBFC Romans (© AREVA, Geoffray Yann).

10.4.3 Fuel assemblies

Fuel rods are arranged in fuel assemblies, which are placed in the reactor core. Such fuel assemblies for light-water reactors typically consist of a bundle of fuel rods and some structural elements. The fuel rods are arranged in 14×14 to 18×18 pin lattices for Western PWRs or in a hexagonal structure for Russian VVER PWRs, in an 8×8 to 10×10 structure surrounded by a fuel assembly channel for BWRs or in a concentrical bundle arrangement for CANDU PHWRs. The main structural elements are, for example, for a PWR fuel assembly a top nozzle and a bottom nozzle, and some spacer and mixing grids with mixing vanes to keep fuel rods in a defined position and to ensure a stable cooling of the cladding by water, cf. Fig. 10.5. Some positions in the assembly are occupied by control rods, which can slide in and out of guide tubes. Some fuel rods can also be replaced by shim rods, which contain burnable poisons[18] to compensate for excess initial reactivity of the core after start-up of the reactor.

---

15 Contains the stable isotope ^{157}Gd (15.,65 % natural abundance) with an extremely large thermal neutron capture cross section of 255 000 barn.

16 It is necessary to have an excess initial reactivity as during reactor operation ^{235}U depletion and the buildup of fission products occur, both reducing the reactivity of the fuel.

17 The free space is needed because of a certain swelling of the fuel and the production of gaseous fission products during operation. These gaseous fission products, mostly noble gases, build up a pressure of up to 180–200 bar inside the fuel rod during operation.

18 Such neutron poisons (e.g., boron carbide (B_4C) pellets in a zircaloy cladding) burn out during operation and thus contribute with an increasing positive reactivity, while the fuel depletes and

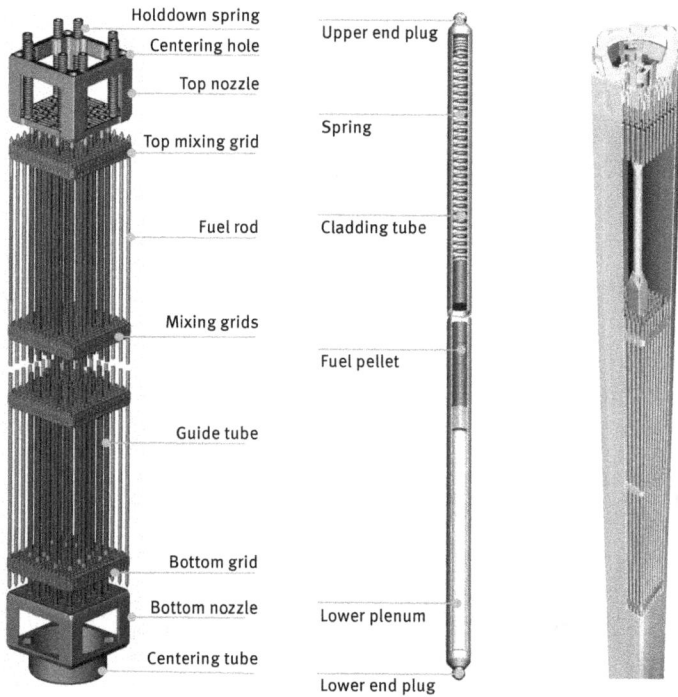

Fig. 10.5: HTP™ fuel element and single fuel rod for a PWR (left) and for a BWR (right) (© AREVA).

10.4.4 Control rods

Control rods are used to control the reaction rate in the fuel or to shut down the reactor. Physicochemical properties of these materials are listed in Tab. 10.6.

They may be inserted into the core or withdrawn from it. Control rods are made of neutron absorbing material, e.g., cadmium (Cd), silver (Ag), indium (In), hafnium (Hf), or boron (B). Typical materials are Ag–In–Cd (AIC) alloys or boron carbide (B_4C). Control rods can be arranged between fuel assemblies (e.g., BWR) or be part of the fuel assembly (e.g., PWR, cf. Fig. 10.6).

---

contributes with a negative reactivity due to the decreasing amount of fissile material and the increasing amount of neutron-absorbing fission products.

Fig. 10.6: PWR fuel assembly mock-up with the rod cluster control assembly on top (© FBFC, Romans).

10.4.5 Moderator

The probability for neutron capture by fissile material and subsequent fission is increased if the neutrons have a low energy of less than 0.1 eV.[19] Therefore, the initially fast neutrons have to be thermalized to initiate further fissions of ^{235}U, ^{239}Pu, and other fissile isotopes. The moderator does this slowdown. Neutrons collide with the moderator nuclei and lose their kinetic energy. Moderators consisting of light elements and with low neutron absorption, such as light water (H_2O), heavy water (D_2O), or graphite (C), are appropriate for this. Table 10.7 gives an overview of some important properties of typical moderators.

Listed are
- the neutron capture cross section σ_c,
- the scattering cross section for epithermal neutrons σ_s,
- a quantity called "lethargy" ξ, which is a dimensionless number describing the energy loss after a collision, and the bigger it is, the better,

---

19 Neutrons released after fission have a high kinetic energy in the range of more than 10 MeV. Those fast neutrons can directly induce further fission reactions if an appropriate fissile material (e.g., ^{238}U) is able to absorb them for further fission. This is accomplished by so-called fast reactors, like the fast breeding reactor (FBR).

Tab. 10.6: Some properties of absorbing materials (for control rods or burnable absorbers) under normal conditions (298 K, 0.1 MPa).

Unit	B	B_4C	BN	Hf	Ag–In–Cd (AIC) alloy	Dy_2O_3	Gd_2O_3
Density ρ in g/cm	2.33	2.51	2.25	13.09	10.17	8.10	7.07
Melting point T_{melt}							
in K	2348	2723	3270	2493	1073	2613	2612
(in °C)	(2075)	(2450)	(3000)	(2220)	(800)	(2340)	(2339)
Heat capacity c_P in J/(kg K)	387	960	848	363	230	290	550
Thermal conductivity λ in W/(m K)	27	92	28.7	22.3	60	2.3	
Linear expansion coefficient α in 10^{-6}/K	5	4.5	2	5.9	22.5	8.3	
Absorption cross section of isotopes in b (natural abundances in%)	3990 (^{10}B, 19.9%)			1500 (^{177}Hf, 18.4%) 75 (^{178}Hf, 27.0%) 65 (^{179}Hf, 13.8%) 14 (^{180}Hf, 35.4%)	37.6 (^{107}Ag, 51.8%) 91 (^{109}Ag, 48.2%) 202 (^{115}In, 95.8%) 20,600 (^{113}Cd, 12.36%)	585 (^{161}Dy, 19.0%) 180 (^{162}Dy, 25.5%) 130 (^{163}Dy, 24.9%) 2700 (^{164}Dy, 28.1%)	61,000 (^{155}Gd, 14.8%) 254,000 (^{157}Gd, 15.65%)
Mean absorption cross section of all stable isotopes in b	790 (B)			105 (Hf)	63 (Ag) 194 (In) 2550 (Cd)	948 (Dy)	48,900 (Gd)

Tab. 10.7: Some properties of typical moderators.

Moderator	σ_c (barn)	σ_s (barn)	ξ	$\xi\,\sigma_s/\sigma_a$	z	L (cm)
H_2O	0.66	49	0.927	62	19	2.73
D_2O	0.00092	10.6	0.51	5860	35	171
Be	0.009	5.9	0.209	138	88	20.8
C	0.0045	4.7	0.158	166	115	50.8

- a combined quantity which describes the "quality" of a moderator ($\xi\,\sigma_s/\sigma_a$),
- the average number z of collisions, needed to slowdown a fission neutron to thermal energy, and
- the linear distance L between neutron source (fissile nucleus) and absorber.

Technical consequences for the choice of a moderator are, for example:
(1) since the neutron capture cross section of H_2O is high, in light-water reactors only enriched uranium can be used to sustain a chain reaction, or
(2) if D_2O, Be, or C is used as moderator, the dimension of the moderator tank has to be increased, compared to H_2O, for the same power level.

10.4.6 Coolant

A coolant flowing through the reactor core is necessary to transfer and transport the transformation heat from it. This can be a liquid (light and heavy water, liquid metals such as sodium or lead alloy) or a gas (helium, carbon dioxide). A coolant should have specific properties like high heat capacity, high thermal conductivity, radiation resistance, low activation by (n,γ)-reactions, and not explosive or burnable. It should also have good moderating properties if it is also used as a moderator. The coolant is pumped by powerful reactor coolant pumps (RCP) around the circuit or, in some alternative reactor designs, naturally circulating.[20] To give an impression, a rough estimation shall be done: Assuming a pressurized water reactor with $1000\,MW_{el}$ and $3000\,MW_{el}$, a temperature increase in the reactor core of 35 K, and H_2O as coolant,

---

[20] The coolant can be modified by chemical additives to influence certain parameters. A common procedure is the addition of boric acid (H_3BO_3) for long-term reactivity control in the coolant of a PWR and to minimize the operation of control rods, especially at start-up or shutdown of the reactor. Boric acid can also be added to the coolant as a neutron poison to terminate the chain reaction and subsequently keep the reactor subcritical in case of an emergency. Another common method is the addition of ^{7}LiOH or KOH to the coolant in PWRs to alkalize the water and to reduce the corrosion rate.

the coolant mass flow rate can be estimated with the heat capacity of water (4.185 kJ/(kg K)) by

$$\dot{m}_{H_2O} = \frac{E}{c\Delta T} = \frac{3\cdot10^9\,J/s\cdot3600\,s/h}{4185\,J/(kg\,K)\cdot35\,K} = \frac{10.8\cdot10^{12}\,kg}{1.465\cdot10^5\,h} \approx 73,000\,t/h = 20\,t/s.$$

This simple estimate is close to the values given in Tab. 10.8.

Tab. 10.8: Main characteristics of the 1400 MW$_{el}$ PWR Isar 2 (Germany), exemplary for a modern PWR.

Type of reactor		Pressurized water reactor (PWR)
Net output		1410 MW$_{el}$
Reactor specific data		
Containment	Inner diameter	56 m
	Wall thickness	38–60 mm
Reactor pressure vessel	Inner diameter	5 m
	Total height	12.01 m
	Wall thickness at cylinder	250 mm
	Weight without internals	ca. 507 t
Reactor	Main coolant loops	4
	RPV inlet temperature	293 °C
	RPV outlet temperature	328 °C
	Coolant pressure	15.8 MPa (158 bar)
Reactor core	Number of fuel assemblies in core	193
	Total uranium mass in initial core	ca. 103 t
	Number of control rod assemblies in core	61 bundles
Steam generator	Number of steam generators	4
	Height	21.5 m
	Mass of each steam generator	440 tca. 5400 m²
	Heat exchange area per steam generator	4700 kg/s
	Coolant flow rate per steam generator	
Plant specific data		
Turbine	Turbine speed	1500 rpm
	Steam pressure at inlet of high-pressure turbine	64.3 bar
	Steam pressure at outlet of high-pressure turbine	11.3 bar
	Mass flow of saturated steam	2200 kg/s

Tab. 10.8 (continued)

Type of reactor		Pressurized water reactor (PWR)
Net output		1410 MW$_{el}$
Plant specific data		
Generator	Rounds per minute	1500 rpm
	Apparent power	1640 MVA
	Rated current	35 kA
Cooling water system	Cooling surface of condenser	96 000 m^2
	Mass flow of cooling water	60 000 kg/s
Cooling tower	Basic diameter	ca. 145 m
	Height	165 m
	Upper diameter	ca. 86 m

10.4.7 Reactor pressure vessel

The reactor pressure vessel (RPV) is the central part of the reactor coolant circuit. This massive steel vessel houses the reactor core consisting of fuel assemblies with control rods, the coolant and moderator, and other core and vessel internals, cf. Fig. 10.7. The inlet and outlet nozzles are above the upper end of the reactor core thus enabling to keep the core flooded even in if one of the coolant pipes breaks. The RPV has to withstand high pressure inside the vessel, high heat fluxes, neutron irradiation, and corrosion. In BWRs, steam separators and steam dryers are also placed in the RPV. Some reactor types such as RBMK and CANDU reactors have a series of pressure tubes instead of a single large pressure vessel.

10.4.8 Steam generator

Steam generators (SG) are heat exchangers and part of the cooling system, cf. Fig. 10.8. The temperature of the coolant in the primary circuit is high enough to boil water in the connected heat exchanger and to produce steam there. Steam generators are often upright U-tube bundle heat exchangers, straight-tube heat exchangers, or horizontal heat exchangers in case of the Russian VVER PWRs. The steam produced is used to drive a turbine to generate electricity. Steam generators are the interconnections of primary (coolant) and secondary (water-steam) circuit. In BWRs, the coolant evaporates in the reactor pressure vessel and the dry steam produced directly drives the turbine. Thus, a steam generator and a secondary circuit are not needed for BWR plants.

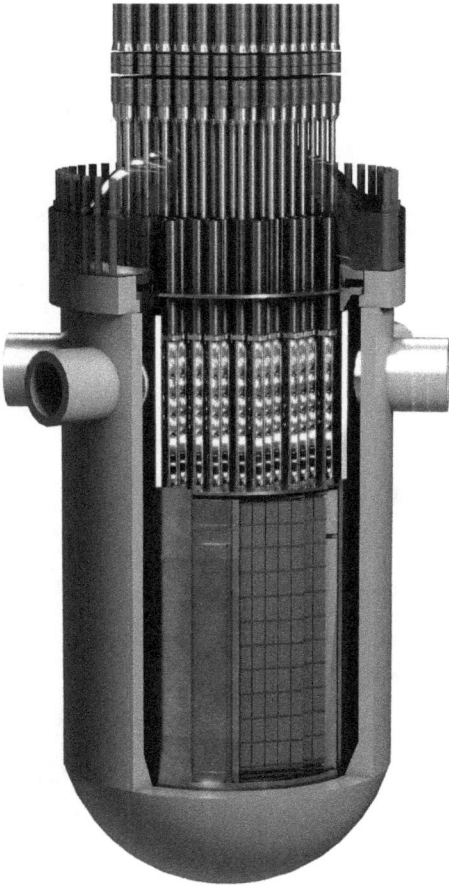

Fig. 10.7: Sketch of the RPV of the EPR PWR (© AREVA, IMAGE and PROCESS).

10.4.9 Nuclear power plant

A nuclear reactor is the main part of a nuclear power plant (NPP). An NPP consists of two main parts: the "nuclear island" and the "conventional island". The word "nuclear island" comprises several installations: nuclear reactor, reactor building, spent fuel pit, auxiliary nuclear systems, control room, and also emergency diesel generator engines. "Conventional island" comprises the complete nonnuclear power generation system with turbine and generator, grid connection, condenser, and turbine hall. Figure 10.9 shows a photo of an NPP.

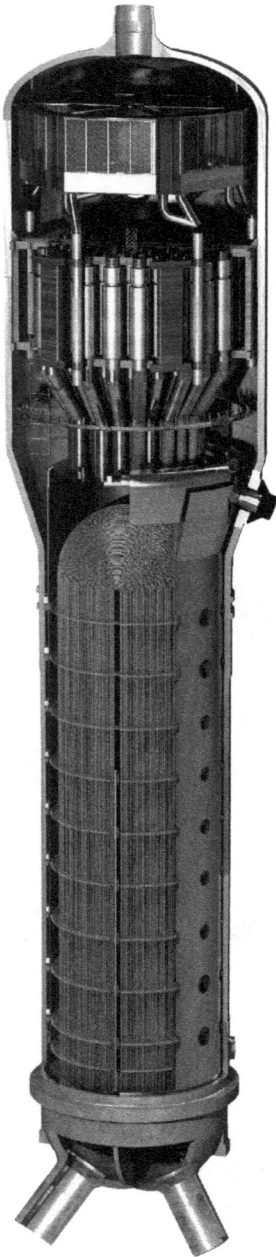

Fig. 10.8: Sketch of a steam generator for an EPR PWR reactor (© AREVA, IMAGE, and PROCESS).

Fig. 10.9: Paluel Nuclear Power Plant with four 1300 MW$_{el}$ PWR units in Normandy (Northern France) (© AREVA, Laurence Godart).

10.5 Common power reactor types

The general characteristics of the most common reactor types PWR, BWR, and PHWR are described in the following section.[21]

10.5.1 Pressurized water reactors

Pressurized water reactors (PWRs) are the most common reactors in the world, cf. Tab. 10.4. They use nonboiling light water (H_2O) for neutron moderation and heat removal. Their basic functionality is shown in Fig. 10.10. The heat produced in the reactor core is removed by the coolant (H_2O) in the closed primary circuit to steam generator(s) and transferred to a closed water-steam circuit (secondary circuit). The saturated steam produced in the steam generator at a pressure of about 65 bar and a temperature of about 280 °C is directed to the turbines which drive the generator. The separation of both circuits confines radionuclides in the primary circuit and keeps the secondary steam-water circuit and the turbines free of radioactivity.

The coolant in the primary circuit is pressurized to about 160 bar and has a temperature between about 290 °C (reactor core *inlet* temperature) and about 325 °C (reactor core *outlet* temperature). The water is kept nonboiling under these conditions.

---

21 Since the existing reactors of one type can strongly vary in detailed features and properties, the given information can only be exemplary.

The circuit is driven by reactor coolant pumps (RCP). Parts of the primary circuit are also the main coolant lines and the electrically heated pressurizer, which ensures a stable pressure in the circuit. A typical pressurized water reactor has 2–4 separate loops, each with a steam generator and a reactor coolant pump. All components of the primary circuit are surrounded by a so-called "containment" with thick steel and concrete walls, which acts as a safety barrier against the release of radionuclides to the environment in case of an accident.

Reactivity control in PWRs is mainly achieved by adjusting the boron concentration in the primary circuit. This allows controlling the long-term reactivity. Initial excess reactivity is compensated by burnable absorbers. Control rods are mainly used for quick shutdowns of the reactor or for fast variation of the reactor power.

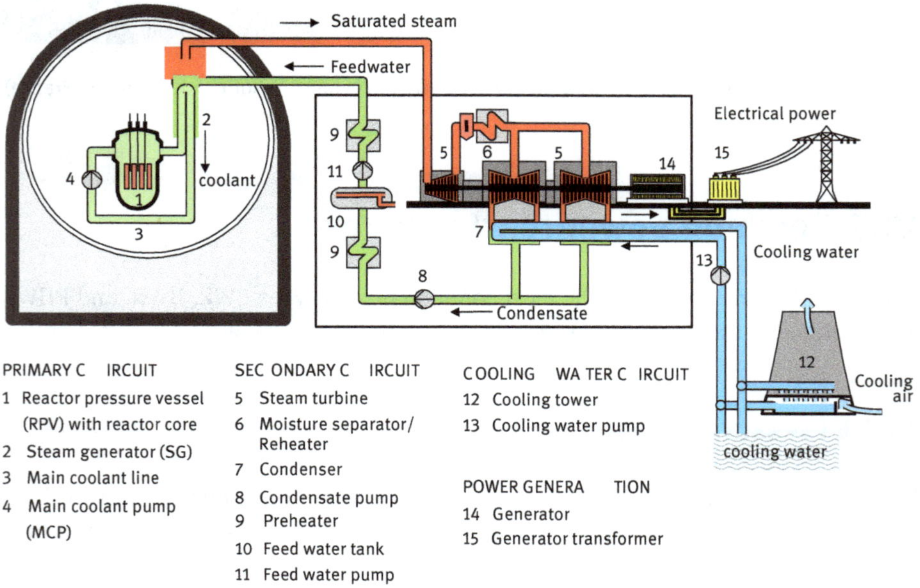

PRIMARY CIRCUIT
1 Reactor pressure vessel (RPV) with reactor core
2 Steam generator (SG)
3 Main coolant line
4 Main coolant pump (MCP)

SECONDARY CIRCUIT
5 Steam turbine
6 Moisture separator/ Reheater
7 Condenser
8 Condensate pump
9 Preheater
10 Feed water tank
11 Feed water pump

COOLING WATER CIRCUIT
12 Cooling tower
13 Cooling water pump

POWER GENERATION
14 Generator
15 Generator transformer

Fig. 10.10: Schematic diagram of a pressurized water reactor (PWR).

10.5.2 Boiling water reactors

Boiling water reactors (BWRs) are also light-water cooled and moderated reactors. The basic difference to PWRs is that the coolant already starts to boil in the reactor and the steam produced is directly – i.e., without interconnected steam generators – directed to the turbine. Steam generators and pressurizers are not needed. Thus, the configuration of a BWR circuit is less complex. However, the RPV of a BWR has to be larger than in a PWR. The reason is that due to the boiling process in the RPV there is a lower power density (about 50–60 MW/m^3 compared to about 100 MW/m^3

in a PWR) and in the upper part of the RPV a water-steam separator and a steam dryer are placed, whereas the control rods are located (contrarily to a PWR) at the bottom, cf. Fig. 10.11. BWRs are operated at lower temperatures than PWRs, typically about 285 °C and about 70 bar.

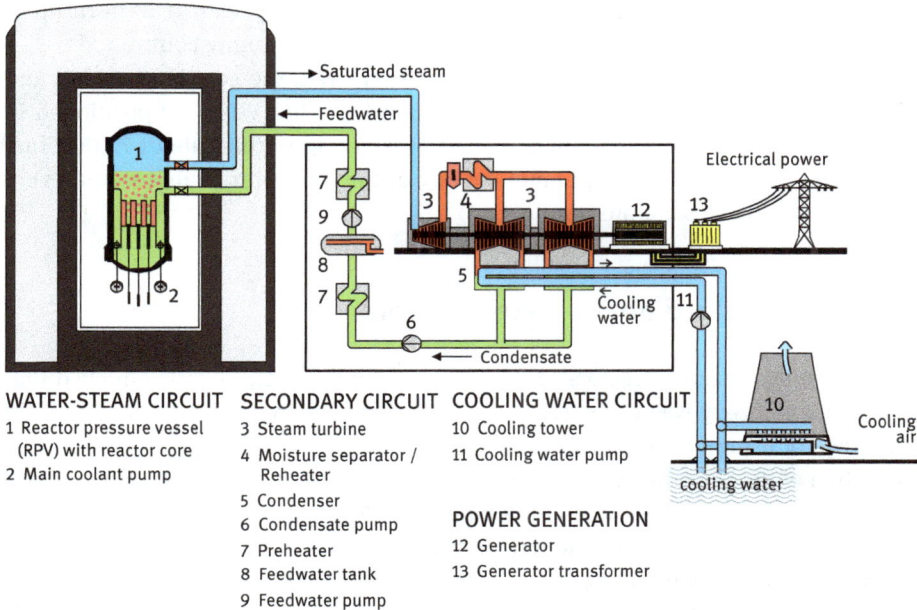

WATER-STEAM CIRCUIT
1 Reactor pressure vessel
  (RPV) with reactor core
2 Main coolant pump

SECONDARY CIRCUIT
3 Steam turbine
4 Moisture separator /
  Reheater
5 Condenser
6 Condensate pump
7 Preheater
8 Feedwater tank
9 Feedwater pump

COOLING WATER CIRCUIT
10 Cooling tower
11 Cooling water pump

POWER GENERATION
12 Generator
13 Generator transformer

Fig. 10.11: RPV of a boiling water reactor.

The steam, as it arrives directly from the reactor, contains radioactive nuclides which lead to turbine contamination and higher γ-radiation at the turbine hall. Most of the γ-radiation is due to the short-lived isotope ^{16}N ($t_{1/2} = 7.13$ s), which is produced by the activation of the coolant H_2O in the reaction $^{16}O(n,\gamma)^{16}N$. Also other activation products of water, such as ^{13}N ($^{16}O(p,\alpha)^{13}N$) and ^{18}F ($^{18}O(p,n)^{18}F$), as well as fission products (noble gases and iodine from uranium contamination of core components) are to some extent volatile and contribute to dose rate. Volatile and dissolved radionuclides in aerosols carried over by steam lead also to radionuclide contamination. The radioactive contamination is not only restricted to the turbine but to all components of the water-steam circuit which are in contact with the coolant.

Reactivity control is performed with control rods and by varying the recirculation water flow through the reactor core by varying the speed of coolant recirculation pumps. By reducing the water flow through the reactor core the void fraction in the core increases, giving worse neutron moderation and reducing quickly fission within the fuel. The change of void fraction between the fuel rods by varying the speed of recirculation pumps enables quick power variation of BWR. The control rods are cruciform blades with neutron absorber rods (e.g., BN, B_4C inside) and are placed between four fuel assemblies, which are arranged in fuel assembly channels. At the beginning of an operation cycle, criticality starts with still some of these rods remaining in core positions. As BWR are operated without any addition of boric acid in water, the decreasing reactivity during operation cycle, due to burn up of fuel, is compensated by a stepwise pulling out of these remaining control rods.

Due to the lower power density achievable in BWR, the core is extended and around 800 fuel assemblies (compared to around 200 in PWRs) are positioned in the core. Each of these has around 90 fuel rods giving in total about 72,000 fuel rods in the core. To achieve optimum cooling of the rods with boiling water, every fuel assembly is in a separate can.

10.5.3 Pressurized heavy-water reactors

Pressurized heavy-water reactors generally use natural or only slightly enriched uranium fuel and heavy water (deuterium oxide, D_2O) for neutron moderation. The main advantage of using heavy water instead of light water for moderation is the reduced absorption of neutrons by the moderator (cf. Tab. 10.7). This allows a sustained chain reaction using nonenriched natural uranium fuel. Most common heavy-water reactors are of the CANDU type, a pressure tube heavy-water-moderated reactor.[22] The CANDU design enables on-line refueling, a necessity to maintain criticality for long-term operation – a unique feature compared to common light-water reactors. Individual channels can be refueled during power operation when new fuel is inserted by one refueling machine while the opposite refueling machine receives the discarded burned up fuel. Figure 10.12 shows the bundle and Fig. 10.13 illustrates the layout for CANDU reactors.

---

22 Those reactors consist of an array of some hundred fixed horizontal pressure tubes installed in a vessel ("calandria") containing D_2O at low temperature (about 70 °C). Each pressure tube is filled with a stack of cylindrical fuel assemblies, each a bundle of 28, 37, or 43 half-meter-long fuel tubes with 10 cm diameter. The generated heat within the pressure tubes is transported in a separated pressurized primary cooling loop by heavy water to steam generators. The gap between the calandria tank and a pressure tube is thermally isolated by calandria tubes, filled with CO_2 to avoid boiling of heavy water within the moderator tank during power operation. The coolant is also D_2O, with about 265 °C inlet temperature, about 310 °C outlet temperature and about 110 bar pressure. The proposed, but finally not realized reactor design of the ACR-1000, a generation III CANDU reactor, switched to light water as coolant. This would have resulted in a negative void reactivity, considerably lower tritium production by neutron activation and reduced costs for the heavy water, but also in the use of more expensive enriched uranium instead of natural uranium.

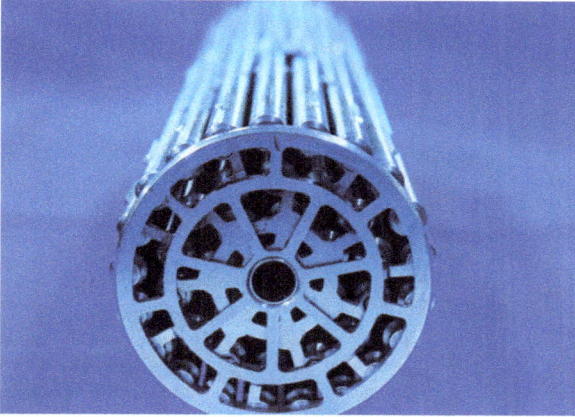

Fig. 10.12: Prototype fuel bundle for CANDU reactors. Photo provided courtesy of Canadian Nuclear Laboratories.

10.6 The fuel cycle

In nature, uranium is found in about 230 minerals, in ore deposits or as trace impurity in soil, rock, and water. In these chemical forms, it is not useable directly for nuclear reactors and processing is needed. The processing comprises uranium mining, chemical conversion, enrichment of the ^{235}U fraction, and the production of the fuel. Because after utilization in the reactor, the burned up fuel can be finally disposed of as waste or recycled for further use, a cycle appears. This "life cycle" of uranium for power plants is often called the nuclear fuel cycle, cf. Fig. 10.14. The cycle that includes recycling of used fuel is called "closed cycle", and the cycle without recycling is called "once-through-cycle".

10.6.1 Mining and refining

As of 2020, the world production of uranium was 47,731 tons worldwide, of which Kazakhstan had a share of 41% (19,477 t), Australia 13% (6203 t), Namibia 11% (5413 t), Canada 8% (3885 t), and Uzbekistan 7% (3500 t), cf. Fig. 10.15. The uranium ore is recovered by mining (open pit or underground mining) or by in-situ techniques.

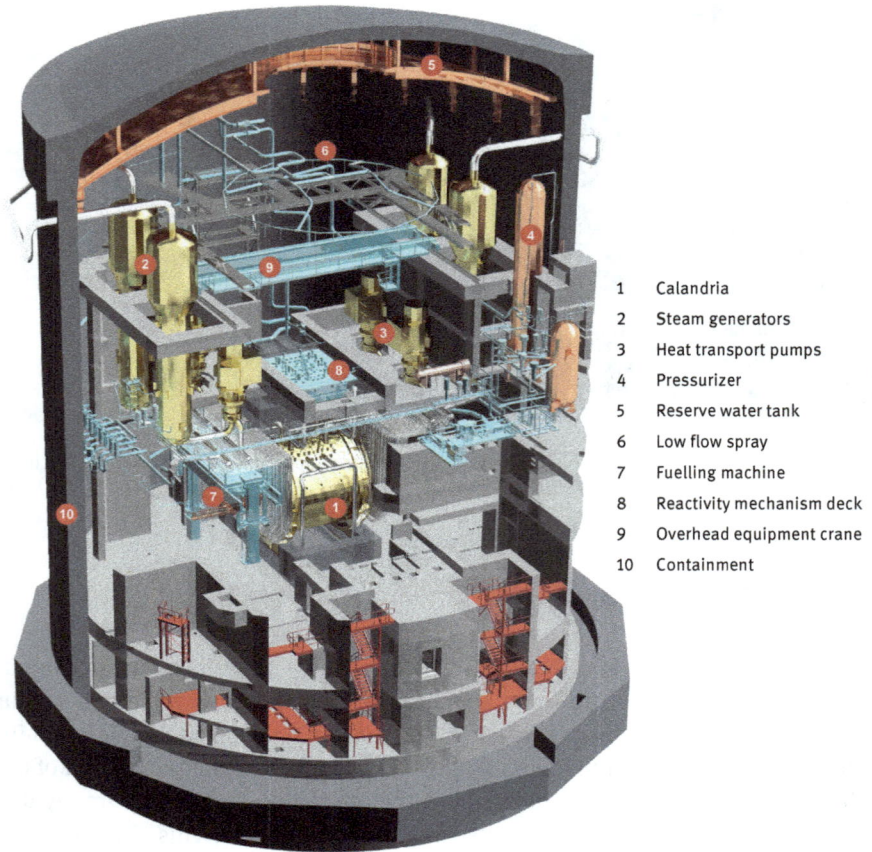

1 Calandria
2 Steam generators
3 Heat transport pumps
4 Pressurizer
5 Reserve water tank
6 Low flow spray
7 Fuelling machine
8 Reactivity mechanism deck
9 Overhead equipment crane
10 Containment

Fig. 10.13: Layout for CANDU reactors. Courtesy of Candu Energy Inc., a Member of the SNC-Lavalin Group.

10.6.1.1 Mining

The uranium ore, obtained in crushed form after milling, is separated from its ore companions with classical procedures such as further milling, flotation, or sorting by the aid of radiation or fluorescence detectors. This concentrates the raw material to about 5–30%. The uranium ore concentrates (UOC) are treated with sulfuric acid (H_2SO_4) or soda (Na_2CO_3) to leach out soluble uranium compounds ($UO_2(SO_4)_2^{2-}$, $UO_2(CO_3)_2^{2-}$, $UO_2(CO_3)_3^{4-}$, and others). These are separated typically by the use of anion exchangers or a solvent extraction process and precipitated with ammonia to form ammonium diuranate (($NH_4)_2U_2O_7$, ADU). After chemical treatment, the precipitate is washed, filtered, and roasted to form a solid, yellow or dark-green concentrate, often called "yellowcake" (cf. Fig. 10.16), with a content of 75–90% U_3O_8 (urania).

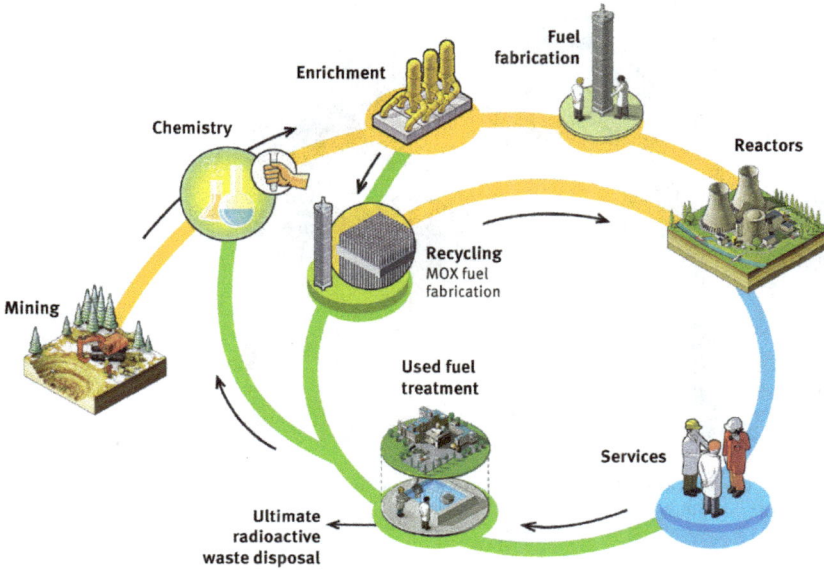

Fig. 10.14: Scheme of the nuclear fuel cycle.

World uranium production in 2020
(in tU)

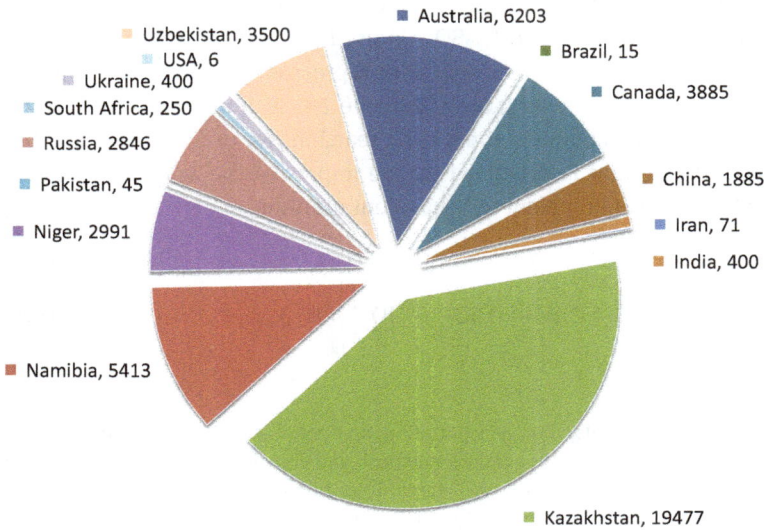

- Uzbekistan, 3500
- USA, 6
- Ukraine, 400
- South Africa, 250
- Russia, 2846
- Pakistan, 45
- Niger, 2991
- Namibia, 5413
- Australia, 6203
- Brazil, 15
- Canada, 3885
- China, 1885
- Iran, 71
- India, 400
- Kazakhstan, 19477

Fig. 10.15: World production of uranium (in tons). Source: Word Nuclear Association, 2021.

Fig. 10.16: Somaïr ore processing plant, Arlit, Niger. Yellowcake on band filter (© AREVA, Jean-Marie Taillat).

10.6.1.2 *In situ* leaching

In situ leaching (ISL, sometimes also *in situ* recovery, ISR) became a considerable alternative to mining. For ISL, porous rock or permeable sand is leached with an acid (H_2SO_4, HNO_3, HCl) or carbonate (Na_2CO_3, $NaHCO_3$, $(NH_4)_2CO_3$, NH_4HCO_3) solution[23] to dissolve uranium from the ore minerals and the solution is brought to the surface by pumping. Most often sulfuric acid (H_2SO_4) is used for ISL. Alkaline (carbonate) leaching is used if for some specific reasons the acid process does not apply, e.g., for a relatively high Ca content in the rock or sand above 2%. Although the chemical reactions during ISL are rather complex, some basic reactions for uranium can be rationalized as follows.[24] Uranium is dissolved by the impact of the acid and possible oxidizing agents (to oxidize tetravalent to hexavalent uranium):

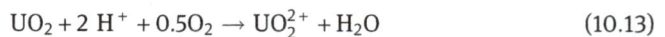

$$UO_3 + 2\,H^+ \rightarrow UO_2^{2+} + H_2O \tag{10.12}$$

$$UO_2 + 2\,H^+ + 0.5O_2 \rightarrow UO_2^{2+} + H_2O \tag{10.13}$$

---

23 The choice of the leaching agents, their concentration, and possible co-agents is made by considering various criteria. These include economic, geological, and chemical considerations such as: Which composition do the rocks and ores have? Which other nonuranium compounds would be leached? How much uranium can be recovered? What is the impact on the environment, e.g., possible groundwater contamination with sulfate or uranium, and how can nature be restored after leaching?
24 This and other more detailed information on the ISL process can be found in the IAEA report "Manual of acid *in situ* leach uranium mining technology" (IAEA-TECDOC-1239, 2001) and other publicly available resources.

$$UO_2 + 2\,Fe^{3+} \rightarrow UO_2^{2+} + 2\,Fe^{2+} \tag{10.14}$$

$$3\,UO_2 + ClO_3^- + 6\,H^+ \rightarrow 3\,UO_2^{2+} + Cl^- + 3\,H_2O \tag{10.15}$$

$$UO_2 + 2\,H_2SO_4 \rightarrow U(SO_4)_2 + 2\,H_2O \tag{10.16}$$

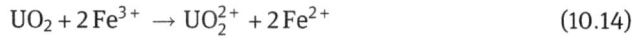

Hexavalent uranium gives with sulfuric acid the sulfates UO_2SO_4, $UO_2(SO_4)_2^{2-}$, and $UO_2(SO_4)_3^{4-}$ which are in equilibrium with each other, with the uranyl trisulfate $UO_2(SO_4)_3^{4-}$ as predominant form. Similar to this, $UO_2(CO_3)_3^{4-}$ is the predominant form in carbonate leaching. Typical oxidants used in carbonate leaching are hydrogen peroxide (H_2O_2) or oxygen (O_2). A typical reaction for alkaline carbonate leaching is

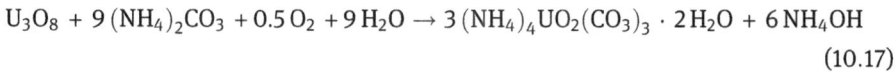

$$U_3O_8 + 9\,(NH_4)_2CO_3 + 0.5\,O_2 + 9\,H_2O \rightarrow 3\,(NH_4)_4 UO_2(CO_3)_3 \cdot 2\,H_2O + 6\,NH_4OH \tag{10.17}$$

Depending on the uranium concentration in the leaching solution, the uranium can then be recovered by ion exchange, solvent extraction or precipitation with processes similar to mining. Also other valuable by-products such as rhenium, molybdenum, vanadium, and rare-earth elements can be recovered.

10.6.2 Chemical conversion

After shipment from the mine to a conversion plant, "yellowcake" is converted chemically into uranium hexafluoride (UF_6), a form suitable for enrichment, or to uranium dioxide (UO_2), if used in reactors working with natural uranium. The conversion involves several steps. In each of these steps, existing impurities are being continuously removed.

In the common "wet process", the yellowcake is dissolved in nitric acid. The aqueous solution containing uranyl nitrate hexahydrate ($UO_2(NO_3)_2 \cdot 6\,H_2O$, UNH) is then subject to a solvent-solvent extraction process with n-tributyl phosphate (TBP) in kerosene or dodecane. In a following step, it is re-extracted by nitric acid from the organic phase. After evaporation of the water phase, the residue is heated in a bed reactor to produce UO_3. It is then reduced to UO_2 by hydrogen following the reaction:

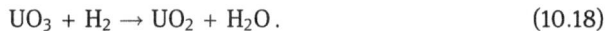

$$UO_3 + H_2 \rightarrow UO_2 + H_2O. \tag{10.18}$$

Uranium dioxide is then treated with hydrofluoric acid (HF) to form uranium tetrafluoride (UF_4), a green salt, which can be used for the production of metallic uranium:

$$UO_2 + 4\,HF \rightarrow UF_4 + 2\,H_2O. \tag{10.19}$$

UF_4 reacts with F_2 or other fluorinating agents to form UF_6:

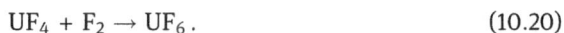

$$UF_4 + F_2 \rightarrow UF_6. \tag{10.20}$$

UF_6 is a white solid at room temperature and sublimates at 56 °C. It is condensed and stored in special steel cylinders.

10.6.3 Isotopic enrichment

The concentration of the fissile isotope ^{235}U of 0.72% in natural uranium is too low to sustain a chain reaction in a light-water reactor. Therefore, the fraction of ^{235}U in the fuel must be increased to about 3–5%; the isotope ^{235}U must be "enriched". The two relevant industrial enrichment processes are the gaseous diffusion process and the centrifuge process. Both start from the chemical species UF_6.

10.6.3.1 Diffusion

The diffusion process uses a series of porous membranes or diaphragms through which both $^{235}UF_6$ and $^{238}UF_6$ can diffuse, cf. Fig. 10.16. The lighter molecule, $^{235}UF_6$, passes the membrane faster so that two gas streams, a ^{235}U-enriched ("product") and a ^{235}U-depleted ("tail") gas stream, remain. The diffusion through the membrane is supported by pressure provided by compressors. The diffusion velocity of the molecule is indirectly proportional to the root of their molar mass. Hence, the maximum separation factor α_{max} for a single separation can be calculated

$$\alpha_{max} = \sqrt{\frac{M_{\,^{238}UF_6}}{M_{\,^{235}UF_6}}} = \sqrt{\frac{352\,g/mol}{349\,g/mol}} = 1.0043. \tag{10.21}$$

Thus, many enrichment steps are necessary to achieve the specified enrichment.[25]

10.6.3.2 Centrifugation

Gas centrifuges (cf. Fig. 10.18) work with cylinders spinning at very high speed into which the UF_6 gas is fed. The strong centrifugal force causes enrichment of the heavier $^{238}UF_6$ molecules towards the inner wall of the cylinder and an enrichment of the lighter $^{235}UF_6$ towards the center, cf. Fig. 10.16. Heating of the lower part and cooling of the upper part of the centrifuge can support the flow inside the centrifuge. This thermal gradient sets up a countercurrent flow inside the centrifuge so that the light $^{235}UF_6$ can be removed axially from the upper end and the heavier $^{238}UF_6$ from

---

25 The industrial use of the diffusion process is currently being phased out because it is an expensive and energy-intensive process, mainly due to the use of many compressors.

the lower end of the centrifuge. The separation factor is about 1.1. Therefore, centrifuges are interconnected in cascades to reach the specified enrichment. Typically, about 1000 centrifuges are interconnected representing one line of stepwise enrichment, cf. Fig. 10.17. Less energy is consumed during enrichment with gas centrifuges by a factor of about 60 compared to diffusion. Again, the tiny difference in mass of the gaseous $^{238}UF_6$ and $^{235}UF_6$ is utilized.

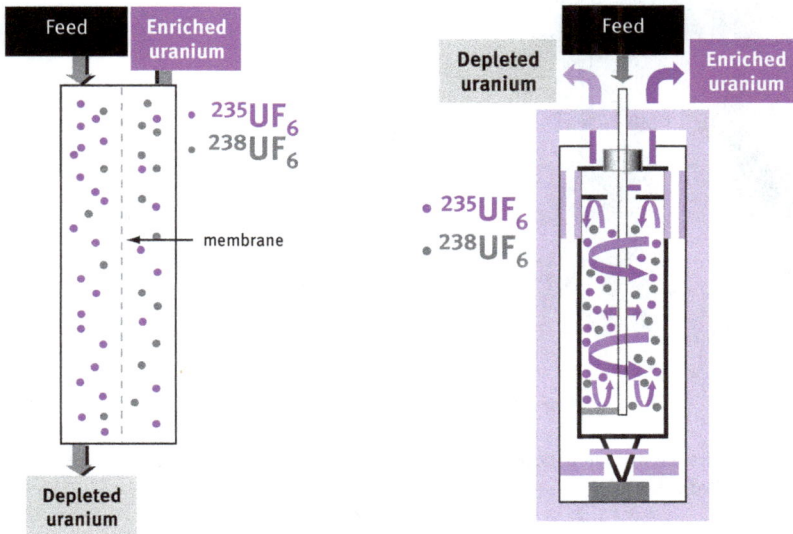

Fig. 10.17: Concepts of uranium-235 isotope enrichment via diffusion (left) and centrifugation (right).

While the enriched uranium hexafluoride is later converted to UO_2 fuel, the depleted uranium (DU) is often stored long-term at the site as UF_6. It may also be converted to uranium trioxide (U_3O_8), recovering the used hydrofluoric acid (HF) for further conversion according to the two reactions

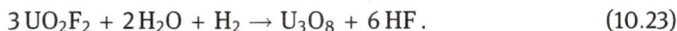

$$UF_6 + 2\,H_2O \rightarrow UO_2F_2 + 4\,HF, \tag{10.22}$$

$$3\,UO_2F_2 + 2\,H_2O + H_2 \rightarrow U_3O_8 + 6\,HF. \tag{10.23}$$

The depleted uranium can be finally stored or used for other applications, e.g., for shielding (due to its high atomic mass and its high density) of highly radioactive material, for fabrication of MOX fuel by blending with fissile reprocessed plutonium, or to produce DU metal for armor-piercing ammunition for the military.

Fig. 10.18: Rotation of first centrifuge cascade. George Besse II site unit 1. December 9, 2009 (© AREVA, Nicolas Petitot).

10.6.4 Fabrication of fuel assemblies

The fuel for light-water reactors is typically UO_2. It is obtained after enrichment by defluorination of UF_6 as a black UO_2 powder either by a "dry conversion" (DC) process according to

$$UF_6 + 2\,H_2O + H_2 \rightarrow UO_2 + 6\,HF \tag{10.24}$$

or by "wet conversion" processes. Two wet processes have been established. In the ADU process, an intermediate product is ammonium diuranate $((NH_4)_2U_2O_7$, ADU), that is obtained after hydrolysis of UF_6 and precipitation with ammonia. UO_2 is then produced by reduction of ADU with hydrogen (H_2)

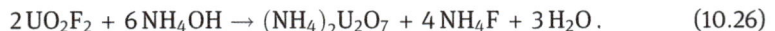

$$UF_6 + 2\,H_2O \rightarrow UO_2F_2 + 4\,HF \tag{10.25}$$

$$2\,UO_2F_2 + 6\,NH_4OH \rightarrow (NH_4)_2U_2O_7 + 4\,NH_4F + 3\,H_2O\,. \tag{10.26}$$

In an alternative process, the AUC process, ammonium uranyl carbonate $((NH_4)_4UO_2$ $(CO_3)_3$, AUC) is formed as intermediate after treatment of UF_6 with water, ammonia, and carbon dioxide. It is then decomposed and reduced to UO_2 in a reaction with a mixture of water, nitrogen, and steam

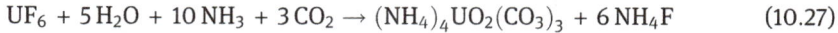

$$UF_6 + 5\,H_2O + 10\,NH_3 + 3\,CO_2 \rightarrow (NH_4)_4UO_2(CO_3)_3 + 6\,NH_4F \qquad (10.27)$$

$$(NH_4)_4UO_2(CO_3)_3 + H_2 \rightarrow UO_2 + 3\,CO_2 + 4\,NH_3 + 3\,H_2O. \qquad (10.28)$$

The UO_2 powder is pressed to small, green pellets which are then sintered in a high temperature furnace at about 1780 °C in a hydrogen atmosphere to yield dense and hard "ceramic" UO_2 pellets, cf. Fig. 10.19. Ceramic UO_2 has some advantageous properties (e.g., a high melting point of 2865 °C) and is resistant to water, hydrogen, or radiation. Because of its limited thermal conductivity, the diameter of the UO_2 pellets is limited. The pellets are precisely ground to a specified diameter (about 1 cm) and stacked in a fuel cladding tube, which is then pressurized and sealed. These fuel rods are then arranged in a fuel assembly.

Fig. 10.19: Raw pellets before entering sintering furnace, Melox MOX fuel fabrication plant, Bagnols-sur-Cèze, France (© AREVA, Patrick Lefevre).

10.6.5 Treatment and recycling of spent fuel

After burning up the fuel in the reactor, the fuel assemblies, still emitting radiation and heat released from fission products, are unloaded from the reactor and put to an interim underwater storage pond nearby the reactor. The water shields the radiation and cools the fuel assemblies. After months or years in the water pool, the assemblies are transferred to a dry storage. The used fuel can then be prepared for storage in a final waste repository or utilized for reprocessing. Aim of the reprocessing of spent fuel is the recovery of fissile and fertile isotopes for fresh fuel and the reduction of highly active waste volume. This leads to a more economic use of fuel, less mining and enrichment, a significant reduction of high-level active waste (but increase in low-level waste), the recovery and consumption of fissile plutonium in MOX fuel, and, hence, a better nonproliferation of Pu for military weapons.[26]

Used fuel still consists of about 95% uranium, depending on the initial enrichment and the time operated in the reactor, and up to about 1% plutonium. Thereof, about 0.4–1% is still ^{235}U, about 0.5% is ^{236}U, a strong neutron absorber, and a small amount is ^{232}U, which is a strong gamma emitter. Spent fuel also contains about 3% stable fission products, about 0.1% of minor actinides, and low amounts of long-lived fission products (e.g., Cs, Sr, I, Tc).[27] The separation of the fissile material from the spent fuel is a challenging technical and chemical task. It involves mechanical and chemical processing steps.

10.6.5.1 Head-end process

A typical process for the treatment of fuel assemblies is the aqueous "head-end process", cf. Fig. 10.20. The fuel assembly is sheared into short segments or chopped and de-cladded.[28] Already during shearing, some of the gases contained in the fuel are released into off-gas treatment systems. The sheared fuel segments are dissolved in hot nitric acid. The dissolution of the uranium oxide fuel can be expressed by the formation of uranyl nitrate and nitrous oxides (NO_x)

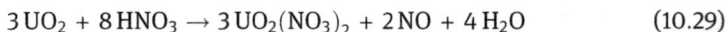

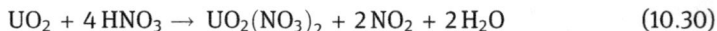

$$3\,UO_2 + 8\,HNO_3 \rightarrow 3\,UO_2(NO_3)_2 + 2\,NO + 4\,H_2O \tag{10.29}$$

$$UO_2 + 4\,HNO_3 \rightarrow UO_2(NO_3)_2 + 2\,NO_2 + 2\,H_2O \tag{10.30}$$

As a rule of thumb, eq. (10.29), with the formation of NO, is applicable when the HNO_3 concentration is below 10 mol/L, the latter equation for concentrations above 10 mol/L. By adding oxygen (O_2), the highly corrosive nitrous oxides NO_x can recombine to HNO_3 and the dissolution of oxide fuel can be written as eq. (10.31). The dissolution is typically operated batchwise, although several designs for continuous processes have been developed and deployed

$$2\,UO_2 + 4\,HNO_3 + O_2 \rightarrow 2\,UO_2(NO_3)_2 + 2\,NO_2 \tag{10.31}$$

During dissolution, a number of volatile isotopes and compounds are released into "off-gas" systems, e.g., ^3H, $^{14}CO_2$, ^{14}CO, ^{129}I, ^{85}Kr. The off-gas treatment involves

---

26 On the other hand, it should be mentioned that – historically – reprocessing was already developed in the 1940s for military purposes – to recover the formed fissile plutonium for weapons.
27 These long-lived fission products mostly contribute to the heat production of spent fuel, especially Sr and Cs isotopes.
28 After moving a spent fuel assembly from the storage site, it can then be disassembled to a certain degree, allowing further treatment. The end sections of LWR fuel assemblies are typically removed and disposed as metal waste. The cladding fragments are not completely dissolved in nitric acids and are removed, together with other undissolved or precipitated solids, and disposed as waste.

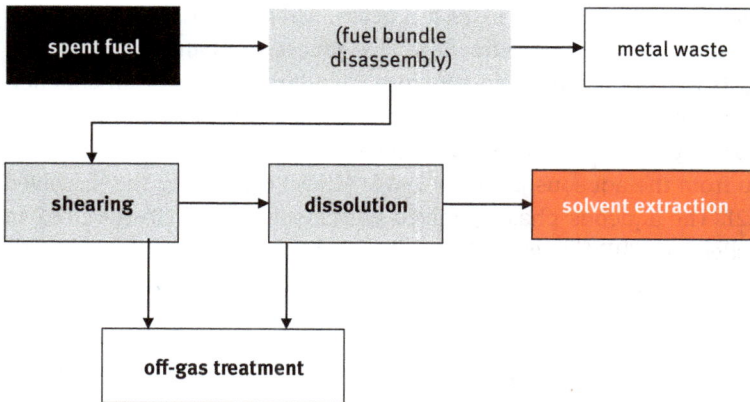

Fig. 10.20: Block diagram for the aqueous head-end process.

several physical or chemical processes to retain activity to match regulatory requirements for the limited release of activity to the environment. Tritium, formed by ternary fission[29] in the fuel, may be removed from the off-gas by the use of molecular sieves, desiccants, or anhydrous $CaSO_4$. Iodine, mostly ^{129}I, may be removed by various adsorbents or by wet scrubbing. Typical adsorbents are porous materials (e.g., zeolite, alumina, silica gel), sometimes loaded with metals (e.g., Ag) or metal compounds (e.g., $AgNO_3$). Wet scrubbing by caustic substances (e.g., NaOH) retains much of the iodine activity and also some other off-gas components (e.g., gaseous Ru compounds, ^{14}C, or NO_x). While the decontamination factors are often high enough for an effective depletion of those components, a disadvantage of caustic scrubbing is the partial decomposition of the solution by some reactants. The chemically inert noble-gas isotopes, Kr and Xe isotopes, can only be recovered by physical separation. Primary processes are cryogenic distillation, adsorption in fluorocarbons (due to their high solubility), and adsorption on adsorbents like molecular sieves or charcoal. But more often, those noble gases are released. ^{14}C, mainly occurring as $^{14}CO_2$, is removed from the off-gas streams by, e.g., wet scrubbing and molecular sieves or other adsorbents.

10.6.5.2 Solvent extraction

Several processes have been developed to recover material from spent fuel, depending on the nature of the fuel and fuel assemblies and on the purpose of the reprocessing, based on solvent extraction or ion exchange.

---

29 Cf. Chapter 10 of Vol. I.

The most common process is the PUREX process, where PUREX is an acronym for plutonium uranium recovery by extraction. This process was already developed in the 1940s and 1950s and has become a mature, standard process in all commercial reprocessing plants. It is used for the recovery of plutonium and uranium from spent fuel. The basic principle of the PUREX process is the solvent–solvent extraction of U and Pu from the aqueous phase (≈ 3–$4\,M\,HNO_3$) containing the dissolved fuel, cf. Fig. 10.21. The aqueous phase is mixed with a solution of 20–30% TBP in kerosene or n-dodecane. TBP forms soluble and neutral complexes with uranium^{VI} nitrate and plutonium^{IV} nitrate (and also other metal ions in the +4 and +6 oxidation states), cf. Chapter 7, Section 7.3 on radiochemical separations:

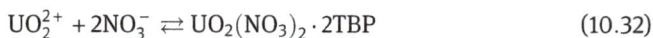

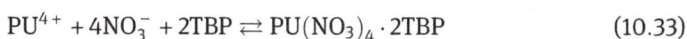

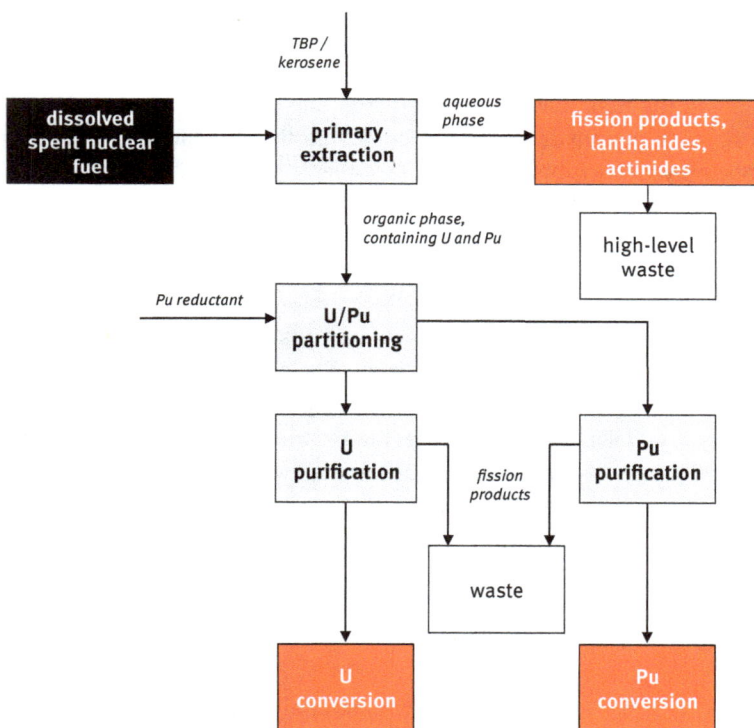

$$UO_2^{2+} + 2NO_3^- \rightleftharpoons UO_2(NO_3)_2 \cdot 2TBP \tag{10.32}$$

$$PU^{4+} + 4NO_3^- + 2TBP \rightleftharpoons PU(NO_3)_4 \cdot 2TBP \tag{10.33}$$

Fig. 10.21: Block diagram of the PUREX process.

Under these highly acidic conditions, uranium and plutonium go to the organic phase while most of the fission products and metals remain in the aqueous phase. The distribution coefficients determining the distribution between the organic and aqueous phase of U, Pu, and other companions in the fuel depend on the HNO_3 concentration and are best for U^{VI} and Pu^{IV} at high HNO_3 concentrations, cf. Tab. 10.9.

Tab. 10.9: Distribution coefficients K_D of uranium, plutonium, and various fission products without Zr, Ru, and Ce with 30% TBP and 1 M HNO_3 at 25 °C.[30]

Species	UO_2^{2+}	Pu^{4+}	PuO_2^{2+}	HNO_3	Zr	Ce^{3+}	Ru	Pu^{3+}	Nb	Rare earths	Combined beta emitters
K_D	8.1	1.55	0.62	0.07	0.02	0.01	0.01	0.008	0.005	0.002	<0.0001

Pu^{4+} and Pu^{6+} are then reduced to Pu^{3+} by a reducing agent such as ferrous sulfamate ($Fe(NH_2SO_3)_2$) in nitric acid or U^{IV} nitrate in nitric acid and back extracted from the organic phase to an aqueous phase ("plutonium stripping"), cf. Tab. 10.8 and eq. (10.34):

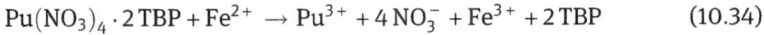

$$Pu(NO_3)_4 \cdot 2\,TBP + Fe^{2+} \rightarrow Pu^{3+} + 4\,NO_3^- + Fe^{3+} + 2\,TBP \qquad (10.34)$$

UO_2^{2+} is unaffected by this process and remains in the organic phase. Since Pu^{3+} is unstable in solutions of nitric acid due to the existence of nitrous acid HNO_2 (eqs. (10.35), (10.36)), it can be stabilized by adding hydrazine (N_2H_4) as hydrazine nitrate ($N_2H_5NO_3$) as nitrous acid scavenger, eq. (10.37)

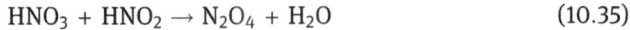

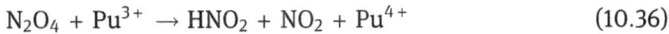

$$HNO_3 + HNO_2 \rightarrow N_2O_4 + H_2O \qquad (10.35)$$

$$N_2O_4 + Pu^{3+} \rightarrow HNO_2 + NO_2 + Pu^{4+} \qquad (10.36)$$

$$N_2H_5NO_3 + HNO_2 \rightarrow HN_3 + 2\,H_2O + HNO_3 \qquad (10.37)$$

Uranium is stripped from the organic phase by a 0.01 M HNO_3 aqueous phase. The obtained product and waste fractions are further purified by subsequent back-extraction and cleaning steps.

The extraction efficiency is increased by multiple stages of each single step. On a technical scale, the PUREX process is realized in continuously, countercurrent operated mixer settlers, pulsed columns, or centrifugal contactors.[31] The uranyl nitrate

---

30 According to Danesi, Solvent Extraction in the Nuclear Industry, 1984.

31 The continuous operation allows a high throughput and the multiple stages a high purity of the products. The recycling of the solvent is possible, thus minimizing waste, but the inherent disadvantages of the PUREX process are a certain solvent degradation due to radiolysis and hydrolysis, the necessity of large-volume tankage as this process is a dilute process and the production of high-level and low-level active waste which needs to be stored, and the release of radioactive gases.

produced is later converted to UO_3 by denitration at higher temperatures, and plutonium nitrate is precipitated and calcined to PuO_2.

Several variations of the PUREX process with other extractants and solvents have been developed with the aim to further purify the products or to separate waste fractions, cf. Fig. 10.22. For example, in the TRUEX (transuranic extraction) process a second extracting agent – CMPO (octyl(phenyl)-*N*, *N*-dibutyl carbamoylmethyl phosphine oxide) – is added to TBP. With this agent, the transuranium elements americium (Am) and curium (Cm) can be removed from waste, thus reducing the alpha activity of waste. In the DIAMEX (diamide extraction) process, malondiamide is used to remove americium, curium, and lanthanides from highly active waste. Americium and curium may then be separated by a SANEX (selective actinide extraction) process from the lanthanides. The separated actinides can then be reused.

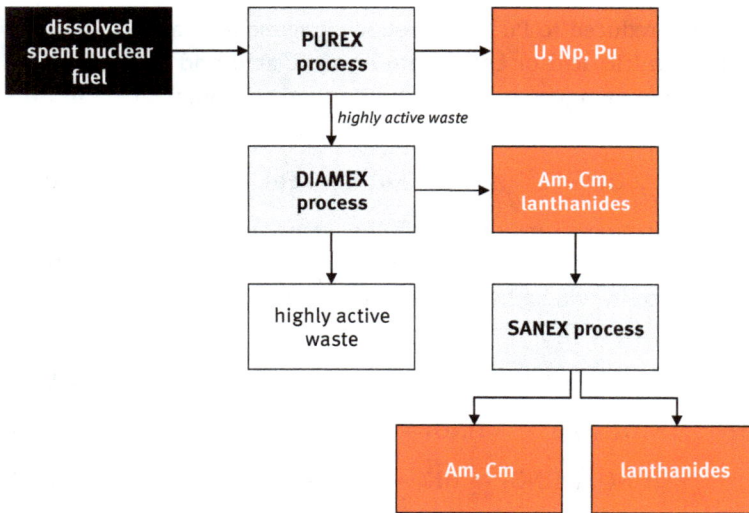

Fig. 10.22: Block diagram of partitioning processes for the separation of minor actinides from spent nuclear fuel.

Such a sequence of separation processes is often called partitioning. By "partitioning", uranium, plutonium, neptunium, lanthanides, and minor actinides can be separated from the waste. This is favorable for final waste disposal as minor actinides[32] and plutonium determine, due to their long half-life times, the radiotoxicity of spent fuel in final

---

32 Referring to uranium and plutonium as "major actinides" in (spent) fuel, the elements neptunium, americium, curium (and berkelium, californium, einsteinium, and fermium) are often called "minor actinides". Most relevant isotopes are ^{237}Np, ^{241}Am, ^{242m}Am, ^{243}Am, and $^{242-246}$Cm, formed in the reactor during neutron irradiation.

disposal sites for geological periods, cf. Fig. 10.23. Partitioning reduces the amount of long-lived and highly active (often called "toxic") radionuclides in the waste.

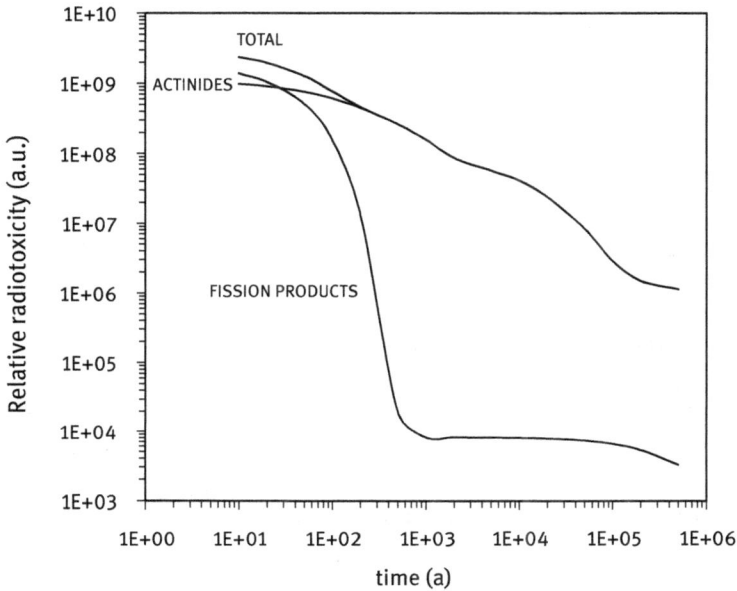

Fig. 10.23: Radiotoxicity of spent nuclear fuel and contribution of fission products and actinides to the radiotoxicity.

10.6.6 Management of waste

The use of nuclear technology for civil and military applications produces a steadily increasing amount of radioactive waste, mainly spent nuclear fuel. It is still a widely unresolved issue to manage the large amounts of highly radioactive waste produced in the past decades.[33]

10.6.6.1 Categories of waste

The storage and disposal policy depends on the classification of the radioactive waste. The classification into different levels is based on the concentration and

---

33 It is consensus that the generations which produce the radioactive waste have the responsibility to develop safe and economically and ecologically acceptable solutions for the long-term management of waste and its isolation from biosphere by multiple, natural, or artificial barriers.

half-life of radionuclides and the generation of heat from their radioactive transformations. A common classification (cf. Fig. 10.24) divides waste into "exempted waste"
(EW), "very low-level waste" (VLLW), "low-level waste" (LLW), "intermediate-level
waste" (ILW), and "high-level waste" (HLW).[34] Nuclear power plants produce about
200,000 m^3 low- and intermediate-level waste and about 10,000 m^3 high-level waste
per year. Nearly 400,000 m^3 of high-level waste have been produced by 2016.[35]

10.6.6.2 Storage of waste

So far, solutions for the storage of short-lived and low-level waste are available, but
the disposal of long-lived and intermediate or high-level waste is still at a conceptual and demonstration stage. First final repositories for long-lived waste have been
selected in some countries and are anticipated to put into operation in the 2020s.
Very short-lived waste (VSLW) with very short half-lives up to about some months
(typically radionuclides used for research or medical applications) can be stored for
transformation and then released as nonradioactive waste. Its storage site is usually
where the activity is applied. Very low-level waste (VLLW), i.e., waste with a low
activity or activity concentration close to naturally occurring radioactivity, can be
stored in landfill facilities, also with other hazardous waste.

Low- and intermediate-level waste (LILW) is produced, e.g., during routine operation and decommissioning of NPP. It is produced by the contamination of various materials by fission or activation products in the reactor. Main parts of the
waste are operational wastes (e.g., filters and ion-exchange resins collecting contaminants from the coolant) and components containing activated solids (e.g., ^{60}Co
and ^{63}Ni in stainless steel components). That waste can be solid or liquid. It needs
to be treated to reduce the volume of the active waste and to be conditioned to im-
mobilize it for disposal.[36] Depending on the half-lives of the radionuclides, the disposal of treated and conditioned LILW requires different strategies, which have
been established during the last decades. Short-lived LILW, with half-lives up to

---

34 This classification is a practical one – it does not say anything about the chemical and physical
nature of the waste, i.e., the kind of radionuclides, physical and chemical properties of the waste.
Exact definitions of these waste levels vary from country to country and are therefore not given
here.
35 This sounds impressive, but for comparison, one 1000 MW$_{el}$ coal plant produces about 300,000
tons of ash per year, containing not least radionuclides and metals, mainly from natural decay
chains. While these end up in the atmosphere or in landfill sites, the radioactive waste from nuclear
technology application is subject to much more stringent and tight regulations, of course.
36 Liquid waste treatment is typically done by evaporation, precipitation, ion exchange, or solid-
phase extraction. Solid waste may be directly conditioned or, if it combustible, combustioned and
compacted, if possible. Immobilization may be done by embedding or pouring waste into materials
such as concrete or bitumen.

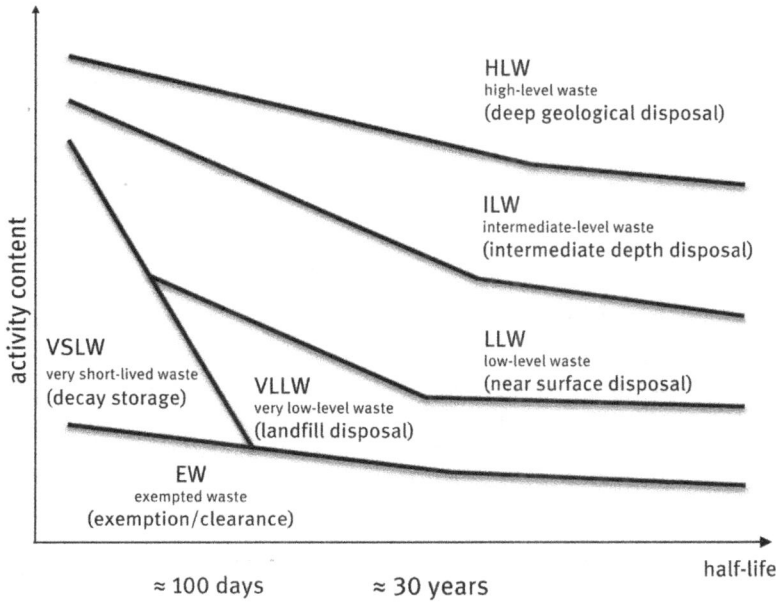

Fig. 10.24: General classification scheme for radioactive waste and proposed storage options, according to IAEA.[37]

about 30 years, is contained and isolated in near surface repositories with engineered barriers and surface structures or in geologic repositories at a depth of between a few ten and a few hundred meters for isolation for up to about 300 years. Long-lived LILW, with half-lives above, requires a much stronger isolation from biosphere typically provided by geologic repositories.

HLW is typically spent fuel or waste from reprocessing. It contains much higher concentrations (about 10^{16}–10^{18} Bq/m^3 and/or with thermal power >2 kW/m^3) of short- and long-lived radionuclides with half-lives of up to several million years and generates remarkable amounts[38] of heat for centuries. At first, it is stored in interim storage facilities for decades to cool down. The typical activity concentration is then on a level of 10^{16} Bq/m^3. HLW has to be contained and isolated to a much higher degree than ILW. Deep geological repositories are considered, equipped with engineered and

---

37 Classification of Radioactive Waste, General Safety Guide, IAEA, 2009.
38 The transformation heat is mainly due to the transformation of a number of fission products for the first years after removing the fuel from the reactor, e.g., 200 kW/t U after 1 day, 100 kW/t U after 1 week and 10 kW/t U after about 1 year. After 20 years, the transformation (1 kW/t U) heat is mainly attributable to the long-lived fission products ^{90}Sr, ^{137}Cs, ^{137m}Ba, and the actinide ^{241}Pu. After 1000 years (60 W/t U), it is mainly due to ^{239}Pu, ^{240}Pu, and ^{241}Am and after 20,000 years (10 W/t U) mainly due to ^{239}Pu and ^{240}Pu.

natural barriers. Such geological repositories have got to be stable geological formations several hundred meters below the surface. Intensive research on such repositories
has been performed. Tests in underground research laboratories and test repositories
demonstrated that this approach could hold for salt, plastic clay, clay stone, and crystalline rock (e.g., granite) formations.[39]

The high-level waste, produced during reprocessing, is composed mainly of fission products and minor actinides, cf. Fig. 10.23. For long-term storage, it needs to
be brought to an insoluble and solid waste form that will be stable for many thousand years. This is accomplished by vitrification. The waste, existing as metal nitrates, is calcined to evaporate water and denitrate the nitrate, and fused with
borosilicate, silicate, or phosphate glass at about 1100 °C to form a new glass in
which the waste is chemically bound in a glass matrix. The molten glass mixture
with about 25% waste content is poured into steel cylinders and cooled down for
many years. The waste products are immobilized in the water-resistant glass matrix
allowing safe storage for many thousands of years. Without reprocessing, high-
level waste from NPP operation can also be directly disposed in a final repository.

10.6.6.3 Transmutation

Over the last few years, research on the conversion of long-lived actinides to short-
lived or stable isotopes is being conducted. This "transmutation" is based on the
reaction of fast neutrons (e.g., from accelerators, spallation sources, or reactors)
with long-lived actinides to produce shorter-lived fission products. It also demands
a proper separation ("partitioning") of actinides before their transmutation and is
also therefore technically challenging. So far, it is not clear when this method can
be applied on a technical scale.

10.7 Radionuclide generation in nuclear reactors

During operation of nuclear reactors, large quantities of different radionuclides are
formed. This extensive generation of radionuclides cannot be avoided during the
operation of nuclear power plants. The radionuclides produced can be divided into

---

39 The detailed concepts for the use of repositories depend on the specific properties of the hosting
rock. Properties being considered are the viscoplasticity, the possibility of water entry to the repository by leakage or diffusion processes, the transport of radionuclides through the rock, and the chemical environment in the repository which can have an impact on the stability of the waste packages,
e.g., by corrosion processes or leaching out of waste by entering water. To date, good progress in the
development of repositories for HLW and spent fuel has been achieved in Finland, Sweden, and
France. Yet, many countries are still at the initial stage of defining their disposal strategies.

fission and activation products whereby fission products are the main part of radionuclides formed in nuclear reactors.

10.7.1 Fission products

A lot of radionuclides with half-lives from split seconds to 10^{24} years are formed by
fission of heavy atomic nuclei during irradiation of nuclear fuel. Most of them remain trapped in the crystal matrix of the fuel. Fission products cover a large area of
chemical elements. Figure 10.25 gives a distribution of fragments of different isobars for the fission of ^{235}U and ^{239}Pu induced by thermal neutrons. The isobaric fission yield of each mass number can cover a lot of different isobaric elements,
typically beta-emitting radionuclides with short half-lives. In spent fuel, only long-
lived radionuclides remain, which constitute part of the radioactive inventory that
has to be disposed of safely. Important information regarding the condition of the
fuel elements can be deduced during operation from the measured activity concentrations of the individual nuclides in reactor coolant and off-gas if the underlying
mechanisms and behavior are understood.

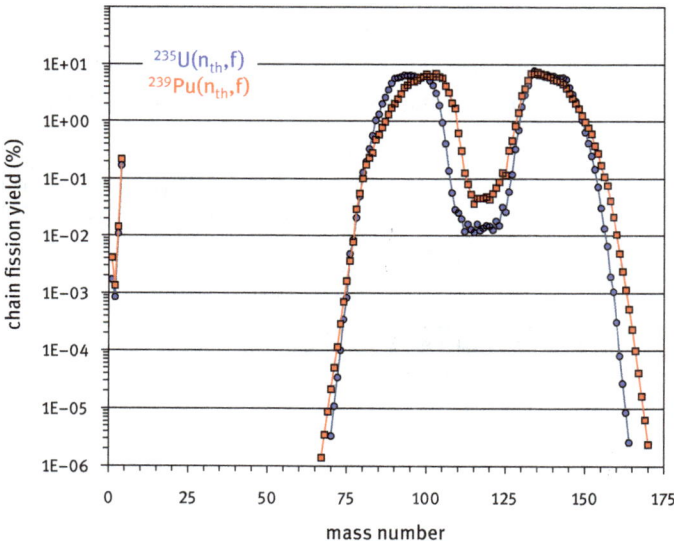

Fig. 10.25: Mass distribution of primary fission fragments for fission of ^{235}U and ^{239}Pu induced by
thermal neutrons (see also Koning et al. 2006).

10.7.2 Activation products

Activation reactions can be divided in different groups. A common classification is based on the components activated by neutrons:
- activation of fuel
- activation of water, its additions, and impurities
- activation of primary loop materials

10.7.2.1 Activation of fuel

Fuel can not only be fissioned, it can also undergo neutron capture reactions, mainly (n,γ) reactions. Substantial amounts of transuranium elements are produced in the nuclear fuel by a complex series of neutron captures and β-transformations. Most of these transuranium nuclides are α-emitters and their radiotoxicity is therefore significantly higher than the radiotoxicity of β-emitting fission products. Figure 10.26 shows schematically the most important of these radionuclides and their formation reactions. In reality, this structure is much more complex because of a multitude of side and competing reactions. In general, the concentrations of the transuranium nuclides decrease rapidly with increasing atomic numbers. This, however, does not say anything about the level of the total activity concentrations. For example, ^{242}Cm, though at low elemental concentration but because of its (relative) short half-life of 163 days, is frequently one of the most intense alpha emitters in the irradiated nuclear fuel.

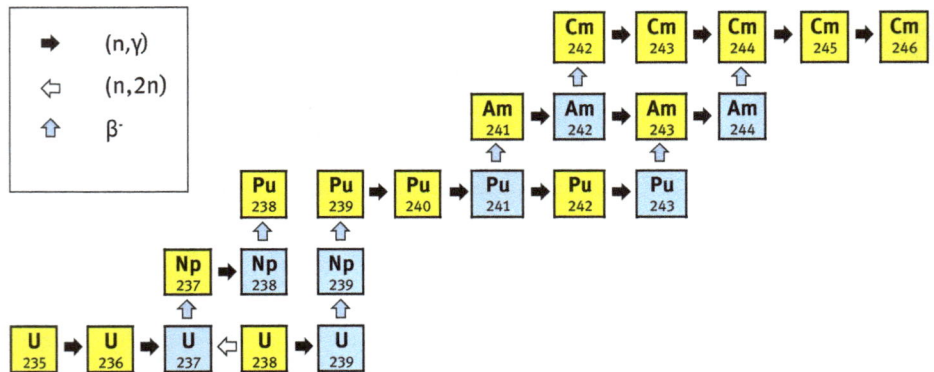

Fig. 10.26: Formation of transuranium nuclides in UO$_2$ fuel.

The inventory of transuranium nuclides in nuclear fuel and its further buildup during irradiation depends on the type of fuel used. There are considerable differences between UO$_2$ fuel and MOX fuel. Due to the content of plutonium nuclides in MOX fuel, the inventory of alpha emitters is considerably higher from the beginning of burn-up in comparison to UO_2 fuel. With increasing burn-up in both types of fuel the transuranium inventory increases but the alpha inventory in MOX fuel even at high burn-up remains significantly above the alpha inventory of UO_2 fuel, cf. Fig. 10.27. The activation of already formed fission products is one possibility.[40]

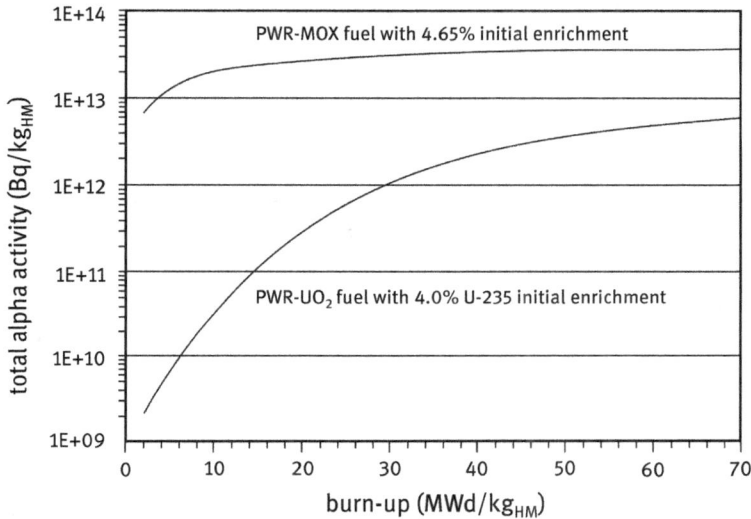

Fig. 10.27: Total alpha activity of transuranium nuclides in irradiated MOX and UO_2 fuel in a PWR.

10.7.2.2 Activation products of water, its additives, and impurities

The reactor coolant in light-water reactors, H_2O, is exposed to intensive neutron radiation while passing through the reactor core. The neutrons are not only moderated, they may also initiate nuclear reactions with the atoms of the water molecules. By elastic scattering of neutrons with hydrogen atoms of the water, i.e., $H(n,n')p$, a proton flow of similar magnitude is created. Thus, proton-induced nuclear reactions may also occur in water.

The boron added as boric acid (H_3BO_3) to the coolant in PWRs for a long-term control of the reactivity is also a reactant to neutrons. Neutron capture leads to the formation of ^7Li via $^{10}B(n,\alpha)^7Li$. Activation of boron also produces very energy-rich tritium, mainly due to the reactions $^{10}B(n,2\alpha)^3H$ and $^{10}B(n,\alpha)^7Li(n,n+\alpha)^3H$ that can

---

40 Such fission products may also act as neutron absorbers. Therefore, with increasing burn-up of fuel during reactor cycle, the amount of reactivity controlling neutron absorbers (boron concentration in PWR or number of inserted control rods) has to be reduced to keep the power level constant.

further trigger nuclear reactions. The latter sequence also shows a further path for the tritium formation, because isotopic pure ^{7}LiOH is often used to adjust the pH in the coolant and reduce the corrosion rate in the primary circuit of a PWR. Further details of tritium formation are discussed in a separate chapter.

Important nuclear reactions taking place are listed in Tab. 10.10 indicating prominent activation product radionuclides. (Apart from these nuclides, there are also a number of different nuclides formed from impurities such as ^{7}Be, ^{72}Ga, ^{123m}Te, and ^{187}W.)

Tab. 10.10: Activation products in the water circuit of nuclear reactors and their source.

Formation reaction	$t\frac{1}{2}$ of the product	Source
^{16}O(n,p)^{16}N	7.13 s	Oxygen of the H_2O molecule
^{16}O(p,α)^{13}N	9.97 min	
^{18}O(p,n)^{18}F	109.8 min	
^{16}O(t,n)^{18}F	109.8 min	
^{17}O(n,α)^{14}C	5730 a	
^{14}N(n,p)^{14}C	5730 a	Dissolved nitrogen, ammonia
^{2}H(n,γ)^{3}H	12.33 a	Deuterium in light water or coolant in heavy-water reactor
^{10}B(n,2α)^{3}H	12.33 a	Boric acid for reactivity control in PWR
^{10}B(n,α)^{7}Li	stable	
^{7}Li(n,nα)^{3}H	12.33 a	Boric acid, Li as alkalizing agent in PWR
^{23}Na(n,γ)^{24}Na	14.96 h	Na traces in reactor coolant
^{40}Ar(n,γ)^{41}Ar	109.3 min	Ar traces in reactor coolant

^{16}N: ^{16}N is of decisive importance for the dose rates during reactor operation because of its very high-energy gamma emission (>6 MeV). It has a short half-life of 7.1 s and transforms immediately after a reactor shutdown. ^{16}N measuring points in the main steam system are part of the reactor protection system to detect reactor-sided steam generator leakages. In BWR's, the turbine building is part of the controlled area in which ^{16}N dominates the dose rate during power operation.

^{18}F: The ^{18}F activity concentrations are also largely predetermined by the design of the reactor core and the reactor recirculation system. Both ^{13}N ($t\frac{1}{2}$ = 9.96 min) and ^{18}F ($t\frac{1}{2}$ = 109.7 min) are β^+-emitters emitting annihilation photons at 511 keV energy. The very high concentration of both nuclides in the reactor coolant (order of magnitude 10^6 to 10^7 Bq/kg) may considerably interfere with the gamma measurement, in particular in the presence of a defect-free core with a low fission product level in the coolant.

^{24}Na: Responsible for the formation of ^{24}Na in the coolant are primarily faint Na traces in injection agents such as boric acid, LiOH, or KOH and also in the make-up water (Na slip in make-up water). Another possible source could also be the resins used for coolant purification.

^{41}Ar: Trace amounts of argon in the primary coolant are unavoidable due to its good solubility in water and its remarkable content of 1% in air. Therefore, ^{41}Ar is produced by neutron activation at a relative high cross section of ^{40}Ar for thermal neutrons. Elevated levels of ^{41}Ar may be considered as an indicator for air ingress into the reactor loop.

10.7.2.3 Activation products of primary loop materials

There are essentially two mechanisms responsible for the formation of activation products from constituents of the primary loop: Corrosion products formed outside of the neutron field (e.g., steam generator) are transported into the neutron field with the reactor coolant and mainly deposited and activated on the fuel assemblies. They are carried away again by the reactor coolant after a certain residence time and redeposited on the primary loop surfaces. Activated corrosion products from highly activated materials (e.g., fuel element spacers, top or bottom of fuel assemblies, and core internals) located in the neutron field are released into the reactor coolant and also deposited on the surfaces of the primary system.

Activation products formed in the neutron field are transported by the reactor coolant along the primary loop, cf. Fig. 10.28. The kinetics of their incorporation into the surface layers under the prevailing water chemistry conditions is decisive for their deposition on the loop surfaces and thus for the contamination buildup in the system. Apart from this, mechanical deposits at irregular flow path locations play a certain role. The activity inventory of the activation products is determined by metal release rates of the primary system materials and also by the corrosion product exposure time in the neutron field, whereby the water chemistry (e.g., Eh- and pH-control, alkalizing agent) is of substantial influence. The lowest metal release rates are caused by alkalization of the coolant when maintaining the required specification values, and transport of activation products from higher to lower temperatures and vice versa are reduced to a minimum. The most important activation products, their half-lives, the nuclear reactions leading to their formation, and also the sources are listed in Tab. 10.11.

Regarding the long-term dose rates of contaminated plant components, ^{60}Co is of fundamental importance. But ^{58}Co (from nickel activation) and ^{124}Sb can also substantially contribute to the dose rate. Limiting the level of Co impurities in steel alloys and the substitution of stellite (Co alloy) of primary loop components are demands to reduce outage dose rate.

REACTOR CORE   **PRIMARY CIRCUIT**

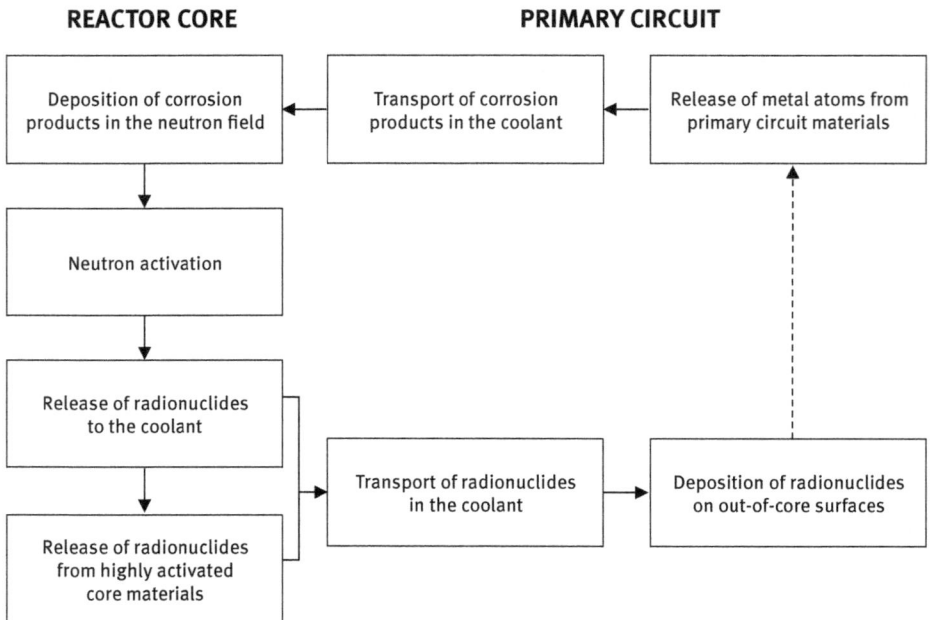

Fig. 10.28: Formation and behavior of radioactive corrosion products in the coolant of a PWR.

Tab. 10.11: Activation products in the primary circuit of a PWR and their sources.

Nuclear reaction	$t\frac{1}{2}$ of the product	Source
$^{54}Fe(n,\gamma)^{55}Fe$	2.73 a	Incoloy 800, stainless steels
$^{58}Fe(n,\gamma)^{59}Fe$	44.5 d	
$^{54}Fe(n,p)^{54}Mn$	312.3 d	
$^{56}Fe(n,p)^{56}Mn$	2.58 h	
$^{50}Cr(n,\gamma)^{51}Cr$	27.7 d	
$^{62}Ni(n,\gamma)^{63}Ni$	100.1 a	Incoloy 800, stainless steels, stellite, Co impurities in
$^{64}Ni(n,\gamma)^{65}Ni$	2.52 h	incoloy and stainless steels
$^{58}Ni(n,p)^{58}Co$	70.86 d	
$^{59}Co(n,\gamma)^{60}Co$	5.27 a	
$^{54}Zn(n,\gamma)^{65}Zn$	244.26 d	Zn addition
$^{94}Zr(n,\gamma)^{95}Zr \xrightarrow{\beta^-} {}^{95}Nb$	34.98 d	Zircaloy cladding, fission products
$^{96}Zr(n,\gamma)^{97}Zr \xrightarrow{\beta^-} {}^{97}Nb$	72.1 min	
$^{98}Mo(n,\gamma)^{99}Mo \xrightarrow{\beta^-} {}^{99}Tc$	$2.1 \cdot 10^5$ a	Fuel element spacer, fission products
$^{109}Ag(n,\gamma)^{110m}Ag$	249.8 d	Silver-containing seals

Tab. 10.11 (continued)

Nuclear reaction	$t\frac{1}{2}$ of the product	Source
$^{114}Cd(n,\gamma)^{115m}Cd$	44.6 d	Control rods
$^{121}Sb(n,\gamma)^{122}Sb$	2.72 d	Antimony-containing seals
$^{123}Sb(n,\gamma)^{124}Sb$	60.2 d	Pump bearings

Other nuclides make either negligible or no contribution to long-term dose rate because of their negligible formation rate (^{54}Mn, ^{59}Fe), low gamma energy (^{51}Cr), short half-life (^{56}Mn, ^{65}Ni), or because of a transformation without associated second-order transition (^{55}Fe, ^{63}Ni).

10.7.2.4 Tritium

Formation of ^3H in nuclear reactors is caused by fission as well as by different nuclear activation reactions. Specific differences between light-water reactors of PWR and BWR type as well as heavy-water reactors (HWR) are seen. In heavy-water reactors, a large buildup of tritium is observed due to activation of deuterium in heavy water.[41] The main part of the formed ^3H in LWR originates from ternary fission of uranium and plutonium, cf. Tab. 10.12. Because of the retention capability of the zircaloy cladding, only a small portion, which is estimated at 0.1%, is transferred into the reactor coolant. For BWR, this diffusion is one of the main tritium sources as no additives of boron or lithium are present in the reactor coolant. In PWR, the main tritium source in the reactor coolant is the impact of fast neutrons on ^{10}B used for reactivity control with boric acid. The parallel buildup of tritium by activation of ^6Li, ^7Li, and ^3H is of minor importance.

The ^3H atoms are quickly integrated into the water molecules by isotopic exchange so that ^3H is completely present as HTO, resp. DTO in HWR. Therefore, reduction of the ^3H content is not possible by coolant purification or degasification. In order to prevent a very high tritium concentration in the coolant water, part of the reactor coolant is regularly discarded.

---

41 The same reaction occurs in light-water reactors, but the deuterium content in water is almost 4 magnitudes lower there.

Tab. 10.12: Formation of tritium in a PWR and BWR light-water reactor per year (1300 MW$_{el}$).

Formation reaction	Source	PWR (Bq/a)	BWR (Bq/a)
$^{235}U(n,f)^3H$ $^{239}Pu(n,f)^3H$	Ternary fission in fuel	$6.7 \cdot 10^{14}$	$6.7 \cdot 10^{14}$
$^{10}B(n,2\alpha)^3H$	Boric acid for reactivity control	$1.9 \cdot 10^{13}$	–
$^{10}B(n,2\alpha)^3H$	Boron in control rods	–	$3.6 \cdot 10^{14}$
$^7Li(n,n\alpha)^3H$	Alkalization agent ^7LiOH and product of $^{10}B(n,\alpha)$	$6.3 \cdot 10^{11}$	–
$^6Li(n,\alpha)^3H$	^6Li impurities in ^7LiOH	$6.3 \cdot 10^{11}$	–
$^2H(n,\gamma)^3H$	Deuterium in light water	$1.0 \cdot 10^{12}$	$9.3 \cdot 10^{10}$
	Total	$\approx 7.5 \cdot 10^{14}$	$\approx 1.3 \cdot 10^{15}$
	reactor coolant content	$\approx 2.2 \cdot 10^{13}$	$\approx 8 \cdot 10^{11}$

10.8 Some chemical aspects in nuclear reactors

10.8.1 Minimization of corrosion

To minimize corrosion of nuclear reactor components, highly corrosion-resistant components are selected. Steel components of reactor vessel, steam generator, and loops have austenitic (stainless steel) plating on their inner surface. Nickel-based alloys are used for steam generator tubes. Permanent control of pH and redox potential (Eh) in the water circuit is a feature to achieve lowest corrosion rates.

10.8.2 pH control

For BWR, pH control is not an option. Due to the evaporation process on the surface of the fuel rods, it is essential that the boiling water has highest purity to avoid any depositions on fuel rods. In PWRs and HWRs, the primary circuit is conditioned slightly alkaline. Western-type reactors use isotopically pure ^7Li-lithium hydroxide. Eastern-type reactors (PWRs of the VVER type) mostly use KOH/NH_3 for alkalization. The use of these chemicals has the disadvantage that the production of long-lived ^{40}K and ^{14}C is enhanced due to the reactions $^{39}K(n,\gamma)^{40}K$ and $^{14}N(n,p)^{14}C$, respectively.

Lowest solubilities of steel alloy constituents (mainly Fe, Cr, Ni) are achieved at high temperature conditions of around 300 °C at a pH of 7.4. Yet, the adjustment of such pH has some restrictions. At the beginning of the operation cycle, high boron concentration is present in the reactor water to compensate the high initial core reactivity. A typical boron concentration at reactor start-up is 1200 ppm B (when using natural boron) or 800 ppm B (when using ^{10}B enriched boric acid). Such high concentrations of boric acid would need LiOH concentrations in the range of 10 ppm. Those LiOH concentrations are out of the manufacturers' specifications of fuel rod cladding, as high Li concentrations bring the risk of enhanced corrosion and Li deposition on fuel rods. Upper Li concentrations are specified for each plant specifically and are typically in the range of 2 to 6 ppm.

The link between boric acid and Li concentration is illustrated in Fig. 10.29 for a whole operation cycle with pH control. At the first stage of the cycle, optimum pH cannot be achieved due to maximum Li concentrations specified. With ongoing operation, the core reactivity decreases due to the decreasing content of fissile material ^{235}U and the increasing content of fission products (which also act as neutron absorbers). Lowering the boron concentration in reactor water counteracts the loss of reactivity and keeps the thermal power constant. Therefore, less and less LiOH is necessary during an ongoing cycle to keep the pH at the optimum.[42]

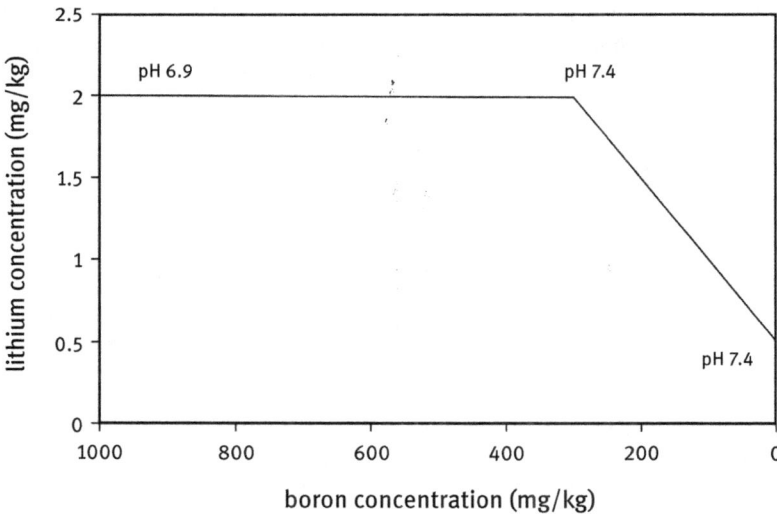

Fig. 10.29: Lithium vs. boron concentration along a typical reactor cycle.

---

42 Additionally, ^{7}Li is formed from boron activation (^{10}B(n,α)^{7}Li) increasing the alkalinity. To counteract this, the Li concentration has to be monitored and the excess of Li has to be removed by a suitable filter system. Filters with protonated cation-exchange resin are used for this task.

10.8.3 Control of redox potential

Lowest corrosion rates are achieved under reducing, oxygen-free water conditions. In BWRs, the water-steam circuit is held nearly gas-free (except for steam), but there is still a low oxygen content in the reactor water due to the steady production from water radiolysis.[43] Part of the BWR plants operates the water-steam cycle with no further reduction of oxygen. This means pure water without any additive is used for operation. The main focus of chemistry in these plants is to guarantee and to monitor water of highest purity. Some BWR plants further reduce the redox potential in water by adding hydrogen. In the radiation field of the core, this dissolved H_2 reacts with traces of oxygen leading to oxygen-free, reducing water conditions. Under such reducing conditions, the growth rate for unwelcome "stress corrosion cracking" (SCC)[44] can be substantially reduced.[45]

All PWRs are operated under reducing conditions as otherwise the oxygen content in the closed primary cycle would be above material specifications, giving not tolerable corrosion rates. Most plants add hydrogen into the primary circuit. Some Eastern-type reactors still use hydrazine addition to keep the oxygen level in primary circuit low, though this mode enhances the production rate of ^{14}C.

10.8.4 Coolant purification

One aim of reactor chemistry is to keep the coolant as clean as possible and to remove unwanted impurities and corrosion products. The level of corrosive anions such as chloride or sulfate and corrosion products from primary loop components should remained as low as possible. To achieve this, part of the flow is directed in

---

43 Radiolytic reactions are generally important to consider in nuclear technology. Radiolysis is the dissociation (in Greek "lysis") of molecules by cleavage of chemical bonds due to the influence of a highly energetic radiation environment. The energy in the environment is described by the dose D (absorbed energy per mass in Gray, Gy, with Gy = J/kg) and the dose rate $\dot{D}$ or DR (dose per time unit in Gy/s or Gy/h). If the absorbed energy is high enough, chemical bonds break and fragments and radicals are primarily formed. The extent of radiolytic reactions depends on the nature of the species undergoing radiolysis and the nature of the radiation field.
44 Stress corrosion cracking describes a failure mechanism of normally inert materials (mostly metals or alloys) that are subject to both tensile stress and a corrosive environment (e.g., high-alloy steel in chloride solutions, steel in water with O_2 dissolved at high temperatures) leading to crack formation and growth. Such cracks may lead to an abrupt and difficult-to-predict failure of the material.
45 This method is applied when the selected material of reactor components is sensitive for SCC. But this application has the disadvantage that the dose rate in the turbine building is elevated. This fact is driven by the enhanced formation of ammonia leading to stronger volatility of the water activation product ^{16}N, usually remaining dissolved as nitrate in water. As ^{16}N is a high gamma energy emitter (up to 7 MeV), this brings the dose rate up for compartments where only steam is present.

bypass to a purification system. This flow is cooled down to approx. 50 °C, a temperature which allows the use of ion-exchange resins (Li-loaded cation- and borate-loaded anion-exchange resin). Most of the impurities are removed. For the removal of lithium, a separate filter system is used on demand, which contains protonated cation-exchange resins instead of Li loaded ones.

10.8.5 Dose rate reduction

The selection of optimized material is of major importance to guarantee a low dose rate in NPPs. In particular, the content of cobalt in steel alloys should be low as its activation product ^{60}Co is a key nuclide dominating the dose rate exposition of nuclear power plant personnel due to its half-life (5.3 years) and its high gamma energy emission (1.17 and 1.33 MeV). The use of stellite, a cobalt/chromium-based containing alloy, should be avoided whenever possible.

For operating plants, the status of selected materials cannot be changed easily. To avoid the buildup of dose rate at the loop surface, it is necessary to avoid the uptake of ^{60}Co into its oxide layer. This can be done by dosing zinc (depleted in ^{64}Zn, as zinc acetate) into the primary circuit aiming at zinc concentrations of 5 ppb. Zn has a higher tendency than Co to be bound in spinel-type oxide and can occupy the available bounding positions. By this, the migrating activated cobalt is less absorbed in the surface oxide layer and remains in the water until it is removed by the coolant purification system. The zinc incorporated does not lead to buildup of dose rate at primary circuit components outside the core as activated Zn isotopes are stable or have short half-lives (^{69}Zn: 13.8 h; ^{71}Zn: 3.9 h).

10.9 Research reactors

10.9.1 Technical characteristics

Research reactors are neutron sources for research and many related applications. They are not operated for power production, although some have thermal powers of up to 200 MW_{th}. More than 750 research reactors have been constructed worldwide, whereof 247 were operational as of 2015.[46] These reactors are typically used, according to an IAEA classification, for radioisotope production for medicine and industry, nuclear data measurement, neutron scattering, neutron radiography, material irradiation, transmutation, teaching and training, neutron activation analysis (NAA),

---

46 According to IAEA's Research Reactor Database.

geochronology, boron neutron capture therapy (BNCT), research on ultracold neutrons (UCN), and other purposes, cf. Tab. 10.13.

Tab. 10.13: Main applications of research reactors.

Application	Basic research	Technical applications	Medical applications
Radioisotope production	×	×	×
Nuclear data measurement	×		
Neutron scattering	×	×	
Neutron radiography	×	×	
Material irradiation	×	×	
Transmutation	×		
Neutron activation analysis (NAA)	×	×	
Geochronology	×	×	
Boron neutron capture therapy (BNCT)	×		×
Ultracold neutron measurements	×		
Teaching and training	×	×	

Since research reactors have a different function compared to power reactors, they also have different designs. In simple terms, it can be said that they are smaller and simpler than power reactors. Research reactors operate with far less fuel and at lower temperatures, but to achieve a high power density and high neutron fluxes, the fuel is more highly enriched, typically up to 20% ^{235}U (low-enriched uranium, LEU), but also up to 93% ^{235}U (highly enriched uranium, HEU). Although the use of HEU is sometimes favorable as it enables to get more compact reactor cores with higher power density and higher neutron fluxes (up to $2.5 \cdot 10^{15}$ cm^{-2}/s in the 85 MW high flux isotope reactor (HFIR) at Oak Ridge, USA), the proportion of research reactors working with HEU is decreasing.[47]

Research reactors vary strongly in their design. They can be, similarly to power reactors, distinguished based on their specific combination of fuel, moderator, and cooling, and their configuration. Most common are pool and tank reactors. In pool reactors, the mostly compact reactor core consisting of a number of fuel elements is immersed in a large pool of light water (H_2O). The water shields the radiation efficiently and provides moderation and cooling. In most pool-type research reactors, the water pool is sufficient for heat exchange by natural convection. The reactor core of such light-water reactors is commonly surrounded by a neutron reflector (e.g., graphite, beryllium) to minimize neutron losses and to reduce the critical

---

47 The major reason are concerns over proliferation of HEU for nuclear weapons. Therefore, many reactors have already been converted to LEU cores.

mass. In tank-type reactors, the reactor core is placed in a tank, typically filled with heavy or light water. The use of heavy water is advantageous for the use of highly compact cores with high power densities and, thus, high neutron fluxes. A coolant, often light water, surrounds the tank.

The reactors also contain control rods and empty spaces for further instrumentation at different places on the core. Those spaces provide the possibility to place samples to be irradiated directly in or near the core. The samples are placed manually or, e.g., by a pneumatic tube in their irradiation position.[48]

Beside these irradiation channels, it is also possible to extract neutron beams from the core by using beam tubes. They guide the neutrons to external experimental facilities or setups. The tubes are typically made of aluminum to minimize neutron losses during passage. Depending on the design of the reactor core and the position of the beam tube tips, the extracted neutrons have different energies (often cold, thermal, and fast neutrons are needed) and fluxes.

Other designs are less common, such as fast reactors (with U/Pu fuel and no moderator) or homogeneous reactors with a liquid solution of uranium in the tank, thus having a homogeneous solution of fuel, moderator, and coolant.

With more than 60 reactors constructed, the TRIGA reactors are the most popular research reactors. This reactor type was developed in the 1950s and since then adapted to increasing demands, with power levels from 20 kW to 16 MW. TRIGA reactors offer a unique feature, making them inherently safe reactors: The fuel is made of a uranium zirconium hydride (U-$ZrH_{1.65}$) alloy with enriched uranium (standard 20% enriched, sometimes also 30% or 70%). Zirconium hydride acts as a moderator making the fuel self-moderated. The bound hydride has some unusual moderating properties to thermal neutrons, compared to hydrogen in water. If the fuel (hydride) temperature is increased, the cold neutrons in the fuel gain energy from the hydride and become warmer ("warm neutron principle") than neutrons in the surrounding cold water. Since warmer neutrons cause less fissions than colder neutrons and tend to escape to the water, the moderation becomes worse and the reactor power is quickly (within a few milliseconds) reduced. The result of this specific fuel design is a prompt negative temperature coefficient of reactivity of the reactor. Moreover, fission products are very effectively retained in the fuel matrix. Another feature of TRIGA reactors is the possibility of transient operation ("pulsing"). After removing the standard control rods from the core, the reactor is critical at low power, and only a transient rod remains in the core. By rapidly removing the transient rod from the core, excess reactivity is inserted and the reactor power increases quickly. Since the fuel temperature also increases, the

---

48 The cover of Vol. I shows a photograph of the core of the research reactor TRIGA Mark II of the Johannes Gutenberg University, Mainz, Germany. TRIGA stands for training, research, and isotope production (GA indicates the producer, general atomics).

excess reactivity is being compensated. After compensation, the reactor power decreases while the temperature is still increasing. The thereby produced pulse generates
for a period of approx. 30–50 ms a very high neutron flux, about 1000 times higher
than the flux at steady operation.

Figure 10.30 is a photo of the TRIGA Mainz research reactor, looking through
the water tank at the fuel elements. In addition to inherent reactor equipment,
there are several tubing systems to introduce samples for irradiation or to deliver
neutrons to peripheral experimental setups.

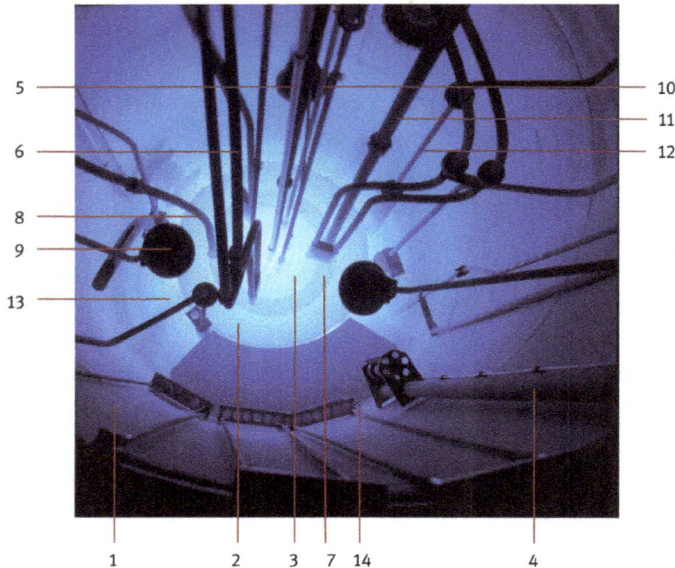

Fig. 10.30: TRIGA Mainz research reactor core, looking through the water tank at the fuel elements.
In addition to inherent reactor equipment, there are several tubing systems to introduce samples
for irradiation or to deliver neutrons to peripheral experimental setups: (1) reactor tank
(18 m^3 water at atmospheric pressure); (2) outer graphite reflector; (3) reactor core; (4) fuel element
inspection unit; (5) central irradiation tube; (6) rabbit system for fast sample transport (1 of 3);
(7) carousel irradiation unit with 40 positions; (8) loading tube for "lazy Susan"; (9) underwater
lamp; (10) regulation rod; (11) "lazy Susan" drive shaft; (12) neutron flux measuring chamber
(1 of 4); (13) beam tube for larger experimental setups close to the reactor core (1 of 4); and
(14) thermal neutron column for experimental setups at greater distance to the reactor core.

10.9.2 Production of medical isotopes

Research reactors may also be used for the production of medically relevant radionuclides. The most important one produced in research reactors is ^{99}Mo ($t^{1/2} = 66.0$ h). It is
produced either by neutron irradiation of ^{98}Mo or by irradiation of a ^{235}U target inducing

fission. ^{99}Mo yields the metastable isotope ^{99m}Tc ($t\frac{1}{2}$ = 6.0 h) in transient radionuclide generator systems.[49] ^{99m}Tc is used for nuclear medicine diagnostics and accounts for about 80% of all radiodiagnostic examinations. Here, the uranium fission-derived ^{99}Mo gained upmost importance to design ^{99}Mo/^{99m}Tc radionuclide generators to guarantee ca. 30 million diagnostic procedures worldwide each year, cf. Chapter 12.

The medically relevant isotope ^{131}I is obtained during ^{99}Mo processing. It is also dissolved in the NaOH solution to form mainly iodide (^{131}I$^-$) and iodate (^{131}IO$_3^-$). Iodine is removed from the solution by means of an anion-exchange resin and purified by distillation. It is then typically stored in NaOH solution. Another production route is by using TeO$_2$ targets. Other typical "medical" radionuclides produced directly at research reactors are ^{153}Sm, ^{166}Ho, ^{186}Re, ^{188}Re, ^{60}Co, ^{32}P, ^{35}S, ^{67}Cu, ^{89}Sr, ^{90}Sr, and ^{177}Lu for diagnostic or therapeutic purposes, cf. Chapters 12 and 13. Also actinides could be produced.

Suggested reading

Neeb K. The radiochemistry of nuclear power plants with light water reactors. Boston, De Gruyter, Berlin, 1997.

Loveland WD, Morrissey DJ, Seaborg GT. Modern nuclear chemistry. Hoboken, New Jersey, John Wiley & Sons, Inc., 2006.

Stacey WM. Nuclear reactor physics. Weinheim, Wiley-VCH, 2007.

Prussin SG. Nuclear physics for applications: A model approach. Weinheim, Wiley-VCH, 2007.

Oka Y. ed. Nuclear reactor design. Japan, Springer, 2014.

Joonhong A, Carson C, Jensen M, Juraku K, Nagasaki S, Tanaka S. Reflections on the Fukushima Daiichi Nuclear Accident. Japan, Springer, 2015, ISBN 978-3-319-12090-4 Online, Open Access.

Morss LR, Edelstein NM, Fuger J. eds. The chemistry of the actinide and transactinide elements. Dordrecht, Springer, 2010.

Nash KL, Lumetta GJ. Advanced separation techniques for nuclear fuel reprocessing and radioactive waste treatment. Cambridge, Woodhead Publishing, 2011.

Kok KD. ed. Nuclear engineering handbook. Boca Raton, CRC Press, 2009.

Lin CC. Radiochemical technology in nuclear power plants. La Grange Park, American Nuclear Society, 2013, ISBN 978-0-89448-045-4.

References

Classification of radioactive waste : safety guide. – Vienna : International Atomic Energy Agency, 2009. ISBN 978-92-0-109209-0.

Danesi PR. Solvent extraction in the nuclear industry. 1984. CONF-8410131--2-Abst. DE84 016321, International School of Solvent Extraction, Barcelona, Spain.

---

49 Cf. Chapter 5 of Vol. I.

IAEA. Technology roadmap for small modular reactor deployment. IAEA, 2021, ISBN 978–92–0–110021–4

JEFF-3.1: Joint Evaluated Fission and Fusion File, Incident-neutron data, http://www-nds.iaea.org/exfor/endf00.htm, 2 October 2006.

Koning A, Forrest R, Kellett M, Mills R, Henriksson H, Rugama Y The JEFF-3.1 Nuclear Data Library, JEFF Report 21, OECD/ NEA, Paris, France, 2006.

Thomas Moenius

11 Life sciences: Isotope labeling with tritium and carbon-14

Aim: Pharmaceutical research and development requires a broad variety of study data, making the application of radioactivity not only useful, but in distinct cases absolutely necessary. Suitable tracers require study-specific specifications such as type and position of label, specific activity, and radiochemical purity. For low-weight molecules tracers involve tritium and carbon-14. Tritium and carbon-14 have different, sometimes complementary advantages and disadvantages. Tritium not only has a significantly higher specific activity, but can also be synthetically incorporated more easily; carbon-14 is considered to be metabolically more stable.

With the "derivatization", the "construction", and the "reconstitution" approaches there are three different strategic opportunities to incorporate isotopes into the drug substance. While proton/tritium exchange, halogen/tritium exchange, or tritiations of unsaturated systems are of fundamental importance to tritium chemistry, ^{14}C-chemistry mostly makes use of ^{14}C-labeled building blocks to synthesize the isotope analog by means of a construction approach. The identification of most straightforward, labeled building blocks, the minimization of the number of radioactive steps, the optimization of the radiochemical yield, and prevention of any risk of contamination and/or incorporation for the operator determine the quality of a radiosynthesis.

11.1 Introduction

Research and development (R&D) in life sciences (i.e., pharmaceuticals, animal health, agrochemicals, and crop protection) requires a comprehensive understanding of a substance and its behavior to ensure its safety, tolerability, and effectiveness. The highly regulated area of drug development is an illustrative example that supports the importance of this goal.[1] Despite increasing development of nonradiolabeled analytical methods (e.g., fluorescent tags, mass spectroscopy techniques) the application of radiolabeled isotope analogs remains a key technology. The advantages of radiolabeled

---

[1] Although most of the following examples are taken from this area, application of radioactivity is not restricted to this field.

Acknowledgment: Ulrike Glaenzel, Esther van de Kerkhof and Robert Nufer for providing illustrative study data. Joel Krauser for careful and patient corrections and helpful discussions.

https://doi.org/10.1515/9783110742701-011

tracers in life sciences are not only the opportunity to detect any "drug-related substance", but also their extremely high detectability compared to other methods.[2] Their application combines key aspects of organic chemistry and pharmacology, and in many cases utilizes a specific terminology. The main wordings are listed in Tab. 11.1.

Tab. 11.1: Terminology used in the present chapter "Isotope labeling with tritium and carbon-14".

Terminology	Abbr.	Explanation
Absorption, Distribution, Metabolism, Excretion study	ADME study	Type of study required for any registration of pharmaceuticals
Drug-Drug Interaction study	DDI study	Type of study investigating the interaction of drugs
Good Laboratory (or Manufacturing) Practice	GLP/GMP	Set of regulations established to guarantee the quality of processes
Isotopomer		Molecules containing the same type and number of isotopes at different positions
(kinetic) Isotope effect	(k)IE	Isotope related differences in the physical and/or chemical behavior of molecules (such as their kinetics)
Metabolic Switching		Isotope related effects in metabolic pathway
Metabolite		Drug-related compounds derived from metabolism
Pharmacodynamics	PD	Impact of the drug substance on the organism
Pharmacokinetics	PK	Impact of the organism on the drug substance
(Quantitative) Whole Body Autoradiography	(Q)WBA	Technique to determine the tissue distribution of radiolabeled materials
Radioimmunoassay	RIA	Assay that uses the competition of radiolabeled and unlabeled material in an antigen-antibody reaction for concentration determination
Specific activity	A_{spec}	Radioactivity per mass (Bq/mg or Bq/mmol)

11.1.1 The tracer concept

The classical concept was to postulate that a radioisotope of a given metal behaves almost identical to a stable isotope of the element and to the chemical element in general. Hevesy first used the radioisotopes ^{212}Pb and ^{212}Bi to quantify the extremely low solubility of salts of lead and bismuth by "spiking" the natural salts with the

---

2 Continuous technical efforts result in an increasing sensitivity in the *on-line* and *off-line* detection mode (see also Kiffe 2008; Bruin 2006).

radioisotopes. Because even extremely low mass of, e.g., $[^{212}\text{Pb}/^{\text{nat}}\text{Pb}]$ oxalate dissolved in water at room temperature still represented a sufficient radioactivity ($\hat{A} = \lambda \cdot \hat{N}$), ratios between dissolved $[^{212}\text{Pb}]^{\text{nat}}\text{PbC}_2\text{O}_4$ could be determined, yielding solubility constants otherwise not measurable. The same idea was transferred to life sciences. However, this rather required "organic" radioisotopes, which only became available later when manmade artificial radioactivity was accessible. HEVESY for example used the β^--emitter ^{32}P in its form of $[^{32}\text{P}]\text{PO}_4^{3-}$ for biological studies on phosphate metabolism in rat and mice.[3] The tracer concept was adopted by YALOW in the 1950s to develop the radioimmunoassay (RIA).[4]

Radiolabeled molecules might serve as "tracers" because of a dichotomy: The labeled compound is usually applied at ultralow concentration (at a "trace" level), and once applied to a biological (or industrial) process, its radioactive emission signal allows "tracing" or following its pathway in a system.[5] The conclusive application of a tracer requires labeled and unlabeled compounds to behave identically (or at least similarly) in the respective study environment. Therefore, the overall idea is to synthesize (radio)labeled molecules with a chemical behavior quite close to that of their native ones. The "isotope tracer" results from substitution of one or several naturally existing isotopes $^A X$, which constitute part of the molecule's chemical structure, by one of its radioisotopes, i.e., ^{A+i}X or ^{A-i}X. For low molecular weight molecules, this means hydrogen and carbon substituted by their long-lived radioisotopes tritium (^3H) and carbon-14 (^{14}C).

11.1.2 Designing a suitable tracer

The goal of pharmaceutical R&D is to discover and develop innovative, efficacious, and safe drugs. Therefore, both efficacy and safety issues must be addressed in animal and human studies. Figure 11.1 summarizes some typical studies in the course of pharmaceutical research and development.

In this aspect, radiotracers (Penner 2012) can qualitatively and quantitatively detect drug substances and their degradation products. Metabolite identification by HPLC-RA (Fig. 11.2) and investigation of drug distribution by Whole Body Autoradiography (Solon 2012) (Fig. 11.3) are two examples illustrating the power of the tracer technique in the field of drug development.

---

3 The Nobel Prize in Chemistry 1943 was awarded to George de HEVESY "for his work on the use of isotopes as tracers in the study of chemical processes".
4 The Nobel Prize in Physiology or Medicine 1977 was divided, one half jointly to Roger GUILLEMIN and Andrew V. SCHALLY "for their discoveries concerning the peptide hormone production of the brain" and the other half to Rosalyn YALOW "for the development of radioimmunoassays of peptide hormones".
5 The term "tracer" denotes the isotopically (radio)labeled drug substance.

| TARGET & HIT IDENTIFICATION, HIT VALIDATION, LEAD SELECTION | LEAD OPTIMIZATION | CANDIDATE SELECTION | EARLY CLINICAL DEVELOPMENT |

³H labelled compounds

| **Support early compound selection:** |
| Establishment of screening assay |
| Characterization of binding sites |
| Characterization of receptor distribution and expression |

³H and ¹⁴C labelled compounds

| **Research:** |
| *in vitro* binding to characterize structure activity relationships |
| *ex vivo* binding studies to determine receptor occupancy |
| **Early ADME:** |
| Metabolite identification, kinetics, *in vivo* / *in vitro* correlation for cross validation, DDI characterization |

³H and ¹⁴C labelled compounds

| **ADME-full support:** |
| Mass Balance, Pharmacokinetics Metabolism, Transporter, PPB |
| Distribution (QWBA) |
| Enviromental compatibility |
| human-ADME (GMP) |
| **Investigative studies for PET-labelling** |

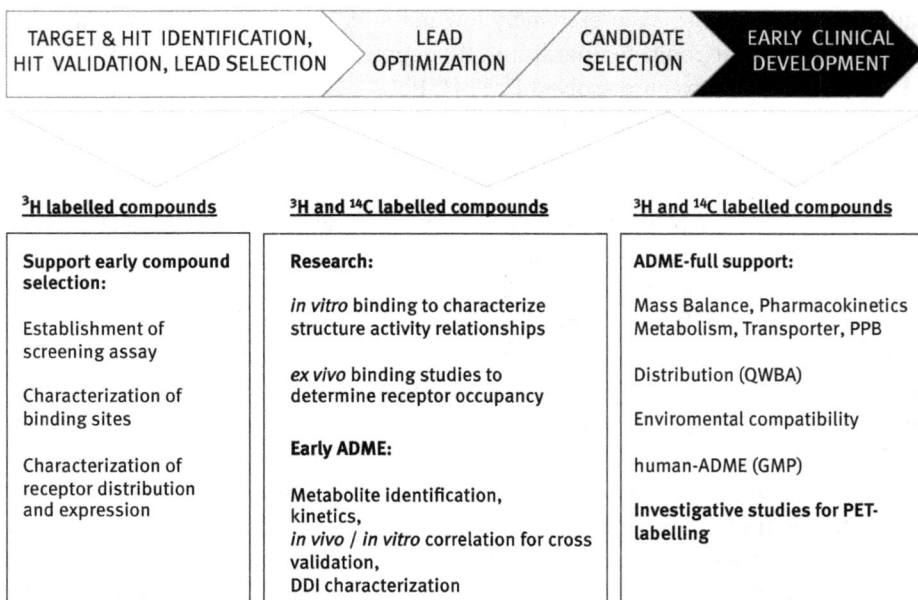

Abbreviations: ADME – Absorption, distribution, metabolism and excretion; DDI – drug drug Interaction; GMP – good manufacturing practices PPB – plasma protein binding; PET – positron emission tomography; QWBA – quantitative whole body autoradiography

Fig. 11.1: Studies requiring tracer support in the course of pharmaceutical R&D (adapted from Krauser 2013).

In order to investigate the pharmacokinetic behavior of any drug substance, circulation (i.e., full blood, plasma) and excreta (i.e., urine, feces, exhaled air) have to be analyzed for drug-related material. In a pharmacokinetic (PK) study, a C-14 labeled isotope analog of a drug candidate was dosed to rats (1 mg/kg bodyweight rat; 3.5 MBq/kg bodyweight rat). Plasma probes (12 µl) were analyzed by HPLC-UV detection (trace a), HPLC-MS detection (trace b), and HPLC-RA detection (trace c) to investigate circulatory metabolites.

Comparison of the detection modes identifies some key advantages of radioactivity (see radioactivity (RA) detection; red line in Fig. 11.2), i.e.,

- extremely low quantities (as low as 20 pmol/injection in the example above), otherwise difficult to detect can easily be analyzed by RA detection
- from a broad variety of endogenous and exogenous compounds (detectable by mass spectrum (MS) evaluation; i.e., black line), radioactivity detection identifies any drug-related compounds
- in contrast to other methods with structure-related sensitivities (such as UV and mass spectrum (MS) detection), the RA detection mode allows straightforward quantification of different degradation compounds

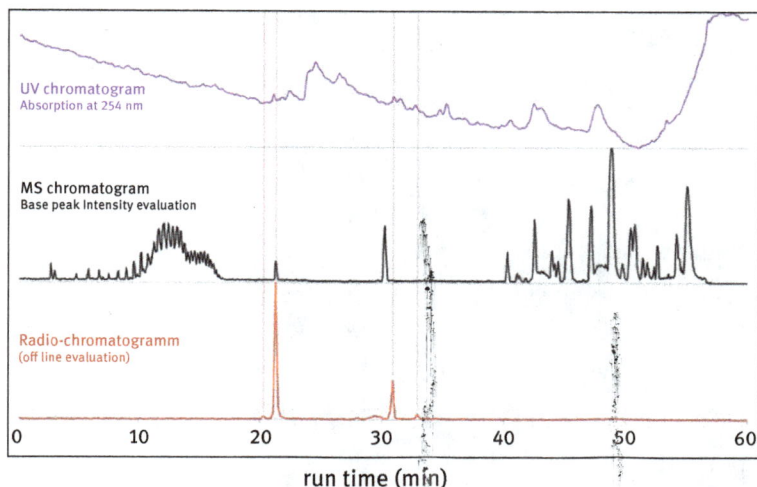

Fig. 11.2: Metabolite identification in plasma samples by HPLC analysis with different detection modes.

Subsequent Mass Spectroscopy (MS) and/or Nuclear Magnetic Resonance (NMR) analysis further allows their structural evaluation. Extension to additional compartments (such as urine, feces and tissue) allows not only comprehensive metabolic profiling but also a complete mass balancing (applied vs. excreted activity).

In addition, Quantitative Whole Body Autoradiography (QWBA) quantitatively "visualizes" the whole body or organ distribution of the radioactivity (i.e., parent drug and/or drug-related compounds) *in vivo*. (Cryo-)Sectioning the carcass of dosed test animals provided slices which were quantitatively evaluated by phosphor imaging technique. The intensity relates to the exposure of radioactivity (the lighter the higher exposure) and is therefore a measure for any drug and/or drug-related material. This insight into the *in vivo* distribution of the radiolabeled tracer allows safety and efficacy-relevant assessments for pharmaceuticals.

Some important study-specific specifications must be fulfilled in order to guarantee full utility of the tracer. For instance:
- type of isotope
- position of the isotope within the molecule structure
- required specific activity [Bq/mmol]
- extent of labeling (i.e., ratio of radioactive isotope to nonradioactive isotope per position)
- chemical and radiochemical purity
- type of synthesis and analysis
- depending on its final application (*in vitro*, animal, human and/or registration studies) GLP/GMP-regulated synthesis and/or release procedures might be required too.

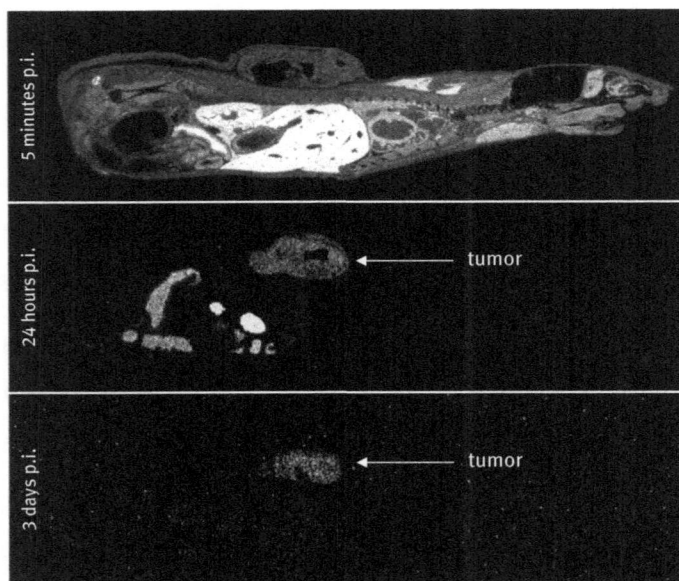

Fig. 11.3: QWBA in tumor-bearing mice at different time-points (p.i. = post intravenous injection) after dosing with a ^{14}C-labeled oncology compound to obtain proof of its uptake in the tumor tissue (upper tissue in all the slices). Comparison of slices at different time-points (0.08 h, 24 h, and 72 h) demonstrated a distinct affinity of the drug compound for the tumor tissue.

These specifications depend not only on the type of study, but also on the test compound itself (see Tab. 11.2). Although a case-to-case approach is required, some general considerations will be given on the "selection of the isotope" and the "selection of the position".

Tab. 11.2: Specifications of different study types.

Type of Study	Isotope	Specific activity	Stability	Radiochemical purity (RCP)	Chemical purity (CP)
Receptor binding	H-3	1.1–3.7 TBq/mmol	Chemical	>98%	>95%
Protein binding	C-14 (H-3)	t.b.d.	Chemical	>99%	>95%
ADME *in vivo*	H-3, C-14	t.b.d.	Chemical and Metabolical	>98%	>95%
hADME *in vivo*	C-14 (H-3)	t.b.d.	Chemical and Metabolical	>98%	>98%

t.b.d.: to be determined on a "case-by-case" basis;
less preferred options are given in parentheses.

11.1.3 Selection of the isotope

Table 11.3 summarizes the relevant decay parameters of the two isotopes. Due to the different kinetic energies of the β^--electrons emitted, ranges in condensed matter differ.[6] Due to the different half-lives, tritium and carbon-14 have significantly different specific activities, cf. eq. (11.1).[7]

Tab. 11.3: Physical data of tritium and carbon-14.

Isotope	Decay mode	Half-life	Specific activity A_{spec}	β^--particle		
				E_{max}	Range in air	Range in water
		(a)	(GBq/mmol)	(keV)	(mm)	(mm)
H-3	β^-	12.3	1080	18.6	6	0.006
C-14	β^-	5730	2.3	156	200	0.28

$$A_{spec} = \ln 2 \cdot N_A / t^{1/2} \tag{11.1}$$

A_{spec} specific activity, e.g., in [Bq/mmol];
N_A AVOGADRO constant [mmol^{-1}];
$t^{1/2}$ half-life of the radioisotope [s^{-1}].

Table 11.3 highlights that complete substitution of one proton (^{1}H) by one tritium atom (^{3}H) results in a maximum specific activity of 1080 GBq/mmol and substitution of an unlabeled carbon (C-12) by a C-14 isotope will result in a maximum specific activity of the tracer of 2.3 GBq/mmol. Since the isotopic purities of H-3 and C-14 labeled compounds will usually be <100%, their molar specific activities emerge proportionally. This means, for example, that [^{14}C]carboxyl-benzoic acid with an isotopic purity of 50% accounts for a specific activity of 1.15 GBq/mmol.

The lower value of E_{max} of the β-particles of tritium also reduces their analytical detectability, i.e., limit of detection. Nonetheless, significantly higher specific activity (factor of about 450) often makes tritium the "isotope of choice" for study compounds administered at lower concentrations. The specific activity required is determined by the chemical amount delivered (dose in mg) and the activity required for its analytical determination.

---

6 The maximum kinetic energies are listed. However, note that the energy spectrum of the emitted β^--particles is a continuous one, and most of the β^--electrons, i.e., those that are more responsible for the detection characteristics, are of lower energy, cf. Chapter 8 in Vol. I.
7 Specific activity is the radioactivity per mass of the nuclide or labeled tracer; see Chapter 3, Vol. I, for details.

11.1.4 Selection of position

Not all tracers allow any positioning of the label. Certain limitations may occur, if
- the drug substance is cleaved (either chemically or metabolically)
- a key interest lies in a specific fragment
- the labeling of a specific position may result in an interfering isotope effect
- the specific position is difficult or impossible to reach by preparative organic synthesis.

A comprehensive understanding of a molecule's biochemical fate, i.e., metabolism and routes of elimination is a necessary step in pharmaceutical R&D. In many cases, drug substances can undergo significant metabolic degradation. Therefore, the optimal labeling position should provide the most information on the process. However, as emphasized in the example of *zibotentan* (1) (Lenz 2011 and Fig. 11.4), such decisions are not always obvious. The initial labeling of the oxadiazole ring ($X = {}^{14}C$) identifies the metabolites (1b, 1c, and 1e), but not the metabolites (1a and 1d) being formed by cleavage of the oxadiazole moiety. Therefore, additional labeling of the pyridine-moiety ($Y = {}^{14}C$) was necessary to obtain comprehensive information.

The introduced label must be stable under the respective experimental conditions. In practice, the label must not be lost under chemical (i.e., pH- and/or temperature-catalyzed exchange reactions) and *in vivo* applications. Loss of the isotope would not only result in loss of the opportunity to detect some potentially important metabolites, but might also lead to misinterpretation of the data. Such an example is the pharmacokinetic characterization of (2) (Shaffer 2006). While the *in vivo* ADME study (Sprague-Dawley rats) of (2b) (Fig. 11.5) gives a terminal half-life[8] of 31 h, the isotopomer (2a) suggests a significantly higher value of 133 h. Further evaluation of the data demonstrated that the tritium label of the position 3 (2a) is metabolically unstable, i.e., oxidatively cleaved off by CYP[9]-enzymes of the organism. The biological half-life of the generated tritium water (HTO) in rats is consistent with the observed higher value of the terminal half-life.

11.1.5 Isotope effects

Isotopes are chemically very similar, but not identical. The difference in their nuclear composition (i.e., different number of neutrons) might result in various mass-dependent (e.g., kinetic, equilibrium IE) and mass-independent (e.g., quadrupole and magnetic moment) discriminations (i.e., isotope effects; Simon 1966). The

---

8 Terminal half-life is an index of drug persistence in the body during the terminal phase.
9 Cytochrome P450 enzymes (CYP) are hemeproteins with oxido-reductive activity.

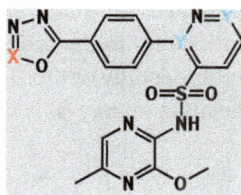

1

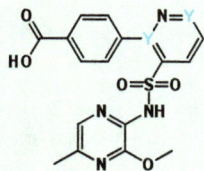

1a

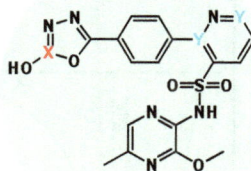

1b

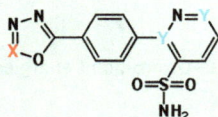

1e

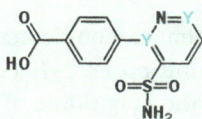

1d

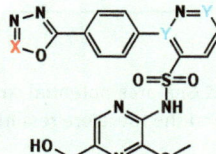

1c

Fig. 11.4: Major metabolites of *zibotentan* **1** in circulation and excreta of rat, dog and human. The initial labeling of the oxadiazole ring (X = ^{14}C) identifies the metabolites (1b, 1c, and 1e), but not the metabolites being formed by cleavage of the oxadiazole moiety, (1a and 1d), which were labeled in the pyridine-moiety (Y = ^{14}C).

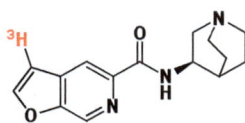

2a

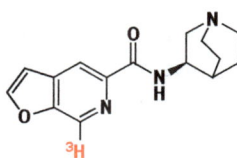

2b

Fig. 11.5: Structures of the isotopomers (2a) and (2b).

mass-dependent isotope effect originates from the differences in zero-point energies (E_0).[10] Due to their inverse proportionality to the reduced mass (m_r) compared to the C–H bond energy (E^H_0), the zero-point energy of the C–D bond (E^D_0) is lower and consequently its bond strength (dissociation energy D) is higher $D^D > D^H$ (cf. Fig. 11.6).

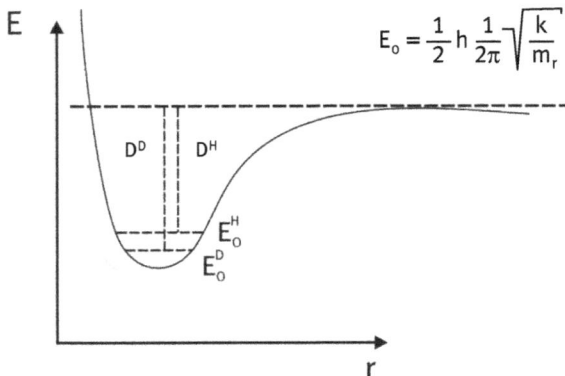

$$E_0 = \frac{1}{2} h \frac{1}{2\pi} \sqrt{\frac{k}{m_r}}$$

Fig. 11.6: Morse potential curve[11] indicating the difference in C–H versus C–D zero-point energies (E_0) and the therefore resulting difference in dissociation energy (D).

Consequently, the carbon-hydrogen, carbon-deuterium, and carbon-tritium binding energies increase in the following order C–H ($338.4 \pm 1.2\,\text{kJ mol}^{-1}$) < C–D ($341.4 \pm 1.2\,\text{kJ mol}^{-1}$) < C–T. In general, the magnitude of the mass-dependent isotope effects increases with the mass ratio of the isotopes (i.e., $^3H/^1H \ggg {}^{14}C/^{12}C$). Exploitation of the kinetic isotope effects (kIE) is often used for mechanistic investigations (Gomez-Gallego 2011).

Isotope effects can translate into significant differences on the biochemical level. Examples of particular interest for tracer studies are isotopic fractionation[12] (Filer 1999) and "metabolic switching" (Miwa 1987). If the C-H cleavage of a metabolic process is rate limiting, its isotope labeling might impede/block its (oxidative) cleavage and therefore favor an alternative metabolism pathway. A distinct impact of isotope labeling on the metabolite pattern was observed in several cases. This phenomenon was strategically used to optimize pharmacodynamic and pharmacokinetic properties (i.e., prevention of critical metabolites). Systematic deuteration of *boceprevir* is an example (Morgan 2011 and Fig. 11.7). Incubation of differently labeled isotopomers (3a–g) with human liver microsomes demonstrated the influence

---

10 Zero-point energy (E_0) is the lowest possible energy that a quantum mechanical physical system may have.

11 A MORSE potential is a model for the potential energy of a diatomic molecule.

12 In rare cases there is the chance to separate chromatographically unlabeled compounds from their isotope analogs. Filer provides examples and proposes explanations (Filer 1999).

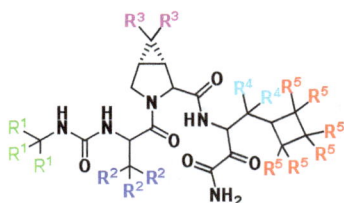

	R¹	R²	R³	R⁴	R⁵	average t½ (min)
3	CH_3	CH_3	CH_3	H	H	19.0±0.87
3a	CH_3	CH_3	CH_3	H	D	26.6±1.70
3b	CH_3	CH_3	CH_3	D	D	28.7±2.19
3c	CH_3	CH_3	CD_3	D	D	31.4±2.36
3d	CH_3	CH_3	CD_3	H	H	19.1±1.19
3e	CD_3	CD_3	CD_3	H	H	23.1±1.31
3f	CD_3	CD_3	CD_3	H	H	25.0±0.68
3g	CD_3	CD_3	CD_3	D	D	37.3±3.12

Fig. 11.7: Structure of *boceprevir* (3), its deuterated analogs (3a–g) and their metabolic stabilities (t½) in human liver microsomes.

of deuterium labeling on their metabolic stabilities. Simultaneous deuteration of the positions R¹, R², R³, R⁴, and R⁵ (3g) resulted in an almost duplication of the metabolic half-life (37.3 vs. 19.0 min).

However, application of such an isotope analog in a tracer study would result in misleading, nonconclusive study data. For example, if the label were subjected to a bond-breaking event, the label might not only be "lost", but there is also a risk that other pathways are favored ("metabolic switching"). In the case of *boceprevir*, labeling in position R³ (3d) shows the least metabolic impact and might be the most suitable position for any tritiated tracers intended for *in vivo* studies.[13]

11.2 Labeling strategies

The goal of radiochemistry is to identify the most straightforward synthesis strategy to the most suitable target, i.e., minimizing the number of radioactive steps, optimizing the radiochemical yield and preventing the operator from any risk of contamination and/or incorporation. A comprehensive synthetic arsenal of different strategies allows identifying optimal opportunities. Most reactions of classic organic chemistry can be applied directly to isotope chemistry. Exceptions, however, exist if inherent physical properties of the labeled intermediates (i.e., volatility) result in

---

13 However, a specific label preference for a given application could be quite costly from a synthetic point of view. Therefore, the decision on the final tracer will also be influenced by economic factors (such as time, money and capacity expenditure), which must be balanced with synthetic feasibility (availability/accessibility of isotopic materials, unlabeled materials, number of steps, complexity of the synthesis), timelines and study needs. Development of *boceprevir* in chronic hepatitis C virus (HCV) infection was therefore performed with the C-14 isotope analog (Ren 2012).

an increased incorporation and/or contamination risk,[14] the intrinsic irradiation favors instability[15] or alternative reaction mechanism. In all cases, however, isotope chemistry requires specific adaption of reaction or work-up conditions. The main purpose of this section is to provide a basic understanding of tritium and carbon-14 chemistry and to discuss the principles of their labeling strategies. In a first step, general approaches are introduced. Selected examples with isotope-specific differences will be given for tritium and carbon-14.

In general, there are three different strategies to incorporate tritium or carbon-14 into molecules, cf. Tab. 11.4.

Tab. 11.4: Strategies in tritium- and ^{14}C-labeling.

Strategy	Approach
Derivatization	Attachment of the label under modification of the molecular structure
Construction	Step-by-step construction of the tracer starting from a labeled building block
Reconstitution	Modification of the final drug substance (or its advanced intermediate) to facilitate introduction of the label

11.2.1 Derivatization approach

The derivatization approach attaches the radioisotope label to the drug substance and changes its molecular structure, thereby accepting modification of its physical and chemical properties. The type and extent of these structural modifications depend on the individual case, but usually exceed the impact of isotope effects. The recently reviewed (Filer 2011) application of dihydro-derivatives as radioligand surrogates provides an example for a successful derivatization approach (Fig. 11.8). All unsaturated drug substances allowing tritiation without significant impact on their molecular structure provide access to highly tritiated tracers, which might be used to investigate the binding characteristics of their unlabeled analogs. An example is dihydro-*picrotoxinin* (5), which serves as tracer for the structurally highly complex and therefore difficult to label *picrotoxinin* (4).

---

14 Due to these risks any handling of radioactivity is strongly regulated and therefore requires special training. Regulations are country- and institution-specific and can therefore not be generalized.
15 Compared to unlabeled compounds, radioactive compounds might be considerably less stable. Due to radiation-induced side reactions their degradation rate might be significantly above the physical half-life (parameters influencing the stability are specific activity and storage conditions).

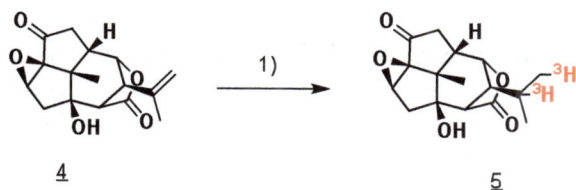

Fig. 11.8: Synthesis of dihydro-*picrotoxinin* (5). Reaction conditions: 1) $^3H_2$, Pd/C 10%, EtOH.

Derivatization as in the reduction of *strychnine* (6) may alter important chemical and biological properties of a molecule. Reduction of the double bond changes the conformation of the piperidine ring, which might explain the significantly decreased affinity of the dihydro-derivative (7) towards the glycine receptor, vs. *strychnine* (6), cf. Filer (2011) and Fig. 11.9.

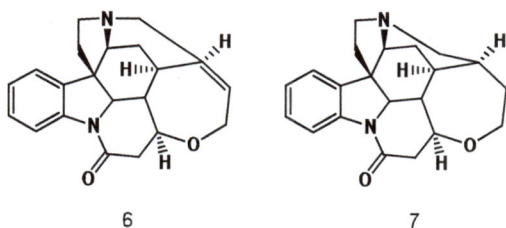

Fig. 11.9: Anticipated molecular structures of *strychnine* (6) and dihydro-*strychnine* (7).

11.2.2 Construction approach

Similar to organic synthesis, retrosynthetic analysis facilitates the synthesis of isotopically labeled compounds. Bond cleavages identify reasonable synthesis steps of isotope chemistry, as well as readily available intermediates and starting materials. The scheme shown in Fig. 11.10 summarizes this approach.

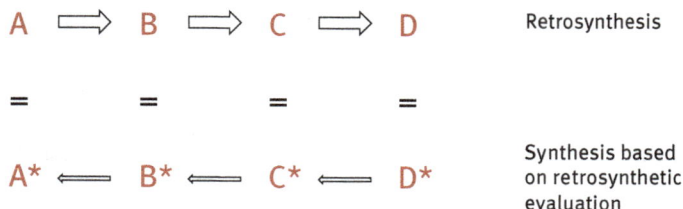

Fig. 11.10: Outline of the "construction approach".

An example is the synthesis of *imitanib* (Salter 2006 and Fig. 11.11). The strategies place the ^{14}C-label in different positions of the molecule: Route A in the amine moiety, route B in the carboxyl moiety of the amide. Both strategies follow the sequence of research chemistry and derive the label from readily available building blocks, i.e., [^{14}C]CO$_2$ and [^{14}C]cyanamide, respectively.

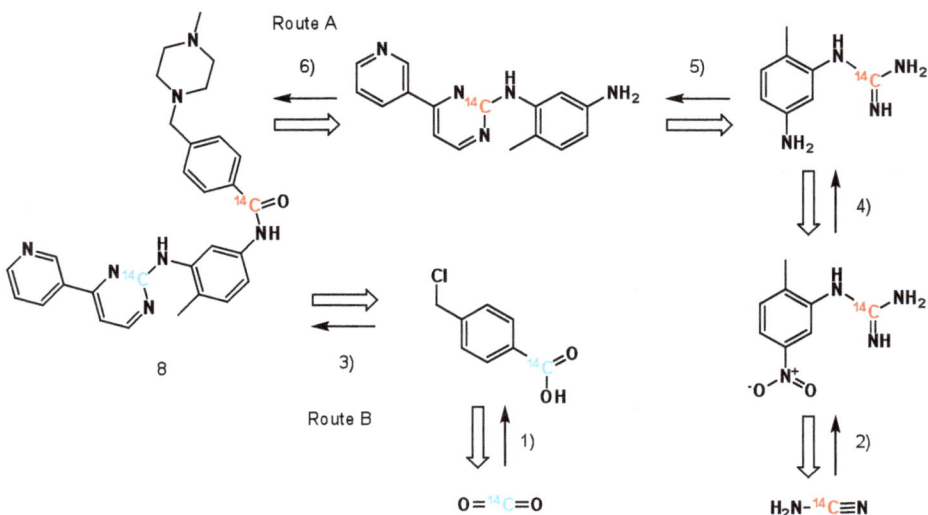

Fig. 11.11: Retrosynthesis/synthesis of [^{14}C]*imitanib* (**8**). Reaction conditions: 1) ClCH$_2$PhMgI, Et$_2$O; 2a) HCl (g), dioxane; b) n-BuOH, 100 °C, 6 h; 3a) SOCl$_2$, DMF (catalytic amount); b) THF, r.t., 16 h; 4) Pd/C, H$_2$, n-BuOH; 5) n-BuOH, 130 °C, 3.5 h; 6) 1-methylpiperazine, EtOH, 45 °C, 5 h.

11.2.3 Reconstitution approach

The most common strategy for isotope labeling is the "reconstitution" approach, which includes the "direct replacement exchange", "indirect replacement" and "disconnection/reconnection" approaches. All strategies start from the unlabeled drug substance (A), but differentiate in the structural modification of their intermediates (A, A′, B), cf. Fig. 11.12.

While direct replacement ("exchange") results in the isotope analog, "indirect replacement" requires a structurally slightly modified intermediate (A′), which then allows incorporation of the isotope, cf. Fig. 11.13. An example for this type of labeling is the halogen/tritium exchange.

A first synthetic step introduces the halogen, which will then selectively be exchanged for the tritium isotope. Not only the regio-selective introduction of the halogen is critical, but also selective exchange for tritium. Based on this strategy, highly

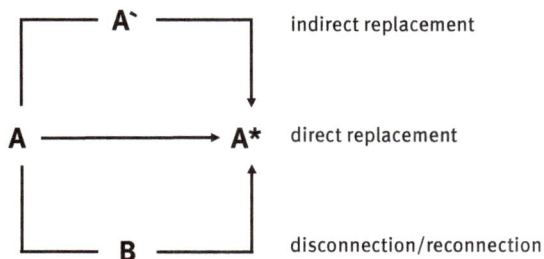

Fig. 11.12: Outline of the "reconstitution approach".

Fig. 11.13: Br/T-exchange labeling – an example for indirect replacement. Reaction conditions: 1) Br$_2$, CH$_3$COOH, r.t., 4 h; 2) ^{3}H$_2$, Pd/C 5%, HCl, r.t., 4 h.

tritiated dopamine (9a) could be prepared at a specific activity of 1.6 TBq/mmol,[16] a suitable tracer for binding and ADME studies (Filer 2006).

Strategically similar is the "disconnection/reconnection approach", but it differs with some modification of intermediate (B). This strategy substitutes part of the drug substance by its isotope analog. An example is the selective synthesis of ω-D$_6$-*tocotrienol* (11a)[17] (West 2008). The modified parts of the molecule are highlighted (Fig. 11.14).

Oxidative cleavage of the terminal isopropenyl moiety (steps 1b–e) of the *tert*-. butyl(dimethylsilyl)-protected ω-D$_6$-*tocotrienol* (11) results in the aldehyde derivative (12), which was reacted with per-deuterated phosphonium ylide to obtain the protected ω-D$_6$-tocotrienol (13). Deprotection allowed access to the isotope analog of the ω-D$_6$-tocotrienol (11a). Overall, this synthetic sequence substitutes the terminal unlabeled isopropyl moiety by the corresponding D$_6$-labeled moiety. The advantage of this strategy is obvious. While the major part of the drug substance remains unaffected, only the isotopically labeled substituents will be subject to synthetic manipulations.

---

16 Competing exchange processes with the solvent (HCl) might explain that only 50% of the theoretical specific activity was achieved.
17 This example makes use of deuterium (^{2}H) – a nonradioactive hydrogen isotope. The resulting ω-D$_6$-*tocotrienol* (11a) is used as internal standard for mass-spectroscopic quantification of ω-*tocotrienol* (11).

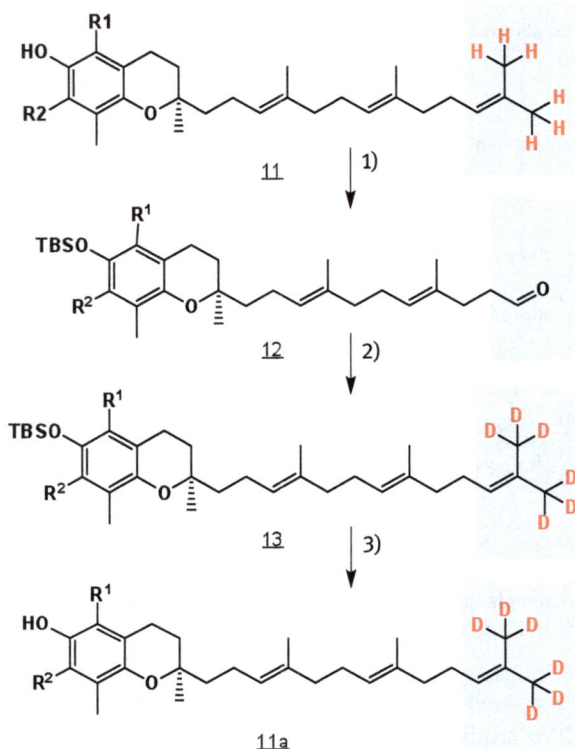

Fig. 11.14: Synthesis from ω-^{2}H$_6$-*tocotrienol* (11a). Reaction conditions: 1a) TBSOTf, imidazole, DMF; b) NBS, THF, H$_2$O; c) K$_2$CO$_3$, CH$_2$Cl$_2$, MeOH; d) HClO$_4$, THF, water e) NaIO$_4$, THF, water 2) Ph$_3$PC^{2}H (C^{2}H$_3$)$_2$ Br, LiHDMS, THF; 3) TBAF, THF.

11.3 Labeling with tritium

The different valencies of carbon and hydrogen explain their individual molecular behaviors. While carbon's polyvalence allows it to constitute the structure of organic molecules, the monovalent hydrogen saturates free valences. Therefore, hydrogen located in the periphery of the molecule is chemically less stable and more easily accessible to enzyme-catalyzed degradation. Any application of tritium in tracer experiments must therefore ensure adequate label stability. The labeled compound should be evaluated in this aspect and, if necessary, tested experimentally to confirm its suitability. In practical terms, this includes testing for chemical stability (i.e., pH stability at various temperatures) as well as *in vivo* stability (i.e., *in vivo* rat stability study) (Krauser 2013).

11.3.1 The source of tritium

Commercially available tritium is produced in form of tritium gas ($^3H_2$ or T_2) by nuclear reaction $^6Li(n,^4He)^3H$. Since $^3H_2$ is the origin of all tritium labeling, safe and efficient handling is a key prerequisite. In contrast to the previous "Toeppler Pumpen" technique, the uranium trap technique is current state of the art. The underlying principle is the equilibrium between uranium tritide (UT_3) and tritium gas (T_2), cf. eq. (11.2).

$$2\,UT_3 + energy \rightarrow 2\,U + 3\,T_2\,. \tag{11.2}$$

Heating of uranium tritide releases tritium gas (T_2-pressure, 1 atm/760 torr at 436 °C), which will be nearly completely reabsorbed ($1.3 \cdot 10^{-3}$ Pa (10^{-5} torr) at 25 °C) after cooling. Many years of research development of this technology resulted in a series of well-engineered commercially available systems. An example is the system engineered by TRITEC AG (Switzerland) given in Fig. 11.15.

11.3.2 Tritiated building blocks

Tritium gas is the common starting material for tritiations of unsaturated bonds. In addition, derivatives of tritium are synthesized from T_2 to serve as so-called building blocks. These are small, easy to synthesize species, which subsequently provide alternative synthesis routes for tritium labeling. An example is CT_3 I, tritiated methyl iodide, which is applied to methylation reactions. Figure 11.16 shows the most relevant T-building blocks. Along with tritium gas, tritides and methyl iodide three building blocks will be discussed with their specific applications.

Even if a given reaction is considered trivial in an organic chemistry laboratory, synthesis and safe handling of radioactivity is not a trivial matter. An example is generation of tritiated water (HTO), as another building block. Major challenges in HTO production are high radiotoxicity, autoradiolytic decomposition and working on an extremely small mass scale.[18] One protocol uses oxidation of T_2 with PdO in THF producing HTO at specific activities of 370–1500 GBq/mmol (i.e., 17–70% of the theoretical activity). HTO synthesized under these conditions is commercially available.

11.3.2.1 Tritium gas as a building block

Exchange labeling ("direct replacement") using T_2 or tritiated water (HTO) as source, provides the shortest and most direct access to isotope analogs. The original way of

---

18 At its maximum theoretical specific activity 37 GBq of tritiated water would correlate to a volume of 0.3 µl.

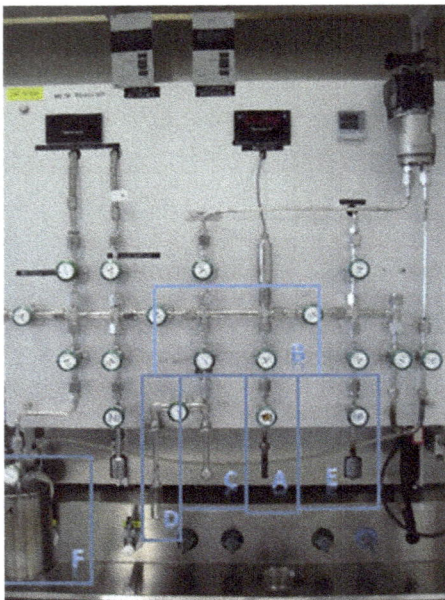

Fig. 11.15: TRITEC AG (Switzerland) uranium trap system allowing safe handling of tritium gas. Careful heating of the storage vessel (A) containing depleted uranium releases tritium gas (T_2) into the compartment (B). Controlled release into the reaction compartment (C) allows exact determination of the added amount of tritium gas. From the pressure difference (before and after release into the reaction compartment), the volume of compartment B and the specific activity of the tritium gas the exact amount of activity (tritium gas) released to compartment C can be calculated. After completed reaction, there is reabsorption of the unreacted amount of tritium gas to the waste vessel (E). As a safety measure, an additional vessel (F) allows the complete retention of the volatile activity avoiding any release to the environment.

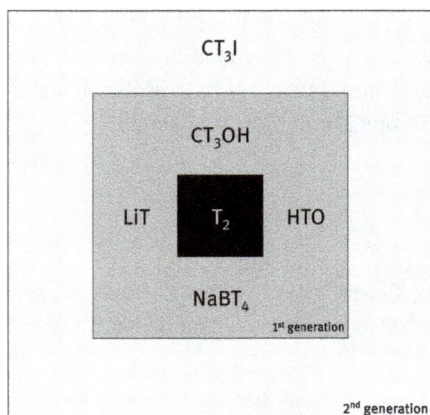

Fig. 11.16: Tritium labeled building blocks of different generations from T_2 gas.

exposing the undissolved drug substance to an atmosphere of tritium gas (WILZBACH method) made exclusive use of the disintegration energy.[19] In the following decades, activation was optimized (temperature, acid/base, and metal catalysis). A wide range of available catalyzed methodologies provides not only increased incorporation (i.e., increased specific activities), but also distinct regioselectivity for certain cases.[20] Iridium(I)-phosphine catalyzed exchange reactions will be discussed as the most recent type of this approach (Salter 2010).

Certain iridium(I) phosphine complexes catalyze the exchange of aromatic hydrogens with tritium gas, whereby suitable coordination sites at the drug substance are essential for their activation. In case of the labeling of *tolmetin* (Hickey 2007 and Fig. 11.17), both carbonyl groups (benzoyl- and carboxyl-group) allow the incorporation of T_2 into the aromatic *ortho*-positions via 5- and 6-metallacyclic intermediates. The individual incorporation (i.e., degree of labeling per position) adds up to a total specific activity of 1.5 TBq/mmol.[21]

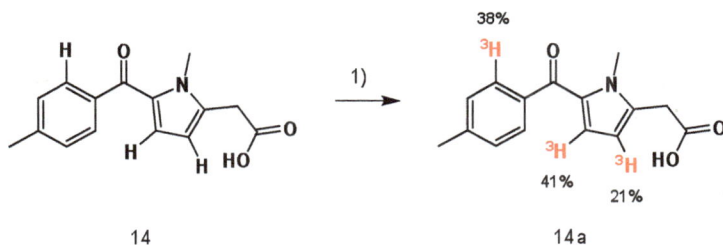

Fig. 11.17: Iridium(I)-catalyzed exchange labeling of *tolmetin* (14). Reaction conditions: 1) solvated polystyrene-bound triphenylphosphine complex of cyclooctadienyl iridium(I) hexafluorophosphate (1.1 equivalents), CH_2Cl_2, $^3H_2$, 3 h.

The polymer-supported iridium(I) exchange catalyst applied is the result of intensive, systematic catalyst development. Its ability to form not only 5- but also 6-metallacylic intermediates, its compatibility with functional groups such as free carboxylic acids and its immobilization, which affords easier work-up, provide differentiation from the recent Crabtree catalyst ([codIr PCy$_3$ py]PF$_6$).

Tritium gas, is also directly applied for the reduction of unsaturated bonds and for halogen/tritium exchange reactions. Despite numerous incompatibilities with other reducible functionalities, these reduction reactions are still most commonly used[22] in

---

19 The energy formed during radioactive decay is called disintegration energy.

20 For a review of this broad field see Atzrodt (2007), Lockley (2010).

21 Estimation of the molecular specific activity: $(0.21 + 0.41 + 0.38 + 0.38 = 1.38) \cdot 1.08$ TBq/mmol $= 1.5$ TBq/mmol.

22 For 2009 a statistical analysis suggested that 45% of the tritiations were accomplished by heterogeneously catalyzed reductions (Meisterhans 2010).

tritium chemistry. Although these reactions are experimentally very straightforward, the synthesis of the suitable precursor ("labeling precursor") might become a major problem. This is typically the case if the respective functionalization (i.e., introduction of a double bond or a halogen) of the desired labeling position is synthetically not easily accessible.

An example for difficulties encountered with heterogeneous catalyzed tritiations is the labeling of the phospholipase A2 inhibitor 16 (given in Culf 1996 and Fig. 11.18). Since its application in ^{3}H-NMR studies required exclusive labeling of the positions 3 and 4 of the decanecarboxylic acid, the synthetically accessible olefin 15 was considered a suitable precursor. In an initial experiment, the heterogeneous catalyzed reduction (method 1a: Pd/C 10%, T_2, CH_3OH) resulted in a product (16a) with an unexpected high specific activity (found: 3.15 TBq/mmol; expected 2.16 TBq/mmol). ^{3}H-NMR spectroscopy data showed distribution of the label over the complete alkyl chain, which indicated a combination of double bond migration, saturation and nonspecific exchange processes. This reaction product did not meet required specifications and was not suitable for study purposes. Alternatively, the reduction was performed under homogenous conditions (method 1b: $RhCl(PPh_3)_3$, T_2, Benzene/EtOH 3/2) providing the reaction product (16b) at the expected specific activity of 2.1 TBq/mmol. (16b) is labeled in the positions 3 and 4, exclusively, cf. Table 11.5. This example demonstrates not only that isotopes can visualize mechanisms (side reactions) not detectable with hydrogen, but also underlines that a complete analytical characterization of isotopically labeled compounds must include structural characterization of the labeled positions (i.e., for tritiated compounds ^{3}H-NMR, for C-14 labeled compounds a ^{13}C-NMR analysis).

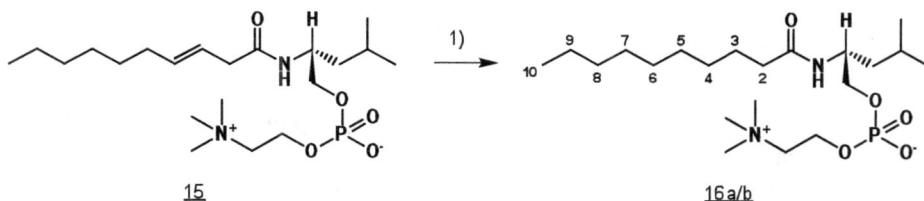

Fig. 11.18: Method-dependent distribution of tritium on the example of phospholipase A2 inhibitor 16. Synthesis of the tritium labeled phospholipase A2 inhibitor 16.

In addition, electrophilic aromatic halogenations via T_2 allow direct functionalization of drug substances. In general, chlorine, bromine, and iodine can be used for halogen/tritium exchange reactions. An example – as given in Thurkauf (1997) and Fig. 11.19 – is the multiple tritiation of *NGD 94–1* (18), a dopamine D_4 neurotransmitter analog resulting in a tracer (18a) of specific activity of 2.1 TBq/mmol. The specific activities experimentally achievable in the halogen exchange are increasing from chlorine to iodine (Tab. 11.6).

Tab. 11.5: Reaction conditions applied to the reduction of the alkene (15).

Method	Reaction conditions	Position of label	Distribution (%)
1a)	T_2, Pd/C, EtOH	2	1.9
		3	29.7
		4–9	67.0
		10	1.4
1b)	T_2, RhCl(PPh$_3$)$_3$, benzene/EtOH 3/2	3	49.0
		4	51.0

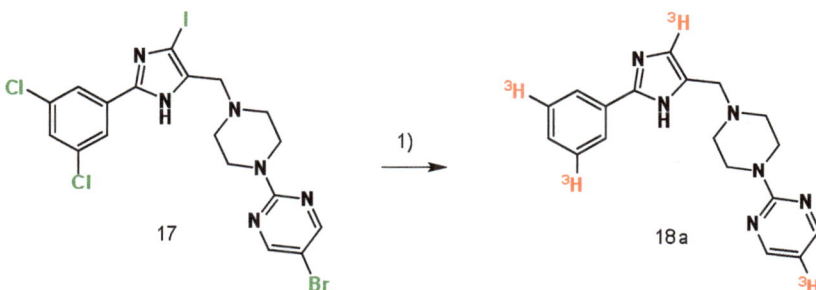

Fig. 11.19: Multiple tritiation of the poly-halogenated precursor (17) to yield the dopamine D$_4$-affine tracer *NGD 94-1* (18a). Reaction conditions: 1) $^3H_2$, Pd/C 10%, MeOH, r.t., 1 h.

Tab. 11.6: Achievable specific activities by halogen/tritium-exchange reactions.

Halogen substituent X	Range of specific activities (GBq/mmol)
iodo	740–1030
bromo	520–920
chloro	290–590

However, the chemical stabilities of the halogenated precursor are decreasing from chloro- to iodo-compounds. Bromo-derivatives are often considered a suitable compromise. Reaction additives allow not only interhalogen selectivity (reaction rate of X/T-exchange: iodo > bromine > chlorine), but also tritiodehalogenation in the presence of other reducible functionalities (–NO$_2$, double bonds). Selection of the solvent, the type of the catalyst and type and stoichiometry of additives further influence the reaction rates and therefore allow selectivity.

In cases of neighbor group activation, tritiodehalogenation reactions are also observed at sp^3-carbons. An example is the labeling of (21) (Szammer 2005) as given in Fig. 11.20.

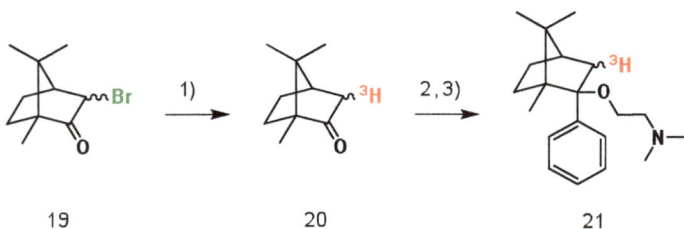

Fig. 11.20: Br/T exchange reaction at an activated sp^3-center with subsequent refunctionalization. Reaction conditions: 1) $^3H_2$, Pd/C (10%), dioxane, 3 h; 2) PhMgBr, diethyl ether; r.t., 1 h; 3a) NaH, toluene, reflux, 1 h; b) $(CH_3)_2NCH_2CH_2Cl$, reflux, 3 h.

Reduction of the α-bromo ketone (19) results in the α-^3H-camphor (20). Since its label is highly susceptible to acid/base-catalyzed enolization, reaction with phenylmagnesium bromide and subsequent etherification of the tertiary alcohol is performed to produce [^3H]deramciclane (21). This approach, the use of functionality to incorporate the label followed by subsequent derivatization to avoid any re-exchange, is of strategic importance.

11.3.2.2 Tritide

Tritides[23] were initially accessible by thermal exchange with tritium gas. In case of LiT, the exchange (330 °C, 16 h) of 3.0 mmol LiH with 1.9 mmol carrier-free[24] T_2-gas results in a product with a specific activity of 4.2 TBq/mmol (39% of the theoretical specific activity). NaBT$_4$ at 2.2–3.0 TBq/mmol (i.e., 50–70% of the theoretical specific activity) was one of the few tritides commercially available.[25] Most important is the highly reactive LiH, cf. eq. (11.3).[26]

$$n\text{-BuLi} + T_2 \xrightarrow{\text{(TMEDA, THF)}} n\text{-BuT} + \text{LiT}. \tag{11.3}$$

---

23 Comparable to hydrides, tritides are compounds of tritium with more electropositive elements.
24 Carrier-free in this context means undiluted, i.e., not diluted with hydrogen.
25 Starting from NaBT$_4$ derivatizations to other tritides (such as NaBT$_3$CN and Na/NMe$_4$ B^3H (OAc)$_3$) with different stabilities and selectivities were reported.
26 For instance, a carefully degassed solution of n-BuLi and TMEDA (= N,N,N′,N′-tetramethylethylenediamine) is reacted with T_2 on the tritium handling manifold at 50–90 kPa for 0.5–2 h to form a white, creamy suspension of LiT. After reabsorption of excess T_2 and lyophilization of all volatile constituents (i.e., n-hexane, TMEDA), LiT is suspended in THF and combined with respective Lewis acids (such as BR$_3$, BF$_3$ OEt$_2$, AlBr$_3$, R$_3$SnX, CP$_2$ZrCl$_2$).

The reaction is applicable to practical handling levels of 0.05 to 2.0 mmol. This procedure provides access to a broad series of tritides, cf. Andres (2001) and Fig. 11.21.

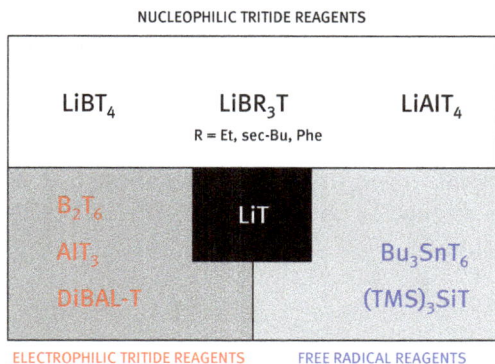

Fig. 11.21: Variety of tritide reagents accessible from highly reactive LiT.

With these reagents available, several protocols of organic chemistry can be applied to isotope labeling chemistry. For instance, as given in Fig. 11.22, the "Oxidation Reduction Strategy" using [^{3}H]-L-*selectride* (Li(sec-Bu$_3$)B^{3}H) produces [35–^{3}H]*sanglifehrin A* (22a) at a specific activity of 943 GBq/mmol (Wagner 2005). After selective protection of the C$_{31}$-hydroxy group as its trichloroacetate, Dess–Martin periodinane oxidizes the free C$_{35}$-OH-moiety into the respective ketone. Careful deprotection produces the immediate precursor (23), which is stereoselectively reduced to *sanglifehrin A* by [^{3}H]L-selectride.

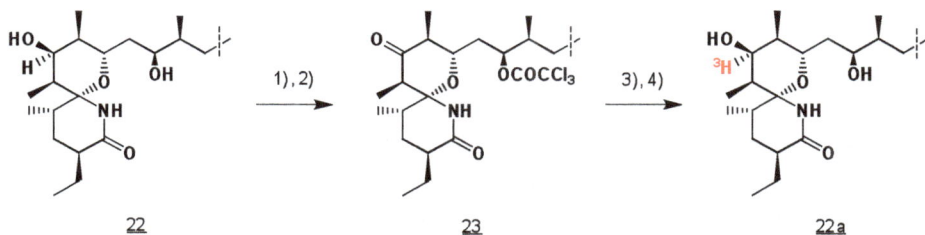

Fig. 11.22: Labeling of *sanglifehrin A* in position 35 (only part of the complete molecule is shown). Reaction conditions: 1a) Ac$_2$O, DMAP, CH$_2$Cl$_2$/pyridine 27:1; b) (Cl$_3$CO)$_2$O, DMAP (catalytic); 2) Dess–Martin periodinane, CH$_2$Cl$_2$; 3) 1.0 M aqueous NaHCO$_3$/MeOH 10:1; 4) Li(sec-Bu$_3$)B^{3}H ([^{3}H]L-selectride), THF, – 20 °C.

11.3.2.3 Methyl iodide (CT$_3$I)

Another important building block of tritium chemistry is tri-tritiated methyl iodide (CT$_3$I). Since it allows simultaneous incorporation of three tritium atoms, it provides access to highly tritiated isotopo-analogs at specific activities 2.9–3.1 TBq/mmol.[27] Ndash-, O-, and S-methyl groups provide the potential for the "reconstitution approach". Demethylation of the original drug substance first provides the respective des-methyl compound, which next allows access to the tritiated isotope analog by re-methylation. Even if there is a clear tendency for their *in vivo* instability, they represent valuable tracers.

The main disadvantage of tritiated methyl iodide, however, is not only its chemical instability, but also its high volatility associated with its low vapor pressure (bp. 43 °C). This volatility is a radioprotection risk.[28] Therefore, development of a less volatile tritium-containing reagent such as methyl nosylate offers a practical alternative, cf. Pounds (2004) and Fig. 11.23.

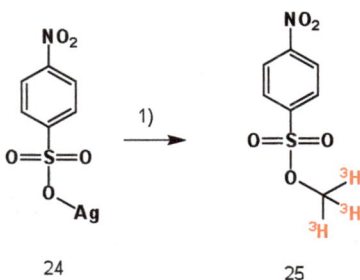

Fig. 11.23: Radiolabeled [^{3}H]-methyl nosylate (25) synthesized via [^{3}H]methyl iodide. Reaction conditions: 1) C^{3}H$_3$I, CH$_3$CN, 80 °C, 12 h.

With comparable reactivity methyl nosylate demonstrates not only increased stability, but also a significantly decreased volatility. Therefore, handling of methyl nosylate is much more convenient than that of methyl iodide.

11.3.2.4 Learning example

Labeling of *AZ8349* (26), cf. Fig. 11.24, at a high specific activity (>1 TBq/mmol; i.e., >1 T/molecule) summarizes some of the strategies described above (Malmquist 2012). Interestingly, the publication introduces not only the solution, but also discusses some failures. The following strategies were tested:

---

27 Commercially available CT$_3$I is produced by catalytic reduction of CO$_2$ with T$_2$ and subsequent reaction with HI. Tritiodehalogenation of 4-trichloromethoxybiphenyl with subsequent HI-mediated ether cleavage provides an alternative for the small scale (Voges 2009).
28 It is a general goal of isotope chemistry to design in a way to minimize the volatility of the labeling reagents and intermediates (and therefore to avoid any risk of contamination and incorporation), known as the "devolatilization concept".

- **Option 1:** H/^{3}H-exchange labeling of *AZ8349* (26) with iridium(I) complex: Successful H/^{3}H-exchanges with iridium(I)-type catalysts reported for comparable structural features prompted the authors to test this approach. Application of an iridium (I) N-heterocyclic carbene (NHC) complex, however, was unsuccessful in this case.
- **Option 2:** Methylation of the O-desmethyl compound: The respective O-desmethyl derivative of *AZ8349* (26) reacted with methyl iodide under exclusive N-alkylation. Previous protection of the aniline moiety with N-*boc.* (N-*tert.* butyloxycarbonyl) and N-SEM (β-(trimethylsilyl)ethoxymethyl) did not give the expected product.
- **Option 3:** Acylation of (28) with [^{3}H]acetic anhydride: Since downscaling of the N-acylation to a scale reasonable for tritium chemistry was not successful, the following strategy was selected. N-Acylation with tribromoacetyl chloride gave the precursor (28). Heterogeneous catalyzed Br/T-exchange reaction in the activated acyl position produced [^{3}H]-*AZ8349* in a μmole scale.

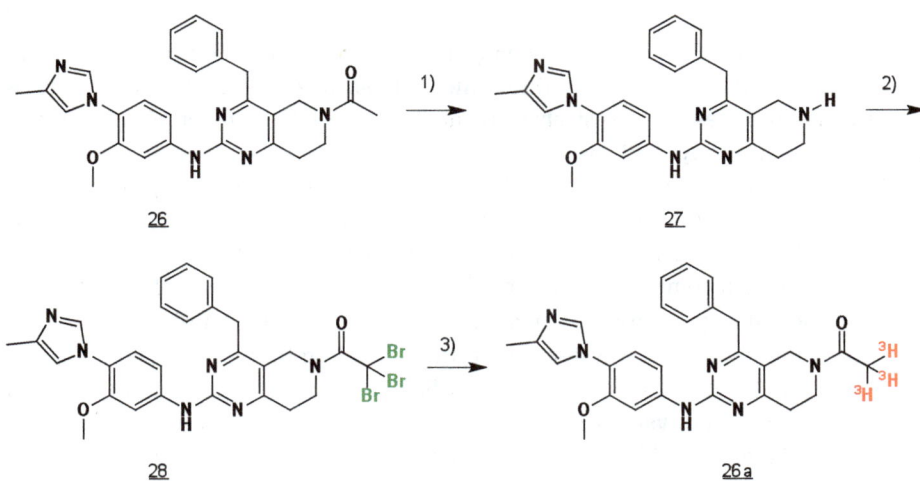

Fig. 11.24: ^{3}H-labeling of *AZ8349* (26). Reaction conditions: 2) 2,2,2-tribromoacetyl chloride, NEt$_3$, THF, room temperature, 150 min; 3) ^{3}H$_2$, Pd/C 10%, EtOH, r.t., 12 h.

The above reaction conditions resulted in ^{3}H-labeled *AZ8349* at a specific activity of 1.2 TBq/mmol (= 38% of the theory). The relatively low specific activity might be due to competing exchange processes with exchangeable protons of the solvent (EtOH). Therefore the authors hypothesized that optimization (= lowering) of the reaction time might increase the specific activity. Positioning the label in the α-position of an amide, however, has to take into consideration a distinct risk of limited chemical and *in vivo* stability.

11.4 Labeling with carbon-14

11.4.1 The source of carbon-14

Carbon-14 is produced in a nuclear reaction by irradiating nitrogen-14 with neutrons, i.e., $^{14}N(n,p)^{14}C$, and subsequently oxidized to $[^{14}C]CO_2$. Therefore, $Ba[^{14}C]CO_3$ (alternatively $Ca[^{14}C]CO_3$) are the forms for commercialization and storage. Due to the ubiquitous presence of carbon, the theoretical specific activity of C-14 (2.3 GBq/ mmol) is difficult to achieve. Common isotopic purities of commercially available $Ba[^{14}C]CO_3$ are in the range from 80% to 90%, correlating to specific activities from 1.8 to 2.1 GBq/mmol.

11.4.2 ^{14}C-precursors

The versatility of C-14 chemistry is dependent on the availability of differently functionalized, ^{14}C-labeled building blocks. Due to their fundamental importance, the syntheses of early intermediates are intensively elaborated and available[29] in optimized protocols (cf. for example Murray-Williams 1958). Figure 11.26 gives an overview.

The key intermediate is $Ba[^{14}C]CO_3$. It provides access to $[^{14}C]CO_2$, $BaN[^{14}C]CN$, $[^{14}C]C_2H_2$, and $M[^{14}C]CN$ (M = Na, K). These "first generation" building blocks allow synthesis of further, more sophisticated building blocks. Therefore, Fig. 11.25 might be considered as a combination of the different synthesis branches: the **carbon dioxide** branch, the **cyanamide** branch, the **acetylene**, and the **cyanide** branch.

According to the above system, the third-generation building block $[2-^{14}C]$acetone (i.e., $(CH_3)_2\,[^{14}C]CO$) is accessible from $Ba[^{14}C]CO_3$ via $[^{14}C]CO_2$ and $CH_3\,^{14}COOH$ with subsequent pyrolysis of Li-, resp. $Ba[1-^{14}C]$acetate. Alternatively to this sequence, reaction of $[1-^{14}C]$acetyl chloride with Me_2Cd or direct reaction of $[^{14}C]CO_2$ with MeLi might be used.[30]

Different isotopomeric forms of small molecules are distinct entities. For example, three different isotopomeres of C-14 labeled acetic acid: $[1-^{14}C]$acetic acid (29a), $[2-^{14}C]$acetic acid (29b) and $[1,2-^{14}C]$acetic acid (29c), cf. Fig. 11.26. While $[1-^{14}C]$acetic acid (29a) is synthesized in one step by carboxylation of CH_3MgX (X = Br, I) with $[^{14}C]CO_2$, $[2-^{14}C]$acetic acid (29b) requires a four step sequence: reduction of $[^{14}C]$ CO_2 to $[^{14}C]CH_3OH$, iodination to $[^{14}C]CH_3I$, nucleophilic substitution to $[^{14}C]CH_3CN$ and subsequent hydrolysis to the title compound (29b). Doubly labeled $[1,2-^{14}C]$acetic acid (29c) requires either combination of both strategies (carboxylation of $[^{14}C]$

---

29 The synthesis of building blocks is usually outsourced to external (specialized) suppliers, while the isotope laboratories focus more on the labeling of later stage intermediates and/or final steps.
30 Since both of the reaction products produced by this approach require thorough purification by fractionated distillation, this alternative is considered less attractive.

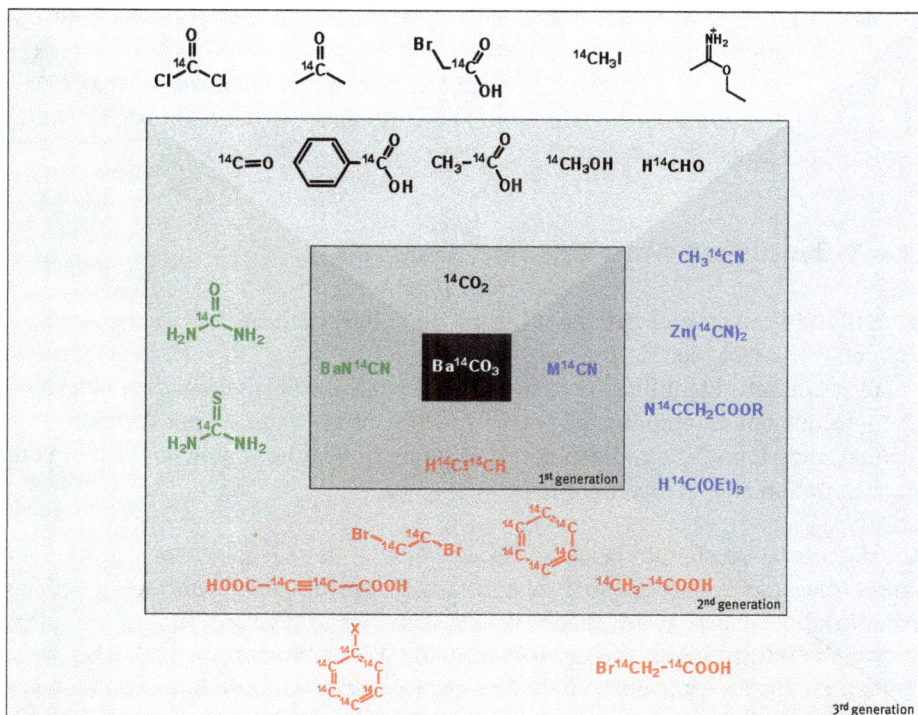

Fig. 11.25: Step-by-step synthesis of key building blocks of C-14 chemistry.

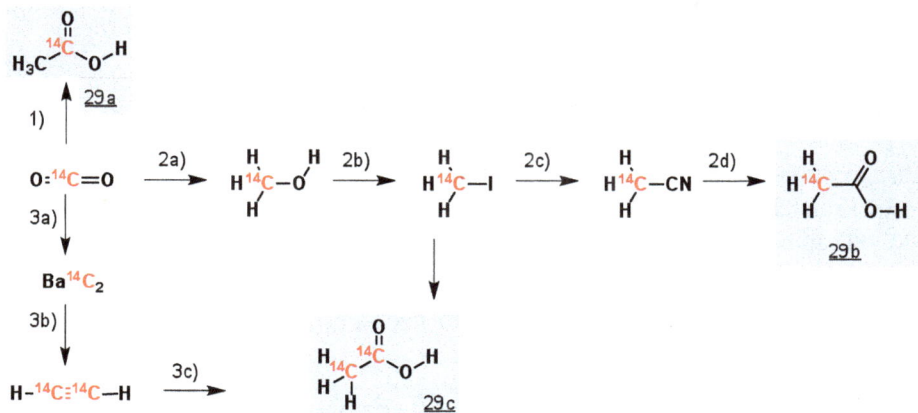

Fig. 11.26: Syntheses of differently labeled isotopomeres of ^{14}C-labeled acetic acid. Reaction conditions: 1a) CH$_3$MgI, ether, − 78 °C, 5 min; b) NaOH, Ag$_2$SO$_4$, H$_2$O; c) H$_2$SO$_4$ 50%, steam distillation; 2a) LiAlH$_4$, diethylene glycol dimethyl ether; b) HI, 40–90 °C; c) KCN, H$_2$O, r.t., 16 h; d) 20% NaOH, 80 °C, 4 h; H$_2$SO$_4$, AgSO$_4$, steam distillation; 3a) Ba, 800–830 °C; b) HClO$_4$/H$_2$O 1/1, reflux; expulsion with hydrogen; c) HgO, (NH$_4$)$_2$S$_2$O$_8$, aq. H$_2$SO$_4$, r.t., 16 h; Cu$_2$O, 40 °C, steam distillation.

CH$_3$MgI with [^{14}C]CO$_2$) or HgO-mediated hydrolysis of [^{14}C$_2$]acetylene with subsequent oxidation by ammonium persulfate.

This underlines that isotope labeling of individual isotopomers, such as (29a), (29b), and (29c), requires specific synthetic strategies (with usually significantly varying labor costs).[31]

11.4.3 Syntheses strategies for ^{14}C-tracers

More than in the case of tritium, ^{14}C-labeling is the result of *de novo* syntheses.[32] Main objectives for their design are
– retrosynthetic identification of straightforward, easy to label building blocks
– development of reproducible syntheses with optimal radiochemical yields
– minimization of the radiochemical steps and avoiding any major risk of contamination and/or incorporation

Not all cases allow simultaneous accomplishment of these requirements. Most often short radiosyntheses are economical and capacity saving. Short syntheses give high overall radiochemical yields, keep specific activities at high levels, and provide a responsible utilization of the radioactivity and a low number of radioactive steps limits the capacity-consuming handling of radioactivity to a minimum. Three options for designing short radiosyntheses are highlighted below. They are
– linear synthesis with late-stage introduction of the label
– convergent synthesis with an advanced unlabeled building block
– degradation-reconstitution approach.

11.4.3.1 Linear synthesis with late-stage introduction of the label

Linear syntheses with the introduction of the label in a very late stage are most convenient. Examples are the C-14 labeling of *PKC412* (30) and *midazolam* (36).[33] Structurally, PKC412 can be considered as the amide of the benzoic acid with the secondary amine of *staurosporine* (32). The *in vivo* stability of this functionality[34] allowed the opportunity to place the C-14 label not only in the structurally easily accessible benzoyl moiety and

---

31 Synthetically more complex examples are discussed for arene isotopomers (Gregson 2011).
32 The term "*de novo* synthesis" refers to the synthesis of structurally complex molecules from simple starting materials.
33 C-14 labeled *PKC412* (N-benzoyl-*staurosporine*) (30a) was requested for ADME studies requiring its label in a metabolically stable position (Burtscher 2010).
34 Environmental risk assessment studies, however, required an isotopomer labeled in the *staurosporine* moiety itself. Due to the chemical complexity of this molecule a fermentative approach was

but also to introduce it in the last step of the synthesis. Thus, labeling of PKC412 could be performed in a two-step synthesis, as given in Fig. 11.27: carboxylation of phenyl-magnesium iodide with $^{14}CO_2$ and reaction of the [carboxyl-^{14}C]benzoic acid (31) with unlabeled *staurosporine* (32), which provided ^{14}C-*PKC412* (30a) in an overall radiochemical yield of 62% (related to Ba[^{14}C]CO_3).

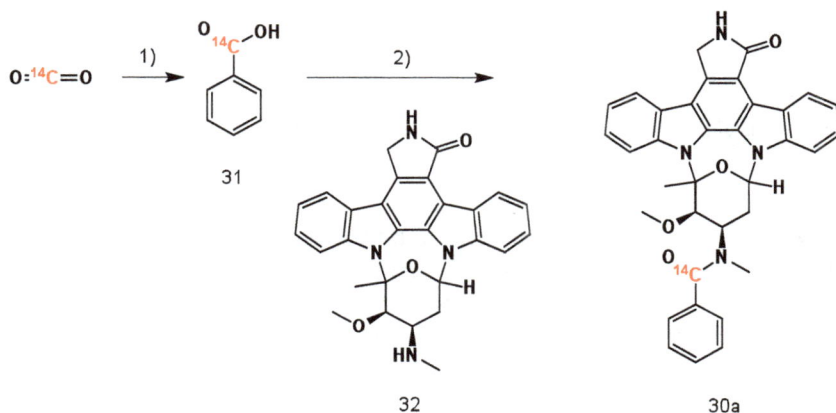

Fig. 11.27: Synthesis of C-14 labeled PKC412 (30a). Reaction conditions: 1) PhMgI; 2) 2-chloro-1-methyl pyridinium iodide, triethylamine, CH_2Cl_2, 12 h, r.t., flash chromatography.

Another example is the labeling of *midazolam* in its imidazole moiety as given in Fig. 11.28. Retrosynthetic analysis identifies *midazolam* as a cyclic amidine derivative (35), which suggests its construction by reaction of the bisamine derivative (33) with the respective acid equivalent (34). (2-amino-5-chlorophenyl)(2-fluorophenyl)methanone (33) becomes a key intermediate that allows reaction with ethyl [1–^{14}C]acetimidate hydrochloride (34)[35] to the immediate precursor (35), which is subsequently oxidized to [^{14}C]*midazolam* (36a) (Easter et al., 2011).

11.4.3.2 Convergent synthesis with an advanced unlabeled building block

If late-stage introduction of the label is not feasible, a convergent synthesis design might help to minimize the number of radioactive steps. Instead of constructing the isotope analog step-by-step, combination with an advanced building block provides an opportunity, cf. Fig. 11.29 for the ^{14}C-labeling of *NVP MTH958*. In contrast to the linear

---

applied. Addition of [3–^{14}C]tryptophan to the fermentation broth resulted in doubly labeled [4c,7a-$^{14}C_2$]*staurosporine* with a specific activity of 700 to 1300 MBq/mmol (Burtscher 2010).

35 In contrast to the orthoacetate used in the nonlabeled synthesis, ethyl [1–^{14}C]acetimidate hydrochloride (34) is more easily accessible and more reactive.

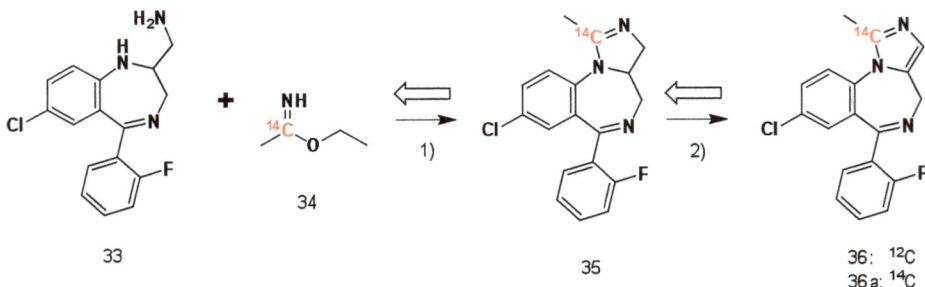

Fig. 11.28: Synthesis of [¹⁴C]*midazolam* (36a). Reaction conditions: 1) EtOH/THF, 0 °C; 2) MnO₂, toluene, reflux.

synthesis, which required six synthetic steps to transform the alanine derivative (37) into the final drug substance (39), the convergent synthesis making use of the advanced intermediate (38) shortened the radiosynthesis significantly. Cutting down the number of radiolabeled steps from 6 to 2 increased the overall radiochemical yield from 1 to 9% (Voges 2001).

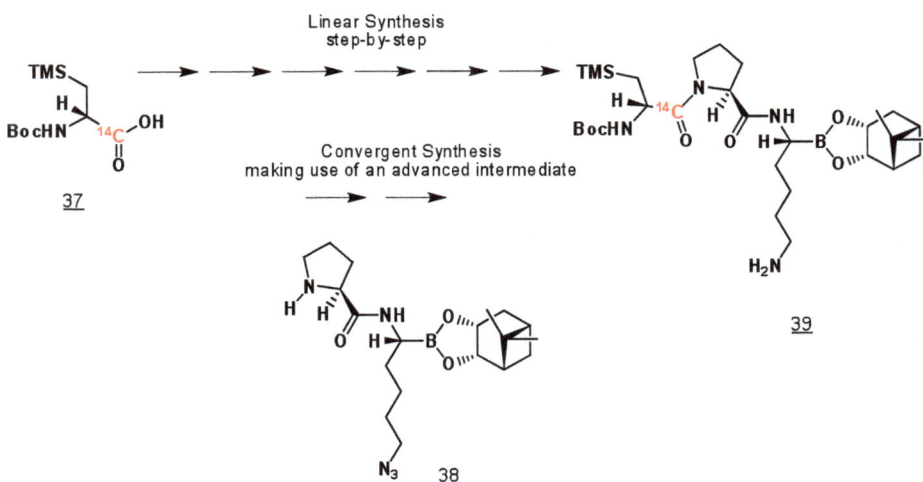

Fig. 11.29: Convergent synthesis of [¹⁴C]*NVP MTH958* (39).

11.4.3.3 Disconnection-reconstitution approach

Allowing access to structurally complex starting materials the "degradation/re-constitution" approach is of particular interest as already indicated in Fig. 11.12. The trimethylsilyl-labeling (-Si[¹⁴C]CH₃(CH₃)₂) of *JNZO92* (43) requires the bromo octahydrobenzoquinoline derivative (41) as intermediate. This compound is accessible via a long and low-yielding synthesis with diastereo- and enantio-selective purifications. Alternatively (Fig. 11.30), the availability of unlabeled *JNZ092* (43) itself allows access to (41) by an easy and straightforward "degradation" approach (Moenius 2007). Acid/catalyzed desilylation, regio-selective bromination, and bromo/silyl exchange with $Cl_2Si(CH_3)_2$ produced the immediate, stereoselective precursor (41), which reacted with $^{14}CH_3Li$ undergoing a chloro-/methyl-exchange reaction to give [^{14}C]*JNZ092* (43a).[36]

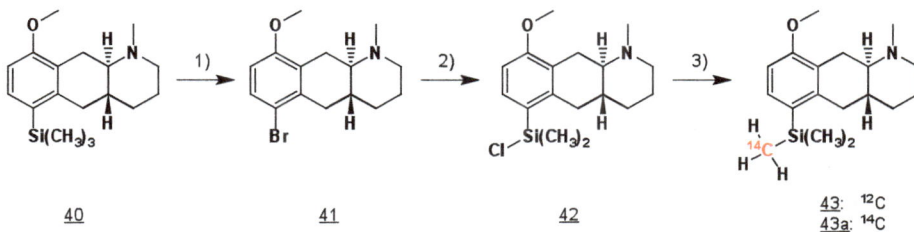

Fig. 11.30: Trimethylsilyl-labeling of *JNZ092* (43). Reaction conditions: 1a) 0.2N HCl, 80 °C; b) Br$_2$, CCl4; 2) *tert*-BuLi, THF, – 78 °C, 1 h, 0 °C, Cl$_2$Si(CH$_3$)$_2$, 2 h; 3a) synthesis of ^{14}CH$_3$Li, *tert*-BuLi, – 78 °C, Et$_2$O, 15 min, RT; b) addition.

Unfortunately, ADME studies demonstrated a distinct *in vivo* instability of (43a) and therefore initiated labeling in alternative position. Since O- and N-Me groups are known to be less metabolically stable, the interest concentrated on the labeling of the ring system. An optimized synthesis strategy (elaborated from research chemistry and chemical development, Bänziger 2003) available at that time (see Fig. 11.31) constructed ring C (= piperidine ring) by functionalization of the naphtaline moiety (47) with the ethylcyano acetate derivative (46). ^{14}C-labeled ethylcyano acetate (44) therefore became a suitable starting material for an isotope synthesis. In a first step the different isotopomers (44a, b, c, d) had to be assessed for their suitability.

A comparison of the isotopomeres (Tab. 11.7) identifies ethyl[^{14}C]cyano acetate (44d) as the most suitable one, since it is not only readily accessible (from K^{14}CN and BrCH$_2$COOEt), but also carries the label in a most suitable position. In contrast to the carboxyl moiety, which is lost during the decarboxylation step, the cyano moiety of (44d) will be incorporated in the piperidine moiety of the target molecule (43b).

---

36 The shortness of the synthesis (3 unlabeled steps, 2 radiolabeled steps), however, is balanced by the risk of handling volatile radioactive starting material (^{14}CH$_3$I; bp. 43 °C) and the possible formation of volatile side products (^{14}CH$_4$; bp. – 161.5 °C). Therefore, the technical execution required careful optimization.

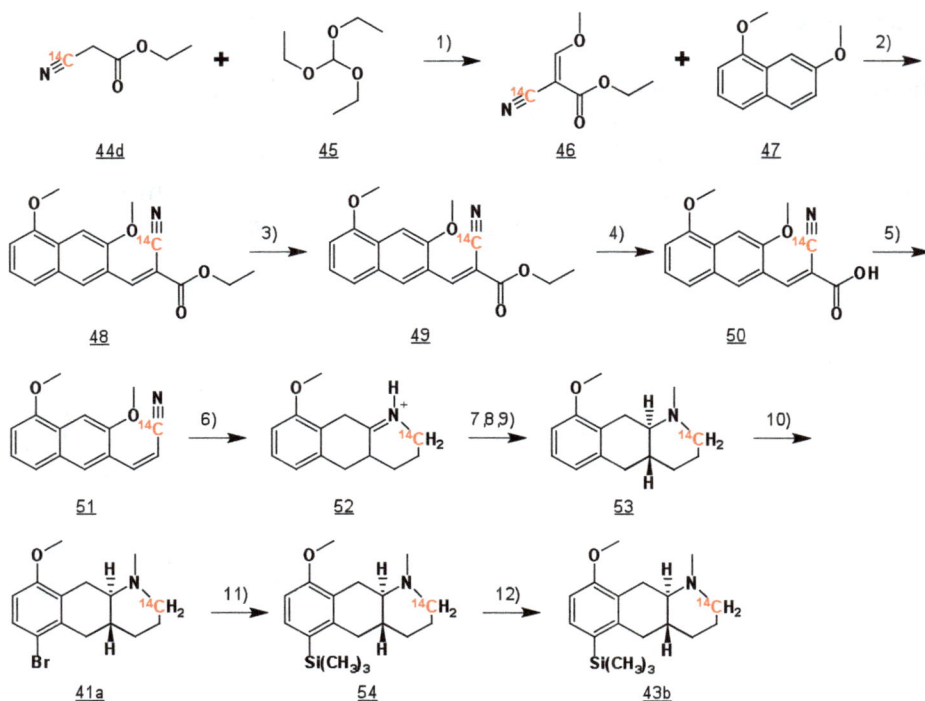

Fig. 11.31: *De novo* ^{14}C-labeling of *JNZ092* (43). Reaction conditions: 1) acetic anhydride, 145 °C, 5 h, flash chromatography (74%); 2) n-BuLi, THF, – 20 °C, – 70 °C, 1 h, – 25 °C, H$_2$SO$_4$, (not isolated); 3) Pd/C 5%, THF/EtOH, H$_2$, 25%, 22 h, flash chromatography (46%); 4) NaOH, EtOH, 90 °C, 3 h (96%); 5) N,N,-dimethylacetamide, NaCl, H$_2$O, 150 °C, 2.5 h (not isolated); 6) Na-metal, n-butanol, 90 °C, 3 h (not isolated); 7) NaBH$_4$, n-butanol, EtOH, acetic acid, water (not isolated); 8) formaldehyde, NaCNBH$_3$, EtOH, 25 °C, 0.5 h; 9) separation of *cis/trans* isomers (42%); 10) HBr, HBrO$_3$, acetic acid/water 1 : 1, flash chromatography (94%); 11) n-BuLi, trimethylsilyl chloride, THF, flash chromatography (88%); 12) separation of antipodes by preparative HPLC-chromatography (48%).

Tab. 11.7: Selection of the most suitable isotopomer as precursor for the synthesis shown in Fig. 11.31.

Compound	Position of label (within the precursor)	Loss of label during synthesis	Synthetic accessibility	Suitable position of label (product)
44a	methoxy-	yes	easy	n.a.
44b	carboxyl-	yes	easy	n.a.
44c	position2	no	difficult	yes
44d	nitril-	no	easy	yes

Reaction of (44d) with triethylorthoformiate (45) results in the Michael acceptor (46), which allows functionalization of the methoxynaphthalene moiety and therewith preconfiguration of the ring C. Reduction, ring formation and methylation form the labeled basic structure, which has to undergo bromination and silylation to obtain (54). The length of the synthesis (12 steps), the required diastereo-selective (*cis/trans* separation – step 9) and enantio-selective (step 12) purification explain the moderate radiochemical yield of 2.5% (related to K[^{14}C]CN). Nevertheless, due to well-elaborated, efficient protocols the synthesis could be executed quite fast without any synthetic or technical problems. Therefore, in cases where optimized protocols, unlabeled intermediates and reference materials are available, linear, multistep synthesis might also be suitable for isotope syntheses.

11.4.3.4 Concluding learning example

A final example (Elmore 2011) summarizes some of the strategies discussed for labeling with tritium and carbon-14. For drug development, the NK3 receptor agonist *AZD2624* (51) was required for

– receptor occupancy studies at high specific activity of >1.1 TBq/mmol) (^{3}H/labeling) and for
– preclinical and clinical ADME studies at a specific activity of >1.8 GBq/mmol (^{14}C-labeling).

^{3}H-labeling of *AZD2624*: Receptor studies require specific activities >1.1 TBq/mmol in chemically stabile form. This value can only be achieved by incorporation of two tritium atoms. Making use of a halogen/tritium exchange reaction, bis-bromo- (preferably a bis-iodo) precursors are required. Direct electrophilic iodination of (55) produced the bis-iodo derivative (56). Heterogeneous catalyzed tritiation gave [^{3}H$_2$] AZ2624 (55a) at a specific activity of 1.3 TBq/mmol, cf. Figure 11.32.

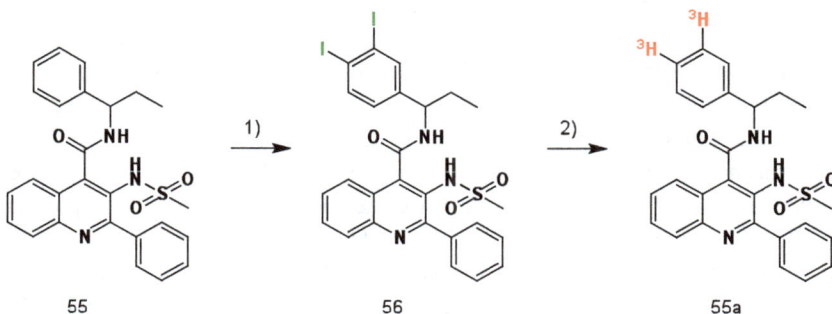

Fig. 11.32: Synthesis of [^{3}H$_2$]*AZD2624* (55a). Reaction conditions: 1) NIS, TFA, r.t., 12 h; 2) ^{3}H$_2$, Pd/C 5%, NEt$_3$, DMF.

^{14}C-labeling of *AZD2624*: The required specific activity necessitates incorporation of one ^{14}C-atom in at least 80% (i.e., 50 mCi/mmol; 62.4 mCi/mmol · 0.8) isotopic purity in a chemically and metabolically stable position. Labeling concentrated on the quinoline moiety. Based on the synthetic route of medicinal chemistry (Fig. 11.33), the key intermediate (57) is synthesized from the fragments (58) and (59). Their retrosynthetic analysis identified [^{14}C]benzoic acid (31) and diethyl [^{14}C]oxalate (61) as starting materials.

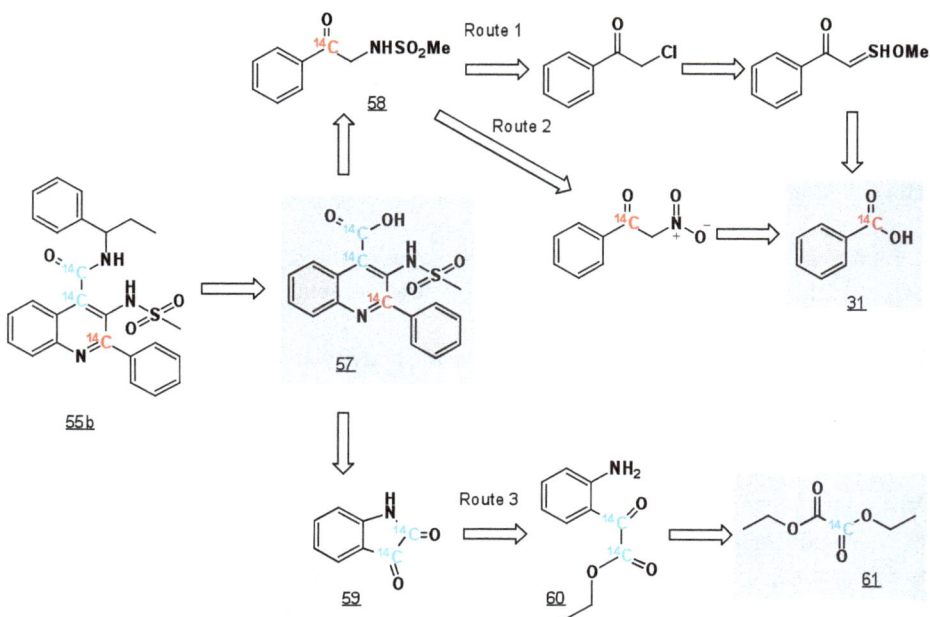

Fig. 11.33: Retrosynthesis of *AZD2624* with the labeling positions.

Route 1 and route 2 (fragment 58): Although [^{14}C-carboxyl]benzoic acid (31) as starting material appears to be very straightforward, experimental execution identified both routes to be less attractive. Low yields and difficulties in purifying the product encouraged the authors to concentrate on route 3. Route 3 (fragment 59): Precursor of (57) is the isatin (59) accessible from diethyl [^{14}C]oxylate (61).

Reaction of monolabeled diethyl [^{14}C]oxalate (61) with the *ortho*-lithiated N-*boc*-aniline gave the ethyl [^{14}C]oxoacetate (60). Due to the symmetry of the deployed diethyl [^{14}C]oxalates, the ^{14}C-label is equally distributed between both carboxyl groups of the product. Deprotection of the aniline with subsequent cyclization results in the isotopomeric mixture of the isatines (59). Reaction of unlabeled (58) and amide formation with (S)-1-phenylpropyl amine gave [^{14}C]*AZD2624* (55b) at the requested specific activity of 53 mCi/mmol (Fig. 11.34). Thus, the respective tracers required for drug development could be made available.

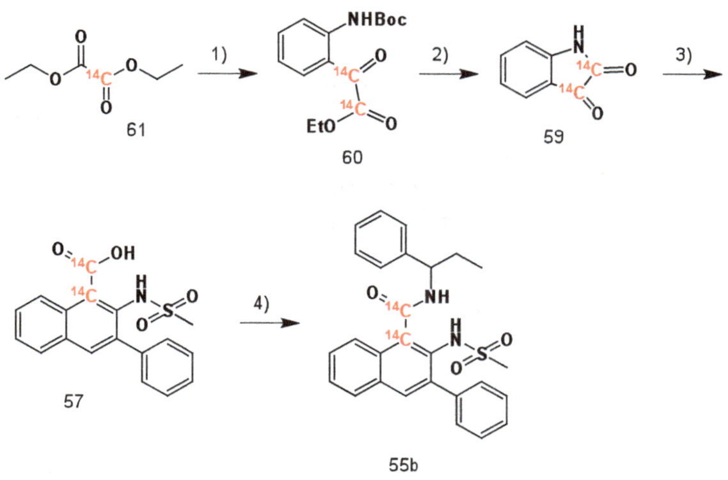

Fig. 11.34: ^{14}C-labeling of *AZD2624* (55b). Reaction Conditions: 2) HCl, DME, 80 °C; 3) PhCH$_2$NHSO$_2$Me, EtOH; (S)-PhCHNH$_2$CH$_2$CH$_3$, HOBT, NMM, CH$_2$Cl$_2$.

11.5 Outlook

Radiochemistry is a critical support function for pharmaceutical R&D that continuously faces increased demands and requirements from end users. A glance at the scientific programs and abstracts from the most recent national and international isotope congresses provides a glimpse of current research interests and possible future developments. Some examples are given from the fields of "synthesis" and "alternative methods".

11.5.1 Synthesis

(a) The application of iridium(I)-catalysts for selective H/T-exchange to synthetically complex molecules has stimulated an intensive interest on catalyst optimization to increase their selectivity, efficiency and functional compatibility. Although numerous improvements have been achieved, the vast number of complex targets are still challenging.

(b) Due to technical difficulties, [^{14}C]carbon monoxide did not earn the same synthetic relevance as unlabeled carbon monoxide. Solutions recently reported (Elmore 2011, Lindhardt 2012) to control them, encourage an application in late-stage labeling.

(c) Identification and development of new classes of compounds also require labeling for the development process. Antisense compounds, antibodies, or antibody-drug-conjugates are such examples needing new labeling strategies and methods.

11.5.2 Alternative methods

To improve analysis of tracer molecules in pharmaceutical R&D study, efforts are continuously taken to establish alternative and more "comfortable" detection methods to support studies. These might be either nonradioactive ones, or ones working at a considerably lower activity level. Convincing examples for the latter are solid phase scintillation technology and accelerator mass spectroscopy, cf. Chapter 4.

In conclusion, radiochemistry is constantly evolving in an ever-changing environment to meet the expectations of its customers.

Further Reading

Lockley WJS, McEwen A, Cooke R. Tritium: A coming of age for drug discovery and development ADME studies. JLCR 2012, 55, 235–257.

Penner N, Xu L, Prakash C. Radiolabeled absorption, distribution, metabolism, and excretion studies in drug development: Why, when, and how?. Chem Res Toxicol 2012, 25, 513–531.

Rick NG. Drugs from discovery to approval. Hoboken, John Wiley & Sons, 2004.

Voges R, Heys JR, Moenius T. Preparation of compounds labeled with tritium and carbon-14. Chicester, John Wiley & Sons, 2009.

References

Andres H. New developments in tritium labelling. In: Pleiss U, Voges R. eds. Synthesis and applications of isotopically labeled compounds, Vol. 7. Chicester, John Wiley & Sons Ltd., 2001, 49–62.

Atzrodt J, Derdau V, Fey T, Zimmermann J. The renaissance of H/D-exchange. Angew Chem Int Ed 2007, 46, 7744–7765.

Bänziger M, Küsters E, La Vecchia L, Marterer W, Nozulak J. A new practical route for the manufacture of (4aR, 10aR)-9-Methoxy-1-methyl-6-trimethylsilanyl-1,2,3,4,4a,5,10,10a-octahydrobenzo[g]quinoline. Org Process Res Dev 2003, 7, 904–912.

Bruin G, Waldmeier F, Boernsen O, Pfaar U, Gross G, Zollinger M. A microplate solid scintillation counter as a radioactivity detector for high performance liquid chromatography in drug metabolism: Validation and applications. J Chromatogr A 2006, 1133, 184–194.

Burtscher P, Haecker H, Moenius T. Labeling of PKC412 – a fermentative radiolabeling. JLCR 2010, 53, 613–615.

Culf AS, Morimoto H, Williams PG, Lockley WJS, Primrose WU, Jones JR. Synthesis of a tritium labeled phospholipase A2 inhibitor: A ligand for macromolecular ^{3}H NMR spectroscopy. JLCR 1996, 38, 373–384.

Easter JA, Burrell RC, Bonacorsi SJ, Balasubramanian B. Synthesis of carbon-14 labeled midazolam. JLCR 2011, 52, 419–421.

Elmore CS. The synthesis and use of [^{14}C]carbon monoxide in Pd-catalyzed carbonylation reactions. JLCR 2011, 54, 59–64.

Filer CN, Orphanos D, Seguin RJ. Synthesis of [ring-³H] dopamine and fluoro analogues at high specific activity. Synth Commun 2006, 36, 975–978.

Filer CN. Tritiated dihydro compounds employed as radioligand surrogates. JLCR 2011, 54, 731–742.

Filer CN. Isotopic fractionation of organic compounds in chromatography. J Label Compd Radiopharm 1999, 42, 169–197.

Gomez-Gallego M, Sierra M. Kinetic isotope effects in the study of organometallic reactions. Chem Rev 2011, 111, 4857–4963.

Gregson TJ, Herbert JM, Row EC. Synthetic approaches to regiospecifically mono- and dilabeled arenes. JLCR 2011, 54, 1–32.

Hickey MJ, Kingston LP, Lockley WJS, Allen P, Mather A, Wilkinson DJ. Tritium-labeling via an iridium-based solid-phase catalyst. JLCR 2007, 50, 286–289.

Kiffe M, Schmid DG, Bruin GJM. Radioactivity detectors for High-Performance Liquid Chromatography in drug metabolism studies. J Liquid Chromat & Related Technol 2008, 31, 1593–1619.

Krauser J. A perspective on tritium versus carbon-14: Ensuring optimal label selection in pharmaceutical R&D. JLCR 2013, 56, 441–446.

Lenz EM, Kenyon A, Martin S, Temesi D, Clarkson-Jones J, Tomkinson H. The metabolism of [¹⁴C]-zibotentan (ZD4054) in rat, dog and human, the loss of the radiolabel and the identification of an anomalous peak, derived from the animal feed. J Pharmaceut Biomed Anal 2011, 55, 500–517.

Lindhardt AT, Simonsson R, Taaning R, Gogsig TM, Elmore CS, Skrystrup T. ¹⁴C-Carbon monoxide made simple – novel approach to the generation, utilization, and scrubbing of ¹⁴C-carbon monoxide. JLCR 2012, 55, 411–418.

Lockley WJS, Heys JR. eds. Metal-catalysed hydrogen isotope exchange. JLCR 2010 special issue, 53, 635–752.

Malmquist J, Bernlind A, Sandell J, Ström P, Waldman M. Tritium labeling of a secretase inhibitor and two modulators as *in vitro* imaging agents. JLCR 2012, 55, 80–83.

Meisterhans C, Nückel S, Zeller A. Survey of different tritium-labeling methods used at RC TRITEC. JLCR 2010, 53, 253.

Miwa GT, Lu AY. Kinetic isotope effects and "Metabolic Switching" in cytochrome P450-catalyzed reactions. BioEssays 1987, 7, 215.

Moenius T, Andres H, Nozulak J, Salter R, Ray T, Burtscher P, Schnelli P, Zueger C, Voges R. Labeling strategies of selected subtypes of the hexahydro-napht[2,3-b]-1,4-oxazine- and octahydrobenzo[g]quinolone type. JLCR 2007, 50, 616–619.

Morgan AJ, Nguyen S, Uttamsingh V, Bridson G, Harbeson S, Tung R, Masse CE. Design and synthesis of deuterated boceprevir analogs with enhanced pharmacokinetic properties. JLCR 2011, 54, 613–624.

Murray A, Williams DL. Organic syntheses with isotopes. New York, Interscience, 1958.

Pounds S. Synthesis and Applications of Isotopically Labelled Compounds In: Dean DC, Filer CN, McCarthy KE. eds. Synthesis and applications of isotopically labeled compounds, Vol. 8. Chicester, John Wiley & Sons, Ltd, 2004, 469–472.

Ren S, Royster P, Lavey C, Hesk D, McNamara P, Kpoharski D, Truoung V, Borges S. Synthesis of [¹⁴C]boceprevir, [¹³C₃]boceprevir, and [D₉]boceprevir, a hepatitis C virus protease inhibitor. JLCR 2012, 55, 108–114.

Salter R, Bordeaux K, Burtscher P, Metz Y, Moenius T, Rodriguez I, Ruetsch R, Voges R, Zueger C. Isotopic labeling of STI571 and its metabolites in the development of gleevec. JLCR 2006, 49, 208–210.

Salter R. The development and use of iridium(I) phosphine systems for ortho-directed hydrogen-isotope exchange. JLCR 2010, 53, 645–657.

Shaffer CL, Gunduz M, Thornburgh BA, Fate GD. Using a tritiated compound to elucidate its preclinical metabolic and excretory pathways in vivo: Exploring tritium exchange risk. Drug Metab Dispos 2006, 34, 1615–1623.

Simon H, Palm D. Isotope effects in organic chemistry and biochemistry. Angew Chem Intern Ed 1966, 5, 920–933.

Solon E. Use of radioactive material and autoradiography to determine drug tissue distribution. Chem Res Toxicol 2012, 25, 543–545.

Szammer J, Simon-Trompler E, Banka Z, Szunyog J, Klebovich I. Synthesis of deramciclane labeled with tritium in various positions. JLCR 2005, 48, 693–700.

Thurkauf A. The synthesis of tritiated 2-phenyl-4-[4-(2-pyrimidyl)piperazinyl]methyl-imidazole ([^{3}H] NGD 94-1), a ligand selective for dopamine D_4 receptor subtype. JLCR 1997, 39, 123–128.

Voges R, Kohler B, Metz Y, Riss B. Linear vs. convergent synthesis of [D-TMS[1-^{14}C]Ala]NVP MTH958. In: Pleiss U, Voges R. eds. Synthesis and applications of isotopically labeled compounds, Vol. 7. Chicester, John Wiley & Sons Ltd, 2001, 295–299.

Wagner J, Andres H, Rohrbach S, Wagner D, Oberer L, France J. Efficient synthesis of [^{3}H]/ Sanglifehrin A via selective oxidation/reduction of alcohols at C_{31} and C_{35}. J Org Chem 2005, 70, 9588–9590.

West R, Wang Y, Atkinson J. Di-(trideuteromethyl)-tocotrienols as probes for membrane orientation and dynamics of tocotrienols. JLCR 2008, 51, 413–418.

Tobias L. Ross and Frank Roesch

12 Life sciences: Nuclear medicine diagnosis

Aim: Molecular imaging in nuclear medicine uses radionuclides attached to relevant molecules in order to illuminate ongoing biological, biochemical, or physiological processes on-line, noninvasively and as quantitatively as possible. In order to register the radiation originating from the radiolabeled molecule (then named "tracer"), electromagnetic emissions are needed – in this case photons. The photons are created either by de-excitation of excited nuclear states as a consequence of a primary transition in terms of emission of a single photon or by following a primary positron emission with subsequent annihilation. This defines the tomographic technology used for imaging, i.e., *single photon emission* computed tomography (SPECT) or *positron emission* tomography (PET), respectively.

This chapter introduces the concept of a radiotracer. It identifies the photon-emitting radionuclides of interest. The great opportunities provided by molecular imaging using radiopharmaceuticals are exemplified for the quantitative measurement of energy consumption in the human body, which is the glucose consumption in terms of g per ml and minute. This is achieved by synthesizing and utilizing an ^{18}F-labeled glucose analog: 2-[^{18}F]fluoro-2-deoxy-D-glucose, 2-[^{18}F]FDG.

Next, the different chemical approaches to attach these radionuclides to different biomolecules are introduced for covalent ^{11}C-, ^{18}F-, and ^{123}I-labeling as well as for coordinative chemistry needed, e.g., ^{99m}Tc and ^{68}Ga, ^{111}In, radiometal labeling.

Finally, the most relevant clinical concepts are highlighted for noninvasive diagnosis in patients.

12.1 Introduction

Molecular imaging is a general approach contributing to addressing two fundamental question of life: What does a certain molecule "do" in the human body, i.e., what kind of biochemical process is it involved in? This question significantly contributes to the principle understanding of how life "works". One of the most impressive aspects is to "see" how signal transduction proceeds in the living human body or to "see" the pharmacology of a pharmaceutical. Idioms such as "seeing is believing" and "an image is worth a thousand words" hold true both in a scientific and a medical sense.

Once the role of a certain molecule is understood: Could a radiolabeled version of the molecule be used to report on the current status of a biochemical and physiological process for an individual human? If the current status is affected by a disease, the approach will provide diagnostic information. Finally, this will hopefully have consequences for the treatment of that disease.

https://doi.org/10.1515/9783110742701-012

Molecular imaging in nuclear medicine uses radionuclides (the "probe") attached to relevant molecules, thereby yielding radiolabeled "tracers". In order to register the radiation originating from the radiolabeled molecule, the emission type must be photons. This emission makes the trace of the molecule visible and the human body (or any other object or subject) "transparent". The way the molecule is selected and how the radionuclide is attached are issues in radiopharmaceutical chemistry. The way the different photons are registered by detectors, how the registered radioactivity is turned in tomographic images of the human body, and how these are interpreted in a diagnostic context are the domain of nuclear medicine.

This chapter introduces the concept of a radiotracer. The great opportunities provided by molecular imaging using radiopharmaceuticals are exemplified for the quantitative measurement of energy consumption in the human body, which is the glucose consumption in terms of g per ml and minute. This is achieved by synthesizing and utilizing an ^{18}F-labeled glucose analog: 2-[^{18}F]fluoro-2-deoxy-D-glucose, 2-[^{18}F]FDG.

According to the chemical element the photon-emitting radioisotopes belong to, individual chemical approaches must be utilized to attach these radionuclides to different biomolecules. This is different for covalent and for coordinative binding. The most relevant examples are covalent ^{11}C-, ^{18}F-, and ^{123}I-labeling as well as coordination chemistry needed for radiometal labeling with ^{99m}Tc and trivalent metal radioisotopes such as ^{68}Ga, ^{111}In, and ^{177}Lu.

Finally, the most relevant clinical concepts are highlighted for noninvasive diagnostics of soft-tissue and bone tumors, brain diseases, and dysfunctions of other organs such as thyroid, kidneys, heart, and lung. Some highlights are referred to the impressive diagnostic value of the corresponding SPECT or PET radiopharmaceuticals.

Another focus is on the new era of positron emitting ^{68}Ga and other metal isotopes, increasingly used for detecting various kinds of cancer such as neuroendocrine tumors and prostate cancer. This class of PET/CT tracers is becoming more and more important in the context of theranostics, i.e., linking diagnosis and treatment within one molecule.

12.2 The tracer principle

Radiolabeled molecules are called "tracers", which has a dual meaning: The labeled compound is usually applied at ultralow concentration (at "trace" levels), and once applied to a biological (or industrial) process, its radioactivity allows its pathway in the system to be "traced".[1]

---

1 The Nobel Prize in Chemistry 1943 was awarded to GEORGE de HEVESY "for his work on the use of isotopes as tracers in the study of chemical processes".

12.2.1 The tracer molecule's structure

If the application of a tracer is to be conclusive, the radiolabeled and the unlabeled (i.e., original biomolecule) compound should behave identically (or at least similarly) in the respective study environment. For the case of "identical behavior", this means substitution of an appropriate atom in the original compound by its respective radioisotope. The most conclusive examples are the replacement of hydrogen and carbon atoms in the hydrocarbon-structure of biomolecules with their long-lived radioisotopes tritium (^{3}H) and carbon-14 (^{14}C). Indeed, the introduction of these unstable isotopes of hydrogen and carbon is a matter for radiochemistry and radiopharmaceutical chemistry. The ^{3}H- or ^{14}C-labeled compound is a chemical analog of the H- and C-compounds and only shows very minor changes in physical and chemical profiles, and most importantly, almost no differences in its biochemical and physiological profiles. This makes the labeled molecule and its biological fate "visible" through measurement of the radioactivity distribution. It turns a radiolabeled molecule into a "tracer", which in the context of, e.g., biology, "traces" the pathways of the original biomolecule in terms of, e.g., organ uptake, route and velocity of excretion characteristics, metabolism.[2,3]

12.2.2 The tracer molecule's concentration

The second unique feature of the tracer concept takes us back to the relationship between radioactivity and the number of atoms behind that radioactivity, cf. Chapter 4 of Vol. I. A radioactivity A sufficient for biological evaluations (which is in most cases at MBq levels) represents an extremely low *number of radioactive atoms Ñ*, and correspondingly, an extremely low *amount* of the radiolabeled compound ($Â = \lambda Ñ$). This low amount transfers to an extremely low, in many cases even negligible *concentration* of the tracer compound in the whole object, e.g., the human body. Figure 12.1 illustrates how a radioactivity of 10 mCi = 370 MBq of a diagnostically relevant ^{18}F-PET tracer is translated into the number of atoms, labeled molecules, molarity, and mass.

With a molecular mass of the diagnostic tracer of, e.g., 500 g/mol, overall mass of the ^{18}F-labeled compound is ca. 3 ng. If diluted into an, e.g., 7 liter blood pool, the concentration is ca. 0.5 pg/ml.

---

2 The same principle has been applied in "pure" chemistry, such as measuring physicochemical properties of inorganic salts and organic molecules, of "tracing" the retention profile of chemicals in industrial processes. HEVESY first used the radioisotopes ^{212}Pb and ^{212}Bi to quantify the extremely low solubility of salts of lead and bismuth by "spiking" the natural salts with the radioisotopes.
3 The exciting information obtainable by this approach applied to ^{14}C- and ^{3}H-labeled compounds used for *in vitro* and *ex vivo* protocols is illustrated in Chapter 11.

Radioactivity proportional
to transformation constant λ and number Ń of the radionuclide

$$\acute{A} = \lambda \cdot \acute{N}$$

Á

370 MBq = 10 mCi
of a ^{18}F-labeled molecule
= tracer

Ń

$\approx 4 \cdot 10^{12}$ atoms
$\approx 6.6 \cdot 10^{-12}$ mol
≈ 3.3 ng

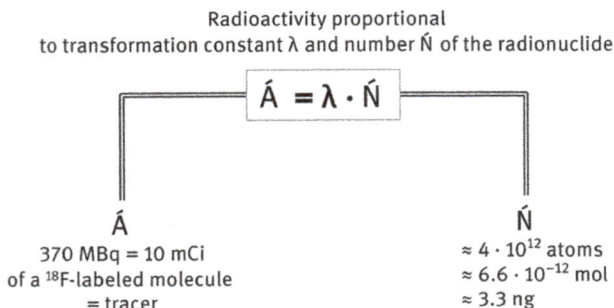

Fig. 12.1: A radioactivity of 10 mCi = 370 MBq of a diagnostically relevant ^{18}F-PET tracer of a molecular weight of 500 g/mol is translated into the number of atoms, labeled molecules, molarity, and mass. Note that the real mass of the injected tracer is larger than this theoretical one due to the presence of "cold", i.e., stable fluorine.

Consequently, the low amount of radiolabeled tracer does not disturb the biochemical process it is involved in and generally is not toxic. The feature is illustrated in Fig. 12.2, comparing the mass of an ^{18}F-labeled neurotransmitter analog functioning as a molecular imaging tracer with a structurally analog neuroleptic acting as a drug. Thus, it is exclusively acting as a signaling probe.

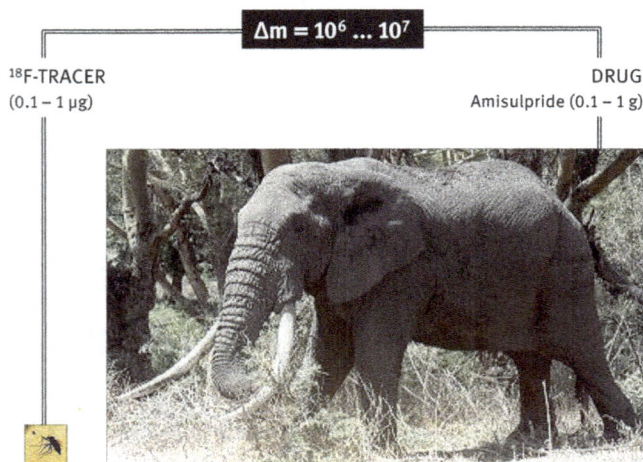

$$\Delta m = 10^6 \dots 10^7$$

^{18}F-TRACER
(0.1 – 1 µg)

DRUG
Amisulpride (0.1 – 1 g)

Fig. 12.2: Comparison of the mass of an ^{18}F-labeled neurotransmitter analog functioning as a molecular imaging tracer (e.g., ^{18}F-FP, see Section 12.7.8) with a structurally analog neuroleptic (e.g., amisulpride) acting as a drug. The difference in mass Δm and concentration is 6 to 7 orders of magnitude. This is the ratio in mass between elephant (drug) and mosquito (tracer). When injecting these tracers into humans, animals or plants, this characteristic is similar to what is generally referred to as "homeopathic".

12.2.3 *In vitro* and *ex vivo* tracers vs. *in vivo* tracers

Although ^{3}H- and ^{14}C-labeled tracers are perfect because there is no difference between the original molecule and the labeled version due to ^{3}H vs. ^{1}H- and ^{14}C vs. ^{12}C-substitutions, there is a fundamental limitation inherent to ^{3}H and ^{14}C. Both are β^--emitting radioisotopes: they offer a relatively low kinetic energy of the β-particles and a lack of accompanying electromagnetic (γ) emission. Consequently, the radioactive emission, the β-particle, cannot penetrate the object under study. Thus, ^{3}H- and ^{14}C-labeled tracers are used for *ex vivo* investigations only. The animal studied must be sacrificed, "cut" into very thin slices (of µm thickness), and those slices are finally exposed to β^--sensitive detectors by means of quantitative autoradiography, cf. Chapter 11. In other words, ^{3}H- and ^{14}C-labeled tracers, although representing the ideal for the tracer concept, cannot be applied to noninvasive imaging investigations on a living object, because they are invisible from outside the object.[4]

12.2.4 Designing a suitable tracer for noninvasive (*in vivo*) molecular imaging

With the impressive advantages of the tracer concept as applied in life sciences, there are consequences and requirements for radiopharmaceutical chemistry: The radioisotope selected must not only ideally fit the requirements of an "ideal" labeling such as by substituting ^{14}C for stable carbon – it must also fit the intention of its application. If this is *noninvasive* molecular imaging, the radioisotope MUST provide a character of radioactive emissions that (i) is able to penetrate the object and (ii) can be registered outside the object's body by adequate detector systems in a quantitative way.

This is not the case for ^{14}C – so what? Let us stay within the world of radiocarbon labeled compounds. The challenge is to find another radioisotope that, like ^{14}C, substitutes stable carbon leaving the original chemical structure of the biomolecule of interest unchanged – and that emits photons instead of β^--particles! The choice is ^{11}C, a neutron-deficient isotope of carbon. In contrast to the neutron-rich ^{14}C, which transforms by β^--emission, ^{11}C transforms through the primary β^+ pathway.[5] The positrons released annihilate with surrounding electrons and induce emission of two

---

4 Of course, some very important information can be obtained by taking blood and organ tissue samples at certain time-points after the injection of the radiolabeled compound and by subsequent, off-line, measurement of the radioactivity they represent – typically referred to as the percent fraction of the injected activity. This kind of the tracer concept was adopted by Rosalyn YALOW in the 1950s to develop the radioimmunoassay (RIA): The Nobel Prize in Physiology or Medicine 1977 was awarded (to one half) to YALOW *"for the development of radioimmuno-assays of peptide hormones"*.
5 Cf. Chapter 7 of Vol. I on the β-transformation pathways.

511 keV photons.[6] This type of radioactive emission easily traverses water and tissue, and is effectively registered by scintillation detectors.[7] Thus, the same radiocarbon-labeled tracer molecule (with ^{11}C instead of ^{14}C) makes the tracer "visible", i.e., noninvasively "traceable" in an object. The object itself becomes "transparent".

12.3 Radionuclides for noninvasive molecular imaging

Thus, the 511 keV photon-emitting ^{11}C is the choice for substituting stable carbon in the hydrocarbons – i.e., in many biomolecules. Could more radionuclides be useful for noninvasive molecular imaging? This takes us back to the question of which radioactive transformations release photons. There are two main sources of photons, illustrated in Fig. 12.3.

The first one is the post-effect of positron/electron annihilation, which is a logical consequence of the primary β^+-transformations as for ^{11}C. Here, only neutron-deficient unstable nuclides are important. However, one may expect a high branching of the β^+ option relative to the electron capture alternative. While the branching of annihilation photons per single transformation event of *K1 may vary, the energy of the photons emitted is always the same, namely exactly 511 keV. Accordingly, the noninvasive molecular imaging technology based on this particular effect is named *positron emission tomography* (PET).

The second source is the pathway of de-excitation of excited nuclear states[8] of the transformation product nuclei $^\circ$K2 as populated by unstable nuclides $^\circ$K1, which underwent primary transformation pathways (for molecular imaging: mainly β- and α-processes). High branching of these photons within the various transformation options of one and the same unstable nuclide could be expected. First, the primary transformation must populate a defined excited level to a very high percentage relative to the transformation to the ground state of the transformation product, and (second) this particular level de-excites toward the final ground state of ^{g}K2. In total, ideally speaking, this particular photon emission should occur in a fraction of close to 1 per each transformation of $^\circ$K1. In most cases, this is realized for just one or two prominent photon(s), and the noninvasive molecular imaging technology is named after them: *Single Photon Emission* Computed Tomography (SPECT). The energy of these photons may vary from about 0.1 to about 2 MeV. In order to meet the optimum energy window, e.g., NaI(Tl) scintillation, photon energies for molecular energies should be between 100 and 250 keV for SPECT imaging.

---

6 Cf. Chapter 11 of Vol. I on the origin of 511 keV annihilation radiation.
7 Cf. Chapter 1 in this volume on detection of radioactivity and radiation detectors.
8 Cf. Chapter 11 in Vol. I on the origin and fate of excited nuclear states.

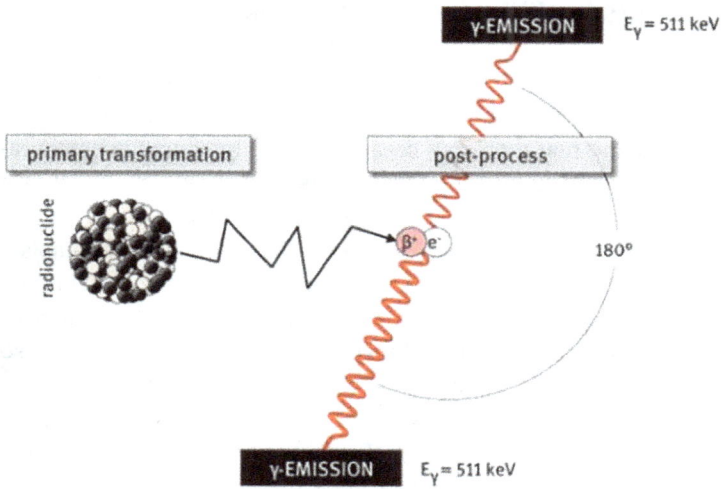

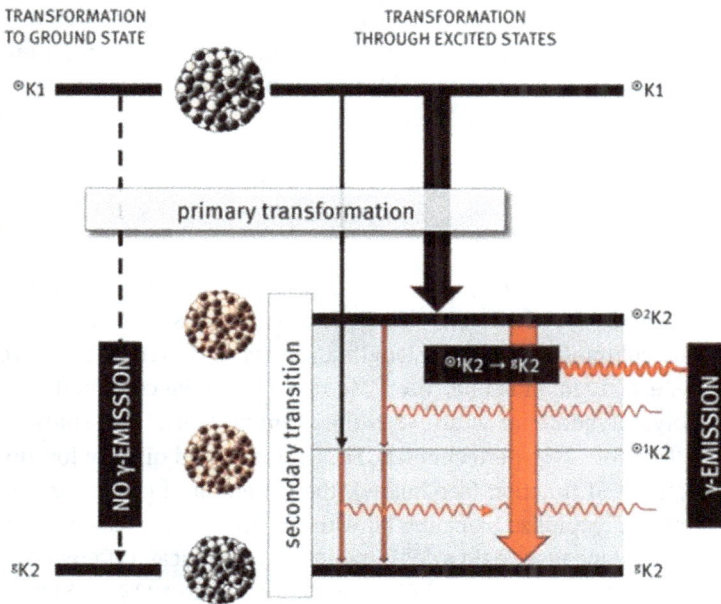

Fig. 12.3: The two sources of photons in radioactive transformation processes. The upper scheme shows the principle of PET based on the annihilation of an electron (e^-) and a positron (β^+) and the resulting two 511 keV photons. The lower case shows both kinds of primary transformation, with and without γ-emission. On the left, the direct transformation to the ground state of the daughter nuclide ($^gK2$) proceeds without γ-emission. On the right, the transformation proceeds into different excited states of the daughter nuclide ($^{o1}K2$ and $^{o2}K2$) and the subsequent secondary transition with the corresponding γ-emissions.

12.3.1 Photon emitters for SPECT and planar imaging

A number of γ-ray emitting radionuclides have found application in diagnostic nuclear medicine using either planar scintillation-γ-cameras or SPECT. Their number is not unlimited (there are about half a dozen), because among the many unstable radionuclides only a small fraction fulfils the requirements mentioned, i.e., (i) offers a highly populated excited nuclear level $^\circ$K2 with (ii) a prominent and highly abundant (single) photon emitted at (iii) useful kinetic energy. Note that (iv) the radioisotope´s availability must also be guaranteed. The most relevant single-photon-emitting radionuclides for noninvasive molecular imaging are listed in Tab. 12.1. It summarizes the nuclear data (half-life, energy, and abundance of the single photon) and also adds the main production pathways. While all the single photon emitters listed can be produced "directly" at particle accelerators, ^{99m}Tc is routinely obtained via a radionuclide generator ^{99}Mo/^{99m}Tc. (The question of producing the radionuclide then refers to the production of the generator parent nuclide ^{99}Mo.) However, also the direct production of ^{99m}Tc is possible and currently under intense investigation.

^{99m}Tc is by far the most often used SPECT radionuclide. It covers almost 85% of all radiotracer applications in nuclear medicine. It is commercially available on a routine basis via the ^{99}Mo/^{99m}Tc generator, the ^{99}Mo being produced using a nuclear reactor. The other major SPECT radionuclides are applied only in cases where analog ^{99m}Tc labeled tracers are not available.

12.3.1.1 ^{99m}Tc

The pair ^{99}Mo ($t^{1}\!/_{2}$ = 66 h)/^{99m}Tc ($t^{1}\!/_{2}$ = 6 h) represents a transient equilibrium system.[9] The radiochemical general design of a ^{99}Mo/^{99m}Tc generator is to absorb ^{99}Mo as polymolybdate on alumina columns of e.g., 7 cm length and 1 cm diameter.[10] The ^{99m}Tc generated is eluted by, e.g., 10 ml of saline. The ^{99}Mo remains on the column due to formation of heteropolymolybdate structure with the alumina matrix. It separates ^{99m}Tc in the chemical form of ^{99m}Tc-pertechnetate TcO_4^-. If not used directly for molecular imaging of the thyroid function (see below), the generator eluate is directly added to a vial containing a lyophilized labeling kit with the appropriate labeling precursor to more or less instantaneously yield a ^{99m}Tc-radiopharmaceutical. Different elution vials (evacuated and sterile) are commercially available, cf. Figures 12.4 and 12.5.

---

9 Cf. Chapter 5 in Vol. I on the mathematics of the radionuclide generator systems.
10 There are various commercial products and Fig. 12.4 reproduces the Mallinckrodt design.

Tab. 12.1: Most commonly used single-photon emitters: Transformation characteristics and routine methods of their production.

Radionuclide	t½	Transformation mode (%)	Main γ-ray energy (%)	Production facility	Main nuclear reaction[a]	Production data		
						Energy range (MeV)	Thick target yield (MBq μA^{-1} h^{-1})	
^{99m}Tc	6.0 h	IT (100)	141 keV (98.6)	reactor/generator	generator ^{99}Mo/^{99m}Tc	–	–	
				cyclotron	^{100}Mo(p,2n)	19 → 16	130	
^{123}I	13.2 h	EC (100)	159 keV (83)	cyclotron	^{123}Te(p,n)	14 → 10	130	
					^{124}Te(p,2n)	26 → 23	392	
					^{127}I(p,5n)^{123}Xe[b]	65 → 45	777[c]	
					^{124}Xe(p,x)^{123}Xe[b]	29 → 23	414[c]	
^{67}Ga	3.26 d	EC (100)	93 keV (37)	cyclotron	^{68}Zn(p,2n)	26 → 18	185	
^{111}In	2.8 d	EC (100)	173 keV (91)	cyclotron	^{112}Cd(p,2n)	25 → 18	166	
^{201}Tl	3.06 d	EC (100)	69–82 keV (X-rays) 166 keV (10.2)	cyclotron	^{201}Tl(p,3 n)^{201}Pb[d]	28 → 20	18[e]	

a preferred for most commercial generator designs.
b ^{123}Xe transformations by EC (87%) and β⁺-emission (13%) to ^{123}I.
c ^{123}I yield expected from the transformation of ^{123}Xe over an optimum time of about 7 h.
d ^{201}Pb transforms through EC (100%) to ^{201}Tl.
e ^{201}Tl yield expected from the transformation of ^{201}Pb over an optimum time of 32 h.

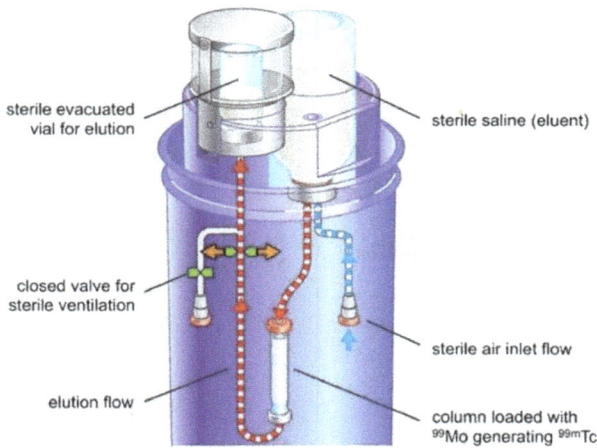

Fig. 12.4: Scheme of the ^{99}Mo/^{99m}Tc generator system (left) and picture of the Mallinckrodt ^{99}Mo/^{99m}Tc generator *Ultra-Technekow™* (right). For elution, simply a sterile, evacuated elution vial (in shielding) is pressed on the elution needle. The reduced pressure automatically sucks the eluent (sterile saline) from the eluent bottle through the column into the sterile elution vial. The elution valve is opened by the lowering of the elution vial in the shielding. The elution process can be stopped by turning the vial by 90°, which in parallel opens the ventilation valve for filling the elution vial with sterile air. Similarly, the pressure balance in the eluent bottle is guaranteed by a sterile air inlet filter.

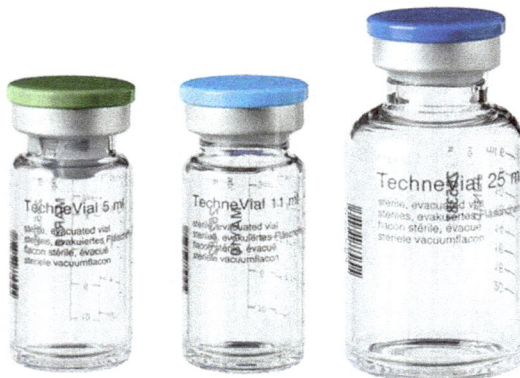

Fig. 12.5: Ready-to-use sterile, evacuated vials (*Techne*Vial™, Mallinckrodt) for elution of the ^{99}Mo/^{99m}Tc generator. Each color of the color-coded vials represents different elution volumes, controlled by different levels of reduced pressure in the vials. Green caps for elution volumes of up to 5 ml; light blue caps for elution volumes of up to 11 ml; and dark blue caps for elution volumes of up to 25 ml.

12.3.1.2 ^{67}Ga and ^{111}In

The two trivalent metallic radionuclides ^{67}Ga and ^{111}In are produced via (p,2n) reactions, the former on ^{68}Zn and the latter on ^{112}Cd. For production on a smaller scale, the (p,n) reactions on ^{67}Zn and ^{111}Cd have also been utilized. Nuclear reactions use isotopically enriched target materials. Chemical separation of the no-carrier-added trivalent Ga and In isotopes from the two-valent target metals uses aqueous chemistry and ion-exchange processes. ^{67}Ga and ^{111}In are finally obtained as M^{3+} cations, ready for synthesizing M^{III}-ligand complexes of medical relevance.

12.3.1.3 ^{123}I

Its nuclear properties are also almost ideal for SPECT studies. In parallel, the radionuclide ^{123}I is a halogen and thus a very useful analog label for creating radiotracers. About 25 different nuclear processes have been investigated for the production of ^{123}I. The development of an optimum process is a good example of the changing demands on the quality of medically important radionuclides. Among them only the four listed in Tab. 12.1 are significant. At low-energy cyclotrons the ^{123}Te(p,n) reaction over $E_p = 14.5 \rightarrow 10$ MeV is applied, and at medium-energy cyclotrons the ^{124}Te(p,2n) process over $E_p = 26 \rightarrow 23$ MeV. In either case highly enriched target material is used. The isolation of ^{123}I is carried out via dry distillation from the irradiated TeO_2 target. The third route of ^{123}I-production consists of the ^{127}I(p,5n)^{123}Xe $\rightarrow$ (EC,β^+) $\rightarrow$ ^{123}I process. The optimum energy range for this nuclear reaction is $E_p = 65 \rightarrow 45$ MeV, needing a higher-energy cyclotron. Interestingly, an iodine target (NaI) is used to produce a radioiodine product. Here, first the proton-rich short-lived radionuclide ^{123}Xe ($t\frac{1}{2} = 2.1$ h) is formed. It instantaneously generates ^{123}I via electron capture and β^+ routes. The radioxenon is collected into a separate vessel and is allowed to transform for about 7 h during which period the maximum growth of ^{123}I occurs. The ^{123}I formed is removed from the vessel, containing radioxenon. This is a relatively high-yield process and the only impurity observed is ^{125}I ($\approx 0.25\%$).

The fourth route is even more complex. It consists of the ^{124}Xe(p,x)-process. Highly enriched and very expensive ^{124}Xe gas is needed,[11] but it requires only a medium-sized cyclotron. This process leads to ^{123}I of the highest radioisotopic purity.

---

11 The natural abundance of ^{124}Xe is only 0.1%.

12.3.1.4 ^{201}Tl

For ^{201}Tl, among several methods the most common production pathway consists of the ^{203}Tl(p,3n)^{201}Pb → EC ($t\frac{1}{2} = 9.4$ h) → ^{201}Tl process. The optimum energy range is $E_p = 28 \rightarrow 20$ MeV. Similar to the ^{127}I(p,5n)^{123}Xe → (EC, β^+) → ^{123}I process, stable thallium is irradiated in order to obtain – following radiochemicals separation steps – a thallium radioisotope.[12] The ^{201}Tl is selectively isolated via the ^{201}Pb intermediate. Utilizing elaborate wet chemical processing, first ^{201}Pb is separated from the irradiated target by co-precipitation. Thereafter, it is allowed to transform (for an optimum time of 32 h) whereby ^{201}Tl grows in. This is removed from ^{201}Pb via anion-exchange chromatography. The ^{201}Tl isolated must be a ^{201}Tl$^+$ species for its sole medical application as a myocardial blood flow tracer.

12.3.2 Positron emitters for PET and PET/CT

12.3.2.1 Organic vs. metallic PET nuclides

Only a few positron-emitting radionuclides "make it" to radionuclides for clinical applications, cf. Tab. 12.2. There are two different groups if structured according to labeling chemistry:

(i) ^{11}C ($t\frac{1}{2} = 20.4$ min), ^{13}N ($t\frac{1}{2} = 10.0$ min), ^{15}O ($t\frac{1}{2} = 2.0$ min), ^{18}F ($t\frac{1}{2} = 109.8$ min), and ^{124}I ($t\frac{1}{2} = 4.2$ d) form a group commonly referred to as "organic" positron emitters, because the elements they originate from constitute the backbone of the hydrocarbon-based organic and biologically relevant compounds, and because their labeling chemistry belongs to the field of organic chemistry. While the latter aspect definitely is true for ^{18}F and ^{124}I representing the halogens, the first may not be. However, because the fluorine atom is in many ways similar to H (atom radius) and OH (electronic influences), it may replace these moieties yielding the ^{18}F-analogs of the corresponding biomolecules, the so-called analog tracers. In addition, it may easily be used in terms of halogenation chemistry, and here it is similar in many ways to the iodine positron emitter ^{124}I.

(ii) All other positron emitters listed in Tab. 12.2 are metals: ^{44}Sc, ^{62}Cu, ^{64}Cu, ^{68}Ga, ^{82}Rb, ^{89}Zr. Consequently, a different approach must be applied, which is the coordinating complex formation between the central metal and adequate ligand structure.

---

12 Thallium itself is a toxic heavy metal. The final ^{201}Tl should therefore not contain any measurable quantities of the target material. The demands on the chemical, radiochemical, and radionuclidic purity of the final product are thus stringent.

Tab. 12.2: Most commonly used positron emitters for PET: Transformation characteristics and routine methods of their production.

Nuclide	Group	Half-life $t\frac{1}{2}$	Transformation mode	Production route	Production	Product	Maximum specific activity (GBq/μmol)
^{15}O	"organic"	2.04 min	β^+ (99.9%)	cyclotrons	$^{14}N(d,n)^{15}O$ $^{15}N(p,n)^{15}O$	$[^{15}O]O_2$ $[^{15}O]O_2$	$3.4 \cdot 10^6$ $3.4 \cdot 10^6$
^{11}C		20.39 min	β^+ (99.8%)		$^{14}N(p,\alpha)^{11}C$	$[^{11}C]CO_2$ $[^{11}C]CH_4$	$3.4 \cdot 10^5$
^{13}N		9.97 min	β^+ (99.8%)		$^{16}O(p,\alpha)^{13}N$	$[^{13}N]NH_3$	$7.0 \cdot 10^5$
^{18}F		109.8 min	β^+ (96.7%)		$^{18}O(p,n)^{18}F$ $^{20}Ne(d,\alpha)^{18}F$	$[^{18}F]F^-_{aq}$ $[^{18}F]F_2$	$6.3 \cdot 10^4$ –
^{124}I		4.17 d	β^+ (22.8%)		$^{124}Te(p,n)^{124}I$	$[^{124}I]I^-$	$1.2 \cdot 10^3$
^{61}Cu	metallic	3.32 h	β^+ (61.5%)	cyclotrons	$^{64}Zn(p,\alpha)^{61}Cu$	$[^{61}Cu]Cu^{2+}$	$3.5 \cdot 10^4$
^{62}Cu		9.74 min	β^+ (98.0%)	generators	$^{62}Zn\ (t\frac{1}{2}=9.2\ h)/^{62}Cu$	$[^{62}Cu]Cu^{2+}$	$7.1 \cdot 10^5$
^{64}Cu		12.7 h	β^+ (54.0%)	cyclotrons	$^{64}Ni(p,n)^{64}Cu$	$[^{64}Cu]Cu^{2+}$	$9.1 \cdot 10^3$
^{44}Sc		3.97 h	β^+ (94.3%)	generators cyclotrons	$^{44}Ti\ (t\frac{1}{2}=60\ a)/^{44}Sc$ $^{44}Ca(p,n)^{44}Sc$	$[^{44}Sc]Sc^{3+}$ $[^{44}Sc]Sc^{3+}$	$2.9 \cdot 10^4$ $2.9 \cdot 10^4$
^{68}Ga		67.7 min	β^+ (54.0%)	generators cyclotrons	$^{68}Ge\ (t\frac{1}{2}=271\ d)/^{68}Ga$ $^{68}Zn(p,n)^{68}Ga$	$[^{68}Ga]Ga^{3+}$ $[^{68}Ga]Ga^{3+}$	$1.0 \cdot 10^5$ $1.0 \cdot 10^5$
^{82}Rb		1.3 min	β^+ (54.0%)		$^{82}Sr\ (t\frac{1}{2}=25.34\ d)/^{82}Rb$	$[^{82}Rb]Rb^+$	$5.35 \cdot 10^6$
^{89}Zr		78.41 h	β^+ (54.0%)	cyclotrons	$^{89}Y(p,n)^{89}Zr$	$[^{89}Zr]Zr^{4+}$	$1.5 \cdot 10^3$

12.3.2.2 Production of "organic" positron emitters

Due to their half-lives, the radionuclides ^{11}C, ^{13}N, ^{15}O, and ^{18}F are generally used at the site of production. ^{18}F can, in addition to on-site use, be transported to medical PET units without a cyclotron. In general, low-energy nuclear reactions like (p,n), (p,α), (d,n), and (d,α) are used and a small-sized cyclotron is adequate for the production of these radionuclides. The excitation functions have been measured systematically. The thick target yield over the optimum energy range for the production of each radionuclide is well known. The calculated values are given in Tab. 12.1.

Gas targets: Considerable effort has also been devoted to targetry designs. For production of ^{11}C and ^{15}O, generally pressurized gas targets in batch mode are used. The product activity is removed from the target by simple expansion and led to vessels where conversion to other chemical forms suitable for labeling of organic compounds is done. The chemical form of the activity leaving the target depends upon the additive given to the gas in the target and, in the case of ^{11}C also on the radiation dose effective in the target. The resulting ^{15}O from an $N_2(O_2)$ target, for example, is $[^{15}O]O_2$. The most reliable production of ^{15}O demands a deuteron beam at the cyclotron. If it is not available, the (p,n) reaction on highly enriched ^{15}N is utilized. An efficient recovery system for the enriched target gas must then be incorporated in the target system. The production of electrophilic ^{18}F is also carried out using a gas target, this time neon. However, the removal of the activity from the target demands addition of some F_2 carrier. In the case of $^{20}Ne(d,α)^{18}F$ reaction, an F_2 (0.1–0.2%) carrier amount is added directly to the neon gas target.

Water targets: The radionuclides ^{13}N and ^{18}F are commonly produced using water targets, and particularly for ^{18}F production, several types of water targets have been developed. A typical target consists of a titanium body, electron-beam welded to two titanium foils (75 μm thick), which act as front and back windows. The target takes 1.3 ml of water with no expansion space and the thickness of the water filling amounts to 3.5 mm. During irradiation, the back window is water-cooled, typically to 8–10 °C, and the front window is helium-cooled to −7 °C. It withstands pressures up to 7 bar and proton beam currents of up to 20 μA. A polyethylene-polypropylene copolymer tube with an i.d. of 0.8 mm and a He drive pressure of 1.3 bar is considered a reliable transfer system. Due to great importance of ^{18}F in PET, several modified designs of high-pressure (>10 bar) water targets have been developed. They can withstand beam currents of up to about 100 μA. A novel spherical niobium target has proved to be very effective. The resulting batch yield of ^{18}F amounts to about 150 GBq, and even more. In the case of ^{18}F, enriched water ($[^{18}O]H_2O$) is employed and the effective projectile energy is $E_p = 16 \rightarrow 3$ MeV. Recovery of both ^{18}F and the enriched water is essential. This is commonly achieved by transferring the irradiated water to an ion-exchange bed whereby the ^{18}F activity is adsorbed and the enriched water passes through and is collected. The ^{18}F activity is removed from the column, i.e., by washing it with a 0.1 M solution of K_2CO_3. The same target can be used for the production of ^{13}N, where natural water ($[^{16}O]H_2O$) is used in high current irradiations. The resulting product is $[^{13}N]NO_3^-$, or $[^{13}N]NH_4^+$ in case of ethanol addition to the target water.

12.3.2.3 Generator produced metallic positron emitters

For generator pairs, the parent nuclides are produced at particle accelerators as well, but it is the convenient availability which makes radionuclide generators attractive.[13] The most relevant system is the $^{99}Mo/^{99m}Tc$ generator as discussed already – but other generators providing positron emitting daughters are emerging.

$^{68}Ge/^{68}Ga$: The parent ^{68}Ge ($t\frac{1}{2} = 278$ d) of the $^{68}Ge/^{68}Ga$ ($t\frac{1}{2} = 67.7$ min) pair is produced at a very limited number of cyclotrons only. Production is based either on $^{69}Ga(p,2n)^{68}Ge$ reactions or on (p,xn) reactions on germanium targets. Though having relatively high cross sections, production batch yields are determined by the long half-life of ^{68}Ge and require cyclotrons of high proton beam intensity.

The $^{68}Ge/^{68}Ga$ and the $^{82}Sr/^{82}Rb$ generators represent secular equilibria. It is a radiochemical challenge to design a separation concept which both guarantees high elution efficacy for the generator daughters and low breakthrough of the parent nuclide. For the $^{68}Ge/^{68}Ga$ system, solid-phase-based chromatographic generators are commercially available utilizing either inorganic (TiO_2, SnO_2, ZrO_2) materials to absorb tetravalent ^{68}Ge or organic resins. The ^{68}Ga is eluted in all systems with hydrochloric acid (of 0.05–0.6 N HCl), which provides the ^{68}Ga as a $^{68}Ga^{3+}$ cation. From this species ^{68}Ga is ready for subsequent synthesis of radiopharmaceuticals via coordination chemistry based on tailored bifunctional chelates. For the $^{82}Sr/^{82}Rb$ generators, the column consists of tin oxide (SnO_2). The daughter is eluted with saline yielding $^{82}Rb^+$.

$^{44}Ti/^{44}Sc$: This generator is comparable in its chemical structure with $^{68}Ge/^{68}Ga$ generator in several ways: tetravalent parent Ti^{IV} / trivalent daughter Sc^{III}, and positron emission of the daughter ^{44}Sc. Yet, the half-lives differ significantly: $t\frac{1}{2} = 60$ years for ^{44}Ti and $t\frac{1}{2} = 3.97$ h for ^{44}Sc. It is exactly this longer half-life of ^{44}Sc compared to ^{68}Ga, which makes this generator attractive. However, while a 5 mCi prototype is demonstrating the clinical value of the system, the availability of ^{44}Ti is problematic.

---

13 For the various features of radionuclide generators cf. Chapter 5 in Vol. I on the mathematics of the radionuclide generator equilibria and Rösch F, Knapp FF, Radionuclide Generators, in Handbook of Nuclear Chemistry, Vértes A, Nagy S, Klencsár Z, Lovas RG, Rösch F (eds.) Vol. 4, pp 1935–1976, Second edition, Springer, Berlin-Heidelberg, 2011.

^{82}Sr/^{82}Rb: The possible production methods for the parent ^{82}Sr ($t^{1/2} = 25.34$ d) include the ^{3}He- and α-particle induced reactions on natural krypton, the (p,4n) reaction on ^{85}Rb and the spallation of molybdenum with high-energy protons. In practice, however, the ^{nat}Rb(p,xn)^{82}Sr process is commonly used, and major quantities of ^{82}Sr are presently produced at intermediate energy accelerators. Again, only a few production facilities exist, this time due to the need for relatively high-energy protons of $65 \rightarrow 45$ MeV.

12.3.2.4 Cyclotron-based production of "metallic" positron emitters

^{64}Cu and ^{89}Zr are produced via irradiation at small- to medium-sized cyclotrons. In both cases, solid metallic targets[14] in form of foils are employed. After irradiation, such target foils need extensive chemical workup to isolate the desired radionuclides. ^{64}Cu is available from the bombardment of enriched ^{64}Ni targets via the ^{64}Ni(p,n)^{64}Cu reaction. The optimal energy range is $E_p = 20 \rightarrow 5$ MeV. The irradiated target foil is dissolved in hydrochloric acid and ^{64}Cu is isolated by ion-exchange chromatography. The expensive enriched ^{64}Ni is efficiently recovered in the same process. In case of ^{89}Zr, the target material is 100% naturally abundant and stable ^{89}Y, which is the only stable isotope of yttrium and thus, very cost-efficient. Again the (p,n)-reaction (^{89}Y(p,n)^{89}Zr) is used and targets are yttrium foils or a sputtered yttrium layer on gold foils. Ideally, targets are irradiated with protons in the energy range of $E_p = 20 \rightarrow 5$ MeV. The workup is identical to the procedure described for ^{64}Cu, the irradiated target material is dissolved in hydrochloric acid, and the desired ^{89}Zr is separated by ion-exchange chromatography.

Also short-lived metallic positrons emitter such as ^{68}Ga and ^{44}Sc or ^{43}Sc and ^{61}Cu can be produced at low-energy cyclotrons, in most cases utilizing (p,n) reactions on isotopically enriched target material. This is in particular relevant when (commercial) ^{68}Ge/^{68}Ga generators cannot satisfy the increasing clinical demand of ^{68}Ga or when ^{44}Ti/^{44}Sc generators are not available. This production is either organized by irradiating "liquid targets" such as zinc nitrate solutions to produce ^{68}Ga or "solid" targets. The advantage of the liquid target strategy is that target designs can be modified already installed for the ^{18}F-production and that the irradiated solution can immediately transferred to synthesis modules including a radiochemical purification step (i.e., isolating ^{68}Ga from the target solution prior to labeling reactions. The advantage of solid targets lies in their much higher production yield, yet the disconnection of the irradiated target from the cyclotron and its transfer is a logistic challenge as well as target dissolution and radiochemical purification. All

---

14 Other metallic positron emitters are produced with solid targets utilizing salts instead of metal target materials, such as ^{86}Y as produced via ^{86}Sr(p,n) reaction on ^{86}Sr strontium oxide, cf. Chapter 7 for radiochemical separations.

those aspects are currently covered both by academia and by cyclotron industry. For effective radioisotope purification, mainly dedicated solid-phase-based ion exchange cartridges are used; see Chapter 7.

12.4 Molecular imaging: Tomography

One major advantage of using radioactivity as a probe is the extremely high sensitivity of detection. In molecular imaging the radioactivity permits the noninvasive *in vivo* detection and tracing of radiolabeled compounds at ultralow concentration levels. Today, this "tracer principle" is the base for the two noninvasive molecular imaging technologies in nuclear medicine, the single photon emission computed tomography (SPECT) and positron emission tomography (PET). There is, actually, only one further (nonradioactive) imaging modality which also fits under the generic term *molecular* imaging, which is optical imaging or near infrared imaging (NIR imaging). NIR imaging is based on molecules carrying fluorescence dyes, which emit light at certain wavelengths after excitation. NIR imaging also provides a very high sensitivity, but due to the poor tissue penetration of these wavelengths, its clinical practicability is very limited. It is frequently used and applied in research. In contrast, other noninvasive imaging techniques such as X-ray, computed tomography (CT) and magnetic resonance imaging (MRI) provide morphological information of the human body. The imaging information from X-ray and CT is based on the absorption or transmission of photons (X-rays) in different tissues and thus shows mostly density differences of tissues and organs. An overview of the different imaging modalities, their sensitivity and other characteristics is given in Fig. 12.6.

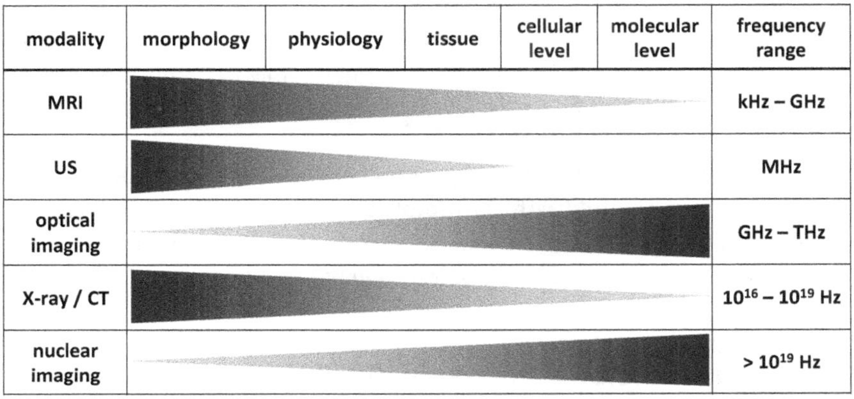

modality	morphology	physiology	tissue	cellular level	molecular level	frequency range
MRI						kHz – GHz
US						MHz
optical imaging						GHz – THz
X-ray / CT						$10^{16} - 10^{19}$ Hz
nuclear imaging						$> 10^{19}$ Hz

Fig. 12.6: Sensitivity and range of applications of various imaging modalities.

Of the above-mentioned imaging modalities, only SPECT and PET allow the visualization and detection of biochemical processes by tracing molecules *in vivo*. They offer only a lower spatial resolution compared to MRI and CT, but due to the functional aspect of imaging biochemistry, their sensitivity is excellent. In clinical routine, modern hybrid systems are used to provide optimal information by combining molecular imaging with morphological imaging. Systems combining SPECT or PET with CT are typical. Furthermore, PET/MRI systems have been installed in clinics.

12.4.1 Single Photon Emission Computed Tomography (SPECT)

SPECT is based on nuclides emitting γ-rays (photons). The detection system consists of a collimator to direct and parallelize the photons for their detection in a scintillator crystal, usually made of NaI (Tl doted). Noteworthy is that the highest efficiency of detection for the Na(Tl)I crystals is in the range of the 140 keV-photons of ^{99m}Tc, the most used radionuclide in nuclear medicine. At the back of the scintillation crystal, a photomultiplier amplifies the signal and passes it on to the electronics.

The simplest imaging systems are planar gamma cameras, which are frequently used for non-tomographic (planar) images (i.e., thyroid imaging). In planar gamma cameras, the patient is placed in front of the collimator of the camera head. The camera head can be orientated either horizontally for patients in lying position, or vertically for patients sitting or staying.

In tomographic systems, generally two planar camera heads are moved around the patient in circles and generate the 3D image. Figure 12.7 shows a typical SPECT/CT (Siemens Symbia T2) in the clinical routine of nuclear medicine diagnostics. In latest SPECT cameras, a revival of three-, four- or even more-headed detection systems for increased sensitivity and shorter scan duration has been seen, but the compromise between spatial resolution, detector distance and geometry as well as collimator design is still challenging. Semiconductor detector technology such as *cadmium zinc telluride* (CZT) allow more compact detectors and enable detector ring systems for SPECT. Furthermore, CZT detectors generally provide superior energy resolution and provide higher sensitivity and contrast.

Developments of SPECT/MRI hybrid imaging machines are still in their infancy and the efforts toward the first clinical prototype are in progress. However, the major challenge is to combine both modalities without interfering with each other, which might even compromise with safety and compatibility issues.

12.4.2 Positron emission tomography (PET)

The emitted positron is able to interact with electrons of the shells of surrounding atoms. As a result, both particles annihilate and give two γ-rays with a total energy of 1.022 MeV, the sum of the masses of positron and electron, 511 keV each. Both γ-rays

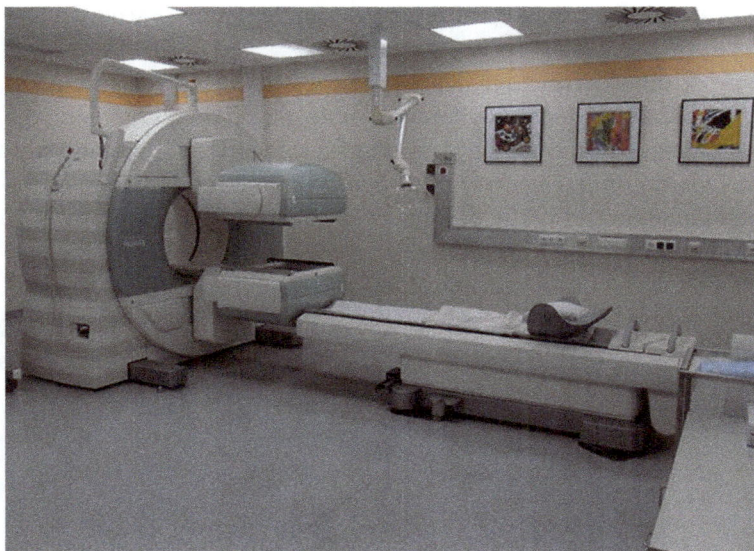

Fig. 12.7: A two-headed SPECT/CT scanner (Siemens Symbia T2) at the Department of Nuclear Medicine of the Hannover Medical School (Germany). The Symbia T2 provides two rotating gamma detector units ("heads") for SPECT and a two lined CT system.

show a nearly 180° distribution. Hence, the transformation of positron emitters results in two body-penetrating photons which can be detected by an appropriate PET scanner. A major aspect in PET is the specific β^+-energy of the nuclides, which determines the distance the positron travels before annihilation. Consequently, the β^+-energy of each nuclide sets limits for the spatial resolution and adds a certain amount of distance between the positron's origin and the place of annihilation.

In PET systems, one or more circular ring(s) of detector pairs, which only record coincidence events in a very short time window of a few nanoseconds, register the pairs of photons. The detectors are constructed of scintillation crystals connected with a photomultiplier which transfers an amplified signal to the coincidence electronics. Due to the basic characteristics of the transformation and the mode of detection, the spatial resolution and sensitivity in PET is much better than in SPECT. Furthermore, PET allows exact quantification of the radioactivity concentration *in vivo*. Appropriate computer-aided data acquisition provides PET images with information about the *in vivo* distribution and levels of accumulation of the radionuclide and the radiopharmaceutical, respectively. As a result, biochemical processes can be visualized and dynamic data acquisition further allows for registration of a temporal component such as pharmacokinetics of a certain radiolabeled molecule. In combination with biomathematical models and individual corrections of attenuation, transmission and scatter effects, physiological and pharmacological processes can be precisely acquired and quantified.

State-of-the-art scanner technology enable *time-of-flight* (TOF) registration of the two coincidence photons, which provides further details on their origin and the position of their annihilation. In combination with new software solutions and suitable algorithms, even the inherent inaccuracy deriving from the traveling of the positron before annihilation can partly be compensated. As a result, modern PET/CT scanners provide higher spatial resolutions and much shorter scanning times. Figure 12.8 shows a modern PET/CT for clinical PET routine.

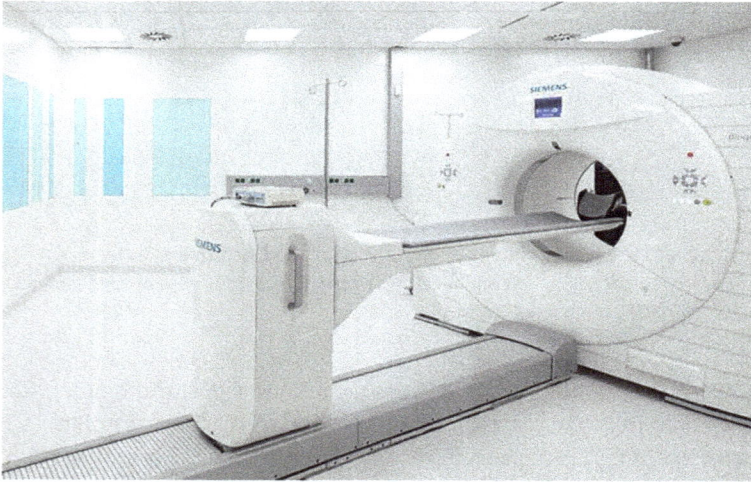

Fig. 12.8: A TOF PET/CT scanner (Siemens Biograph mCT 128 Flow) at the Department of Nuclear Medicine of the Hannover Medical School (Germany). The Biograph mCT 128 Flow consists of a 128 lined CT and a PET including TOF registration. The 128 lined CT allows full body CT scans within only a few minutes. The flow technology enables scans with continuous moving of the patient/bed instead of stepwise protocols.

Latest developments in PET scanner technology increase the number of detector rings to enlarge the *field-of-view* (FOW) with a longer axial length. Scanners with an axial field of up to 100 cm and longer are available. This increased axial length has several advantages such as strongly increased sensitivity and reduced scanning duration, which in turn provides better imaging quality while reduced injected activity lowers radiation burden to the patient. Shorter scanning times increase patient throughput and simplify patient management. An extreme example is the total-body PET/CT (United Imaging) which has an axial FOW of 194 cm. A full-body PET/CT using 2-[^{18}F]FDG can be performed in only one minute by administration of a normal dose. On the other hand, with longer scanning times in a total-body PET/CT new areas in PET diagnostics can be explored, such as simultaneous dynamic imaging of multiple organs with new insights into pharmacokinetics and biodistribution.

Novel PET/MRI hybrid imaging systems have already arrived in clinical routine, but only in limited numbers. Generally, clinical key applications leveraging the advantages of PET/MRI compared to PET/CT would help to install more PET/MRI systems. Development in PET imaging will benefit from larger axial FOW, but also TOF in combination with sophisticated reconstruction algorithms and motion correction will further PET imaging quality and broaden the range of applications. Not surprisingly, AI has also found its way into the field, and will surely make major contributions to PET imaging and how PET data will be processed in future.

12.5 Molecular imaging: Concepts

In the following, some of the most commonly applied concepts in molecular imaging are discussed in more detail.

12.5.1 Blood flow and perfusion: Heart and brain

The concept of blood flow and perfusion imaging is one of the most commonly applied molecular imaging concepts in clinical routine. It is frequently used in cardiology questions such as tissue blood supply post myocardial infarct. The most "natural" blood flow and perfusion tracer is $[^{15}O]H_2O$, but due to the very short half-life of ^{15}O ($t^{1/2} = 2$ min), it is limited to sites with a cyclotron. The single cations $[^{201}Tl]Tl^+$ and $[^{82}Rb]Rb^+$ function as surrogates of endogenous potassium cations K^+. As alternatives for myocardial perfusion imaging, the ^{99m}Tc-labeled tracers *sestamibi* (hexakis(2-methoxy-2-methyl-propylisonitrile)) and *tetrofosmin* (1,2-*bis*[di-(2-ethoxyethyl)phosphino]ethane) are well established. These alternatives even provide cell viability information, as they show uptake into mitochondria of healthy myocardial cells.

As a standard tracer for perfusion, $[^{15}O]H_2O$ is further applied for blood flow and perfusion imaging of other organs such as kidneys or in different malignancies. Regarding blood flow and perfusion of the brain, a tracer with very high lipophilicity is ^{99m}Tc-HMPAO (hexamethylpropyleneamine oxime). With the high lipophilicity, ^{99m}Tc-HMPAO is able to pass the blood-brain barrier, and thus, provide information about perfusion of brain tissue. An example is given in Fig. 12.9, where stroke patients were scanned using ^{99m}Tc-HMPAO before and after noninvasive treatment with a stimulation technique (*focal repetitive muscle vibration*, fMV) to promote the post-stroke motor recovery. The ^{99m}Tc-HMPAO scans show increased cortical perfusion after treatment, and confirm a positive treatment effect.

before treatment   after treatment

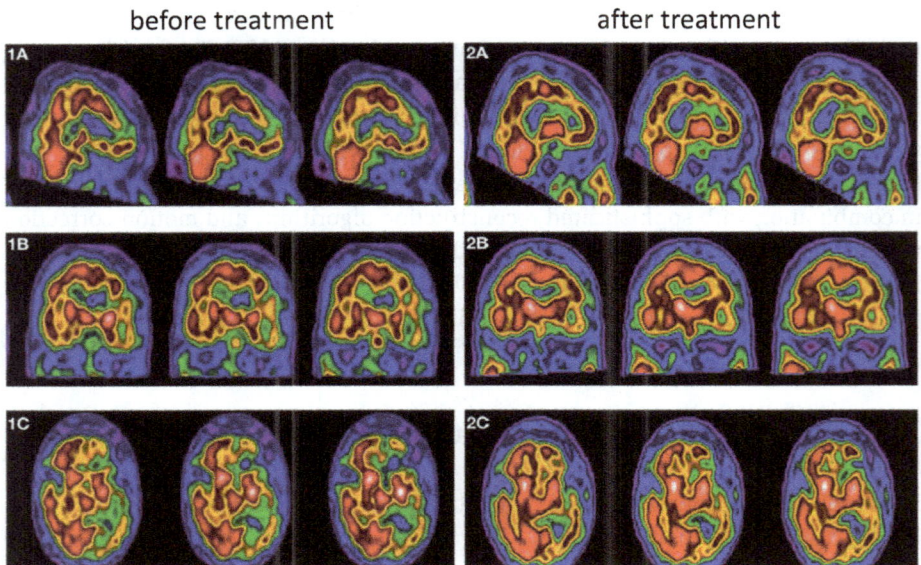

Fig. 12.9: ^{99m}Tc-HMPAO brain perfusion SPECT imaging in stroke patients before (baseline) (left column) and after a noninvasive stimulation treatment (right column). The three rows (A-C) represent images in sagittal (A), coronal (B), and transaxial (C) plane. ^{99m}Tc-HMPAO scans show increased perfusion after treatment in left frontal region (2A), in left parietal region (2B) and in left occipital region (2C).[15]

12.5.2 Trapping (direct vs. metabolic)

The concept "trapping" is one of the most crucial concepts in molecular imaging. In general, trapping (in contrast to "flow") covers the effects and mechanisms leading to prolonged retention of the radioactivity in the target region. Two basic concepts of trapping can be considered, direct trapping and metabolic trapping. In direct trapping, the radiotracer is bound or fixed directly by a mechanism (i.e., cell internalization, uptake) or an interaction with endogenous structures (i.e., receptor binding, protein binding).

A good example of direct trapping is the binding of a somatostatin (sst) receptor ligand (i.e., the peptidic tracer [^{68}Ga]Ga-DOTA-TOC) to the somatostatin receptor. After intravenous administration, the radiolabeled sst-ligand binds to the transmembrane sst receptors on the tumor cells, which are highly overexpressed by neuroendocrine tumors. The sst receptors are G-protein coupled receptors (GPCR) and

---

15 Motor recovery after stroke: From a Vespa Scooter ride over the Roman *Sampietrini* to focal muscle vibration (fMV) treatment. A ^{99m}Tc-HMPAO SPECT and neurophysiological case study. 2020.

may undergo endocytosis after the ligand has bound. As a result, the radiotracer is not only bound to the receptor, but is actively internalized into the cell and thus trapped inside the cell, cf. Section 12.6.3.

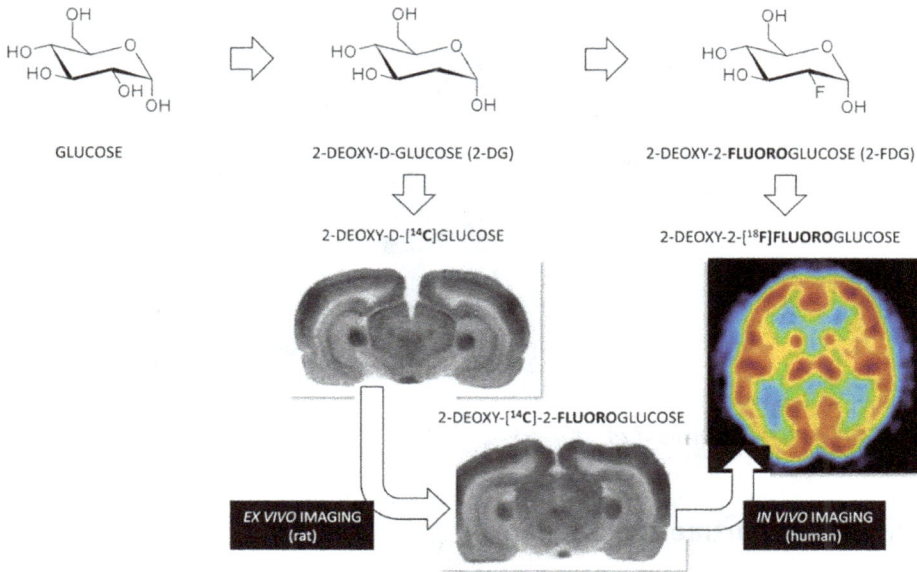

Fig. 12.10: From *in vitro/ex vivo* autoradiography of brain functions applying 2-[^{14}C]DG in animals to noninvasive PET neuroimaging in humans using 2-[^{18}F]FDG.

The best and most prominent example for the metabolic trapping concept in molecular imaging is 2-[^{18}F]FDG. This metabolic trapping concept goes back to the first *in vitro/ex vivo* investigations of L. SOKOLOFF on the metabolism of 2-deoxy-D-glucose in the brain. 2-deoxy-D-[^{14}C]glucose (2-[^{14}C]DG) behaves identically to D-glucose in the first metabolic step (cf. Figs. 12.10 and 12.11). 2-[^{14}C]DG is taken up by the cells through glucose transporters (GLUT). In the cells, 2-[^{14}C]DG is phosphorylated to 2-[^{14}C]DG-6-phosphate by the enzyme hexokinase. While D-glucose-6-phosphate is further metabolized by glucose phosphate isomerase and several following enzymes to CO_2 and H_2O, 2-[^{14}C]DG-6-phosphate is not a substrate for glucose phosphate isomerase and is trapped in this form for a certain time period. This concept of metabolic trapping allowed the study of brain functions in animals by using *ex vivo* autoradiography.

Realizing that 2-[^{14}C]DG is a very attractive tool for studying brain (patho)physiology, the radiotracer was further developed to 2-[^{18}F]FDG to enable noninvasive molecular imaging using PET. The 2-[^{18}F]FDG-6-phosphate stays for a prolonged time in the cells. 2-[^{18}F]FDG is the most frequently employed radiotracer for PET worldwide and used in oncology, cardiology, neuroimaging, and several other diseases or medical questions.

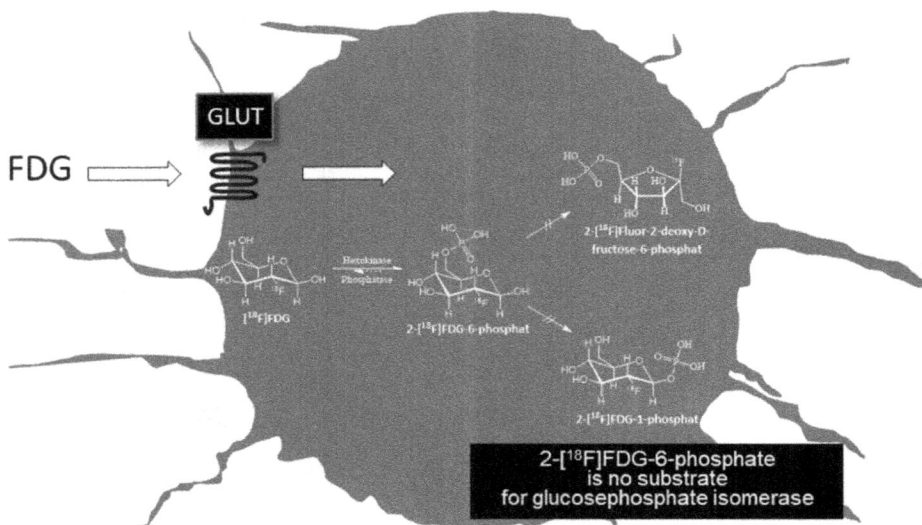

Fig. 12.11: Metabolic trapping of 2-[^{18}F]FDG as 2-[^{18}F]FDG-6-phosphate. GLUT = glucose transporters.

12.5.3 Compartment modeling

The concept of compartment modeling is a tool to transform data from molecular PET imaging into a quantitative interpretation of the biochemistry of a certain radiotracer. Compartment models simplify the description of the biodistribution of a (radio)tracer and make the pharmacology and *in vivo* biochemistry mathematically accessible. The compartments represent different distribution volumes, sections, and regions of the body in which the molecule or its metabolite is uniformly and instantly distributed. Applying the compartment modeling concept to 2-[^{18}F]FDG, the result is a two-tissue-compartment model, cf. Fig. 12.12.

2-[^{18}F]FDG is administrated intravenously and distributed in the compartment C_s representing the fraction of 2-[^{18}F]FDG in serum/blood. 2-[^{18}F]FDG is taken up by the cells into compartment C_f with the rate K_1. K_2 represents the efflux rate of free 2-[^{18}F]FDG from the cell back into serum. Free 2-[^{18}F]FDG in the cell is phosphorylated by the hexokinase to 2-[^{18}F]FDG-6-phosphate with the rate K_3 into the compartment C_m. 2-[^{18}F]FDG-6-phosphate is trapped in C_m and there is only very slow dephosphorylation (K_4) back into C_f. As a result, K_4 is neglected and the compartments C_f and C_m can be combined to C_t. Furthermore, from molecular imaging data of a 2-[^{18}F]FDG scan, C_f and C_m could not be distinguished. In practice, C_s is determined via blood-sampling and C_t from PET scans.

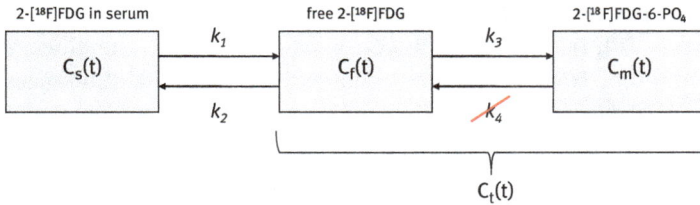

Fig. 12.12: Two-tissue-compartment model of 2-[^{18}F]FDG. Compartment C_s = serum/blood, compartment C_f = free 2-[^{18}F]FDG in the cell, compartment C_m = metabolized 2-[^{18}F]FDG-6-phosphate in the cell, compartment C_t = trapped 2-[^{18}F]FDG-6-phosphate (combined from C_f and C_m). K_1–K_4 = different influx and efflux rates between the compartments. K_4 is very slow and negligible.

12.5.4 Quantification

There are different levels of quantification methods which can be applied. In daily clinical routine, a precise and complex quantitative analysis including multiple blood sampling is too time-consuming and not of special benefit for, i.e., assessment of tumors and metastases. A qualitative analysis by visual inspection of the images is then applied. The major disadvantage is a variability between different people interpreting the images. The most commonly used method of quantification is based on the so-called standard uptake value (SUV), see eq. (12.1). The semi-quantitative method of SUV is a compromise between accuracy and practicability and is commonly applied as semi-quantitative value for accessing PET images in regions of interest (ROI).

$$\text{SUV} = \frac{\text{activity concentration in ROI} \left[\dfrac{\text{MBq}}{\text{ml}}\right] \cdot \text{injected dose}[\text{MBq}]}{\text{body weight [g]}}. \tag{12.1}$$

An SUV of 1 means a uniform distribution of the radioactivity in the whole body; values greater than 1 show accumulation of activity. Typical SUVs of healthy tissues (background) are 1–2, while tumor or metastases can reach SUVs of 15 to 20 and even higher. In some cases, the SUV is related to the body surface area or lean body weight instead of the body weight.

For a more precise and absolute analysis, such as the quantitative assessment of the glucose metabolism, more complex mathematical and practical methods are required. Beside dynamic PET scans, blood sampling at several time-points over the scan period is needed. Based on the compartment modeling described in Section 12.5.3, the glucose metabolic rate (MR^{glu}) can be calculated from the following equation:

$$\text{MR}^{\text{glu}} = \frac{C_s}{\text{LC}} \cdot \frac{K_1 \cdot k_3}{k_2 + k_3} = \frac{C_s}{\text{LC}} \cdot K_i. \tag{12.2}$$

The calculation is based on two-tissue-compartment modeling and the fact that the dephosphorylation of 2-[^{18}F]FDG-6-phosphate (K_4) is neglected. C_s is the glucose blood (serum) concentration, LC is the lumped constant correcting the differences in kinetics between 2-[^{18}F]FDG and glucose, K_1 is the rate constant of the uptake of 2-[^{18}F]FDG, K_2 is the rate constant of the 2-[^{18}F]FDG efflux, K_3 is the rate constant of the phosphorylation of 2-[^{18}F]FDG by hexokinase. K_i is the net rate of the 2-[^{18}F]FDG intake. However, assessing K_i from time activity curves of dynamic PET scans by nonlinear regression comes with inaccuracies such as partial volume effects, varying signal-to-noise ratios, different image quality, or just from the assumptions in the modeling by setting K_4 as zero.

Another similarly complex approach with high accuracy is Patlak graphical analysis. In a 2-[^{18}F]FDG scan, the radioactivity concentration at the time t can be described as:

$$c(t) = \lambda c_s(t) + K_i \int_0^t c_s(\tau)\mathrm{d}\tau. \tag{12.3}$$

(t) is the radioactivity in the region of interest (ROI) at the time t derived from the dynamic PET scan. λ is the distribution volume of 2-[^{18}F]FDG, $c_s(t)$ is the blood concentration of 2-[^{18}F]FDG at the given time t, K_i is the net influx of 2-[^{18}F]FDG into the ROI, and τ is an integration variable. The method of Patlak graphical analysis determines K_i as the slope in a simple plot and is a bit more robust than the method based on nonlinear regression.

12.6 Radiopharmaceutical chemistry for molecular imaging

The introduction of the radionuclide into a biomolecule or a structure of (patho) physiological interest is obviously a crucial step for radiopharmaceuticals and the key competence in radiopharmaceutical chemistry. Furthermore, the decision for a certain radionuclide as the label for a structure defines the whole development path of a radiopharmaceutical. Radiopharmaceutical chemistry has gained systematic expertise in the rapidly growing field. The reader is referred to excellent reviews on topics on the progress of labeling chemistry such as *Chemistry for Positron Emission Tomography: Recent advances in ^{11}C-, ^{18}F-, ^{13}N- and ^{15}O-labeling reactions.* Deng X et al., Angew. Chem. Int. Ed. 2019; *Fluorine-18 radiochemistry, labeling strategies and synthetic routes.* Jacobson O et al., Bioconjugate Chem. 2015; *Fluorine-18 labelled building blocks for PET tracer synthesis.* van der Born D et al., Chem. Soc. Rev., 2017;

[18]F-Labeling of sensitive biomolecules for Positron Emission Tomography. Krishnan HS et al., Chemistry. 2017; *Reaction of [^{18}F]Fluoride at heteroatoms and metals for imaging of peptides and proteins by Positron Emission Tomography*. Scroggie KR et al., Front. Chem. 2021; *Matching chelators to radiometals for radiopharmaceuticals*. Price EW et al., Chem Soc Rev. 2014; and recent publications on related to neuroscience and oncology, such as *Radionuclide Imaging for Neuroscience: Current Opinion and Future Directions*. Gee et al., Molecular Imaging 2020; *Novel tracers and radionuclides in PET imaging*, 2021; *Single Photon Emission Computed Tomography tracer*. Pietzsch et al., Recent results in cancer research, 2020; and *Novel Tracers and Radionuclides in PET Imaging*. Mason et al., Radiol Clin North Am. 2021 (see "Further reading").

The most favorable strategy in labeling chemistry of a molecule is so-called authentic labeling, which means an atom in the lead structure is exchanged by one of its radioisotopes suitable for molecular imaging purposes. An authentic radiolabel eliminates any changes in (bio)chemical and physiological behavior of the radiolabeled molecule compared to the nonradioactive lead. As biomolecules mainly consist of only carbon, hydrogen, oxygen, nitrogen, sulfur and phosphorous, the range of suitable radionuclides for authentic labeling is quite narrow. The "organic" radionuclides oxygen-15, nitrogen-13, and carbon-11 allow such authentic labels. However, these isotopes are very short-lived with half-lives from 2 min to 20 min which strongly limits their applicability. In clinical routine, ^{15}O ($t^{1/2} = 2\,min$) and ^{13}N ($t^{1/2} = 10\,min$) are only employed as [^{15}O]water and [^{13}N]ammonia for perfusion imaging. Both are available from quick on-line processes after the radionuclides have been delivered from a cyclotron. Only the half-life of 20 min of ^{11}C provides more flexibility for radiolabeling and radiosynthesis.

Alternatively, the so-called analog radionuclides with longer half-lives are commonly introduced into biomolecules. Frequently, labeling with analog radionuclides makes use of the concept of bioisosteres. Similarities in steric demand and/or in electronic character of the substituted atoms or functional groups in the lead structure are utilized. Generally, the analog radionuclides cause only insignificant structural differences, but differences in electronic characters and in chemical reactivity can be critical. In each individual case, the pharmacological behavior and properties of analog radiotracers have to be tested for changes in biochemical behavior. With a rapidly increasing number of new pharmaceuticals and bioactive molecules, more and more substances are originally carrying fluorine, bromine, iodine, or even metals. As a result, the advantages of authentic radiolabeling and longer half-lives mount up and simplify the development of corresponding radiopharmaceuticals for molecular imaging.

Several metallic radionuclides enable molecular imaging. The half-lives of metallic radionuclides for molecular imaging vary from minutes to days and offer a broad range of applicability. Some radiometals can be used directly in their free cationic form such as rubidium-82 and thallium-201 as monovalent cations.[16] However, in contrast to organic or analog radionuclides, the many other important metallic radionuclides (^{99mc}Tc, divalent ^{64}Cu, trivalent ^{68}Ga, ^{111}In, ^{90}Y, and the various trivalent lanthanide and actinide radioisotopes) are incorporated into suitable chelators to form metal-chelate complexes which are stable *in vivo*. The most prominent examples are technetium-99m which is an artificial transition metal of versatile redox-chemistry, and trivalent ^{68}Ga. On the one hand, several of these chelators required to stabilize the metal are designed to play an active biochemical/physiological role following the radiometal-chelate complex formation. For many other applications, the metal-chelate complex itself is of no physiological value and is attached covalently to biologically active targeting vectors (e.g., peptides) – and in this case the complex should be as inert as possible. Figure 12.13 compares a single ^{99m}Tc-chelate complex used for imaging kidney function and a ^{99m}Tc-chelate-tropane vector used for imaging brain functions to illustrate the two basic concepts. The complex ^{99m}Tc-MAG3 allows assessment of the kidney function in a dynamic and quantitative manner (see section 12.7.8 kidney function).

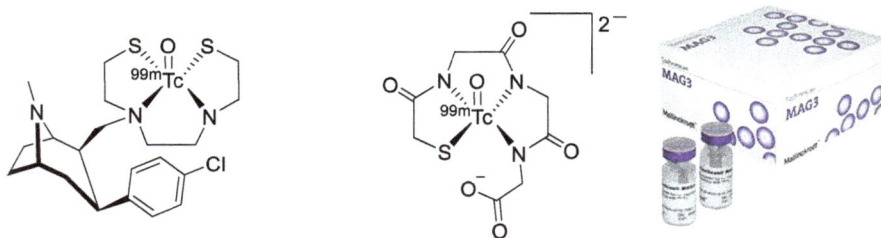

Fig. 12.13: Single radiometal-chelate complex vs. radiometal-chelate-targeting vector concepts: ^{99m}Tc-MAG3 used for visualizing kidney function representing the concept of a single metal-chelate complex possessing physiological properties (right) and ^{99m}Tc-(N2S2-type) chelate-tropane, TRODAT-1, for quantifying dopamine transporter function in the human brain (left). In contrast to the ^{99m}Tc-MAG3 complex, the ^{99m}Tc-chelate complex TRODAT-1 should be biologically inert in order to not ruin the properties of the cocaine-analog tropane moiety binding to the dopamine transporters (DAT). The photo shows how the ligand mercaptoacetyl triglycine is available as commercial, lyophilized kit containing a reducing agent, ready for the addition of the ^{99m}Tc generator eluate. For its application in imaging kidney function cf. 12.7.9.

12.6.1 Radiopharmaceuticals without radiosynthesis

In a few cases, the "naked" radionuclide already displays an important biochemical or pharmacological behavior. Consequently, there is no need for radiosynthesis.

---

16 or strontium-89 and radium-223 as divalent cations for therapeutic purposes, cf. Chapter 13.

The corresponding radionuclides are only purified and formulated for their *in vivo* application. Table 12.3 gives an overview of the most important ones of these radionuclides and their central parameters for the applicability in molecular imaging.

Tab. 12.3: The most important radionuclides achievable without synthetic steps and their central parameters and characteristics. These nuclides are applied in molecular imaging as pure anions or cations. Pertechnetate is the stable and native form of technetium in aqueous solutions.

Radionuclide	Species	Target organ	Imaging modality
^{123}I	iodide anion	thyroid	SPECT
^{131}I			scintigraphy
^{124}I			PET
^{99m}Tc	pertechnetate anion		SPECT
^{18}F	fluoride anion	bone metastases	PET
^{201}Tl	thallium(I) cation	myocard	SPECT
^{82}Rb	rubidium(I) cation		PET

12.6.1.1 Radioiodide [*I]I⁻

Radioiodine is one of the most prominent radionuclides which can be used directly in form of their ions, here it is the anion iodide. Identical to nonradioactive iodide, it is accumulated in the thyroid. After administration, radioiodide is quickly taken up by the thyroid via the *sodium iodide symporter* (NIS) and excess radioiodide is excreted renally. For administration, they all are formulated as sodium iodide in slightly basic aqueous solutions. It can be taken orally or given by injection. The scintigraphic images of the thyroid allow identification of various diseases and dysfunctions of the organ such as local foci of hyperthyroidism or hypothyroidism. Due to their long half-lives, these iodine isotopes are typically provided by external suppliers for clinical routine. For example, [^{131}I]iodide is usually delivered as capsules with the radioactivity already calibrated for certain patients. As shown in Tabs. 12.1 and 12.2, several radioisotopes of iodine are suitable for molecular imaging and, furthermore, enable PET (^{124}I) and SPECT (^{123}I, ^{131}I), respectively.

12.6.1.2 ^{99m}Tc-pertechnetate [^{99m}Tc]TcO$_4^-$

This is the native form of technetium^{VII} in aqueous solutions as eluted from ^{99}Mo/^{99m}Tc radionuclide generators simply by physiological saline solution under sterile conditions and available as readily injectable solution. $[^{99m}Tc]TcO_4^-$ has tetrahedron geometry and a similar size to the iodide anion (volumes of $4.05 \cdot 10^{-23}$ and $4.22 \cdot 10^{-23}$ cm^3, respectively). As a result, $[^{99m}Tc]TcO_4^-$ is accepted by the NIS in the same manner as iodide and thus taken up into the thyroid. Typically, $[^{99m}Tc]TcO_4^-$ is employed for initial diagnostics of thyroid dysfunctions and stratification of patients for a radioiodide $[^{131}I]$ therapy.

12.6.1.3 $[^{18}F]$fluoride $[^{18}F]F^-$

$[^{18}F]$fluoride $[^{18}F]F^-$ in aqueous solution is the form of ^{18}F that comes directly from the target of the cyclotron after applying the $^{18}O(p,n)^{18}F$ nuclear reaction to $[^{18}O]$water. $[^{18}F]F^-$ is separated from the target filling by trapping on an anion ion exchanger cartridge. For formulation, the anion exchanger is washed with sterile water and the $[^{18}F]F^-$ is eluted with sterile saline solution to provide the final injectable solution of the radiopharmaceutical $[^{18}F]NaF$. The $[^{18}F]$fluoride anion is a very specific and efficient bone-seeking agent. The fluoride anion has a high affinity to the hydroxyapatite in bone. In hydroxyapatite $(Ca_{10}(PO_4)_6(OH)_2)$ the hydroxyl anion (OH^-) is rapidly exchanged by the fluoride anion to form fluoroapatite $(Ca_{10}(PO_4)_6F_2)$, cf. molecular imaging of bone metastases.

12.6.1.4 $[^{201}Tl]Tl^+$

^{201}Tl is provided as $[^{201}Tl]TlCl$ in sterile saline solution ready for injection. The $[^{201}Tl]Tl^+$ cation is recognized by the Na^+/K^+-ion exchange pump of cardiac cells and actively pumped into the cells (cf. Fig. 12.40).

12.6.1.5 $[^{82}Rb]Rb^+$

^{82}Rb is available from the $^{82}Sr/^{82}Rb$ radionuclide generator. The $^{82}Sr/^{82}Rb$ radionuclide generator is FDA approved (1989) and on the market as CardioGen-82$^©$ (Bracco Diagnostics, USA). ^{82}Sr is absorbed on tin^{IV} oxide and ^{82}Rb is eluted as $[^{82}Rb]RbCl$ under sterile conditions by use of sterile saline solution and obtained as readily injectable solution. ^{82}Rb emits positrons and enables PET imaging. Similar to $[^{201}Tl]Tl^+$, $[^{82}Rb]Rb^+$ is an analog of K^+ and used for cardiac PET imaging.

12.6.2 Organic radiolabeling chemistry

12.6.2.1 Carbon-11

Carbon-11 is obviously the best choice for authentic radiolabeling of biomolecules and it is one of the most commonly used positron emitters for PET radiopharmaceuticals. Although the short half-life of only 20.4 min impedes time-consuming radiosynthesis or transport of ^{11}C-labeled radiopharmaceuticals, several important ^{11}C-tracers are routinely employed in clinical routine. ^{11}C-chemistry requires an on-site cyclotron for the production via the $^{14}N(p,\alpha)^{11}C$ nuclear reaction. The two major product forms, depending on the target gas composition, and thus the two primary ^{11}C-labeling synthons are $[^{11}C]CH_4$ and $[^{11}C]CO_2$. $[^{11}C]CO_2$ can be used to react directly with Grignard reagents (R-MgBr) giving the corresponding direct ^{11}C-carboxylation product. A commonly used ^{11}C-labeled radiopharmaceutical is $[1-^{11}C]$acetate deriving from the direct ^{11}C-carboxylation of CH_3MgBr. Generally, the short half-life allows only reactions and conversions with very fast kinetics. However, several secondary and further ^{11}C-labeling synthons deriving from $[^{11}C]CH_4$ and $[^{11}C]CO_2$ offer a broad and flexible range of ^{11}C-labeling strategies (cf. Fig. 12.14), even with the short half-life.

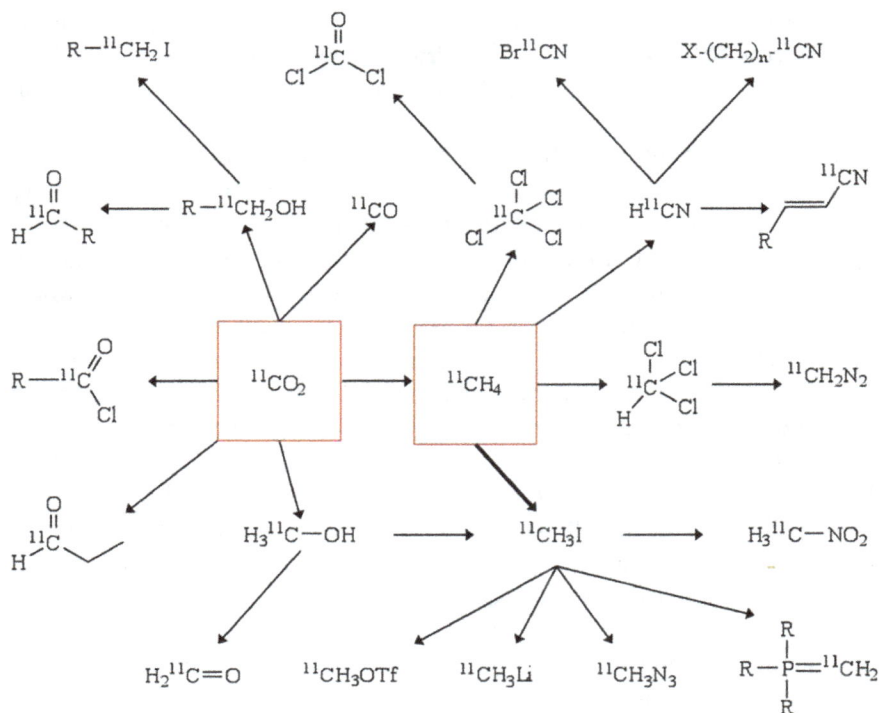

Fig. 12.14: Secondary and further ^{11}C-labeling synthons derived from the primary $[^{11}C]CO_2$ and $[^{11}C]CH_4$.

Along with all the possible pathways, the ones using [^{11}C]CH$_3$I are the preferred routes applied for ^{11}C-labeling. [^{11}C]CH$_3$I can be used for very fast and efficient ^{11}C-methylations of biomolecules. Generally, ^{11}C-labeling via methylation is performed as N-, O- or S-heteroatom ^{11}C-methylation using the desmethyl precursors. [^{11}C]CH$_3$I is produced via two different methods, the "wet" and the "dry" process (cf. Fig. 12.15).

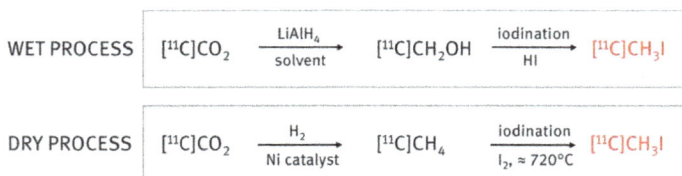

WET PROCESS   [^{11}C]CO$_2$ $\xrightarrow[\text{solvent}]{\text{LiAlH}_4}$ [^{11}C]CH$_2$OH $\xrightarrow[\text{HI}]{\text{iodination}}$ [^{11}C]CH$_3$I

DRY PROCESS   [^{11}C]CO$_2$ $\xrightarrow[\text{Ni catalyst}]{\text{H}_2}$ [^{11}C]CH$_4$ $\xrightarrow[\text{I}_2, \approx 720°C]{\text{iodination}}$ [^{11}C]CH$_3$I

Fig. 12.15: Radiosyntheses of [^{11}C]CH$_3$I according to the "wet" and "dry" method starting from primary [^{11}C]CO$_2$.

In the wet method, [^{11}C]CO$_2$ is reduced to [^{11}C]CH$_3$OH by means of lithium aluminium hydride in organic solvents (diethyl ether). In a second step, [^{11}C]CH$_3$OH is iodinated to the final [^{11}C]CH$_3$I using hydroiodic acid. [^{11}C]CH$_3$I is distilled into a solution of a precursor molecule, where the ^{11}C-methylation takes place. A more recent approach is the "dry" process. Starting from [^{11}C]CO$_2$, hydrogen reduction in presence of a nickel catalyst provides [^{11}C]CH$_4$ which is passed through a heated glass tube (≈ 720 °C) with iodine vapor for radical iodination. The product [^{11}C]CH$_3$I is trapped on a Porapak cartridge. For the subsequent ^{11}C-methylation, [^{11}C]CH$_3$I is released by heating the Porapak cartridge and is transferred by a gas stream into the reaction vessel. The dry method is the more preferred method as it generally provides much higher specific activities which can be very crucial for PET studies of receptor systems in the CNS. In the wet method, the use of LiAlH$_4$ is a source of nonradioactive carbon dioxide which in turn brings in isotopic carrier (carbon-12) and thus dramatically reduces the specific radioactivity of the product. Moreover, the dry process allows several consecutive productions of [^{11}C]CH$_3$I without cleaning or opening the synthesis apparatus. After use of the wet process, vessels and tubings need to be cleaned or exchanged. Automatic radiosynthesis modules using the dry process are commercially available and offer a broad flexibility and a convenient access to a variety of ^{11}C-methylated radiopharmaceuticals, such as L-[S-methyl-^{11}C]methionine or [N-methyl-^{11}C]flumazenil enabling PET imaging of brain tumors or the translocator protein (TSPO, first described as peripheral benzodiazepine receptor), respectively (cf. Fig. 12.16).

In some cases, the reactivity of the synthon [^{11}C]CH$_3$I might not be high enough, and by a simple on-line transformation the more reactive ^{11}C-methylation agent [^{11}C]CH$_3$OTf can be obtained. [^{11}C]CH$_3$OTf is formed by passing the [^{11}C]CH$_3$I in a gas stream over a cartridge containing silver triflate. The final gas stream contains the product [^{11}C]CH$_3$OTf and can directly be transferred into a precursor solution to perform

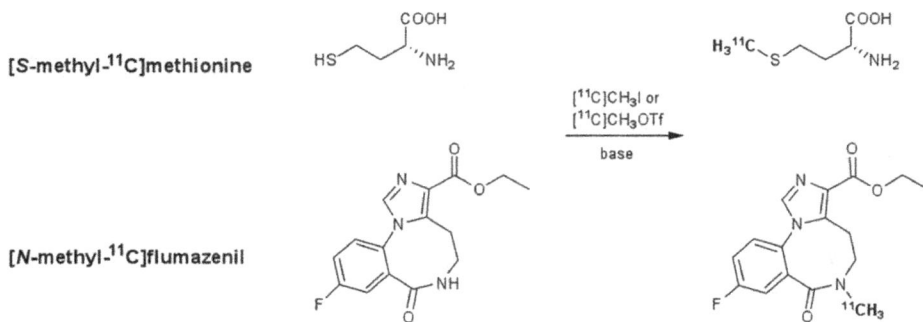

Fig. 12.16: N- and S-heteroatom ^{11}C-methylation reactions based on $[^{11}C]CH_3I$ or $[^{11}C]CH_3OTf$.

the ^{11}C-methylation. ^{11}C-methylation reactions using $[^{11}C]CH_3OTf$ show extremely fast kinetics and allow efficient ^{11}C-methylations already at room temperature, which is very helpful in case of heat sensitive biomolecules.

12.6.2.2 Fluorine-18

Fluorine-18 is the most important organic radionuclide in PET. Consequently, methods for its introduction into biomolecules are of paramount interest and already a broad variety of procedures has been developed. In general, three basic principles for ^{18}F-fluorination reactions can be defined. (1) direct nucleophilic substitutions; (2) direct electrophilic substitution and (3) indirect ^{18}F-labeling by synthons/prosthetic groups. Which procedure is applicable and used depends on the biomolecule to be labeled and on the intended position in its molecular structure. By far the most preferred method is the direct nucleophilic substitution using no-carrier-added $[^{18}F]$fluoride. Furthermore, it is also the method of choice to label small synthons and prosthetic groups with ^{18}F for subsequent indirect labeling reactions. The electrophilic pathway uses carrier-added $[^{18}F]F_2$ gas, which is the primary product from the target when applying the $^{20}Ne(p,\alpha)^{18}F$ nuclear reaction. C.a. $[^{18}F]F_2$ comes with low specific radioactivities and the production routes are less efficient than for n.c.a. $[^{18}F]$fluoride. Accordingly, the use of the direct electrophilic substitution via c.a. $[^{18}F]F_2$ is decreasing.

Direct electrophilic ^{18}F-fluorination: A very popular example of electrophilic radio-fluorinations using c.a. $[^{18}F]F_2$ is the original method to produce 2-$[^{18}F]FDG$ in 1978. $[^{18}F]F_2$ was used in a electrophilic addition to the double bond of triacetoxyglucal to yield 2-$[^{18}F]FDG$ in a radiochemical yield of only 8% and the radioactive side product 2-$[^{18}F]$fluoro-2-deoxy-mannose ($[^{18}F]FDM$) with 3%.

In the 1980s, the new synthesis route of 2-[^{18}F]FDG via nucleophilic substitution with n.c.a. [^{18}F]fluoride became available and replaced the electrophilic method. However, c.a. [^{18}F]F$_2$ was still necessary for the radiolabeling of molecules which are not available via n.c.a. [^{18}F]fluoride. The most prominent and most important radiopharmaceutical in this respect is 6-[^{18}F]fluoro-L-DOPA ([^{18}F]DOPA). The molecular structure of [^{18}F]DOPA does not offer convenient access via n.c.a. [^{18}F]fluoride and the electron-rich aromatic ring calls for electrophilic methods for the ^{18}F-introduction (Fig. 12.17).

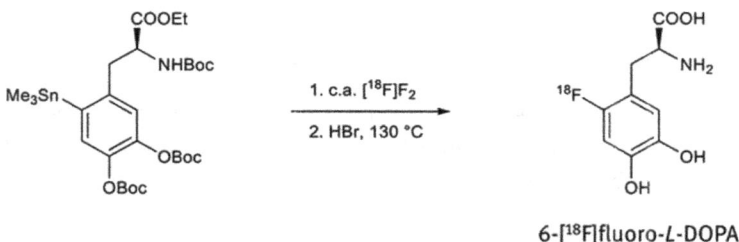

Fig. 12.17: Direct electrophilic ^{18}F-labeling of [^{18}F]DOPA.

The importance of [^{18}F]DOPA for the diagnosis of Parkinson's disease has helped to keep the direct electrophilic ^{18}F-labeling methods alive. Furthermore, commercial radiosynthesis modules are available for [^{18}F]DOPA productions in clinical routine. Several attempts have been reported to find a nucleophilic pathway to [^{18}F]DOPA. The most promising nucleophilic approaches are based on precursor molecules carrying electron-withdrawing groups to activate the aromatic ring for nucleophilic substitution reaction using ^{18}F-for-NO$_3$ exchange or even an ^{18}F/^{19}F-isotopic exchange, cf. Fig. 12.20. In a second step, auxiliary and protection groups are removed and/or modified (oxidized) to yield the desired [^{18}F]DOPA.

Direct nucleophilic ^{18}F-fluorination: As mentioned before, the nucleophilic substitution of ^{18}F as [^{18}F]fluoride is by far the most preferred method for ^{18}F-introduction into molecules. [^{18}F]Fluoride is obtained from the cyclotron in an aqueous solution, wherein it is present as [^{18}F]HF with a large and tight hydration shell. In this form, the ^{18}F is very unreactive and not useful for organic chemistry reactions. To activate the ^{18}F and remove the target water ([^{18}O]water), three major tricks are used. First, the target filling is passed over an anionic exchanger resin, where [^{18}F]fluoride is trapped and the target water runs through. This step removes most of the water and enables recycling of the enriched [^{18}O]water. In the next step, the [^{18}F]fluoride is eluted from the resin by a mixture of acetonitrile and a base, frequently potassium carbonate. Furthermore, a phase transfer catalyst is added. The mixture is heated and the acetonitrile is removed under reduced pressure and a gas stream. The acetonitrile further forms an azeotropic mixture with residual water traces, which are removed by distilling this azeotropic mixture.

For complete removal of all water traces, dry acetonitrile is added twice and distilled off. Dry [18F]fluoride is left and as counter ion either large and soft cations such as cesium or tetrabutylammonium are used, or a large cationic complex of potassium and an aminopolyether (Kryptofix2.2.2®) is formed. As a result, a "naked" and activated [18F]fluoride is available for nucleophilic reactions. To keep this activation, the following labeling steps need to be free of water, protons, metal cations, and everything which would attract the [18F]fluoride anion.

The most important 18F-labeled molecule is 2-[18F]fluoro-2-deoxy-D-glucose (2-[18F]FDG), as already mentioned before. The replacement of the original electrophilic 18F-labeling of 2-[18F]FDG was driven by a new precursor, the 1,3,4,6-tetraacetate 2-(trifluoromethane-sulfonate), and the use of an aminopolyether (Kryptofix2.2.2®) as phase-transfer catalyst (cf. Fig. 12.18). The major advantages of the new method are very high and reliable radiochemical yields, no-carrier-added conditions as well as further benefits from the starting material n.c.a. [18F]fluoride. Today, 2-[18F]FDG is produced in automated synthesis modules (cf. Fig. 12.19) in very large batches of up to 100 GBq and even more. The automated synthesis has been improved and ready-to-use sterile, single-use cassettes (all chemicals and materials are included) are applied (cf. Fig. 12.18).

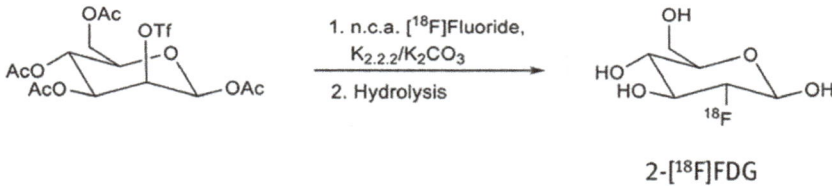

Fig. 12.18: Direct nucleophilic 18F-labeling of 2-[18F]FDG based on n.c.a. [18F]fluoride and the phase-transfer-catalyst Kryptofix2.2.2®.

Several commercial suppliers provide 2-[18F]FDG. The product batches are dispensed in vials according to customer requests and shipped to different clinics and institutes for PET imaging.

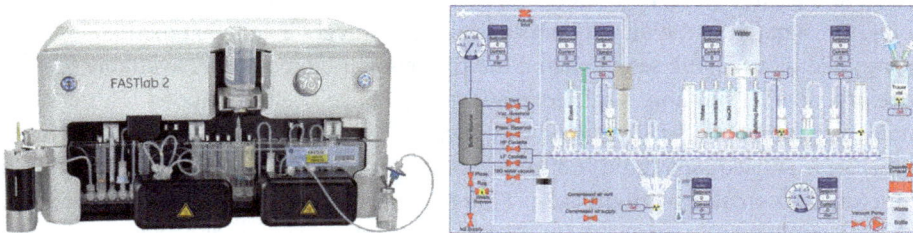

Fig. 12.19: Automated radiosynthesis module (GE healthcare FASTlab2®) for direct nucleophilic 18F-labeling reactions such as for 2-[18F]FDG, and the corresponding scheme of the Fastlab2-based [18F]FDG radiosynthesis.

Besides 2-[^{18}F]FDG, several biomolecules and structures have been labeled with ^{18}F via nucleophilic substitution. Important examples are the developed nucleophilic radiosyntheses toward [^{18}F]DOPA, Fig. 12.20. The new radiosyntheses are based on nucleophilic aromatic ^{18}F-labeling using n.c.a. [^{18}F]fluoride. The major challenges in a nucleophilic n.c.a. approach toward [^{18}F]DOPA are the reduction of electron density in the aromatic ring enabling nucleophilic reactions, suitable protection group chemistry allowing ^{18}F-labeling and retaining the right stereochemistry (L-configuration of the amino acid), and a convenient and efficient radiosynthesis suitable for automation. The first two precursors fulfilling these requirements lead to multi-step procedures, which are available for different automated radiosynthesis modules. Recently, the introduction of copper-mediated ^{18}F-fluorinations using arylboronic acids and boronic pinacol esters as precursors enabled the direct nucleophilic aromatic ^{18}F-fluorination of a wide range of (novel) molecules with relatively high electron density. By applying such a boronic pinacol ester precursor (cf. Fig. 12.20) for the ^{18}F-labeling of [^{18}F]DOPA, two-step automated procedures became available, yielding [^{18}F]DOPA in high specific activities.

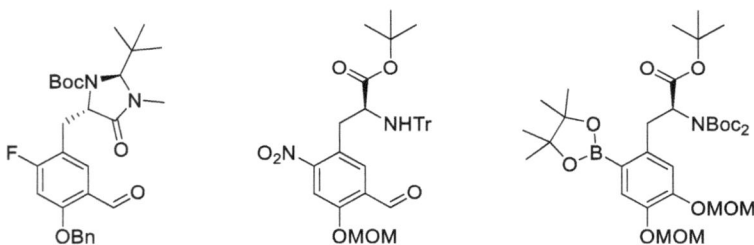

Fig. 12.20: Precursors for direct nucleophilic ^{18}F-labeling of [^{18}F]DOPA. The first two molecules use the auxiliary aldehyde function to activate a nucleophilic attack of [^{18}F]fluoride on the aromatic ring (C6), while the third precursor is based on the novel direct copper-mediated ^{18}F-labeling using a boronic pinacol ester precursors avoiding auxiliary moieties.

^{18}F-Labeling synthons and prosthetic groups: The range of biomolecules is very broad and a large fraction of them has complex and often sensitive molecular structures. Consequently, the quite harsh conditions of direct ^{18}F-labeling methods cannot be applied. For making sensitive and multifunctional biomolecules available for ^{18}F-labeling, special synthons and prosthetic groups have been developed. These small molecules can be efficiently labeled with n.c.a. [^{18}F]fluoride and offer additional functions for a subsequent (mild) coupling to the desired biomolecule. A very simple structure and prosthetic group is 2-[^{18}F]fluoroethyl tosylate ([^{18}F]FETos), cf. Fig 12.21, which is available by nucleophilic ^{18}F-labeling of the ethylene ditosylate precursor. [^{18}F]FETos is a versatile ^{18}F-labeling synthon and can conveniently be coupled to molecules bearing nucleophilic functions such as NH$_2$ and OH groups. O-(2-[^{18}F]fluoroethyl)-L-tyrosine ([^{18}F]FET) is a widely used ^{18}F-labeled amino acid in PET imaging of brain tumors and for which the ^{18}F-labeling via [^{18}F]FETos is one of the common radiosynthesis pathways.

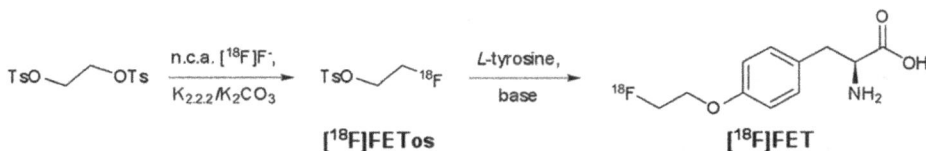

Fig. 12.21: Direct nucleophilic ^{18}F-labeling of the ^{18}F-synthon [^{18}F]FETos and subsequent coupling to the amino acid *L*-tyrosine toward the PET tracer [^{18}F]FET.

In the case of peptides, the complexity of the molecular structure and their multifunctionality are obvious. Peptides are quite sensitive and only rarely offer options for standard methods of direct ^{18}F-labeling. As a consequence, several different prosthetic groups for ^{18}F-labeling of peptides have been developed. The diversity of their functionalities for subsequent coupling reactions with the peptides allow various strategies such as alkylation, acylation, amidation, imidation, thiol coupling, oxime formation, photochemical conjugation, thiourea formation, or click chemistry. One of the first and commonly applied prosthetic groups for ^{18}F-labeling of peptides is the activated succinimidyl ester of *para*-[^{18}F]fluorobenzoate ([^{18}F]SFB, *N*-succinimidyl-4-[^{18}F]fluorobenzoate). [^{18}F]SFB is available from a fast two-step radiosynthesis in high yields and enables coupling to an amino functionality in the desired biomolecule, Fig. 12.22.

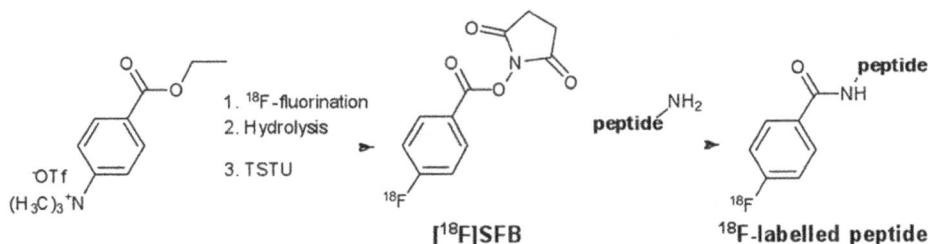

Fig. 12.22: Direct nucleophilic ^{18}F-labeling of [^{18}F]SFB and subsequent coupling to free amino functionalities for ^{18}F-acylation of peptides. TSTU: *O*-(*N*-succinimidyl)-*N*-*N*,*N'*,*N'*-tetramethyluronium tetrafluoroborate.

12.6.2.3 Radioiodine

Radioiodine offers very versatile applications in molecular imaging and nuclear medicine. The radioisotopes ^{120}I and ^{124}I are suitable for PET and ^{123}I for SPECT. Furthermore, ^{131}I is commonly applied as β^--emitter for endoradiotherapy of malignancies of the thyroid and other tumors, but it also offers well-detectable photons of 364 keV for SPECT imaging.

Direct electrophilic radioiodination: A major route and very convenient way to label with radioiodine is direct electrophilic substitution. Generally, direct electrophilic iodinations can be performed at room temperature and provide good to very high radiochemical yields, already after a few minutes reaction times. To achieve reactive electrophilic species an *in situ* oxidation of radioiodide is applied. The most commonly used oxidants are chloramine-T (CAT; *para*-tosylchloramide sodium) and Iodogen™ (1,3,4,6-tetrachloro-3α,6α-diphenylglycouril). CAT allows oxidations in homogeneous aqueous solutions, whereas Iodogen™ is insoluble in water, and thus it is the proper substance for heterogenic reaction routes, which are advantageous for oxidation sensible precursors where the oxidant has to be removed quickly. Furthermore, so-called (radio)iodination tubes are commercially available in which Iodogen is coated on the walls of test tubes. After the radioiodination, the reaction mixture is removed from the (radio)iodination tube and instantly separated from the oxidant. Figure 12.23 depicts the direct electrophilic radioiodination of *N*-succinimidyl-3-(4-hydroxyphenyl)propionate (SHPP), the so-called Bolton–Hunter reagent, for radioiodination of antibodies and peptides via prosthetic group labeling. As a drawback, the direct electrophilic radioiodination is not a regiospecific method; as a result, isomeric derivatives may occur.

Electrophilic demetalation: In contrast to the direct electrophilic procedure, electrophilic demetalation almost provides regiospecificity. Particularly in automated syntheses it offers simple purification and isolation of the radiotracer. On the other hand, the syntheses of the organometallic precursors might be complex and extensive. As precursors for the electrophilic demetalation radioiodinations, organotin compounds in particular are used and often show excellent results. Consequently, radioiodo-destannylation is the most appropriate radioiodination procedure and thus it is the most commonly employed method.

Radioiodination via prosthetic group labeling: In cases of biomolecules that are sensitive to the oxidative reagents or if functional sites for direct radioiodinations are lacking, direct radioiodination methods fail. Alternatively, small molecules can be radioiodinated as prosthetic groups and subsequently coupled with the desired compound. Principally it is the same strategy as for ^{18}F-labeling via prosthetic groups described above. The first approach on prosthetic groups for radioiodination was the above mentioned Bolton–Hunter reagent, *N*-succinimidyl-3-(4-hydroxyphenyl)propionate (SHPP), an activated ester as prosthetic group for peptides and proteins via coupling with a free amino function (cf. Fig. 12.23). Several further coupling methods of prosthetic groups for radioiodinations have been developed. Other common examples are molecules carrying a maleimide function which can be coupled to thiol groups in biomolecules.

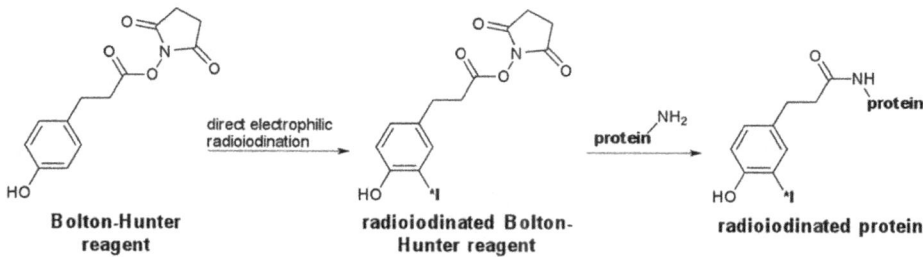

Fig. 12.23: Direct electrophilic radioiodination of SHPP, the so-called Bolton–Hunter reagent, for radioiodination of antibodies and other proteins via prosthetic group labeling.

12.6.3 Coordination chemistry

Several metallic radionuclides present suitable and attractive characteristics for molecular imaging. Certainly, the transition metal ^{99m}Tc is the most widely applied radionuclide in molecular imaging and nuclear medicine. The most important radiometals today are listed in Tab. 12.4.

Tab. 12.4: Clinically applied metallic radionuclides and their characteristics and applications.

Radiometal valency	Nuclide	Half-life	Coordination geometry/number	Imaging modality
Me^{II}	^{62}Cu	9.67 min	/6	PET
	^{64}Cu	12.7 h		PET
	^{67}Cu	61.8 h		SPECT
Me^{III}	^{44}Sc	3.93 h	/8	PET
	^{68}Ga	67.7 min	octahedron/6	PET
	^{67}Ga	3.26 d	octahedron/6	SPECT
	^{111}In	2.80 d	7/6	SPECT
Me^{VII}	^{99m}Tc	6.01 h		SPECT

Except in cases where pure metallic cations can directly be applied for molecular imaging, the availability and applicability of radiometal-labeled molecules depend on the stable incorporation of the metal into molecular structures. For such a stable complexation of the radiometal, certain ligands or chelating agents are required, which additionally enable a linkage to a biomolecule. Alternatively, the chelating molecule already exhibits interesting biological functionality and can be applied for molecular imaging. The best examples for such cases are many of the ^{99m}Tc-radiopharmaceuticals such as the example ^{99m}Tc-MAG3 in Fig. 12.12.

In terms of labeling, the general concept with metallic nuclides is based on a bifunctional chelating system which forms stable complexes with the radiometal and is coupled to the biomolecule, Fig. 12.24. Frequently, linking structures and spacers are introduced between chelating structure and the biomolecule. Spacer are just used to get some distance between the two bulky molecules and avoid interference of the biomolecule binding through the chelating systems. On the other hand, spacer molecules can be important as pharmacokinetic modulators and influence the pharmacokinetic profile such as the increase or decrease in lipophilicity of the whole construct.

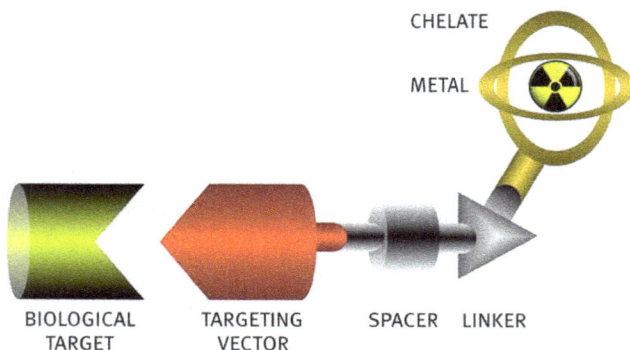

Fig. 12.24: Concept of radiolabeling of biomolecules with radiometals using a bifunctional chelating system covalently attached via chemical linkers and spacer units to a targeting vector (the "magic bullet").

Prominent examples of chelating systems are DOTA (1,4,7,10-tetraazacyclododecane-*N,N′,N″,N‴*-tetraacetic acid), NOTA (1,4,7-triazacyclononane-*N,N′,N″*-triacetic acid) and DTPA (diethyltriamine-*N,N,N′,N″,N″*-pentaacetic acid), Fig. 12.25. Of these, DTPA is the only noncyclic chelating system, while DOTA and NOTA are cyclic chelating systems. Due to the open structure of the polyamino polycarboxy ligand DTPA, complexes with various radiometals are rapidly formed, already at low temperatures. Therefore, DTPA is quite often used for radiolabeling of antibodies and sensitive proteins with radiometals.

Among the commonly used chelating systems, DOTA is by far the most versatile one. The cage size fits a broad range of radiometals from small ones such as radiogallium to very large atoms of actinides. The flexibility offered by DOTA is the major advantage and enables the development of so-called *theranostics* without alternating the molecular structure and just changing from a diagnostic radiometal to a therapeutic radiometal in the radiolabeling.

Concerning radiogallium, NOTA has the perfect cage size and forms very strong octahedron complexes with gallium[III] already at mild conditions. Noteworthy, in an

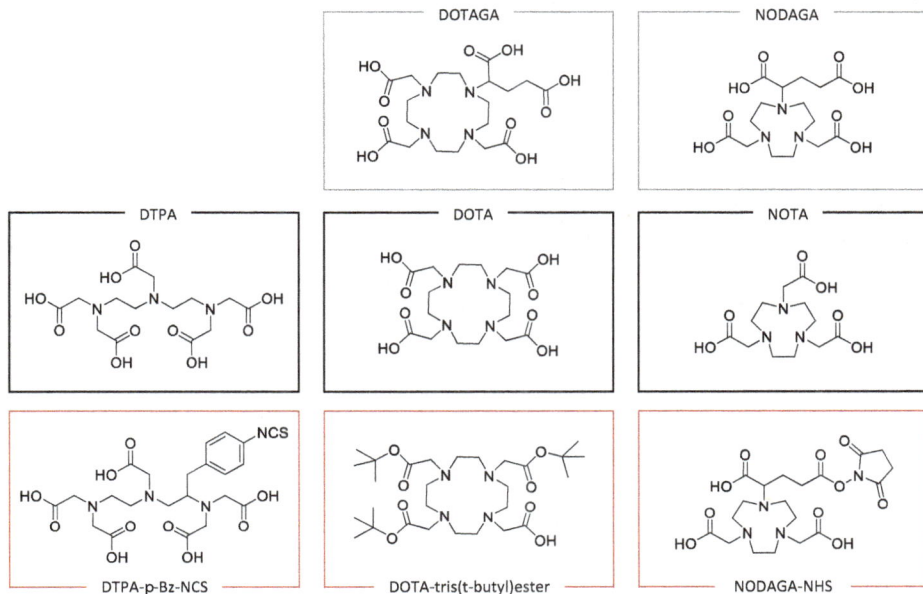

Fig. 12.25: Molecular structures of prominent chelating systems for radiometals. Middle row represent the initial chelators DTPA (acyclic), DOTA and NOTA (cyclic); upper row indicates extensions of DOTA and NOTA toward DOTAGA and NODAGA, respectively, adding one extra COOH-group; lower row shows selected examples of bifunctional derivatives ready for coupling the chelator to targeting vectors.

octahedron complex all functionalities of NOTA are occupied and there is no further possibility for conjugation to a biomolecule. Alternatively, derivatives of NOTA, such as NODAGA (1,4,7-triazacyclononane-*N,N*′-diacetic acid-*N*″-glutaric acid, Fig. 12.25) are available which offer additional coupling options while all coordinating ligands are retained for stable complexation of radiometals with an octahedron coordination. For DOTA, there is a similar strategy, yielding DOTAGA.

In all cases, the coupling of the chelator to a targeting vector is the next step, which, however, requires transforming the initial chelator into a bifunctional one. "Bifuntional" refers to the two tasks of the chelator, namely keep the coordination sites available for complex formation with the radiometal, and offer one functionality (mainly -COOH) for covalent coupling to the targeting vector. Examples of are given in Fig. 12.25, lower row, with three different coupling moieties.

12.7 Radiopharmaceuticals for molecular imaging

12.7.1 Biological targets

In oncology the most important target, the enzyme hexokinase, was already introduced above (cf. Section 12.5.2). The corresponding radioligand is 2-[^{18}F]FDG. Imaging hexokinase activity correlates with the glycolysis rate of cells as an indicator of their metabolism. Consequently, the drastically increased glycolysis rate of tumor cells enables an almost universal oncological imaging tool using 2-[^{18}F]FDG. On the other hand, tumor cells own specific targets, which further allows selective addressing of these cells by suitable radioligands, Fig. 12.26.

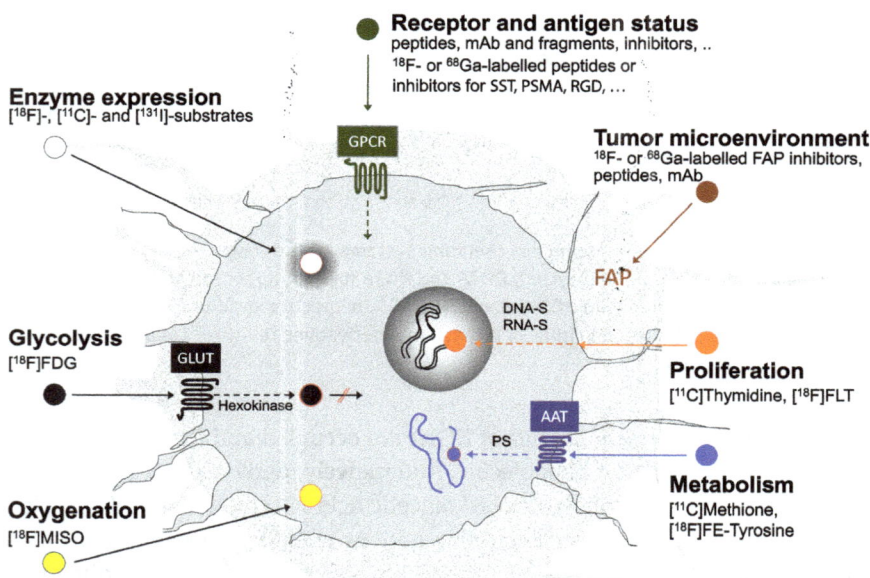

Fig. 12.26: Various targets for PET imaging in oncology and corresponding radiotracers: 3′-deoxy-3′-[^{18}F]fluoro-L-thymidine ([^{18}F]FLT), [^{11}C-*methyl*]methionine ([^{11}C]Methionine), [^{18}F]fluoromisonidazol ([^{18}F]MISO), 2-deoxy-[^{18}F]fluoroglucose (2-[^{18}F]FDG). GLUT = glucose transporters, AAT = amino acid transporters, GPCR = G-protein coupled receptors, SST = somatostatin, mAb = monoclonal antibodies, PSMA = prostate specific membrane antigene, FAP = fibroblast activation protein, RDG = Arg-Gly-Asp, PS = protein synthesis, DNA-S/RNA-S = DNA and RNA synthesis.

Many targets are based on the enhanced metabolism and need for substances and building blocks for the high proliferation rate of tumor cells. Radiotracers as analogs of building blocks for DNA and RNA are compounds such as 5-[^{18}F]fluorouracil ([^{18}F]FU), [^{11}C]thymidine or 3′-[^{18}F]fluoro-3′-deoxy-L-thymidine ([^{18}F]FLT), which allow determination of the proliferation status of tumor cells by PET imaging.

Another important mechanism for oncological targeting is the amplified protein synthesis of tumor cells. O-(2-[^{18}F]fluoroethyl)-L-tyrosine ([^{18}F]FET) and [^{11}C-*methyl*] methionine ([^{11}C]MET) are the most commonly applied radiolabeled amino acid analogs and will be discussed in detail in Section 12.7.4.

Transmembrane and surface receptors are another important class of target structures. In the last decade, the most successful of such receptor systems in molecular imaging has been the family of somatostatin receptors (hSSTR). hSSTRs provide a very specific target on neuroendocrine tumors and in combination with the highly specific radioligands [^{68}Ga]Ga-DOTA-TOC and -TATE, this target-ligand system almost ideally matches the magic bullet concept of EHRLICH.[17]

In addition, as a consequence of their generally enhanced metabolism and fast proliferation, tumor cells quickly become hypoxic and substances like [^{18}F]fluoromisonidazole are intracellularly reduced and metabolically trapped. Several other specific targets are expressed by tumor cells and different tumor entities show different expression patterns.

The basis for molecular imaging of certain (patho)physiology and functions is the availability of a suitable radiotracer which takes part in the relevant processes. With an appropriate radiotracer in hand, molecular imaging of almost any biochemical process or function is possible. However, the development of new and specific radiotracers is not trivial and besides target specificity, further characteristics such as pharmacokinetics, metabolic stability or excretion pathways must be taken into account. In the following section a selection of targets and organ functions are presented, and the most commonly applied corresponding radiotracers are discussed.

12.7.2 Oncology: Glycolysis

In relation to healthy tissues, tumor cells exhibit strongly enhanced energy consumption, and thus, their glucose uptake and metabolism (glycolysis) is drastically increased. These findings go back to 1956, when the Nobel laureate Otto H. WARBURG investigated tumor cells and identified a very high rate of glycolysis as major energy source of these cells. This phenomenon is called the (oncological) Warburg effect.[18] Glucose, the major energy source for eukaryotic cells, is transported into cells via the glucose transporters (GLUT1-6, mainly GLUT-1), phosphorylated to glucose-6-phosphate by the enzyme hexokinase and then either converted to glycogen or enzymatically degraded to carbon dioxide and water (cf. Fig. 12.11).

---

[17] PAUL EHRLICH, Nobel Prize in Physiology or Medicine, 1908, "in recognition of their work on immunity" and related to the concept of antibodies (carried out together with EMIL VAN BEHRING).
[18] OTTO WARBURG. On the origin of cancer cells. Science 1956, 123, 309–314. **Note:** *The (oncological) WARBURG effect should not be confused with another phenomenon also named WARBURG effect in plant physiology describing a decrease of photosynthesis and an increase in photorespiration.*

Like glucose, 2-[^{18}F]FDG is transported into the cell via the GLUT and then also phosphorylated to 2-[^{18}F]FDG-6-phosphate. However, the ^{18}F in 2-position prevents further conversion through the enzyme glucose-6-phosphate isomerase and due to the phosphorylation 2-[^{18}F]FDG-6-phosphate is no longer a substrate for GLUT. The phosphorylated tracer is essentially *trapped* (metabolic trapping) in the cell and becomes an ideal tracer for the *glycolysis rate* of cells. This is exemplified in Fig. 12.27 for the visualization of the glucose consumption of the living brain – the organ utilizing ca. 90% of the human glucose reservoir.

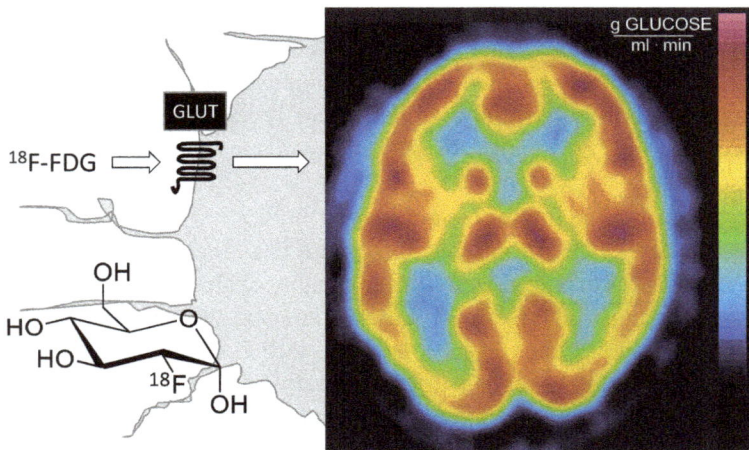

Fig. 12.27: Transport of 2-[^{18}F]FDG into healthy (brain) cells, and the corresponding PET image (axial/transversal plane) showing the high physiological glycolysis rate of normal brain cells.[19] The color-coded column on the right translates radioactivity measured into biochemistry in terms of glucose consumption in gram per ml and minute.

Today, 2-[^{18}F]FDG is the most important PET tracer worldwide. It is conveniently producible in very high yields or can be obtained from other sites and commercial suppliers shipping 2-[^{18}F]FDG even over distances of several hundred kilometers. Its main application is in tumor diagnosis. Based on the WARBURG effect, 2-[^{18}F]FDG is predominantly applied for oncological imaging such as primary identification of malignancies, staging of tumors and metastases and therapy control and monitoring, Fig. 12.28. Besides, several other diseases and pathophysiological processes cause changes in cells' glycolysis rate. Accordingly, 2-[^{18}F]FDG is a very helpful tool to identify and follow these.

---

19 The image represents the glucose consumption of the brain of one of the authors of this chapter, FRANK ROESCH.

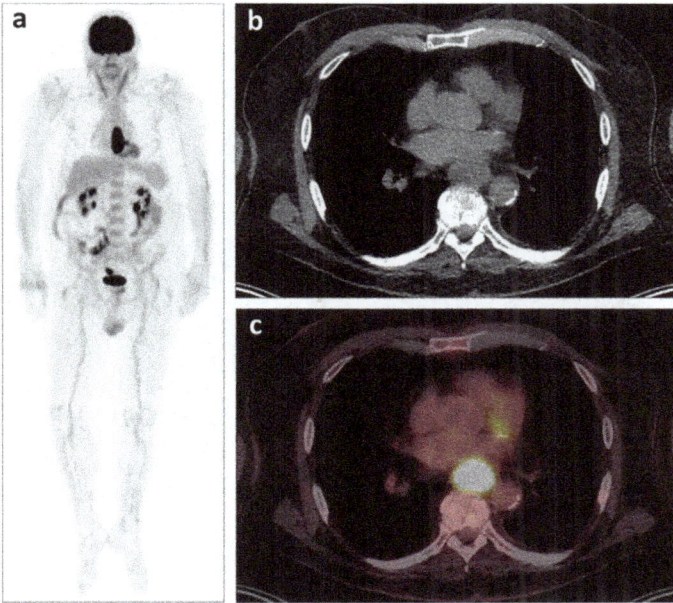

Fig. 12.28: 2-[¹⁸F]FDG anterior maximum intensity projection in a patient with advanced esophageal cancer (a), a transversal CT of the tumor (b), and the transversal fused PET/CT image of the tumor (c). Courtesy of T. Derlin, Nuclear Medicine, Hannover Medical School, Germany.

12.7.3 Oncology: Protein synthesis

Besides the increased glycolysis rate, tumor cells are generally in a status of enhanced metabolism and proliferation. As a result, protein synthesis is also amplified and the required building blocks, amino acids, are increasingly taken up by tumor cells, Fig. 12.29. Radiolabeled amino acids are therefore important radiotracers in tumor diagnostics.

The most frequently applied radiolabeled amino acids in clinical routine are [^{11}C-*methyl*]methionine ([^{11}C]MET) and O-(2-[^{18}F]fluoroethyl)-L-tyrosine ([^{18}F]FET), Figs. 12.29 and 12.30. Both enable molecular imaging via PET, but with different half-lives. Another important aspect of the difference in the ^{11}C- and ^{18}F-label in this particular case is the type of labeling or the structural changes. The ^{11}C-label of [^{11}C] MET is an authentic labeling via ^{11}C-methylation while [^{18}F]FET is an analog of tyrosine bearing the 2-[^{18}F]fluoroethyl group additionally. As a consequence, they exhibit slightly different biochemistry. Due to the authentic label, the metabolism of [^{11}C]MET is identical to native methionine. Hence, [^{11}C]MET is taken up into tumor cells via the amino acid transporters (AAT) and further incorporated into proteins in the protein biosynthesis which provides the desired metabolic trapping. [^{18}F]FET

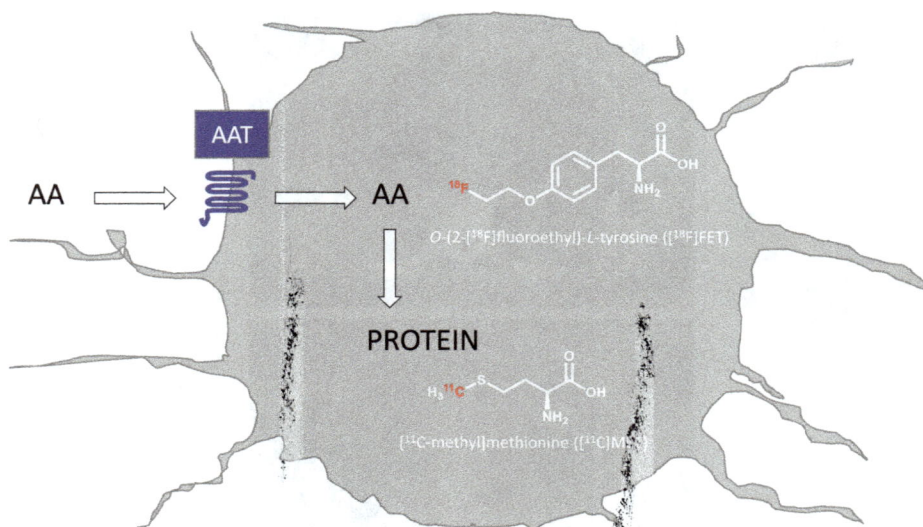

Fig. 12.29: Cellular uptake of radiolabeled amino acids (AA) via amino acid transporters (AAT) and further use in protein biosynthesis. Molecular structures of [^{11}C-*methyl*]methionine ([^{11}C]MET) and O-(2-[^{18}F]fluoroethyl)- L-tyrosine ([^{18}F]FET).

is a substrate for the AATs, and thus transported into tumor cells like native amino acids, but the structural differences of the ^{18}F-labeling moiety prevent the incorporation into proteins. Although [^{18}F]FET is not metabolically trapped via the protein biosynthesis, neither does it show efflux from the tumor cells. Consequently, there is a trapping effect of [^{18}F]FET, but the detailed mechanism still awaits clarification. [^{11}C]MET represents a direct tracer for the protein biosynthesis rate, while [^{18}F]FET reflects the activity of the amino acid transporters.

On the other hand, healthy brain cells with a normal protein synthesis rate show low uptake levels for amino acids. In contrast, tumor cells have strongly elevated protein synthesis rates and a high need for amino acids. Unsurprisingly, the radiolabeled amino acids are ideal tracers to visualize brain tumors. Figure 12.30 shows the schematic transport of [^{18}F]FET into a (brain) tumor cell and the corresponding PET image of a patient with a glioma.

Is there a significant difference between imaging the protein synthesis rate with [^{11}C]MET and the amino acid transporters' activity with [^{18}F]FET? A comparative study on brain glioma or brain metastases found comparable diagnostic information on gliomas and brain metastases was observed from both tracers. Figure 12.31 shows PET images of [^{11}C]MET and [^{18}F]FET, and the corresponding gadolinium-enhanced MRT images in the same patient with recurrent brain metastases. For both tracers the diagnostic information seems identical. Obviously, there is no relevant difference between targeting the amino acid transporters or the protein synthesis rate. [^{11}C]MET as well as

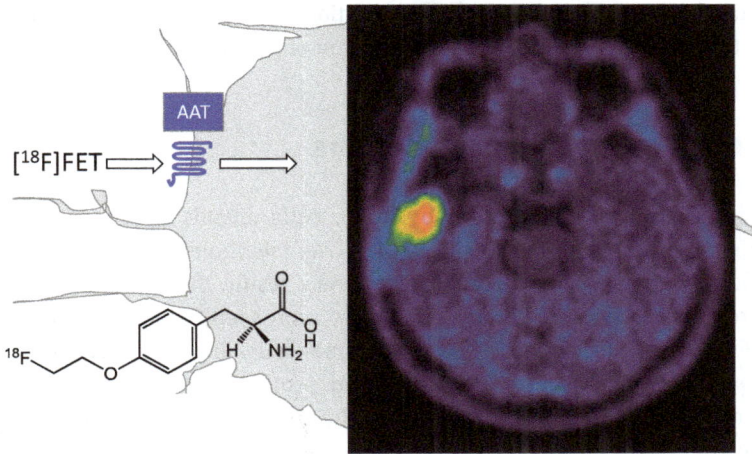

Fig. 12.30: Transport of [^{18}F]FET into brain tumor cells, and the corresponding PET image (axial/transversal plane) showing the hot spot of the brain tumor (anaplastic oligoastrocytoma) with a high contrast to healthy brain areas. Courtesy of G. Berding, Nuclear Medicine, Hannover Medical School, Germany.

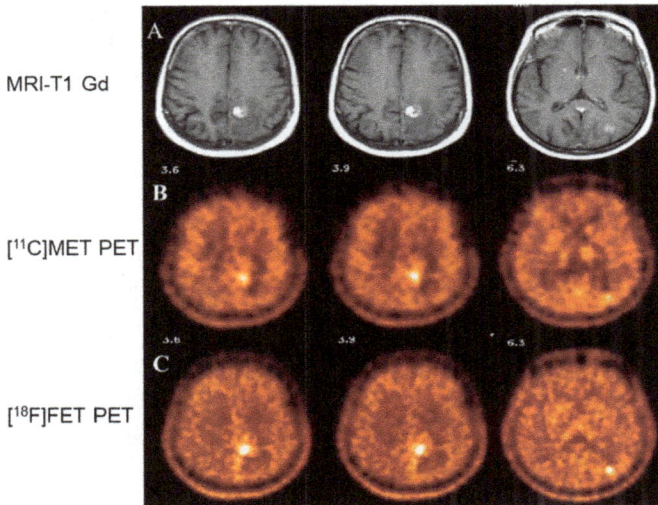

Fig. 12.31: Transversal images of recurrent brain metastases in the same patient using gadolinium-enhanced MRT imaging (A), [^{11}C]MET (B), and [^{18}F]FET (C) PET imaging.[20]

20 An interindividual comparison of O-(2-[^{18}F]fluoroethyl)-L-tyrosine (FET)- and L-[*methyl*-^{11}C]methionine (MET)-PET in patients with brain gliomas and metastases. 2011.

[^{18}F]FET can be used for visualization and differentiation of brain tumors, and especially important, they are equally suitable for delineation of gliomas prior to surgery.

12.7.4 Oncology: Transmembrane receptors

Looking at the surface proteins expressed by tumor cells reveals an astonishing number and variety of potential targeting structures. The metabolism and other cell functions of tumor cells appear utterly unregulated. As a result, the expression levels of surface proteins on tumor cells are drastically increased, which can easily reach a million-fold overexpression in relation to normal cells. Furthermore, several tumor entities (over)express (specific) transmembrane proteins which show almost no or only very restricted expression levels in healthy tissues. Transmembrane receptors are large protein structures located in the cell membrane. These receptors have three major domains: the transmembrane, the extra- and the intracellular domain. The extracellular domain is the part of the cell surface recognized by specific ligands. In contrast, the intracellular domain is the part which generates a signal inside the cell by activation of (*effector*) proteins or enzymes.

One prominent family are the G-protein coupled receptors (GPCR), where the intracellular domain is coupled to the so-called G-protein which is responsible for further signal transduction in the cell. Some GPCRs perform endocytosis after ligand binding, which is a great advantage as it provides effective trapping of the radiotracer inside the tumor cell, Fig. 12.32.

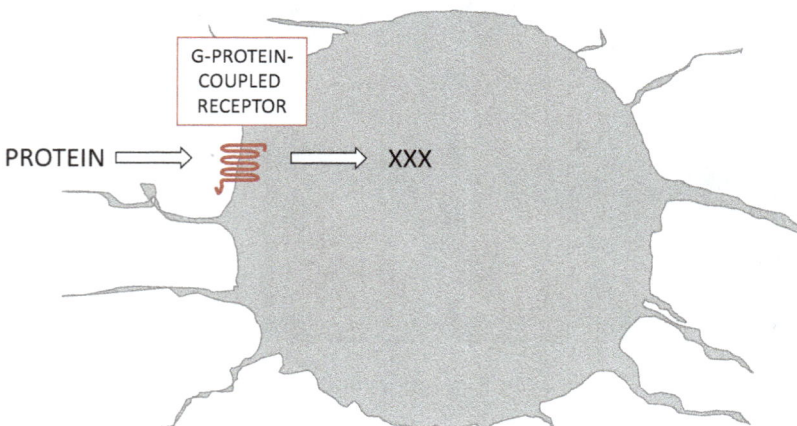

Fig. 12.32: General principle of ligand binding to G-protein coupled receptors (GPCR) with subsequent signal transduction and/or endocytosis.

Obviously, these surface structures can generally be considered as potential targets in diagnostics and therapy of tumors. The most important transmembrane receptor class of the last decade in nuclear medicine is the family of the human somatostatin receptors (hSSTR), Fig. 12.33. Of five subtypes, the most relevant subtype is hSSTR2, and to a much lesser extent hSSTR1, 3, and 5. hSSTRs are strongly overexpressed on neuroendocrine tumors and some other kinds of human tumors. Expression levels of hSSTR2 on tumor cells can reach a 2.5 million-fold overexpression. The physiological expression of hSSTR2 in healthy tissue is very limited and is mainly found in spleen. Consequently, hSSTR2 provides ideal characteristics as a targeting structure for tumor imaging (and therapy). Moreover, the hSSTR family performs endocytosis after ligand binding offering a very efficient trapping of the ligand in the addressed cell.

The native ligand, somatostatin, is a peptide hormone of 14 or 28 amino acids, respectively (two isoforms). Somatostatin has only a plasma half-life of 2 min, and thus, it is inappropriate for molecular imaging purposes. The truncated derivative octreotide, a much smaller cyclic peptide, was developed as a somatostatin analog for the treatment of acromegaly and the treatment of neuroendocrine tumors. In a first approach, octreotide was conjugated with DTPA as chelating system for radiolabeling with indium-111. ^{111}In-labeled DTPA-octreotide is available as *Octreoscan*$^©$ for SPECT imaging. Further developments provided a corresponding derivative, [DOTA^0-D-Phe^1-Tyr^3]-octreotide (DOTA-TOC), with the chelating moiety DOTA for radiolabeling with various radiometals and a strongly increased affinity to hSSTR2. Together with another variation, [DOTA^0-D-Phe^1-Tyr^3]-octreotade (DOTA-TATE), DOTA-TOC is the most important precursors for radiolabeled somatostatin analogs. [^{68}Ga]Ga-DOTA-TOC/-TATE are the commonly applied tracers in clinical routine for PET imaging of hSSTR2 positive tumors and metastases, Fig. 12.33.

The prostate-specific membrane antigen (PSMA) has been identified as a suitable target for prostate cancer diagnosis and therapy. The corresponding binding motif is a very small peptidometic structure of Glu-urea-Lys (KuE). Several radiolabeled derivatives have been developed for molecular imaging using, e.g., fluorine-18 and gallium-68. The most successful and clinically employed diagnostic one is the ^{68}Ga-labeled ligand from DKFZ Heidelberg, utilizing an HBED chelate: [(EuK)-Ahx-HBED-CC], PSMA-11, Fig. 12.34. Figure 12.35 represents three of the most common KuE-based radiopharmaceuticals routinely applied to diagnose prostate cancer via PET/CT. The [^{68}Ga]Ga-DOTA-PSMA-617 is a very important PSMA-targeting molecule, which enables via the DOTA chelate the radiolabeling with therapeutic radionuclides such as lutetium-177. [^{177}Lu]Lu-DOTA-PSMA-617 has already passed a clinical trial phase III (VISION trial, and its market authorization as *[^{177}Lu] vipivotide tetraxetan*$^©$ can be expected soon. As prostate cancer is the second-most common cancer worldwide and is the fourth-leading cause of cancer-based death in men, the affected patient population is very large. With increasing implementation of PET/CT for diagnostics of prostate cancer in clinical routine, several nuclear

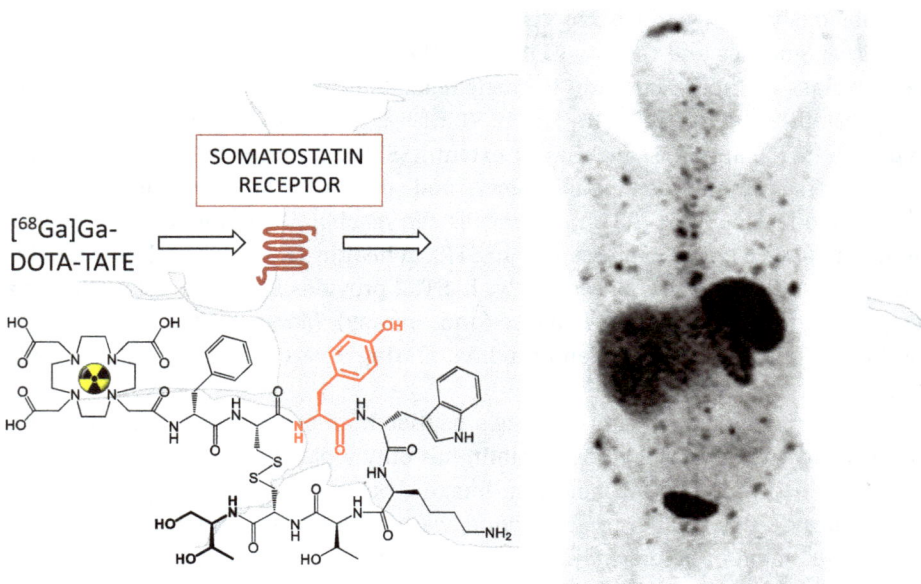

Fig. 12.33: Molecular structure of the hSSTR2 addressing ^{68}Ga-labeled DOTA-octreotate derivative [^{68}Ga]Ga-DOTA-[d-Phe[1]-Tyr[3]]-octreotate (^{68}Ga-DOTA-TATE). Principle of radioligand binding to hSSTRs, endocytosis and a corresponding PET image of a patient with multiple liver and bone metastases from neuroendocrine tumors: (1) bone metastases, (2) spine metastases, (3) skull metastases, (4) pituitary gland, (5) liver and liver metastases, (6) spleen, (7) kidney, (8) bladder.

medicine facilities needed to switch from ^{68}Ga-labeled PSMA-ligands to a ^{18}F-labeled variant, such as [^{18}F]F-PSMA-1007 (cf. Fig. 12.35), to manage increased patient numbers. [^{18}F]F-PSMA-1007 is one of the clinically most employed ^{18}F-labeled PSMA-ligand, and at some clinics PSMA-based PET/CT scans already reach levels of [^{18}F]FDG PET/CT.[21] Very recently, the first ^{99m}Tc-labeled PSMA-ligands for SPECT imaging have been developed and made their way into the clinics. Clinical trials are ongoing and the first market authorization of the corresponding ^{99m}Tc-labeling kits can be expected soon. Besides PSMA-based SPECT imaging, such ^{99m}Tc-labeled radiopharmaceuticals are of special interest for *radio-guided-surgery* (RGS). For a RGS, the radiopharmaceutical is administrated before the surgery, accumulating in malignant tissue. At the time point of the surgery, the surgeon is guided either by a hand-held gamma probe or in case of an endoscopic system by a miniaturized gamma probe at the endoscope, identifying tumor tissue and metastases by their radiation.

21 For a recent review cf. Radiolabeled PSMA inhibitors, Neels OC et al., 2021, 13, 6255. doi.org/10.3390/cancers13246255.

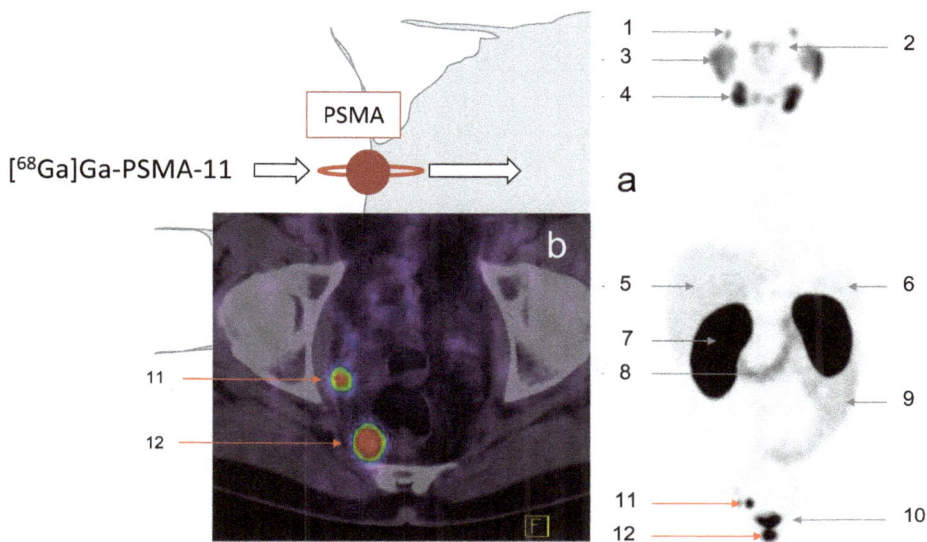

Fig. 12.34: [^{68}Ga]Ga-PSMA-11 ([^{68}Ga]Ga-HBED-CC-PSMA): Molecular imaging of prostate cancer (PC) as well as bone metastases, disseminated from primary PC, and lymph nodes. (a) Maximum Intensity Projection (MIP) of a PET/CT at 1 h post injection. Physiological uptake of [^{68}Ga]Ga-PSMA -11 is seen in lacrimal glands (1), nasal mucosa (2), parotid glands (3), sublingual and submandibular glands (4), liver (5), spleen (6), kidneys (7), proximal parts of the small intestines (8) and in some parts of the colon (9). The surplus tracer is excreted via the urinary tract including the bladder (10). This patient presents with a primary prostate cancer (12) and two lymph nodal metastases (11). [^{68}Ga]Ga-PSMA-11 is able to show prostate cancer lesions with a contrast that has never been achieved before. (b) Fusion of PET and CT. Courtesy by Afshar-Oromieh, DKFZ Heidelberg, Germany.

In addition to the key/lock systems SSTR/octreotide derivatives for neuroendocrine tumors and PSMA/PSMA inhibitors for prostate cancer, there are other examples validated for diagnosing more types of cancer in nuclear medicine. Important cases are bombesin (BBN) analogs for the gastrin-releasing peptide receptor (GRPR) in PSMA-negative prostate cancer, neurotensin antagonists for neurotensin receptors for various human cancers including pancreatic adenocarcinoma, peptidic vectors for the C-X-C motif chemokine receptor (i.e., CXCR4), which is overexpressed by many different tumor entities, folic acid derivatives for the folate receptor system for different endothelial human carcinoma including ovarian carcinomas.

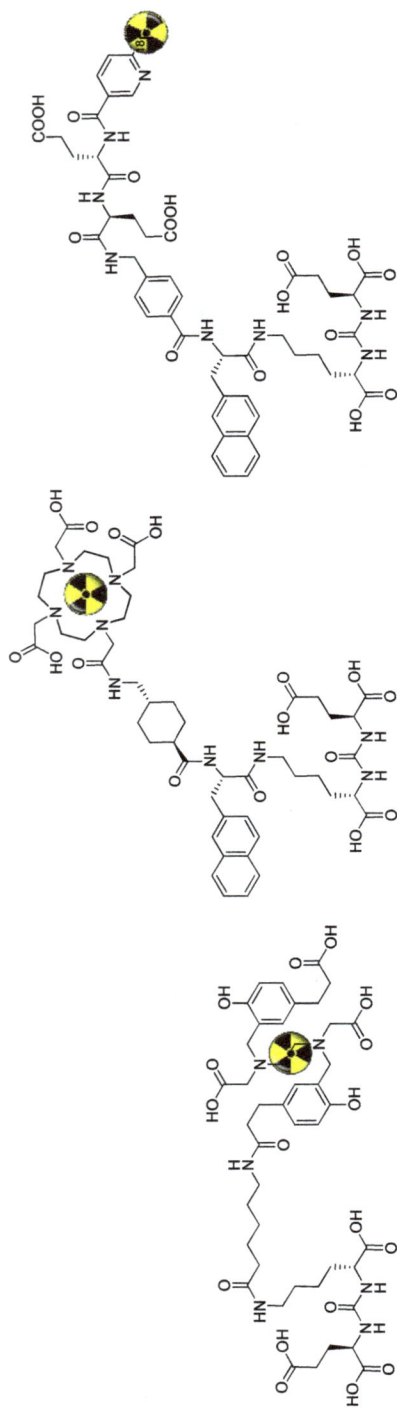

Fig. 12.35: Chemical structures of [⁶⁸Ga]Ga-PSMA-11 (left), [⁶⁸Ga]Ga-DOTA-PSMA-617 (middle), and [¹⁸F]F-PSMA-1007 (right).

12.7.5 Oncology: Tumor microenvironment

Different to specific molecular tumor targets such as the somatostatin receptor of neuroendocrine tumors and the prostate-specific membrane antigen of prostate cancer more recently the tumor microenvironment has been identified as a suitable target for cancer diagnosis and therapy. Cancer is a heterogeneous disease formed within an extremely complex microenvironment. Malignant tumors do not only consist of cancerous cells but also a vast majority of endogenous host stromal cells (e.g., fibroblasts, vascular and immune cells) and extracellular matrix (ECM) components, collectively known as tumor microenvironment (TME). The stromal cells in the tumor (tumor stroma) deem to be the largest portion of the total tumor mass (over 90%) connected through desmoplastic reaction. Among all cells within the TME matrix, fibroblasts are dominant cells that have a strong association of their biological functions to all stages of cancer progression and metastasis. Cancer-associated fibroblasts (CAFs) have been implicated to have a strong tumor-modulating effect and are commonly found in most solid tumors. Generally, CAFs account for up to 80% of all fibroblasts in the TME. CAFs are identified by the expression of various specific biomarkers on their surface such as α-smooth muscle actin (α-SMA), fibroblast-specific protein 1 (S100A4 or FSP-1), platelet-derived growth factor receptors (PDGFRα/β), and fibroblast activation protein (FAP). Fibroblast activation protein alpha (FAPα or FAP), also known as prolyl endopeptidase FAP or seprase, is a type II membrane-bound serine protease associated with fibrosis, tissue repair, inflammation, and ECM degradation. FAP contains two types of enzymatic activity: dipeptidyl peptidase and endopeptidase. The endopeptidase allows FAP to mediate the proteolytic processing of matrix metalloproteinase-cleaved collagen I, leading to the prevention of morphogenesis, tissue remodeling, and repair. FAP appears to be a promising target in oncology due to its nonexpression in normal fibroblasts and the stroma of benign epithelial tumors compared to its significantly high accumulation mainly on the stromal compartments of a variety of malignant tumors. The challenge to radiopharmaceutical chemistry is to develop molecular targeting vectors of high affinity to FAP und high selectivity to related proteases. Recently, molecular antibodies, small peptides, and inhibitors have been turned into molecular imaging probes, with inhibitors utilizing the (4-quinolinoyl)glycyl-cyanopyrrolidine scaffold representing the most exciting class (nanomolar affinity to FAP compared to micromolar affinity to PREP such as the difluoro-version UAMC1110).

Radiopharmaceutical chemists from the German Cancer Research Centre Heidelberg, Germany, pioneered in translating this FAP inhibitor into potent diagnostic radiopharmaceuticals by coupling chelators to the amine of the quinoline-part of the inhibitor introducing a number of linker and spacer motifs. The University of Mainz, Germany, utilized a squaric acid (SA)-based linker bridging between several bifunctional chelators (DATA, DOTA, and DOTAGA) to the inhibitor. Currently, ^{68}Ga labeled derivatives such as DOTA-FAPi-04 and DOTA-FAPi-46 and DATA^{5m}.SA.FAPi or DOTA.SA.FAPi, etc. (cf. Figs. 12.36A and 12.36B) are showing promising value in diagnosing many different kinds of tumors yielding new information even when compared to the ^{18}F-FDG glycolysis tracer or PET tracers specific to neuroendocrine tumors or prostate cancer, cf. Fig. 12.7.4. ^{99m}Tc-and ^{18}F-labeled FAPi tracers for SPECT or PET imaging were also developed.[22]

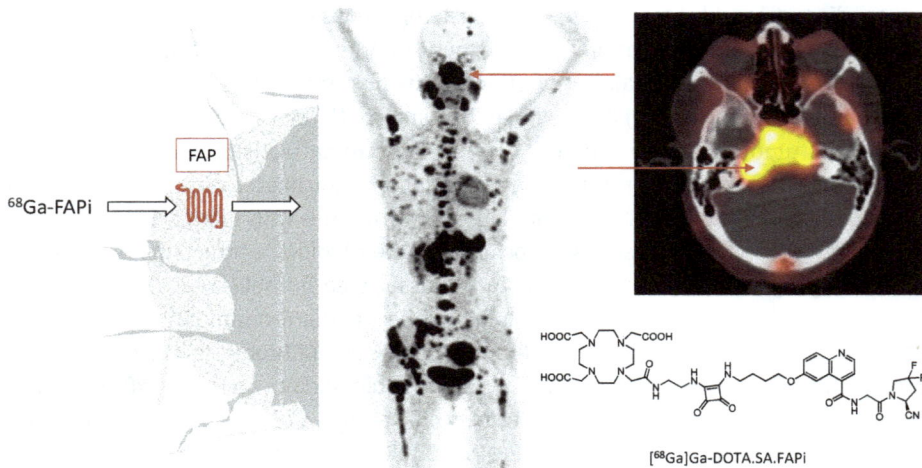

Fig. 12.36A: [^{68}Ga]Ga-DOTA.SA.FAPi PET/CT MIP: A 39-year-old woman was diagnosed with widely invasive follicular thyroid cancer. A significantly high tumor to brain background uptake (arrows) was observed with [^{68}Ga]Ga-DOTA.SA.FAPi. Intense accumulation of radiotracer in the soft tissue mass and multiple skeletal sites. Courtesy by CS Bal, AIIMS, New Delhi, India.

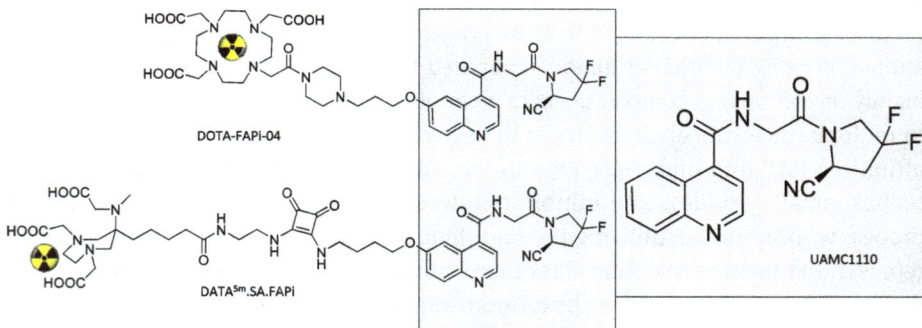

Fig. 12.36B: Chemical structures of the FAP inhibitor UAMC1110, [^{68}Ga]Ga-DOTA-FAPi-04 and [^{68}Ga]Ga-DATA.SAFAPi.

22 For a recent review cf. New frontiers in cancer imaging and therapy based on radiolabeled fibroblast activation protein inhibitors: A rational review and current progress. Imlimthan S et al. 2021.

12.7.6 Oncology: Bone metastases

Malignant tumors of prostate, breast, and others at an advanced stage often lead to the development of painful metastases of the bone. The disease relates to the action of osteoblasts and osteoclasts, constituting the active components of bone growth. The inorganic bone matrix itself is basically made of calcium hydroxyl apatite. Conventional drugs as well as diagnostic and therapeutic radiopharmaceuticals all have in common a high binding affinity to the inorganic matrix and/or the enzymes regulating bone growth. Figure 12.37 illustrates the structure of bone and the various constituents of hydroxyl apatite, which are targeted by radiopharmaceuticals. Pharmaceuticals for therapy of osteoporosis are often used as bisphosphonate moiety. Bisphosphonates are similar to pyrophosphate, but due to the substitution of the central oxygen atom by a carbon atom, they are metabolically stable. Furthermore, this carbon atom enables the attachment of two functional groups. By using a hydroxyl group as a substituent the affinity to hydroxyl apatite can be increased.

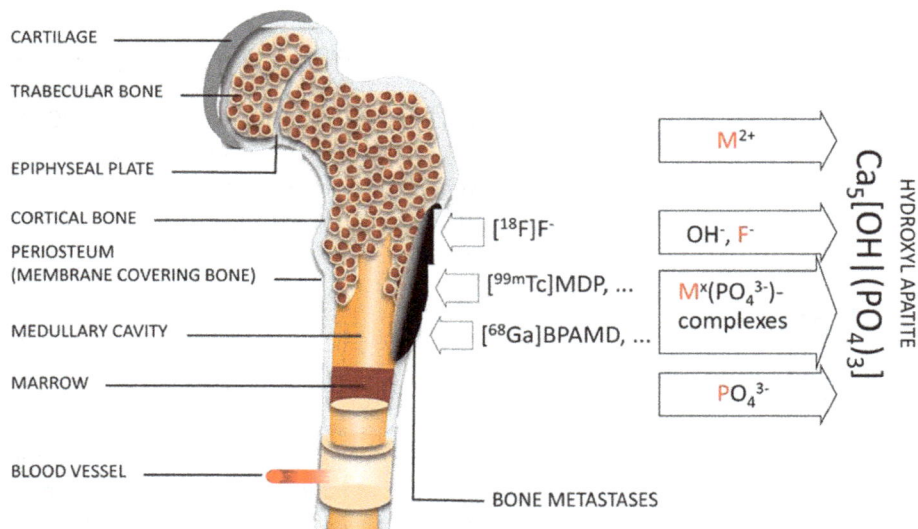

Fig. 12.37: Structure of bone and the various constituents of hydroxyl apatite. The metastasis is schematically drawn, targeted by three diagnostic radiopharmaceuticals.

There are a number of radiopharmaceuticals available for diagnosis that either substitute the calcium, the hydroxyl group (e.g., ^{18}F-fluoride), or the phosphate (bisphosphonate-containing complexes of relevant radio-metals such as ^{99m}Tc or ^{68}Ga). Accordingly, the increased metabolism of the bone material can be visualized via both SPECT and PET. ^{99m}Tc is predominantly used in terms of scintigraphy utilizing [^{99m}Tc]bisphosphonate complexes. For PET, ^{18}F in form of [^{18}F]NaF is used. [^{18}F]NaF

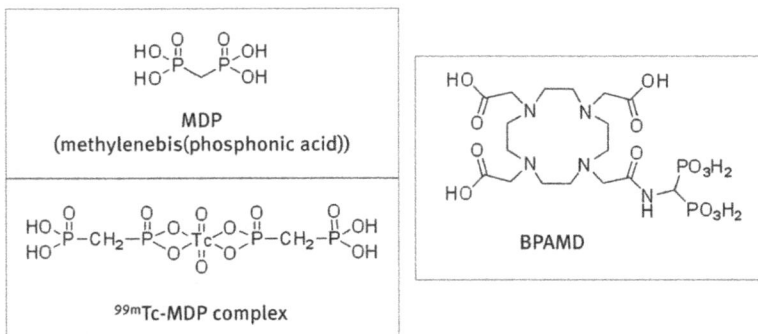

Fig. 12.38: Molecular structures of a methylene-bridged bisphosphonate (MDP) and the structure of its ^{99m}Tc complex for SPECT and a DOTA-conjugated bisphosphonate for ^{68}Ga-labeling and PET imaging of bone metastases.

has almost no background and fast renal clearance from blood. As a result, 45–60 min after injection PET imaging using [^{18}F]NaF provides clear-cut bone images with high signal-to-noise ratios to identify certain bone disorders, malignancies and bone metastases. Recently, new developments indicate a significant potential for ^{68}Ga-bisphosphonate complexes. A common bisphosphonate for ^{99m}Tc scintigraphy (MDP) and a DOTA-based bisphosphonate are illustrated in Fig. 12.38. Figure 12.39 shows molecular imaging of a patient suffering from disseminated bone metastases with three representative tracers. For therapy, ^{68}Ga may be substituted by, e.g., ^{177}Lu, see Chapter 13.

12.7.7 Heart function

Cardiac molecular imaging is one of the most important disciplines in nuclear imaging. To access the heart function by molecular imaging two major questions are of interest. First, information about perfusion is essential to provide information on blood supply to myocardial regions. The second major question is about the viability of myocardial cells, especially after myocardial infarct. The currently available radiotracers are either solely cations, i.e., [^{201}Tl]Tl$^+$ or [^{82}Rb]Rb$^+$, which mimic the K$^+$-cation; or lipophilic complexes such as [^{99m}Tc]sestamibi and [^{99m}Tc]tetrofosmin, Fig. 12.40.

The [^{201}Tl]Tl$^+$ cation is recognized by the Na$^+$/K$^+$-ion 3 : 2-exchange pump (Na$^+$/K$^+$-ATPase pump) of cardiac cells and actively pumped into the cells. In the so-called cardiac stress test or thallium stress test, the uptake of ^{201}Tl in cardiac tissue correlates with the blood supply to this region. Furthermore, only viable cardiac cells show normal uptake of ^{201}Tl. The test involves two SPECT scans, one under cardiac stress (exercise or medication) and a second scan at rest. The images enable

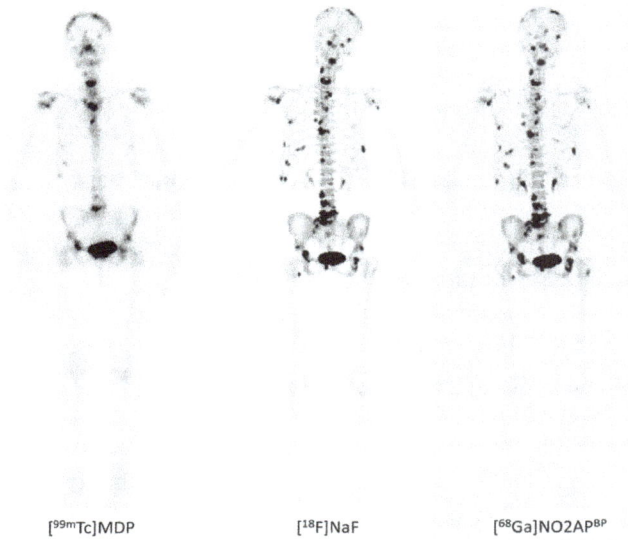

Fig. 12.39: Diagnostic clinical imaging of a patient suffering from disseminated bone metastases with three representative tracers: ^{99m}Tc-MDP (scintigraphy), ^{18}F-fluoride (PET) and [^{68}Ga]Ga-BPAMD (PET). Note that the quality of the images is different both due to the different imaging technologies and the affinity of the tracers. Courtesy of C Bal, AIIMS, New Delhi, 2014.

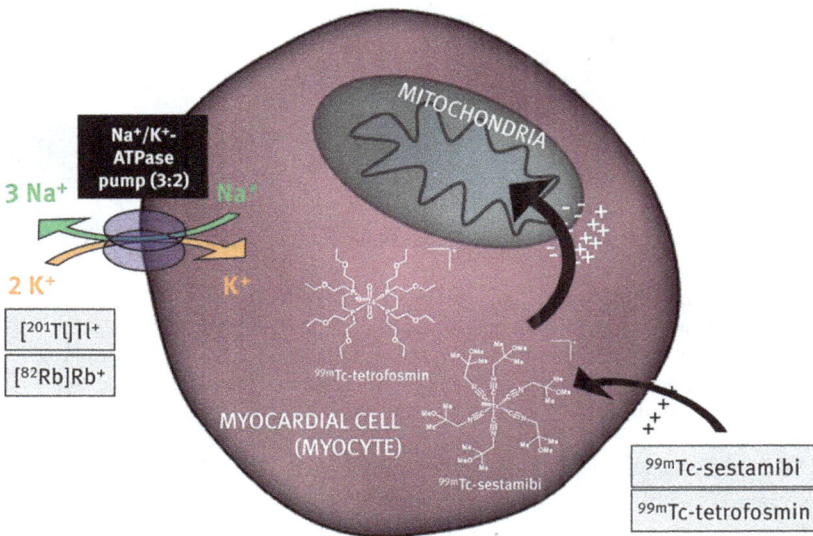

Fig. 12.40: Principle of cardiac perfusion imaging using either the actively transported cations [^{201}Tl]Tl$^+$ and [^{82}Rb]Rb$^+$ as surrogates of K$^+$, or the lipophilic ^{99m}Tc-labeled complexes sestamibi and tetrofosmin.

identification of myocardial regions which would benefit from a revascularization after myocardial infarct, and the identification of ischemic coronary artery disease.

Similar to $[^{201}Tl]Tl^+$, $[^{82}Rb]Rb^+$ is an analog of K^+ and accepted by the Na^+/K^+-ion ATPase-pump on cardiac cells. $[^{82}Rb]RbCl$ is rapidly cleared from blood and scans can start within the first minutes after administration. As a result, $[^{82}Rb]RbCl$ enables myocardial perfusion imaging using PET. The very short half-life of ^{82}Rb allows quick repetition of scans and due to the radionuclide generator the availability is given almost at any time. Scanning protocols as described above for a thallium cardiac stress test can be facilitated within much shorter time periods.

Besides the simple cations, lipophilic radiometal complexes such as $[^{99m}Tc]$sestamibi and $[^{99m}Tc]$tetrofosmin with a single positive charge are also routinely applied for cardiac imaging. $[^{99m}Tc]$Sestamibi carries six (sesta) methoxyiso-butylisonitrile (MIBI) ligands, while $[^{99m}Tc]$tetrofosmin is a P4 complex ([6,9-bis(2-ethoxyethyl)-3,12-dioxa-6,9-diphosphatetradecane]). Both radiotracers diffuse passively into myocardial cells with no active uptake. They accumulate in mitochondria of the myocardial cells by electrostatic interactions due to a higher negative membrane potential of the mitochondria. $[^{99m}Tc]$Sestamibi and $[^{99m}Tc]$tetrofosmin are trapped in myocardial cells. Both give almost identical results and are considered equal in clinical use. They are available as standard kit formulations. Figure 12.41 shows a $[^{99m}Tc]$sestamibi kit and corresponding SPECT images.

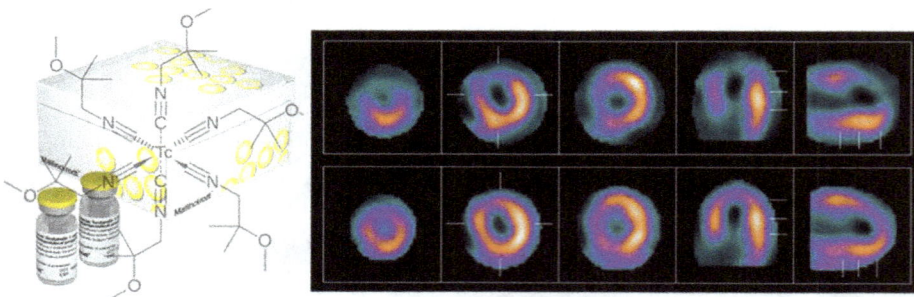

Fig. 12.41: $[^{99m}Tc]$Sestamibi and *Cardiolite*© kit, Lantheus Medical Imaging, and $[^{99m}Tc]$sestamibi images of a patient after myocardial infarct under stress (right, upper row) and at rest (right, lower row). The lower or missing uptake in the "stress row" in comparison to the "rest row" shows regions of reduced perfusion. $[^{99m}Tc]$Sestamibi scans courtesy of G Berding, Nuclear Medicine, Hannover Medical School, Germany.

The novel FAP inhibitors for PET imaging described above (cf. 12.7.5) are also able to detect activated fibroblasts in cardiac fibrosis.[23] The process of fibrosis in general is an ongoing deposition of extra cellular matrix (ECM) , mainly collagen-I. Fibrosis

---

23 For an overview cf. Emerging imaging targets for infiltrative cardiomyopathy: Inflammation and fibrosis. Bengel FM et al., 2019.

usually occurs in wound healing and scar formation for repair and regeneration. The same holds for organs, where an injury triggers an inflammatory response, which in turn stimulates fibrosis for repair and regeneration. But, what happens if this process is not adequately controlled and does not stop? A positive feedback loop between the inflammation and fibrosis can lead to an overload of ECM and finally to an impairment of the organ function. Organ fibrosis can occur in any organ and is responsible for more than 45% of all deaths in the industrialized countries.

Myocardial infarction or other cardiac events (i.e., ischemia, stress, tissue injury, etc.) trigger an inflammatory response, which in turn activates resident (quiescent) fibroblast differentiating into myofibroblasts expressing FAP and secreting ECM. Hence, using radiolabeled FAP inhibitors enable visualization of activated fibroblasts/myofibroblasts and can help identifying the extend of fibrogenesis and fibrosis. In Fig. 12.42, [68Ga]Ga-DOTA-FAPi-46 was used to assess the extent of

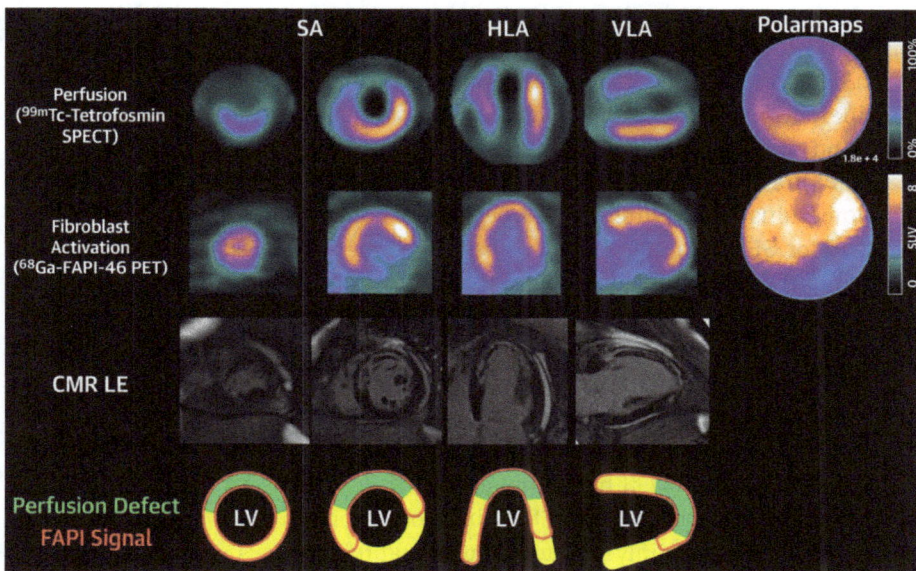

Fig. 12.42: PET imaging using [68Ga]Ga-DOTA-FAPi-46 (left) and SPECT imaging using [99mTc] tetrofosmin in a patient after myocardial infarction. In addition, the patient also underwent cardiac magnetic resonance imaging (CMR). The fibroblast activation strongly exceeds the region of the actual myocardial infarction. CMR = cardiac magnetic resonance imaging, HLA = horizontal long axis, LE = gadolinium contrast late enhancement, LV = left ventricle, SA = short axis.[24]

24 Diekmann J, Koenig T, Zwadlo C, Derlin T, Neuser J, Thackeray JT, Schäfer A, Ross TL, Bauersachs J, Bengel FM. Molecular Imaging Identifies Fibroblast Activation Beyond the Infarct Region After Acute Myocardial Infarction. *Journal of the American College of Cardiology* **2021**, *77*, 1835. https://doi.org/10.1016/j.jacc.2021.02.019.

cardiac fibrosis after myocardial infarction, and compared with perfusion imaging using [^{99m}Tc]tetrofosmin to define the infarcted area. In this example, the fibroblast activation extensively exceeded the area of the actual myocardial infarct.

12.7.8 Kidney function

The kidneys are our major excretion organs for hydrophilic substances, and their unhampered functionality is essential. Renally excreted substances are either filtered by the glomerulus (glomerular filtration) and/or actively secreted by tubular cells from peritubular capillary into tubular lumen; both mechanisms take place in the cortical nephron. Subsequently, the filtered or secreted substances follow the system of renal tubules, where substances are either reabsorbed or they travel further into the collecting ducts in the renal medulla. From the collecting ducts the urine leaves via papillary ducts and renal papilla into the ureter and finally into the urinary bladder; see Fig. 12.43.

In renal imaging and investigating kidney function it is of interest to determine the different renal functions reflected by the parameters *glomerular filtration rate* (GFR) and *effective renal plasma flow* (ERPF). The GFR is difficult to determine as an ideal radiotracer must be 100% free (no plasma protein binding), fully filtered in the glomerulus and neither reabsorbed nor secreted. The nonradiolabeled gold standard for GFK is inulin, a fructose-based polymer of ca. 5 kDa, which is rapidly filtrated in the glomerulus and neither reabsorbed nor secreted by tubular cells. Of several approaches to develop a GFK-radiotracer, different radiometal complexes of dieethylenetriaminopentaacetic acid (DTPA) showed the most promising results. ^{99m}Tc-DTPA meets the requirements of neither reabsorption nor tubular secretion. In spite of a low plasma protein bound fraction of 5–10%, ^{99m}Tc-DTPA is the radiotracer of choice to determine GFR by renal scintigraphy and SPECT.

The ERPF is the flow of blood that is effectively separated from plasma by renal extraction. The gold standard for ERPF, but not radiolabeled, is *para*-aminohippuric acid (PAH), which is mainly secreted by proximal tubular cells and which shows almost complete renal extraction and no retention in kidneys. Its radiolabeled analog, *ortho*-[^{123}I]iodohippuric acid (OIH) is a radiotracer with similar characteristics (20% glomerular filtration, 80% tubular secretion). OIH reaches about 90% of the PAH plasma clearance and is the radiotracer to determine the ERPF with renal scintigraphy and SPECT.

The development of [^{99m}Tc]mercaptoacetyl triglycine (^{99m}Tc-MAG3) offered a ^{99m}Tc-based imaging agent for ERPF and tubular function as an alternative for OIH. ^{99m}Tc-MAG3 combines the imaging qualities of OIH with the availability and convenience of ^{99m}Tc and a kit preparation, Fig. 12.44. ^{99m}Tc-MAG3 exhibits renal clearance by both mechanisms, glomerular filtration and tubular excretion, whereby ^{99m}Tc-MAG3 reaches a renal extraction ratio of about 70% of OIH. ^{99m}Tc-MAG3 has

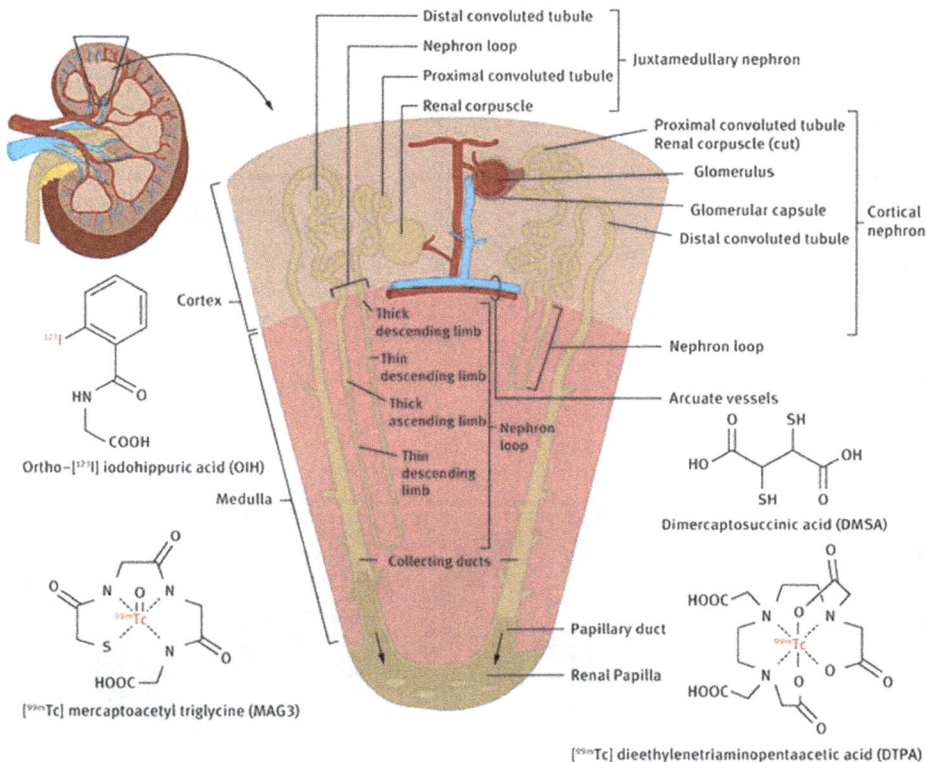

Fig. 12.43: Anatomy of kidneys and detailed view on the nephron. Molecular structures of selected radiotracers for molecular imaging of renal functions. Effective renal plasma flow (ERPF): ortho-[^{123}I]iodohippuric acid (OIH) and [^{99m}Tc]mercaptoacetyl triglycine (^{99m}Tc-MAG3). Glomerular filtration rate: [^{99m}Tc]diethylenetriaminopentaacetic acid (DTPA). Tubule secretion: dimercaptosuccinic acid (DMSA) in form of ^{99m}Tc-DMSA$_2$. *(Note: the exact molecular structure of ^{99m}Tc-DMSA$_2$ is still unresolved).*

a higher plasma protein binding and its glomerular filtration rate is smaller than OIH. On the other hand, the higher plasma protein bound fraction allows an increased tubular secretion rate and thus a higher renal extraction. ^{99m}Tc-MAG3 is currently the standard radiotracer for studying the ERPF and tubular cell functions by scintigraphy and SPECT.

12.7.9 Thyroid function

Imaging the function of the thyroid gland has a long tradition in nuclear medicine. Functional imaging of the thyroid is based on the uptake of iodide and iodide analogs for the biosynthesis of thyroid hormones, namely triiodothyronine (T3) and thyroxine (3,5,3′,5′-tetraiodothyronine, T4). T3 and T4 are synthesized in the thyroid follicular cells (thyroid epithelial cells) and the follicle colloid. The required iodide anions are actively pumped into the cells by the sodium-iodide (Na/I) symporter (NIS), Fig. 12.45.

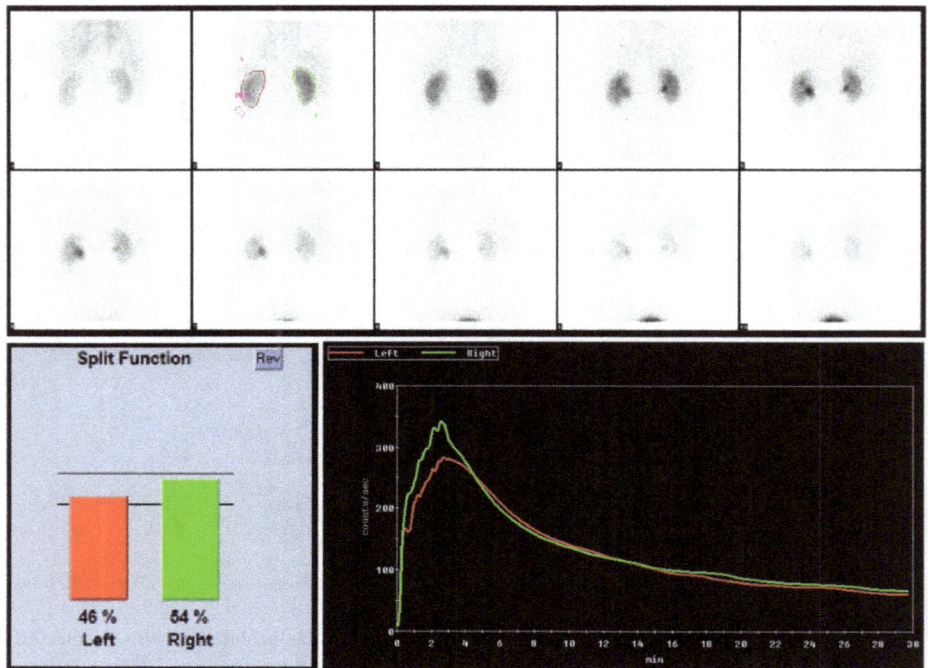

Fig. 12.44: Radioisotope renography using ^{99m}Tc-MAG3 to assess kidney function, showing a normal kidney functionality. Typical renogram with 8 sequential images of the kidneys over a period of 30 min, red and green cycle in the second image define the region of interest (ROI) for subsequent quantitative analysis (upper row). The split function gives an overview of the functional distribution between both kidneys (lower row, left). The dynamic renographic curves show the course of the radioactivity in the kidneys over time, red line = left kidney, green line = right kidney. ^{99m}Tc-MAG3 renography scans by courtesy of T. Brunkhorst, Nuclear Medicine, Hannover Medical School, Germany.

Molecular imaging of the thyroid is based on the specificity of the NIS toward iodide anions and analogs such as the pertechnetate anion, which has an almost identical volume/size as well as identical negative charge. As a result, radioiodide and [^{99m}Tc]pertechnetate [^{99m}Tc]TcO$_4^-$ are used for thyroid imaging. Radioiodide follows the identical metabolism as cold iodide, and is eventually metabolically trapped by incorporation into the biosynthesis of T3 and T4. However, there is no significant efflux of [^{99m}Tc]TcO$_4^-$ from the thyroid gland and it even has some advantages for thyroid scintigraphy over radioiodide. [^{99m}Tc]TcO$_4^-$ enables SPECT

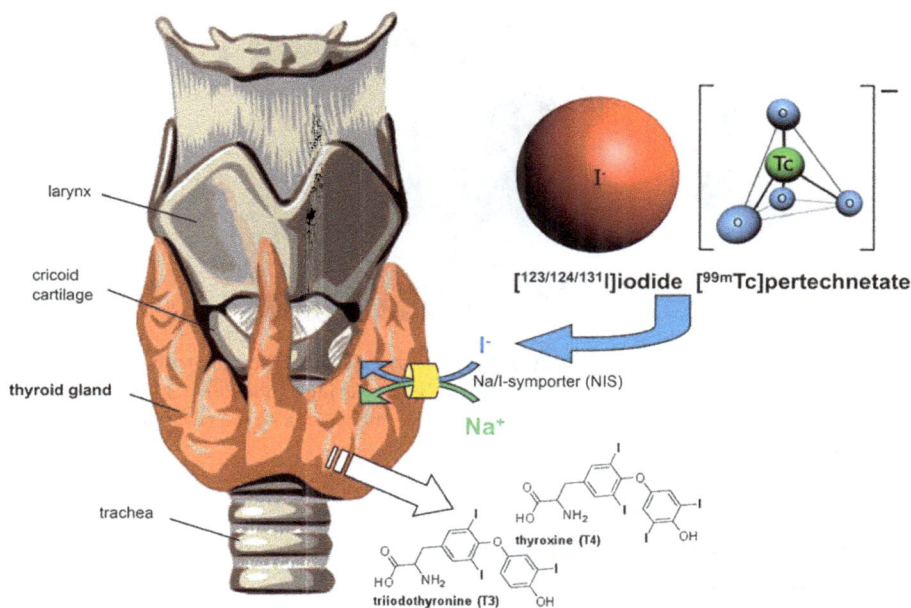

Fig. 12.45: Anatomy of thyroid gland showing the active transport of iodide ions via the Na/I-symporter (NIS) and the release of the thyroid hormones triiodothyronine (T3) and thyroxine (T4). Radiotracers for molecular imaging of the thyroid functions: [$^{123/131}$I]iodide for scintigraphy and SPECT; [^{124}I]iodide for PET, and [^{99m}Tc]pertechnetate as iodide analog for scintigraphy and SPECT. Both anions are similar in dimension and charge, and thus substrates for the NIS.

imaging and scintigraphy of the thyroid by the convenient use of ^{99m}Tc. Due to the shorter half-life, usually [^{99m}Tc]TcO$_4^-$ allows administration of higher activities and thus enables faster scanning protocols. Moreover, the time window between administration and scan is generally much shorter than with radioiodide, making patient handling and organization in clinical routine much simpler.

Molecular imaging of the thyroid is an important tool in the management of patients with benign or malignant thyroid disease. In case of malignancies of the thyroid, endoradiotherapy using ^{131}I is the primary therapy. For exact dose and therapy planning, thyroid scintigraphy is performed using radioiodide ($^{123/131}$I) prior to therapy. Additionally, thyroid scintigraphy is also the method of choice for differentiation and posttherapy controls, cf. Fig. 12.46.

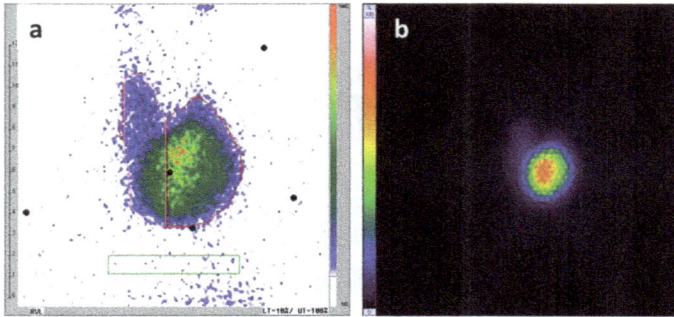

Fig. 12.46: Pretherapeutic scintigraphy of the thyroid showing large unifocal autonomy. (a) using [^{99m}Tc]pertechnetate, and (b) the corresponding scintigraphy using [^{131}I]iodide. Courtesy of T. Derlin, Nuclear Medicine, Hannover Medical School, Germany. Cf. also the cover of this volume for a [^{99m}Tc]pertechnetate thyroid image.

12.7.10 Neurosignal transduction and molecular imaging of the brain

Figure 12.47 shows the correlation between morphological and functional imaging of the living human brain, highlighting the basic aim of molecular imaging using nuclear medicine and radiopharmacy technologies. PET and SPECT molecular imaging have contributed to significant developments in addressing, among other important physiologies, the functionality of several important neurotransmitter systems.

An example is the dopaminergic system. Because of its relevance in several very key neurological disorders such as Parkinson's disease and because of the variety of very potent radiolabeled PET and SPECT tracers used for basic research as well as for patient diagnosis, the different approaches to imaging the dopaminergic system are illustrated in the following. Figure 12.48 illustrates the biological structure of a dopaminergic synapse and identifies the key steps in presynaptic dopamine (DA) synthesis, presynaptic dopamine re-uptake by dopamine transporters (DAT), postsynaptic binding at various subtypes of dopaminergic receptors (D$_1$–D$_5$) and the enzymatic degradation by monoamine oxidases (MAO) of the dopamine released into the synaptic cleft.

The highest density of dopaminergic neurons in the human brain is in the basal ganglia. There are small midbrain areas called substancia nigra projecting to the dorsal striatum, as well as ventral tegmental areas projecting to prefrontal cortex and nucleus accumbens, to amygdala, cingulate gyrus, hippocampus and olfactory bulb. The regions most relevant for molecular imaging of dopamine neurotransmission are illustrated in Fig. 12.49.

Both the enzymes (AADC, MAO) and transmembrane receptors (DAT, DR) are the biologically relevant targets for radiolabeled tracers. Using the appropriate

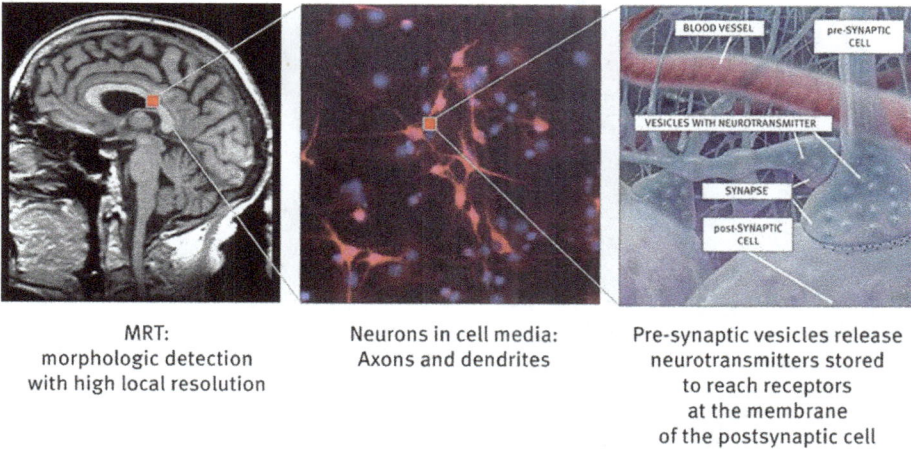

MRT:	Neurons in cell media:	Pre-synaptic vesicles release
morphologic detection	Axons and dendrites	neurotransmitters stored
with high local resolution		to reach receptors
		at the membrane
		of the postsynaptic cell

Fig. 12.47: Concepts of medical imaging: morphology vs. function. The left image shows an MRT scan of a slice of human brain, perfectly revealing tissue structure. However, single neurons as imaged in cell media (middle) remain morphologically invisible. Going into even further detail, the right image in a very simplified way illustrates the physiology of neurotransmission within and between neuronal compartments. Presynaptic vesicles release stored neurotransmitters to reach receptors at the membrane of the postsynaptic cell. Although morphologically invisible, this neurotransmission can be visualized by nuclear medicine imaging.

molecules, synthesis, binding, re-uptake, and metabolic turnover of dopamine can be covered by means of *in vivo* molecular imaging. This allows to detect (i) the molecular, i.e., chemical principles of biological and physiological processes and (ii) the molecular origin of malfunctions of the human brain, (iii) to trace back treatment strategies or to transfer information about anatomical research results to suitably challenge paradigms to better understand pharmacological manipulations of target structures *in vivo* and to evaluate neurobiological vulnerability factors of diseases.

12.7.10.1 Dopamine synthesis capacity

In the nerve cells, endogenous L-dopa is a substrate of the enzyme "aromatic L-amino acid decarboxylase" (AADC). AADC thus is converting endogenous L-dopa into endogenous dopamine[25] (the dopamine is formed by enzymatically removing the carboxy moiety from L-dopa) and does the same to a similar extent with $6\text{-}[^{18}F]$ fluoro-L-DOPA, a fluorinated derivative of L-dopa: it turns $6\text{-}[^{18}F]$fluoro-L-DOPA

---

25 Dopamine = 4-(2-aminoethyl)benzene-1,2-diol; L-dopa = (*S*)-2-amino-3-(3,4-dihydroxyphenyl) propanoic acid (also: L-3,4-dihydroxyphenylalanine).

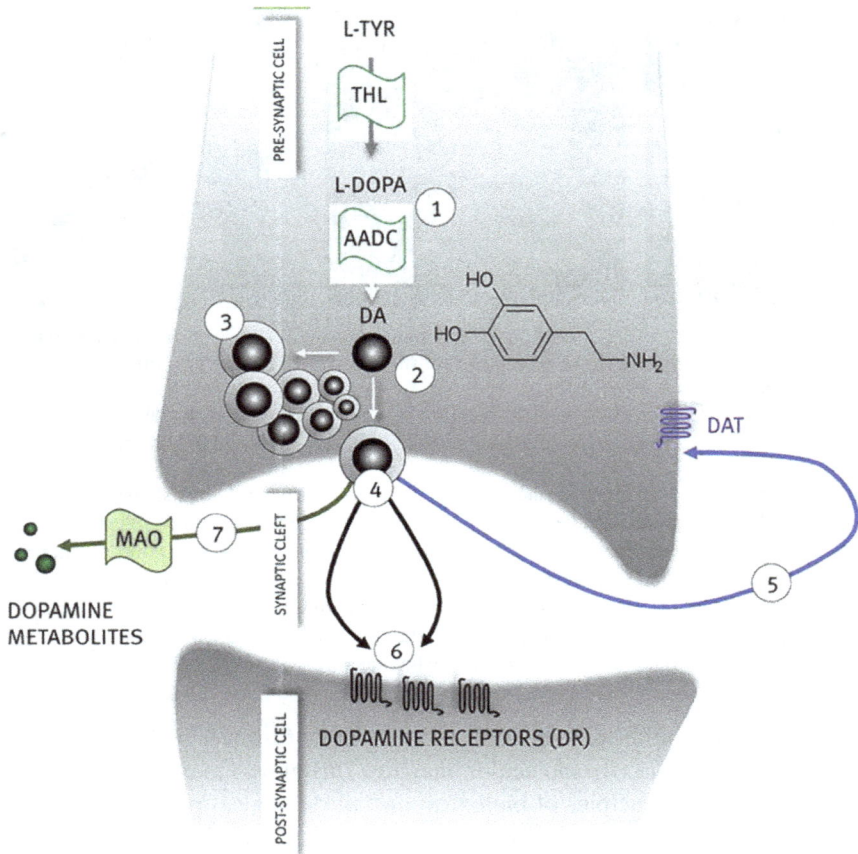

Fig. 12.48: Diagrammatic representation of a dopaminergic synapse, composed of presynaptic cell, synaptic cleft and postsynaptic cell, showing relevant steps in dopamine synthesis and dopaminergic neurotransmission. Dopamine synthesis starts from L-tyrosine (L-TYR), which is enzymatically converted by tyrosine hydroxylase (THL) to L-dopa (also referred to as "levidopa"), which is (1) enzymatically converted by aromatic L-amino acid decarboxylase (AADC) to dopamine (DA). DA (2) is stored in synaptic vesicles (3), which – following activation signals – release DA into the synaptic cleft (4). The fate of DA is manifold. Key steps in the context of molecular imaging are: (5) presynaptic L-dopa (re)uptake via transmembrane proteins called dopamine transporters (DAT); (6) postsynaptic binding to another class of transmembrane proteins called dopamine receptors (DR) (the essential part of neurotransmission); (7) oxidative deamination of dopamine by the enzyme monoamine oxidase (MAO).

([^{18}F]DOPA) into 6-[^{18}F]fluoro-dopamine ([^{18}F]DA). [^{18}F]DOPA, if not degraded by peripheral O-methylation, is cleared from blood to the brain and is shuttled into the neuronal cells by the same amino acid transporter that usually transports the L-dopa precursor L-tyrosine as well as several other amino acids. Afterward, 6-[^{18}F]

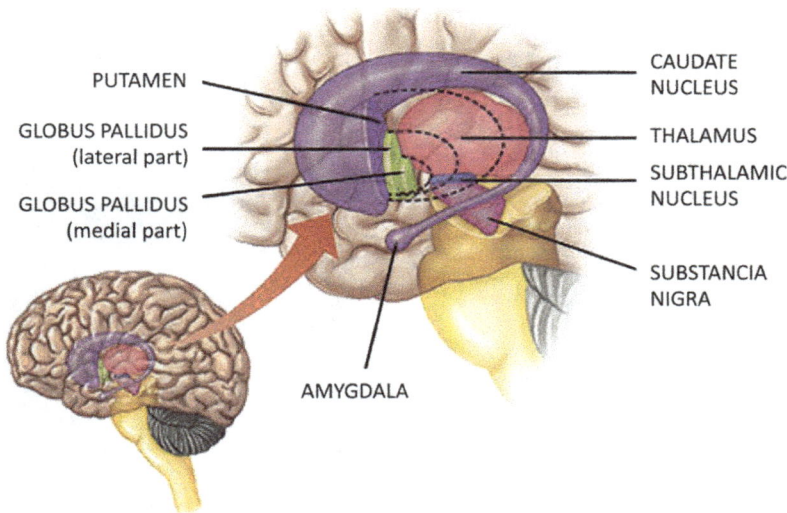

Fig. 12.49: Diagrammatic representation of the small area of dopaminergic neurotransmission located in the basal ganglia of the human brain. Modified from http://brainmind.com/images/ba salgangliadetail.gif.

DA is stored in the vesicles; cf. Fig. 12.50. Molecular imaging utilizes the trapping[26] of this enzymatic conversion product in the presynaptic cell.

Based on early autoradiographic analysis using tritiated L-dopa, the synthesis and application of [^{18}F]DOPA was one of the first developments for clinical and scientific investigations of the brain by means of PET. Still, [^{18}F]DOPA is a very important radiotracer – in spite of its complex kinetic modeling, the low signal-to-noise ratio, and its difficult neurophysiological interpretation.[27]

Using the radiotracer and its outcome parameters, a vast amount of clinically and scientifically important findings have been reported. In neurology, [^{18}F]FDOPA served as an important tool to detect the pathologies in the nigrostriatal dopamine system, in particular, for the diagnosis of Parkinson's disease. Today, the use of [^{18}F]FDOPA is no longer a standard routine in neurology, due to the availability of

---

26 This is metabolic trapping similar as described for 2-[^{18}F]FDG.
27 The interpretation of [^{18}F]FDOPA PET scans is complicated since AADC is an enzyme of high capacity and clearly not the rate limiting step of the dopamine synthesis. Usually the rate of blood/brain clearance is the major outcome parameter (K). Its neurophysiologic meaning is usually carefully described as *dopamine synthesis capacity*: this parameter mirrors the anatomical and structural density of dopaminergic neurons (a rather stable influencing factor) but also some regulatory processes on enzyme activity under environmental or pharmacological challenges. A high dopamine synthesis capacity does not necessarily depict a high rate of dopamine transmission but rather enables the system to provide a high capacity if needed.

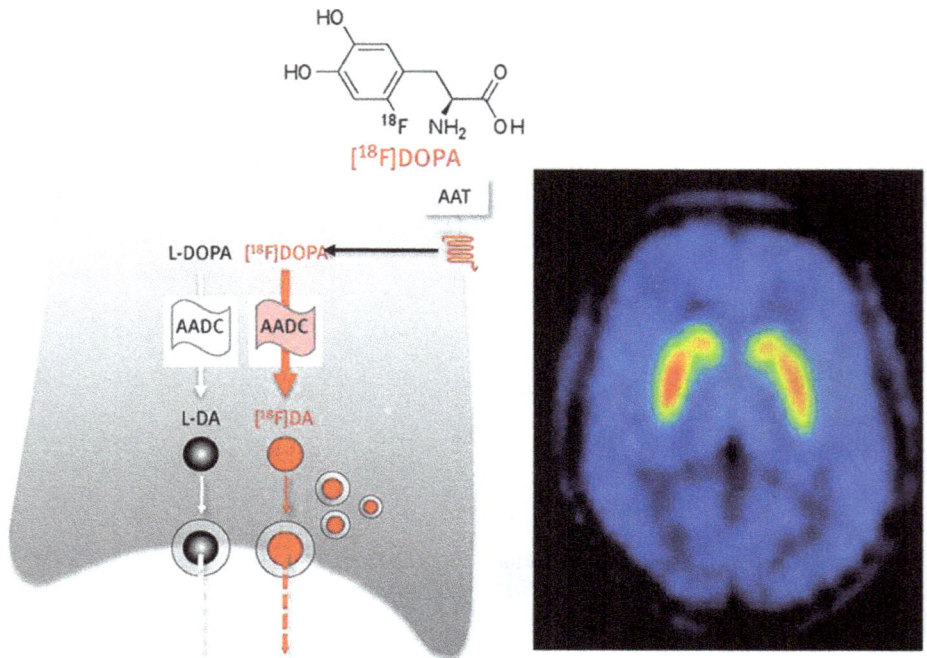

Fig. 12.50: Schematic illustration of transfer of [^{18}F]DOPA into the presynaptic neuron, conversion to [^{18}F]DA by AADC and terminal storage in vesicles (left). The right image shows [^{18}F]DA uptake in dopaminergic neurons of a healthy brain (sagittal view).

SPECT/PET presynaptic dopamine transporter ligands – whose targets are less subject to substantial upregulation (see below). Recently, a 10–20% increase in the dopamine synthesis capacity in schizophrenia was shown, which in some studies was reported to be associated with the extent of psychotic symptoms.

12.7.10.2 Fate of dopamine

Dopamine, once released from the vesicles, is emitted into the synaptic cleft. Here, it is involved in three fundamental pathways. The first one is degradation: Enzymes such as monoamine oxidases are present and "waiting" for biochemical cascades "deactivating" the neurotransmitter – see below. The second one is the re-uptake into the presynaptic cell. This is organized by specific "transporters". The final – and in terms of *signal transduction* most important step – is the binding to specific receptors, expressed at the postsynaptic cell.

Fate of dopamine (I) – Studies of the presynaptic dopamine re-uptake system:
The dopamine emitted into the synaptic cleft exists in this compartment to a certain
concentration. A fraction of it is transported "backward" into the presynaptic cell.
Here it functions like a control of the dopamine synthesis rate: A certain amount of
L-dopa re-transported indicates that there is – at the current moment – a sufficient
concentration in the synaptic cleft. Consequently, the original dopamine synthesis
rate (via AADC) is lowered – or vice versa. This particular transmembrane transport
is managed by transmembrane proteins, so-called dopamine (re-uptake) transporters
(DAT). How is this pathway visualized? There are small molecules other than L-dopa
that can also bind to the dopamine transporters. However, these molecules just
"bind", i.e., they are not "transported" into the cell. Still they compete with dopamine
for one DAT.[28] One DAT protein is able to accept only one molecule at a time, i.e.,
either dopamine or a competitor.[29] Among those small molecules, the most important
one is cocaine.[30] To visualize its binding to a DAT, ^{11}C-labeled cocaine analogs have
been developed. Also other β^+ (^{18}F) or single photon (^{123}I and ^{99m}Tc) emitters have
been attached to the cocaine basic structure (cf. Fig. 12.51).

The corresponding radiolabeled tracers such as [^{18}F]FE-β-CIT and [^{18}F]PRO4.MZ
selectively bind to the DAT. Some others are even commercially available: [^{123}I]
tropane (DATSCAN) and [^{99m}Tc]TRODAT. According to the isotope, the tracers are
applied for PET molecular imaging (^{11}C, ^{18}F) or SPECT (^{123}I, ^{99m}Tc). Figure 12.52 (left)
summarizes the most potent derivatives, and Fig. 12.52 (right) and Fig. 12.53 illustrates
the visualization by means of [^{18}F]PRO4.MZ.

Fate of dopamine (II) – PET studies on postsynaptic dopamine receptors: Dopamine receptors are transmembrane proteins, localized at the surface of the postsynaptic
cell. There are several subtypes of this receptor (D_1–D_5), which are different in terms
of the amino acid sequences of the protein, and in terms of action. All are G-protein
coupled receptors (GPCR) in as much as they induce an action on the G-protein within
the cell. Stimuli (which is binding of the corresponding endogenous agonistic

---

28 Note that this competition will happen only if the radiotracer is able to cross the blood-brain
barrier and in the case it shows a sufficient binding affinity (cf. Tab. 12.5) to the dopamine transporter. The way this is achieved belongs to the traditional domain of medicinal chemistry and drug
development – it requires systematic organic synthesis and biological evaluations in terms of struc-
ture/affinity relationships. Minor structural changes may have a dramatic impact on whether the
molecule "functions" – i.e., binds to the DAT or not.
29 In other words, binding of another small molecule to a particular DAT "blocks" the transporter –
it cannot transport the dopamine. This is one of the fundamentals of drug addiction, see below.
30 This explains the rational of the drug cocaine: By blocking the DAT, there is no incoming signal
"telling" the presynaptic dopamine "machinery" that there is already a sufficient dopamine concentration in the cleft. Consequently, more and more dopamine is synthesized and released. The
increasing concentration of the neurotransmitter induces a higher signal transduction via the postsynaptic dopamine receptors – creating the "high".

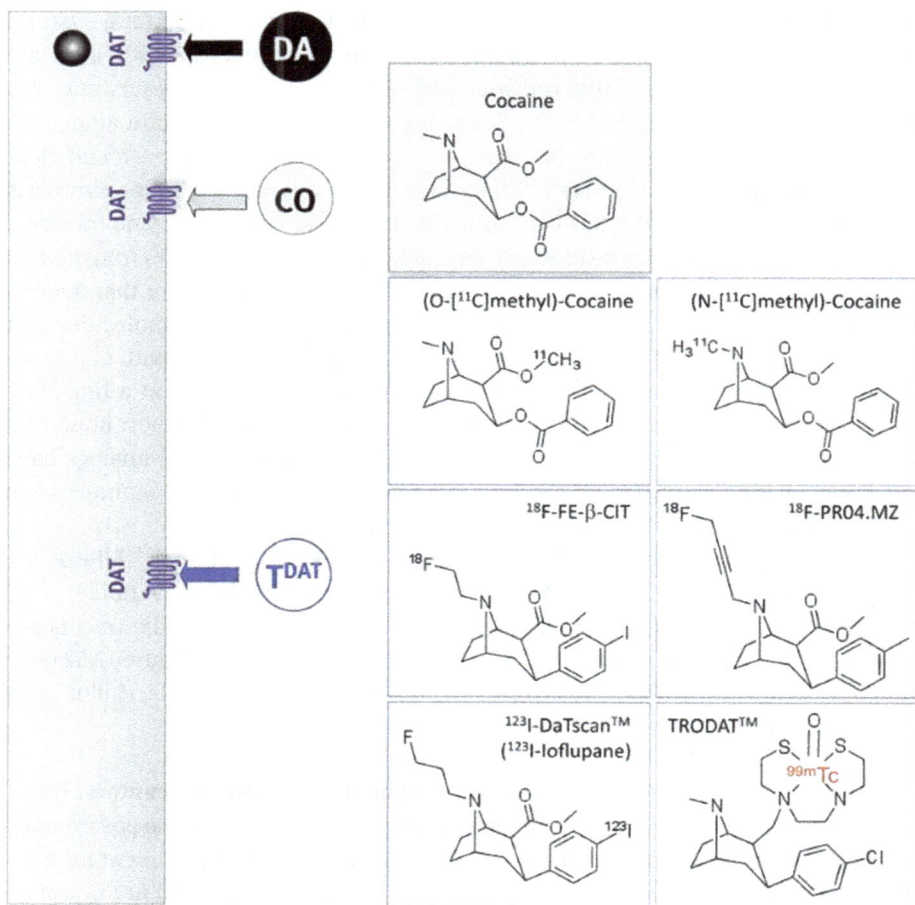

Fig. 12.51: Molecules addressing the DAT: dopamine (DA), which is transported into the presynaptic neuron; cocaine (CO), which is bound by the DAT but not transported (thus "blocking" DAT binding sites for DA and other molecules); and radiolabeled analogs of cocaine: (O-[^{11}C]methyl}-cocaine, (N-[^{11}C]methyl}-cocaine, ^{18}F-β-CIT, ^{18}F-PRO4.MZ, ^{123}I-tropane (DATSCAN), and ^{99m}Tc-TRODAT.

neurotransmitter dopamine[31]) at D_1 and D_5 receptors ("D1-like family") endogenous to the G protein G_{sg}, in turn *activates* the intracellular enzyme *adenylyl cyclase*, thereby increasing the intracellular concentration of the second messenger *cyclic*

---

31 The Nobel Prize in Physiology or Medicine 2000 was awarded jointly to A. CARLSON, P. GREENGARD and E.R. KANDEL "for their discoveries concerning signal transduction in the nervous system". CARLSON discovered that dopamine is a transmitter, GREENGARD found that dopamine and other "slow" transmitters act by protein phosphorylation, and KANDEL showed that phosphorylation is necessary for the formation of short-and long-term memory.

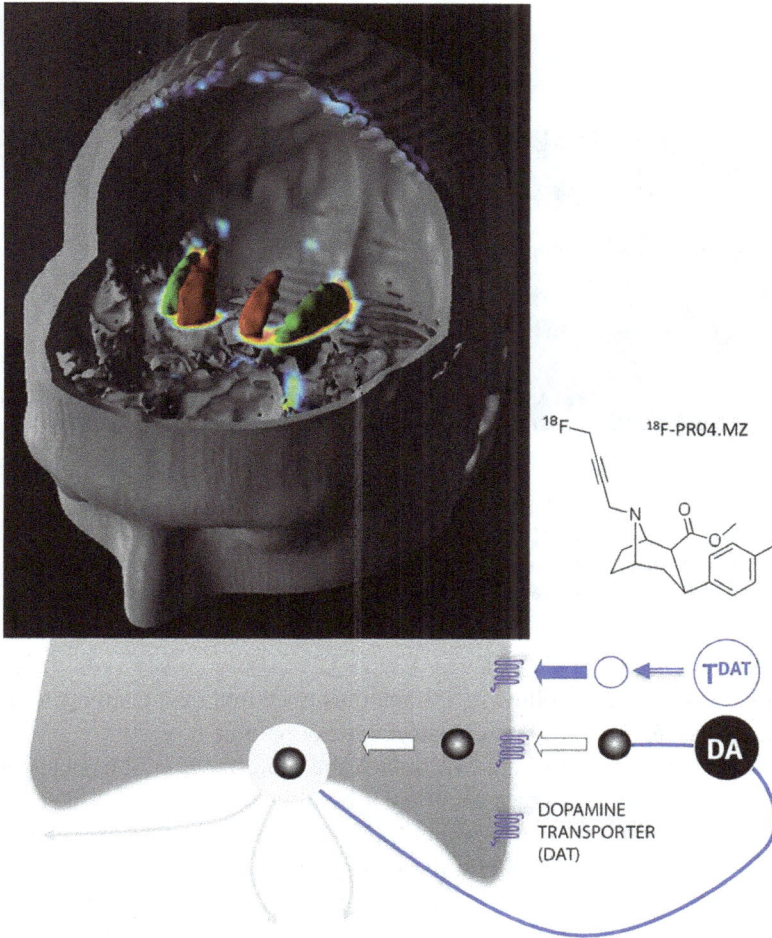

Fig. 12.52: Schematic binding of a radiolabeled tracer T^{DAT} to the dopamine transmembrane (re-uptake) transporter (DAT) in analogy to dopamine (left) and visualization of [^{18}F]PRO4.MZ (see also Fig. 12.53) uptake in healthy human brain. The PET-tracer uptake in DAT-rich regions of the striatum (red) and putamen (green) are overlaid with the morphology of the head as recorded by simultaneously MRI. Courtesy H Amaral and V Kramer, PositronPharma, Santiago de Chile, 2014.

adenosine monophosphate cAMP. In contrast, D_2, D_3, and D_5 receptors ("D_2-like family") are linked to the G protein G_{ig}, *inhibiting* the *adenylyl cyclase*, and consequently, the formation of cAMP.

Among these subtypes, D_2 and D_3 are of main interest. They are involved in very important and clinically relevant diseases such as M. Parkinson and other neurological disorders, they serve as targets to treat many diseases such as schizophrenia, depression, addiction, and obesity; they are understood best – and they can be

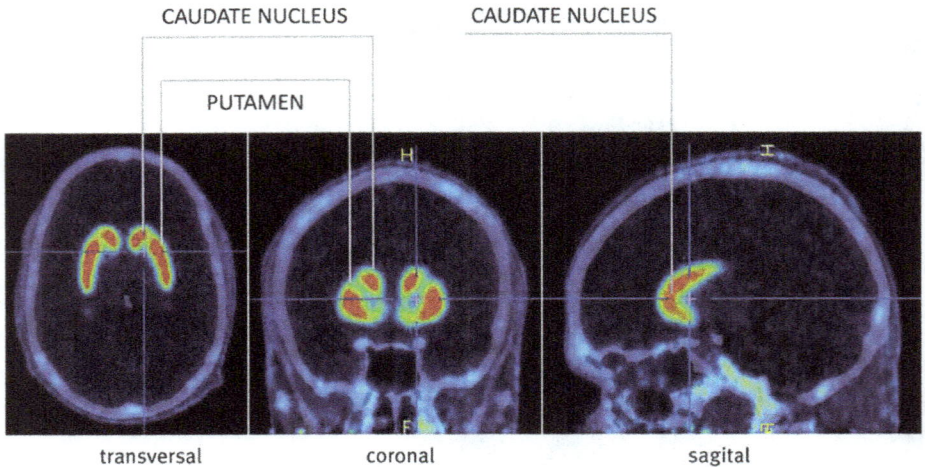

Fig. 12.53: [^{18}F]PR04.MZ uptake in DAT-rich regions of the striatum, putamen and amygdala visualized as sagittal, coronal and transversal views of a healthy brain. Fusion PET/CT; color table cold; transversal, sagittal and coronal view in slices of 2 mm thickness 120 min after tracer injection summed image from 120–135 min. p. Courtesy H Amaral and V Kramer, PositronPharma, Santiago de Chile, 2014.

visualized by adequate radiolabeled tracers. Figure 12.54 shows how a radiotracer for the dopamine receptor (T^{DR}) following intravenous injection, may compete with endogenous dopamine for the binding to postsynaptic dopamine receptors.

Note that this competition will happen only if the radiotracer is able to cross the blood-brain barrier and if it shows a sufficient binding affinity to the target receptor (or its subtype). Among the tracers that fulfill these requirements are several classes of compounds, such as initially (in the 1980s) radiolabeled spiperones (e.g., [^{76}Br]bromospiperone) and later, substituted benzamides, the most important of which was [^{11}C]raclopride with moderate affinity at $D_{2/3}$ receptors, but high selectivity. Regions outside the striatum were not detectable with [^{11}C]raclopride. Important ^{18}F-labeled benzamides are [^{18}F]desmethoxy-fallypride ([^{18}F]DMFP) and [^{18}F]fallypride ([^{18}F]FP). The latter is extremely affine to the $D_{2/3}$ receptors and is capable of even visualizing extrastriatal dopamine receptors. Figure 12.54 shows the structure of the tracers mentioned, and Tab. 12.5 compares binding affinities of the two fluorinated benzamides DMFP and FP to dopamine and other receptors.

The differences in binding affinity have consequences for the ability to address different clinical applications – i.e., either for fundamental brain research, drug development or rather "routine" patient diagnosis. [^{18}F]DMFP is a very useful tracer in routine clinical application to quantify $D_{2/3}$ receptor in the striatal regions. Yet, the one order of magnitude higher affinity of [^{18}F]FP compared to its structural analog [^{18}F]DMFP allows to visualize even the extrastriatal regions of $D_{2/3}$ receptors, where

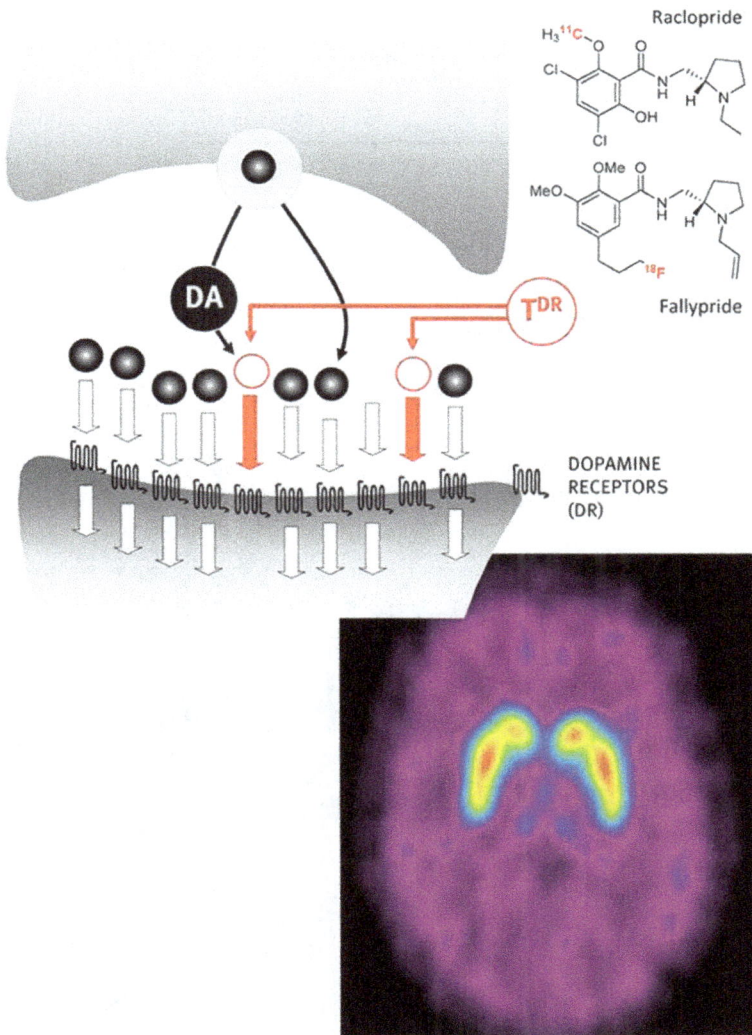

Fig. 12.54: Simplified concept of how a radiotracer T^{DR} (examples show benzamides labeled with either ^{11}C (raclopride) or ^{18}F (fallypride, FP)) for the postsynaptic dopamine receptor DR may compete with endogenous dopamine (DA). [^{18}F]DMFP sagittal PET images: Binding to striatal regions of the healthy human brain. Color scaling parallels increasing levels of radioactivity per volume element (voxel), which in turn parallels the density of dopamine receptors available.

D$_{2/3}$ receptors exist, but at a much lower concentration compared to the striatum, cf. Fig. 12.55.

It is interesting to recall the basic principle of *in vivo* molecular imaging: The distribution of the PET tracer [^{18}F]FP in the living human brain allows to quantitatively

Tab. 12.5: Binding affinities of DMFP and FP to selected human neuroreceptors.

Receptor	Subtype	K_i (nM)					
		DMFP			FP		
Dopaminergic	pD1	29 500	±	1 500	18 000	±	2 000
	hD2$_{short}$	30	±	4	2.2	±	0.1
	hD2$_{long}$	30	±	5	2.2	±	0.1
	hD3	32.8	±	7.8	1.7	±	0.3
	hD4.4	3 000	±	400	1 300	±	250
Serotonergic	p5-HT$_{1A}$	2 250	±	750	1 080	±	120
	p5-HT$_{2A}$	4 450	±	50	3 700	±	300
Alpha	pα1	1 900	±	100	970	±	330

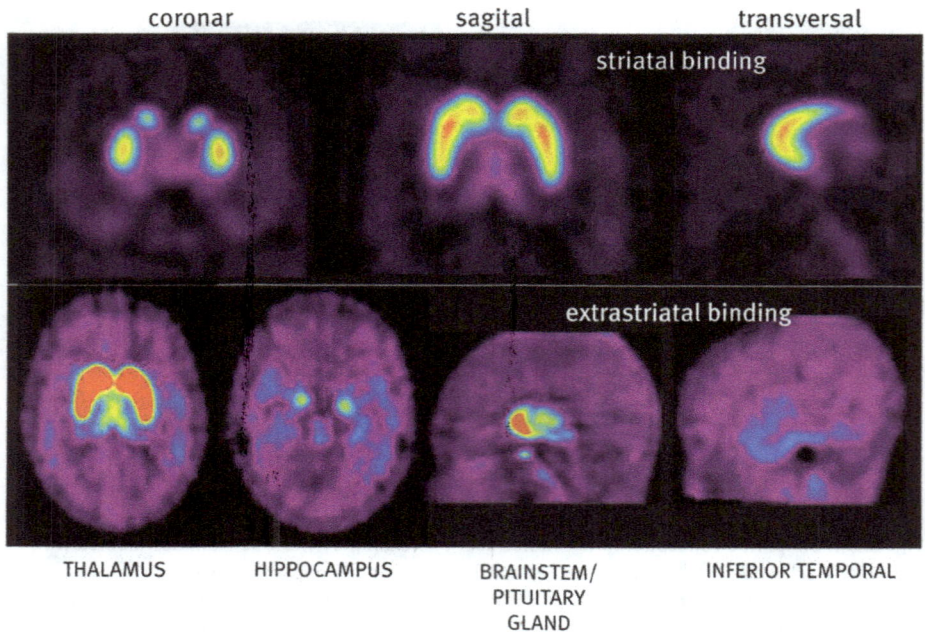

Fig. 12.55: [^{18}F]FP: Binding to striatal and to extrastriatal regions of the healthy human brain.

determine the availability of postsynaptic dopamine receptors. It fits well with the distribution of these receptors as analyzed in postmortem brain slices by quantitative autoradiography as measured with the same, but tritiated tracer [^{3}H]FP, cf. Fig. 12.56.

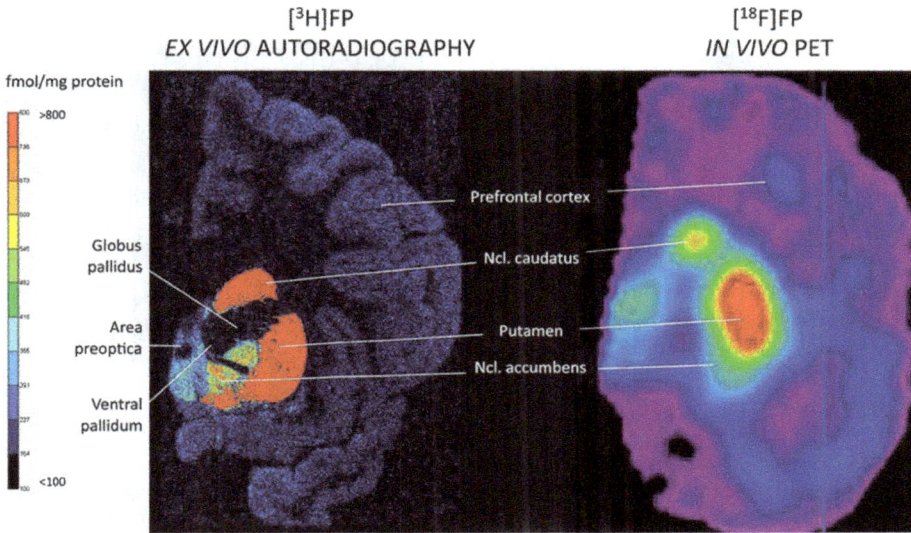

Fig. 12.56: *Postmortem* quantitative auroradiography (transversal) with tritiated fallypride [^{3}H]FP vs. *in vivo* PET with [^{18}F]FP: Spatial resolution as obtained by means of PET is inferior to quantitative autoradiography, but uptake reflects the real distribution of the tracers (in the slices selected, the putamen) and – following data processing – results in correlations matching the DR densities similar to autoradiography. Color scaling for the *postmortem* data parallels the (protein) density of dopamine receptors as indicated by <100 to >800 pmol cm^{-3}; and the same can be achieved for *in vivo* PET.

$D_{2/3}$ tracers are clinically most relevant for the noninvasive diagnosis of Morbus Parkinson. The biological origin of the disease is the loss of striatal neurons, causing a reduced dopamine synthesis and consequently, a reduced dopaminergic signal transduction. Suppose a presynaptic dopaminergic neuron is not (or to a lesser degree) emitting dopamine, the concentration of the neurotransmitter in the synaptic cleft will be low. Consequently, more postsynaptic dopamine receptors are available for the radiotracer binding.[32] A PET or SPECT image will detect a higher radioactivity in the negrostriatal regions. This is exemplified in Fig. 12.57, showing the brain of a Parkinson patient. The radioactivity concentration of the PET tracer injected ([^{18}F] DMFP in the present case) in the left striatum is normal, while the activity is increased in the right striatum. This is a typical finding for this disease.

---

32 The same effect occurs for presynaptic transporters, which is not shown in the figure. However, the principle remains the same: Low concentration of dopamine (due to a disease such as M. Parkinson or due to other neurological disorders) in the synaptic cleft means that less DAT is occupied by dopamine and consequently, a DAT-radiotracer such as ^{11}C-cocaine, or the various tropane derivatives labeled with ^{123}I or ^{99m}Tc for SPECT or ^{18}F for PET binds to a higher extent (as revealed by higher radioactivity concentration).

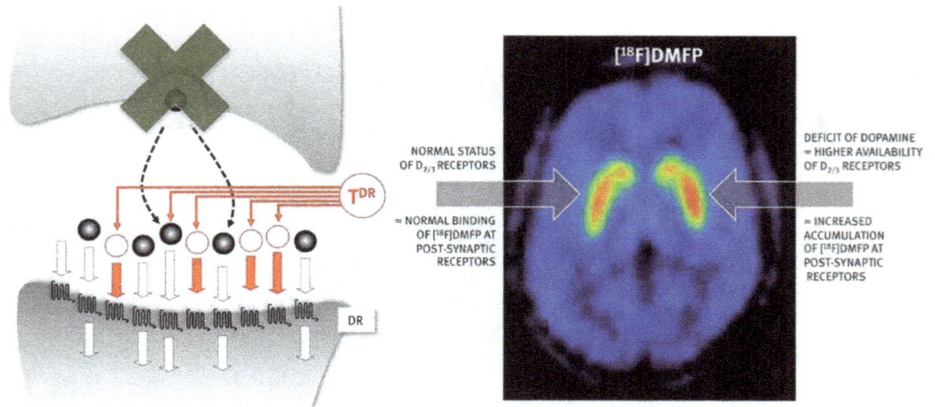

Fig. 12.57: Simplified concept of how a radiotracer T^{DR} for the postsynaptic dopamine receptor DR accumulates more intensely in the case where a deficit of endogenous dopamine keeps far more postsynaptic receptors available for a T^{DR} ligand such as the radioligand [^{18}F]DMFP (left). [^{18}F]DMFP PET sagittal image representing a M. Parkinson patient (right). The radioactivity concentration of the PET tracer in the left striatum is normal, while the activity is increased in the right striatum due to the increased availability of postsynaptic $D_{2/3}$ receptors.

Fate of dopamine (III) – Enzymatic degradation: Mainly involved in the complex fate of [^{18}F]DA is the monoamine oxidase (MAO), which is a cytosolic enzyme. Given that [^{18}F]DA is protected against degradation within the vesicles, the dopaminergic release (and subsequent re-uptake) makes [^{18}F]DA a substrate of MAO in the synaptic cleft. Following this argumentation, the occurrence of ^{18}F-containing degradation products, i.e., the radioactive metabolites of [^{18}F]DA, is a surrogate parameter of the presynaptic dopamine turnover, cf. Fig. 12.58. Using arterial plasma withdrawal (for arterial plasma activity and detection of metabolite fractions) and a 2-hour PET scan, these can be reliably determined. Although quantitative modeling is very complex, the physiological impact is highly relevant and yields astonishing clinical results including those in schizophrenia.[33]

12.7.10.3 More radioligands for more neurotransmitter systems

The concept of molecular imaging of neurotransmission was described here for the dopaminergic synapse. The various approaches apply, though to a different extent, to many other brain receptors. Figure 12.59 depicts the most relevant and clinically

---

33 Usually, dopamine synthesis capacity and turnover rates under resting conditions are in equilibrium; in schizophrenia, however, the turnover rate is doubled though the synthesis capacity is increased by 10–20% as mentioned above.

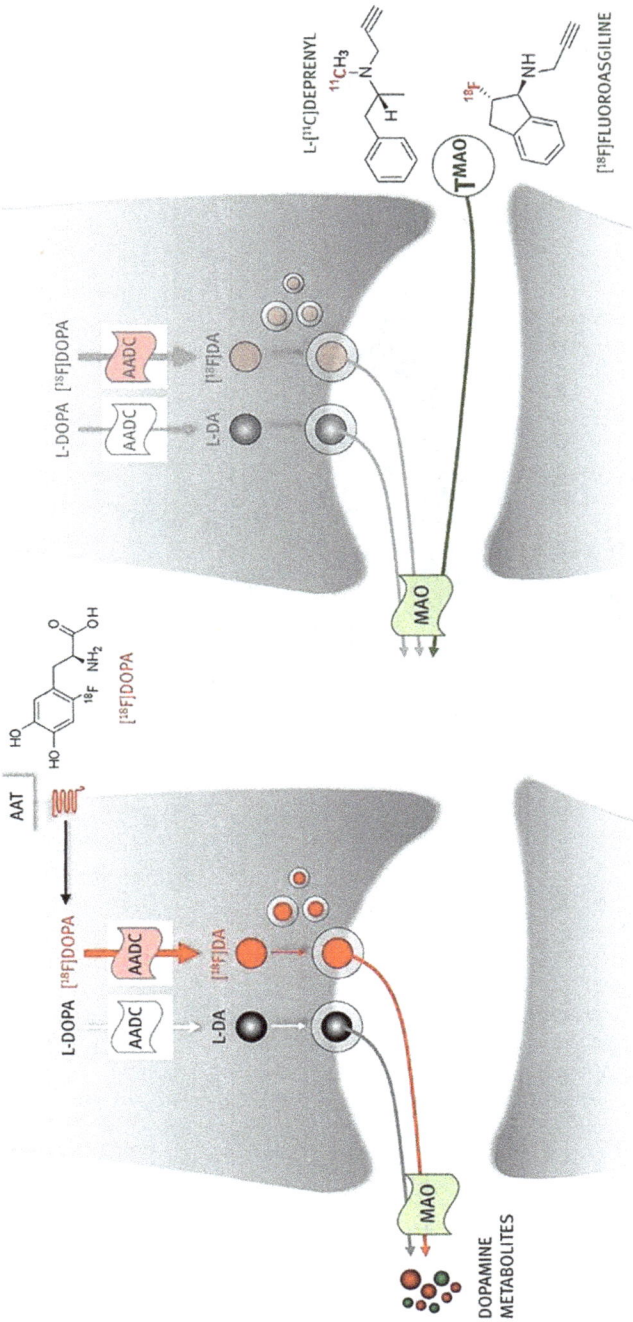

Fig. 12.58: Illustration of the metabolic fate of [18F]DA once released into the synaptic cleft. Monoamine oxidase (MAO) starts a catalytic degradation chain similar to the pathway of endogenous dopamine. The level of 18F-containing metabolites can be quantified from blood samples and is a measure of the enzyme activity (left). In parallel, radiolabeled substrates of MAO can be applied. Examples shown are 11C-labeled L-deprenyl and 18F-labeled asgiline (right).

established neuroreceptors, which are addressed by well-studied and highly selective PET analogs. Each of the radioligand selectively targets "its" receptor, and each receptor system is specifically involved in important physiological processes and as such is a tool to diagnose the related diseases.

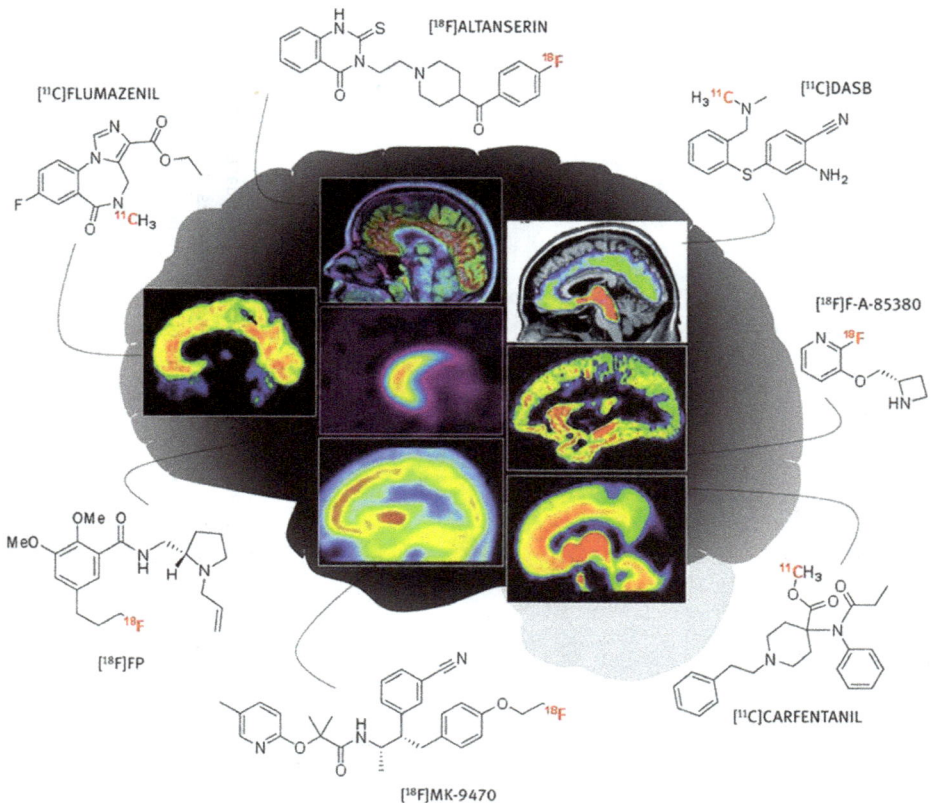

Fig. 12.59: Most relevant and clinically established neurotransmitter analogs for PET imaging of a variety of neuroreceptors. Selected neurotransmitter analogs for PET imaging of clinically relevant neuroreceptor systems: Serotonin receptors system: [^{18}F]Altanserin; serotonin receptor transporter system: [^{11}C]DASB; nicotinic acetylcholine receptor system: [^{18}F]F-A-85,380; opiate receptor system: [^{11}C]carfentanil; cannabinoid receptor system: [^{18}F]MK-9470; dopamine receptor system: [^{18}F]FP; denzodiazepine receptor system: [^{11}C]Flumazenil.

12.7.11 Other molecular imaging of the brain

12.7.11.1 Radioligands in imaging of Alzheimer's disease

Alzheimer's disease (AD) is the most relevant form of dementia and an early and accurate diagnosis is not possible by clinical symptoms. Only at a late stage of the disease do symptoms allow a rough differentiation from other neurodegenerative diseases. A confirmation of AD can only be given by postmortem histopathological examination of brain tissue (Aβ plaques deposit). At a cellular level, pathophysiological changes occur years before the first clinical symptoms are recognized. These changes are generally deposits of β-amyloid plaques (Aβ plaques) and neurofibrillary tangles of hyperphosphorylated tau protein; see Fig. 12.60. These deposits and accumulations affect the neurons and finally lead to the loss of neurons and synapses in the cerebral cortex. The cause for AD is still unknown. One of the major hypotheses is the amyloid hypothesis. Other hypotheses are discussed, of which some already have been disproved. The

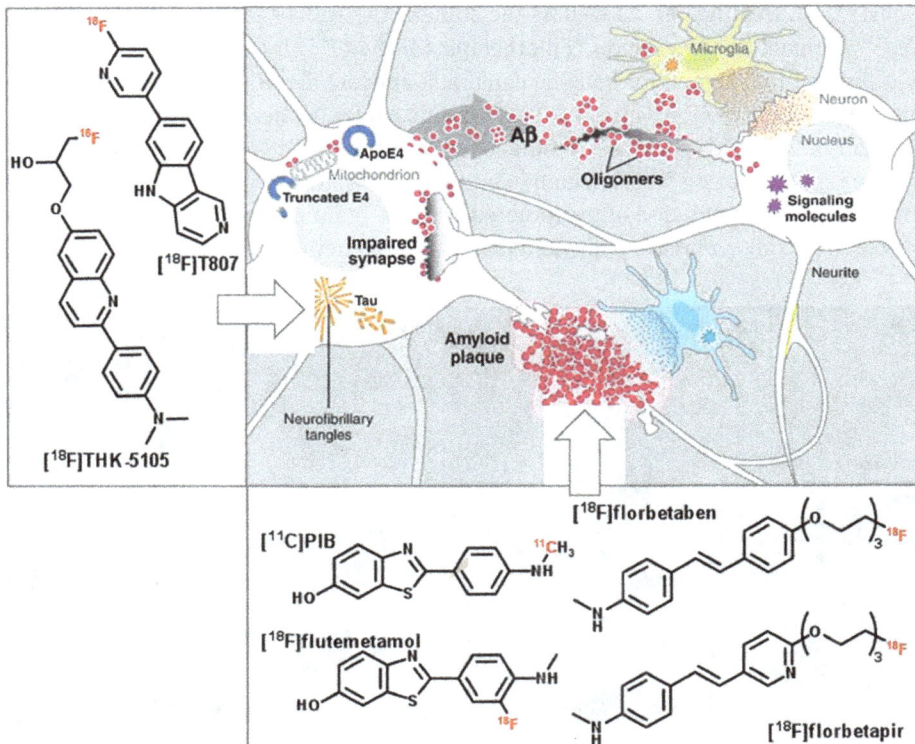

Fig. 12.60: Cellular and molecular processes of the neuropathology of Alzheimer's disease. The most relevant and clinically used radioligands for imaging of Aβ plaques in AD and dementia (lower box); and new radioligands for specific PET imaging of tau protein and neurofibrillary tangles (left box).

amyloid hypothesis postulates the extracellular β-amyloid peptide deposit as major cause of AD. The abnormal accumulation of these β-amyloid peptides is produced by a mutation of the amyloid precursor protein (APP). The accumulation or reduced clearance of β-amyloid peptides first produces β-amyloid oligomers which already impair synaptic functions. Further accumulation leads to β-amyloid plaques. β-amyloid plaques provoke an immune response, and thus oxidative stress as well as neuroinflammation. As a result, neuronal kinases and phosphatases are activated, which are assumed to cause the hyperphosphorylation of tau proteins leading to intracellular neurofibrillary tangles. These tangles further damage neurons and synapses, and contribute to their loss.

For early diagnosis of AD, molecular imaging of the pathophysiological changes is the method of choice. Numerous different structures and radioligands have been developed in the last decade. Most of them are ligands for Aβ plaques derived from the lead structures of dyes used in histopathological staining of Aβ plaques postmortem. The most prominent radioligands for Aβ plaque imaging are the thioflavin T-based "Pittsburgh compound B" [^{11}C]PiB and its ^{18}F-labeled analog [^{18}F]flutemetamol (*Vizamyl*™, GE Healthcare); as well as the stilbene derivatives [^{18}F]florbetaben (*NeuraCeq*™, Piramal Imaging) and [^{18}F]florbetapir (*Amyvid*™, Eli Lilly), Fig. 12.57. [^{11}C]PiB was the first radioligand for imaging cerebral β-amyloid deposits and today is still the most studied tracer for Aβ imaging. It has been used for further development of new β-amyloid radioligands such as [^{18}F]flutemetamol. Figure 12.61 shows PET imaging of β-amyloid plaques using [^{18}F]florbetaben in a healthy volunteer and an AD patient. The much higher accumulation of the radiotracer in the brain of the AD patient are clearly visible, reflecting a massive β-amyloid plaque deposit in the diseased brain.

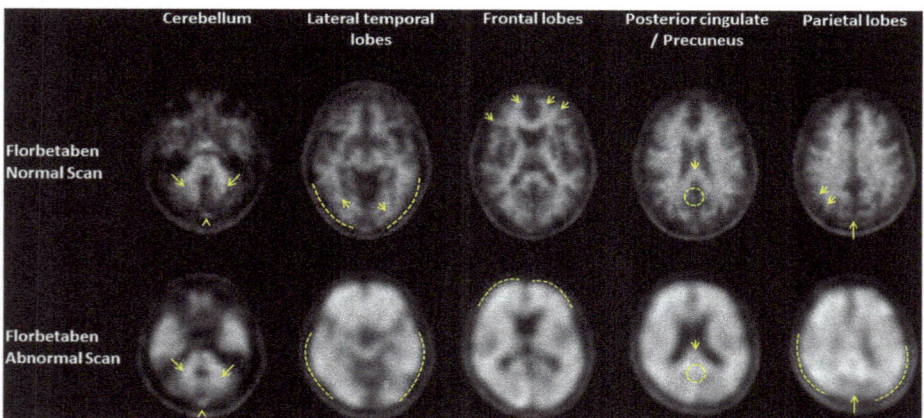

Fig. 12.61: Molecular PET imaging using [^{18}F]Florbetaben showing an example of a normal scan (i.e., negative for beta amyloid plaques) (top row) and an abnormal scan (i.e., positive for β-amyloid plaques) (bottom row).

Besides β-amyloid ligands, specific tau protein ligands are of particular interest. Two examples are the benzimidazole pyrimidine [^{18}F]T807 and the 2-arylquinoline [^{18}F]THK5117, Fig. 12.61. Both show high selectivity to tau over β-amyloid and represent specific tau radioligands. One reason tau imaging is assumed to be the next frontier in AD imaging is that there are indications that the level of tau deposition in post-mortem immune-staining correlates with the severity of disease.

12.7.11.2 2-[^{18}F]FDG in brain imaging

Today, 2-[^{18}F]FDG is known as the most commonly applied PET radiotracer in oncology, but originally 2-[^{18}F]FDG was intended as a brain imaging agent for noninvasive molecular imaging of brain functions (cf. Section 12.5.2 on "trapping"). Still it is a useful radiotracer for brain imaging. In several pathologies of the CNS, cells are damaged and show decreased glycolysis, or even complete cell loss. In PET imaging, such brain areas show reduced uptake of 2-[^{18}F]FDG. Incidentally, in imaging of neurodegenerative diseases 2-[^{18}F]FDG is often used as reference, and for correlation of pathophysiological processes with basic metabolism (glycolysis).

12.7.11.3 Brain perfusion imaging

As mentioned in Section 12.5.1 on "blood flow and perfusion", the ideal perfusion tracer is [^{15}O]H$_2$O. The half-life of 2 min of oxygen-15 limits its availability and common use, but allows several consecutive PET scans right after another. In contrast, the [^{99m}Tc]labeled perfusion tracers [^{99m}Tc]HMPAO (hexamethylpropyleneamine oxime) and [^{99m}Tc]ECD (ethylene dicysteine diethyl ester, [^{99m}Tc]bicisate) offer the convenience of technetium-99m with simple kit preparations and availability everywhere through the ^{99}Mo/^{99m}Tc generator. [^{99m}Tc]HMPAO is applied for blood flow and brain perfusion imaging. Its relatively high lipophilicity allows [^{99m}Tc]HMPAO to penetrate the blood-brain barrier and it provides information about brain perfusion. [^{99m}Tc]ECD is applied for cerebral blood flow measurements. Another clinically employed SPECT tracer for brain perfusion is [^{123}I]IMP (N-isopropyl-p-[^{123}I]iodoamphetamine). The oldest brain perfusion imaging agent is ^{133}Xe gas. The inert gas is inhaled and shortly afterward SPECT scans of the brain are performed to determine the cerebral blood flow. Nowadays, ^{133}Xe is less commonly applied, the [^{99m}Tc]labeled alternatives are much cheaper and easier in handling and use.

12.8 Outlook

12.8.1 More nuclides, improved chemistry, more radiopharmaceuticals, new targets

The chapter focused on the most relevant and clinically applied nuclides in molecular imaging of PET and SPECT, based on the most relevant radionuclides. However, there are numerous "exotic" radionuclides feasible for medical imaging, which show interesting results and applications. Generally, the availability of "exotic" radionuclides is limited to certain research centers. However, the development of new methods enabling the use of certain radionuclides in medical applications is of great importance and help to bring medical needs and radiopharmaceuticals closer together.

Additional efforts are being put into the optimization of known radiopharmaceuticals regarding their labeling strategies and automation as well as improvement of their pharmacological characteristics.

Developing novel radiopharmaceuticals is by far the most active research area in the field. This area is driven by aims of research groups and networks to enable molecular imaging of pathophysiology, where reliable and simple diagnostic methods are still lacking.

Before the development of new radiopharmaceuticals can start, suitable targets have to be identified. The beginning of radiopharmaceutical research is usually the discovery or identification of suitable targeting structures. Besides new transmembrane receptors such as the chemokine receptor 4 or structures like matrix metalloproteases, developments are even aiming for radiotracers addressing kinases and signaling pathways as general targets for multiple pathologies.

An excellent example with outstanding importance and impact in molecular imaging is the so-called prostate-specific membrane antigen (PSMA). The enormous impact ^{68}Ga-PSMA radiopharmaceuticals have had since about 2013 is a combination of the factors mentioned: the availability of the ^{68}Ge/^{68}Ga generator (available since ca. 2005), the application of a new chelate for ^{68}Ga (HBED), the adoption of the PSMA-targeting motif. PET imaging of prostate cancer using the ^{68}Ga- and ^{18}F-radiotracers is today established in several clinics as well as the corresponding therapeutic approaches using DOTA chelates labeled with ^{177}Lu, cf. Chapter 13.

The latest developments in radiopharmaceuticals, which have already proven great potential, are the novel FAP-targeting radiopharmaceuticals for PET. Besides the first successful reports on FAP-based *theranostic* approaches using PET/CT and radionuclide therapy (cf. Chapter 13) in various human cancers, molecular imaging of activated fibroblast is of further interest in diseases such as organ fibrosis.

12.8.2 Molecular imaging vs. therapy

Lively and broad research on novel radiopharmaceuticals is the solid base for molecular imaging and nuclear medicine. Especially the combination of diagnostics and therapy in so-called *theranostics* is a very powerful weapon against cancer and can make a substantial contribution to society's health. The noninvasive visualization of the organ distribution of radiotracers is primarily utilized for the diagnosis of a variety of diseases in patients: a radiotracer in a characteristic behavior reaches its biological target (such as a tumor, an organ of physiological malfunction, or regions of, e.g., the brain corresponding to the expression of specific neurotransmitters involved in a certain disease). However, this "diagnostic" information also provides upmost important information in the context of subsequent therapeutic protocols. In its simplest form, the quantitative identification of the tumor-specific uptake of an adequate tumor-visualizing radiotracer identifies the spatial expression of the tumor – which may next be resected by surgery.

Moreover, a radiotracer identifies that a certain biological target of a particular tumor is expressed (e.g., a transmembrane tumor receptor such as the somatostatin receptor for neuroendocrine tumors addressed by receptor antagonists as drugs) which consequently allows to apply the *clinically established* corresponding drug specific to this particular target. In addition, a diagnostic radiopharmaceutical represents the combination of a targeting moiety and a photon-emitting radionuclide. Supposing the targeting moiety remains the same, but a "therapeutic" one substitutes the "diagnostic" radionuclide, radiopharmaceuticals may directly be used to treat tumors and other diseases. This aspect is covered in detail in Chapter 13.

12.8.3 Drug development

Significant improvements in human longevity and health have been provided by the development of new pharmaceuticals for the treatment of various diseases in the last decades. During this time the research and development (R&D) of new drugs was also subject to significant alterations, especially by the introduction of good manufacturing practice (GMP) at the end of the 1930s. Later these guidelines were expanded to further improve quality management by the introduction of good laboratory practice (GLP) and good clinical practice (GCP), which have also increased the expenditure (time and money) required for the development of new pharmaceuticals. Although new methods for an efficient R&D have been established, such as databases for an improved identification of lead compounds, combinatorial chemistry, which drastically increased the number of compounds a chemist can synthesize per year, or high-throughput screening, which allowed a faster testing of compounds, the number of new molecular entities gaining approval steadily declined over the last decades.

Higher R&D efficiency is desirable, since today only 1–2 of every 10 000 substances synthesized in the basic research phase will successfully pass all stages of the critical path and finally achieve approval. Even in the clinical phases there is a high rate of failure, resulting in high financial expenditures until this point.[34] Molecular imaging is one of the key technologies to make drug development more efficient and economic. PET can provide fast access to *in vivo* data in the preclinical phase, to verify the proof of concept, or in the clinical phases, for example to determine the dosage of drugs or to determine the efficacy of the drug. This holds true for all biological targets and indications, in particular for oncology. The concept, however, is illustrated best for neuroleptics.

Commercial R&D had created generations of neuroleptic drugs with more and more improved action and fewer side effects. There are, among others, several key questions to answer in this development, such as:

(i) Does the new drug for a given orally administered amount (in mg or g) give a sufficient blood plasma concentration?

(ii) Is the drug metabolized and if yes, what is the pharmacology of the metabolites?

(iii) Does the drug penetrate the blood-brain barrier to a sufficient extent?

(iv) Does the new drug reach the biological target it was aimed at, such as a particular neuroreceptor?

Whereas points (i) and (ii) can directly be analyzed by taking blood samples and analyzing the concentration of the drug by HPLC or other technologies in laboratory animals and in patients, point (iii) can be determined in a similar way, but not in humans. Point (iv) is typically investigated through behavior experiments in laboratory animals by analyzing brain sections of those animals *ex vivo*, but direct access to the binding of the drug molecule to a brain receptor is impossible by means of invasive technologies in humans. This is the moment where noninvasive and quantitative molecular imaging with appropriate radiolabeled neurotransmitter analogs contributes in a unique way. Figure 12.62 schematically illustrates the concept. After crossing the blood-brain barrier a neuroleptic drug (NLD), also referred to as an antipsychotic (AP), reaches the postsynaptic (in other cases presynaptic) receptors. According to both its concentration and its receptor binding affinity, it competes with endogenous dopamine for binding to the dopamine receptors. With both dopamine and NLD occupying the receptors, the number of potential binding places for the radiolabeled neurotransmitter analog decreases. Consequently, less radioactivity is measured in that particular area.

In particular for antipsychotics, the entire *in vivo* properties at central nervous receptors should be investigated in order to predict better clinical effects. Figure 12.63

---

34 For example, only 12% of the compounds tested in the clinical phase II finally get approval, but to this point about 40% of the total R&D expenditures to develop a new drug have already been spent.

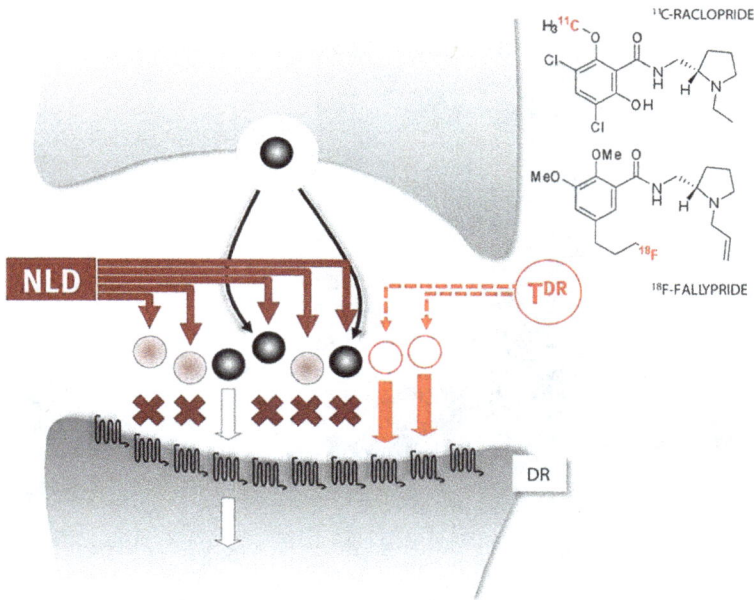

Fig. 12.62: Concept of how a radiolabeled neurotransmitter analog is used in R&D. A neuroleptic drug (NDL), after crossing the blood-brain barrier, reaches the postsynaptic receptors. According to both its concentration and its receptor binding affinity, it competes with endogeneous dopamine for binding to the dopamine receptors.[35] With both dopamine and NLD occupying the receptors, the number of potential binding places for the radiolabeled neurotransmitter analog decreases. Thus, whether the radiolabeled neurotransmitter analog's uptake is changed or not (or to what degree) is a measure of the potential of the drug candidate.

shows in a real study how the $D_{2/3}$ receptor-specific PET-tracer [^{18}F]FP is used to quantify *in vivo*, whether a certain NLD (clozapine, ziprasidone, quetiapine, amisulpride) targets $D_{2/3}$ receptors in healthy humans.

Future R&D focuses on the attempt to produce effective antipsychotics with less extra-pyramidal side effects (EPMS). For drug development of suitable agents, it was and is necessary, for example, to understand whether the absence of EPMS is simply a question of correct dosage or depends on additional properties. This requires additional studies as shown in Fig. 12.62 for $D_{2/3}$ receptors also for other clinically relevant neurotransmitter systems, such as serotonin, acetylcholine, and muscarinic. In each case, the corresponding PET tracer should be applied.

However, a suitable antipsychotic-to-EPMS ratio does not necessarily depend on those multireceptor binding profiles. The frequently used amisulpride was systematically evaluated by $D_{2/3}$ receptor PET including the use of high-affinity ligands. These

---

35 The same is true for competitive binding to presynaptic DAT.

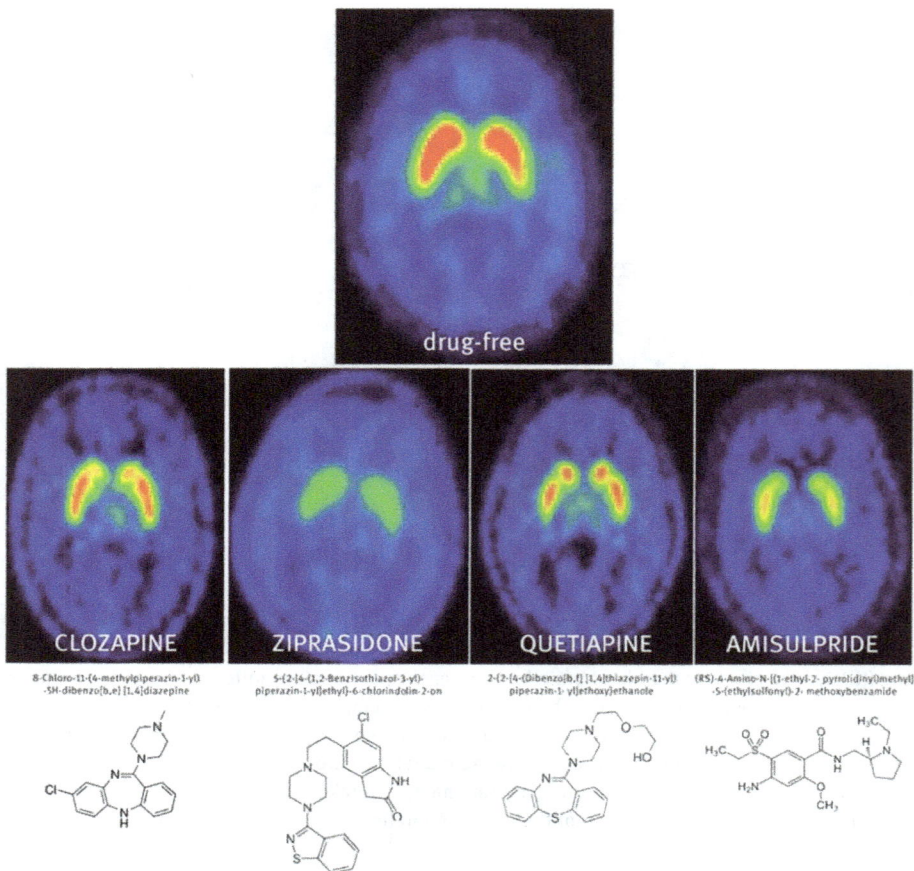

Fig. 12.63: Sagittal views of the striatum of healthy volunteers injected with the $D_{2/3}$ receptor-specific PET-tracer $[^{18}F]FP$. Upper image: distribution of the PET-tracer illustrating endogenous levels of receptor availability. Lower row: Volunteers following oral application of clinically relevant amounts of four neuroleptics: clozapine, ziprasidone, quetiapine, amisulpride. Efficient accumulation of the drug at $D_{2/3}$ receptors is detected by decreased radioactivity in striatal areas due to the blocking of $D_{2/3}$ receptors by the drug molecule, thereby minimizing the number of $D_{2/3}$ receptors available for $[^{18}F]FP$. Obviously, this is best achieved by ziprasidone while, e.g., clozapine has no effect on $[^{18}F]FP$ uptake. It thus seems to preferentially bind to other targets than $D_{2/3}$ receptors (which may be analyzed using PET tracers selective for those receptors). Courtesy G. Gründer, RWTH Aachen, Germany.

investigations were of highest interest since amisulpride shares atypical properties but is only able to act at $D_{2/3}$ receptors; thus, it does not depend on any 5-HT$_{2A}$ or M$_1$-antagonism. PET studies revealed that two mechanisms might contribute to this effect. Mainly based on the moderate affinity of amisulpride at the target receptors, the striatal plasma concentration vs. occupancy curve in the striatum was less steep in comparison to high-affinity drugs. In extrastriatal regions, the occupancy reached a profound intensity earlier; meanwhile, striatal occupancies were rather low.[36] In particular for amisuplride, ziprasidone, quetiapine, and clozapine profound differences between the regions could be detected. The highest differences were found for clozapine, which was found to bind early only moderately in the striatum. Nevertheless, it showed much higher (>20%) occupancies in extrastriatal regions (i.e., thalamus and cortex). Similar results could be obtained for quetiapine questioning the most important target site for AP action. A moderate to low affinity appears to be a relevant factor for "atypicality" even in absence of additional receptor binding properties. PET studies demonstrate that antipsychotics with moderate affinity show a broad range of plasma concentrations that provides the 65–80% occupancy window whereas high-affinity agents usually provide only a narrow range which in the individual case can hardly be met.

12.8.4 Therapeutic drug monitoring

Even with a well-studied receptor binding profile of a drug, the optimum, patient-individual dose should be determined. Therapeutic success is finally correlated with the receptor occupancy by the drug. Clinical dose finding for lurasidone, for example, was performed by [^{11}C]raclopride PET, revealing that more than 40 mg are necessary to provide effective $D_{2/3}$ receptor occupancies. Another PET investigation on lurasidone including a large group of schizophrenia patients, however, discovered that daily doses only modestly correlate with receptor occupancies, whereas plasma levels of the compound including its active metabolites did. It is well known in therapeutic drug monitoring (TDM) that daily doses may fail to reliably predict plasma levels as well as clinical response or side effects. Thus, the orally administered dose is not necessarily predictive for the concentration of the drug at its site of action, i.e., at "its" neuroreceptor target in the brain. Several patient-characteristic properties such as metabolism and penetrability of the blood-brain barrier make the final concentration "act" much lower than the totally administered one. This may be even less than required to achieve the therapeutic effect desired. In contrast, doses higher than needed cause significant side effects.

Molecular imaging with adequate radiolabeled neurotransmitter analogs is indeed able to quantify the receptor occupancy depending on the dose taken. Figure 12.64 illustrates the increasing occupation of postsynaptic $D_{2/3}$ receptors as analyzed by [^{18}F]DMFP PET. This in turn may contribute to modulate the daily dose to the receptor occupancy level needed.

---

36 Extrastriatal (cf. Fig. 12.55 for [^{18}F]FP) binding properties are reported for many second generation antipsychotics.

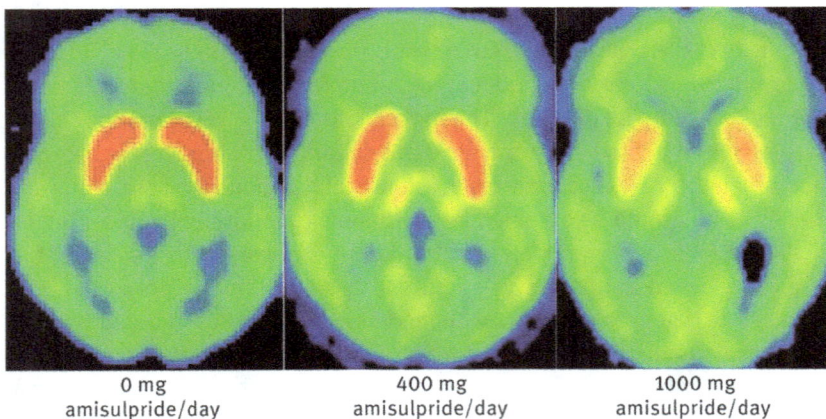

| 0 mg amisulpride/day | 400 mg amisulpride/day | 1000 mg amisulpride/day |

Fig. 12.64: Increasing occupation of postsynaptic $D_{2/3}$ receptors as analyzed by [^{18}F]DMFP PET depending on the daily dose of the neuroleptic drug amilsulpride.

Further reading

Bengel FM, Ross TL. Emerging imaging targets for infiltrative cardiomyopathy: Inflammation and fibrosis. J Nucl Cardiol 2019, 26, 208. https://doi.org/10.1007/s12350-018-1356-y

Gee AD, Herth MM, James ML, Korde A, Scott PJH, Vasdev N. Radionuclide imaging for neuroscience: Current opinion and future directions. Mol Imaging 2020, 19, 1536012120936397.

Deng X, Rong J, Wang L, Vasdev N, Zhang L, Josephson L, Liang SH. Chemistry for positron emission tomography: Recent advances in ^{11}C-, ^{18}F-, ^{13}N- and ^{15}O-labeling reactions. Angew Chem Int Ed 2019, 58, 2580–2605. 10.1002/anie.201805501

Grosu AL, Astner ST, Riedel E, et al. An interindividual comparison of O-(2-[^{18}F]fluoroethyl)-L-tyrosine (FET)- and L-[methyl-^{11}C]methionine (MET)-PET in patients with brain gliomas and metastases. Int J Radiat Oncol Biol Phys 2011, 81, 1049–1058.

Histed SH, Lindenberg ML, Mena E, Turkbey B, Choyke PL, Kurdziel KA. Review of functional/anatomical imaging in oncology. Nucl Med Commun 2012, 33, 349–361.

Imlimthan S, Moon ES, Rathke H, Afshar-Oromieh A, Rösch F, Rominger A, Gourni E. New frontiers in cancer imaging and therapy based on radiolabeled fibroblast activation protein inhibitors: A rational review and current progress. Pharmaceuticals 2021, 14, 1023. https://doi.org/10.3390/ph14101023

Jacobson O, Kiesewetter DO, Chen X. Fluorine-18 radiochemistry, labeling strategies and synthetic routes. Bioconjugate Chem 2015, 26, 1–18. x.doi.org/10.1021/bc500475e

Krishnan HS, Ma L, Vasdev N, Liang SH. ^{18}F-Labeling of sensitive biomolecules for Positron Emission Tomography. chemistry 2017, 23(62), 15553–15577. doi:10.1002/chem.201701581

Matthews PM, Rabiner EA, Passchier J, Gunn RG. Positron Emission Tomography for drug development. Br J Clin Pharmacol 2011, 73, 175–186.

Mason C, Gimblet GR, Lapi SE, Lewis JS. Novel tracers and radionuclides in PET imaging. Radiol Clin North Am 2021, 59(5), 887–918.

Neels OC, Kopka K, Liolios C, Afshar-Oromieh A. Radiolabeled PSMA inhibitors. Cancers 2021, 13, 6255. doi.org/10.3390/cancers13246255

Piel M, Rösch F. Radiopharmaceutical chemistry. In Gründer G. ed.. Molecular imaging in the clinical neurosciences. New York, Heidelberg, Dordrecht, London, Humana Press/ Springer, Springer, 2012, 41–73.

Pietzsch HJ, Mamat C, Müller C, Schibli R. Single photon emission computed tomography tracer. Recent Results Cancer Res 2020, 216, 227–282.

Price EW, Orvig C. Matching chelators to radiometals for radiopharmaceuticals. Chem Soc Rev 2014, 43(1), 260–290. doi:10.1039/c3cs60304k

Rösch F (ed.). Radiochemistry and radiopharmaceutical chemistry in life sciences. In: Vértes A, Nagy S, Klencsár Z, Lovas RG, Rösch F. eds. Vol. 4 of: Handbook of nuclear chemistry. 2nd ed. Berlin-Heidelberg, Springer, 2011.

Piel M, Vernaleken I, Rösch F. Positron Emission Tomography in CNS drug discovery and drug monitoring. J Med Chem 2014, 57(22), 9232–9258.

Ramogida CF, Orvig C. Tumour targeting with radiometals for diagnosis and therapy. Chem Commun 2013, 49, 4720–4739.

Ross TL, Ametamey SM. PET chemistry: An introduction. In Khalil MM. ed.. Basic sciences of nuclear medicine. Switzerland AG, Springer Nature, 2021, 131–176.

Ross TL, Ametamey SM, Chemistry: PET. Radiopharmaceuticals. In Khalil MM. ed.. Basic sciences of nuclear medicine. Switzerland AG, Springer Nature, 2021, 177–199.

Scroggie KR, Perkins MV, Chalker JM. Reaction of [^{18}F] fluoride at heteroatoms and metals for imaging of peptides and proteins by Positron Emission Tomography. Front Chem 2021, 9, 687678. doi:10.3389/fchem.2021.687678

Toscano M, Ricci M, Celletti C, Paoloni M, Ruggiero M, Viganò A, Jannini TB, Altarocca A, Liberatore M, Camerota F, Di Piero V. Motor recovery after stroke: From a Vespa Scooter ride over the Roman Sampietrini to focal muscle vibration (fMV) treatment. A 99mTc-HMPAO SPECT and neurophysiological case study. Front Neurol 2020, 11, 567833. https://doi.org/10.3389/fneur.2020.567833

van der Born D, Pees A, Poot AJ, Orru RVA, Windhorst AD, Vugts DJ. Fluorine-18 labelled building blocks for PET tracer synthesis. Chem Soc Rev 2017, 46, 4709.

Wong DF, Gruender G, Brasic JR. Brain imaging research: Does the science serve clinical practice?. Int Rev Psychiatry 2007, 19, 541–558.

Cathy S. Cutler

13 Life sciences: Therapy

Aim: The aim of this chapter is to explain the basic theory that underlies the use of radioactivity for therapeutic applications. Radiotherapy has taken on many forms from the use of external beams, to brachytherapy, use of small molecules to the more advanced methods of targeted radiotherapy and the emerging field of nanotechnology. The strategy of all these methods is to deliver an optimum radiation dose (ionization) to the diseased site while minimizing the dose encountered by normal tissues. The methods used for systemic "endoradiotherapy" have been built on the concepts of systematically probing a biological target through the chemical and radiochemical design of the appropriate radiopharmaceutical originally developed for noninvasive imaging. However, to turn a "radiodiagnostic" into a "radiotherapeutic", the "diagnostic" radionuclide must be substituted by a "therapeutic" one. The selection criteria thus are the particle emission (electron and/or α-particles) rather than electromagnetic radiation (γ-photons). However, the constraints for radiotherapy are far greater than those for imaging which requires merely a high concentration of the radionuclide in an area of interest for a period of time long enough for imaging. Due to the more destructive nature of radionuclides used for therapy and the higher doses administered it is necessary to achieve a high uptake at the diseased site with little uptake in normal tissues. Dosimetry is more critical and often is done via imaging and the emerging field of theranostics. Here we'll discuss what is important to consider when inducing a biological effect with radioactivity, what therapeutic radionuclides are used, and how they are delivered selectively to the disease site.

13.1 Introduction

Today radiopharmaceuticals are largely used for both diagnosis and therapy. The largest use of radiopharmaceuticals is for diagnosis where it is estimated that over 30 million procedures are performed annually worldwide. In diagnosis, the radionuclide normally transforms giving off a photon with energy that is energetic enough to escape the body and be detected by an external detector array. In diagnosis, most radiopharmaceuticals are used as tracers to noninvasively image physiological function on a molecular level. The tracer principle was developed by George de HEVESEY who recognized that radioactivity could be detected at very low levels and that radioactive isotopes do not differ chemically from cold or nonradioactive isotopes. Thus, small amounts of molecules (nano to picomolar) that have been conjugated with radioactive isotopes can be injected that do not interfere or perturb biochemical reactions and do not elicit a pharmacological effect. The second principal commonly

https://doi.org/10.1515/9783110742701-013

used was that developed by P. EHRLICH termed the "magic bullet". EHRLICH[1] recognized that most drugs failed due to their negative effects or toxicity to normal tissues. Thus, if drugs could be formulated that were highly selective for diseased tissue over normal tissue, they would be highly effective at ablating the disease. He termed this approach the *magic bullet*, which is now a commonly used mechanism for delivering toxins to diseased tissue. The toxin, in this case the radionuclide, is attached (stably) to the targeting vector that will selectively deliver its cargo to the diseased tissue. A number of vectors have been utilized including small molecules such as proteins, and large peptides such as monoclonal antibodies to the newly emerging field of nanoparticles.

13.2 Historical background

A number of "mysterious discoveries" lead to the utilization of radioactivity in medicine. The first was the discovery of X-rays by RÖNTGEN at the University of Würzburg in 1895. He discovered that a new kind of rays given off by a gas discharge tube could blacken photographic film. In 1896 he publicly demonstrated their use by imaging a hand showing how X-rays could be used to view the underlying bone structure. In 1896 a physics professor at Vanderbilt University, J. DANIELS, demonstrated the first biological effect of ionizing radiation by removing hair on the scalp using X-ray irradiation. The first published therapeutic use of X-rays was in 1896 by L. FREUND at the Medical University who demonstrated the eradication of a hairy mole after irradiation by X-rays.

H. BECQUEREL in 1896 was the first to discover that uranium compounds could give off "radioactivity". This discovery prompted M. and P. CURIE to investigate these minerals and led to the eventual isolation of the radionuclide's polonium and radium in 1898. It was not long after the discovery of radioactivity that its use for treating malignant diseases such as cancer was understood and actively being pursued. Madame CURIE spent a great deal of time trying to garner a large supply of radium which she believed would aid in the eradication of certain types of cancer. The first tumor therapy occurred in 1899 in Sweden by STENBECK and SJOGREN who removed a tumor from a patient's nose with X-rays.

Initially, however, radioactivity was not applied systematically by injecting a well-defined chemical and pharmacologically "active" species into the blood stream of a patient – it was rather used as an "external" source to treat tumors, cf. Fig. 13.1.

G. BELL then suggested radium could be used for the treatment of cancer by direct implantation into the tumor. In 1903 in the *New York Journal* a physician from

---

1 Nobel Prize in Physiology or Medicine 1908 awarded jointly to Ilya Ilyich MECHNIKOV and Paul EHRLICH "in recognition of their work on immunity".

Fig. 13.1: Cancer treatment with a "radium bomb" in 1917. The α-emitting radium isotopes used to treat tumors by external irradiation. Science Photo Library www.sciencephoto.com/media/ 300040/view/N060/0042.

Chicago reported on the use of X-rays to ameliorate leukemia. In 1937 J. LAWRENCE performed the first ^{32}P treatment of leukemia. It was the first use of artificially produced radioactivity to treat a patient. J. HAMILTON followed this in 1931, with the use of iodine-131 for the diagnosis of disease. In 1946 S. SEIDLIN was able to cure all metastases associated with thyroid cancer using ^{131}I for therapy. In the same year C. PECHER used ^{89}Sr to treat metastatic bone cancer.

13.3 Classifications

Radiotherapy can be used alone as a primary cure for cancer or in combination with surgery and chemotherapy or it may be used as a palliative treatment. There are three ways that radiation is used to treat cancer: external beam, brachytherapy, and use of systemic radioisotope treatments.

13.3.1 External radiation therapy

External radiotherapy involves a beam of radiation directed outside the body to irradiate the cancerous mass with a precisely controlled, targeted beam of electromagnetic radiation (X-rays or γ-photons) generated through machines or originating from radionuclide sources (such as historically shown in Fig. 13.1); or particles (electrons, protons, and large ions) produced from accelerators. Much of external radiation therapy is still done with megavoltage-energy beams of electromagnetic radiation. Skin cancer and structures close to the surface are treated by kilovoltage electrons, whereas tumors located deep inside the body are treated by megavoltage X-rays. For photons, a targeted tumor site within the body would receive a certain dose, but both the pre- and post-tumor ("up- and downstream"), i.e., healthy tissue, additionally receive radiation doses. Changing the angular distribution of the "beam", a second and a third external irradiation would create cumulative radiation doses within the tumor tissue. Proton therapy uses a synchrotron or a cyclotron to create the protons, which are then guided by magnets to the patient tumor site, with tumors within the eye being the main targets. In addition, external radiation therapy is very common for brain tumors – a target which is rather difficult to remove surgically. The advantage of accelerated charged particles compared to electromagnetic radiation or neutrons is the improved dosimetric profile due to the BRAGG effect: Charged particles deposit a maximum of interaction with condensed matter when approaching rest energy – and this parallels the degree of ionization and radiation dose, cf. Fig. 13.2. In addition, changing the angular distribution of the "beam", a second and a third external irradiation would create cumulative radiation doses.

External beam therapy is normally given in small doses over several cycles requiring normally daily or weekly visits. A challenge for external beam therapy is that every tissue that lies along the path of the beam also receives a dose, which may require some time to heal between treatments to reduce toxic side effects. Multifield or intensity modulated *teletherapy* can result in superficial dose sparing with a deep dose to tumor that is high. This is accomplished by using multiple fields, which intersect at the tumor thereby sparing dose to normal tissues and can be used to shape the dose deposition profile. Use of a special filtering system allows the effective energy of the beam to vary across the surface of the beam enabling a more accurate shaping of the dose volume. This requires specialized computer planning and specialized equipment and requires more time.

13.3.2 Brachytherapy

Another common form of radiation treatment called "brachytherapy" aims to reduce the dose to normal or surrounding tissues while at the same time delivering a high dose to tumor. Brachytherapy, often called "internal source of radiation",

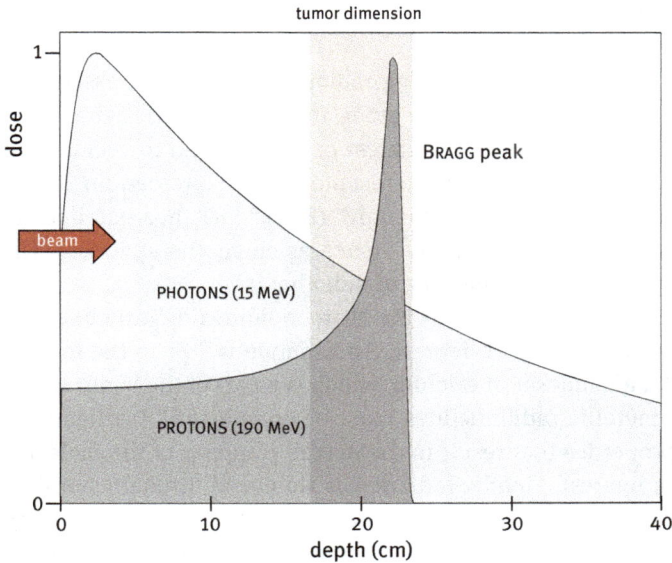

Fig. 13.2: Distribution of radiation doses along the path of photons and protons (starting from a virtual 100% level) with increasing penetration depth in tissue. For photons, a tumor positioned within a depth of, e.g., 17–23 cm would receive a certain dose, but both the pre- and post-tumor ("up- and downstream"), i.e., healthy tissue, receive radiation doses. For protons, the relative radiation dose in tumor tissue would be larger than that in the healthy tissue because of the Bragg peak, and the maximum depth is 25 cm for this particular kinetic energy.

involves placing radioactive material, often referred to as *seeds*, either directly into the tumor or very close to it. Widely used in clinical practice brachytherapy results in the tumor receiving a very high dose and the normal tissues being protected as the dose drops off by the inverse square and the dose margin can be as small as 5 mm. Brachytherapy involves using sealed sources that can be made of a variety of radionuclides. Typically, they are made of radionuclides such as ^{60}Co, ^{137}Cs, ^{192}Ir, ^{125}I, or ^{103}Pd. Most often the effect is usually from the β^--particles, although γ-emissions have been used.

Total dose is delivered faster than teletherapy and normally in one to a few sessions requiring fewer visits and therefore is much more convenient for the patient. Some disadvantages are that the tumor needs to be well localized, only applicable for small tumors and placement can be very labor intensive. Treatments can range from a single treatment to multiple. Dose calculation tends to be more difficult and sources may move and/or be lost.

13.3.3 Endoradiotherapy

The third type of radiotherapy involves injecting radiopharmaceuticals, essentially tumor-avid "drugs" that contain a radioactive atom, that are either designed such to be accumulated in the tissue or organ of interest or are attached to vectors that selectively target for instance transmembrane receptors overexpressed on cancer cells. This is often referred to as "endoradiotherapy" (ERT).[2] ERT thus parallels radiodiagnostic protocols, as both are "systemic" (such as chemotherapeutics). The rest of the chapter will concentrate on this type of radiotherapy.

Only a few compounds can be injected as the "bare radionuclide" and be selectively trapped in the tissue or organ of interest. An example is ^{89}Sr in the form of strontium chloride, $SrCl_2$, a mimicker of calcium, which is a part of the hydroxyapatite bone matrix. Most metallic radionuclides must be complexed by a ligand to form a molecule with properties that result in the overall trapping of the molecule in the tissue or organ of interest. Finally, radiometals stabilized through coordination chemistry by adequate chelate-type bifunctional ligands can be attached to biological targeting vectors such as peptides and monoclonal antibodies. In turn, those "targeting vectors" systemically deliver the radionuclide to the target site similar to EHRLICH's *magic bullets*; see also below.

13.4 "Therapeutic" particles and corresponding radionuclides

13.4.1 Particles: Types of particles and their origin

The biological basis of radiotherapy is that cancer cells are dividing more rapidly than normal cells and that they are less able to tolerate DNA strand breaks and will thereby lose their ability to replicate. This local ionization causing the DNA strand breaks is most effectively achieved by particles emitted from radionuclei. In order to focus the therapeutic action of the radiation to the biological target, two parameters are most relevant: (1) suitable range of the radioactive emission and (2) a high linear energy transfer. The appropriate range refers to the dimension of the biological target, representing a tumor down to single cells. This criterion is exclusively met by particular radiations in contrast to electromagnetic radiation (photons, X-rays). Nonelectromagnetic emissions in radioactive transformations are the electron

---

2 "Endo" refers to the systemic approach of delivering the radiotherapeutical and its action *in vivo*. In parallel to ERT, wordings such as PRRT (peptide receptor radiation therapy) are common.

(β⁻-particles, IC electrons, AUGER electrons) and the α-particle.[3] Those are the weapons applied to fight cancer by means of ERT.

What is the origin of these two sorts of particles? For α-particles, obviously this is the primary process of α-transformation, cf. Chapter 9 of Vol. I. The therapeutically relevant radionuclides releasing the α-particles are typically[4] isotopes of elements starting from lead and bismuth ($Z = 82$ and 83) and range to actinium ($Z = 89$). They are members of the naturally occurring transformation chains, cf. Chapter 5 of Vol. I, but can also be produced artificially through particle-induced nuclear reactions, cf. Chapter 13 of Vol. I. Discrete α-particle energies are between ca. 4 and 8 MeV.

In the case of electrons, in contrast, there are several sources. The first one is the primary process of β⁻-transformation of neutron-rich unstable nuclei, cf. Chapter 8 of Vol. I. Those nuclides represent the largest group of therapeutically relevant radioisotopes. The β⁻-particles emitted have a continuous distribution of their kinetic energy with a certain maximum value, which is in the range of ca. 0.5 to ca. 2.2 MeV. Another source of electrons are those released as conversion electrons within the secondary transformation processes, cf. Chapter 11 of Vol. I. Those IC electrons are emitted from mostly K- and L-shell orbitals and have a discrete kinetic energy. The absolute value of E_{IC} is lower than $E^{max} \beta^-$ values by ca. 2 orders of magnitude (ca. 10–100 keV). The final source of electrons is due to the group of post-processes of radioactive transformations. It starts with a vacancy in the K- or L-electron shell as created either by the secondary IC process mentioned above or by the primary transformation process of electron capture in proton-rich unstable nuclei. The complex events of re-filling the initial hole in the inner electron shell is accompanied by the release of a cascade of AUGER electrons as described in Chapter 12 of Vol. I. Several individual electrons are ejected, each of discrete kinetic energy ranging from ca. 10 keV down to several eV.

13.4.2 Energy of particles and their range in water

The electron and the α-particle differ in charge, mass, and kinetic energy. Concerning mass, despite its high kinetic energy of ca. 5 ± 2 MeV, the "heavy" α-particle shows a much more intense interaction with surrounding matter, and consequently, its range is short. Electrons of similar energy would traverse the same matter by distances that are orders of magnitude larger.

There is even a difference in the "travel profile". Beta particles differ from alpha particles in that they do not travel in straight lines but have treacherous paths and

---

3 The neutrons released in the course of the spontaneous fission are not relevant in a medical context.

4 There are just a very low number of lighter or heavier α-emitters, such as ¹⁴⁹Tb.

travel much farther than α-particles. They are considered low *linear energy transfer* or low LET as they transfer low levels of energy along their path depositing most of their energy at the end of their path length. Whereas α-particles travel 80–100 μm, beta particles can travel from 1–12 mm in tissue. Shown in Fig. 13.3 is a comparison of the path taken by an α-particle to that of a path traveled by a β-particle.

Fig. 13.3: Schematic illustration of the path of an α-particle and a β-particle.

In addition, electrons used in ERT have lower kinetic energies that are dependent on the radioactive transformation processes they are generated from. Consequently, there are different ranges among the same group as well as between the three different types of electrons. Figure 13.4 lists maximum β⁻-energies for relevant therapeutic β⁻-emitters. Figures 13.5 and 13.6 schematically compare the range of the various particles in water, given either as "real" distance or in relation to the dimension of a cell.

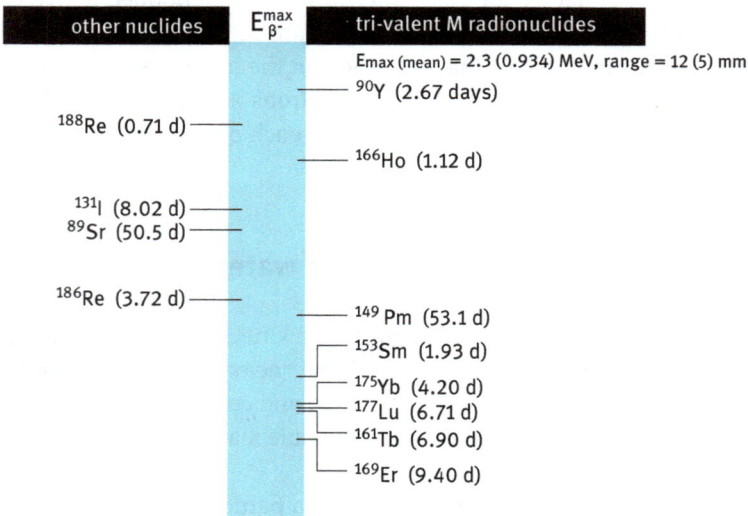

Fig. 13.4: Maximum β⁻-energies for relevant therapeutic β⁻-emitters, catalogued by their chemistry: Trivalent metals right, others left. Half-lives are in brackets.

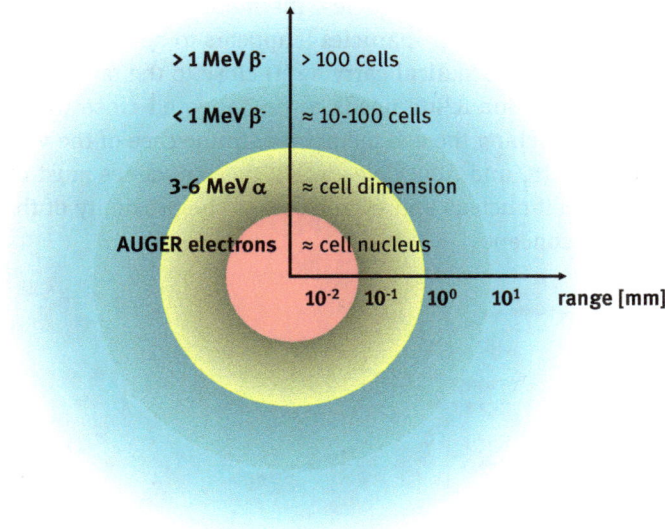

Fig. 13.5: Particle radiative emissions used in radiotherapy and their range in water dependent on type and energy of the particle. The range is also given in terms of cell diameters, assuming an average cell diameter of 20 µm.

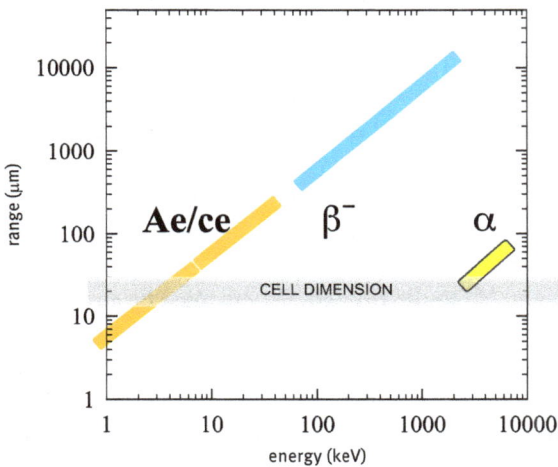

Fig. 13.6: Electron and α-particle ranges in water are dependent on the type and energy of the particle (ce = conversion electrons, Ae = Auger electrons). Assuming an average cell diameter of 20 µm, both low-energetic electron and α-particle ranges may have ranges less than one cell diameter.[5]

---

5 For β⁻-particles note that the energy given refers to the maximum kinetic energy. Average values or generally 1/3 of it.

With this varying range for therapeutic particles, the mode of delivery must vary as well to be effective. For long-range β^--particles it appears to be sufficient to deliver the radionuclide (or the pharmaceutical carrying it) close to the target cell. For short-range α-particles, the radionuclide needs to be delivered close to the tumor cell nucleus, at least approaching the cell membrane. In the case of the very short-range low-energy electrons (IC and AUGER electrons), those nuclides must be delivered directly to the tumor cell nucleus and even closer to the proximity of the DNA. Figure 13.7 illustrates this concept.

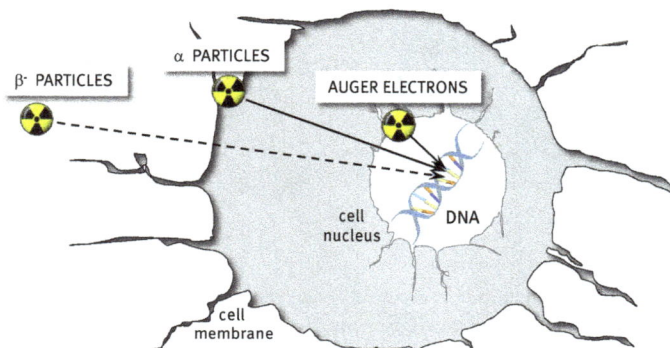

Fig. 13.7: Different ranges of the particles require different specific biochemical/biological targeting strategies.

13.4.3 Radionuclides: Their availability and their chemistry

Another factor that can dictate which radioisotope to use is the chemistry of the radioisotope and how to complex it or stably bind it. Certain radionuclides can be administered close to their native form and will show uptake to the desired site naturally such as ^{89}Sr to the bone and ^{131}I to the thyroid. Others must be complexed using chelates or prosthetic groups and others may need to be complexed in microspheres or nanoparticles to provide high *in vivo* stability. Some metals have very rich chemistry such as rhenium with a large number of oxidation states that allows for a wide variety of complexes to be formed and with redox chemistry that if not controlled may be accessible to proteins *in vivo* and cause uptake in normal tissues. Rhenium, for instance, prefers to be in higher oxidation states and if not stably complexed can oxidize up to perrhenate which then can result in lower dose to the intended site. In certain applications such as palliation of bone pain this mechanism has been shown to lead to lower normal bone uptake and dose over time. Certain radioisotopes require very bulky groups that when conjugated to a biomolecule will dramatically change its *in vivo* behavior and thus are not effective. Others can be metabolized *in vivo* resulting in release and trapping in normal organs. Iodine for instance is often introduced by complexing it to functional groups of normal amino acids in proteins and peptides. These peptides and proteins can be degraded by enzymes to the individual amino acids, which are then released back into the body. Due to this when labeling peptides and proteins it is important to use attachment mechanisms that are residualizing or known to stay in the cell.

Finally, the cost and availability of the radionuclide must be considered. Many therapeutic isotopes can be produced most economically at medium and high-flux reactors (neutron flux $> 4 \cdot 10^{14}$ n/cm^{2}s^{-1}) in neutron-induced nuclear reactions. A growing number of therapeutic radionuclides, charged-particle-induced nuclear reaction processes are needed, which are performed at dedicated accelerator sites.

Despite the direct production routes, radionuclide generators are ideal for radionuclide production as they provide a portable radionuclide as needed, at a relatively low cost, and most have a long shelf-life. Few therapeutic radionuclides are available in generator form. The most notable generators of a therapeutic radionuclide are the ^{90}Sr/^{90}Y generator and the ^{188}W/^{188}Re generator. Another therapeutic radionuclide ^{213}Bi is obtained from its parent nuclide ^{225}Ac through a radionuclide generator.

In the following, the most relevant radionuclides used in ERT are discussed. They are grouped within β^-/electron emitters and α-emitters. Only those nuclides clinically used are listed.

13.4.4 β^-/electron emitters

Selected examples are listed in Tab. 13.1 listed by mass number. Prominent examples of radionuclides that transform via β^--transformation are ^{131}I, ^{177}Lu, and ^{90}Y. These and more are and will be discussed below in terms of nuclear transformation parameters, production pathways and other relevant aspects.

13.4.4.1 Iodine-131

^{131}I was originally used for diagnosis but now is more commonly used for radiotherapy. It has a half-life of 8.1 days and transforms 89% of the time giving off a maximum β^--particle energy of 0.61 MeV (average of 0.19 MeV) and several γ-rays such as 362 keV.[6] Its average β^--range in tissue is 0.8 mm. ^{131}I transforms to stable ^{131}Xe.

---

6 For the β^--transformation scheme and a γ-spectrum of ^{131}I see Chapters 8 and 11 (Figs. 11.11 and 11.12) of Vol. I, respectively.

Tab. 13.1: Radionuclides that transform by β⁻-emission commonly used in radiotherapy (arranged by increasing mass number).

Radionuclide	t½ (h)	E^max(β⁻) (MeV)	Accompanying γ-emission (keV (%))	Mean tissue range (mm)
^{64}Cu	12.7	0.58	1346 (47.5)	0.19
^{67}Cu	61.9	0.58	184.6 (46.7), 93.3 (16.6)	0.19
^{90}Y	64.1	2.28	–	1.1
^{105}Rh	35.4	0.57	319 (19.6)	0.19
^{131}I	192	0.61	364 (81)	0.21
^{149}Pm	53.1	1.10	286 (3)	0.43
^{153}Sm	46.3	0.81	103 (28.3)	0.30
^{161}Tb	165.8	0.59	48.9 (17), 74.6 (10.2)	0.20
^{166}Ho	26.8	1.86	80.6 (6.2), 1379 (1.1)	0.84
^{177}Lu	161	0.50	208 (11), 113 (6.6)	0.16
^{186}Re	89.2	1.10	137 (9)	0.43
^{188}Re	16.9	2.10	155 (15)	0.98
^{198}Au	64.7	0.96	412 (95.6)	0.38
^{199}Au	75.4	0.45	208 (9.1), 158 (40)	0.14

There are two production pathways to obtain ^{131}I. The first is from the neutron-induced fission of ^{235}U, which results in a 3% production rate. However, it is accompanied by the production of other iodine isotopes particularly stable forms, thus contains carrier. This is not an issue when used for the treatment of thyroid cancer; however, it is an issue in the use of targeted therapy in which the carrier requires additional targeting vector and saturation of the receptor sites maybe an issue. A second method of production is the neutron irradiation of tellurium-130, which then forms ^{131}Te (t½ = 25 min), which further transforms to ^{131}I. This route of production has a very low cross section and thus results in a lower amount of activity, however upon separation from the tellurium target material it results in a no-carrier-added ^{131}I.

The use of ^{131}I for medical applications was started back in the 1930s and ^{131}I is still one of the most highly used radionuclides for radiotherapy today. Radioactive isotopes of iodine are often used to both diagnose and treat hyperthyroidism and different types of thyroid cancer. An initial use of radioiodine was for the treatment of hyperthyroidism caused by Graves's disease. Therapy normally consists of oral administration of ^{131}I in the form of sodium iodide. In this case, radioiodine is one of the few radionuclides that can be administered as the bare radionuclide. It was the first and probably only bare radionuclide that behaves as a true "magic bullet" in that it is taken up selectively in the hyperthyroid. Another common iodine therapy treatment is metaiodobenzylguanidine (MIBG) typically used to deliver targeted therapy to treat high-risk neuroblastoma (called neuroendocrine tumors) in children and infants. The iodinated MIBG is selectively taken up by the neuroblastoma cells resulting in selective dose delivery to the cancerous cells.

13.4.4.2 Phosphorous-32

^{32}P is a pure β^--emitting radionuclide, which has a half-life of 13.29 days and maximum beta energy of 1.71 MeV and mean beta energy of 0.70 MeV. The range in tissue is 3–8 mm. It is readily available and cheaply produced in medium flux reactors by neutron irradiation of sulfur within a ^{32}S(n,p)^{32}P process, and thus allows for chemical isolation. It has been used for over 50 years for radiotherapy.

13.4.4.3 Strontium-89

^{89}Sr is a pure beta emitter with a relatively long half-life of 50.5 days and maximum beta energy of 1.46 MeV and range in tissue of 2.4–6.7 mm. There are two routes of production, one from the thermal neutron irradiation of ^{88}Sr to form low specific activity ^{89}Sr, the second involves the irradiation of ^{89}Y with fast neutrons in a fast flux reactor:

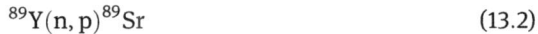

$$^{88}\text{Sr}(n, \gamma)^{89}\text{Sr} \tag{13.1}$$

$$^{89}\text{Y}(n, p)^{89}\text{Sr} \tag{13.2}$$

Both reactions have fairly low cross sections of 0.005 thermal and 0.06 resonance for eq. (13.1) and $1.7 \cdot 10^{-14}$ barns for eq. (13.2). Strontium is an alkaline earth and is used for bone palliation, as it is chemically and biologically very similar to calcium and is administered as the chloride salt (SrCl_2). It exchanges with calcium in the hydroxyapatite of bone.

13.4.4.4 Yttrium-90

^{90}Y is a pure beta emitter with an average beta range of 0.94 MeV (maximum of 2.28 MeV). It transforms with a half-life of 64.2 h to stable ^{90}Zr. It has an average range in tissue of 2.5 mm (maximum range 11 mm). It transfers 95% of its energy to tissue within 5.94 mm. It has been calculated that 27 mCi of ^{90}Y results in a dose of 50 Gy/kg.

There are currently two ways in which ^{90}Y can be produced resulting in high or low specific activity material. The low specific activity route is by neutron irradiation of ^{89}Y by thermal neutrons in a nuclear reactor, where ^{90}Y cannot be isolated from ^{89}Y target material:[7]

---

[7] This is the route commonly used to produce ^{90}Y for Theraspheres®, a glass microsphere that contains ^{90}Y that is used to treat inoperable liver cancer.

$$^{89}\text{Y}(n,\gamma)^{90}\text{Y} \tag{13.3}$$

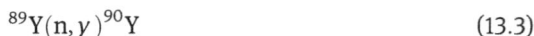

The second route, which produces high specific activity ^{90}Y, is produced by β^--transformation of ^{90}Sr. The generator parent radionuclide is obtained from ^{235}U fission. The ^{90}Sr has a long half-life of 28.8 a, and the system $^{90}\text{Sr}/^{90}\text{Y}$ represents a secular radionuclide generator. A variety of generator designs have been developed including liquid-liquid extraction generators, ion exchange-based generators, single stage or two stage supported liquid membrane (SLM) generators and an electrochemical generator. Most ^{90}Y is provided via a vendor that elutes the generator and then supplies the doses of ^{90}Y to central radiopharmacies and hospitals for formulation of the ^{90}Y radiopharmaceuticals.

13.4.4.5 Copper-64 and copper-67

There are two major radionuclides of Cu that are of interest for radiotherapy applications: ^{64}Cu and ^{67}Cu. Copper-67 with a half-life of 2.58 days has long been of interest because of its intermediate energy β^--particles (141 keV average energy), and imageable photons of 184.6 (46.7%), 91.27 (7.3%), and 93.31 (16.6%) keV for both antibodies and peptides. Its half-life is sufficient for the uptake kinetics of many monoclonal antibodies and other carriers administered *in vivo*. It also has easy chemistry for labeling. It is produced on a high-energy accelerator by bombarding an enriched ^{68}Zn target with 193 MeV high-energy protons by the following nuclear reaction:

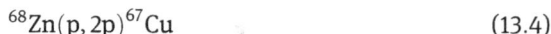

$$^{68}\text{Zn}(p,2p)^{67}\text{Cu} \tag{13.4}$$

This reaction also results in the production of ^{64}Cu with a half-life of 12.7 h and ^{61}Cu with a half-life of 3.4 h and other radionuclides, which cannot be removed by purification. The ^{61}Cu rapidly transforms away and is not a major concern. Although there is significant interest in ^{67}Cu, its use has been curtailed by its low availability. Currently production is limited to only a few sites worldwide that make approximately 3700 MBq (100 mCi) per month.

Recently production has been performed through the photonuclear reaction using bremsstrahlung photons with an electron accelerator by bombarding an enriched ^{68}Zn target with an e-LINAC at 27 MeV or higher by the following nuclear reaction:

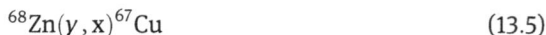

$$^{68}\text{Zn}(\gamma,x)^{67}\text{Cu} \tag{13.5}$$

This reaction does not result in the co-production of ^{64}Cu observed by the high-energy proton route. For the photonuclear route, the use of enriched target material is required to avoid the co-production of ^{64}Cu and zinc radionuclide impurities. This route of production can provide curie quantities of material in very high specific activity and is being used to support clinical trials of a ^{67}Cu radiopharmaceutical for treating neuroendocrine and prostate tumors. A challenge with this route is the need for large-enriched zinc-68 targets requiring the target material to be recycled to make this route cost-effective.

^{67}Cu can also be produced in a reactor by irradiation of enriched ^{67}Zn oxide by the nuclear reaction:

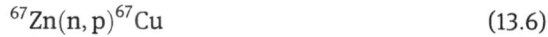

$$^{67}Zn(n, p)^{67}Cu \qquad (13.6)$$

This reaction, however, is low yielding and results in the production of the long-lived ^{65}Zn, resulting in expensive waste disposal.

^{64}Cu transforms to 19% by positron emission (0.653 MeV), 40% by β^- emission (0.579 MeV), and 41% by electron capture, and can be used for both imaging and therapeutic applications. It has a half-life of 12.7 h and can be produced by proton irradiation of an enriched nickel target following the nuclear reaction:

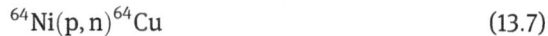

$$^{64}Ni(p, n)^{64}Cu \qquad (13.7)$$

Copper has two oxidation states of importance to radiopharmaceuticals: Cu^I and Cu^{II}, with the latter being more prevalent as it is more kinetically inert. For *in vivo* studies, kinetic inertness is more important than thermodynamic stability. Copper has been shown to have greater kinetic inertness and *in vivo* stability when complexed with macrocyclic chelators than with linear polyamino carboxylates. Copper forms compounds with coordination numbers ranging from four to six and prefers binding to nitrogen- and oxygen containing functional groups. A number of Cu compounds result in a reduction of Cu^{II} to Cu^I *in vivo* and subsequent loss of the Cu, which can then be complexed by proteins such as superoxide dismutase (SOD) and result in accumulation of Cu in the liver.

13.4.4.6 Lutetium-177

^{177}Lu transforms via β^- emission to stable ^{177}Hf. Due to its long half-life, availability and low-energy β^- particles (max. 0.49 MeV) as well as imageable gamma rays (208 (11%) and 113 (6.6%) keV), ^{177}Lu is being investigated for a number of therapeutic applications. The low energy of the β^- particle results in the majority of the dose remaining localized in small areas, which has been shown to be ideal for metastases and in minimizing kidney dose. The 6.64-day half-life is long enough that it can be attached to biomolecules with short or long biological half-lives and allows for distribution to researchers or hospitals in remote regions.

Although there are methods that can produce ^{177}Lu via a cyclotron, it is most commonly produced in a nuclear reactor. There are currently two methods available to produce ^{177}Lu: direct and indirect. Due to the large cross section of ^{176}Lu (2100 b) for the (n,γ) reaction, direct neutron activation of isotopically enriched[8] ^{176}Lu:

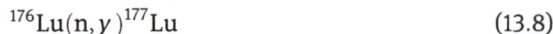

$$^{176}Lu(n, \gamma)^{177}Lu \tag{13.8}$$

The nuclear reaction in a medium flux reactor results in 20–30% of the ^{176}Lu atoms being converted to ^{177}Lu, yielding a specific activity of 740–1110 GBq/mg (20–30 Ci/mg). Higher specific activities can be achieved by irradiation in higher flux reactors,[9] achieving specific activities of 1850–2405 GBq/mg (50–85 Ci/mg). A disadvantage of the direct approach is the production of the long-lived impurity ^{177m}Lu with a half-life of 160 days.

^{177}Lu can also be produced by an indirect method: neutron capture followed by beta transformation of a parent radioisotope to the desired daughter radioisotope. For example, neutron activation of enriched ^{177}Yb produces ^{177}Yb (1.9 h half-life), which is followed by beta transformation to produce ^{177}Lu.

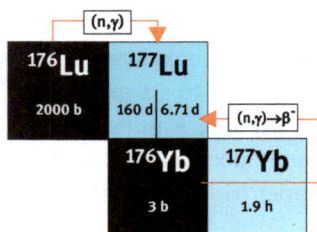

$$^{176}Yb(n, \gamma)^{177}Yb \overset{(\beta^-)}{\rightarrow} 177Lu \tag{13.9}$$

Fig. 13.8: Comparison of the two production pathways leading to ^{177}Lu. Whereas ^{177}Lu cannot be isolated from the target lutetium isotope and the metastable and long-lived ^{177m}Lu is co-produced at low level, neutron capture on ^{176}Yb forms ^{177}Yb of short half-life, instantaneously transforming via β⁻transformation to the ground state of ^{177}Lu exclusively. The latter process requires (and allows) chemical isolation of almost carrier-free ^{177}Lu from ^{176}Yb.

---

8 The natural abundance of ^{176}Lu is low, only 2.6%.
9 Such as the HIFR reactor at Oak Ridge National Laboratory, USA, or at reactors in Dmitrovgrad, Russian Federation.

Separation of the daughter material from the parent is possible (cf. Chapter 7 for radiochemical separations) and comes with the distinct advantage that nearly all of the lutetium atoms produced are radioactive. Further, the indirect production route does not produce the long-lived ^{177m}Lu impurity because the beta transformation has no branching through the metastable isomer. Separation from the Yb target produces a high specific activity ^{177}Lu product, on the order of 4107 GBq/mg (111 Ci/mg), which gives a carrier-free product requiring less biolocalization agent to achieve a given patient dose (because there are fewer nonradioactive lutetium atoms competing with ^{177}Lu). Figure 13.8 compares the two neutron-capture pathways.

13.4.4.7 Samarium-153

Samarium, like lutetium, is also a member of the lanthanide series. Radioisotopes of Sm can be produced via reactor or accelerator. The most common radioisotope, ^{153}Sm, has a half-life of 46.27 h and emits β^- particles with a maximum energy of 0.8 MeV and a mean energy of 0.23 MeV, with an average soft-tissue range of 0.3 mm. It emits an imageable gamma ray of 103 keV (28.3%), which can be used for monitoring distribution as well as calculating dosimetry. It is produced by neutron irradiation of an enriched ^{152}Sm target.

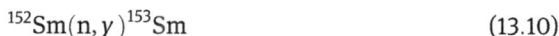

$$^{152}\text{Sm}(n, \gamma)^{153}\text{Sm} \tag{13.10}$$

The target is normally irradiated as the oxide or nitrate and then dissolved in dilute hydrochloric acid. The cross section is relatively high (206 b) and produces ^{153}Sm with a specific activity of 222 GBq/mg (6 Ci/mg) when irradiated at a flux of $1.2 \cdot 10^{14}$ n/(cm^2 s) for approximately 155 h.

13.4.4.8 Holmium-166

^{166}Ho is also a lanthanide that has a radionuclide with favorable properties for radiotherapy. ^{166}Ho transforms with a half-life of 26.8 h and emits two β^- particles; one with energy of 1.85 MeV (51%) and a second at 1.77 MeV (48%) and gamma rays with energies of 80.6 (6.2%) and 1379 keV (1.13%). The mean range of ^{166}Ho in soft tissue is 4 mm (maximum range 8.7 mm). The 80 keV photon is used for imaging. Holmium-166 is routinely made by direct neutron irradiation of ^{165}Ho, which is 100% abundant.

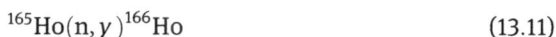

$$^{165}\text{Ho}(n, \gamma)^{166}\text{Ho} \tag{13.11}$$

Production via this route results in a low specific activity material as only a very small portion, 0.31%, is actually converted to ^{166}Ho. The radionuclide thus produced has seen applicability in nuclear medicine, but its agents contain macroscopic quantities of nonradioactive ^{166}Ho isotopes, and thus have comparatively low specific activities. Another mode of ^{166}Ho production exists via an indirect route, cf. Fig. 13.9. In this method, an enriched ^{164}Dy target undergoes a (2n,γ) reaction to produce ^{166}Dy that subsequently transforms by β$^-$emission to the desired ^{166}Ho product:

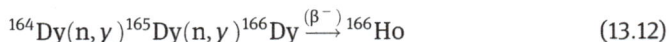

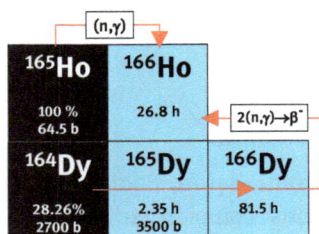

$$^{164}\text{Dy}(n,\gamma)^{165}\text{Dy}(n,\gamma)^{166}\text{Dy} \xrightarrow{(\beta^-)} {}^{166}\text{Ho} \tag{13.12}$$

(n,γ)		
^{165}Ho	**^{166}Ho**	
100 % 64.5 b	26.8 h	← 2(n,γ)→β$^-$
^{164}Dy	**^{165}Dy**	**^{166}Dy**
28.26% 2700 b	2.35 h 3500 b	81.5 h

Fig. 13.9: ^{166}Ho production via direct and indirect route. In the direct route, ^{165}Ho is irradiated; in the indirect method, an enriched ^{164}Dy target undergoes a (2n,γ) reaction to produce short-lived (2.35 h half-life) ^{166}Dy that immediately transforms further by β$^-$emission to ^{166}Ho.

Double neutron capture reactions typically are not favorable; however, in this case the neutron absorption cross sections of ^{164}Dy and ^{165}Dy have been experimentally shown to produce up to one Ci of ^{166}Ho per mg of ^{164}Dy. The specific activity of ^{166}Ho produced using this route can be much higher than that of ^{166}Ho produced by direct irradiation, if a chemical separation of ^{166}Ho from the ^{164}Dy target material is achieved. The first neutron capture cross sections are effectively 2731 barns thermal and 932 barns epithermal to form ^{165}Dy with a half-life of 2.33 h. Although the half-life of this intermediate is relatively short, ^{165}Dy has thermal and epithermal neutron capture cross sections of 3600 and 22 000 barns, respectively, and thus a significant percentage of ^{165}Dy atoms capture a second neutron to form ^{166}Dy.[10]

13.4.4.9 Terbium-161

Terbium-161 is also a lanthanide and combined with Terbium-155 makes an ideal theranostic radionuclide pair. Terbium-155 has a 5.32-day half-life and two imageable γ-rays (86.6 keV (32.0% and 105 keV (25.1%)) and can serve as a potential imaging and dosimetry surrogate for the β$^-$ emitting ^{161}Tb. Terbium-161 [t½ = 6.89 d, β$^-$ 100%, $E_{\beta max}$ 594 keV, γ 48.9 keV (17.0%) and 74.6 keV (10.2%)] has physical and chemical properties that are comparable to the clinically established ^{177}Lu. However,

---

10 One mg of enriched ^{164}Dy irradiated over 155 h at a thermal flux of $4 \cdot 10^{14}$ neutrons cm^{-2} s^{-1} thermal and an epithermal flux of $1.6 \cdot 10^{13}$ neutrons cm^{-2} s^{-1} at the University of Missouri research reactor (MURR) has been shown to produce close to the theoretical yield of 1.2 Ci of ^{165}Dy.

unlike ^{177}Lu, ^{161}Tb has a true theranostic isotope counterpart. Also in contrast with ^{177}Lu, ^{161}Tb decays with co-emission of ~ 12 Auger and conversion electrons (≤50 keV, vs. ~ 1 for ^{177}Lu) along with its low-energy β^- particle, resulting in a higher total electron-to-photon dose ratio for ^{161}Tb than ^{177}Lu. This higher percentage of low-energy Auger/conversion electrons for ^{161}Tb may enhance tumor cell killing, particularly if the targeting ligand is internalized.

Terbium-161 can be made in a reactor indirectly by neutron irradiation of enriched ^{160}Gd to form ^{161}Gd that transforms to ^{161}Tb by the following nuclear reaction:

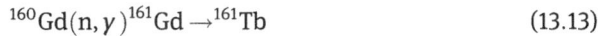

$$^{160}\text{Gd}(n,\gamma)^{161}\text{Gd} \rightarrow {}^{161}\text{Tb} \qquad (13.13)$$

With the true theranostic isotope pair of ^{155}Tb and ^{161}Tb, new chemically identical radiopharmaceuticals can be realized.

13.4.4.10 Rhenium-186 and rhenium-188

Fortuitously, ^{186}Re and ^{188}Re have potential to stand alone as theranostic radionuclides, due to their transformation by both particle and gamma emissions. Rhenium-186 has a 3.72-day half-life, a 1.07 MeV β^- emission, and a 137 keV (9%) gamma emission. It can be reactor-produced by neutron capture on ^{185}Re to yield a low specific activity product through the ^{185}Re(n,γ)^{186}Re nuclear reaction. Alternatively, ^{186}Re can be accelerator-produced with higher specific activity via proton/deuteron bombardment of tungsten targets, such as ^{186}W(p,n)^{186}Re using enriched ^{186}W target material. Various accelerator approaches are being explored, but regular productions have not yet been established. Routine availability of high specific activity ^{186}Re would have a great impact in theranostic research, as its longer half-life would allow the study of slower *in vivo* processes through labeling of biomolecules with longer *in vivo* circulation.

Rhenium-188 has a 17.0-h half-life, a 2.12 MeV β^- emission, and a 155 keV (15%) gamma emission. It is obtained from an ^{188}W/^{188}Re generator with very high specific activity. The ^{188}W parent radionuclide (69.4-day half-life) is produced by double neutron capture on ^{186}W [^{186}W(n,γ)^{187}W(n,γ)^{188}W] in a high-flux nuclear reactor, cf. Fig. 13.10, and is then loaded onto an alumina-based generator. Following β^-/ transformation of the ^{188}W parent radionuclide, the ^{188}Re daughter radionuclide is eluted as radioperrhenate (Na^{188}ReO$_4$) in normal saline. This separation of the parent and daughter radionuclides results in near theoretical levels of specific activity for ^{188}Re.

Rhenium is a group 7 congener of Tc and therefore, in many cases, shares remarkably similar chemical behavior to Tc. Technetium-99m is a diagnostic radionuclidic workhorse in nuclear medicine, and a wealth of information exists on its

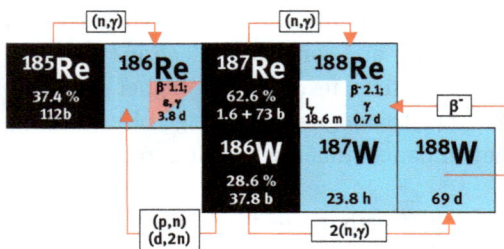

Fig. 13.10: Production pathways of ^{186}Re through ^{185}Re(n,γ) at reactors or via p- and d-induced nuclear reactions at cyclotrons and ^{188}Re; production pathways of ^{188}Re through ^{187}Re(n,γ) at reactors or double neutron capture on ^{186}W followed by β⁻ transformation of the short-lived ^{188}W. The ^{188}W/^{188}Re pair represents a valuable radionuclide generator system.

incorporation into biomolecules.[11] Thus, ^{99m}Tc radiocomplexation knowledge can often be applied to the ^{186}Re and ^{188}Re radionuclides; some even describe ^{99m}Tc and $^{186/188}$Re as a matched pair for radiodiagnostics/radiotherapeutics.[12] In nuclear medicine applications, Re complexes are typically reported in the + 5 and + 1 oxidation states, and to a lesser degree + 3. With two thermodynamic sinks (ReO$_4^-$ and ReO$_2$) and a proclivity toward oxidation on high dilution, design of a stable chelate for Re is critical. Should oxidation occur, the body quickly clears ReO$_4^-$ through the renal-urinary system and avoids unwanted accumulation in most nontargeted tissues; blocking agents such as sodium perchlorate or potassium iodide can be administered to prevent the transient uptake of ReO$_4^-$ in thyroid, salivary gland, and stomach tissues caused by their expression of the sodium-iodide symporter for which ReO$_4^-$ is a substrate. Nonetheless, highly stable and selectively targeted $^{186/188}$Re complexes are desired, to maximize therapeutic efficacy. A brief summary of Re^V/Re^{VII} chemistry and example applications are presented here; more detailed reviews can be found in the literature (e.g., for rhenium radiopharmaceutical chemistry see Liu and Hnatowich 2007; Liu 2008; Donelly 2011; and Blower 1999).

The + 5-oxidation state is the most easily achieved from reduction of the Re^{VII} starting material ReO$_4^-$, which leads to the majority of complexes with Re as Re^V. These complexes typically have square pyramidal geometry, consisting of an N$_x$S$_{4-x}$ tetradentate chelate surrounding a monooxo Re^V metal center. The oxorhenium^V

---

11 Cf. Chapter 12 on the diagnostic applications of radionuclides in life sciences.
12 However, significant differences between ^{99}Tc and $^{186/188}$Re complexes do occur, most often involving dissimilarities in their substitution kinetics and redox chemistry. Rhenium is generally slower to substitute than Tc, which can result in different coordination geometries and ligand patterns. Further, Re is more difficult to reduce and, conversely, easier to oxidize. This difference in redox behavior both reduces the number of readily accessible oxidation states for Re relative to Tc, and requires a more reducing environment to prevent oxidative loss of the radiometal from the complex.

core can be incorporated through the use of bifunctional chelating agents, like di-amine dithiols (DADT) and monoamine monoamide dithiols (MAMA), or by direct integration into molecules, such as α-melanocyte-stimulating hormone and so-matostatin peptide analogs, with the bifunctional chelate approach far more widely used. In either case, the nitrogen donors include either or both amines and amides, while the sulfur donors are typically thiols. Stability against reoxidation is key, and chelates containing two or more sulfur donors (sometimes phosphine donors, also reducing in nature) are used to protect against instability at the ra-diotracer level *in vivo*. However, the lipophilicity and charge changes imparted to these complexes by additional thiol/phosphine incorporation must be considered during the selection process, to prevent undesirable effects on the pharmacokinet-ics (e.g., increased hepatobiliary clearance).

The Re^I tricarbonyl monocationic core has a low-spin d^6 electronic configura-tion, making it more kinetically inert than Re^V and therefore attractive for nuclear medicine applications. This stable core can be obtained as the tricarbonyl triaqua form $[Re(CO)_3(OH_2)_3^+]$ through reduction of perrhenate using a two-step kit formu-lation though the conditions are harsher than those used in *IsoLink* kits (Mallinck-rodt) for the analogous ^{99m}Tc synthon. Substitutions of the coordinated water ligands are facile and are typically carried out with tridentate chelates containing amines, imines, carboxylates, phosphines, thiols, and/or thioethers. The *fac*-[Re $(CO)_3]^+$ organometallic synthon is small, which allows its addition to biomolecules of various sizes, but also adds considerable lipophilicity to the biomolecule.

Rhenium radiopharmaceuticals have been used in a variety of clinical applica-tions; these include clinical trials for the palliation of pain from bone metastases and the treatment of inoperable hepatocellular carcinomas, advanced lung cancer, and rheumatoid arthritis; see below.

13.4.5 α-Emitter

There is significant interest in using α-emitters as their short range of ca. 10^{-2} mm results in low toxicity to normal cells and their high linear energy transfer results in very high cell kill with a very low amount of activity. It is estimated that only 1–3 α-transformations are needed to result in cell death compared to 1000 s of β^-.[13]

The most common radionuclides are ^{223}Ra, ^{212}Pb, ^{213}Bi, ^{225}Ac, and ^{211}At. Except for ^{223}Ra, which is used as the bare radiocation for the therapy of bone metastases, all the other radiometals require adequate chelates, which are attached to small

13 SCHEINBERG was able to demonstrate the eradication of tumors in cells and in tumor-bearing mice using ^{225}Ac complexed to monoclonal antibodies in picocurie quantities in cell culture and nanocurie quantities ablated 40–50% of tumors in mice for greater than 6 months.

molecules, peptides and antibodies for eradicating tumors, cf. Tab. 13.2. Due to their high recoil energy, it has been challenging to develop ligands that remain complexed to the daughter upon transformation of the parent. This can result in the daughter, often times radioactive as well, to cause damage to normal tissues.[14]

Potential applications of α-emitters for radiotherapy are in micro metastases, radioresistant tumors such as tumors that are known to be hypoxic, and in tumors that are located near radiation sensitive areas, such as in the brain and located near the spinal cord or the heart. The use of α-particles can result in a boost in therapy at the primary site for conventional treatments. Alpha emitters are being investigated for the treatment of minimal residual disease (e.g., surgery margins) and metastatic disease. They are being evaluated for treating cancers that are isolated to body compartments such as ovarian and pancreatic metastases in the peritoneal cavity, leukemia, metastatic breast and prostate cancer, and glioma in the brain. The α-emitter ^{223}Ra has been investigated and approved in the form of ^{223}RaCl$_2$ for the palliation of bone pain. Additionally, α-emitters are being investigated for the use of drug-resistant bacterial, fungal, and viral infections such as HIV. They have been in evaluation for the conditioning for hematopoietic (stem) cell transplantation.

Tab. 13.2: List of α-emitters being used clinically (gray) or investigated for targeted α-therapy, arranged by increasing mass number. ^{avg} considers an average energy covering not a single, but several successive α-transformations in a transformation chain until terminating at the final transformation product; 75% represents the main branching.

Radionuclide	t½	E$_\alpha$ (keV)	accompanying γ-emission: energy (keV) and branching (%)
^{149}Tb	4.12 h	4077	β$^+$, γ = 352
^{211}At	7.21 h	5867	γ = 79
^{212}Bi	60.6 min	8785	γ = 727
^{213}Bi	45.7 min	8378	γ = 440
^{223}Ra	11.4 d	5348^{avg}	γ = 269
^{224}Ra	3.62 d	5094^{avg}	γ = 241
^{225}Ac	10.0 d	5450^{avg}	γ = 86
^{228}Th	30.9 min	6338$^{(75\%)}$	γ = 111 (3%)
^{227}Th	18.7 d	5562^{avg}	γ = 236 (11.5%)
^{255}Fm	20.1 h	7022	γ = 16

---

14 Actinium-225 (t½ = 9.92 d) has been suggested as an *in vivo* "nano-generator" since it transforms via α-emission through 3 atoms, each of which also emits α-particles, ^{221}Fr (t½ = 4.8 min), ^{217}At (t½ = 32.3 ms), and ^{213}Bi (t½ = 47 min). For a review on radionuclide generators in nuclear medicine see Roesch and Knapp (2011).

13.4.5.1 Astatine-211

^{211}At is an alpha emitter with a half-life of 7.2 h. Astatine-211 simultaneously transforms by two routes; 58% by electron capture to ^{211}Po and by α-transformation (42%) to ^{207}Bi, cf. Fig. 13.11. Both transformation products are unstable and transform further: ^{211}Po with a half-life of 0.5 s via alpha emission to stable ^{207}Pb and ^{207}Bi with a half-life of 31.5 a transforms by electron capture to the same stable ^{207}Pb. This gives two α-particles per each transformation of ^{211}At. The energy of the two alpha particles is 4.87 and 7.5 MeV, respectively.

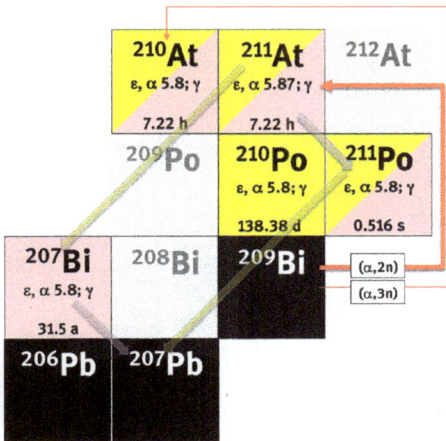

Fig. 13.11: Astatine-211 simultaneously transforms by α-transformation (42%) to ^{207}Bi and by electron capture to ^{211}Po (58%). Following on there is a second α-emission to stable ^{207}Pb.

It is mainly produced by bombarding natural bismuth with 28 MeV α-particles. It is critical that the energy be kept at 28 MeV as a lower energy bombardment will result in the production of ^{210}At which transforms to ^{210}Po that is a long-lived alpha emitter and highly toxic. Astatine can be purified either by liquid-liquid extraction using organic solvents or by using dry distillation in a tubular oven. Dry distillation is the more preferred route as it results in high recoveries of 75–85% of the ^{211}At in a very small volume, cf. Chapter 7.

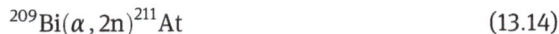

$$^{209}\text{Bi}(\alpha, 2n)^{211}\text{At} \tag{13.14}$$

Astatine is the heaviest halogen X. Its most relevant oxidation state is − I, resembling the typical halogenide X$^-$, but according to the increasing metallic character within a group of the PSE, also the astatine cation At$^+$ is known. Due to its chemical similarity to iodine, it was postulated that ^{211}At could be incorporated into proteins using the same prosthetic groups that have been successfully used with iodine and could be directed against a variety of targets to ablate bacteria, viruses, and tumor cells. Incorporation of [211]At was investigated, e.g., with labeling of N-succinimidyl *para*-tri-*n*-butylstannylbenzoate and its derivatives. Briefly, the [211]At is added to a solution containing the N-succinimidyl *para*-tri-*n*-butylstannylbenzoate and by destannylation gives the N-succinimidyl-astobenzoate (SAB) shown in Fig. 13.12. *In vitro* stability studies including serum showed high stability and high retention of the [211]At with the protein. However, *in vivo* studies showed a high degree of instability, which is thought to be due to enzyme catabolism and the lower pH in lysosomes resulting in release of the [211]At. To stabilize [211]At to proteins *in vivo*, several new approaches are being investigated.

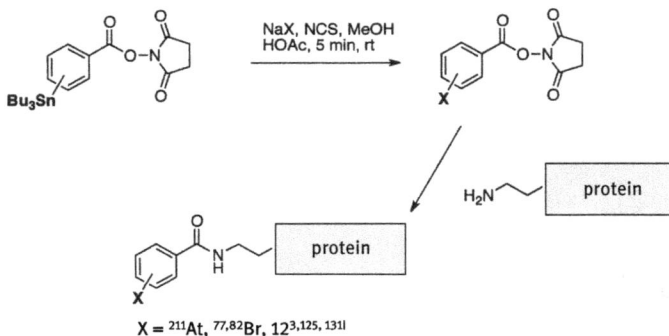

$X = {}^{211}At, {}^{77,82}Br, {}^{123,125, 131}I$

Fig. 13.12: Halogenation reaction of N-succinimidyl-benzoate used for [211]At labeling.

13.4.5.2 Radium-223

[223]Ra is an α-emitting radioisotope with a half-life of 11.43 days. It transforms by four successive α-emissions and two β⁻-emissions to stable [207]Pb. It is a member of the natural [235]U transformation chain.[15] Within that chain, it can be obtained from [227]Ac with a half-life of 21.8 years, representing a radionuclide generator system, cf. Fig. 13.13.

---

15 Cf. Chapter 4 of Vol. I.

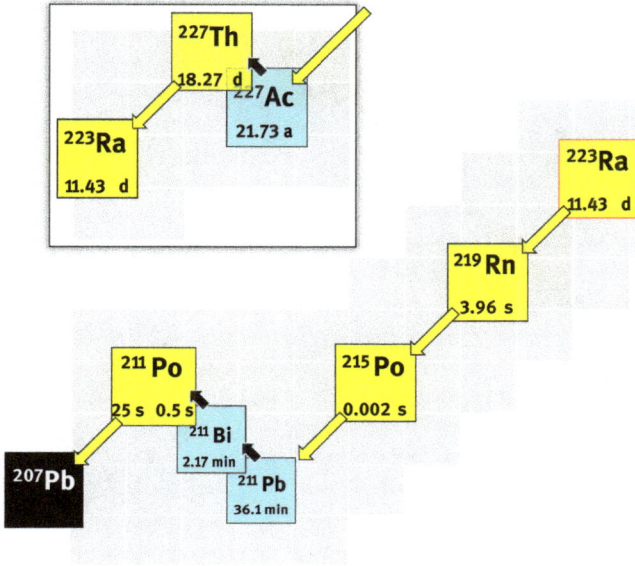

Fig. 13.13: Transformation chain generating ^{223}Ra (inserted) and subsequent ^{223}Ra transformations.

13.4.5.3 Actinium-225 and Bismuth-213

^{225}Ac and ^{213}Bi are trivalent radiometals and transform by α-emission. The two nuclides are linked to each other as they are daughter nuclides of another natural transformation chain, this time starting from ^{237}Np/^{233}U.

^{225}Ac transforms with a half-life of 10 days through successive α-emissions, thus releasing 4 α-particles of individual energy per transformation, cf. Fig. 13.14. The daughters involved are relatively short-lived. One of those daughters is ^{213}Bi. This radionuclide represents the last steps of the ^{225}Ac transformation chain. Once radiochemically separated from ^{225}Ac in a radionuclide generator system, it is used as a therapeutic radionuclide despite its short half-life of 45.6 min.

Actinium-225 has traditionally been obtained from a long-lived ^{229}Th (t½ = 7340 a) of which ^{233}U has been the only viable source. This route provides 44.4–62.9 GBq (1.2–1.7 Ci) of ^{225}Ac world-wide. Recently high-energy proton ^{225}Ac through the spallation of natural Th through the nuclear reaction:

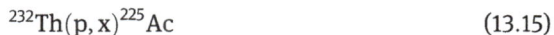

$$^{232}\text{Th}(p, x)^{225}\text{Ac} \tag{13.15}$$

This route can provide high quantities of ^{225}Ac in short irradiations meeting the amounts required for clinical evaluations.

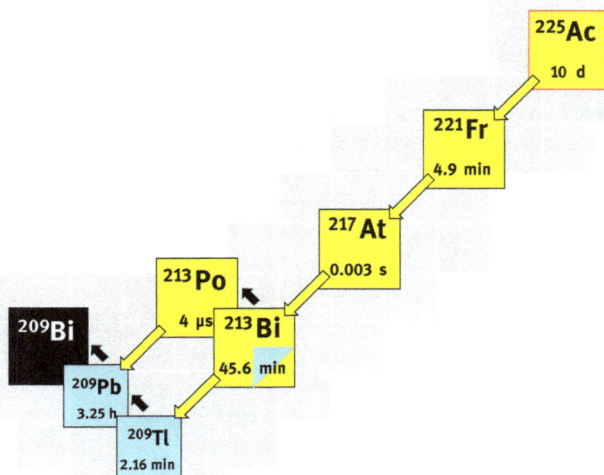

Fig. 13.14: Transformation chain of ^{225}Ac including the ^{213}Bi segment.

The advantage of ^{225}Ac and ^{213}Bi relative to, e.g., ^{223}Ra or ^{211}At is their coordination chemistry. Being trivalent metals, it is rather straightforward to radiolabel the two α-emitters using chelate ligands (such as DOTA) developed for other trivalent metals such as ^{68}Ga or ^{111}In for imaging or ^{90}Y or ^{177}Lu for therapy. A rather problematic issue is the stability of those chelates, as the radiation dose may destroy chemical bonds even within the radiopharmaceutical compound itself. Moreover, the individual daughter nuclides may be released from the initially labeled compound and develop their own pharmacology *in vivo*.

13.5 LET and ionization

Linear energy transfer (LET) is the amount of energy released by a radioactive particle or photon over the length of its transformation track. Specific ionization refers to the number of ion pairs produced along the path of transformation. LET is measured by the ionization density along the track of transformation (ion pairs/cm of tissue). High LET radiation (α- and β-particles) ionizes H_2O into H^+ and OH^- radicals over a very short track. These radicals then combine forming peroxide H_2O_2, which can result in oxidative damage in the cell. Low LET radiation (photons or β-particles) also ionizes water molecules, but over a longer track. The radicals H^+ and OH^- formed in these reactions often unite and form H_2O.

13.5.1 From radiation to ionization

Radiation in biological tissue may lead to either excitation or ionization. If the absorption leads to the raising of an electron to a higher level without removal it is called excitation. If enough energy is imparted to eject an electron from an atom or molecule it is termed ionizing radiation. The energy released per ionizing event is 33 eV, which is enough to break a typical bond. For instance, the energy associated with a carbon double bond is 4.9 eV. Ionizing radiations are normally classified as either electromagnetic or particulate: the two most common forms of ionizing radiation. What occurs when radiation interacts with matter depends on which type of radiation it is, the energy of the radiation; the makeup of the matter such as its composition and density. The type of radiation can impact such things as the rate at which energy is lost, average path length and whether the path is continuous with a gradual loss of energy or sporadic. Particulate radiation passing through matter results in the breakage of chemical and biological bonds. Electromagnetic radiations are indirectly ionizing and do not directly result in the breakage of bonds or chemicals, but the energy they release to produce fast molecules may in turn produce damage. What occurs is dependent on the energy of the photon and the composition and density of the material it is interacting with.

13.5.2 From ionization to energy

The concept of radiation dosimetry was introduced in Chapter 2. Accordingly, the radiation dose delivered to a biological target is expressed in terms of Gray (Gy) and/or Sievert (Sv). Dose is reported in units of Gray (Gy) for mass, and dose equivalent is reported in units of Sievert (Sv) for biological tissue, where 1 Gy or 1 Sv is equal to 1 Joule per kilogram. Radiation dosimetry is the calculation of the amount of energy deposited and its effect on living tissue. Exposure to a radioactive source will give a dose which is dependent on the activity, time of exposure, energy of the radioactivity emitted, distance from the source and shielding. The dose equivalent is dependent upon the additional assignment of weighting factors describing biological effects for different kinds of radiation on different organs.

Alpha emitters are classified as high linear energy transfer (high LET) as their paths tend to be very straight and highly charged resulting in significant damage. Due to their high LET, α-particles are very toxic over very short distances, leading to highly localized radiation, which can kill individual cells. Unlike other forms of radiation, they are independent of the oxygen concentration of the tumor and thus can function just as well in hypoxic cells (lacking oxygen) as in well-oxygenated cells. Their effect is independent of dose rate and unlike chemotherapy or inhibitory agents that block specific pathways, cancer cells cannot develop resistance. Furthermore, they do not need to get inside of cells to be effective.

Newer dosimetry techniques allow for the separation of organs and use of individualized patient data to calculate the dose such as in MIRDOSE3. Even newer dosimetry versions such as OLINDA allow for the calculation of the absorbed dose of each organ and a calculation of the effective dose of the radiopharmaceutical. To obtain the effective dose, the absorbed dose is first corrected for the radiation type to give the equivalent dose, and then corrected for the tissue receiving the radiation. Some tissues like bone marrow are particularly sensitive to radiation, so they are given a weighting factor that is disproportionally large relative to the fraction of body mass they represent. Other tissues like the hard bone surface are particularly insensitive to radiation and are assigned a disproportionally low weighting factor.

To further improve dose calculation imaging is often used to measure the activity in a given region of interest with time. Planar images can be taken over time generating counts, which can then be converted to activity. This is the technique that has often been used by doing an initial planar scan of a radiopharmaceutical labeled with a gamma emitter, generating information on how that specific patient retains and clears the radiopharmaceutical over time. This data can then be used to determine how much of a therapeutic radiopharmaceutical should be given to result in the optimal dose. Due to its ability to be quantitative, PET imaging is often used to evaluate new drugs particularly in clinical trials to gain more specific patient quantification and better efficacy.

13.5.3 From energy to the breaking of chemical bonds

The main goal of radiotherapy is malignant cell destruction, which is normally through DNA double-strand breaks. This leads to the cell's inability to reproduce and thus eventual death. Additionally, radiation can directly interfere with cellular processes necessary for growth such as cell senescence (process of aging) and cell apoptosis (programmed cell death). When radionuclides transform in biological tissues, they give off either energy or particles, which leads to excitation or ionization of molecules and atoms. It is these two events which result in the actual breakage of the chemical bonds.

13.5.4 Relative biological effectiveness

Different kinds of radiation create different biological effects. This is defined as the "relative biological effectiveness" (RBE), which is derived by multiplying the dose by the weighting factor (W_T). The higher the LET, the greater the biological impact and thus is assigned a higher weighting factor. For example, for a photon or an electron the W_T is defined as 1, whereas for a highly charged α-particle the W_T is 20. For details see Chapter 2.

The radiation doses required for tumor cell killing are also dependent on the specific cancer to be treated (e.g., its radiation sensitivity), the tumor size and location, and the nontarget radiation dose delivered by the radiopharmaceutical. Alpha particles are high LET radiation and produce very high-localized ionization densities in cells that minimize the ability of the cells to repair radiation damage. Alpha particles are therefore most effective in cell killing (per Gy delivered); however, their range is short (approximately one to three cell diameters) and best suited for small tumors; α-particles deposited in tumors deliver minimal irradiation to nearby tissues. AUGER emitters produce very low-energy electrons that have very short ranges and are considered a form of high LET radiation.

The presence of oxygen is very critical for most forms of therapy including the use of β-particles, however, α-particles due to their high energy have the highest probability of producing double DNA strand breaks and are not dependent on the presence of oxygen. The average separation between ionizing events is on the order of 20 Å at 100 keV/μm, this ionization density has the highest probability of causing a double-strand break from a single charged particle. In comparison, the cell damage based on mass and energy by an α-particle compared to a β⁻-particle is 7400 times higher it can be illustrated by visualizing a β-particle as a light 12-pound bowling bowl vs. a fully loaded semi-truck and trailer weighing 36 tons.

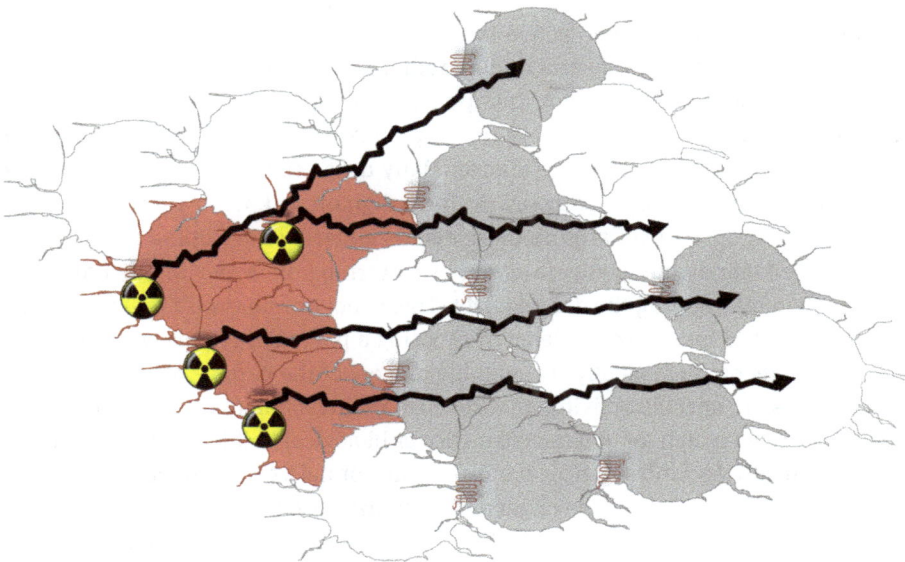

Fig. 13.15: A radiotherapeutic compound that is selectively delivered to a cell (brown) with a specific target at the cell membrane. Black lines indicate the particles' pathway originating from the radionuclide "traveling" through a number of cell diameters thus delivering a dose to a tumor cell (gray) that has not been selectively targeted, but also to healthy cells (white).

Some tumors are known to have low blood flow or blood flow that is compromised to some of the cells due to the rapid growth of the tumor. For systemically delivered drugs this greatly minimizes their effect, as they are only effective on those cells they are directly delivered too. In the case of radiotherapy, it is possible to select radionuclides that upon transformation have particles that travel long distances thus enabling an effect on cells that the radioactivity has not been directly delivered to; this is called the cross-fire effect and is illustrated in Fig. 13.15.

13.6 Selected radiotherapeutics

Radiotherapy involves the administration of a radioactive drug that is selectively accumulated in cancerous or diseased tissue versus normal tissue and either ablates or damages the diseased tissue through the emission of an energetic particle. Because these particle emissions result in damage to tissue, it is imperative that the drug accumulates selectively in the diseased tissue, as any uptake in normal tissue will result in unwanted dose to the patient and injury to normal tissue.

Radiotherapy has many clear advantages over the use of other drugs in treating disease. Radionuclides used for medical applications have a short physical half-life. The importance of this is that the toxicity of radioactivity is only for a short time. Most drugs will continue to be active for long periods of time capable of traveling around in the body causing destruction. In radiopharmaceuticals, the radionuclides have relatively short half-lives, for therapy normally lasts days to weeks and upon transformation the radionuclides are no longer toxic. Another clear advantage of radiotherapy is that there is no resistance. Many drugs such as chemotherapeutics work by inhibiting a specific pathway in the malignant cell. Cells are very adept at quickly learning how to get around these agents either by developing resistance in which the cell merely pumps the drug out or by turning off or moving to a different pathway thereby negating the drug's effect. This is not possible with radioactivity.

Several factors must be considered in choosing a particular radioisotope for therapeutic applications, such as physical half-life, energy of the particle emission, type of particle emission, specific activity, and the cost and availability of the radioisotope. The half-life must match the pharmacokinetics or biological half-life of the radioactive drug, specifically for uptake and clearance from normal versus targeted tissues, in order to maximize the dose to the target and minimize dose to normal tissue.

In choosing a therapeutic isotope for monoclonal antibodies, it is important to consider that it normally takes at least four days for monoclonal antibodies to clear the bloodstream and reach maximum uptake at the target site. Labeling monoclonal antibodies with an isotope with too short of a half-life will result in most of the dose being administered to normal tissues, with little to none reaching the target site. On the other hand, faster-clearing small molecules and peptides can be labeled with short-lived isotopes and achieve maximum dose delivery in the target tissue. Half-life also affects how easily the isotope can be transported from the site of production to the final user. Short-lived isotopes require either on-site production, which can be prohibitively expensive, or development of a generator system.

Specific activity – a measure of the radioactivity per unit mass of the compound – is an indicator of potency; the higher the specific activity of a radionuclide, the higher both the percentage of radioactive atoms and the deliverable dose. Specific activity may or may not be important depending on the number of sites available for targeting. For instance, bone is considered a large capacity site and does not require radioisotopes with high specific activity. Low specific activity radioisotopes such as ^{153}Sm in Quadramet® and ^{166}Ho in the Skeletal Targeted Radiotherapy agent, STR, have been used for metastatic bone pain palliation and bone marrow ablation, respectively (Cutler et al. 2000). Low-capacity sites such as receptor sites, which are present in low numbers, require high specific activity radioisotopes. Radioisotopes produced in reactors are normally available in what is called carrier-added form as only a small portion of the target material is converted to the desired radioisotope. This results in a high portion of stable isotope being present in the sample. There are indirect methods for making radioisotopes which upon separation from the target material result in a no-carrier-added form in that no stable isotope is present, only the radioactive isotope. No-carrier-added radioisotopes are preferred as they result in the highest delivery of dose to the diseased tissue with minimal targeting material.

The choice of type and energy of the particle emission is largely determined by the size of the lesion or tumor being treated, site of delivery, whether the tumor is homogeneous, and whether the dose can be delivered uniformly to each cell. For example, smaller tumors may respond better to lower β^--energies, such as for ^{177}Lu, whereas the higher energy β^--emitter ^{166}Ho may be required for larger tumors. In certain cases, the elimination or minimization of toxic side effects determines which energy is optimal.

Another factor that can dictate which radioisotope to use is the chemistry of the radioisotope and how to complex it or stably bind it. Certain radionuclides can be administered close to their native form and will show uptake to the desired site naturally such as ^{89}Sr to the bone and ^{131}I to the thyroid. Others must be complexed using chelates or prosthetic groups and others may need to be complexed in microspheres or nanoparticles to provide high *in vivo* stability. Some metals have very rich chemistry such as rhenium with a large number of oxidation states that allows for a wide variety of complexes to be formed and with redox chemistry that if not controlled may be accessible to proteins *in vivo* and cause uptake in normal tissues. Rhenium for instance prefers to be in higher oxidation states and if not stably complexed can oxidize up to perrhenate which then can result in lower dose to the intended site. In certain applications such as palliation of bone pain this mechanism has been shown to lead to lower normal bone uptake and dose over time. Certain radioisotopes require very bulky groups for stable complexation that when conjugated to a biomolecule will dramatically change its *in vivo* behavior and thus are not effective. Others can be metabolized *in vivo* resulting in release and trapping in normal organs. Iodine for instance is often introduced by complexing it to functional groups of normal amino acids in proteins and peptides. These peptides and proteins can be degraded by enzymes to the individual amino acids, which are then released back into the body. Consequently, when labeling peptides and proteins it is important to use attachment mechanisms that are residualizing or known to stay in the cell.

Many of the radiotherapies being delivered today are utilizing selective targeting methods, which aid in the radioactivity being delivered only to malignant cells. This greatly enhances the activity that can be delivered to the tumor sites and minimizes what is observed by normal cells. In the following, the most prominent directions of endoradiotherapy are discussed. This is not structured according to a certain radionuclide. Instead, the main diseases are listed and the various therapeutic radionuclides and their chemical species are mentioned.

13.6.1 Thyroid cancer

Iodide accumulates in cells of the thyroid due to the biochemical pathways mentioned in Chapter 12 in the context of molecular imaging diseases of the thyroid. The strategy of imaging can easily be directed towards the treatment of thyroid disease – by just changing the radioisotope of iodine.

For the treatment of thyroid cancer, ^{131}I is normally administered in NaOH solution containing 0.02 M Na_2SO_4 at a pH of 9–12 and a radioactivity concentration between 7.4–37 TBq (200–1000 Ci/ml).[16] The amount delivered varies depending on the application. It is often given after a thyroidectomy in the range of 2.75–5.50 GBq (75–150 mCi) to ablate any remaining thyroid cells. The amount given is dependent on the radioactive iodine uptake (abbreviated RAIU) as well as the biological half-life of the radioiodine in the thyroid gland, which has been determined to be patient dependent and to vary widely. To treat distant metastases higher than 7.4 GBq (200 mCi) is given. Shown in Fig. 13.16 is an image of a patient undergoing thyroid ablation.

An important consideration for radioiodine therapy is the calculation of dose to both the thyroid and normal tissues to ensure efficacy and to minimize side effects. Normally, the dose is calculated by utilizing imaging and either calculating the dose to regions of interest (ROI) or utilizing SPECT and CT to derive the integrated activity on a voxel (volumetric pixel) by voxel basis. Monte Carlo computation methods are then used to determine the absorbed dose from both the beta particle and the gamma radiation. Dosimetry is critical to ensure the absorbed dose to the targeted areas is

---

16 Patients treated in Europe are normally kept in the hospital to minimize dose to the public. In the United States the Nuclear Regulatory Commission requires that a determination be made to ensure the dose to the public does not exceed 5 mSv from the patient. If below this, the patient is released with a series of instructions to minimize exposure to others.

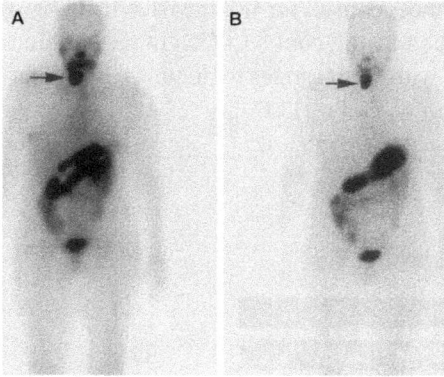

Fig. 13.16: Focal radioiodine uptake in the thyroid bed (arrows) noted on the anterior view of a 370 MBq ^{123}I pre-treatment scan (A) and a 5550 MBq ^{131}I post-treatment scan (B).

optimal and without overcoming the maximum tolerated dose in the red marrow and the lungs, the two dose limiting organs. Pre-therapy imaging with low dose $[^{131}$I]I$^-$ can cause "stunning" (reduced uptake of the therapeutic administration) causing resistance to further treatment.[17] Imaging post-therapy is also done for "staging" (measurement of extent of disease and response). Patients are normally imaged by planar scintigraphy 7–14 days post treatment for staging purposes.

The main limiting organ is the bone marrow in which the radioiodine is known to accumulate which can result in bone marrow ablation if not tightly controlled. Further acerbating this issue is that in roughly 20–30% of cases of thyroid cancer radioiodine is not accumulated in the cancer resulting in poor diagnosis and treatment.

Astatine is a halogen and shares similar chemical properties to the other halogens, most notably iodine. The astatide anion behaves similarly in the body in that it shows high accumulation in the thyroid gland and it has been suggested that it may be useful in the treatment of hyperthyroidism, utilizing the α-particles emitted by ^{211}At.

13.6.2 Bone tumors and metastases

It has been estimated that 60–75% of patients diagnosed with breast and prostate cancer will eventually develop bone metastases. These metastases are extremely painful and result in a low quality of life. Radionuclides that mimic calcium or that are bound to bone-seeking phosphorous chelators have been developed and used to palliate the pain associated with metastatic bone disease, see e.g. Paes and Serafini

---

17 Imaging with $[^{123}$I]I$^-$ may prevent stunning.

2010. Chapter 12 introduces the targeting concepts used for noninvasive imaging of bone metastases. The same concept functions in the context of therapy: Radionuclides chosen need to have β- or α-emissions with short ranges to minimize bone marrow ablation, which can lead to loss of immunity, cf. Fig. 13.17.

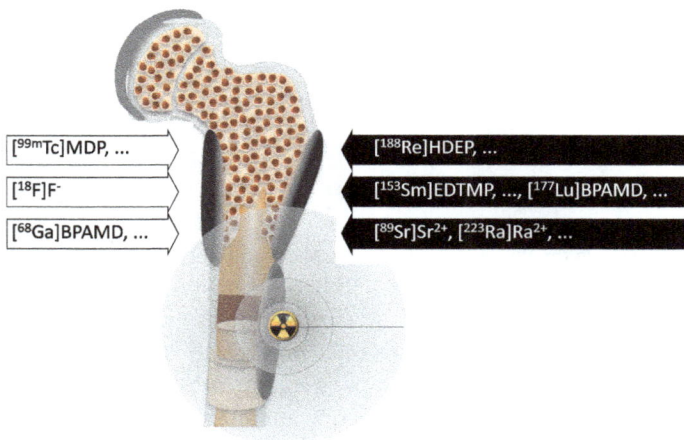

[^{99m}Tc]MDP, ...

[^{18}F]F-

[^{68}Ga]BPAMD, ...

[^{188}Re]HDEP, ...

[^{153}Sm]EDTMP, ..., [^{177}Lu]BPAMD, ...

[^{89}Sr]Sr^{2+}, [^{223}Ra]Ra^{2+}, ...

Fig. 13.17: Therapy of bone metastases with endoradiotherapeutics targeting hydroxyapatite. For comparison, the diagnostic approach is illustrated on the left-hand side.

The initial use of ^{32}P was for bone pain palliation. Phosphorous-32 has been used in a variety of forms from sodium orthosphosphate, as (Na$_6$O$_{18}$P$_6$), pyrophosphate (P$_2$O$_7$) and hydroxyethylidene diphosphonate (HEDP), with orthosphosphate being the most prevalent. It can be administered either by intravenous injection or orally. It localizes by replacing phosphate ions in the hydroxyapatite of the bone. It additionally localizes in the liver, spleen, GI tract and the kidneys. Roughly 85% of the administered dose localizes in the skeleton dependent on metabolic activity, with another 15% localizing in the liver, muscle, and spleen and then 5–10% is excreted via the urine and GI tract. Doses are normally from 370–777 MBq (10–21 mCi) fractionated for the treatment of metastatic bone pain. A major obstacle is the bone marrow suppression, which can last up to 36 months. Relief begins 2–4 weeks post administration and can last from 1.5–11 months.

Strontium-89 is the most commonly used radionuclide for bone pain palliation and is administered intravenously. Approximately 50% localizes in the bone, in normal bone it is quickly cleared but is highly taken up and retained in osteoblastic metastases. The remaining quantity clears 80% via the kidneys and 20% via the gut. The normal dose is 1.5 MBq/kg (0.04–0.06 mCi/kg). Mild transient bone marrow suppression is noted (less than that observed for ^{32}P) lasting approximately 12 weeks. Pain relief begins 2–6 weeks post treatment and lasts for 4–15 months. Retreatment with ^{89}Sr is possible.

^{223}Ra is, like ^{89}Sr, an alkaline rare earth and its bone uptake is similar to that of calcium in the bone. Comparative studies between ^{89}Sr chloride and ^{223}Ra chloride show they have similar bone uptake and target areas of metabolically active bone such as osteoblastic metastases. Consequently, ^{223}Ra is used for radioimmunotherapy of bone metastases that arise from prostate cancer. Due to the short distance traveled by α-particles the dose to the bone marrow is minimal. ^{223}Ra delivers a very high dose to a very small area. The dose is cumulative as four successive α-transformations occur following one transformation of ^{223}Ra, cf. Fig. 13.13. Unlike the previous pain palliation agents described above, which are excreted very quickly normally within four hours via the kidneys, $RaCl_2$ is cleared more slowly primarily via the intestines. In a large, randomized phase III trial, ^{223}Ra has demonstrated improvements in median overall survival of 3.6 months, a delay of skeletal related events (pathological fractures, spine cord compression, and requirements for surgery or radiation to the bone) and biochemical parameters, with a remarkably tolerant adverse event profile in men with castration-resistant prostate cancer. This has been the first agent to demonstrate overall survival, although it should be pointed out this was never evaluated for the other agents. Doses range from 50–80 kBq per kg of body weight. Dosimetry models have estimated that a 50 kBq dose per kg of body weight results in a dose of 13.05 Sv to the bone surfaces. The bone marrow to bone surface dose ratio was estimated to be 10.3. Doses to the liver, large intestine, and colon were equivalent and estimated to be 0.635, 0.367, and 0.254 Sv. Tumor absorbed dose was estimated to be 243 cGy/MBq and 12.15 Sv/MBq in tumors. The treatment is given on an outpatient basis. Treatment is well tolerated and more preferred by patients as there is minimal to no risk of contamination due to GI clearance unlike ^{89}SrCl$_2$ and ^{153}Sm-EDTMP (see below).

Skeletal targeted radiotherapy (STR) offered a promising and more focused alternative to total body radiation and chemotherapy regimens.[18] It targets multiple myeloma, a cancer of the plasma cells (a type of white blood cell), that presents in the bone marrow. Several trivalent radiometals bind to bone metastases in a "pure" cationic form, but clinically complexes of those radionuclides with chelate ligands, which include phosphonate moieties, are used. This targeted radiotherapy treatment places the radiation where it is the most effective, which results in a minimum amount of damage to normal healthy cells and maximum damage to the bone marrow. The remaining amount is quickly excreted from the body via the kidneys into the urine.

One example uses the β$^-$-emitting radionuclide ^{166}Ho coupled to the small-molecule biolocalization agent DOTMP (1,4,7,10-tetraazacyclododecane-1,4,7,10-tetramethylenephosphonic acid), cf. Fig. 13.18. In STR treatment using ^{166}Ho-DOTMP, the patient would initially receive a small dose 1110 MBq (30 mCi) and be imaged to

---

18 Multiple myeloma destroys bone structure, causes anemia, and often results in kidney failure, and affects 500 000 people in the US alone, claiming 11,000 lives each year. The conventional chemotherapeutic and/or total body radiation treatment regimens are only marginally effective, with a five-year survival rate for myeloma patients of only 30%.

determine how that patient handled the drug (uptake in the skeleton and clearance) to calculate dosimetry to the bone and kidneys to determine a patient-specific dose. Imaging methods were shown to be more reliable for assessing patient pharmacokinetics, clearance, and dose. A week later, the patient received a higher dose up to 74 kBq (2 Ci), calculated based on the imaging of STR, during which the [166]Ho-DOTMP adduct accumulates in the bone.[19]

Samarium *lexidronam*, which is [153]Sm-complexed with ethylene diamine tetramethylene phosphonate (EDTMP), trade name (*Quadramet®*), shown in Fig. 13.18, was approved by the FDA in 1997 for the palliation of bone pain associated with metastatic disease derived from breast, prostate and lung cancer. The dose is 18–37 MBq/kg (1 mCi). As with most bone agents, it exhibits a much higher accumulation in the metastatic lesions (five times higher compared to normal bone). *Quadramet®* exhibits a high clearance from the blood and reaches a maximum bone uptake and blood clearance at six hours post injection. The main excretion path is through the kidneys into the urine. Pain relief is normally observed within one to two weeks and persists from four to 35 weeks. Patients may experience a flair response (transient increase in pain) 72 h post injection. The main side effects are thrombocytopenia, transient myelosuppression, and a drop in platelet counts, which return to normal typically 10 weeks post treatment. The skeletal uptake averages around $65.5 \pm 15.5\%$ of the administered dose dependent on the level of metastases. Patients receiving up to six cycles of low dose [153]Sm-EDTMP have remained pain-free for years. Calculations using MIRDOSE indicate the skeletal lesions are receiving a range of 23 to 24 mGy/MBq. The marrow doses ranged from 1.2 to 2.0 mGy/MBq and urinary bladder doses ranged from 0.83 to 0.12 mGy/MBq. Nonskeletal sites received negligible doses.[20]

Due to its low specific activity, [153]Sm has not been evaluated with many of the new molecular selective targeting agents. In contrast, ^{177}Lu is a promising radionuclide for bone pain palliation due to its low-energy β⁻-emission, long half-life, and ability to be produced in large quantities and almost carrier-free at most medium flux reactors. [177]Lu complexed to both EDTMP and DOTMP is being evaluated for bone pain palliation. A study comparing ^{177}Lu-DOTMP to [153]Sm-EDTMP showed that little to no toxicity was observed for the ^{177}Lu agent when administered to give the same skeletal dose as the [53]Sm agent, indicating larger ^{177}Lu doses could be given that may result in longer

---

19 After approximately one week, the previously withdrawn bone marrow, having been cleansed of cancer cells, is returned to the patient to restore disease-free marrow function. It was calculated that the dose the bone marrow received was 0.54 mGy/MBq (0.02 Gy/mCi), the bladder received 0.54 mGy/MBq (0.02 Gy/mCi), the bone surface received 0.81 mGy/MBq (0.03 Gy/mCi), kidneys received 0.0135 mGy/MBq (0.0005 Gy/mCi), and the whole body received an absorbed dose of 0.054 mGy/MBq (0.002 Gy/mCi).

20 Clinical trials evaluating the combination of *Quadramet®* with chemotherapy of *docetaxol* and *estramustine* have been performed in castration-resistant prostate cancer patients and have shown an increase in the progression-free survival of the patients. Side effects were minimal and results indicated this might be a viable option for patients that should be explored further.

EDTMP
(ethylene diamine tetramethylene phosphonate)

^{153}Sm-EDTMP
(Quadramet)

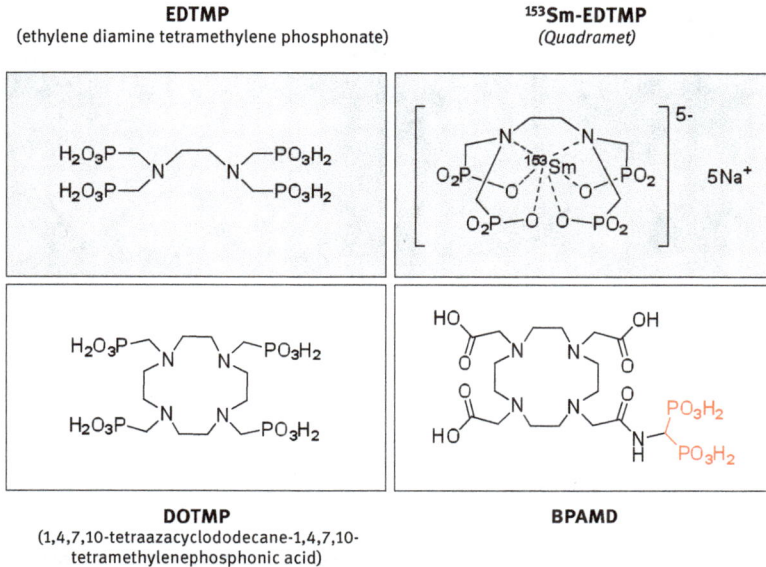

DOTMP
(1,4,7,10-tetraazacyclododecane-1,4,7,10-
tetramethylenephosphonic acid)

BPAMD

Fig. 13.18: Structures of EDTMP and the ^{153}Sm-EDTMP complex, DOTMP and BPAMD (with the bisphosphonate moiety in red: Here, the targeting vector is not involved in the M-chelate coordination chemistry). The DOTA in this case becomes a "bifunctional chelator".

remission times, Bryan et al. 2009. More recently, ^{177}Lu-DOTA-based bisphosphonates, i.e., the trivalent radiometal coordinated to a bifunctional macrocyclic chelator attached to a bone-seeking bisphosphonate, cf. Fig. 13.18, have shown significant uptake at bone metastases and appear to be promising for treatment, cf. Fig. 13.19.

13.6.3 Lung and liver and synoviorthesis

One of the most widely used medical applications of ^{90}Y is as microspheres for the treatment of inoperable liver cancer. There have been two types of spheres developed and used clinically. The first was *Therasphere®* FDA approved in 1999 which are spheres comprised of highly pure glass sized at 20–30 µm to be selectively trapped in the capillary bed. The glass contains ^{89}Y, which is irradiated in a reactor for approximately a week to produce the ^{90}Y, cf. eq. (13.3). These spheres contain a higher activity, roughly 2.5 MBq (67.6 µCi) per sphere, and there are roughly 2–8 million spheres per dose. The second type of microsphere is called *Sir-Spheres®* for selected internal radiotherapy. The spheres were approved in 2002 and comprise a resin of 20–40 µm that has the ^{90}Y loaded onto it. These spheres contain 50 Bq (1.35 nCi) per sphere and roughly 40–80 million spheres per dose.

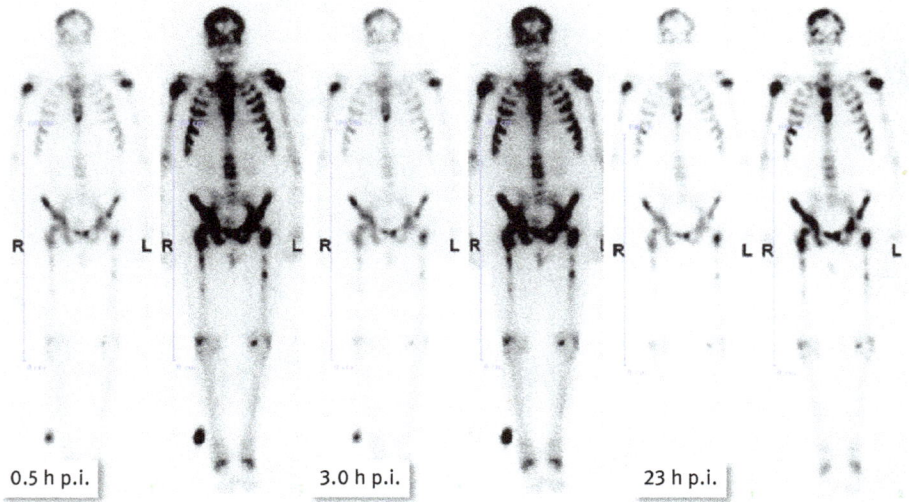

Fig. 13.19: Skeletal uptake of ^{177}Lu-BPAMD, a DOTA-amide-conjugated bisphosphonate, at three time-points post injection of 5 GBq ^{177}Lu, in a patient suffering from bone metastases.

Both sets of spheres are administered via a microcatheter located in the femoral artery, which delivers the arterial supply to the liver. A microcatheter is advanced into the hepatic artery or segment branch, cf. Fig. 13.20.

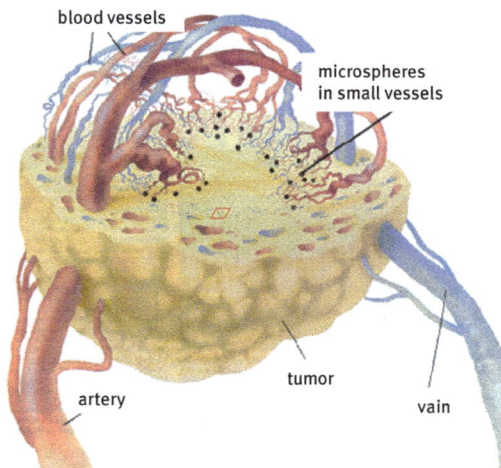

Fig. 13.20: Spheres are administered via catheter through the hepatic artery (Nordion Physician information pamphlet) and microspheres being trapped in the capillary bed feeding a tumor.

Tumors larger than 3 mm receive 80–100% of their blood supply from the hepatic artery thus this results in high delivery of the spheres to hypervascularized tumor capillary beds. Normal liver tissue receives blood from the portal vein. The spheres are used in cases of hepatic neoplasia and metastases as well as in metastatic colon cancer. The spheres are trapped in the precapillary alveolar level of the tumors. Thus, the highest dose is on the boundary of the tumors and falls off toward the interior of the tumor. This delivers the beta radiation to the most active growing site of the tumor. Very few spheres are delivered to the normal liver. Figure 13.21 is a scan of a patient undergoing a ^{99m}Tc macroaggregated human albumin (MAA) nuclear medicine scan to determine the extent of arteriovenous shunting to the lungs and to confirm the absence of gastric and duodenal flow. This is a pretest to ensure the microspheres are delivered selectively to the tumor.

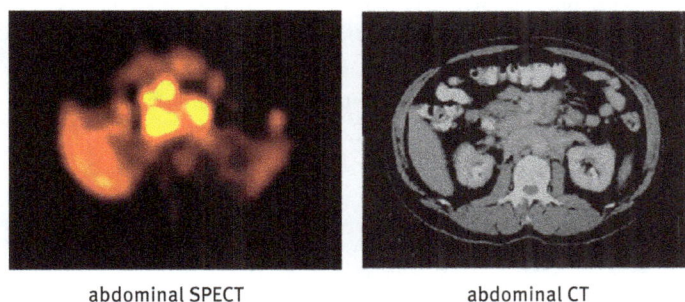

abdominal SPECT abdominal CT

Fig. 13.21: Planar image of ^{99m}Tc-MAA to determine shunting of doses of ^{90}Y microspheres to lungs or gastrointestinal tract. Planar anterior and posterior views of the chest and abdomen.

13.6.4 Soft tissue cancer overexpressing transmembrane receptors

Many types of cancer are characterized by an overexpression of specific transmembrane receptors, acting as G-protein coupled receptors. Those proteins are recognized at their extracellular site by small peptides in the classical sense of a *magic bullet*, cf. Fig. 13.22.

One of the most studied G-PCR has been for the treatment of neuroendocrine tumors. These tumors tend to grow at a slow rate and upon diagnosis tend to present as multiple metastases that are spread throughout the liver and abdomen. The tumor cells overexpress somatostatin receptors. Octreotide (OC) peptides have been designed as a drug to inhibit the somatostatin system. Radiolabeled versions for diagnostic imaging, such as ^{68}Ga-DOTA-TOC for PET/CT, are described in Chapter 12.

The chemical design of these diagnostic targeting vectors, when macrocyclic chelates such as DOTA are used, perfectly meets the requirements for introducing

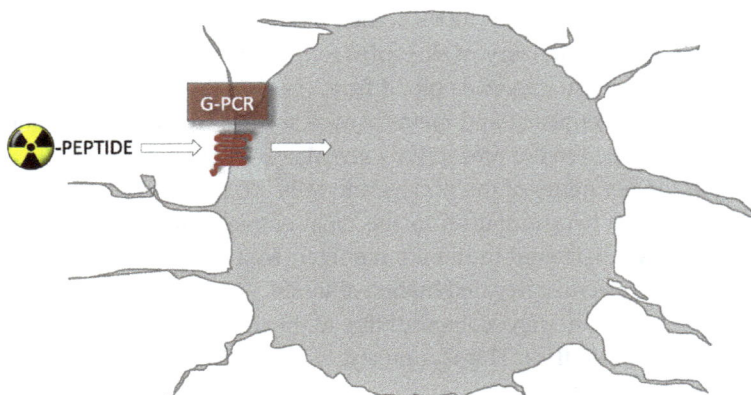

Fig. 13.22: Targeting tumor transmembrane receptors, typical G-protein coupled receptors (G-PCR), to deliver therapeutic radionuclides attached to receptor-specific peptides.

also therapeutic radionuclides – in particular in the case of trivalent therapeutic radiometals such as ^{90}Y, ^{177}Lu, ^{67}Cu, ^{225}Ac and ^{213}Bi, cf. Fig. 13.23.

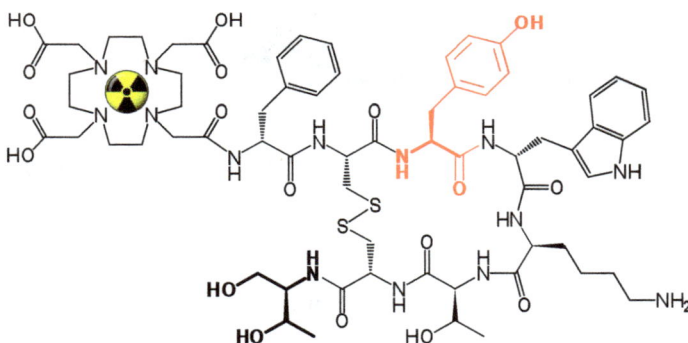

Fig. 13.23: Prototype of the radiopharmaceutical design of a tumor-affinity peptide (D)Phe^1-Tyr^3-octreotide, conjugated covalently through an amide bond at the N-terminus to the macrocycle DOTA. The DOTA will coordinatively bind the trivalent therapeutic radiometal. Red: The tyrosine part (T) of the peptide amino acid sequence makes the octreotide (OC) conjugate the TOC. The C-terminus makes the difference between a TOC and a TATE-derivative, cf. also Chapter 12.

Yttrium-90 has been attached to octreotide and used for somatostatin peptide receptor radiation therapy (PRRT) of neuroendocrine tumors. ^{177}Lu labeled octreotide analogs were investigated after patients treated with ^{90}Y-DOTA-TOC began showing signs of kidney toxicity with the expectation that a lower energy beta emitter such as ^{177}Lu may result in lower kidney doses. This led to a 30% objective response with a survival benefit of 40–72 months. Patients are typically given 4 cycles of 7.4 GBq (200 mCi) of ^{177}Lu-DOTA-TATE over a 6–10-week period.

Shown in Fig. 13.24 is the response of neuroendocrine tumors for one patient in a series of therapies with first ^{90}Y-DOTA-TOC and next ^{177}Lu-DOTA-TOC analogs. PET images shown are obtained from ^{68}Ga-DOTA-TOC as an analog. This imaging allows the physician to evaluate if the tumor expresses the somatostatin receptor and if the level of expression is high enough to allow for targeted therapy. The ^{68}Ga derivative has some differences in distribution, however the uptake in the spleen, liver, kidneys, and clearance closely follows that of ^{90}Y or ^{177}Lu, giving the physician a valuable tool to tailor the dose specifically for each patient to give ultimate dose to the tumor and minimize dose to normal tissues. The highest absorbed doses are to the spleen, the urinary bladder wall, and the kidneys. Due to the observed kidney toxicity co-infusion with amino acids is done.

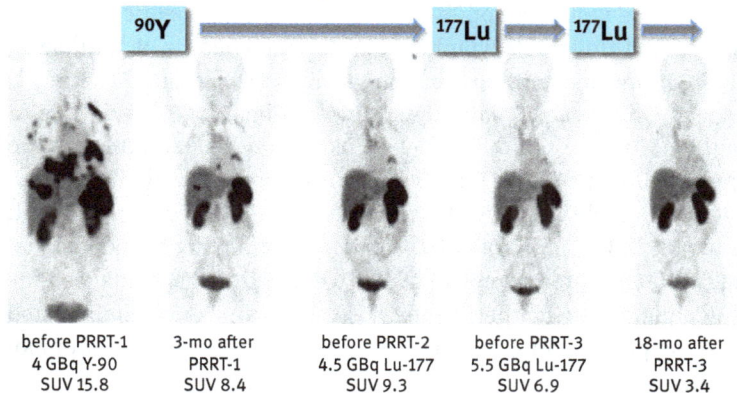

before PRRT-1	3-mo after	before PRRT-2	before PRRT-3	18-mo after
4 GBq Y-90	PRRT-1	4.5 GBq Lu-177	5.5 GBq Lu-177	PRRT-3
SUV 15.8	SUV 8.4	SUV 9.3	SUV 6.9	SUV 3.4

Fig. 13.24: PET images of DOTA-(D)Phe]-Tyr^3-octreotide (DOTA-TOC) labeled with ^{68}Ga for pre/therapeutic PET/CT-imaging and for validating the success of the PRRT (peptide receptor radiation therapy) treatment. ^{68}Ga-DOTA-TOC PET images of a patient prior and post treatment. For PRRT first 4 GBq of ^{90}Y-DOTA-TOC was applied. Three months later, 4.5 GBq of ^{177}Lu-DOTA-TOC was used and 18 months later another ^{177}Lu-DOTA-TOC treatment with 5.5 GBq was performed (courtesy of RP Baum).

Based on the promising results observed for these agents and the low toxicity profiles observed, particularly with ^{177}Lu, a number of peptides are being evaluated with ^{177}Lu and more are expected to enter the clinic. Several analogs of bombesin (BBN) have been radiolabeled with ^{177}Lu and evaluated for therapy of prostate cancer. The one that has been taken the farthest in the clinic for therapy is ^{177}Lu-AMBA. A problem encountered with these peptides is toxic side effects if given in large enough doses; thus, labeling in high specific activity appears to be a requirement with these peptides. New BBN analogs have been developed that show better tumor targeting with less normal tissue retention, especially for the antagonists.

13.6.5 Soft tissue cancer overexpressing antigens

Besides thyroid cancer there has been significant interest in attaching ^{131}I-radioiodine to targeting vectors such as peptides and antibodies to selectively deliver radioactivity to cancerous tissues. In order to accomplish this, radioiodine must be stably bound to the organic targeting vectors. Various methods have been developed to attach radioiodine to these vectors. Two main routes have been developed: direct labeling that involves attachment of the radioiodine to an amino acid such as tyrosine, or indirect labeling which uses the attachment of radioiodine via a prosthetic group. The synthetic strategies of radioiodine radiopharmaceutical chemistry introduced in Chapter 12 apply. In both cases, a covalent bond is obtained. A typical application is the radiotherapy with ^{131}I-labeled monoclonal antibodies (mab).

Bexxar® was one of the first FDA approved radioimmunotherapeutic agents for use in the treatment of non-Hodgkin's lymphoma (NHL) and consists of ^{131}I-labeled tositumomab, an IgG G2A murine antibody that targets the CD20 antigen. Therapy is based on patient-specific pharmacokinetics determined by imaging 7–14 days prior to treatment. Initially 450 mg of nonradiolabeled tositumomab is administered to block CD20 sites present on circulating B-cells and B-cells present in the bone marrow, and spleen. It additionally improves tumor uptake of the radiolabeled agent. Next a small amount 185 MBq (5 mCi) of ^{131}I-Tositumomab on 35 mg of monoclonal antibody is administered and three scans are taken to determine the residence time and to calculate the patient-specific therapeutic dose. Then a dose is administered to result in 75 cGy to the patient, adjusted to 65 cGy if concerns regarding thrombocytopenia are present.

The advantages of ^{131}I lie in the concept of attaching iodine covalently to mainly tyrosine residues of the amino acid sequences, which does not alter the pharmacology of the radioiodinated monoclonal antibody, and in the relatively low range of its β^--particles compared to, e.g., ^{90}Y, additionally it can be imaged; see Fig. 13.25.

[90]Y is also used for the treatment of non-Hodgkin's lymphoma. Zevalin™ is a targeted radioimmunotherapy treatment comprised of ^{90}Y labeled ibritumomab, a murine IgG1 monoclonal antibody that targets the CD20 antigen overexpressed on the surface of B-cells or non-Hodgkin's lymphoma cells. The CD20 antigen is not shed and does not internalize or modulate upon binding to the antibody. Different to radio-iodination of proteins, the trivalent radiometal does not bind to the amino acid sequence of proteins. Like for peptides, bifunctional chelators are needed to attach the metal to the antibody, cf. Fig. 13.26.

The ^{90}Y is stably complexed to the antibody also by use of Tiuxetan [N-[2-bis (carboxymethyl)amino]-3-(p-isothiocyanatophenyl)-propyl]-[N-[2-bis(carboxymethyl)amino]-2-(methyl)-ethyl]glycine, a modified version of DTPA whose carbon backbone contains an isothiocyanatobenzyl functionality, which forms a strong urea-type bond with the antibody. Imaging is done initially with a 185 MBq (5 mCi) (250 mg/m^2) dose of ^{111}In-Zevalin™ as a surrogate tracer from 2 h out to 120 h post

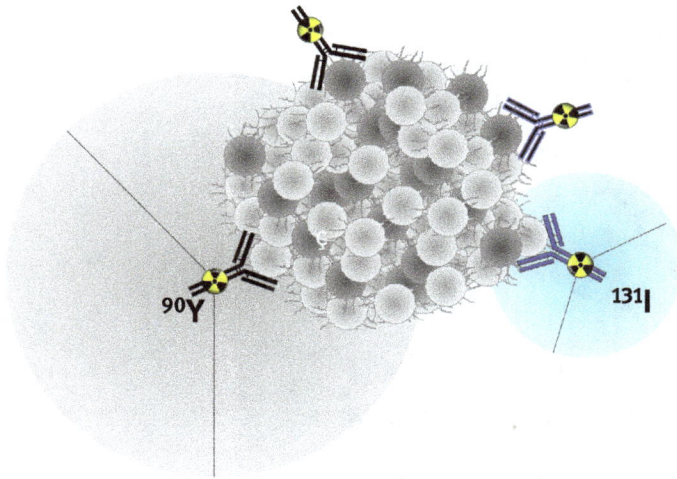

Fig. 13.25: Different track ranges of the β^--particles released from ^{131}I relative to ^{90}Y, both attached to antibodies targeting antigens.

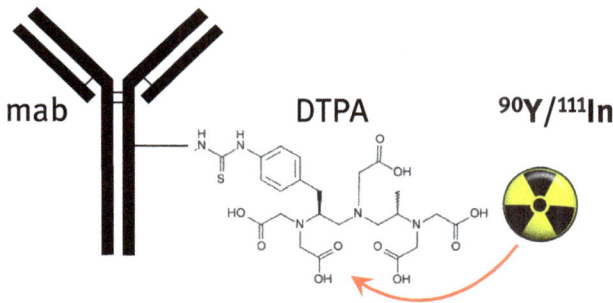

Fig. 13.26: Examples of attachment of bifunctional chelators (in this case: DTPA), an amine function, to a monoclonal antibody (mab).

injection by planar scintigraphy and SPECT cameras.[21] Figure 13.27 shows a scintigraphy of ^{111}In-*Zevalin*™.

---

[21] The organ uptake and clearance of ^{111}In is measured by regional analysis of 5 to 8 serial gamma scans over one week. Blood draws are performed concurrently with the imaging to determine the activity in the blood. Residence times are calculated from area under the curve (fraction of injected activity versus time). Radiation absorbed doses are estimated by entering the residence times into the MIRDOSE 3.1 computer program (ORISE). Normal distribution of ^{111}In-*Zevalin*™ includes uptake in the blood pool; moderately high uptake in the liver, spleen; and low uptake in the kidneys, bladder and bowel. Tumor uptake may or may not be seen and is not necessary for therapy to proceed. Altered biodistribution results in higher uptake in organs such as the lungs, kidneys, or bowels that is higher than that observed in the liver and indicates the patient should not proceed with therapy.

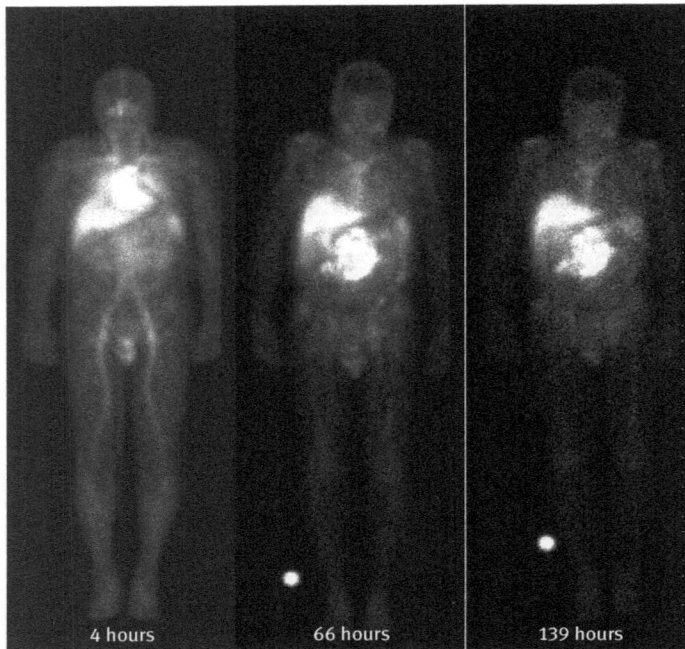

Fig. 13.27: Scintigraphy of ^{111}In-*Zevalin*™.

The first clinical evaluation of ^{177}Lu was for the treatment of ovarian cancer. Lutetium-177 was attached to the CC49 antibody via the PA-DOTA chelator {α-[2;(4-aminophenyl)ethyl-1,4,7,10-tetraazacyclodecane-1,4,7,10-tetraacetic acid} via reaction with a primary amine to form a thiourea linkage. The CC49 used in these initial trials was a murine monoclonal antibody that targets the tumor-associated glycoprotein TAG-72 found overexpressed in various carcinomas. Initial studies in mice bearing human colon carcinoma cells showed the potential therapeutic efficacy of ^{177}Lu labeled CC49, particularly the advantage gained by administering in fractionated doses and its ability to ablate large human colon xenografts in mice. Based on these studies a phase 1 study was performed in nine patients with carcinoma that expressed Tag-72. At the initial dose three patients were treated with 370 MBq/m^2 (10 mCi/m^2), which developed only grade one thrombocytopenia.

There has been a fair amount of interest in using ^{211}At radiolabeled antibodies for treating radioresistant tumors, treatment of minimal residual disease and metastatic diseases, conditioning for hematopoietic cell transplantation, bone pain palliation and treatment of cancer in isolated body compartments such as ovarian or glioma cancers. Astatine-211 attached to MX35 F(ab')$_2$, an antibody fragment, was evaluated for intraperitoneal α-radioimmunotherapy of disseminated ovarian cancer initially in tumor bearing mice and then in nine patients receiving 22.4–104.1 MBq/l that was infused via dialysis into the peritoneal with a catheter. No significant toxicity was observed and the dose to the peritoneum was estimated to be 15.6 ± 1.0 mGy, 0.14 ± 0.04 mGy to the red bone marrow, 0.77 ± 0.19 mGy to the urinary bladder wall, 24.7 ± 11.1 mGy to the unblocked thyroid and 1.4 ± 1.6 mGy to the blocked thyroid. Imaging was used along with measuring biological parameters to determine the optimal dose for each patient.

13.7 Outlook

13.7.1 More nuclides

Currently, there is a relatively large number of β^--emitters routinely used for ERT. The number of α-emitters is low – but growing and indeed there are routinely, and commercially applied radiotherapeutics labeled with α-emitters. In addition, radio-therapeutics labeled with Auger electron-emitting radionuclides are in the clinics, due to significant expectations related to those low-range radionuclides. However, the de-livery of the labeled targeting vectors close to tumor DNA is a challenge: It requires a targeting vector that selectively accumulates at the tumor cell, an internalization into the cytosol, and another moiety to deliver the nuclide close or into the cell nucleus.[22]

13.7.2 More radiotherapeutics

Over the last decades, there has been a steady increase in the development of selec-tive tumor targeting vectors. R&D (cf. Chapter 11) is continuing to contribute ever more candidates that can serve as lead structures for radiotherapeutics. An exciting development concerns the identification of an increasing number of small peptides with high affinity and selectivity for certain tumor G-PCRs. The recent development of PSMA motifs to target prostate cancer is an excellent example: The ^{68}Ga-PSMA structures have shown great success over the last few years, and since 2014 those structures are being translated into DOTA-PSMA analogs which are labeled with ^{177}Lu and ^{225}Ac. There are expectations that this strategy will apply to breast, ovar-ian, and other cancers.

---

22 For a list of candidates, cf. Chapter 12 in Vol. I, Tab. 12.5.

13.7.3 Theranostics

One of the problems with the use of ^{90}Y is that since it is a pure beta emitter it cannot be followed by either SPECT or quantitative PET imaging, although *bremsstrahlung*[23] and an extremely low fraction of annihilation radiation can be detected. This raises concerns due to incorrect delivery and accurate calculation of dose to tumor and normal tissues. The patient treated with such an "ideal" therapeutic radionuclide is a black box, cf. Fig. 13.28. This holds true for the concept of ERT in general: Whenever radionuclides are needed to emit radiation within a very short range in order to deposit a maximum radiation dose to the desired target and keep the surrounding tissue almost unaffected, there should be no (or very little) penetrating radiation which escapes from the human body that can be used for qualitative or quantitative imaging. In reality, most of the therapeutic radionuclides do emit photons useful for imaging and yielding qualitative information, but mostly at low branching, which prevents direct access to quantitative information.

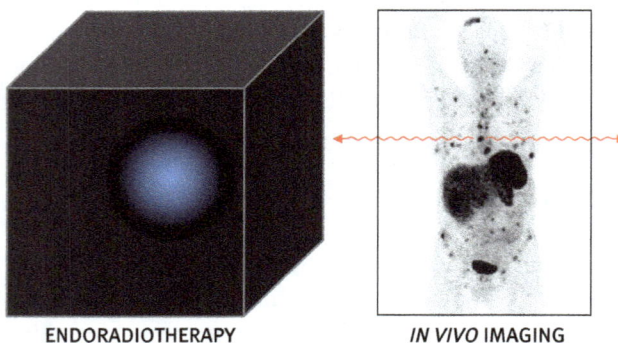

ENDORADIOTHERAPY   *IN VIVO* IMAGING

Fig. 13.28: The patient treated with an "ideal" therapeutic radionuclide is like a black box. This is exactly opposite to the diagnostic approach with radionuclides emitting photons. Here, the patient is "transparent".

Obviously, the ability to image the agent would be a benefit. By imaging the agent, it is possible to determine how a specific patient is handling the drug, such as its uptake at the site, how fast it is being cleared, how fast it is being metabolized. It also aids in estimating the dose that is delivered to normal tissues. This capability allows the doctor to deliver an amount that is more effective at treating the patient than if imaging is not possible. If imaging is not possible the doctor is restricted to giving an amount of the drug that in the typical patient will not result in toxicity. It has been estimated that this results in doses that are not effective 25–65% of the time.

---

23 Cf. Chapter 11 in Vol. I.

Theranostics is a combination of the words "Therapy" and "Diagnosis". It refers to the increasingly close interlocking of diagnosis and therapy. The aim of Theranostics is to allow for the right therapy for the right person at the right time with the right drug. Theranostics thus is a modern therapy-accompanying diagnosis with the ultimate aim of a patient-specific treatment. It is also described by *personalized medicine* and *predictive medicine*. In the context of nuclear medicine and radiopharmaceutical chemistry it basically includes the determination of the efficacy of a therapeutic drug for a given disease for an individual patient and evaluation of the toxicity to normal tissues. What is needed for a given radiotherapeutic compound is a positron emitter nuclide analogue or at least an imaging nuclide with similar chemistry *in vivo*, cf. Fig. 13.29.

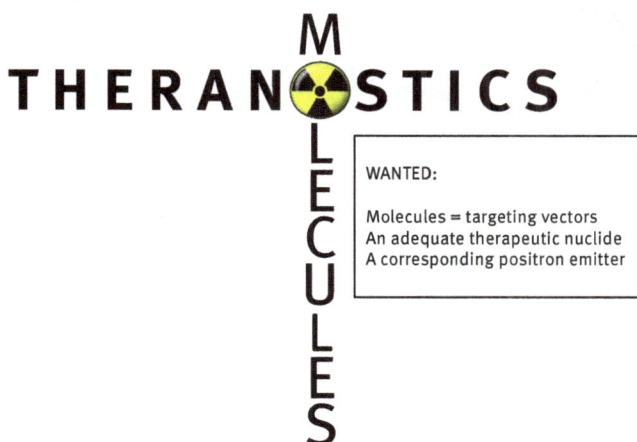

M
THERAN☢STICS
L
E
C
U
L
E
S

WANTED:

Molecules = targeting vectors
An adequate therapeutic nuclide
A corresponding positron emitter

Fig. 13.29: Very simple scheme of the theranostic concept applied to nuclear medicine. A radiotherapeutic contains a therapeutic radionuclide within its molecular structure. Keeping that molecule the same, a positron emitter is needed with similar chemistry which will fit into the same "labeling" position as the therapeutic nuclide. Changes in chemical characteristics due to the new nuclide should be minimal – to ensure that changes in pharmacology between the therapeutic and the diagnostic derivative are negligible.

An ideal option is to apply a diagnostic isotope of the same chemical element as the therapeutic nuclide, called "matched pairs". There are several options to pair a positron emitting nuclide with the therapeutic nuclide, such as ^{124}I for ^{131}I, ^{86}Y for ^{90}Y, ^{64}Cu for ^{67}Cu. The positron emitting yttrium isotope ^{86}Y (t½ = 13.6 h, 36% β⁺ branch) may be used as a surrogate and PET may be applied for pre-therapeutic patient-individual diagnosis. Figure 13.30 illustrates how quantitative PET imaging allows both monitoring of the uptake and pharmacology of a ^{90}Y-radiotherapeutic and calculation of organ-specific radiation doses for the therapeutic isotope. In this case it

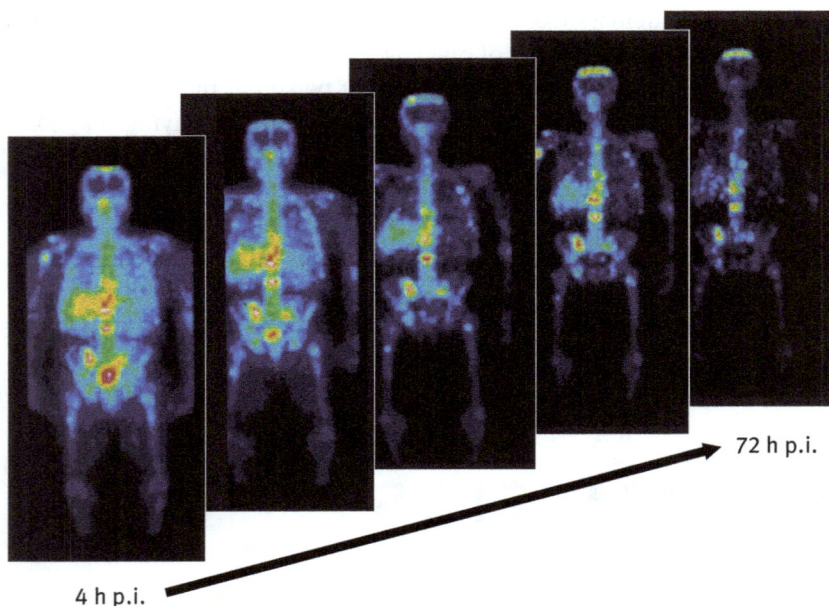

72 h p.i.

4 h p.i.

Fig. 13.30: PET image of the whole-body distribution and pharmacology of ^{86}Y citrate as a surrogate of ^{90}Y-citrate for bone metastases treatment, see also Section 13.6.3 (courtesy Roesch et al. 1996).

even allows for calculation of the radiation dose ^{90}Y-citrate delivered to the bone metastases, the red bone marrow and the healthy organs such as liver.

This was one of the first applications performed in 1996, when quantitative PET imaging of the pharmacology of two different ^{90}Y-compounds considered for the palliative treatment of bone metastases, namely ^{90}Y-citrate and ^{90}Y-EDTMP, was used to determine radiation dose, comparing ^{86}Y-citrate and ^{86}Y-EDTMP. It allowed for calculation of the radiation dose values delivered in terms of [mGy/MBq ^{90}Y injected]. The values for the bone metastases, the red bone marrow and the healthy organs such as liver were 25.5, 2.5, and 1.8 for ^{90}Y-citrate and 17.8, 1.8, and 0 for ^{90}Y-EDTMP. This enabled the physician to choose the most effective radiotherapeutic and the most appropriate amount of radioactivity to administer to the individual patient to obtain optimal dosing. Similar approaches were used when the dosimetry of ^{90}Y-DOTA-TOC was investigated using ^{86}Y-DOTA-TOC prior to its clinical application. Fortunately, other several "matched pairs" exist. Ideal examples are the even isotopic pairs ^{124}I for ^{131}I, ^{86}Y for ^{90}Y, ^{64}Cu for ^{67}Cu.

If matched pairs are not available, surrogate radionuclides may be used that do not represent another isotope of the therapeutic nuclide, but a chemically similar nuclide of a different chemical element. The "surrogate" versions are for example the positron emitters ^{68}Ga and ^{44}Sc to "match" the trivalent radiometals ^{90}Y, the radiolanthanides ^{153}Sm, ^{161}Tb, ^{177}Lu, etc., and also for ^{213}Bi or ^{225}Ac.

A proof-of-principle application is SPECT imaging using ^{111}In-labeled pharmaceuticals or PET/CT imaging using ^{68}Ga-labeled pharmaceuticals in a pre- or post-therapeutic strategy for [90]Y- or ^{177}Lu-labeled analogue targeting vectors. A prominent application is the endoradiotherapy of neuroendocrine tumors with [177]Lu- or ^{90}Y-labeled DOTA-octreotide derivatives with [68]Ga- or [44]Sc-labeled analogue molecules, cf. Fig. 13.24.

More recently, the prostate specific membrane antigen (PSMA) has been identified as a suitable target for prostate cancer diagnosis and therapy. The corresponding binding motif is a very small peptidometic structure of Glu-urea-Lys (KuE). Several radiolabeled derivatives have been developed for molecular imaging using, e.g., fluorine-18 and gallium-68, cf. Chapter 12 for the concept and for the structure of those radiopharmaceuticals. Overexpressed in a majority of prostate cancer cells, PSMA inhibitors can allow effective use not only as a molecular imaging agent, but also for targeted therapy of castration resistant prostate cancer. A recent example is the use of theranostics for metastatic castration-resistant prostate cancer evaluating the potential of prostate-specific membrane antigen (PSMA)-targeted radioligand therapy with initial PET imaging with ^{68}Ga-labeled inhibitors followed by therapy with ^{177}Lu and/or [225]Ac-labeled analogs. A beta emitter ^{177}Lu or alpha emitter [225]Ac binding with a PSMA inhibitor was demonstrated to be effective as a theranostics agent. This theranostic approach, using [^{68}Ga]Ga-PSMA-11 for patient selection and decision-making informing therapy and monitoring has led to favorable outcomes with personalized medicine. Clinical evaluation of [^{177}Lu]Lu-DOTA-PSMA-617 has already passed phase III (VISION trial, and its market authorization as *[^{177}Lu]vipivotide tetraxetan[©]* is expected soon.

Shown in Fig. 13.31 is a patient in which all convential treatments were exhausted. The patient entered a trial for [^{177}Lu]Lu-PSMA-617 and as demonstrated by the initial [^{68}Ga]Ga-PSMA image tumors showed high avidity for ^{68}Ga-PSMA. The second image taken after two cycles of [^{177}Lu]Lu-PSMA administration (7.4 GBq per cycle) showed tumor progression as seen in the ^{68}Ga-PSMA image and the PSA level that rose from 294 ng/mL to 419 ng/mL. The patient was switched to [[225]Ac]Ac-PSMA-617 and received two administrations (6.4 MBq, 100 kBq per kilogram of body weight). A third round of treatment was given and the [^{68}Ga]Ga-PSMA-11 PET/CT image indicated a complete remission and a drop in PSA to < 0.1 ng/mL, highlighting how theranostics effectively changes patient treatment and implements choice of radionuclide and personalized medicine optimizing patient outcome.

In addition to the key/lock systems SSTR/octreotide derivatives for neuroendocrine tumors and PSMA/PSMA inhibitors for prostate cancer, there are other examples validated for diagnosing more types of cancer in nuclear medicine which can be turned into theranostics. Important cases are bombesin (BBN) analogues for the gastrin-releasing peptide (GRPR) in PSMA-negative prostate cancer, neurotensin antagonists for neurotensin receptors for various human cancers including pancreatic adenocarcinoma, peptidic vectors for the C-X-C motif chemokine receptor

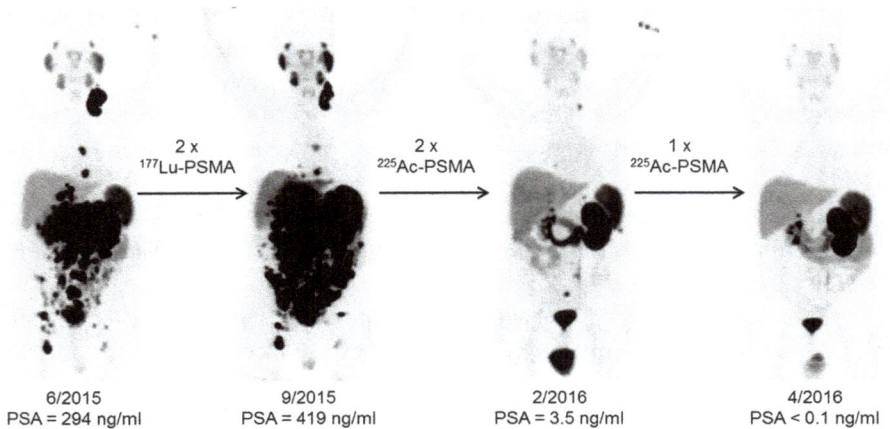

(i.e., CXCR4), which is overexpressed by many different tumor entities, folic acid derivatives for the folate receptor system for different endothelial human carcinoma including ovarian carcinomas.

Another example is the use of the same DOTA-conjugated bisphosphonate for the treatment of disseminated bone metastases, cf. Fig. 13.32.

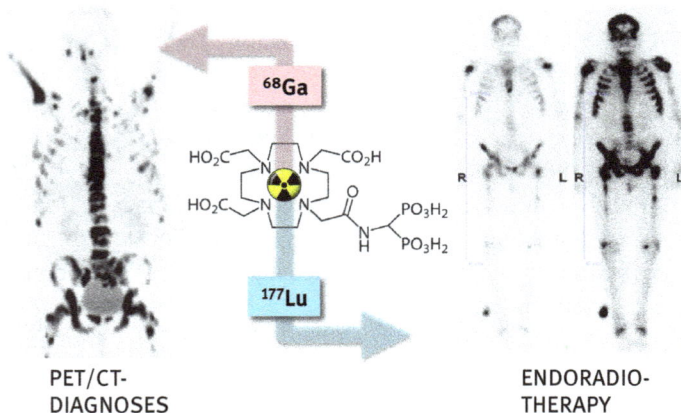

PET/CT-
DIAGNOSES

ENDORADIO-
THERAPY

Fig. 13.32: Example of a theranostic approach in nuclear medicine: The same targeting vector, the DOTA-conjugate bisphosphonate BPAMD, is labeled first with ^{68}Ga for quantitative PET to verify the bone metastases expression and the intensity of the tracer uptake (in SUV). Next the same patient undergoes therapy with the ^{177}Lu-BPAMD analog (courtesy Roesch 2011).

Different to specific molecular tumor targets such as the somatostatin receptor of neuroendocrine tumors and the prostate specific membrane antigen of prostate cancer more recently the tumor microenvironment has been identified as a suitable target for cancer diagnosis and therapy. Fibroblast activation protein (FAP), a type II transmembrane serine protease is highly expressed in the microenvireonnment of various cancers and is being evaluated as a target for diagnosis and therapy of many cancer types, cf. Chapter 12; or a recent review cf. *New frontiers in cancer imaging and therapy based on radiolabeled fibroblast activation protein inhibitors: A rational review and current progress* (Imlimthan et al. (2021)). It promotes tumor growth, proliferation, and angiogenesis. This expression of FAP in a broad spectrum of cancers offers an optimum target for theranostics and personalized medicine. Recently, several fibroblast activation protein specific radiopharmaceuticals have been evaluated for theranostics of numerous malignant tumors using radiolabeled FAPI species such as

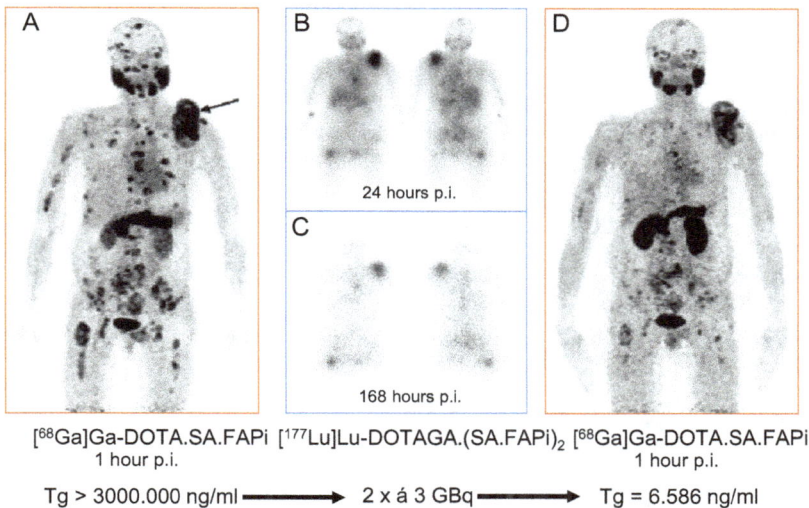

[^{68}Ga]Ga-DOTA.SA.FAPi [^{177}Lu]Lu-DOTAGA.(SA.FAPi)$_2$ [^{68}Ga]Ga-DOTA.SA.FAPi
  1 hour p.i.   1 hour p.i.

Tg > 3000.000 ng/ml ⟶ 2 x á 3 GBq ⟶ Tg = 6.586 ng/ml

Fig. 13.33: FAP inhibitor-based theranostic approach demonstrated for a 50-year-old woman with follicular variant of papillary carcinoma failed radioiodine therapy, Soranib and Lenvatinib. (A) Baseline [^{68}Ga]Ga-DOTA.SA.FAPi PET/CT MIP, normal biodistribution in the oral mucosa, salivary glands, liver, pancreas, gall bladder, colon, kidneys. Intense accumulation of radiotracer in the soft tissue mass (black arrow) and multiple skeletal sites. (B,C) [^{177}Lu]Lu-DOTAGA.(SA.FAPi)$_2$ whole body scintigraphic images, after intravenous injection of 40 mCi of [^{177}Lu]Lu-DOTAGA.(SA.FAPi)$_2$, radiotracer retention in the metastatic sites till 168-h delayed images. Patient received two cycles of [^{177}Lu]Lu-DOTAGA.(SA.FAPi)$_2$ therapy, resulting in significant clinical improvement with a decrease of thyroglobulin levels. (D) follow-up MIP [^{68}Ga]Ga-DOTA.SA.FAPi PET/CT demonstrating reduction in the SUVmax values after treatment; significant decrease in the VASmax score from 10 to 4 in a follow-up of 9 months. Ballal S, Yadav MP, Moon ES, Roesch F, Kumari S, Agarwal S, Tripathi M, Sahoo RK, Mangu BS, Tupalli A, Bal C. Novel Fibroblast Activation Protein Inhibitor-Based Targeted Theranostics for Radioiodine-Refractory Differentiated Thyroid Cancer Patients: A Pilot Study. Thyroid. 2022, 32(1), 65–77. doi: 10.1089/thy.2021.0412. With permission from the authors.

peptides or inhibitors comprising the DOTA and DOTAGA chelator covalently attached to derivatives of the FAP targeting vectors, cf. Chapter 12. These conjugates offer the possibility to use imaging radionuclides such as ^{68}Ga as well as therapeutic radionuclides such as ^{225}Ac and ^{177}Lu. Recent work demonstrated that the tumor retention could be increased by going from a monomer to a dimer and that the use of the DOTAGA chelator radiolabeled with ^{177}Lu attached to the FAPi dimer [^{177}Lu]Lu-DOTAGA.(SA.FAPi)$_2$ resulted in an agent that was well tolerated in patients up to 3 cycles (1.48 GBq) and should be further evaluated in clinical trials, Fig. 13.33.

References

Blower PJ, Prakash S. The chemistry of rhenium in nuclear medicine. Perspect Bioinorg Chem 1999, 4, 91–143.

Bryan JN, Bommarito D, Kim DY, Berent LM, Bryan ME, Lattimer JC, et al. Comparison of systemic toxicities of ^{177}Lu-DOTMP and ^{153}Sm-EDTMP administered intravenously at equivalent skeletal doses to normal doses. J Nucl Med Technol 2009, 37, 45–52.

Cutler CS, Smith CJ, Ehrhardt GJ, Tyler TT, Jurisson SS, Deutsch E. Current and potential therapeutic uses of lanthanide radioisotopes. Cancer Biother Radiopharm 2000, 15, 531–545.

Donnelly PS. The role of coordination chemistry in the development of copper and rhenium radiopharmaceuticals. Dalton Trans 2011, 40, 999–1010.

Imlimthan S, Moon ES, Rathke H, Afshar-Oromieh A, Rösch F, Rominger A, Gourni E. New frontiers in cancer imaging and therapy based on radiolabeled fibroblast activation protein inhibitors: A rational review and current progress. Pharmaceuticals 2021, 14, 1023. https://doi.org/10.3390/ph14101023

Liu S. Bifunctional coupling agents for radiolabeling of biomolecules and target-specific delivery of metallic radionuclides. Adv Drug Deliv Rev 2008, 60, 1347–1370.

Liu G, Hnatowich DJ. Labeling biomolecules with radiorhenium – a review of the bifunctional chelators. Anti-Cancer Agents Med Chem 2007, 7, 367–377.

Paes FM, Serafini AN. Systemic metabolic radiopharmaceutical therapy in the treatment of metastatic bone pain. Semin Nucl Med 2010, 40, 89–104.

Roesch F, Baum RP. Generator-based PET radiopharmaceuticals for molecular imaging of tumours: On the way to THERANOSTICS. Dalton Trans 2011, 40, 6104–11.

Roesch F, Herzog H, Qaim SM. The beginning and development of the theranostic approach in nuclear medicine, as exemplified by the radionuclide pair ^{86}Y and ^{90}Y. Pharmaceuticals 2017, 10, 1–28.

Roesch F, Knapp FF. Radionuclide generators. In: Vértes A, Nagy S, Klencsár Z, Lovas RG, Roesch F. eds. Handbook of nuclear chemistry, Vol. 4, 2nd ed. Berlin-Heidelberg, Springer, 2011, 1935–1976.

Symbols and abbreviations

Symbol/ Abbreviation	Meaning	Chapter
A		
A	mass number	1
A	area of neighboring mass peaks	4
$\acute{A}$	(absolute) radioactivity	1, 3, 5, 12
Å	Ångström	6
A_{abs}	absorption	6
$\acute{A}_{cal}$	known activity	1
$\acute{A}_{corrected}$	transformation-corrected activity	3
$\acute{A}_{sat}$	saturation activity	3
A_{spec}	specific activity	11
$A_S(t)$	activity in source organ S at a certain time t	2
AAT	amino acid transporters	12
AADC	aromatic L-amino acid decarboxylase	12
AD	Alzheimer's disease	12
ADME	absorption, distribution, metabolism, excretion	11
AGR	advanced gas-cooled reactor	10
AIDA	detection system for alpha spectroscopy	7
AMS	accelerator mass spectrometry	4, 5
AP	antipsychotic	12
APP	amyloid precursor protein	12
ARCA	automated rapid chemistry apparatus	7, 9
AS	abundance sensitivity	4
a_x	effective attenuation depth	5
B		
B	magnetic field (strength)	4, 6
BP	"before present"	5
BWR	boiling light-water-cooled and moderated reactor	10
b	barn	3, 10
C		
C	count rates	3
C	actual detection rate (or count rate)	1
C^*	compound nucleus	9
C_s, C_m, C_f, C_t	compartments	12
C_{cal}	known count rate	1
CANDU	Canada deuterium uranium reactor	10
CAT	para-tosylchloramide sodium	12
CE	capillary electrophoresis	6
CIC	constant initial concentration	5
CNS	central nervous system	12
CT	computed tomography	1, 2
CPAA	charged particle activation analysis	3
c	speed of light	9

https://doi.org/10.1515/9783110742701-014

(continued)

Symbol/ Abbreviation	Meaning	Chapter
c	concentration	6
c_{aq}	concentration of a solute in the aqueous phase	7
c_{org}	concentration of a solute in the organic phase	7
c_p	heat capacity	10
c_{res}	concentration of an ion in the ion exchange resin	7
c_{sample}	concentration of an element in a sample	3
cps	counts-per-seconds	1
chi	magnetic susceptibility	8
D		
D	absorbed dose	1, 2
D	diffusion coefficient	6
D	dissociation energy	11
D_γ	gamma-radiation absorbed dose	2
D_n	neutron absorbed dose	2
D^{IE}	ratio between the concentration of an ion in the organic phase and its concentration in the aqueous phase	7
$D_{R,T}$	mean absorbed dose from radiation of type R in tissue T	2
D^{SE}	ratio between the concentration of a solute in the organic phase and its concentration in the aqueous phase	7
D_T	mean absorbed dose	2
$D(sf)$	spontaneous fission track densities	5
$D(n,if)$	induced fission track densities	5
D_1-D_5	dopaminergic receptors	12
DA	dopamine	12
DAT	dopamine (re-uptake) transporters	12
DD	daughter deficient situation	5
DDI	drug-drug interaction	11
DE	daughter excess situation	5
DGNAA	delayed gamma neutron activation analysis	3
DNNA	delayed neutron activation analysis	3
DOTA	1,4,7,10-tetraazacyclododecane-N,N',N'',N'''-tetraacetic acid	12
DSB	double-strand break	2
DTPA	diethyltriamine-N,N,N',N'',N''-pentaacetic acid	12
d	thickness of the sample	6
dE	energy loss	1
dx	unit length	1
E		
E	energy	2
E	electric field	6
E	electrical field strength	6
$\bar{E}_{kin}$	mean kinetic energy	10
E_0	absorption edge energy	6
E_0	zero-point energies	11

(continued)

Symbol/ Abbreviation	Meaning	Chapter
E_a	electron affinity	4
$\bar{E}_B$	neutron binding energy	9
$^{\star}E$	excitation energies	9
E_γ	gamma energy	1
E_{bind}	binding energy	1, 6
E_c	critical energy	1
E_{eject}	kinetic energy of the ejected electron (photo) electron	1
E_i	energy of n_i particles	1
E_{IC}	kinetic energy of inner conversion electrons	13
E_{kin}	kinetic energy	1, 4, 6
$E^{max}\beta^-$	maximum kinetic energy of β^- particles	13
$E^{scatter}$	scatter energy	1
E_{tr}	total energy transferred by radiation in gas detectors	1
$E^{X\text{-ray}}$	X-ray energy	6
ED	electro-deposition	7
EM	electromigration	7
EPCARD	European program package for calculation of aviation route doses	2
EPMS	extra-pyramidal side effects	12
EPR	electron paramagnetic resonance	8
ERR	excess relative risk	2
ERPF	effective renal plasma flow	12
ESI	electrospray ionization	6
ESI-MS	electrospray ionization mass spectrometry	6
EXAFS	extended X-ray absorption fine structure	6
e	absolute value of the electron charge ($1.6 \cdot 10^{-19}$ C)	1
e	elementary electric charge	4, 9
e_0	elementary charge of the electron	6
F		
F_{centri}	centripetal force	4
F_{Lor}	LORENTZ force for spatial separation	4
FBR	fast breeder reactor	10
FDA	federal drug administration (USA)	13
FDG	2-deoxy-2-[^{18}F]fluoro-D-glucose	7
FFF	Field Flow Fractionation	6
FID	Free Induction Decay	6
f	branching ratio	5
$f(v)$	probability density functions	7
$f(k)$	scattering parameters	6
G		
$G(E)$	true value of the dosimetric quantity	1
Gy	Gray, unit of the absorbed dose	1, 2
GEANT4	particle transport code	2
GCP	good clinical practice	12

(continued)

Symbol/ Abbreviation	Meaning	Chapter
GCR	gas-cooled, graphite-moderated reactor	10
GDMS	glow discharge mass spectrometry	4
GFR	glomerular filtration rate	12
GLP	good laboratory practice	11, 12
GLUT	glucose transporters	12
GMP	good manufacturing practice	11, 12
GM	Geiger–Müller counters	1
GPCR	G-protein coupled receptors	12
g_+	statistical weights	4
g_0	statistical weights	4
g_I	nucleon g-factor	6
H		
H	abundance of the stable target isotope	3
H	height of neighboring mass peaks	4
H^*	ambient dose equivalent	1
H_p	personal dose equivalent	1
H_T	equivalent dose	2
$H^*(10)$	ambient dose equivalent	2
$H'(d, \Omega)$	directional dose equivalent	2
$H_p(d)$	personal dose equivalent	2
$H(T \leftarrow S, t)$	equivalent dose absorbed in organ T	2
HR-RIMS	high-resolution RIMS	
HTGR	high-temperature gas-cooled reactor	10
h	PLANCK constant	1
$\hbar$	reduced PLANCK constant	6, 9
$h\upsilon_F$	fluorescence light	6
$h\upsilon_{abs}$	stimulating light	6
I		
I	intensity	6
I	ionic strength	6
I_0	intensity of the incident beam	6
I_p	(first) ionization potential	4
IAEA	international atomic energy agency	3
ICP-MS	Inductively Coupled Plasma Mass Spectrometry	4, 5
ICRU	International Committee on Radiation Units	1, 2
ICRP	International Commission on Radiological Protection	1
INAA	Instrumental Neutron Activation Analysis	3
Iodogen™	1,3,4,6-tetrachloro-3α,6α-diphenylglycouril	12
IR	Infrared Spectroscopy	6
IRMM	Institute for Reference Measurements and Materials	4
i.v.	intravenously	11
J		
J	diffusion flux	6

(continued)

Symbol/ Abbreviation	Meaning	Chapter
K		
K_i	net rate of 2-[^{18}F]FDG intake	12
k	wave number	6
k_B	BOLTZMANN constant	1, 4, 7, 10
k_F	apparent fluorescence rate	6
k_{eff}	effective neutron multiplication factor	10
k_I	inherent rate	6
k_Q	quenching rate	6
k_1	2-[^{18}F]FDG uptake rate	12
k_2	2-[^{18}F]FDG efflux rate	12
k_3	2-[^{18}F]FDG phosphorylation rate	12
k_4	de-phosphorylation rate	12
(k)IE	isotope effect	11
L		
L	linear distance	10
LC	lumped constant	12
LD$_{50}$	dose that will be lethal for 50% of an exposed group of individuals	2
LET	linear energy transfer	2
LDM	liquid drop model	9
LIBD	laser-induced breakdown detection	6
LNT	linear-no-threshold	2
LCH	Large Hadron Collider	1
LOD	limit of detection	3, 4, 6
LPAS	laser photoacoustic spectroscopy	6
LSC	liquid scintillation counting	3, 5
LSS	life span study	2
LWGR	light-water-cooled, graphite-moderated reactor	10
M		
M	molar mass of the element	3
M	total number of energies available in the field	1
$M(E)$	actual dosimeter reading	1
MAG3	[^{99m}Tc]mercaptoacetyl triglycine	12
MAO	monoamine oxidases	12
MC	Monte Carlo codes	2
MCNPX	particle transport code	2
MP	Morbus Parkinson	12
MOX	UO_2 mixed oxide	10
$M1$	molar mass	1
m	mass	2, 3
m	cycles	10
m_e	electron mass	6
m_e	mass of the electron	9
m_l	magnetic quantum number	6

(continued)

Symbol/ Abbreviation	Meaning	Chapter
m_M	molecular mass	7
m_a	mass of the projectile	9
m_A	mass of the target	9
m_r	reduced mass	11
$m_{standard}$	mass of the standard	3
m_{sample}	mass of the sample	3
$\dot{m}_{H_2O}$	coolant mass flow rate	10
N		
N	number of particles passing	1
N	number of neighboring atoms	6
N_A	Avogadro constant/number	1, 3
N_d	number of the radioactive nuclides	1, 3, 5
N	number of stable daughter nuclei	5
N_p	number of target atoms	3
N	number of radioactive parent nuclides	5
N_{conc}	number concentration	6
N_{r1}	atomic density in the material of a detector	1
NDP	neutron depth profiling	3
NIS	sodium iodide symporter	12
NLD	neuroleptic drug	12
NMR	nuclear magnetic resonance spectroscopy	6
NODAGA	1,4,7-triazacyclononane-*N,N'*-diacetic acid-*N''*-glutaric acid	12
NOTA	1,4,7-triazacyclononane-*N,N',N''*-triacetic acid	12
NPP	nuclear power plant	10
n	electron shell	6
n_i	number of particles having energy E_i	1
n_{refr}	index of refraction	6
O		
OIH	*ortho*-[^{123}I]iodohippuric acid	12
P		
P	point in the field	1
$P_{0,x}$	production rate of species x at the surface	5
PAA	photon activation analysis	3
PD	pharmacodynamics	11
PET	positron emission tomography	1, 12
PGNAA	prompt gamma neutron activation analysis	3
PHITS	particle transport code	2
PHWR	pressurized heavy-water-moderated and cooled reactor	10
PK	pharmacokinetics	11
PUREX	plutonium uranium extraction (plutonium uranium redox extraction)	7, 8
PWR	pressurized light-water-moderated and cooled reactor	10
p	protons	1

(continued)

Symbol/ Abbreviation	Meaning	Chapter
p	impulse	7
p_C^*	fusion probability	9
pMC	percent Modern Carbon	5
Q		
Q	total electric charge	1
Q	quality factor	2
Q	adsorption heat ($J\ mol^{-1}$)	7
(Q)WBA	(quantitative) whole body autoradiography	11
R		
R	molar gas constant ($\approx 8.314\ J\ mol^{-1}\ K^{-1}$)	7
$*R$	nuclear reaction rate	3
R	radiation type	2
R	total kinetic energy of N particles	1
R	distance to neighbor atoms	6
R_{en}	(intrinsic) energy resolution	1
R_G	energy response of a detector	1
RBE	relative biological effectiveness	2
RIA	radioimmunoassay	11
RIMS	resonance ionization mass spectrometry	4
RMS	radioisotope mass spectrometry	4
RNAA	radiochemical neutron activation analysis	3
ROI	regions of interest	13
RSC	relative sensitivity coefficient	4
RSD	relative standard deviation	4
$\mathfrak{R}$	mass resolution	4
R&D	research and development	12
R_6, R_7	concordia parameters	5
r	radius	4, 6
r	orbital radius	9
r	bond distance	6
r	distance	1
r_{hd}	hydrodynamic ionic radius	6
r_f	retention fractions	7
r_t	retention times	7
S		
S	stopping power	1
S	source organ	2
S_c	COULOMB collisions	1
S_r	radiative losses	1
S_0	electronic ground state	6
S_1	electronic excited state	6
SI	international system of units	2
SEE	specific effective energy	2

(continued)

Symbol/ Abbreviation	Meaning	Chapter
SEM	secondary electron multiplier	4
SF_{AB}	separation factor	7
SHPP	N-succinimidyl-3-(4-hydroxyphenyl)propionate	12
SIMS	secondary ion mass spectrometry	4
SNAP	space nuclear auxiliary power	8
SPECT	single photon emission computed tomography	12
SSB	single-strand break	2
SUV	standard uptake value	12
Sv	Sievert, unit of the equivalent dose	1, 2
Sv/h	dose equivalent rate	1
s	surface	7
s	cross-sectional area of a sphere	1

T		
T	absolute temperature (K)	4, 7
T	reactor period	10
T	temperature	10
T_a	adsorption temperature	9
T_a	adsorption temperature of the substance	7
T_{boil}	boiling point	10
T_{melt}	melting point	10
T_s	starting temperature	7
T3	triiodothyronine	12
T4	thyroxine	12
TBP	tri-n-butyl phosphate	7
TC	thermochromatography	7
TCC	thermochromatographic columns	7
THTR	thorium high temperature reactors	10
TIMS	thermal ionization mass spectrometry	4
TOF	time-of-flight	12
TOF-MS	time-of-flight mass spectrometer	4
TRLFS	time-resolved laser-induced fluorescence spectroscopy	6
TRLFS	time-resolved laser-fluorescence	6
TSTU	O-(N-succinimidyl)-N-N,N′,N′-tetramethyluronium tetrafluoroborate	12
t	time	1, 5, 6
t_{int}	interaction times	4
t^{irr}	irradiation for the time period	3
t^{origin}	year of origin of the sample	5
τ	integration variable	12
τ	lifetime of each neutron generation	10

U		
U_B	electrostatic acceleration potential	4
UNSCEAR	United Nations Scientific Committee on the Effects of Atomic Radiation	2

(continued)

Symbol/ Abbreviation	Meaning	Chapter
UV-VIS	absorption spectroscopy	6
u	electromigration velocity	7
V		
v	velocity, speed	4, 7, 9, 10
$\hat{v}$	most probable speed	7
v	quantum number of vibration levels	6
v_d	fraction of delayed neutrons	10
$\bar{v}^2$	root mean speed	7
W		
W_e	average energy required to produce an electron-ion pair in the detector gas	1
W_e	work function (of the surface material)	4
W_{sur}	probability for the system to survive the evaporation cascade	9
w_T	tissue weighting factors	2
w_R	radiation weighting factor	2
X		
X	thickness	1
X	exposure	1
X	X-rays	1
X	ratio of a specific emission of the particle	1
XANES	X-ray Absorption Near Edge Structure	6
XPS	X-ray Photoelectron Spectroscopy	6
x	position	6
Z		
Z	atomic number	6
Z_a	atomic number of the target material	1
Z_p	atomic number of the primary particle	1
z	effective charge of the ion	6
Z	average number of collisions	10
Greek		
α	alpha particles	1
α	linear expansion coefficient	10
α	polarizability	6
α	fine structure constant	9
α_{max}	maximum separation factor	10
α-HIB	α-hydroxy isobutyric acid	7, 8, 9
β	beta particles	1
$\beta = v/c$	particle velocity	1
β	relative fraction of delayed neutrons	10
Γ	exposure rate constant	1
γ	gamma radiation	1
$-\Delta H_{ads}$	adsorption enthalpy	9

(continued)

Symbol/ Abbreviation	Meaning	Chapter
$\Delta\mu_0(E)$	jump in the absorption	6
Δm	width of the mass peak at mass m	4
Δr	distance from the molecule's equilibrium geometry	6
Δt	time interval	1
Δt	age	5
δ	delta particles	1
$\delta(k)$	scattering parameters	6
δE_B	partial binding energy	10
δ	phase	8
E	photon energy $\varepsilon = h\nu$	1
ε_0	dielectric (field) constant	6, 9
ε_{abs}	absolute efficiency	1
ε_{edge}	edge of COMPTON continuum	1
ε_{emis}	relative abundances (e.g., of γ-emissions)	1
ε_{ero}	erosion rate	5
ε_{geom}	geometric efficiency of a detector	1
ε_{int}	intrinsic efficiency of a detector	1
ε_{mol}	molar absorptivity	6
ε_r	dielectricity constant	6
ζ	zeta potential	6
η	dynamic viscosity	6
θ	scattering angle	1
θ_c	critical angle	1
λ	transformation (decay) constant	3, 5
λ	distribution volume of 2-[^{18}F]FDG	12
λ	thermal conductivity	10
μ	linear attenuation coefficient	1
$\mu, \mu(E)$	absorption coefficient	6
$\mu_0(E)$	background function	6
μ_{EF}	electrophoretic mobility	6
μ_{EOF}	electroosmotic flow	6
μ_N	nuclear magneton	6
μ_{tot}	total mobility	6
N	vibration frequency	6
ν_{in}	frequency of incoming light	6
ξ	lethargy: dimensionless number describing the energy loss after a collision	10
ρ	density	6, 10
ρ	(mass) density of the atoms for the chemical species	1
ρ	material density	1
ρ	reactivity	10
σ	atomic or nuclear cross section	1, 3
σ	shielding constant	6
σ^2	factor that describes the disorder in the neighbor distance	6
σ_c	neutron capture cross section	10

(continued)

Symbol/ Abbreviation	Meaning	Chapter
σ_{cap}	capture cross section	9
σ_{EVR}	evaporation residue cross section	9
σ_{fus}	fusion cross section	9
σ_s	scattering cross section for epithermal neutrons	10
σ_{tot}	total cross section	1
τ	mean lifetime	5
τ	fluorescence lifetime	6
τ_0	period of oscillation of the species in the adsorbed state	7
$x(E)$	EXAFS fine structure function	6
$x()$	fine structure	6
$x_0(E)$	energy spectrum	10
Φ	particle fluence (or flux)	1
Φ	neutron flux in $cm^{-2}\,s^{-1}$	3
Φ	concentration	6
Φ_n	flux of thermal neutrons	
Φ_s	work function	6
ϕ	fluence rate (or flux density)	1
Ψ	energy fluence	1
ψ	energy fluence rate	1
Ω	solid angle	1
ω_0	specific frequency	6
ω	angular frequency	6
Number		
$100 \cdot F^{14}C$	reference isotopic ratio of modern radiocarbon	5

Index

https://doi.org/10.1515/9783110742701-015

www.ingramcontent.com/pod-product-compliance
Lightning Source LLC
Chambersburg PA
CBHW080347220326
41598CB00030B/4629